Handbook of
INORGANIC
COMPOUNDS

SECOND EDITION

Handbook of INORGANIC COMPOUNDS

DALE L. PERRY

CRC Press
Taylor & Francis Group
Boca Raton   London   New York

CRC Press is an imprint of the
Taylor & Francis Group, an **informa** business

CRC Press
Taylor & Francis Group
6000 Broken Sound Parkway NW, Suite 300
Boca Raton, FL 33487-2742

© 2011 by Taylor and Francis Group, LLC
CRC Press is an imprint of Taylor & Francis Group, an Informa business

No claim to original U.S. Government works

Printed in the United States of America on acid-free paper
10 9 8 7 6 5 4 3 2 1

International Standard Book Number: 978-1-4398-1461-1 (Hardback)

Library of Congress Cataloging-in-Publication Data

Perry, Dale L.
  Handbook of inorganic compounds / Dale L. Perry. -- 2nd ed.
    p. cm.
  Rev. ed. of: Handbook of inorganic compounds / edited by Dale L. Perry, Sidney L. Phillips.
  Includes bibliographical references and index.
  ISBN 978-1-4398-1461-1 (alk. paper)
  1. Inorganic compounds--Handbooks, manuals, etc. I. Handbook of inorganic compounds. II. Title.

QD155.5.H36 2010
546.02'12--dc22   2010039438

Visit the Taylor & Francis Web site at
http://www.taylorandfrancis.com

and the CRC Press Web site at
http://www.crcpress.com

During the past several years, inorganic compounds have played a greater role in not only inorganic chemistry but related disciplines as well. These include materials chemistry, with its own subdisciplines such as catalysts, solar materials, superconductors, photonic materials, nuclear radiation detection materials, and inorganic thin films, among many others. This book provides a base set of inorganic compounds that are both inherently important themselves and important for a wide variety of applications that already have been realized and for many that have not, at least to this date.

There are several criteria for the selection of the compounds. Many are very significant with respect to their applications in both basic and applied research; this is true not only for chemical research but also for related fields such as biochemistry and materials science. Other compounds have been included because they are or can be precursors in processes for preparing important materials. Examples are the use of the precursors in sol–gel, vacuum-deposition, and hydrothermal crystallization as steps in a larger preparative procedure. A third reason for inclusion is to provide rapid access to basic compounds such as oxides, halides, sulfides, and other commonly used chemical compounds for a large variety of needs and applications.

As was true with the first edition of this handbook, it was not possible to include every inorganic compound or all the important data associated with them. The *Handbook of Inorganic Compounds* consists of data for 3326 selected gas, liquid, and solid compounds, with an attempt to include representative compounds of several different classes of compounds. Choices of compounds were based on criteria such as inclusion of the compounds in various handbooks of laboratory chemicals, discussion in recent research publications, compounds important to inorganic materials chemistry, and comments of the Advisory Committee guiding the production of the first edition of the handbook.

Compounds included in this book are mainly the chemical elements, binary compounds of the elements with anions such as sulfate and chloride, and metal salts of some simple organic acids. If a compound has more than one form, then each form may be listed individually. A typical example is that of separate listings for an anhydrous compound and its hydrates. Another example is the separate listing of the three calcium carbonates. With some exceptions, minerals (i.e., listed by their mineral names; many exist in nature as inorganic compounds and may be tabulated as such here in the *Handbook*), organometallic compounds, metallic alloys, noncrystalline materials, coordination complexes, and nonstoichiometric compounds (such as many naturally occurring minerals) are not included in this handbook.

The format for presenting information has both numerical data and descriptive information. The data are solubility, melting point, boiling point, density, thermal conductivity, and thermal expansion coefficient. Other data may also be included, for example, vapor pressure, viscosity, hardness, lattice parameters, electrical resistivity, Poisson's ratio, and dielectric constant. There may also be thermodynamic values, mainly enthalpy of vaporization, fusion, and sublimation. However, thermodynamic values for the individual compounds such as enthalpy of formation are not covered in this handbook. Descriptive information for the various compounds is organized into three categories: form, for example, color and particle size; preparation or manufacturing procedure; and commercial or other uses. Also, because of constraints on space and interest in compactness needed for a one volume treatise, no detailed structural data are presented.

Much effort has been made to obtain as many significant numerical values related to physical and chemical constants for each compound as possible. Thus, the reader is saved considerable time in looking for the basic properties from many sources. This handbook is intended to be useful to chemists, chemical engineers, materials scientists, and other scientists who need

1. Basic, essential property data for compounds that they wish to use in their database compilations, research, and applications work.
2. American Chemical Society (ACS) *Chemical Abstract Registry Numbers* (*RN's, or CAS numbers*) for computer and other searches. An effort has been made to include *CAS numbers* for both hydrated chemical compounds and their parent anhydrous compounds.
3. A tabulation of molecular weights for calculations. In this handbook, molecular weights have been calculated to three decimal places in all cases.
4. The synthesis of inorganic compounds and materials.
5. Vendor information for obtaining commercial samples and batches of inorganic compounds.

Complementary physical and chemical data for inorganic compounds can be found in a number of additional reference books, including the *Handbook of Chemistry and Physics* (91st Edition), *Perry's Chemical Engineers' Handbook* (8th Edition), the *Kirk-Othmer Encyclopedia of Chemical Technology* (5th Edition), the *Merck Index* (14th Edition), *Comprehensive Inorganic Chemistry*, *Gmelin Handbook of Inorganic and Organometallic Chemistry* (8th Edition), *Lange's Handbook of Chemistry* (70th Edition), and

Hawley's Condensed Chemical Dictionary (15th Edition). Increasingly, many of them are available online, both as electronic volumes and as journals in virtual libraries. Specialty data source references to the elements also are available, including *Thermochemical Data of Elements and Compounds, Transition Metal Chemistry, Polyhedron* (and older issues of its predecessor, *Journal of Inorganic and Nuclear Chemistry*), *Coordination Chemistry Reviews,* and *Inorganic Chemistry.* More recent information and data can be found in open research journals such as the *Journal of Chemical and Engineering Data, Journal of Solution Chemistry, Journal of Material Research, Journal of Organometallic Chemistry, Journal of the American Ceramic Society, Chemistry of Materials, Chemical Reviews, Material Research Society [MRS] Bulletin, Solid-State Ionics,* and *Journal of the Electrochemical Society.*

And, there have been several upgrades in this second edition relative to the first edition. In line with the dramatic growth of the Internet since the first edition was published in 1995, a number of aspects have been upgraded with respect to Web sites for vendors in the references. The major vendors were used for compound data in the first edition for the data themselves and as intentional sources for commercial venues from which to purchase compounds in a variety of forms and purities. Additionally, a separate section has been added that lists major reference volumes concerning the field of inorganic chemistry itself.

I wish to thank Sidney Phillips, my coeditor of the first edition of this handbook, for his hard work on the first edition. His professional background and scientific judgment were invaluable for producing this book. Additionally, I wish to thank David Lide for his tremendous assistance in both the first edition of this handbook and the present one. Conversations with him have been extremely useful with respect to chemical databases in general and the production of the present volume.

And, finally, I would like to thank Hilary Rowe and Fiona Mcdonald at CRC for their really huge role in directing this effort. Without their editorial guidance, this volume would not be possible.

Dale L. Perry received his PhD in inorganic chemistry from the University of Houston, Houston, Texas. He was a Welch Postdoctoral Fellow and a National Science Postdoctoral Fellow at Rice University. He was a Miller Fellow in chemistry at the University of California, Berkeley, California. He has been on the scientific staff in chemistry at Lawrence Berkeley National Laboratory, University of California, being appointed a senior scientist in chemistry at the same institution.

His research interests are in solid-state inorganic synthesis and spectroscopy, inorganic systems that include those of transition, main group, lanthanide, and actinide metal ions. The classes of compounds and materials on which his research has focused include metal ion-organic complexes, inorganic thin films, semiconductors, superconductors, mixed metal ion oxide catalysts, inorganic crystals, inorganic scintillation materials, and inorganic polymers. He has authored and coauthored over 300 contributed and invited scientific presentations, refereed journal publications, and numerous invited seminars at universities, national laboratories, and industry. He has edited and authored several books, including *Instrumental Surface Analysis of Geologic Materials*, *Applications of Analytical Techniques to the Characterization of Materials*, and *Applications of Synchrotron Radiation Techniques to Materials Science*. He has conducted workshops related to the characterization of inorganic materials and analysis using x-ray photoelectron, Auger, infrared, Raman, nuclear magnetic resonance, and Mossbauer spectroscopy.

Dr. Perry's honors include a Sigma Xi National Research Award and Traineeship, a Miller Fellowship, a Beyer Award, and a National Science Foundation Fellowship. He is a member of the American Chemical Society, American Association for the Advancement of Science, Materials Research Society, the Society for Applied Spectroscopy, and the Royal Society of Chemistry (London). He is also a fellow of both the Royal Society of Chemistry (London) and American Association for the Advancement of Science. He has been a member of the Committee for Corporate Participation in the Materials Research Society and both a member and a chairman of the Chemistry and Engineering Materials Subdivision in the Industrial & Engineering (I & EC) Division of the American Chemical Society.

In addition to research, he has been a member of several ad hoc panels for the U.S. Department of Energy related to instrumentation needs in both heavy metal chemical research and research as it pertains to heavy metals, their chemistry, and materials science related to them. He is an organizer of symposia concerning the application of spectroscopy to materials research, synthesis and characterization of inorganic materials, and the application of surface spectroscopy to materials studies. He has also been very active as a principal investigator and mentor to Hispanic, Native American, and African American students in the Center for Science and Engineering Education at Lawrence Berkeley National Laboratory. He was the winner of the Outstanding Mentor Award for Undergraduate Research Program, U.S. Department of Energy, in 2002.

Acknowledgments

The author would like to thank the following individuals and organizations for their contributions to this edition as well as the first edition of this handbook.

Solubility Data:

Boris Krumgalz
Israel Oceanographie & Limnological Research
Tel Shikmona
P.O.B. 8030
Haifa, Israel 31080
Phone: 04-515202
FAX: 04-511911

Barium and Cerium Data:

Mariska Scholten
Delft University of Technology
P.O.B. 5045
Julianalaan 136
2628 BL Delft
The Netherlands
Phone: 31 (0) 15-782615
FAX: 31 (0) 15-782655

Sodium Silicates Data:

Barry Schenker
Occidental Chemical Corporation
P.O.B. 344
Niagara Falls, New York 14302-0344
Phone: (800) 733-1165

Lanthanide and Ruthenium Data:

Joseph A. Rard
Lawrence Livermore National Laboratory
University of California
Livermore, California 94550
Phone: (925) 422-1100

Computer Work:

Daniel J. Phillips
Camatx/Basic Chemistry
171 El Toyonal
Orinda, California 94563
Phone: (510) 254-7717
FAX: (510) 253-1358

Barium and Cerium Data:

Joop Schoonman
Delft University of Technology
P.O.B. 5045
Julianalaan 136
2628 BL Delft
The Netherlands
Phone: 31 (0) 15-782615
FAX: 31 (0) 15-782655

Consistency of Data:

David R. Lide
CRC Press, Inc.
13901 Riding Loop Dr.
Gaithersburg, Maryland 20878
Phone: (301) 738-7147
FAX: (301) 738-7147

Oxide Solubility Data:

Steven E. Ziemniak
Knolls Atomic Power Laboratory
P.O.B. 1072
Schnectady, New York 12301
Phone: (518) 395-4000
FAX: (518) 395-4422

Data Collections of Radioactive Inorganic Compounds:

Gerald English
CCIS, Inc.
3785 Highland Road
Lafayette, California 94549
Phone: (925) 283-5708
FAX: (510) 486-5757

Organization of Data for the Compounds

References to sources of the data are given in the form [XXXYY], where XXX represents the principal author or work and YY represents the last two digits of the year of the work.

Compound: Commonly used name of inorganic compound.

Formula: Commonly used chemical formula.

Molecular Formula: Modified Hill system in which carbon is always listed first, followed by hydrogen (if any) and then other elements in alphabetical order. If there is no carbon, then the elements are given in alphabetical order. Stoichiometry is always shown in the usual subscript form.

Molecular Weight: Consistently calculated from the stoichiometry of the formula to three decimal places, using atomic weights from *Pure & Applied Chemistry*, *81*, 2131–2156 (2009). Significant figures are not taken into account.

CAS RN: Chemical Abstracts Service (CAS) Registry Number (RN). Where possible, the CAS RN for the compound with hydrated waters is given, as well as that for the anhydrous compound.

Properties: Consists of all or some of the following basic chemical data: crystalline form with lattice parameters; color; gas, liquid, or solid; vapor pressure; hardness; viscosity; dielectric constant; electrical resistivity; Poisson's ratio; enthalpy of vaporization; enthalpy of fusion; preparation; uses.

Solubility: Concentration of compound in solvent under the stated conditions. The solvent is usually water unless otherwise stated; efforts have been made to include the equilibrium solid phase. Abbreviations for solvents are identical to those used in Reference [CRC10], *CRC Handbook of Chemistry and Physics*, W. M. Haynes, Ed., CRC Press, Inc., 91st edition, Boca Raton, FL (2010–2011).

Density: Density of the solid, liquid, or gas.

Melting Point: Temperature at which the pure solid becomes liquid.

Boiling Point: Temperature at which the pure liquid becomes gas.

Reactions: Limited generally to phase changes, decomposition, and hydrolytic reactions.

Thermal Conductivity: Property of the compound that attributes a numerical value to its capability to transmit heat.

Thermal Expansion Coefficient: Change in volume or length per degree change in temperature. [This is an important property for ceramics.]

Literature Cited

[AES93] *AESAR Catalog*, Johnson Matthey, Ward Hill, MA (1992–1993). http://www.alfa.com/en/ge100w.pgm

[AIR87] Specialty Gases, Air Products and Chemicals, Inc., Allentown, PA (1987). http://www.airproducts.com/index.asp

[ALD93] *Aldrich Catalog Handbook of Fine Chemicals*, Aldrich Chemical Co., St. Louis, MO (1992–1993). http://www.sigmaaldrich.com/chemistry.html

[ALD94] Aldrich Catalog Inorganics, Aldrich Chemical Co., Milwaukee, WI (1994). http://www.sigmaaldrich.com/chemistry.html

[ALF93] Alfa Catalog of Research Chemicals and Accessories, Johnson Matthey Catalog Co., Inc., Ward Hill, MA (1993–1994). http://www.alfa.com/en/ge100w.pgm

[ALF95] Alfa Aesar Chemicals Catalog, Johnson Matthey Catalog Co., Inc., Ward Hill, MA (1995–1996). http://www.alfa.com/en/ge100w.pgm

[ASM93] W. Assmus and W. Schmidbauer, Crystal growth of HTSC materials, *Supercond. Sci. Technol.*, 1993, 6, 555–566.

[AST92] Standard Practice for Use of the SI International System of Units, ASTM E 380-92, American Society for Testing and Materials, Philadelphia, PA (1992).

[BAB85] V.I. Babushkin, G.M. Matveyev, and O.P. Mchedlov-Petrossyan, *Thermodynamics of Silicates*, B.N. Frenkel and V.A. Terentyev (transl.), Springer-Verlag, New York (1985).

[BAN90] N.P. Bansal, Influence of several metal ions on the gelation activation energy of silicon tetraethoxide, *J. Am. Ceram. Soc.*, 1990, 73, 2647–2652.

[BHA78] H. Bhattacharya and B.N. Sammaddar, Formation of nonstoichiometric spinel on heating hydrous magnesium aluminate, *J. Am. Ceram. Soc.*, 1978, 61, 279–280.

[BOU93] W.L. Bourcier, K.G. Knauss, and K.J. Jackson, Aluminum hydrolysis constants to 250°C from boehmite solubility measurements, *Geochim. Cosmochim. Acta*, 1993, 57, 747–762.

[BRO73] D. Brown, The actinides, In *Comprehensive Inorganic Chemistry*, Pergamon Press, New York (1973), p. 278.

[CAB85] KBI Electronic Materials, Potassium and Rubidium, Cabot Corp., Revere, PA (1985).

[CAB93] Cabot Performance Materials, Tantalum, Cabot Corp., Revere, PA (1993).

[CEN92] F.J. Adrian and D.O. Cowan, The new superconductors, *Chem. Eng. News*, 1992, 70.

[CEN94] R. Dagani, New opportunities in materials science draw Eager organometallic chemists, *Chem. Eng. News*, 1994, 72.

[CER91] Cerac, Inc., Milwaukee, WI (1991). http://www.cerac.com/

[CHA90] F. Chaput and J.-P. Boilot, Alkoxide-hydroxide route to synthesize $BaTiO_3$-based powders, *J. Am. Ceram. Soc.*, 1990, 73, 942–948.

[CIC73] J.C. Bailar, H.J. Emeleus, R. Nyholm, and A.F. Trotman-Dickenson, Eds., *Comprehensive Inorganic Chemistry*, Pergamon Press, New York (1973).

[CLA66] S.P. Clark, Ed., *Handbook of Physical Constants*, Memoir 97, Geological Society of America, New York (1966).

[CON87] L.E. Conroy, A.N. Christensen, and J. Bottiger, Preparation and characterization of Hi-T_c oxides. $YBa_2Cu_3O_7$ and $REBa_2Cu_3O_7$, *Acta Chim. Scand.*, 1987, A41, 501–505.

[COT88] F.A. Cotton, and G. Wilkinson, *Advanced Inorganic Chemistry*, 5th edn., John Wiley & Sons, New York (1988).

[CRC10] W.M. Haynes, Ed., *CRC Handbook of Chemistry and Physics*, CRC Press, Inc., 91st edn., Boca Raton, FL (2010–2011).

[DAH90] J.R. Dann, U. von Sacken, C.A. and Michal, Structure and electrochemistry of $Li_{1\pm y}NiO_2$ and a new Li_2NiO_2 phase with the $Ni(OH)_2$ structure, *Solid State Ionics*, 1990, 44, 87–97.

[DES91] P. Descamps, S. Sakaguchi, M. Poorteman, and F. Cambier, High temperature characterization of reaction sintered mullite–zirconia composites, *J. Am. Ceram. Soc.*, 1991, 74, 2476–2481.

[DOU83] B. Douglas, D.H. McDaniel, and J.J. Alexander, *Concepts and Models of Inorganic Chemistry*, 2nd edn., John Wiley & Sons, Inc., New York (1983).

[DRE93] M.S. Dresselhaus, C. Dresselhaus, and P.C. Eklund, Fullerenes, *J. Mater. Res.*, 1993, 8, 2054–2097.

[ERI92] T.E. Eriksen, P. Ndalamba, J. Bruno, and M. Caceci, The solubility of $TcO_2 \cdot nH_2O$ in neutral to alkaline solutions under constant pCO_2, *Radiochim. Acta*, 1992, 58/59, 67–70.

[FIE87] P.E. Fielding, and T.J. White, Crystal chemical incorporation of high level waste species in aluminotitanate-based ceramics, *J. Mater. Res.*, 1987, 2, 387–414.

[FMC93] FMC Corporation, Lithium Division, Lithco Products, Your Source of Quality Lithium Chemicals, Gastonia, NC (1993). http://www.fmclithium.com/

[FRE87] B. Freudenberg and A. Mocellin, Aluminum titanate formation by solid-state reaction of fine Al_2O_3 and TiO_2 powders, *J. Am. Ceram. Soc.*, 1987, 70, 33–38.

[FRI87] J.J. Fritz and E. Luzik, Solubility of copper(I) bromide in aqueous solutions of potassium bromide, *J. Solut. Chem.*, 1987, 16, 79–85.

[GEI92] G. Geiger, Ceramic coatings, *Bull. Am. Ceram. Soc.*, 1992, 1470–1481.

[GME76] *Gmelin Handbuch der Anorganischen Chemie. Seltenerdelemente Teil C3: Sc, Y, La und Lanthanide, Fluoride, Oxidifluoride und zugehorige Alkilidoppelverbindungen*, H. Bergmann, Springer, Berlin (1976). In [SCH93] J. Schölten, and J. Schoonman, Delft University of Technology, the Netherlands (October 18, 1993).

[GME77] *Gmelin Handbuch der Anorganischen Chemie*, Springer-Verlag, New York (1977).

[GUM92] R.J. Gummow, M.M. Thackeray, W.I.F. David, and S. Hull, Structure and electrochemistry of lithium cobalt oxide synthesised at 400°C, *Mater. Res. Bull.*, 1992, 27, 327–337.

[HAW93] R.J. Lewis, Ed., *Hawley's Condensed Chemical Dictionary*, 12th edn., Van Nostrand Reinhold Co., New York (1993).

[HIR87] S.-I. Hirano and K. Kato, Synthesis of LiNbO$_3$ by hydrolysis of metal alkoxides, *Adv. Ceram. Mater.*, 1987, 2, 142–145.

[HO72] C.Y. Ho, R.W. Powell, and P.E. Liley, *J. Phys. Chem. Ref. Data*, 1972, 1, 279–421.

[HOL73] C.E. Holcombe and A.L. Coffey, Calculated X-ray powder diffraction data for β-Al$_2$TiO$_5$, *J. Am. Ceram. Soc.*, 1973, 56, 220–221.

[HOU82] V. Houlding, T. Geiger, U. Kölle, and M. Grätzel, Electrochemical and photochemical investigations of two novel electron relays for hydrogen generation from water, *J. Chem. Soc., Chem. Commun.*, 1982, 681–683.

[HUA91] G. Huan, J.W. Johnson, A.J. Jacobson, and D.P. Goshorn, Hydrothermal synthesis, single-crystal structure, and magnetic properties of VOSeO$_3$–H$_2$O, *Chem. Mater.*, 1991, 3, 539–541.

[IUP92] IUPAC Atomic Weights of the Elements 1991, *Pure Appl. Chem.*, 1992, 64, 1519–1534.

[IUP93] PAC Review, Fullerenes, *Chem. Int.*, 1993, 15, 94.

[JAN71] D.R. Stull and H. Prophet, JANAF thermochemical tables, *Nat. Stand. Ref. Data Ser.*, NIST, 37 (1971).

[JAN82] M.W. Chase, J.L. Curnutt, J.R. Downey, R.A. McDonald, A.N. Syverud, and E.A. Valenzuela, JANAF thermochemical tables, 1982 supplement, *J. Phys. Chem. Ref. Data*, 1982, 11, 695.

[JAN85] M.W. Chase, C.A. Davies, J.R. Downey, D.J. Frurip, R.A. McDonald, and A.N. Syverud, JANAF thermochemical tables, 3rd edn., *J. Phys. Chem. Ref. Data*, 1985, 14(Suppl. 1).

[KAT86] J.J. Katz, G.T. Seaborg, and L.R. Morss, Eds., *The Chemistry of the Actinide Elements*, 2nd edn., Vols. 1 and 2, Chapman & Hall, New York (1986).

[KAZ90] A.M. Kazakos, S. Komarneni, and R. Roy, Sol–gel processing of cordierite: Effect of seeding and optimization of heat treatment, *J. Mater. Res.*, 1990, 5, 1095–1103.

[KIR78], [KIR79], [KIR80], [KIR81], [KIR82], [KIR83], [KIR84] Kirk-Othmer, *Encyclopedia of Chemical Technology*, 3rd edn., various volumes, John Wiley & Sons, New York (1978–1984).

[KIR91] Kirk-Othmer, *Encyclopedia of Chemical Technology*, 4th edn., John Wiley & Sons, New York (1991).

[KLE93] J.D. Klein, R.D. Herrick, D. Palmer, M.J. Sailor, C.J. Brumlik, and C.R. Martin, Electrochemical fabrication of cadmium chalcogenide microdiode arrays, *Chem. Mater.*, 1993, 5, 902–904.

[KNA91] O. Knacke, O. Kubascheski, and K. Hesselmann, *Thermochemical Properties of Inorganic Substances*, Springer-Verlag, Berlin (1991).

[KOH88] S.C.H. Koh and R. McPherson, Sintering of metastable ZrO$_2$–Al$_2$O$_3$, ceramic developments, In *Materials Science Forum*, 1988, Vols. 34–36, p. 117–121, Trans. Tech. Publications, Ltd., Switzerland (1988).

[KOR91] N.E. Korte and Q. Fernando, A review of arsenic(III) in groundwater, *Crit. Rev. Environ. Control*, 1991, 21, 1–39.

[KRE91] A.M. Kressin, V.V. Doan, J.D. Klein, and M.J. Sailor, Synthesis of stoichiometric cadmium selenide films via sequential monolayer electrodeposition, *Chem. Mater.*, 1991, 3, 1015–1020.

[KRU93] B.S. Krumgalz, *Mineral Solubility in Water at Various Temperatures*, Israel Océanographie & Limnological Research, Tel-Shikmona, Haifa, Israel (1993).

[LAN05] J.G. Speight, Ed., *Lange's Handbook of Chemistry*, 16th ed., McGraw-Hill Book Co., New York (2005). http://www.knovel.com/web/portal/browse/display?_EXT_KNOVEL_DISPLAY_bookid=1347

[LAU73] R.A. Laudise, Hydrothermal growth. In P. Hartman, Ed., *An Introduction to Crystal Growth*, North-Holland, Amsterdam, New York (1973).

[LAU87] R.A. Laudise, Hydrothermal crystallization, *Chem. Eng. News*, 1987, 30.

[LID94] D.R. Lide, *Physical Constants of Inorganic Compounds*, CRC Press, Inc., Private Communication (November 1994).

[LOP84] M.A. Lopez-Quintela, W. Knoche, and J. Veith, Kinetics and thermodynamics of complex formation between aluminum(III) and citric acid in aqueous solutions, *J. Chem. Soc., Faraday Trans. 1*, 1984, 80, 2313–2321.

[MAE90] K. Maeda, F. Mizukami, S. Miyashita, S.-I. Niwa, and M. Toba, Synthesis of cordierite by complexing agent assisted sol–gel procedure, *J. Chem. Soc., Chem. Commun.*, 1990, 1268–1269.

[MAK90] A. Makishima, M. Asami, and K. Wada, Preparation and properties of TiO$_2$–CeO$_2$ coatings by the sol–gel process, *J. NonCryst. Solids*, 1990, 121, 310–314.

[MAR58] W.L. Marshall, F.J. Loprest, and C.H. Secoy, The equilibrium Li$_2$CO$_3$+CO$_2$+H$_2$O=2Li$^+$+2HCO$_3^-$ at high temperature and pressure, *J. Am. Chem. Soc.*, 1958, 80, 5646.

[MER52] *The Merck Index*, 14th edn., Merck & Co., Inc., Whitehouse Station, NJ (2006). http://library.dialog.com/bluesheets/html/bl0304.html

[MIT87] M. Mitomo and Y. Yoshioka, Preparation of Si$_3$N$_4$ and A1N powders from alkoxide-derived oxides by carbothermal reduction and nitridation, *Adv. Ceram. Mater.*, 1987, 2, 253–256.

[MIT72] P.W.D. Mitchell, Chemical method for preparing MgAl$_2$O$_4$ spinel, *J. Am. Ceram. Soc.*, 1972, 55, 484.

[MIZ89] M. Mizuno and H. Saito, Preparation of highly pure fine mullite powder, *J. Am. Ceram. Soc.*, 1989, 72, 377–382.

[MM09] M.E. Wieser and M. Burgland, Atomic weights of the elements 2007 (IUPAC Technical Report), *Pure Appl. Chem.*, 2009, 81, 2131–2156.

[MOI86] A. Moini, R. Peascoe, P.R. Rudolf, and A. Clearfield, Hydrothermal synthesis of copper molybdates, *Inorg. Chem.*, 1986, 25, 3782–3785.

[MOY86] J.R. Moyer, A.R. Prunier, N.N. Hughes, and R.C. Winterton, Synthesis of oxide ceramic powders by aqueous coprecipitation, *Materials Research Society Symposium Proceedings*, Vol. 73, Material Research Society, Pittsburgh, PA (1986).

[MR89] Y. Hirata, K. Sakeda, Y. Matsushita, K. Shimada, and Y. Ishihara, Characterization and sintering behavior of alkoxide derived aluminosilicate powders, *J. Am. Ceram. Soc.*, 1989, 72, 995–1002.

[OGU88] Y. Oguri, R.E. Riman, and H.K. Bowen, Processing of anatase prepared from hydrothermally treated alkoxy-derived hydrous titania, *J. Mater. Sci.*, 1988, *23*, 2897–2904.

[OKA91] K. Okada, N. Otsuka, and S. Somiya, Review of mullite synthesis routes in Japan, *Ceram. Bull.*, 1991, *70*, 1633–1640.

[OXY93] B. Schenker, *Sodium Metasilicate*, Occidental Chemical Corp., NY, Private Communication, November 10, 1993.

[OZB80] H. Ozbek and S.L. Phillips, Thermal conductivity of aqueous sodium chloride solutions from 20 to 330°C, *J. Chem. Eng. Data*, 1980, *25*, 263–267.

[PAR90] F.J. Parker, Al_2TiO_5–$ZrTiO_4$–ZrO_2 composites: A new family of low-thermal-expansion ceramics, *J. Am. Ceram. Soc.*, 1990, *73*, 929–932.

[PAS88] J.A. Pask, Phase equilibria in the Al_2O_3–SiO_2 system with emphasis on mullite, *Mater. Sci. Forum*, 1988, *34–36*, 1–8.

[PER89] D.S. Perera, Reaction-sintered aluminum titanate, *J. Mater. Sci. Lett.*, 1989, *8*, 1057–1059.

[PFA93] Pfaltz & Bauer Chemicals Catalog, Aceto Corp., Waterbury, CT (1993). http://www.pfaltzandbauer.com/

[PHI93] S.L. Phillips and F.V. Hale, Hydrolysis constants for $AlOH^{++}$ in high temperature water from thermodynamic calculations, *Proceedings of the 1991 Symposium on High Temperature Chemistry*, Provo, UT (August 1993).

[PHU89] P.P. Phule and S.H. Risbud, Low temperature synthesis and dielectric properties of ceramics derived from amorphous barium titanate gels and crystalline powders, *Mater. Sci. Eng.*, 1989, *B3*, 241–247.

[PLU82] L.N. Plummer and E. Busenberg, The solubilities of calcite, aragonite and vaterite in CO_2–H_2O solutions between 0 and 90°C, and an evaluation of the aqueous model for the system $CaCO_3$–CO_2–H_2O, *Geochim. Cosmochim. Acta*, 1982, *46*, 1011–1040.

[POT78] R.W. Potter and M.A. Clynne, The solubility of the noble gases He, Ne, Ar, Kr, and Xe in water up to the critical point, *J. Solut. Chem.*, 1978, *7*, 837–844.

[PRA92] A.V. Prasadarao, U. Selvaraj, S. Komarneni, A.S. Bhalla, and R. Roy, Enhanced densification by seeding of sol-gel-derived aluminum titanate, *J. Am. Ceram. Soc.*, 1992, *75*, 1529–1533.

[RAR83] J.A. Rard, Critical review of the chemistry and thermodynamics of technetium and some of its inorganic compounds and aqueous species, UCRL-53440, Lawrence Livermore National Laboratory, Livermore, CA (1983). http://www.llnl.gov

[RAR84] J.A. Rard, Solubility of $Eu(NO_3)_3$·$6H_2O$ in water at 298.15 K, *J. Chem. Thermodyn.*, 1984, *16*, 921–925.

[RAR85] J.A. Rard, Chemistry and thermodynamics of ruthenium and some of its inorganic compounds and aqueous species, *Chem. Rev.*, 1985, *85*, 1–39.

[RAR85a] J.R. Rard, Chemistry and thermodynamics of europium and some of its simpler inorganic compounds and aqueous species, *Chem. Rev.*, 1985, *85*, 555–582.

[RAR85b] J.A. Rard, Solubility determinations by the isopiestic method and application to aqueous lanthanide nitrates at 25°C, *J. Solut. Chem.*, 1985, *14*, 457–471.

[RAR87a] J.A. Rard, Osmotic and activity coefficients of aqueous $La(NO_3)_3$ and densities and apparent molal volumes of aqueous $Eu(NO_3)_3$ at 25°C, *J. Chem. Eng. Data*, 1987, *32*, 92–98.

[RAR87b] J.A. Rard, Isopiestic determination of the osmotic and activity coefficients of aqueous $NiCl_2$, $Pr(NO_3)_3$, and $Lu(NO_3)_3$ and solubility of $NiCl_2$ at 25°C, *J. Chem. Eng. Data*, 1987, *32*, 334–341.

[RAR88] J.A. Rard, Aqueous solubilities of praseodymium, europium and lutetium sulfates, *J. Solut. Chem.*, 1988, *17*, 499–517.

[RAR92] J.A. Rard, Isopiestic investigation of water activities of aqueous $NiCl_2$ and $CuCl_2$ solutions and the thermodynamic solubility product of $NiCl_2$–H_2O at 298.15 K, *J. Chem. Eng. Data*, 1992, *37*, 433–442.

[RIO92] A. Riou, A. Lecerf, Y. Gerault, and Y. Cudennec, Etude Structurale de Li_2MnO_3, *Mat. Res. Bull.*, 1992, *27*, 269–275.

[RTT86] J.J. Ritter, R.S. Roth, and J.E. Blendell, Alkoxide precursor synthesis and characterization of phases in the barium–titanium oxide system, *J. Am. Ceram. Soc.*, 1986, *69*, 155–162.

[ROB67] R.A. Robie, P.M. Bethke, and K.M. Beardsley, Selected X-ray crystallographic data, molar volume, and densities of minerals and related substances, U.S. Geological Survey Bulletin 1248, 87pp (1967). In [CRC10].

[ROB78] R.A. Robie, B.S. Hemingway, and J.R. Fisher, Thermodynamic properties of minerals and related substances at 298.15 K and 1 bar (10^5 Pascals) pressure and at higher temperatures, Geological Survey Bulletin 1452, U.S. Geological Survey, U.S. Government Printing Office, Washington, DC (1978).

[ROS91] M.H. Rossouw and M.M. Thackeray, Lithium manganese oxides from Li_2MnO_3 for rechargeable lithium battery applications, *Mater. Res. Bull.*, 1991, *26*, 463–473.

[ROS92] M.H. Rossouw, D.C. Liles, and M.M. Thackeray, Alpha manganese dioxide for lithium batteries: A structural and electrochemical study, *Mater. Res. Bull.*, 1992, *27*, 221–230.

[SAF87] A. Safari, Y.H. Lee, A. Halliyal, and R.E. Newnham, O-3 piezoelectric composites prepared by coprecipitated $PbTiO_3$ powder, *Am. Ceram. Soc. Bull.*, 1987, *66*, 668–670.

[SCH88] G. Schwartz, R. Bennett, and S.A. Prokopovich, Sol-gel processing in the ZrO_2–SiO_2 system. In C.C. Sorrell and B. Ben-Nissen, Eds, *Ceramics Developments, Materials Science Forum*, Vols. 34–36 (1988), pp. 841–843.

[SCH93] M. Schölten and J. Schoonman, Barium and cerium, Delft University of Technology, Delft, the Netherlands, October 18, 1993.

[SIE94] S. Siekierskiand and S.L. Phillips, Eds., *Actinide Nitrates, Solubility Data Series*, Vol. 55, Oxford University Press, U.K. (1994).

[SOM91] S. Somiya and Y. Hirata, Mullite powder technology and applications in Japan, *Ceram. Bull.*, 1991, *70*, 1624–1632.

[STR93] Strem Catalog No. 15, 1993–1994, Strem Chemicals, Inc., Newburyport, MA. http://www.strem.com/

[STR08] Strem Catalog No. 15, 2008–2010, Strem Chemicals, Inc., Newburyport, MA. http://www.strem.com/

[STR94] The Strem Chemiker, April 1994, Vol. XV, No. 1, Strem Chemicals, Inc. Newburyport, MA. http://www.strem.com/

[SUB90] E.C. Subbarao, D.K. Agrawal, H.A. McKinstiy, C.W. Saliese, and R. Roy, Thermal expansion of compounds of zircon structure, *J. Am. Ceram. Soc.*, 1990, *73*, 1246–1252.

[TAY84a] D. Taylor, Thermal expansion data. I. Binary oxides with the sodium chloride and wurtzite structures, MO, *Br. Ceram. Trans. J.*, 1984, *83*, 5–9.

[TAY85] D. Taylor, Thermal expansion data. VIII. Complex oxides, ABO_3, the perovskites, *Br. Ceram. Trans. J.*, 1985, *84*, 181–188.

[TAY86] D. Taylor, Thermal expansion data. X. Complex oxides, ABO_4, *Br. Ceram. Trans. J.*, 1986, *85*, 146–155.

[TAY87] D. Taylor, Thermal expansion data: XI. Complex oxides, A_2BO_5, and the garnets, *Br. Ceram. Trans. J.*, 1987, *86*, 1–6.

[TAY88a] D. Taylor, Thermal expansion data. XII. Complex oxides: AB_2O_6, AB_2O_7, $A_2B_2O_7$, plus complex aluminates, silicates and analogous compounds, *Br. Ceram. Trans. J.*, 1988, *87*, 39–45.

[TAY88b] D. Taylor, Thermal expansion data. XIII. Complex oxides with chain, ring and layer structures and the apatites, *Br. Ceram. Trans. J.*, 1988, *87*, 87–95.

[TAY91a] D. Taylor, Thermal expansion data. XIV. Complex oxides with the sodalite and nasicon framework structure, *Br. Ceram. Trans. J.*, 1991, *90*, 64–69.

[TAY91b] D. Taylor, Thermal expansion data. XV. Complex oxides with the leucite structure and frameworks based on the six-membered ring of tetrahedra, *Br. Ceram. Trans. J.*, 1991, *90*, 197–204.

[THA92] M.M. Thackery, A. deKock, M.H. Rossouw, D. Liles, R. Bittihn, and D. Hoge, Spinel electrodes from the Li–Mn–O system for rechargeable lithium battery applications, *J. Electrochem. Soc.*, 1992, *139*, 363–366.

[TOU77] Y.S. Touloukian, R.K. Kirby, R.E. Taylor, and T.Y.R Lee, *Thermophysical Properties of Matter, Volume 13: Thermal Expansion Nonmetallic Solids*, IFI/Plenum, New York (1977).

[WU 88] J.-M. Wu and H.-W. Wang, Factors affecting the formation of $Ba_2Ti_9O_{20}$, *J. Am. Ceram. Soc.*, 1988, *71*, 869–875.

[VIE91] D.J. Viechnicki, M.J. Slavin, and M.I. Kliman, Development and current status of armor ceramics, *Ceram. Bull.*, 1991, *70*, 1035–1039.

[YAM87] O. Yamaguchi, H. Taguchi, and K. Shimizu, Formation of spinel from metal organic compounds, *Polyhedron*, 1987, *6*, 1791–1796.

[YAM89] O. Yamaguchi and Y. Mukaida, Formation and transformation of TiO_2 (anatase) solid solution in the system TiO_2–Al_2O_3, *J. Am. Ceram. Soc.*, 1989, *72*, 330–333.

[YTN92] Y. Ying and Y. Rudong, Study on the thermal decomposition of tetrahydrated ceric sulfate, *Thermochim. Acta*, 1992, *202*, 301–306.

[ZIE89] S.E. Ziemniak, M.E. Jones, and K.E.S. Combs, Solubility and phase behavior of nickel oxide in aqueous phosphate solutions at elevated temperatures, *J. Solut. Chem.*, 1989, *18*, 1133–1152.

[ZIE92a] S.E. Ziemniak, M.E. Jones, and K.E.S. Combs, Copper(II) oxide solubility behavior in aqueous phosphate solutions at elevated temperatures, *J. Solut. Chem.*, 1992, *21*, 179–200.

[ZIE92b] S.E. Ziemniak, M.E. Jones, and K.E.S. Combs, Zinc(II) oxide solubility and phase behavior in aqueous phosphate solutions at elevated temperatures, *J. Solut. Chem.*, 1992, *21*, 1153–1176.

[ZIE93] S.E. Ziemniak, M.E. Jones, and K.E.S. Combs, Solubility behavior of titanium(iv) oxide in alkaline media at elevated temperatures, *J. Solut. Chem.*, 1993, *22*, 601–623.

Glossary of Terms

Reference: "Abbreviated List of Quantities, Units and Symbols in Physical Chemistry," IUPAC, Blackwell Scientific Publications, Oxford, England (1987). "Standard Practice for Use of the International System of Units," ASTME 380-92, Philadelphia, PA 19103 (1992). *Handbook of Chemistry and Physics*, 91st edition, W. M. Haynes, Editor, CRC Press, Inc., Boca Raton, FL 33431 (2010–2011).

, (comma):	and, and also, in	**liq:**	liquid
→:	giving, yielding	**m:**	meter
<:	below	**micro:**	E − 06
>:	above	**min:**	minute(s)
~:	approximately	**mL:**	milliliter
μ:	micro, E − 06	**mm:**	millimeter
10^{-n}:	exponent to number, n	**mol:**	mole
a, b, c:	lattice parameters	**monocl:**	monoclinic
aq:	aqueous	**mp:**	melting point
atm:	atmosphere, atmospheric	**MPa:**	megapascal
bcc:	body-center(ed) cubic	**mW:**	milliwatt
bluish:	having a tinge of blue; somewhat blue	**nm:**	nanometer
bp:	boiling point	**off-white:**	a yellowish or grayish white
Btu:	British thermal unit	**ohm·cm:**	ohm centimeter
cal:	calorie	**ortho-rhomb:**	orthorhombic
CAS RN:	Chemical Abstracts Service Registry Number	**ortho:**	orthorhombic
cm:	centimeter	**Pa:**	pascal
conc:	concentrated	**powd:**	powder(s)
cp:	centipoise	**reddish:**	having a tinge of red
cryst:	crystal(s), crystalline	**rhomb:**	rhombic
cub:	cubic	**sec:**	second(s)
deliq:	deliquescent	**s:**	soluble
dil:	dilute	**sl:**	slightly
E+or −:	exponent($10^{+ \text{ or} -}$)	**soln:**	solution
eV:	electron volt	**$t_{1/2}$:**	half-life
fcc:	face-centered cubic	**T_C, T_K:**	superconducting transition temperature
fp:	freezing point	**temp:**	temperature
GPa:	gigapascal	**tetr:**	tetragonal
h:	hour(s)	**tric:**	triclinic
hex:	hexagonal	**trig:**	trigonal
hygr:	hygroscopic	**v:**	very
i:	insoluble	**V:**	volt
J:	joule	**W:**	watt
K:	Kelvin	**W/(m·K):**	$W\ m^{-1}\ K^{-1}$
k:	kilo	**[XXXYY]:**	Reference to the source publication; XXX denotes first three letters of work or senior author, YY denotes last two digits of the year of publication
kgf:	kilogram force		
kJ:	kilojoule		

Conversion of Units

References: "Abbreviated List of Quantities, Units and Symbols in Physical Chemistry," IUPAC, Blackwell Scientific Publications, Oxford, England (1987). "Standard Practice for Use of the International System of Units," ASTM E 380-92, Philadelphia, PA 19103 (1992).

Å (angstrom) × E − 01 = nm

atm × 1.013 250 E + 05 = Pa

bar × 1.000 000 E + 05 = Pa

Btu in./(h ft^2 °F) × 1.441 314 E − 01 = W/(m · K)

Btu (thermochemical) × 1.054 35 E + 03 = J

cal/(cm · s · °C) × 4.184 000 E + 02 = W/(m · K)

cal (thermochemical) × 4.184 = J

cp × 1.000 000 E − 03 = Pa · s

g/cm^3 × 1.000 000 E + 03 = kg/m^3

g × 1.000 000 E + 03 = kg

J × E + 03 = kJ

kgf/mm^2 × 9.806 650 E + 06 = Pa

kcal × 4.184 = kJ

m × E × 09 = nm

mm Hg (0°C) × 1.333 22 E + 02 = Pa

ohm · cm × 1.000 000 E − 02 = ohm · m

Pa × E + 09 = GPa

W/(m · K) = 0.1 mW/(cm · K)

W/(m · K) = 100 W/(cm · K)

°C = (°F − 32)/(1.8)

°C = K − 273.15

Table of Atomic Weights of the Elements—2007*

The following 2007 Table of Standard Atomic Weights was prepared at a meeting in Pisa, Italy, July 30–31, 2007, under the auspices of the Commission on Atomic Weights and Isotopic Abundances, Inorganic Chemistry Division, and the chairmanship of Prof. T. P. Ding.

INTRODUCTION

Atomic weights represent the bedrock of calculations for reactions, stoichiometries, and other numerical applications regarding the elements of the periodic table.

The detail and the number of significant figures in the IUPAC Table of Standard Atomic Weights usually exceed the needs and interests of most users, who are more interested in the use of abbreviated, short, accurate tables that have validity to the precision limit of their interests. The Commission on Atomic Weights and Isotopic Abundances in 1987, therefore, decided to prepare for publication a revised and updated version of the 1981 Table of Atomic Weights abridged to five significant figures or fewer where uncertainties do not warrant even five-figure accuracy. Additional upgraded tables were published in 1993, 1995, 1997, 1999, 2001, and 2005.

The complete table is given here with the knowledge that the quoted values for these elements will change slightly from year to year, something that has been shown to occur historically with previous versions of the table. It should be understood that the atomic-weight values for many elements are still uncertain by more than one unit in their last significant figure. Moreover, for additional elements, the indicated uncertainty range in the unabridged table here includes values that, when rounded to five significant figures, would show a change in the fifth figure. For some of these elements, minor changes in their best standard atomic weight to five significant figures could occasionally be required as more accurate values become available as a result of the biennial revision of the unabridged table. Most annotated warnings of anomalous geological occurrences, isotopically altered materials, and variability of radioactive elements are relevant even in the abridged table. There are older tables of atomic weights that are abridged to only five significant figures, based on older table values. These are quite useful for workers who do not need the atomic-weight values to more significant figures. Too, many of the values with respect to five significant figures are in fairly good agreement with the values in this table, especially since most workers do not need the greater precision.

This table may be freely reprinted provided it includes the annotations, the rubric at the head of the table, and the IUPAC primary reference sources are acknowledged.

Atomic weights are here quoted to all significant figures included in the most recent (2007) table of atomic weights. The dependable accuracy is more limited either by the combined uncertainties of the best published atomic-weight determinations or by the variability of isotopic composition in normal terrestrial occurrences (the latter applied to elements associated with **Footnote # 1** in the table). The last significant figure of each tabulated value is considered reliable to ±1 except when a larger single-digit uncertainty is inserted in parentheses following the atomic weight. Neither the highest nor the lowest actual atomic weight of any normal sample is thought likely to differ from the tabulated value by more than the assigned uncertainty. However, the tabulated values do not apply either to samples of highly exceptional isotopic composition arising from most unusual geological occurrences (for elements associated with **Footnotes # 1 and # 2**) or to those who isotopic composition has been artificially altered. Such might even be found in commerce without disclosure of that modification (for elements associated with **Footnotes # 3 and # 4**). Some elements have no stable isotope and are generally represented in this table by just one of the element's commonly known radioisotopes in brackets (**Footnote # 5**), with a corresponding relative atomic mass in the atomic-weight column. However, three elements in this grouping (Th, Pa, and U) do have a characteristic terrestrial isotopic composition, and for these an atomic weight is tabulated. For more detailed information, users should refer to the full IUPAC Table of Standard Atomic Weights, as is found in the biennial reports of the Commission on Atomic Weights and Isotopic Abundances. The most recent table (2007) was published in *Pure Appl. Chem.*, 81, 2131–2156 (2009), with only very slight changes to a few elements since the tabulation that was included in the first edition of this handbook. The new values are incorporated in the following table.

The group of elements with the highest atomic numbers in the periodic table that have been identified only recently in the last few decades have not been extensively studied or even named, Elements 113–118. Element 112, copernicium, has had its name officially, formally accepted in the most recent issue (March 2010) of *Pure and Applied Chemistry* (**Footnote # 6**).

---

* Michael E. Wieser and Michael Berglund, *Pure Appl. Chem.*, 81, 2131–2156 (2009), [MM09] in Literature Cited.

Table of Standard Atomic Weights—2007

	Name	Symbol	Atomic Weight	Footnotes
1	Hydrogen	H	1.00794(7)	1, 2, 3
2	Helium	He	4.002602(2)	1, 2
3	Lithium	Li	6.941(2)	1, 2, 3, 4
4	Beryllium	Be	9.012182(3)	
5	Boron	B	10.811(7)	1, 2, 3
6	Carbon	C	12.0107(8)	1, 2
7	Nitrogen	N	14.0067(2)	1, 2
8	Oxygen	O	15.9994(3)	1, 2
9	Fluorine	F	18.9984032(5)	
10	Neon	Ne	20.1797(6)	1, 3
11	Sodium (Natrium)	Na	22.98976928(2)	
12	Magnesium	Mg	24.3050(6)	
13	Aluminum	Al	26.9815386(8)	
14	Silicon	Si	28.0855(3)	2
15	Phosphorus	P	30.973762(2)	
16	Sulfur	S	32.065(5)	1, 2
17	Chlorine	Cl	35.453(2)	3
18	Argon	Ar	39.948(1)	1, 2
19	Potassium (Kalium)	K	39.0983(1)	1
20	Calcium	Ca	40.078(4)	1
21	Scandium	Sc	44.955912(6)	
22	Titanium	Ti	47.867(1)	
23	Vanadium	V	50.9415(1)	
24	Chromium	Cr	51.9961(6)	
25	Manganese	Mn	54.938045(5)	
26	Iron	Fe	55.845(2)	
27	Cobalt	Co	58.9331	
28	Nickel	Ni	58.69344	
29	Copper	Cu	63.546(3)	2
30	Zinc	Zn	65.38(2)	
31	Gallium	Ga	69.723(1)	
32	Germanium	Ge	72.64(1)	
33	Arsenic	As	74.92160(2)	
34	Selenium	Se	78.96(3)	
35	Bromine	Br	79.904(1)	
36	Krypton	Kr	83.798(2)	1, 3
37	Rubidium	Rb	85.4678((3)	1
38	Strontium	Sr	87.62(1)	1, 2
39	Yttrium	Y	88.90585(2)	
40	Zirconium	Zr	91.224(2)	1
41	Niobium	Nb	92.90638(2)	
42	Molybdenum	Mo	95.96(2)	1
43	Technetium	Tc	[98]	5
44	Ruthenium	Ru	101.07(2)	1
45	Rhodium	Rh	102.90550(2)	
46	Palladium	Pd	106.42(1)	1
47	Silver	Ag	107.8682(2)	1
48	Cadmium	Cd	112.411(8)	1
49	Indium	In	114.818(3)	
50	Tin	Sn	118.710(7)	1

Table of Standard Atomic Weights—2007 (continued)

	Name	Symbol	Atomic Weight	Footnotes
51	Antimony (Stibium)	Sb	121.760(1)	1
52	Tellurium	Te	127.60(3)	1
53	Iodine	I	126.90447(3)	
54	Xenon	Xe	131.29(3)	1, 3
55	Cesium	Cs	132.9094(1)	
56	Barium	Ba	137.327(7)	
57	Lanthanum	U	138.90547(7)	1
58	Cerium	Ce	140.116(1)	1
59	Praseodymium	Pr	140.90765(2)	
60	Neodymium	Nd	144.242(3)	1
61	Promethium	Pm	[145]	5
62	Samarium	Sm	150.36(2)	1
63	Europium	Eu	151.964(1)	1
64	Gadolinium	Gd	157.25(3)	1
65	Terbium	Tb	158.92535(2)	
66	Dysprosium	Dy	162.500(1)	1
67	Holmium	Ho	164.93032(2)	
68	Erbium	Er	167.259(3)	1
69	Thulium	Tm	168.93421(2)	
70	Ytterbium	Yb	173.054(5)	1
71	Lutetium	Lu	174.9668(1)	1
72	Hafnium	Hf	178.49(2)	
73	Tantalum	Ta	180.94788(2)	
74	Tungsten (Wolfram)	W	183.84(1)	
75	Rhenium	Re	186.207(1)	
76	Osmium	Os	190.23(3)	1
77	Iridium	Ir	192.217(3)	
78	Platinum	Pt	195.084(9)	
79	Gold	Au	196.966569(4)	
80	Mercury	Hg	200.59(2)	
81	Thallium	Tl	204.3833(2)	
82	Lead	Pb	207.2(1)	1, 2
83	Bismuth	Bi	208.98040(1)	
84	Polonium	Po	[209]	5
85	Astatine	At	[210]	5
86	Radon	Rn	[222]	5
87	Francium	Fr	[223]	5
88	Radium	Ra	[226]	5
89	Actinium	Ac	[227]	5
90	Thorium	Th	232.03806(2)	1, 5
91	Protactinium	Pa	231.03588(2)	5
92	Uranium	U	238.02891(3)	1, 3, 5
93	Neptunium	Np	[237]	5
94	Plutonium	Pu	[244]	5
95	Americium	Am	[243]	5
96	Curium	Cm	[247]	5
97	Berkelium	Bk	[247]	5
98	Californium	Cf	[251]	5
99	Einsteinium	Es	[252]	5

Table of Standard Atomic Weights—2007 (continued)

	Name	Symbol	Atomic Weight	Footnotes
100	Fermium	Fm	[257]	5
101	Mendelevium	Md	[258]	5
102	Nobelium	No	[259]	5
103	Lawrencium	Lr	[262]	5
104	Rutherfordium	Rf	[265]	5
105	Dubnium	Db	[268]	5
106	Seaborgium	Sg	[271]	5
107	Bohrium	Bh	[272]	5
108	Hassium	Hs	[270]	5
109	Meitnerium	Mt	[276]	5
110	Darmstadtium	Ds	[281]	5
111	Roentgenium	Rg	[280]	5
112	Copernicium	Cn	[285]	5, 6
113	Ununtrium	Uut	[284]	5, 6
114	Ununquadium	Uuq	[289]	5, 6
115	Ununpentium	Uup	[288]	5, 6
116	Ununhexium	Uuh	[293]	5, 6
118	Ununoctium	Uuo	[294]	5, 6

FOOTNOTES

1 Geological specimens are known in which the element has an isotopic composition outside the limits for normal material. The difference between the atomic weight of the element in such specimens and that given in the table may exceed the stated uncertainty.

2 Range in isotopic composition of normal terrestrial material prevents a more precise value being given; the tabulated value should be applicable to any normal material.

3 Modified isotopic compositions may be found in commercially available material because it has been subject to an undisclosed or inadvertent isotopic fractionation. Substantial deviations in atomic weight of the element from that given in the table can occur.

4 Commercially available lithium materials have atomic weights that range between 6.939 and 6.996; if a more accurate value is required, it must be determined for the specific material [range quoted for 1995 Table 6.94 and 6.99].

5 Element has no stable nuclides. The value enclosed in brackets, for example, [209], indicates the mass number of the longest-lived isotope of the element. However, three such elements (Th, Pa, and U) do have a characteristic terrestrial isotopic composition, and for these an atomic weight is tabulated.

6 The names and symbols for elements 113–118 are under review. The temporary system recommended by J. Chatt [*Pure Appl. Chem.*, *51*, 381–384 (1979)] is used above. The name for Element 112, copernicium, has been finally approved by the International Union of Pure and Applied Chemistry (IUPAC) [K. Tatsumi and J. Corish, *Pure Appl. Chem.*, *82*, 753–755 (2010)] http://www.iupac.org/news/archives/2010/Element_112_Press_Release.pdf

Selected Reference Books for Inorganic Chemistry

For in-depth coverage of subtopics in the field, one must go to individual monographs and specialty volumes. The following books have been selected as representative, standard treatises on the subject of inorganic chemistry.

Advanced Inorganic Chemistry (6th Edition), F. Albert Cotton, Carlos A. Murillo, and M. Bachmann, Wiley-Interscience, 1999. [ISBN-13: 978-0471199571]

Chemistry of the Elements (2nd Edition), A. Earnshaw and N. Greenwood, Butterworth-Heinemann, 1997. [ISBN-13: 978-0750633659]

Concepts and Models of Inorganic Chemistry (3rd Edition), B. E. Douglas, D. H. McDaniel, and J. J. Alexander, Wiley, 1994. [ISBN-13: 978-0471629788]

Concise Inorganic Chemistry (5th Edition), J. D. Lee, Wiley-Blackwell, 1999. [ISBN-13: 978-0632052936]

Descriptive Inorganic Chemistry (5th Edition), G. Rayner-Canham and T. Overton, W. H. Freeman, 2009. [ISBN-13: 978-1429218146]

Descriptive Inorganic, Coordination, and Solid State Chemistry (2nd Edition), G. E. Rodgers, Brooks Cole, 2002. [ISBN-13: 978-0125920605]

Inorganic Chemistry (4th Edition), D. Shriver and P. Atkins, Oxford University Press, 2006. [ISBN-13: 978-0716748786]

Inorganic Chemistry (4th Edition), G. L. Miessler and D. A. Tarry, Prentice Hall, 2010. [ISBN-13: 978-0136128663]

Inorganic Chemistry, G. Wulfsberg, University Science Books, 2000. [ISBN-13: 978-1891389016]

Inorganic Chemistry, J. E. House, Academic Press, 2008. [ISBN-13: 978-0123567864]

Inorganic Chemistry: Principles of Structure and Reactivity (4th Edition), J. E. Huheey, E. A. Keiter and R. L. Keiter, Prentice Hall, 1997. [ISBN-13: 978-0060429959]

Inorganic Structural Chemistry (Inorganic Chemistry: A Textbook Series) (2nd Edition), U. Muller, Wiley, 2006. [ISBN-13: 978-0470018651]

Introduction to Modern Inorganic Chemistry (6th Edition), R. A. Mackay and W. Henderson, CRC Press, 2002. [ISBN-13: 978-0748764204]

Contents

1
Compound: Acetylferrocene
Formula: $CH_3COC_5H_4FeC_5H_5$
Molecular Formula: $C_{12}H_{12}FeO$
Molecular Weight: 228.074-
CAS RN: 1271-55-2
Properties: orange cryst [STR93]
Melting Point, °C: 83 [STR93]

2
Compound: Actinium
Formula: Ac
Molecular Formula: Ac
Molecular Weight: 227
CAS RN: 7440-34-8
Properties: silvery white metal; fcc, a = 0.5311 nm; $t_{1/2}$ of ^{227}Ac is 21.8 years; a decay product of ^{235}U; stable, colorless solution for Ac^{+++}; ionic radius Ac^{+++}, 0.1119 nm; enthalpy of vaporization 293 kJ/mol; chemistry closely follows that of lanthanum; first discovered in 1899 by Diebierne; preparation: by transmutation of radium: $^{226}Ra + n \rightarrow ^{227}Ra + \gamma$ $^{227}Ra \rightarrow ^{227}Ac$ [KIR78] [KAT86] [HAW93]
Density, g/cm³: 10.07 (25°C) [KIR91]
Melting Point, °C: 1100 [KIR91]
Boiling Point, °C: ~3300 [MER06]

3
Compound: Actinium bromide
Formula: $AcBr_3$
Molecular Formula: $AcBr_3$
Molecular Weight: 467
CAS RN: 33689-81-5
Properties: white; hex, a = 0.806 nm, c = 0.468 nm; preparation: by reacting Ac_2O_3 with $AlBr_3$ at 750°C [CIC73] [KAT86]
Solubility: s H_2O [CRC10]
Density, g/cm³: 5.85 [KAT86]
Melting Point, °C: sublimes at 800 [CRC10]

4
Compound: Actinium chloride
Formula: $AcCl_3$
Molecular Formula: $AcCl_3$
Molecular Weight: 333
CAS RN: 22986-54-5
Properties: white cryst; hex, a = 0.762 nm, c = 0.455 nm; preparation: by reacting $Ac(OH)_3$ with CCl_4 at 500°C [KAT86] [KIR78]
Density, g/cm³: 4.81 [KIR78]
Melting Point, °C: sublimes at 900 [CRC10]

5
Compound: Actinium fluoride
Formula: AcF_3
Molecular Formula: AcF_3
Molecular Weight: 284
CAS RN: 33689-80-4
Properties: white cryst; hex, a = 0.741 nm, c = 0.755 nm; preparation: from the reaction $Ac^{+++} + 3F^- = AcF_3$ at 25°C [KIR78] [KAT86]
Solubility: i H_2O [CRC10]
Density, g/cm³: 7.88 [KIR78]

6
Compound: Actinium hydride
Formula: AcH_2
Molecular Formula: AcH_2
Molecular Weight: 229
CAS RN: 60936-81-4
Properties: black; cub, fluorite structure, a = 0.5670 nm [CIC73]
Density, g/cm³: 8.35 [CIC73]

7
Compound: Actinium hydroxide
Formula: $Ac(OH)_3$
Molecular Formula: AcH_3O_3
Molecular Weight: 278

CAS RN: 12249-30-8
Properties: white [CRC10]
Solubility: i H_2O [CRC10]

8

Compound: Actinium iodide
Formula: AcI_3
Molecular Formula: AcI_3
Molecular Weight: 608
CAS RN: 33689-82-6
Properties: white [CRC10]
Solubility: s H_2O [CRC10]
Melting Point, °C: sublimes at 700–800 [CRC10]

9

Compound: Actinium oxalate decahydrate
Formula: $Ac_2(C_2O_4)_3 \cdot 10H_2O$
Molecular Formula: $C_6H_{20}Ac_2O_{22}$
Molecular Weight: 898
CAS RN: 12002-61-8
Properties: monocl, a=1.126 nm, b=0.997 nm, c=1.065 nm; obtained by adding soluble oxalate to soluble Ac^{+++} solution at 25°C [KAT86]
Solubility: i H_2O [CRC10]
Density, g/cm³: 2.68 [KAT86]

10

Compound: Actinium oxide
Formula: Ac_2O_3
Molecular Formula: Ac_2O_3
Molecular Weight: 502
CAS RN: 12002-61-8
Properties: white cryst; hex, a=0.407 nm, c=0.629 nm; preparation: from decomposition of actinium oxalate at 1100°C [KIR78] [KAT86]
Solubility: i H_2O [CRC10]
Density, g/cm³: 9.19 [KIR78]

11

Compound: Actinium oxybromide
Formula: AcOBr
Molecular Formula: AcBrO
Molecular Weight: 323
CAS RN: 49848-33-1
Properties: tetr, a=0.427 nm, c=0.740 nm; preparation: by reacting $AcBr_3$ with NH_3 and H_2O at 1000°C [KAT86]
Density, g/cm³: 7.89 [KAT86]

12

Compound: Actinium oxychloride
Formula: AcOCl

Molecular Formula: AcClO
Molecular Weight: 278
CAS RN: 49848-29-5
Properties: white, tetr, a=0.424 nm, c=0.708 nm; preparation: by reacting $AcCl_3$ with H_2O at 1000°C [KAT86] [CIC73]
Density, g/cm³: 7.23 [KAT86]

13

Compound: Actinium oxyfluoride
Formula: AcOF
Molecular Formula: AcFO
Molecular Weight: 262
CAS RN: 49848-24-0
Properties: white; cub, a=0.5931 nm; preparation: by reaction of AcF_3 with NH_3 and H_2O at 900°C–1000°C [KAT86]
Density, g/cm³: 8.28 [KAT86]

14

Compound: Actinium phosphate hemihydrate
Formula: $AcPO_4 \cdot 1/2H_2O$
Molecular Formula: $AcHO_{4.5}P$
Molecular Weight: 331
CAS RN: 7778-39-4
Properties: hex, a=0.721 nm, c=0.664 nm; preparation: by precipitation of a soluble Ac^{+++} salt with a solution of PO_4^- [KAT86]
Density, g/cm³: 5.48 [KAT86]

15

Compound: Actinium sulfide
Formula: Ac_2S_3
Molecular Formula: Ac_2S_3
Molecular Weight: 550
CAS RN: 50647-18-2
Properties: bcc, a=0.897 nm; preparation: by reaction between Ac_2O_3 and H_2S at 1400°C [KAT86] [CIC73]
Density, g/cm³: 6.75 [KAT86]

16

Compound: Aluminum
Formula: Al
Molecular Formula: Al
Molecular Weight: 26.981539
CAS RN: 7429-90-5
Properties: silvery white, ductile, metal; fcc, a=0.40496 nm; forms corrosion-resistant oxide film of ~5 nm thickness in moist air; enthalpy of fusion 10.71 kJ/mol; enthalpy of sublimation 314.0 kJ/mol; enthalpy of vaporization 294 kJ/mol; electrical resistivity 2.6548 µohm·cm; tensile strength 6800 psi; hardness 2.9 Mohs; ionic radius of Al^{+++} 0.050 nm;

used in mirrors, beverage cans, buildings and construction [CIC73] [KIR78] [HAW93] [CER91]

Solubility: i H_2O, conc HNO_3 [HAW93]

Density, g/cm³: 2.70 [MER06]

Melting Point, °C: 660.37 [ALD94]

Boiling Point, °C: 2517.66 [JAN85]

Reactions: reacts with dil HCl, H_2SO_4, KOH, NaOH to evolve hydrogen [MER06]

Thermal Conductivity, W/(m·K): 236 (0°C), 237 (25°C), 240 (100°C) [HO72]

Thermal Expansion Coefficient: linear coefficient (30°C–300°C) 24.9×10^{-6}/°C [CIC73]

17

Compound: Aluminum acetate

Synonym: aluminum triacetate

Formula: $Al(CH_3COO)_3$

Molecular Formula: $C_6H_9AlO_6$

Molecular Weight: 204.115

CAS RN: 139-12-8

Properties: white powd; preparation: by heating aluminum or $AlCl_3$ with an acetic acid solution containing acetic anhydride; uses: as an antiseptic, an astringent, and in antiperspirant applications; there is a hydroxyaluminum diacetate, CAS RN 142-03-0 [CIC73] [HAW93] [ALD94]

Solubility: s H_2O [HAW93]

Melting Point, °C: decomposes with moisture [CRC10]

Reactions: minus acetic anhydride at 120°C–140°C, forming basic acetates [CIC73]

18

Compound: Aluminum acetylacetonate

Synonyms: 2,4-pentanedione, aluminum(III) derivative

Formula: $Al[CH_3COCH=C(O)CH_3]_3$

Molecular Formula: $C_{15}H_{21}AlO_6$

Molecular Weight: 324.310

CAS RN: 13963-57-0

Properties: white powd or monocl cryst; preparation: by reacting $AlCl_3$ and acetylacetone; uses: to vapor deposit aluminum and as a catalyst [HAW93] [CIC73] [STR93]

Solubility: i H_2O; s benzene, alcohol [CIC73] [HAW93]

Density, g/cm³: 1.27 [CRC10]

Melting Point, °C: 189 [HAW93]

Boiling Point, °C: 315 [HAW93]

Reactions: decomposes at 320°C; sublimes at 150°C (1 mm Hg) [STR93]

19

Compound: Aluminum ammonium sulfate dodecahydrate

Formula: $AlNH_4(SO_4)_2 \cdot 12H_2O$

Molecular Formula: $AlH_{28}NO_{20}S_2$

Molecular Weight: 453.329

CAS RN: 7784-26-1

Properties: colorless cryst or powd [CRC10]

Solubility: s H_2O; i EtOH [CRC10]

Density: 1.65 [CRC10]

Melting Point, °C: 94.5 [CRC10]

Boiling Point, °C: decomposes at 280 [CRC10]

20

Compound: Aluminum ammonium sulfate

Formula: $AlNH_4(SO_4)_2$

Molecular Formula: $AlH_4NO_8S_2$

Molecular Weight: 237.146

CAS RN: 7784-25-0

Properties: white powd [CRC10]

Solubility: sl s H_2O; i EtOH [CRC10]

21

Compound: Aluminum antimonide

Formula: AlSb

Molecular Formula: AlSb

Molecular Weight: 148.739

CAS RN: 25152-52-7

Properties: electronic dielectric constant 10.2; enthalpy of fusion 58.6 kJ/mol; cryst, lattice constant 0.61361 nm; band gap 1.68 eV at 0 K and 1.58 eV at 300 K; mobility (300 K) 200 cm²/(V·s) electrons, 420 cm²/(V·s) holes; effective mass 0.12 electrons and 0.98 holes; preparation: by fusion of Al and Sb, followed by purification using zone melting; uses: semiconductor research [CIC73] [MER06] [KIR82]

Density, g/cm³: 4.15 [CIC73]

Melting Point, °C: 1050 [MER06]

Thermal Conductivity, W/(m·K): 60 (25°C) [CRC10]

Thermal Expansion Coefficient: 4.2×10^{-6}/K [CRC10]

22

Compound: Aluminum arsenide

Formula: AlAs

Molecular Formula: AlAs

Molecular Weight: 101.903

CAS RN: 22831-42-1

Properties: −20 mesh powd with 99.5% purity; semiconductor; band gap, 2.13 eV (22°C); electronic dielectric constant, 10.3; lattice constant a = 0.5662 nm; enthalpy of fusion 24.5 kJ/mol; uses: in rectifiers, transistors, and thermistors [CIC73] [ALF93] [HAW93]

Density, g/cm³: 3.81 [CIC73]

Melting Point, °C: 1740 [CIC73]

23
Compound: Aluminum borate
Synonyms: eremeyevite, jeremejevite
Formula: $2Al_2O_3 \cdot B_2O_3$
Molecular Formula: $Al_4B_2O_9$
Molecular Weight: 273.543
CAS RN: 11121-16-7
Properties: white, granular powd or cryst needles; prepared by heating Al_2O_3 with B_2O_3; used in glass and ceramics [HAW93] [MER06]
Solubility: i H_2O with decomposition [HAW93] [MER06]
Melting Point, °C: ~1050 [MER06] [CIC73]
Reactions: forms $2Al_2O_3 \cdot B_2O_3$ at 1000°C, and $9Al_2O_3 \cdot 2B_2O_3$ at 1100° [MER06]

24
Compound: Aluminum borohydride
Synonym: aluminum tetrahydroborate
Formula: $Al(BH_4)_3$
Molecular Formula: AlB_3H_{12}
Molecular Weight: 71.510
CAS RN: 16962-07-5
Properties: volatile; liq; enthalpy of vaporization 30 kJ/mol; ignites spontaneously in air; can be formed by reacting sodium borohydride and aluminum chloride in the presence of small quantity of tributyl phosphate; used as a reducing agent and as a fuel for jet engines and rockets [HAW93] [MER06] [CRC10]
Solubility: reacts vigorously with H_2O and HCl evolving H_2 [MER06]
Melting Point, °C: −64.5 [MER06]
Boiling Point, °C: 44.5 [MER06]

25
Compound: Aluminum bromate nonahydrate
Formula: $Al(BrO_3)_3 \cdot 9H_2O$
Molecular Formula: $AlBr_3H_{18}O_{18}$
Molecular Weight: 572.826
CAS RN: 11126-81-1
Properties: white cryst; hygr; can be obtained from mixing aq solutions of $Al_2(SO_4)_3$ and $Ba(BrO_3)_2$, followed by crystallization [CIC73] [CRC10]
Solubility: s H_2O; sl s acids [CRC10]
Melting Point, °C: 62 [CIC73]
Boiling Point, °C: decomposes at >100 [CIC73]

26
Compound: Aluminum bromide
Formula: $AlBr_3$
Molecular Formula: $AlBr_3$

Molecular Weight: 266.694
CAS RN: 7727-15-3
Properties: white to yellow-red; trig cryst or powd; very hygr; fumes strongly in air; enthalpy of sublimation 35.9 kJ/mol; enthalpy of fusion 11.25 kJ/mol; enthalpy of vaporization 23.5 kJ/mol; can be prepared by heating Al and Br_2; used as an acid catalyst for organic syntheses, similar to $AlCl_3$, but is more reactive and more soluble in organic solvents [CIC73] [MER06] [KIR78] [CRC10]
Solubility: reacts with H_2O violently [KIR78]
Density, g/cm³: 3.01 [KIR78]; 2.64, 100°C (liq) [STR93]
Melting Point, °C: 97.45 [KIR78]
Boiling Point, °C: sublimes at 256 [CIC73]

27
Compound: Aluminum bromide hexahydrate
Formula: $AlBr_3 \cdot 6H_2O$
Molecular Formula: $AlBr_3H_{12}O_6$
Molecular Weight: 374.785
CAS RN: 7784-11-4
Properties: colorless to sl yellow; deliq; may be prepared by dissolution of Al or aluminum hydroxide in HBr, followed by precipitation; used as an acid catalyst [MER06] [KIR78]
Solubility: s H_2O, alcohol [MER06]
Density, g/cm³: 2.54 [HAW93]
Melting Point, °C: 93 [MER06]
Boiling Point, °C: decomposes at 135 [CRC10]

28
Compound: Aluminum carbide
Formula: Al_4C_3
Molecular Formula: C_3Al_4
Molecular Weight: 143.959
CAS RN: 1299-86-1
Properties: yellow hex cryst or olive green powd; can be prepared by reacting stoichiometric amounts of Al and C in the absence of both oxygen and nitrogen at ~1000°C; used to generate methane and to manufacture AlN [MER06] [ALF93] [CIC73]
Solubility: decomposes with the evolution of CH_4 in H_2O [MER06]
Density, g/cm³: 2.36 [MER06]
Melting Point, °C: 2100 [MER06]
Boiling Point, °C: decomposes at >2200 [MER06]

29
Compound: Aluminum chlorate
Formula: $Al(ClO_3)_3$
Molecular Formula: $AlCl_3O_9$
Molecular Weight: 277.332

CAS RN: 15477-33-5
Properties: colorless cryst; deliq; occurs as hexahydrate and nonahydrate; evaporation of an aq solution yields the nonahydrate; used as a disinfectant and to prevent yellowing of acrylic fibers [CIC73] [MER06] [HAW93]
Solubility: s H_2O, alcohol [HAW93]
Melting Point, °C: decomposes [CRC10]

30
Compound: Aluminum chlorate nonahydrate
Synonym: mallebrin
Formula: $Al(ClO_3)_3 \cdot 9H_2O$
Molecular Formula: $AlCl_3H_{18}O_{18}$
Molecular Weight: 439.472
CAS RN: 15477-33-5
Properties: deliq cryst; obtained by mixing $Ba(ClO_4)_2$ and $Al_2(SO_4)_3$ solution, followed by evaporation [CIC73] [MER06]
Solubility: s H_2O, alcohol [MER06]

31
Compound: Aluminum chloride
Formula: $AlCl_3$
Molecular Formula: $AlCl_3$
Molecular Weight: 133.340
CAS RN: 7446-70-0
Properties: white or light yellow; cryst or powd; hygr; hex; odor of HCl; exists as dimer at <327°C; triple point 192.5°C (233 kPa); enthalpy of sublimation of dimer Al_2Cl_6 (25°C) 115.73 kJ/mol; enthalpy of solution at 20°C is −325.1 kJ/mol; enthalpy of fusion 35.40 kJ/mol; can be made by reacting HCl and Al at ~150°C; used as an acid catalyst, in cracking petroleum, and in the manufacture of rubbers and lubricants [CIC73] [KIR78] [ALF93]
Solubility: dissolves violently in H_2O evolving HCl [MER06]; g/100 g soln H_2O: 30.84 ± 0.25 (0°C), 31.10 (25°C), 33.23 (98°C), equilibrium solid phase, $AlCl_3 \cdot 6H_2O$ [KRU93]; s HCl, ether, ethanol [KIR78]
Density, g/cm³: 2.44 [KIR78]
Boiling Point, °C: sublimes at 181.2 [KIR78]
Reactions: Al_2Cl_6 (dimer) → $2AlCl_3$ >327°C [KIR78]

32
Compound: Aluminum chloride hexahydrate
Formula: $AlCl_3 \cdot 6H_2O$
Molecular Formula: $AlCl_3H_{12}O_6$
Molecular Weight: 241.431
CAS RN: 7784-13-6

Properties: cryst powd; white or yellow; deliq; preparation: by dissolution of $Al(OH)_3$ in conc HCl, followed by cooling to 0°C and addition of gaseous HCl; uses: to preserve wood, disinfect stables, and in deodorants and antiperspirants [MER06] [HAW93] [KIR78]
Solubility: 1 g/0.9 mL H_2O; s alcohol [MER06]; gAl_2O_3/100 mL at 35°C in each of the following solvents: 15.91 H_2O, 9.43 methanol, 4.77 ethanol, 6.04 ethylene glycol [OKA91]
Density, g/cm³: 2.398 [STR93]
Melting Point, °C: decomposes at 100 [ALF93]

33
Compound: Aluminum chromate
Synonym: aluminum oxide-chromium oxide
Formula: $Al_2O_3 \cdot Cr_2O_3$
Molecular Formula: $Al_2Cr_2O_6$
Molecular Weight: 253.952
CAS RN: 57921-51-4
Properties: yellow amorphous solid; fused 98 wt% Al_2O_3, 2 wt% Cr_2O_3; −200, +325 mesh and −325 mesh, +10 μm of 99% purity; used in ceramics [KIR78] [CER91]

34
Compound: Aluminum citrate
Formula: $AlC_6H_5O_7$
Molecular Formula: $C_6H_5AlO_7$
Molecular Weight: 216.084
CAS RN: 31142-56-0
Properties: white powd or scales; study of complex formation in [LOP84] [MER06]
Solubility: dissolves slowly in cold H_2O, s hot H_2O, ammonia [MER06]

35
Compound: Aluminum diacetate
Synonym: aluminum subacetate
Formula: $Al(CH_3COO)_2(OH)$
Molecular Formula: $C_4H_7AlO_5$
Molecular Weight: 162.079
CAS RN: 142-03-0
Properties: white, amorphous powd or curdy precipitate; can be produced by reacting sodium aluminate solution with acetic acid; used as a mordant in dyeing, to manufacture color lakes, to waterproof and fireproof fabrics, in antiperspirant formulations, and as a disinfectant by embalmers [MER06] [CIC73]
Solubility: i H_2O when dried at 100°C [MER06]

36
Compound: Aluminum diboride
Formula: AlB_2
Molecular Formula: AlB_2
Molecular Weight: 48.604
CAS RN: 12041-50-8
Properties: powd; made by reaction of the elements above 600°C; high neutron absorption; used as a nuclear shielding material [HAW93] [CIC73]
Solubility: s dil HCl [CIC73]
Density, g/cm³: 3.19 [ALF93]
Melting Point, °C: decomposes to AlB_{12} at >920 [CIC73]

37
Compound: Aluminum distearate
Formula: $Al(OH)[CH_3(CH_2)_{16}COO]_2$
Molecular Formula: $C_{36}H_{71}AlO_5$
Molecular Weight: 610.939
CAS RN: 637-12-7
Properties: white powd; preparation in [KIR78]; used as a thickener for paints, inks, and greases, and as a water repellent and lubricant [HAW93]
Solubility: i H_2O, alcohol, ether; forms gel with aliphatic and aromatic hydrocarbons [HAW93]
Density, g/cm³: 1.009 [HAW93]
Melting Point, °C: 145 [HAW93]

38
Compound: Aluminum dodecaboride
Formula: AlB_{12}
Molecular Formula: AlB_{12}
Molecular Weight: 156.714
CAS RN: 12041-54-2
Properties: 3–8 μm powd; high neutron absorption [ALF93] [HAW93]
Solubility: s hot HNO_3; i acid, alkalies [CRC10] [CIC73]
Density, g/cm³: 2.55 [CRC10]
Melting Point, °C: decomposes to boron and carbon at 1900 [CIC73]

39
Compound: Aluminum ethoxide
Synonym: aluminum ethylate
Formula: $Al(C_2H_5O)_3$
Molecular Formula: $C_6H_{15}AlO_3$
Molecular Weight: 162.165
CAS RN: 555-75-9

Properties: liq that slowly solidifies to a white powd; sensitive to moisture; prepared from a reaction of Al with ethanol in the presence of catalytic amounts of I_2 and $HgCl_2$; used as a polymerization catalyst and to reduce aldehydes and ketones [MER06] [STR93] [HAW93]
Solubility: sl s in high boiling organic solvents [HAW93]
Density, g/cm³: 1.142 (20°C) [CRC10]
Melting Point, °C: 130 [STR93]
Boiling Point, °C: 210 (10 mm Hg) [STR93]

40
Compound: Aluminum fluoride
Synonym: aluminum trifluoride
Formula: AlF_3
Molecular Formula: AlF_3
Molecular Weight: 83.977
CAS RN: 7784-18-1
Properties: hex white powd or 99.5% pure highly dense 3–6 mm sintered pieces; enthalpy of fusion 98.0 kJ/mol; dielectric constant 6; formed by heating $(NH_4)_3AlF_6$ in nitrogen; used in electrolyte for production of Al, in ceramics as a flux in metallurgy and to inhibit fermentation, and as an evaporation material and sputtering target for preparation of low index films [CIC73] [KIR78] [STR93] [MER06] [CER91] [CRC10]
Solubility: g/100 g soln H_2O: 0.25 (0°C), 0.50 (25°C), 1.64 (100°C), equilibrium solid phase $AlF_3 \cdot 3H_2O$ [KRU93]
Density, g/cm³: 3.10 [KIR78]
Melting Point, °C: 1290 [COT88]
Reactions: transition from α to β at 455°C; dissociates at 776°C [ROB78]

41
Compound: Aluminum fluoride monohydrate
Synonym: fluellite
Formula: $AlF_3 \cdot H_2O$
Molecular Formula: AlF_3H_2O
Molecular Weight: 101.992
CAS RN: 32287-65-3
Properties: ortho-rhomb [MER06]
Solubility: sl s H_2O [CRC10]
Density, g/cm³: 2.17 [MER06]

42
Compound: Aluminum fluoride trihydrate
Formula: $AlF_3 \cdot 3H_2O$
Molecular Formula: $AlF_3H_6O_3$
Molecular Weight: 138.023

CAS RN: 15098-87-0
Properties: white hygr; cryst powd [HAW93] [STR93]
Solubility: sl s H_2O [HAW93]
Density, g/cm³: 1.914 [STR93]
Reactions: minus H_2O at both 100°C
  and 200°C [MER06]

43
Compound: Aluminum hexafluorosilicate nonahydrate
Synonym: aluminum silicofluoride
Formula: $Al_2(SiF_6)_3 \cdot 9H_2O$
Molecular Formula: $Al_2F_{18}H_{18}O_9Si_3$
Molecular Weight: 642.329
CAS RN: 17099-70-6
Properties: hex prisms; occurs naturally
  as topaz; used to protect and preserve
  construction materials and in the
  manufacture of glass [MER06] [HAW93]
Solubility: s H_2O, decomposes in hot H_2O [MER06]
Melting Point, °C: decomposes at ~1000 [MER06]
Reactions: minus H_2O at <500°C [MER06]

44
Compound: Aluminum hydride
Formula: AlH_3
Molecular Formula: AlH_3
Molecular Weight: 30.005
CAS RN: 7784-21-6
Properties: colorless, nonvolatile solid; can be
  obtained by reacting an ether solution of
  $AlCl_3$ with LiH; used as a catalyst for organic
  polymerization processes [MER06]
Solubility: evolves H_2 in H_2O [HAW93]
Melting Point, °C: decomposes at 160 [HAW93]

45
Compound: Aluminum hydroxide
Formula: $Al(OH)_3$
Molecular Formula: AlH_3O_3
Molecular Weight: 78.004
CAS RN: 21645-51-2
Properties: white, bulky, amorphous powd; forms
  gels if in prolonged contact with H_2O; absorbs
  CO_2; many uses such as an absorbent and
  emulsifier, in ion-exchange chromatography,
  as a mordant in dyeing [MER06] [ALF93]
Solubility: i H_2O; s acids, alkalies [MER06]
Density, g/cm³: 2.42 [HAW93]
Reactions: minus H_2O at 300°C [CRC10]; forms
  gel on contact with H_2O [MER06]

46
Compound: Aluminum hydroxide(β′)
Synonym: nordstrandite
Formula: $β'-Al(OH)_3$
Molecular Formula: AlH_3O_3
Molecular Weight: 78.004
CAS RN: 12752-71-0
Properties: tric, a=0.875 nm, b=0.507 nm,
  c=1.024 nm [KIR78]

47
Compound: Aluminum hydroxide(α)
Synonym: gibbsite
Formula: $α-Al(OH)_3$
Molecular Formula: AlH_3O_3
Molecular Weight: 78.004
CAS RN: 14762-49-3
Properties: white, pearly vitreous; hardness
  2.5–3.5 Mohs; monocl: a=0.868 nm,
  b=0.507 nm, c=0.972 nm; tricl: a=1.733 nm,
  b=1.008 nm, c=0.973 nm [KIR78]
Density, g/cm³: monocl: 2.441; tricl:
  2.42 [KIR78] [ROB78]
Reactions: transition to boehmite at 103°C [ROB78]

48
Compound: Aluminum hydroxide(β)
Synonym: bayerite
Formula: $β-Al(OH)_3$
Molecular Formula: AlH_3O_3
Molecular Weight: 78.004
CAS RN: 20257-20-9
Properties: monocl, a=0.506 nm,
  b=0.867 nm, c=0.471 nm; used to
  make η-alumina catalyst [KIR78]
Density, g/cm³: 2.53 [KIR78]

49
Compound: Aluminum hydroxychloride
Synonym: aluminum chlorohydroxide
Formula: $Al_2(OH)_5Cl \cdot 2H_2O$
Molecular Formula: $Al_2ClH_9O_7$
Molecular Weight: 210.483
CAS RN: 1327-41-9
Properties: glassy solid; prepared by electrolysis
  of Al solutions; used as an antiperspirant
  and in medicine [MER06]
Solubility: s H_2O, forms sl turbid
  colloidal solution [MER06]

50

Compound: Aluminum hydroxystearate
Formula: $Al(OH)[OOC(CH_2)_{10}CH_2O(CH_2)_5CH_3]_2$
Molecular Formula: $C_{36}H_{71}AlO_7$
Molecular Weight: 642.938
CAS RN: 637-12-7
Properties: white powd; preparation in [KIR78]; used to weatherproof leather and cement, to lubricate plastics and rope, and in paints and inks [HAW93]
Density, g/cm³: 1.045 [HAW93]
Melting Point, °C: 155 [HAW93]

51

Compound: Aluminum hypophosphite
Formula: $Al(H_2PO_2)_3$
Molecular Formula: $AlH_6O_6P_3$
Molecular Weight: 221.948
CAS RN: 7784-22-7
Properties: cryst powd; can be made by precipitation with slow heating from a solution of an aluminum salt with 50% hypophosphorus acid at 80°C–90°C; used in finishes for acrylonitrile polymer fibers [MER06] [CIC73]
Solubility: i H_2O; s warm NaOH; decomposes in H_2SO_4 and HCl [MER06]
Reactions: decomposes at ~220°C evolving phosphine [MER06]

52

Compound: Aluminum iodide
Formula: AlI_3
Molecular Formula: AlI_3
Molecular Weight: 407.695
CAS RN: 7784-23-8
Properties: solid; white leaflets, if pure; yellowish to brownish black lumps; fumes in moist air; strong exothermic reaction with H_2O; enthalpy of sublimation 112.1 kJ/mol; enthalpy of fusion 15.90 kJ/mol; enthalpy of vaporization 32.2 kJ/mol; can be prepared by heating Al and I_2 in a sealed tube; used as a catalyst for organic reactions [CIC73] [MER06] [HAW93] [CRC10]
Solubility: reacts violently with H_2O [KIR78]
Density, g/cm³: 3.98 [KIR78]
Melting Point, °C: 191 [KIR78]
Boiling Point, °C: sublimes at 381–382 [CIC73]

53

Compound: Aluminum iodide hexahydrate
Formula: $AlI_3 \cdot 6H_2O$

Molecular Formula: $AlH_{12}I_3O_6$
Molecular Weight: 515.786
CAS RN: 10090-53-6
Properties: yellowish cryst powd; deliq; can be obtained from a reaction between Al or $Al(OH)_3$ and HI; there is a $AlI_3 \cdot 15H_2O$ [MER06] [KIR78]
Solubility: s H_2O, alcohol, ether [MER06]
Density, g/cm³: 2.63 [CRC10]
Melting Point, °C: decomposes at 185 [CRC10]

54

Compound: Aluminum isopropoxide
Synonym: aluminum isopropylate
Formula: $Al[OCH(CH_3)_2]_3$
Molecular Formula: $C_9H_{21}AlO_3$
Molecular Weight: 204.245
CAS RN: 555-31-7
Properties: hygr white solid; prepared from a reaction between aluminum and isopropyl alcohol with $HgCl_2$ catalyst; used to prepare aluminum soaps, as a waterproofing finish for textiles, and to synthesize aluminum titanate [MER06] [YAM89]
Solubility: decomposed by H_2O; s ethanol, isopropanol, benzene, toluene, chloroform, carbon tetrachloride [MER06]
Density, g/cm³: 1.0346 (20°C) [CRC10]
Melting Point, °C: 119 [MER06]
Boiling Point, °C: 141 [CRC10]

55

Compound: Aluminum lactate
Synonym: aluctyl
Formula: $Al(C_3H_5O_3)_3$
Molecular Formula: $C_9H_{15}AlO_9$
Molecular Weight: 294.194
CAS RN: 18917-91-4
Properties: powd; preparation: lactic acid and $AlCl_3$; used in foam fire extinguishers and in dental impression materials [MER06] [ALF95]
Solubility: v s H_2O [MER06]

56

Compound: Aluminum metaphosphate
Formula: $Al(PO_3)_3$
Molecular Formula: AlO_9P_3
Molecular Weight: 263.898
CAS RN: 32823-06-6
Properties: colorless powd; tetr; used as a component of glazes, enamels, and glasses, and in high-temp insulating cement [CRC10] [HAW93]
Solubility: i H_2O [HAW93]

Density, g/cm³: 2.780 [ALD94]
Melting Point, °C: ~1527 [HAW93]

57
Compound: Aluminum molybdate
Formula: $Al_2(MoO_4)_3$
Molecular Formula: $Al_2Mo_3O_{12}$
Molecular Weight: 533.776
CAS RN: 15123-80-5
Properties: −325 mesh powd with 99% purity [ALF93]

58
Compound: Aluminum monopalmitate
Formula: $Al(OH)_2C_{16}H_{31}O_2$
Molecular Formula: $C_{16}H_{33}AlO_4$
Molecular Weight: 316.418
CAS RN: 555-35-1
Properties: white powd; made by heating aluminum hydroxide with palmitic acid and H_2O, followed by filtration and drying; used to waterproof leather, paper, and textiles, and to thicken lubricating oils, also used in varnishes and as a food additive [HAW93]
Solubility: i H_2O, alcohol; gels in hydrocarbons [HAW93]
Density, g/cm³: 1.072 [HAW93]
Melting Point, °C: 200 [HAW93]

59
Compound: Aluminum monostearate
Formula: $Al(OH)_2[CH_3(CH_2)_{16}COO]$
Molecular Formula: $C_{18}H_{37}AlO_4$
Molecular Weight: 344.472
CAS RN: 7047-84-9
Properties: faint odor; white to yellowish white powd; prepared by mixing solutions of sodium stearate and a soluble aluminum salt; used in paints, inks, greases, waxes, to thicken lubricating oils, for waterproofing [HAW93]
Solubility: i H_2O; forms gel with aliphatic and aromatic hydrocarbons [HAW93]
Density, g/cm³: 1.020 [HAW93]
Melting Point, °C: 155 [PFA93]

60
Compound: Aluminum nitrate nonahydrate
Formula: $Al(NO_3)_3 \cdot 9H_2O$
Molecular Formula: $AlH_{18}N_3O_{18}$
Molecular Weight: 375.134
CAS RN: 7784-27-2

Properties: white, hygr cryst; monocl, a = 1.086 nm, b = 0.959 nm, c = 1.383 nm; prepared by adding lead nitrate solution to aluminum sulfate solution; used in leather tanning, as a corrosion inhibitor and as an antiperspirant [MER06] [CIC73] [STR93]
Solubility: 67.3 g/100 mL H_2O (25°C) [CIC73]; gAl_2O_3/100 mL, 35°C, in the following solvents: methanol 14.45, ethanol 8.63, ethylene glycol 18.32 [OKA91]
Density, g/cm³: 1.72 [CIC73]
Melting Point, °C: 73 [MER06]
Boiling Point, °C: decomposes at 135 [MER06]
Reactions: decomposes to oxides of Al and N_2 at 500°C [KIR78]

61
Compound: Aluminum nitride
Formula: AlN
Molecular Formula: AlN
Molecular Weight: 40.989
CAS RN: 24304-00-5
Properties: powd or bluish white cryst; hex or ortho-rhomb; hex, a = 0.311 nm, c = 0.4975 nm; has odor of NH_3 in moist air; hardness 9–10 Mohs; band gap 4.26 eV; manufactured by heating bauxite in flowing nitrogen at 1500°C for 8 h; semiconductor material, used in steel manufacturing, as a crucible to grow cryst of gallium arsenide, and as a 99.8% sputtering target to prepare diodes and integrated circuits [MER06] [CIC73] [ALF93] [MIT87] [CER91]
Solubility: decomposes in H_2O to $Al(OH)_3 + NH_3$; decomposes in acids and alkalies [MER06] [CRC10]
Density, g/cm³: 3.05 [MER06]
Melting Point, °C: 2150–2200 [MER06]
Reactions: sublimes at 2000°C; decomposes to Al gas and N_2 from 1340°C to 1654°C [CRC10] [JAN85]
Thermal Conductivity, W/(m·K): 30 [KIR81]
Thermal Expansion Coefficient: coefficient is 4.03×10^{-6}/°C [KIR81]

62
Compound: Aluminum oleate
Synonyms: 9-octadecanoic acid, aluminum(III) salt
Formula: $Al[CH_3(CH_2)_7CH=CH(CH_2)_7COO]_3$
Molecular Formula: $C_{54}H_{99}AlO_6$
Molecular Weight: 871.358
CAS RN: 688-37-9
Properties: yellowish mass; formed from freshly precipitated aluminum hydroxide and oleic acid; used as a lacquer for metals in oil or turpentine solutions and as drier and waterproofing agent for paints [MER06]

Solubility: i H_2O; s alcohol, benzene, ethanol, oil, turpentine [MER06]
Density, g/cm³: 1.01 [KIR78]
Melting Point, °C: 120 [KIR78]

63
Compound: Aluminum oxalate monohydrate
Formula: $Al_2(C_2O_4)_3 \cdot H_2O$
Molecular Formula: $C_6H_2Al_2O_{13}$
Molecular Weight: 336.037
CAS RN: 814-87-9
Properties: white powd; used as a mordant to print textiles and to dye cotton [MER06]
Solubility: i H_2O, alcohol; s acids [MER06]

64
Compound: Aluminum oxide
Synonyms: native aluminum oxide, bauxite
Formula: Al_2O_3
Molecular Formula: Al_2O_3
Molecular Weight: 101.961
CAS RN: 1344-28-1
Properties: white powd; hex; hardness 8.8 Mohs; electrical resistivity at 300°C ~ $1.2 \times 10^{+13}$ ohm·cm; used as an adsorbent (see corundum); evaporated material of 99.99% purity is used as a high temp dielectric to protect aluminum mirrors and as a support for specimens in electron diffraction [MER06] [CER91]
Solubility: i H_2O; sl s alkalies [MER06]
Density, g/cm³: 3.965 [ALF93]
Melting Point, °C: 2045 [ALF93]
Boiling Point, °C: 2980 [ALF93]
Thermal Conductivity, W/(m·K): 28.9 (100°C), 21.2 (200°C), 12.5 (400°C), 8.70 (600°C), 6.86 (800°C), 5.86 (1000°C), 5.27 (1200°C) [HO72]

65
Compound: Aluminum oxide(α)
Synonym: corundum
Formula: α-Al_2O_3
Molecular Formula: Al_2O_3
Molecular Weight: 101.961
CAS RN: 1302-74-5
Properties: naturally occurring; white rhomb cryst, a = 0.47591 nm, c = 1.2894 nm; impedance 384–410 MPa·s/m; hardness 9 Mohs; enthalpy of fusion 111.1 kJ/mol; used as an abrasive powd, in grinding wheels, and in crucible form to melt metals; some precious stones are forms of corundum, which contain traces of other metals including ruby (chromium) and sapphire (cobalt) [JAN85] [ROB67] [VIE91] [KIR78] [CER91]

Solubility: i H_2O; v sl s acids, alkalies [HAW93]
Density, g/cm³: 3.987 [ROB78]
Melting Point, °C: 2054 [CRC10]
Boiling Point, °C: 2980 [CRC10]
Thermal Expansion Coefficient: ⊥ c-axis: 100°C (0.044), 200°C (0.112), 400°C (0.278), 600°C (0.455), 800°C (0.632), 1000°C (0.815), 1200°C (0.998) [CLA66]

66
Compound: Aluminum oxide(γ)
Formula: γ-Al_2O_3
Molecular Formula: Al_2O_3
Molecular Weight: 101.961
CAS RN: 1344-28-1
Properties: −60 mesh with 96% purity; white powd; a = 0.562 nm, b = 0.780 nm; enthalpy of fusion 78.49 kJ/mol [JAN85] [STR93] [BHA78] [KIR78] [ALF93]
Solubility: i H_2O; sl s acids, alkalies [CRC10]
Density, g/cm³: 3.97 [STR93]
Melting Point, °C: 2018 [ROB78]
Boiling Point, °C: 2980 [STR93]

67
Compound: Aluminum oxide(δ)
Formula: δ-Al_2O_3
Molecular Formula: Al_2O_3
Molecular Weight: 101.961
CAS RN: 1344-28-1
Properties: ortho-rhomb: a = 0.425 nm, b = 1.275 nm, c = 1.021 nm; tetr: a = 0.796 nm, c = 2.34 nm; enthalpy of fusion 93.3 kJ/mol [JAN85] [KIR78]
Density, g/cm³: 3.2 [KIR78]
Melting Point, °C: 2035 [JAN85]
Reactions: transition from δ to α in two steps: from 800°C to 1100°C and at 1200°C [JAN85]

68
Compound: Aluminum oxide(κ)
Formula: κ-Al_2O_3
Molecular Formula: Al_2O_3
Molecular Weight: 101.961
CAS RN: 1344-28-1
Properties: hex, a = 0.971 nm, c = 1.786 nm; enthalpy of fusion 91.2 kJ/mol [JAN85] [KIR78]
Density, g/cm³: 3.1–3.3 [KIR78]
Melting Point, °C: 2040 (fusion) [JAN85]
Reactions: transition from κ to α ~1200°C [JAN85]

69
Compound: Aluminum oxyhydroxide(α)
Synonym: boehmite

Formula: α-AlO(OH)
Molecular Formula: $AlHO_2$
Molecular Weight: 59.989
CAS RN: 1318-23-6
Properties: white; ortho-rhomb; a = 0.286 nm, b = 1.2227 nm, c = 0.380 nm; hardness is 3.5–4 Mohs; obtained by hydrothermal reaction of hydroxide slurries at 200°C–250°C [KIR78] [ROB67]; solubility data are in [BOU93] and [PHI93]
Solubility: i H_2O; s hot acids, hot alkalies [CRC10]
Density, g/cm³: 3.07 [ROB78]
Reactions: transforms to diaspore at 227°C, dehydrates to corundum at 400°C [LAU73]

70
Compound: Aluminum oxyhydroxide(β)
Synonym: diaspore
Formula: β-AlO(OH)
Molecular Formula: $AlHO_2$
Molecular Weight: 59.989
CAS RN: 14457-84-2
Properties: ortho-rhomb; a = 0.439 nm, b = 0.942 nm, c = 0.284 nm; hardness is 6.5–7 Mohs; stable from 275°C–425°C [KIR78]
Solubility: i H_2O; s hot acids, hot alkalies [CRC10]
Density, g/cm³: 3.44 [KIR78]
Reactions: dehydrates to corundum at 400°C [LAU73]

71
Compound: Aluminum palmitate
Synonyms: hexadecanoic acid, aluminum(III) salt
Formula: $Al[CH_3(CH_2)_{14}COO]_3$
Molecular Formula: $C_{48}H_{93}AlO_6$
Molecular Weight: 793.244
CAS RN: 555-35-1
Properties: white to yellow powd; made by heating aluminum hydroxide, palmitic acid, and water; used to thicken petroleum and lubricants and for water proofing fabrics [MER06] [HAW93]
Solubility: i H_2O, alcohol; s petroleum ether [MER06]
Density, g/cm³: 1.095 (dihydroxy monopalmitate) [CRC10]
Melting Point, °C: 200 (dihydroxy monopalmitate) [CRC10]

72
Compound: Aluminum perchlorate
Formula: $Al(ClO_4)_3$
Molecular Formula: $AlCl_3O_{12}$
Molecular Weight: 325.329
CAS RN: 14452-39-2

Properties: can be prepared by evaporation of solutions of $AlCl_3$ and $AgClO_4$ in methanol or benzene, with subsequent evaporation at 150°C; forms hydrates with 3, 6, 9, and 15 waters of hydration; finds use in studies of cation–ligand interactions, e.g., hydrolysis products [CIC73]
Solubility: g/100 g soln, H_2O: 54.87 (0°C), 64.62 (91.5°C); equilibrium solid phase $Al(ClO_4)_3 \cdot 9H_2O$ [KRU93]

73
Compound: Aluminum perchlorate nonahydrate
Formula: $Al(ClO_4)_3 \cdot 9H_2O$
Molecular Formula: $AlCl_3H_{18}O_{21}$
Molecular Weight: 487.470
CAS RN: 81029-06-3
Properties: white cryst [STR93]
Density, g/cm³: 2.0 [STR93]
Melting Point, °C: 82 [ALF93]

74
Compound: Aluminum phosphate
Synonyms: berlinite, aluminum orthophosphate
Formula: $AlPO_4$
Molecular Formula: AlO_4P
Molecular Weight: 121.953
CAS RN: 7784-30-7
Properties: white; rhomb plates; cryst, a = 0.4942 nm, c = 1.097 nm, isomorphous with quartz; naturally occurring; can be prepared by mixing a solution of aluminum sulfate and sodium phosphate; used as a flux for ceramics, in dental cements, for special glasses [MER06] [CRC10] [HAW93]
Solubility: i H_2O; v sl s HCl, HNO_3 [MER06]
Density, g/cm³: 2.56 [MER06]
Melting Point, °C: >1460 [MER06]
Reactions: α-$AlPO_4$ form is stable below 584°C [LAU87]

75
Compound: Aluminum phosphate dihydrate
Synonym: variscite
Formula: $AlPO_4 \cdot 2H_2O$
Molecular Formula: AlH_4O_6P
Molecular Weight: 157.984
CAS RN: 7784-30-7
Properties: rhomb; green; hardness 3.5–4.5 Mohs; $AlPO_4 \cdot xH_2O$ can be prepared as a gelatinous precipitate by adding a neutral solution of an Al salt to a solution of an alkali metal phosphate [CIC73] [CRC10]
Solubility: i H_2O; sl s HNO_3, HCl [CIC73]
Density, g/cm³: 2.57 [CIC73]
Melting Point, °C: 1850 [CIC73]

76
Compound: Aluminum phosphate trihydroxide
Synonym: angelite
Formula: $Al_2(PO_4)(OH)_3$
Molecular Formula: $Al_2H_3O_7P$
Molecular Weight: 199.954
CAS RN: 12004-29-4
Properties: naturally occurring mineral; colorless, white, yellowish-white or rose; monocl; hardness 4.5–5 Mohs [CRC10] [MER06]
Density, g/cm³: 2.696 [CRC10]

77
Compound: Aluminum phosphide
Synonyms: celphos, detia, phostoxin
Formula: AlP
Molecular Formula: AlP
Molecular Weight: 57.956
CAS RN: 20859-73-8
Properties: dark gray or yellow powd; cub, a = 0.5467 nm; band gap 2.42 eV; reacts readily in moist air to produce phosphine; electronic dielectric constant 8.5; can be prepared from red phosphorus and aluminum powd; used as a fumigant, as a source of phosphine, and in semiconductor work [CIC73] [MER06]
Solubility: reacts with H_2O to produce phosphine [MER06]
Density, g/cm³: 2.40 [MER06]
Melting Point, °C: >1000 [MER06]
Thermal Conductivity, W/(m·K): 92.0 [CRC10]

78
Compound: Aluminum selenide
Formula: Al_2Se_3
Molecular Formula: Al_2Se_3
Molecular Weight: 290.843
CAS RN: 1302-82-5
Properties: 6.5 mm and down black pieces or yellowish to light brown powd; unstable in air; formed by reaction of stoichiometric amounts of Al and Se at 1000°C; used to prepare H_2Se for semiconductor work; a = 0.389 nm, c = 0.630 nm [STR93] [MER06] [CIC73]
Solubility: decomposed by H_2O, acids [MER06]
Density, g/cm³: 3.437 [MER06]

79
Compound: Aluminum silicate
Synonym: metakaolinite

Formula: $Al_2O_3 \cdot 2SiO_2$
Molecular Formula: $Al_2O_7Si_2$
Molecular Weight: 222.128
CAS RN: 1302-76-7
Properties: white powd [STR93]
Density, g/cm³: 2.60 (dickite) [ROB78]
Reactions: forms mullite and SiO_2 at 1200°C [BAB85]; forms amorphous aluminosilicate at 980°C [CHA90]

80
Compound: Aluminum silicate
Synonym: sillimanite
Formula: $Al_2O_3 \cdot SiO_2$
Molecular Formula: Al_2O_5Si
Molecular Weight: 162.041
CAS RN: 12141-45-6
Properties: ortho, a = 0.78483 nm, b = 0.7673 nm, c = 0.57711 nm [ROB67]
Density, g/cm³: 3.247 [ROB78]
Reactions: forms mullite and SiO_2 from 1345°C to 1550°C [CLA66]
Thermal Expansion Coefficient: 100°C (0.088), 200°C (0.215), 400°C (0.531), 1000°C (1.979) [CLA66]

81
Compound: Aluminum silicate
Synonym: andalusite
Formula: $Al_2O_3 \cdot SiO_2$
Molecular Formula: Al_2O_5Si
Molecular Weight: 162.041
CAS RN: 12183-80-1
Properties: gray, greenish, reddish, or bluish; hardness 7–7.5; used in dental cements and the glass industry, in enamels, ceramics, and as a paint filler [MER06] [HAW93]
Density, g/cm³: 3.145 [ROB78]
Reactions: forms mullite and SiO_2 from 1325°C to 1410°C [CLA66]
Thermal Expansion Coefficient: 100°C (0.151), 200°C (0.417), 1000°C (3.606) [CLA66]

82
Compound: Aluminum silicate
Synonym: mullite
Formula: $3Al_2O_3 \cdot 2SiO_2$
Molecular Formula: $Al_6O_{13}Si_2$
Molecular Weight: 426.048
CAS RN: 1302-93-8

Properties: colorless; rhomb, a=0.7557 nm,
   b=0.76876 nm, c=0.28842 nm; hardness: hot
   pressed 13.6 GPa, sintered 12.7 GPa; indentation
   microfracture 2.02 MPa·m$^{1/2}$; submicrometer
   powd can be prepared by hydrolysis of mixed
   alkoxides, followed by drying and calcining
   up to 1600°C; sinter, microstructure [SOM91];
   phases equilibria in mullite [PAS88]; other
   data in [HIR89] [ROB67] [MIZ89]
Solubility: i H_2O, acids, HF [CRC10]
Density, g/cm³: theoretical 3.17 [MIZ89]
Melting Point, °C: 1750 [JAN85]
Thermal Conductivity, W/(m·K): 100°C (5.39),
   200°C (4.89), 4000°C (4.18), 600°C (3.81),
   800°C (3.59), 1000°C (3.43) [HO72]
Thermal Expansion Coefficient: 100°C (0.070),
   200°C (0.188), 400°C (0.471), 600°C (0.786),
   800°C (1.121), 1000°C (1.439) [CLA66]

83

Compound: Aluminum silicate
Synonym: kyanite
Formula: $Al_2O_3 \cdot SiO_2$
Molecular Formula: Al_2O_5Si
Molecular Weight: 162.041
CAS RN: 12141-46-7
Properties: mineral; tricl, a=0.7123 nm, b=0.7848 nm,
   c=0.5564 nm [ROB67] [HAW93]
Density, g/cm³: 3.247 [CRC10]
Reactions: forms mullite and SiO_2 from
   1000°C to 1325°C [CLA66]
Thermal Expansion Coefficient: 100°C (0.127),
   200°C (0.360), 400°C (0.890), 600°C (1.478),
   800°C (2.081), 1000°C (2.687) [CLA66]

84

Compound: Aluminum silicate dihydrate
Synonyms: kaolin, China clay
Formula: $Al_2O_3 \cdot 2SiO_2 \cdot 2H_2O$
Molecular Formula: $Al_2H_4O_9Si_2$
Molecular Weight: 258.161
CAS RN: 1332-58-7
Properties: white to yellowish or grayish fine powd;
   high lubricity, feels slippery to touch; tricl,
   a=0.5055 nm, b=0.8959 nm, c=1.4736 nm;
   used as a filler and coating for paper and
   rubber, in paint [STR93] [HAW93]
Solubility: i H_2O, dil acids and alkali
   hydroxides [HAW93]
Density, g/cm³: 2.594 [ROB78]
Reactions: transforms to metakaolinite
   about 525°C [BAB85]

85

Compound: Aluminum stearate
Formula: $Al(C_{18}H_{35}O_2)_3$
Molecular Formula: $C_{54}H_{105}AlO_6$
Molecular Weight: 877.390
CAS RN: 637-12-7
Properties: white powd [CRC10]
Solubility: i H_2O, EtOH, eth; s alk [CRC10]
Density: 1.070
Melting Point, °C: 115 [CRC10]

86

Compound: Aluminum sulfate
Synonyms: alum, pearl alum
Formula: $Al_2(SO_4)_3$
Molecular Formula: $Al_2O_{12}S_3$
Molecular Weight: 342.154
CAS RN: 10043-01-3
Properties: white, lustrous cryst; can be
   prepared by treating kaolin, aluminum
   hydroxide or bauxite with sulfuric acid,
   followed by filtration and crystallization;
   used in tanning leather as a mordant for
   dyeing, to purify water, and to waterproof
   and fireproof cloth [MER06] [HAW93]
Solubility: g/100 g soln, H_2O: 27.50 (0°C), 27.82
   (25°C), 43.9 (99.2°C); equilibrium solid
   phase, $Al_2(SO_4)_3 \cdot 16H_2O$ [KRU93]
Density, g/cm³: 1.61 [MER06]
Melting Point, °C: decomposes at 770 [ALD94]
Reactions: decomposes to γ-Al_2O_3 and SO_3
   from 580°C to 900°C [KIR78]

87

Compound: Aluminum sulfate octadecahydrate
Synonyms: alunogen, cake alum
Formula: $Al_2(SO_4)_3 \cdot 18H_2O$
Molecular Formula: $Al_2H_{36}O_{30}S_3$
Molecular Weight: 666.429
CAS RN: 7784-31-8
Properties: colorless; monocl; used in paper industry
   and in water treatment [KIR78] [STR93]
Solubility: anhydrous/100 g, H_2O: 27.5 (0°C), 27.8
   (25°C), 46.9 (103.2°C) [KIR78]; gAl_2O_3/100 mL
   at 35°C in the following solvents: methanol 2.89,
   ethanol 0.45, ethylene glycol 8.76 [OKA91]
Density, g/cm³: 1.69 [KIR78]
Melting Point, °C: decomposes at 86.5 [STR93]
Reactions: minus $15H_2O$ from 40°C to 250°C;
   minus $3H_2O$ from 250°C to 400°C [KIR78]

88
Compound: Aluminum sulfide
Formula: Al_2S_3
Molecular Formula: Al_2S_3
Molecular Weight: 150.161
CAS RN: 1302-81-4
Properties: yellowish gray powd; hex, a = 0.642 nm, c = 1.783 nm; H_2S odor; decomposes in moist air; formed by heating stoichiometric amounts of Al and S at 700°C–100°C; used as a semiconductor [CIC73] [MER06]
Solubility: decomposes in H_2O [MER06]
Density, g/cm³: 2.32 [CIC73]
Melting Point, °C: 1100 [MER06]
Reactions: sublimes at 1500°C in N_2 atm; $\alpha \rightarrow \gamma$ transition at 1000°C [CRC10] [JAN85]

89
Compound: Aluminum tartrate
Synonyms: 2,3-dihydroxybutanedioic acid, aluminum(III) salt
Formula: $Al_2(C_4H_4O_6)_3$
Molecular Formula: $C_{12}H_{12}Al_2O_{18}$
Molecular Weight: 498.179
CAS RN: 815-78-1
Properties: odorless granules; used in dyeing textiles [MER06]
Solubility: s H_2O, dissolves faster in hot H_2O; s ammonia [MER06]

90
Compound: Aluminum telluride
Formula: Al_2Te_3
Molecular Formula: Al_2Te_3
Molecular Weight: 436.763
CAS RN: 12043-29-7
Properties: dark gray or black cryst; a = 0.407 nm, c = 0.693 nm; electrical resistivity (27°C) 0.0054 ohm·cm; can be formed by reacting Al and Te at 1000°C; used in semiconductor work [CIC73] [STR93]
Density, g/cm³: 4.5 [CIC73]
Melting Point, °C: decomposes [ALF93]

91
Compound: Aluminum tellurite
Formula: $Al_2(TeO_3)_3$
Molecular Formula: $Al_2O_9Te_3$
Molecular Weight: 580.758
CAS RN: 58500-12-2
Properties: reacted product, –80 mesh particle size; 99% purity [CER91]

92
Compound: Aluminum thiocyanate
Formula: $Al(CNS)_3$
Molecular Formula: $C_3AlN_3S_3$
Molecular Weight: 201.233
CAS RN: 538-17-0
Properties: yellowish powd; aq solution used as mordant in the dye industry and in pottery manufacturing [MER06] [HAW93]
Solubility: s H_2O; i alcohol, ether [HAW93]

93
Compound: Aluminum titanate
Synonym: tielite
Formula: Al_2TiO_5
Molecular Formula: Al_2O_5Ti
Molecular Weight: 181.827
CAS RN: 12004-39-6
Properties: –100 mesh with 99.5% purity; ortho-rhomb, pseudobrookite; preparation: sol-gel [PRA92], by reaction of equimolar amounts of Al_2O_3 and TiO_2 at 1300°C, then sintering at 1400°C for 4 h [PER89]; cryst structure in [FIE87]; other references [ALF93] [FRE87]
Density, g/cm³: 3.73 [ROB78]
Melting Point, °C: 1860 [HOL73]
Reactions: decomposes at 800°C–1300°C to corundum and rutile, which recombine to tielite >1300°C [PAR90]
Thermal Expansion Coefficient: -2×10^{-6}/°C (24°C–1000°C) [PAR90]

94
Compound: Aluminum tristearate
Synonym: aluminum stearate
Formula: $Al[CH_3(CH_2)_{16}COO]_3$
Molecular Formula: $C_{54}H_{105}AlO_6$
Molecular Weight: 877.406
CAS RN: 637-12-7
Properties: white powd; prepared by reacting stearic acid with aluminum salts; used in waterproofing fabrics and ropes, in paint, and varnish driers [MER06] [HAW93]
Solubility: i H_2O, alcohol, and ether; s alkali; forms gel with aliphatic and aromatic hydrocarbons [HAW93]
Density, g/cm³: 1.070 [HAW93]
Melting Point, °C: 115 [HAW93]

95
Compound: Aluminum tungstate
Formula: $Al_2(WO_4)_3$

Molecular Formula: $Al_2O_{12}W_3$
Molecular Weight: 797.476
CAS RN: 15123-82-7
Properties: −325 mesh with 99% purity; white powd [ALF93] [STR93]

96
Compound: Aluminum zirconate
Formula: $Al_2O_3 \cdot 3ZrO_2$
Molecular Formula: $Al_2O_9Zr_3$
Molecular Weight: 471.630
CAS RN: 70692-95-4
Properties: high fracture toughened ceramic material at room temp due to a dispersion of tetr ZrO_2 particles in an alumina matrix; can be prepared by plasma synthesis from the vapor phase by injecting mixtures of $AlCl_3$ and $ZrCl_4$ vapor in an argon–oxygen plasma, resulting in powd consisting of a mixture of δ-Al_2O_3 and tetr ZrO_2; reacted product, −100 mesh; 99% purity; reaction sintering with mullite in [DES91]; other references [CER91] [KOH88], wear coating [GEI92]

97
Compound: Aluminum zirconium
Formula: Al_2Zr
Molecular Formula: Al_2Zr
Molecular Weight: 145.187
CAS RN: 12004-50-1
Properties: −100 mesh with 99% purity; powd [ALF93]
Melting Point, °C: 1645 [ALF93]

98
Compound: Chlorodiethylaluminum
Formula: $AlCl(C_2H_5)_2$
Molecular Formula: $C_4H_{10}AlCl$
Molecular Weight: 120.557
CAS RN: 96-10-6
Properties: col liq [CRC10]
Solubility: reac H_2O [CRC10]
Density, g/cm³: 0.96 [CRC10]

99
Compound: Chlorodiisobutylaluminum
Formula: $AlCl(C_4H_9)_2$
Molecular Formula: $C_6H_{18}AlCl$
Molecular Weight: 176.664
CAS RN: 1779-25-5
Properties: hygr col liq [CRC10]
Solubility: s eth, hx [CRC10]

Density, g/cm³: 0.95 [CRC10]
Melting Point, °C: −40 [CRC10]

100
Compound: Dichloromethylaluminum
Formula: $AlCl_2CH_3$
Molecular Formula: CH_3AlCl_2
Molecular Weight: 112.923
CAS RN: 917-65-7
Properties: cryst [CRC10]
Solubility: Soluble bz, eth, hydrocarbon solvents [CRC10]
Melting Point, °C: 72.7 [CRC10]
Boiling Point, °C: 95 [CRC10]

101
Compound: Americium
Formula: Am
Molecular Formula: Am
Molecular Weight: 243
CAS RN: 7440-35-9
Properties: silvery metal with two forms; α-Am: hex, a=0.3468 nm, c=1.1241 nm; β-Am: cub, a=0.4894 nm; $t_{1/2}$ [241]Am=433 years, $t_{1/2}$ [242]Am=152 h, $t_{1/2}$ [243]Am=7400 years; Am⁺⁺⁺ stable in aq solution; enthalpy of vaporization 230 kJ/mol; enthalpy of fusion 14.4 kJ/mol; enthalpy of sublimation 276 kJ/mol; ionic radius of Am⁺⁺⁺ is 0.0982 nm; can be prepared from [241]Pu; used to diagnose thyroid disorders [KIR78] [CIC73]
Solubility: s dil acids [CRC10]
Density, g/cm³: α-Am: 13.67; β-Am: 13.65 [CIC73]
Melting Point, °C: 1173 [KIR91]
Boiling Point, °C: 2011 [KIR91]
Reactions: α transforms to β at 1079°C [CIC73]

102
Compound: Americium bromide
Formula: $AmBr_3$
Molecular Formula: $AmBr_3$
Molecular Weight: 483
CAS RN: 14933-38-1
Properties: white; ortho-rhomb; a=0.4064 nm, b=1.2661 nm, c=0.9144 nm [CIC73]
Solubility: s H_2O [CRC10]
Density, g/cm³: 6.85 [KIR78]
Melting Point, °C: sublimes [CRC10]

103
Compound: Americium carbonate dihydrate
Formula: $Am_2(CO_3)_3 \cdot 2H_2O$
Molecular Formula: $C_3H_4Am_2O_{11}$

CAS RN: 18421-71-1
Molecular Weight: 702
Properties: a = 1.409 nm [CIC73]

104

Compound: Americium chloride
Formula: $AmCl_3$
Molecular Formula: $AmCl_3$
Molecular Weight: 349
CAS RN: 13464-46-5
Properties: pink; hex, a = 0.7382 nm,
    c = 0.4214 nm [KIR78]
Density, g/cm³: 5.87 [KIR78]
Melting Point, °C: 715 [KIR91]

105

Compound: Americium fluoride
Synonym: americium trifluoride
Formula: AmF_3
Molecular Formula: AmF_3
Molecular Weight: 300
CAS RN: 13708-80-0
Properties: pink; hex, a = 0.7044 nm,
    c = 0.7225 nm [CIC73]
Solubility: i H_2O [CRC10]
Density, g/cm³: 9.53 [KIR78]
Melting Point, °C: 1393 [KIR91]

106

Compound: Americium hydride
Formula: AmH_3
Molecular Formula: AmH_3
Molecular Weight: 246
CAS RN: 13774-24-8
Properties: black; hex, a = 0.377 nm, c = 0.675 nm [CIC73]
Density, g/cm³: 9.76 [CIC73]

107

Compound: Americium hydroxide
Formula: $Am(OH)_3$
Molecular Formula: AmH_3O_3
Molecular Weight: 294
CAS RN: 23323-79-7
Properties: hex, a = 0.6426 nm, c = 0.3745 nm [CIC73]

108

Compound: Americium iodide
Formula: AmI_3
Molecular Formula: AmI_3
Molecular Weight: 624

CAS RN: 13813-47-3
Properties: yellow; ortho-rhomb,
    a = 0.742 nm, c = 2.055 nm [KIR78]
Solubility: s H_2O [CRC10]
Density, g/cm³: 6.9 [CRC10]
Melting Point, °C: ~950 [KIR91]

109

Compound: Americium oxide(α)
Formula: α-Am_2O_3
Molecular Formula: Am_2O_3
Molecular Weight: 534
CAS RN: 12254-64-7
Properties: tan; hex, a = 0.3805 nm,
    c = 0.696 nm [KIR78]
Solubility: s mineral acids [CRC10]
Density, g/cm³: 11.77 [KIR78]

110

Compound: Americium oxide(β)
Formula: β-Am_2O_3
Molecular Formula: Am_2O_3
Molecular Weight: 534
CAS RN: 12254-64-7
Properties: reddish brown; cub, a = 1.103 nm [KIR78]
Density, g/cm³: 10.57 [KIR78]

111

Compound: Americium oxychloride
Formula: AmOCl
Molecular Formula: AmClO
Molecular Weight: 294
CAS RN: 37961-19-6
Properties: cryst; a = 0.400 nm, b = 0.678 nm [BRO73]

112

Compound: Americium phosphate
Formula: $AmPO_4$
Molecular Formula: AmO_4P
Molecular Weight: 338
CAS RN: 14933-41-6
Properties: monocl, a = 0.673 nm,
    b = 0.693 nm, c = 0.641 nm [CIC73]

113

Compound: Americium sulfide
Formula: Am_2S_3
Molecular Formula: Am_2S_3
Molecular Weight: 582
CAS RN: 12446-46-7
Properties: bcc, a = 0.845 nm [CIC73]

114

Compound: Americium(IV) fluoride
Formula: AmF_4
Molecular Formula: AmF_4
Molecular Weight: 319
CAS RN: 15947-41-8
Properties: tan; monocl, a = 1.254 nm,
    b = 1.052 nm, c = 0.820 nm [KIR78]
Density, g/cm³: 7.23 [KIR78]

115

Compound: Americium(IV) oxide
Formula: AmO_2
Molecular Formula: AmO_2
Molecular Weight: 275
CAS RN: 12005-67-3
Properties: black; cub, a = 0.5374 nm [KIR78]
Solubility: s mineral acids [CRC10]
Density, g/cm³: 11.68 [KIR78]

116

Compound: Ammonia
Formula: NH_3
Molecular Formula: H_3N
Molecular Weight: 17.031
CAS RN: 7664-41-7
Properties: colorless gas; very pungent odor; critical
    pressure 111.5 atm; critical temp 132.4°C;
    vapor pressure of liq 8.5 atm (20°C); enthalpy
    of vaporization 23.33 kJ/mol; enthalpy of
    fusion 5.66 kJ/mol; autoignition temp 690°C;
    produced by reaction of steam forced through
    incandescent coke; used in the manufacture
    of nitric acid, explosives, synthetic fibers,
    nitrides, fertilizers, and in electronics and
    refrigeration [MER06] [HAW93] [AIR87]
Solubility: s H_2O: 47% (0°C), 31%
    (25°C), 28% (50°C) [MER06]
Density, g/cm³: 0.5967 (air = 1.0000) [MER06]
Melting Point, °C: −77.7 [MER06]
Boiling Point, °C: −33.35 at 1 atm [MER06]
Reactions: mixtures of NH_3 and air can
    explode when ignited [MER06]

117

Compound: Ammonium 12-molybdophosphate hydrate
Synonym: ammonium phosphomolybdate
Formula: $(NH_4)_3PMo_{12}O_{40} \cdot xH_2O$
Molecular Formula: $H_{12}Mo_{12}N_3O_{40}P$ (anhydrous)
Molecular Weight: 1834.345 (anhydrous)
CAS RN: 12026-66-3

Properties: yellow, cryst powd; prepared by reacting
    ammonium molybdate with phosphoric and nitric
    acids; used as a reagent in ion-exchange columns,
    and as a photographic additive [HAW93]
Solubility: v sl s H_2O; s alkali; i alcohol, acids [HAW93]
Melting Point, °C: decomposes [CRC10]

118

Compound: Ammonium acetate
Synonyms: acetic acid, ammonium salt
Formula: CH_3COONH_4
Molecular Formula: $C_2H_7NO_2$
Molecular Weight: 77.084
CAS RN: 631-61-8
Properties: white cryst; deliq; prepared from acetic
    acid and ammonia by exact neutralization to
    pH 7.0; finds use in analytical chemistry, drugs, and
    textile dyeing; specific gravity of aq solutions in
    % of CH_3COONH_4: 10% (1.022), 20% (1.042), 30%
    (1.062), 40% (1.077), 50% (1.092) [MER06] [KIR78]
Solubility: 148 g/100 g H_2O (4°C) [KIR78]
Density, g/cm³: 1.073 [KIR78]
Melting Point, °C: 114 [MER06]
Boiling Point, °C: decomposes [KIR78]

119

Compound: Ammonium aluminum sulfate
Synonym: burnt ammonium alum
Formula: $NH_4Al(SO_4)_2$
Molecular Formula: $AlH_4NO_8S_2$
Molecular Weight: 237.148
CAS RN: 7784-25-0
Properties: white powd [MER06]
Solubility: g $NH_4Al(SO_4)_2$/100 g H_2O: 2.10
    (0°C), 5.00 (10°), 7.74 (20°C), 10.9 (30°),
    14.9 (40°C), 26.7 (60°C) [LAN05]
Density, g/cm³: 2.45 [CRC10]
Melting Point, °C: decomposes at 280 [KIR78]
Reactions: forms δ-Al_2O_3 at 1000°C–1250°C [KIR78]

120

Compound: Ammonium aluminum
    sulfate dodecahydrate
Synonym: ammonium alum
Formula: $NH_4Al(SO_4)_2 \cdot 12H_2O$
Molecular Formula: $AlH_{28}NO_{20}S_2$
Molecular Weight: 453.331
CAS RN: 7784-26-1
Properties: colorless, cryst, white granules or powd;
    obtained by crystallization from a mixture of
    ammonium and aluminum sulfates; used to purify
    drinking water, in baking powd, for dyeing fabrics,
    and fireproofing [MER06] [HAW93] [KIR78]

Solubility: 1 g/7 mL H_2O; 1 g/0.5 mL hot H_2O
[MER06]; s glycerol; i alcohol [HAW93]
Density, g/cm³: 1.65 [MER06]
Melting Point, °C: 94.5 [MER06]
Boiling Point, °C: decomposes above 280 [MER06]
Reactions: minus $10H_2O$ (250°C) forms γ-Al_2O_3
(1000°C–1250°C) [MER06] [KIR78]

121
Compound: Ammonium arsenate hydrate
Synonym: ammonium orthoarsenate
Formula: $(NH_4)_3AsO_4 \cdot xH_2O$
Molecular Formula: $AsH_{12}N_3O_4$ (anhydrous)
Molecular Weight: 193.035 (anhydrous)
CAS RN: 13462-93-6
Properties: −6 mesh with 99.9% purity;
x = 3: ortho [CER91] [CRC10]
Reactions: minus NH_3 on heating [CRC10]

122
Compound: Ammonium azide
Formula: NH_4N_3
Molecular Formula: H_4N_4
Molecular Weight: 60.059
CAS RN: 12164-94-2
Properties: ortho-rhomb, a = 0.893 nm,
b = 0.864 nm, c = 0.380 nm [CIC73]
Solubility: g NH_4N_3/100 g H_2O: 16.0 (0°C),
25.3 (20°C), 37.1 (40°C) [LAN05]
Density, g/cm³: 1.346 [CIC73]
Melting Point, °C: 160 [CIC73]
Boiling Point, °C: explodes at 134 [CIC73]

123
Compound: Ammonium benzoate
Synonyms: benzoic acid, ammonium salt
Formula: $C_6H_5COONH_4$
Molecular Formula: $C_7H_9NO_2$
Molecular Weight: 139.154
CAS RN: 1863-63-4
Properties: white cryst or powd; manufactured from
benzoic acid and ammonia; used in medicine and
as a preservative for latex [HAW93] [MER06]
Solubility: 1 g/4.7 mL H_2O, 1 g/1.2 mL hot H_2O
[MER06]; s alcohol, glycerol [HAW93]
Density, g/cm³: 1.26 [MER06]
Melting Point, °C: 198 [MER06]
Boiling Point, °C: sublimes at 160 [HAW93]
Reactions: gradually loses NH_3 in air [MER06]

124
Compound: Ammonium bimalate
Synonyms: hydroxybutanedioic acid, monoammonium salt
Formula: $NH_4OOCCH_2CH(OH)COOH$
Molecular Formula: $C_4H_9NO_5$
Molecular Weight: 151.119
CAS RN: 5972-71-4
Properties: ortho-rhomb cryst [MER06]
Solubility: s 3 parts H_2O, sl s alcohol [MER06]
Density, g/cm³: 1.15 [MER06]
Melting Point, °C: 160–161 [MER06]

125
Compound: Ammonium bromide
Formula: NH_4Br
Molecular Formula: BrH_4N
Molecular Weight: 97.943
CAS RN: 12124-97-9
Properties: colorless cryst or yellowish white powd; tetr,
a = 0.4034 nm; vapor pressure, kPa: 7.3 (300°C),
13.3 (320°C), 41.2 (360°C), 73.4 (380°C), 115.4
(400°C); prepared by reacting HBr and NH_4OH;
used to make AgBr salts for photography, in
medicine, engraving, textile finishing, and as a
fire-retardant material [CIC73] [HAW93] [KIR78]
Solubility: g/100 g soln, H_2O: 37.5 (0°C), 43.9 (25°C),
57.4 (100°C) [KRU93]; g/100 g H_2O: 60.6 (0°C); 75.5
(20°C), 145.6 (100°C) [KIR78]; s alcohol [HAW93]
Density, g/cm³: 2.429 [CIC73]
Melting Point, °C: sublimes at 452 [CIC73]
Boiling Point, °C: 235 in vacuum [CIC73]

126
Compound: Ammonium caprylate
Synonym: octanoic acid ammonium salt
Formula: $C_7H_{15}COONH_4$
Molecular Formula: $C_8H_{19}NO_2$
Molecular Weight: 161.245
CAS RN: 5972-76-9
Properties: hygr; monocl cryst; prepared by reacting
caprylic acid and ammonia; used in photographic
emulsions and as an insecticide [MER06]
Solubility: s acetic acid, ethanol [MER06]
Melting Point, °C: 70–85 [MER06]
Reactions: easily hydrolyzed in H_2O [MER06]

127
Compound: Ammonium carbamate
Formula: NH_2COONH_4
Molecular Formula: $CH_6N_2O_2$
Molecular Weight: 78.071
CAS RN: 1111-78-0

Properties: cryst powd; ammonia odor; white; rhomb; very volatile; forms urea when heated; made from liq NH_3 and solid CO_2; used as a fertilizer [HAW93] [MER06]
Solubility: v s H_2O; s alcohol [MER06]
Melting Point, °C: volatilizes ~60 [MER06]
Reactions: gradually evolves NH_3 in air [MER06]

128
Compound: Ammonium carbonate
Synonym: normal ammonium carbonate
Formula: $(NH_4)_2CO_3$
Molecular Formula: $CH_8N_2O_3$
Molecular Weight: 96.086
CAS RN: 506-87-6
Properties: powd or lumps; colorless cryst; prepared by passing CO_2 gas through NH_4OH solution, to crystallize the carbonate [ALF93] [ALD94] [KIR78]
Solubility: 20 g/100 g saturated solution in water (25°C) [MER06]
Melting Point, °C: decomposes at 58 [CIC73]

129
Compound: Ammonium cerium(III) nitrate tetrahydrate
Synonym: cerous ammonium nitrate
Formula: $(NH_4)_2Ce(NO_3)_5 \cdot 4H_2O$
Molecular Formula: $CeH_{16}N_7O_{19}$
Molecular Weight: 558.278
CAS RN: 15318-60-2
Properties: large, colorless, transparent, monocl cryst [CRC10] [MER06]
Solubility: g/100 g anhydrous, H_2O: 242 (10°C), 276 (20°C), 318 (30°), 376 (40°), 681 (60°C) [LAN05]
Melting Point, °C: 74 [CRC10]

130
Compound: Ammonium cerium(III) sulfate tetrahydrate
Synonym: cerous ammonium sulfate
Formula: $(NH_4)Ce(SO_4)_2 \cdot 4H_2O$
Molecular Formula: $CeH_{12}NO_{12}S_2$
Molecular Weight: 422.342
CAS RN: 21995-38-0
Properties: prepared by slow evaporation of a solution containing stoichiometric amounts of cerous and ammonium sulfates; monocl cryst [MER06]
Solubility: g/100 g anhydrous, H_2O: 5.53 (20°C), 4.49 (30°C), 3.48 (40°C), 2.02 (60°C), 1.33 (80°C) [LAN05]
Density, g/cm³: 2.523 (octahydrate) [CRC10]
Reactions: anhydrous at 150°C [CRC10]

131
Compound: Ammonium cerium(IV) nitrate
Synonym: ceric ammonium nitrate
Formula: $(NH_4)_2Ce(NO_3)_6$
Molecular Formula: $CeH_8N_8O_{18}$
Molecular Weight: 548.222
CAS RN: 16774-21-3
Properties: small, reddish orange cryst; precipitates from ceric nitrate solution containing excess nitric acid, on addition of NH_4NO_3; has been used as an oxidizing agent in analytical chemistry and as a catalyst for polymerization of olefins [ALF93] [MER06] [KIR78]
Solubility: g/100 g H_2O: 135 (20°C), 150 (30°C), 169 (40°C), 213 (60°C) [LAN05]; s alcohol, i conc HNO_3 [HAW93]

132
Compound: Ammonium cerium(IV) sulfate dihydrate
Synonym: ceric ammonium sulfate dihydrate
Formula: $(NH_4)_4Ce(SO_4)_4 \cdot 2H_2O$
Molecular Formula: $CeH_{20}N_4O_{18}S_4$
Molecular Weight: 632.556
CAS RN: 10378-47-9
Properties: cryst powd; oxidizing agent; precipitates from a solution of ceric sulfate and ammonium sulfate; anhydrous compound, 13840-04-5, exists [ALF93] [ALD93] [KIR78]

133
Compound: Ammonium chlorate
Formula: NH_4ClO_3
Molecular Formula: ClH_4NO_3
Molecular Weight: 101.490
CAS RN: 10192-29-7
Properties: colorless or white cryst; oxidizing agent; obtained as a product of the reaction of solutions of ammonium chloride and sodium chlorate; can be used as an explosive [HAW93]
Solubility: g/100 g H_2O: 28.7 (0°C), 115 (75°C) [CIC73]
Density, g/cm³: 1.80 [CIC73]
Melting Point, °C: explodes at 102 [CIC73]

134
Compound: Ammonium chloride
Synonyms: ammonium muriate, sal ammoniac
Formula: NH_4Cl
Molecular Formula: ClH_4N
Molecular Weight: 53.492
CAS RN: 12125-02-9

Properties: white; cub cryst, a=0.3866 nm; enthalpy
of formation 317 kJ/mol; enthalpy of sublimation
165.7 kJ/mol; vapor pressure, kPa, at temp
shown: 6.5 (250°C), 17.9 (280°C), 33.5 (300°C),
60.9 (320°C), 101.1 (338°C); prepared by
mixing ammonium sulfate and sodium chloride
solutions; used in batteries, as a soldering flux,
in electroplating [CIC73] [HAW93] [KIR78]
Solubility: g/100 g H$_2$O: 29.4 (0°C), 39.3 (25°C), 77.3
(100°C); equilibrium solid phase, NH$_4$Cl [KRU93]
Density, g/cm³: 1.527 [CIC73]
Melting Point, °C: sublimes without melting [MER06]
Reactions: decomposes at 520°C [CIC73]

135
Compound: Ammonium chromate(VI)
Synonym: ammonium enromate
Formula: (NH$_4$)$_2$CrO$_4$
Molecular Formula: CrH$_8$N$_2$O$_4$
Molecular Weight: 152.071
CAS RN: 7788-98-9
Properties: yellow cryst; obtained by adding NH$_4$OH
to ammonium dichromate solution, followed by
crystallization; used as a mordant in dyeing and
as a corrosion inhibitor [HAW93] [KIR78]
Solubility: g/100 g soln, H$_2$O: 19.9 (0°C), 27.02 (25°C),
41.20 (75°C) [KRU93] [MER06]; i alcohol [HAW93]
Density, g/cm³: 1.90 [KIR78]
Melting Point, °C: decomposes at 185 [MER06]
Reactions: minus some NH$_3$ in air [MER06]

136
Compound: Ammonium chromic sulfate dodecahydrate
Synonym: ammonium chrome alum
Formula: (NH$_4$)Cr(SO$_4$)$_2$ · 12H$_2$O
Molecular Formula: CrH$_{28}$NO$_{20}$S$_2$
Molecular Weight: 478.345
CAS RN: 10022-47-6
Properties: green powd or deep violet cryst; can be
crystallized from a solution of chromic sulfate
and ammonium sulfate; used as a mordant for
dyeing and in tanning [HAW93] [MER06]
Solubility: g/100 g anhydrous, H$_2$O: 3.95 (0°C), 18.8
(30°C), 32.6 (40°C) [LAN05]; sl s alcohol [HAW93]
Density, g/cm³: hydrated form: 1.72 [HAW93]
Melting Point, °C: hydrated form: 94 [HAW93]
Reactions: minus 9H$_2$O on melting, minus
12H$_2$O by 300°C [MER06]

137
Compound: Ammonium citrate tribasic
Formula: H$_4$NOOCCH$_2$C(OH)(COONH$_4$)CH$_2$COONH$_4$

Molecular Formula: C$_6$H$_{17}$N$_3$O$_7$
Molecular Weight: 243.217
CAS RN: 3458-72-8
Properties: structure
Melting Point, °C: decomposes at 185 [ALD94]

138
Compound: Ammonium cobalt(II)
phosphate monohydrate
Synonym: cobaltous ammonium phosphate
Formula: NH$_4$CoPO$_4$ · H$_2$O
Molecular Formula: CoH$_6$NO$_5$P
Molecular Weight: 189.959
CAS RN: 14590-13-7
Properties: red to violet powd or monocl; results from
a reaction between a cobalt(II) salt and ammonium
phosphate; used as a pigment in ceramic glass and
to indicate temp in textile industry [MER06]
Solubility: i H$_2$O; s acids [MER06]

139
Compound: Ammonium cobalt(II) sulfate hexahydrate
Synonym: cobaltous ammonium sulfate
Formula: (NH$_4$)$_2$Co(SO$_4$)$_2$ · 6H$_2$O
Molecular Formula: CoH$_{20}$N$_2$O$_{14}$S$_2$
Molecular Weight: 395.229
CAS RN: 13586-38-4
Properties: red; monocl prismatic cryst [MER06]
Solubility: g (NH$_4$)$_2$Co(SO$_4$)$_2$/100 g H$_2$O: 6.0 (0°C),
9.5 (10°C), 13.0 (20°C), 17.0 (30°C), 22.0
(40°C), 33.5 (60°C), 49.0 (80°C), 58.0 (90°C),
75.1 (100°C) [LAN05]; i alcohol [MER06]
Density, g/cm³: 1.902 [HAW93]

140
Compound: Ammonium copper(II) chloride dihydrate
Synonym: cupric ammonium chloride dihydrate
Formula: 2NH$_4$Cl · CuCl$_2$ · 2H$_2$O
Molecular Formula: Cl$_4$CuH$_{12}$N$_2$O$_2$
Molecular Weight: 277.464
CAS RN: 10060-13-6
Properties: blue to bluish green cryst; tetr; preparation:
by evaporating a solution containing a stoichiometric
amount of NH$_4$Cl and CuCl$_2$; has been used as
an analytical reagent; anhydrous material has
yellow, hygr rhombohedral cryst [MER06]
Solubility: g 2NH$_4$Cl · CuCl$_2$/100 g H$_2$O: 28.2
(0°C), 32.0 (10°C), 35.0 (20°C), 38.3 (30°C),
43.8 (40°C), 56.6 (60°C), 76.5 (80°C), 76.5
(90°C) [LAN05]; s alcohol [MER06]
Density, g/cm³: 1.993 [STR93]

Melting Point, °C: decomposes at
>120 [MER06] [STR93]
Reactions: minus 2H$_2$O over range
110°C–120°C [MER06]

141

Compound: Ammonium cyanide
Formula: NH$_4$CN
Molecular Formula: CH$_4$N$_2$
Molecular Weight: 44.056
CAS RN: 12211-52-8
Properties: colorless cryst solid; readily decomposes;
can be formed by mixing ammonium sulfate
solution with barium cyanide solution [KIR78]
Solubility: v s H$_2$O [KIR78]
Density, g/cm^3: 1.02 (100°C) [CRC10]
Melting Point, °C: decomposes at 36 [CRC10]
Reactions: decomposes to NH$_3$ and
HCN at 36°C [KIR78]

142

Compound: Ammonium dichromate(VI)
Synonym: ammonium dichromate
Formula: (NH$_4$)$_2$Cr$_2$O$_7$
Molecular Formula: Cr$_2$H$_8$N$_2$O$_7$
Molecular Weight: 252.065
CAS RN: 7789-09-5
Properties: reddish orange cryst; monocl; can
be prepared from ammonium sulfate and
sodium dichromate, followed by crystallization
from the solution; used as a mordant for
dyeing, in leather tanning, oil purification,
photography [HAW93] [KIR78]
Solubility: g/100 g H$_2$O: 18.2 (0°C), 25.5 (10°C), 35.6
(20°C), 46.5 (30°C), 58.5 (40°C), 156 (100°C)
[LAN05]; g/100 g soln, H$_2$O: 15.37 (0°C), 28.615
(25°C), 60.89 (100°C) [KRU93]; s alcohol [HAW93]
Density, g/cm^3: 2.155 [KIR78]
Melting Point, °C: decomposes at 180 [KIR78]
Reactions: decomposes with swelling and
evolution of heat and N$_2$ [MER06]

143

Compound: Ammonium dihydrogen arsenate
Formula: NH$_4$H$_2$AsO$_4$
Molecular Formula: AsH$_6$NO$_4$
Molecular Weight: 158.975
CAS RN: 13462-93-6
Properties: cryst [ALF93]
Solubility: g/100 g H$_2$O: 33.74 (0°C), 48.67
(20°C), 122.4 (90°C); equilibrium solid
phase, NH$_4$H$_2$AsO$_4$ [KRU93]

Density, g/cm^3: 2.311 [ALF93]
Melting Point, °C: 300, with decomposition [ALF93]

144

Compound: Ammonium dihydrogen phosphate
Synonym: ammonium phosphate monobasic
Formula: (NH$_4$)H$_2$PO$_4$
Molecular Formula: H$_6$NO$_4$P
Molecular Weight: 115.026
CAS RN: 7722-76-1
Properties: odorless cryst or white powd; made
from ammonia and phosphoric acid; used
with sodium bicarbonate as baking powd
and for fireproofing materials such as paper,
wood, and fiberboard [MER06] [HAW93]
Solubility: g/100 g soln, H$_2$O: 17.8 ± 0.8 (0°C),
28.8 ± 0.4 (25°C), 72.4 ± 7.0 (100°C)
[KRU93]; i alcohol [HAW93]
Density, g/cm^3: 1.803 [HAW93]
Melting Point, °C: 190 [STR93]

145

Compound: Ammonium dimolybdate
Synonym: ammonium molybdenum dioxide
Formula: (NH$_4$)$_2$Mo$_2$N$_2$O$_7$
Molecular Formula: H$_8$Mo$_2$N$_2$O$_7$
Molecular Weight: 339.953
CAS RN: 27546-07-2
Properties: white powd; obtained by crystallization
from a solution of MoO$_3$ containing excess
NH$_3$; used as a high purity source for the
preparation of Mo metal [KIR81] [ALF93]
Density, g/cm^3: 3.1 [ALF95]

146

Compound: Ammonium dithiocarbamate
Synonym: ammonium sulfocarbamate
Formula: NH$_2$CSS(NH$_4$)
Molecular Formula: CH$_6$N$_2$S$_2$
Molecular Weight: 110.204
CAS RN: 513-74-6
Properties: yellow, lustrous; ortho-rhomb; decomposes
in air, odor of H$_2$S; prepared from carbon
disulfide and ammonia; used to precipitate
metals and metal sulfides, and in the synthesis
of heterocyclic compounds [MER06]
Solubility: s H$_2$O [MER06]
Density, g/cm^3: 1.451 [MER06]
Melting Point, °C: decomposes at 99 [MER06]
Reactions: reversible exothermic
transition at 63°C [MER06]

147
Compound: Ammonium ferric chromate
Synonym: ferric ammonium chromate
Formula: $(NH_4)Fe(CrO_4)_2$
Molecular Formula: $Cr_2FeH_4NO_8$
Molecular Weight: 305.871
CAS RN: 7789-08-4
Properties: carmine red microcryst powd; can be prepared by adding ammonia to a solution of CrO_3 and ferric nitrate hexahydrate [MER06]
Solubility: i H_2O [MER06]

148
Compound: Ammonium ferric citrate
Synonym: ferric ammonium citrate
Molecular Formula: $C_6H_{5+4y}Fe_xN_yO_7$
CAS RN: 1185-57-5
Properties: undetermined structure; reddish-brown granules, red scales, or brownish-yellow powd, and green form; deliq; light sensitive; preparation: $Fe(OH)_3$ addition to aq solution of citric acid and ammonia; uses: blueprints, photography, for iron deficiency [MER06]
Solubility: v s H_2O, i alcohol [MER06]
Reactions: light causes reduction to ferrous salt [MER06]

149
Compound: Ammonium ferric oxalate trihydrate
Synonym: ferric ammonium oxalate trihydrate
Formula: $(NH_4)_3Fe(C_2O_4)_3 \cdot 3H_2O$
Molecular Formula: $C_6H_{18}FeN_3O_{15}$
Molecular Weight: 428.065
CAS RN: 13268-42-3
Properties: bright green; monocl, hygr, prismatic cryst; sensitive to light; prepared by addition of ammonium binoxalate and ferric hydroxide; used in blueprint photography, and to color aluminum and aluminum alloys [ALD93] [MER06] [HAW93]
Solubility: v s H_2O; i alcohol [MER06]
Density, g/cm³: 1.780 [ALD93]
Melting Point, °C: decomposes at 160–170 [MER06]
Reactions: minus $3H_2O$ by 100°C [MER06]

150
Compound: Ammonium ferric sulfate dodecahydrate
Synonym: ferric alum
Formula: $(NH_4)Fe(SO_4)_2 \cdot 12H_2O$
Molecular Formula: $FeH_{28}NO_{20}S_2$
Molecular Weight: 482.194
CAS RN: 10138-04-2

Properties: colorless to pale violet; effloresces; octahedral cryst; prepared by mixing solutions of ferric sulfate and ammonium sulfate with subsequent evaporation and crystallization; used in medicine, as a mordant to dye textiles, and as an astringent [HAW93] [MER06] [STR93]
Solubility: v s H_2O; i alcohol [MER06]
Density, g/cm³: 1.71 [MER06]
Melting Point, °C: 39–41 [ALD94]
Reactions: minus $12H_2O$ at 230°C [HAW93]

151
Compound: Ammonium ferricyanide trihydrate
Synonym: ammonium hexacyanoferrate(III) trihydrate
Formula: $(NH_4)_3Fe(CN)_6 \cdot 3H_2O$
Molecular Formula: $C_6H_{18}FeN_9O_3$
Molecular Weight: 320.113
CAS RN: 14221-48-8
Properties: red cryst; sensitive to light [MER06]
Solubility: s H_2O [MER06]

152
Compound: Ammonium ferrous sulfate hexahydrate
Synonyms: Mohr's salt, ferrous ammonium sulfate
Formula: $(NH_4)_2Fe(SO_4)_2 \cdot 6H_2O$
Molecular Formula: $FeH_{20}N_2O_{14}S_2$
Molecular Weight: 392.141
CAS RN: 10045-89-3
Properties: pale bluish green cryst powd; slowly oxidizes and effloresces in air; sensitive to light; prepared from a mixture of ferrous sulfate and ammonium sulfate solutions, with subsequent evaporation and crystallization; used in analytical chemistry and in metallurgy [HAW93] [MER06] [STR93] [ALF93]
Solubility: g/100 g H_2O: 12.5 (0°C), 17.2 (10°C), 26.4 (20°C), 33 (30°C), 46 (40°C); i alcohol [MER06] [LAN05]
Density, g/cm³: 1.865 [HAW93]
Melting Point, °C: decomposes at 100–110 [HAW93]

153
Compound: Ammonium fluoride
Formula: NH_4F
Molecular Formula: FH_4N
Molecular Weight: 37.037
CAS RN: 12125-01-8
Properties: white, deliq cryst; hex, a=0.439 nm, c=0.702 nm; tends to lose NH_3 to form the more stable $NH_4F \cdot HF$; can be obtained by adding ammonia to an ice cold 40% HF solution; NH_4F is used as a laboratory reagent [KIR78] [CIC73]

Solubility: g/100 g soln, H_2O: 41.72 (0°C), 45.5 ± 0.3 (25°C), 54.05 (80°C) [KRU93]
Density, g/cm³: 1.009 [CIC73]
Melting Point, °C: decomposes [CIC73]
Reactions: decomposed by hot water to NH_3 and HF [MER06]

154

Compound: Ammonium fluoroborate
Synonym: ammonium tetrafluoroborate
Formula: NH_4BF_4
Molecular Formula: BF_4H_4N
Molecular Weight: 104.844
CAS RN: 13826-83-0
Properties: white powd; ortho-rhomb below 205°C, a = 0.7278 nm, b = 0.9072 nm, c = 0.5678 nm; cub above 205°C; can be prepared by reacting ammonia gas with fluoroboric acid [KIR78] [ALF93]
Solubility: g/100 mL H_2O: 3.09 (−1.0°C), 5.26 (−1.5°C), 10.85 (−2.7°C), 12.20 (0°C), 25 (16°C), 25.83 (25°C), 44.09 (50°C), 67.50 (75°C), 98.93 (100°C), 113.7 (108.5°C); s HF 19.89% (0°C) [KIR78]
Density, g/cm³: 1.871 [HAW93]
Melting Point, °C: decomposes at 487 [KIR78]; sublimes at 220 [ALF93]

155

Compound: Ammonium fluorosulfonate
Formula: NH_4SO_3F
Molecular Formula: FH_4NO_3S
Molecular Weight: 117.101
CAS RN: 13446-08-7
Properties: long, colorless needles [KIR78]
Solubility: s H_2O, alcohol, methanol [KIR78]
Melting Point, °C: 245 [KIR78]

156

Compound: Ammonium formate
Synonyms: formic acid, ammonium salt
Formula: $HCOONH_4$
Molecular Formula: CH_5NO_2
Molecular Weight: 63.056
CAS RN: 540-69-2
Properties: deliq cryst or white powd; formed by reaction of ammonia and formic acid; has been used to precipitate metals [HAW93] [STR93]
Solubility: g/100 g H_2O: 102 (0°C), 143 (20°C), 204 (40°C), 311 (60°C), 533 (80°C) [LAN05]; s alcohol [HAW93]
Density, g/cm³: 1.26 [HAW93]

Melting Point, °C: 119–121 [STR93]
Boiling Point, °C: decomposes at 180 [CRC10]

157

Compound: Ammonium germanium oxalate hydrate
Synonym: ammonium tris(oxalato)germanate
Formula: $(NH_4)_2Ge(C_2O_4)_3 \cdot xH_2O$
Molecular Formula: $C_6H_8GeN_2O_{12}$ (anhydrous)
Molecular Weight: 372.745 (anhydrous)
CAS RN: 67786-11-2
Properties: hygr [ALD94]

158

Compound: Ammonium heptafluorotantalate
Formula: $(NH_4)_2TaF_7$
Molecular Formula: $F_7H_8N_2Ta$
Molecular Weight: 350.014
CAS RN: 12022-02-5
Properties: hygr [ALD93]

159

Compound: Ammonium hexabromoosmiate(IV)
Formula: $(NH_4)_2OsBr_6$
Molecular Formula: $Br_6H_8N_2Os$
Molecular Weight: 705.731
CAS RN: 24598-62-7
Properties: black powd [ALF93]

160

Compound: Ammonium hexabromoplatinate(IV)
Formula: $(NH_4)_2PtBr_6$
Molecular Formula: $Br_6H_8N_2Pt$
Molecular Weight: 710.581
CAS RN: 17363-02-9
Properties: reddish brown powd [ALF93]
Density, g/cm³: 4.26 [ALD94]
Melting Point, °C: decomposes at 145 [ALF93]

161

Compound: Ammonium hexachloroiridate(III)
Formula: $(NH_4)_3IrCl_6$
Molecular Formula: $Cl_6H_{12}IrN_3$
Molecular Weight: 459.048
CAS RN: 15752-05-3
Properties: olive green powd [ALF93]

162

Compound: Ammonium hexachloroiridate(III) monohydrate
Formula: $(NH_4)_3IrCl_6 \cdot H_2O$

Molecular Formula: $Cl_6H_{14}IrN_3O$
Molecular Weight: 477.063
CAS RN: 29796-57-4
Properties: hygr [ALD93]

163
Compound: Ammonium hexachloroiridate(IV)
Formula: $(NH_4)_2IrCl_6$
Molecular Formula: $Cl_6H_8IrN_2$
Molecular Weight: 441.010
CAS RN: 16940-92-4
Properties: black cryst powd [ALD93] [STR93]
Solubility: g/100 g H_2O: 0.556 (0°C), 0.706 (10°C), 0.77 (20°C), 1.21 (30°C), 1.57 (40°C), 2.46 (60°C), 4.38 (80°C), decomposes (90°C) [LAN05]
Density, g/cm³: 2.856 [ALD93]
Melting Point, °C: decomposes [STR93]

164
Compound: Ammonium hexachloroosmiate(IV)
Synonym: ammonium osmium chloride
Formula: $(NH_4)_2OsCl_6$
Molecular Formula: $Cl_6H_8N_2Os$
Molecular Weight: 439.023
CAS RN: 12125-08-5
Properties: red powd or dark red octahedral cryst [MER06]
Solubility: s H_2O, alcohol [MER06]
Density, g/cm³: 2.930 [ALD93]
Melting Point, °C: sublimes at 170 [ALF93]

165
Compound: Ammonium hexachloropalladate(IV)
Formula: $(NH_4)_2PdCl_6$
Molecular Formula: $Cl_6H_8N_2Pd$
Molecular Weight: 355.213
CAS RN: 19168-23-1
Properties: hygr; reddish brown cryst [ALF93] [STR93]
Density, g/cm³: 2.418 [STR93]
Melting Point, °C: decomposes [STR93]

166
Compound: Ammonium hexachloroplatinate(IV)
Synonym: ammonium chloroplatinate
Formula: $(NH_4)_2PtCl_6$
Molecular Formula: $Cl_6H_8N_2Pt$
Molecular Weight: 443.873
CAS RN: 16919-58-7
Properties: cub, reddish orange cryst or yellow powd [KIR82] [MER06]

Solubility: g/100 g H_2O: 0.289 (0°C), 0.374 (10°C), 0.499 (20°C), 0.637 (30°C), 0.815 (40°C), 1.44 (60°C), 2.16 (80°C), 2.61 (90°C), 3.36 (100°C) [LAN05]; i alcohol [HAW93]
Density, g/cm³: 3.065 [ALD93]
Melting Point, °C: decomposes at >380 [KIR81]

167
Compound: Ammonium hexachlororhodate(III) monohydrate
Formula: $(NH_4)_3RhCl_6 \cdot H_2O$
Molecular Formula: $Cl_6H_{14}N_3ORh$
Molecular Weight: 387.752
CAS RN: 15336-18-2
Properties: red hygr cryst [STR93] [ALF95]

168
Compound: Ammonium hexachlororuthenate(IV)
Formula: $(NH_4)_2RuCl_6$
Molecular Formula: $Cl_6H_8N_2Ru$
Molecular Weight: 349.863
CAS RN: 18746-63-9
Properties: red cryst [ALF93]

169
Compound: Ammonium hexacyanoferrate(II) monohydrate
Synonym: ammonium ferrocyanide
Formula: $(NH_4)_4Fe(CN)_6 \cdot H_2O$
Molecular Formula: $C_6H_{18}FeN_{10}O$
Molecular Weight: 302.120
CAS RN: 14481-29-9
Properties: yellowish green powd; hygr; light sensitive [STR93] [ALD94]; a trihydrate with similar properties is listed in [CRC10] and [MER06]
Solubility: trihydrate s H_2O [MER06]
Melting Point, °C: decomposes [MER06]
Reactions: minus NH_3 on exposure to air and light [MER06]

170
Compound: Ammonium hexafluoroaluminate
Synonym: ammonium aluminum fluoride
Formula: $(NH_4)_3AlF_6$
Molecular Formula: $AlF_6H_{12}N_3$
Molecular Weight: 195.087
CAS RN: 7784-19-2
Properties: white powd; cub; does not attack glass; preparation: NH_4F and $Al(OH)_3$; uses: preparation of pure NH_4F [STR93] [MER06]
Solubility: s H_2O [MER06]

Density, g/cm³: 1.78 [MER06]
Melting Point, °C: thermally stable
to above 100 [MER06]

171
Compound: Ammonium hexafluorogallate
Formula: $(NH_4)_3GaF_6$
Molecular Formula: $F_6GaH_{12}N_3$
Molecular Weight: 237.828
CAS RN: 14639-94-2
Properties: octahedra; preparation: reaction of $Ga(OH)_3$,
HF, NH_4F; uses: preparation of GaF_3 [MER06]
Reactions: heating in air forms Ga_2O_3, forms GaN
if heated in a vacuum at 200°C [MER06]

172
Compound: Ammonium hexafluorogermanate
Formula: $(NH_4)_2GeF_6$
Molecular Formula: $F_6GeH_8N_2$
Molecular Weight: 222.677
CAS RN: 16962-47-3
Properties: white cryst [HAW93]
Solubility: s H_2O; i alcohol [HAW93]
Density, g/cm³: 2.564 [HAW93]
Melting Point, °C: sublimes at 380 [HAW93]

173
Compound: Ammonium hexafluorophosphate
Formula: $(NH_4)PF_6$
Molecular Formula: F_6H_4NP
Molecular Weight: 163.003
CAS RN: 16941-11-0
Properties: white cryst; square leaflets or
tables; cub [MER06] [STR93]
Solubility: 74.8 g/100 mL H_2O (20°C) [MER06]
Density, g/cm³: 2.180 [MER06]
Melting Point, °C: decomposes at 58 [CIC73]

174
Compound: Ammonium hexafluorosilicate
Synonym: cryptohalite
Formula: $(NH_4)_2SiF_6$
Molecular Formula: $F_6H_8N_2Si$
Molecular Weight: 178.153
CAS RN: 16919-19-0
Properties: white, odorless, cryst powd; cub
or trig; used in soldering flux, to etch
glass, and in pesticides [MER06]
Solubility: s H_2O; i alcohol [MER06] [HAW93]
Density, g/cm³: 2.011 [ALD93]
Melting Point, °C: decomposes [ALF93]

175
Compound: Ammonium hexafluorotitanate dihydrate
Formula: $(NH_4)_2TiF_6 \cdot 2H_2O$
Molecular Formula: $F_6H_{12}N_2O_2Ti$
Molecular Weight: 233.964
CAS RN: 16962-40-6
Properties: white cryst [STR93]
Melting Point, °C: decomposes [CRC10]

176
Compound: Ammonium hydrogen acetate
Synonym: ammonium acetate double salt
Formula: $(NH_4)H(CH_3COO)_2$
Molecular Formula: $C_4H_{11}NO_4$
Molecular Weight: 137.136
CAS RN: 631-61-8
Properties: prepared by dissolving ammonium
acetate in hot acetic acid; the product
crystallizes as long, deliq needles [KIR78]
Solubility: v s H_2O [KIR78]
Melting Point, °C: 66 [KIR78]

177
Compound: Ammonium hydrogen arsenate
Formula: $(NH_4)_2HAsO_4$
Molecular Formula: $AsH_9N_2O_4$
Molecular Weight: 176.004
CAS RN: 7784-44-3
Properties: white powd, efflorescing in
air with loss of NH_3 [HAW93]
Solubility: s H_2O, decomposed by hot H_2O [HAW93]
Density, g/cm³: 1.99 [HAW93]
Melting Point, °C: decomposes [CRC10]

178
Compound: Ammonium hydrogen borate trihydrate
Formula: $(NH_4)HB_4O_7 \cdot 3H_2O$
Molecular Formula: $B_4H_{11}NO_{10}$
Molecular Weight: 228.332
CAS RN: 10135-84-9
Properties: colorless cryst; effloresces evolving
NH_3; obtained by the reaction of NH_4OH
and boric acid, then cryst of product; used
to fireproof materials [HAW93]
Solubility: s H_2O [HAW93]
Density, g/cm³: 2.38–2.95 [HAW93]

179
Compound: Ammonium hydrogen carbonate
Synonym: ammonium bicarbonate

Formula: NH_4HCO_3
Molecular Formula: CH_5NO_3
Molecular Weight: 79.056
CAS RN: 1066-33-7
Properties: white cryst; vapor pressure, kPa: 59 (25.4°C), 122 (34.2°C), 201 (40.7°C), 278 (45.0°C), 395 (50.0°C), 72.1 (54.0°C), 108.4 (59.2°C); manufactured by passing CO_2 gas through NH_4OH solution, which evolves heat, followed by crystallization of ammonium bicarbonate; used as a leavening agent for cookies, crackers, and in fire-extinguisher materials [HAW93] [KIR78]
Solubility: g/100 g soln, H_2O: 10.6 (0°C); 19.9 (25°C); 78.0 (100°C); equilibrium solid phase, NH_4HCO_3 [KRU93]; i alcohol [HAW93]
Density, g/cm³: 1.586 [KIR78]
Melting Point, °C: 107.5 (v rapid heating) [MER06]
Boiling Point, °C: sublimes at ~60 with decomposition [MER06]
Reactions: decomposed by hot H_2O [MER06]

180
Compound: Ammonium hydrogen citrate
Synonym: diammonium citrate
Formula: $(NH_4)_2HC_6H_5O_7$
Molecular Formula: $C_6H_{14}N_2O_7$
Molecular Weight: 226.186
CAS RN: 3012-65-5
Properties: granules or cryst; white; stable in air; used in metal cleaning applications [KIR78] [MER06]
Solubility: 100 g/100 mL H_2O at 25°C; sl s alcohol; i ether [KIR78]
Density, g/cm³: 1.48 [MER06]

181
Compound: Ammonium hydrogen fluoride
Synonym: ammonium bifluoride
Formula: NH_4HF_2
Molecular Formula: F_2H_5N
Molecular Weight: 57.044
CAS RN: 1341-49-7
Properties: white, deliq flakes; ortho-rhomb cryst; enthalpy of fusion 19.1 kJ/mol; enthalpy of vaporization 65.3 kJ/mol; enthalpy of solution 20.3 kJ/mol; enthalpy of dissociation to form NH_3 and HF 141.4 kJ/mol; can be prepared by dehydration of NH_4F solutions; can be used as a less hazardous substitute for HF [KIR78] [MER06] [HAW93]
Solubility: 41.5% in H_2O at 25°C; 1.73% in 90% alcohol at 25°C [KIR78] [HAW93]
Density, g/cm³: 1.50 [KIR78]
Melting Point, °C: 126.1 [KIR78]
Boiling Point, °C: 239.5 [KIR78]

182
Compound: Ammonium hydrogen oxalate hemihydrate
Synonym: ammonium binoxalate
Formula: $NH_4HC_2O_4 \cdot 1/2H_2O$
Molecular Formula: $C_2H_6NO_{45}$
Molecular Weight: 116.022
CAS RN: 37541-72-3
Properties: monohydrate: rhomb; uses: removes ink stains [MER06]
Solubility: monohydrate: s 25 parts H_2O; sl s alcohol [MER06]
Melting Point, °C: decomposes at 220 [ALD93]

183
Compound: Ammonium hydrogen oxalate monohydrate
Synonym: ammonium binoxalate monohydrate
Formula: $NH_4OOCCOOH \cdot H_2O$
Molecular Formula: $C_2H_7NO_5$
Molecular Weight: 125.081
CAS RN: 5972-72-5
Properties: colorless rhomb cryst; obtained from solution of NH_4OH and oxalic acid, then cryst; used to remove ink from fabrics [HAW93] [MER06]
Solubility: s 25 parts H_2O [MER06]
Density, g/cm³: 1.56 [MER06]
Melting Point, °C: decomposed by heating [HAW93]

184
Compound: Ammonium hydrogen phosphate
Synonym: ammonium phosphate dibasic
Formula: $(NH_4)_2HPO_4$
Molecular Formula: $H_9N_2O_4P$
Molecular Weight: 132.055
CAS RN: 7783-28-0
Properties: cryst or powd; gradually loses about 8% NH_3 when exposed to air [MER06]
Solubility: g/100 g soln, H_2O: 36.4 (0°C), 41.0 (25°C), 58.6 (100°C); equilibrium solid phase $(NH_4)_2HPO_4$ [KRU93]
Density, g/cm³: 1.619 [ALF93]
Melting Point, °C: decomposes at 155 [ALF93]

185
Compound: Ammonium hydrogen phosphite monohydrate
Formula: $(NH_4)_2HPO_3 \cdot H_2O$
Molecular Formula: $H_{11}N_2O_4P$
Molecular Weight: 134.072
CAS RN: 51503-61-8
Properties: deliq cryst [MER06]
Solubility: s H_2O [MER06]

Density, g/cm³: 1.619 (anhydrous) [CRC10]
Melting Point, °C: decomposes at
155 (anhydrous) [CRC10]

186
Compound: Ammonium hydrogen sulfate
Synonym: ammonium bisulfate
Formula: $(NH_4)HSO_4$
Molecular Formula: H_5NO_4S
Molecular Weight: 115.111
CAS RN: 7803 63 6
Properties: white powd; cryst, deliq; used as
catalyst in organic syntheses and in hair wave
formulations [HAW93] [STR93] [MER06]
Solubility: 100 g/100 g H_2O (0°C) [CIC73];
i acetone, alcohol [HAW93]
Density, g/cm³: 1.78 [CIC73]
Melting Point, °C: 146.9 [CIC73]

187
Compound: Ammonium hydrogen sulfide
Synonyms: ammonium hydrosulfide,
ammonium bisulfide
Formula: NH_4HS
Molecular Formula: H_5NS
Molecular Weight: 51.113
CAS RN: 12124-99-1
Properties: white; tetr or ortho-rhomb; vapor
pressure 99.7 kPa at 32.1°C; readily sublimes;
produced from stoichiometric amounts of
NH_3 and H_2S at 0°C [MER06] [KIR78]
Solubility: 128.1 g/100 g H_2O (0°C), decomposes in
hot H_2O [CIC73]; i ether, benzene [KIR78]
Density, g/cm³: 1.17 [CIC73]
Melting Point, °C: 118 [CIC73]
Reactions: decomposes to H_2S and NH_3
at room temp [MER06]

188
Compound: Ammonium hydrogen sulfite
Synonym: ammonium bisulfite
Formula: $(NH_4)HSO_3$
Molecular Formula: H_5NO_3S
Molecular Weight: 99.111
CAS RN: 10192-30-0
Properties: cryst; available commercially only in
solution form; uses: preservative [MER06]
Solubility: g/100 g soln, H_2O: 72.2 ± 0.4 (0°C),
78.41 (25°C), 85.4 ± 0.7 (60°C); equilibrium
solid phase, $(NH_4)_2S_2O_5$ [KRU93]
Density, g/cm³: 2.03 [CIC73]
Melting Point, °C: sublimes at 150 [CIC73]

189
Compound: Ammonium hydrogen tartrate
Synonym: ammonium bitartrate
Formula: $(NH_4)OOCCH(OH)CH(OH)COOH$
Molecular Formula: $C_4H_9NO_6$
Molecular Weight: 167.118
CAS RN: 3095-65-6
Properties: white, odorless cryst; used in
baking powd [HAW93] [MER06]
Solubility: g/100 g H_2O: 1.00 (0°C), 1.88
(10°C), 2.70 (20°C) [LAN05]; s acids,
alkalies; i alcohol [HAW93]
Density, g/cm³: 1.68 [MER06]
Melting Point, °C: decomposes [CRC10]

190
Compound: Ammonium hydrogen
tetraborate dihydrate
Formula: $(NH_4)HB_4O_7 \cdot 2H_2O$
Molecular Formula: $B_4H_9NO_9$
Molecular Weight: 210.317
CAS RN: 12228-86-3
Properties: cryst [ALF93]
Melting Point, °C: decomposes (tetraborate) [CRC10]

191
Compound: Ammonium hydroxide
Synonym: ammonia solution
Formula: NH_4OH
Molecular Formula: H_5NO
Molecular Weight: 35.046
CAS RN: 1336-21-6
Properties: colorless liq; strong odor of
ammonia; dissolved NH_3 concentration
ranges up to 30% [HAW93]
Density, g/cm³: 0.900 [ALD94]

192
Compound: Ammonium hypophosphite
Formula: $NH_4H_2PO_2$
Molecular Formula: H_6NO_2P
Molecular Weight: 83.028
CAS RN: 7803-65-8
Properties: hygr; deliq cryst or white granules; uses:
catalyst in the manufacture of polyamide [MER06]
Solubility: g/100 mL H_2O: 83 (room temp)
[KRU93]; s alcohol [HAW93]
Density, g/cm³: 1.634 [CRC10]
Melting Point, °C: on heating, decomposes
evolving phosphine gas [MER06]

193
Compound: Ammonium iodate
Formula: NH_4IO_3
Molecular Formula: H_4INO_3
Molecular Weight: 192.941
CAS RN: 13446-09-8
Properties: white, granular powd;
    oxidizing agent [HAW93]
Solubility: g/100 mL H_2O: 2.06 (15°C),
    14.5 (101°C) [CRC10]
Density, g/cm³: 3.309 [CRC10]
Melting Point, °C: decomposes at 150 [CRC10] [ALF93]

194
Compound: Ammonium iodide
Formula: NH_4I
Molecular Formula: H_4IN
Molecular Weight: 144.943
CAS RN: 12027-06-4
Properties: white, odorless, very hygr; tetr; yellow
    to brown on exposure to air due to liberation
    of some iodine; vapor pressure, kPa: 31.3
    (360°C), 54.1 (380°C), 89.8 (400°C), 101.1
    (405°C); enthalpy of fusion 21.00 kJ/mol; formed
    by reacting NH_3 with I_2; finds some use in
    photography [KIR78] [MER06] [CRC10]
Solubility: g/100 g soln, H_2O: 60.55 (0°C), 64.65
    (25°C), 71.3 (100°C); equilibrium solid phase NH_4I
    [KRU93]; g/100 g H_2O: 154.2 (0°C), 172.3 (20°C),
    250.3 (100°C) [KIR78]; s alcohol [HAW93]
Density, g/cm³: 2.514 [CIC73]
Melting Point, °C: partly decomposes and
    sublimes if heated [MER06]
Boiling Point, °C: 220 in vacuum [CIC73]

195
Compound: Ammonium magnesium
    chloride hexahydrate
Synonym: carnallite
Formula: $NH_4Cl \cdot MgCl_2 \cdot 6H_2O$
Molecular Formula: $Cl_3H_{16}MgNO_6$
Molecular Weight: 256.793
CAS RN: 39733-35-2
Properties: deliq cryst [MER06]
Solubility: s in 6 parts H_2O [MER06]
Density, g/cm³: 1.456 [CRC10]
Reactions: minus $2H_2O$, 100°C [CRC10]

196
Compound: Ammonium mercuric chloride dihydrate
Synonym: mercuric ammonium chloride

Formula: $(NH_4)_2HgCl_4 \cdot 2H_2O$
Molecular Formula: $Cl_4H_{12}HgN_2O_2$
Molecular Weight: 440.852
CAS RN: 33445-15-7
Properties: powd; uses: ointment for chronic
    eczema, antifungal [MER06]
Solubility: s H_2O [MER06]

197
Compound: Ammonium metatungstate hexahydrate
Formula: $(NH_4)_6W_7O_{24} \cdot 6H_2O$
Molecular Formula: $H_{36}N_6O_{30}W_7$
Molecular Weight: 1887.188
CAS RN: 12028-48-7
Properties: white cryst; formula also given as
    $(NH_4)_6H_2W_{12}O_{40}$ and $(NH_4)_6W_{12}O_{39} \cdot 4H_2O$;
    preparation: reaction of NH_4OH with tungstic acid;
    uses: to prepare tungsten alloys and ammonium
    phosphotungstate [HAW93] [STR93] [ALD94]
Solubility: s H_2O; i alcohol [HAW93]

198
Compound: Ammonium metavanadate
Synonym: ammonium vanadate
Formula: NH_4VO_3
Molecular Formula: H_4NO_3V
Molecular Weight: 116.979
CAS RN: 7803-55-6
Properties: white or sl yellow cryst powd;
    loses H_2O and NH_3 on heating; obtained
    by precipitation with NH_4Cl from alkaline
    V_2O_5 solutions; used as a catalyst in dyes
    and varnishes [HAW93] [MER06]
Solubility: g/100 g H_2O: 0.48 (20°C), 0.84 (30°C),
    1.32 (40°C), 2.42 (60°C) [LAN05]
Density, g/cm³: 2.326 [HAW93]
Melting Point, °C: decomposes at 210 [HAW93]

199
Compound: Ammonium molybdate tetrahydrate
Synonym: ammonium molybdate(VI)
Formula: $(NH_4)_6Mo_7O_{24} \cdot 4H_2O$
Molecular Formula: $H_{32}Mo_7N_6O_{28}$
Molecular Weight: 1235.857
CAS RN: 12054-85-2
Properties: colorless, sl greenish or yellowish cryst;
    obtained by crystallization from a solution
    of MoO_3 with excess NH_3; used to prepare
    specialty catalysts [KIR81] [MER06]
Solubility: s 2.3 parts H_2O [MER06];
    i alcohol [HAW93]

Density, g/cm³: 2.498 [ALD93]
Melting Point, °C: decomposes [HAW93]

200
Compound: Ammonium nickel chloride hexahydrate
Synonym: nickel ammonium chloride
Formula: $NH_4Cl \cdot NiCl_2 \cdot 6H_2O$
Molecular Formula: $Cl_3H_{16}NNiO_6$
Molecular Weight: 291.181
CAS RN: 16122-03-5
Properties: green cryst; deliq; obtained from salt
   solutions by crystallization; used as a dye mordant
   and for metal finishing [KIR81] [HAW93]
Solubility: s H_2O [HAW93]
Density, g/cm³: 1.65 [HAW93]

201
Compound: Ammonium nickel sulfate hexahydrate
Synonym: nickel ammonium sulfate
Formula: $(NH_4)_2SO_4 \cdot NiSO_4 \cdot 6H_2O$
Molecular Formula: $H_{20}N_2NiO_{14}S_2$
Molecular Weight: 394.989
CAS RN: 7785-20-8
Properties: bluish green cryst; decomposes on heating;
   obtained from an aq solution by crystallization;
   used as a dye mordant and in metal finishing
   [KIR81] [MER06] [HAW93] [ALF93]
Solubility: g $(NH_4)_2Ni(SO_4)_2/100$ g H_2O: 1.00 (0°C),
   4.00 (10°C), 6.50 (20°C), 9.20 (30°C), 12.0 (40°C),
   17.0 (60°C) [LAN05]; i alcohol [HAW93]
Density, g/cm³: 1.923 [MER06]

202
Compound: Ammonium nitrate
Formula: NH_4NO_3
Molecular Formula: $H_4N_2O_3$
Molecular Weight: 80.043
CAS RN: 6484-52-2
Properties: white, transparent, hygr cryst; vapor
   pressure of saturated NH_4NO_3 solutions,
   kPa: 0.85 (10°C), 1.5 (20°C), 2.5 (30°C), 3.9
   (40°C); five cryst forms, stable at: α, < −18°C;
   β, −18°C–32.1°C; γ, 32.1°C–84.2°C; δ,
   84.2°C–125.2°C; ε, 125.7°C–169.6°C; enthalpy
   of fusion 6.40 kJ/mol; enthalpy of neutralization
   51.8 kJ/mol; manufactured by neutralization of
   HNO_3 solutions with NH_3; used as a fertilizer
   and in explosives [MER06] [KIR78] [CRC10]
Solubility: g/100 g soln, H_2O: 54.2 (0°C),
   68.2 (25°C), 90.3 (100°C); equilibrium
   solid phase NH_4NO_3 [KRU93]
Density, g/cm³: 1.725 [CIC73]

Melting Point, °C: 169.6 [CRC10]
Boiling Point, °C: 210 (11 mm Hg) [CIC73]
Reactions: decomposes ~210°C to $H_2O + N_2O$ [MER06]
Thermal Expansion Coefficient: coefficient of expansion
   is 0.000920 (20°C), 0.001113 (100°C) [KIR78]

203
Compound: Ammonium nitrite
Formula: NH_4NO_2
Molecular Formula: $H_4N_2O_2$
Molecular Weight: 64.044
CAS RN: 13446-48-5
Properties: white-yellowish cryst; uncertain stability;
   can be made by adding a solution of barium nitrite
   to ammonium sulfate solution [KIR78] [CRC10]
Solubility: g/100 g soln, H_2O: 56.0 (1.4°C),
   64.3 (19.15°C); equilibrium solid
   phase NH_4NO_2 [KRU93]
Density, g/cm³: 1.69 [CIC73]
Melting Point, °C: explodes at 60–70, producing N_2,
   H_2O, and other products [KIR78] [CIC73]

204
Compound: Ammonium nitroferricyanide
Synonym: ammonium nitroprusside
Formula: $(NH_4)_2Fe(CN)_5NO$
Molecular Formula: $C_5H_8FeN_8O$
Molecular Weight: 252.017
CAS RN: 14402-70-1
Properties: red to brownish red cryst [MER06]
Solubility: s H_2O, alcohol [MER06]

205
Compound: Ammonium O,O-diethyldithiophosphate
Formula: $(C_2H_5O)_2P(S)SNH_4$
Molecular Formula: $C_4H_{14}NO_2PS_2$
Molecular Weight: 203.267
CAS RN: 1068-22-0
Properties: cryst [ALF95]
Melting Point, °C: 164–165 [ALF95]

206
Compound: Ammonium oleate
Synonym: ammonium soap
Formula: $CH_3(CH_2)_7CH=CH(CH_2)_7COONH_4$
Molecular Formula: $C_{18}H_{37}NO_2$
Molecular Weight: 299.498
CAS RN: 544-60-5
Properties: yellowish brown paste, softens at
   10°C–13°C; used to emulsify products,
   and in cosmetics [MER06] [HAW93]

Solubility: s H_2O (27°C); sl s acetone [MER06]
Melting Point, °C: 21–22 [MER06]

207
Compound: Ammonium oxalate
Synonyms: ethanedioic acid, diammonium salt
Formula: $(NH_4)_2C_2O_4$
Molecular Formula: $C_2H_8N_2O_4$
Molecular Weight: 124.097
CAS RN: 1113-38-8
Properties: colorless cryst; used as an analytical
    chemistry reagent, to manufacture oxalates,
    for removing rust and scale [HAW93]
Solubility: g/100 g soln, H_2O: 2.31 (0°C), 4.95
    (25°C), 25.73 (100°C) [KRU93]
Density, g/cm³: 1.5 [ALF93]

208
Compound: Ammonium oxalate monohydrate
Synonyms: ethanedioic acid, diammonium
    salt monohydrate
Formula: $(NH_4)_2C_2O_4 \cdot H_2O$
Molecular Formula: $C_2H_{10}N_2O_5$
Molecular Weight: 142.111
CAS RN: 6009-70-7
Properties: white, granular, odorless cryst; ortho-
    rhomb; used in analytical chemistry and to
    remove rust [MER06] [ALF93] [HAW93]
Solubility: 1 g/20 mL H_2O, 2.6 mL boiling H_2O [MER06]
Density, g/cm³: 1.502 [HAW93]
Melting Point, °C: decomposes on heating [HAW93]

209
Compound: Ammonium palmitate
Synonyms: hexadecanoic acid, ammonium salt
Formula: $CH_3(CH_2)_{14}COONH_4$
Molecular Formula: $C_{16}H_{35}NO_2$
Molecular Weight: 273.460
CAS RN: 593-26-0
Properties: yellowish white powd,
    softens at 3°C–4°C [MER06]
Solubility: s H_2O [MER06]; s hot
    alcohols, benzene [HAW93]
Melting Point, °C: 21–23 [MER06]

210
Compound: Ammonium pentaborate tetrahydrate
Formula: $(NH_4)B_5O_8 \cdot 4H_2O$
Molecular Formula: $B_5H_{12}NO_{12}$
Molecular Weight: 272.150
CAS RN: 12229-12-8

Properties: two cryst forms: ortho-rhomb (α)
    and monocl (β); very stable with respect to
    loss of NH_3, losing less than 1% NH_3 at 50°C
    and only 2% at 200°C; prepared from an
    aq solution of NH_3 and boric acid; used in
    flameproofing formulations [HAW93] [KIR78]
Solubility: % anhydrous by weight, H_2O: 4.00 (0°C),
    8.03 (25°C), 14.4 (50°C), 30.3 (90°C) [KIR78]
Density, g/cm³: 1.58 [KIR78]
Reactions: minus 75% of the H_2O
    content at 50°C [KIR78]

211
Compound: Ammonium pentachlororhodate(III)
    monohydrate
Formula: $(NH_4)_2RhCl_5 \cdot H_2O$
Molecular Formula: $Cl_5H_{10}N_2ORh$
Molecular Weight: 334.261
CAS RN: 63771-33-5
Properties: red cryst [ALF93]
Melting Point, °C: 210–230, decomposes [ALF93]

212
Compound: Ammonium pentachlororuthenate(III)
    monohydrate
Formula: $(NH_4)_2RuCl_5 \cdot H_2O$
Molecular Formula: $Cl_5H_{10}N_2ORu$
Molecular Weight: 332.425
CAS RN: 68133-88-0
Properties: cryst powd [ALF93]

213
Compound: Ammonium pentachlorozincate
Synonym: zinc ammonium chloride
Formula: $(NH_4)_3ZnCl_5$
Molecular Formula: $Cl_5H_{12}N_3Zn$
Molecular Weight: 296.769
CAS RN: 14639-98-6
Properties: ortho-rhomb cryst; hygr; uses:
    manufacture of dry cell batteries, welding
    flux, soldering, galvanizing [MER06]
Solubility: v s H_2O [MER06]
Density, g/cm³: 1.81 [MER06]
Melting Point, °C: sublimes at 340 [MER06]

214
Compound: Ammonium perchlorate
Formula: NH_4ClO_4
Molecular Formula: ClH_4NO_4
Molecular Weight: 117.490
CAS RN: 7790-98-9

Properties: white cryst; oxidizing agent; ortho-rhomb,
a=0.9202 nm, b=0.5816 nm, c=0.7449 nm;
prepared from NH_4OH, HCl and sodium chlorate
with subsequent cryst; cub >240°C, a=0.763 nm;
used as an oxidizer in rocket propellants [CIC73]
[HAW93] [ALD93] [KIR79] [ALF93]

Solubility: g/100 g soln, H_2O: 10.8 (0°C),
19.8 (25°C), 46.9 (100°C); mol/kg H_2O:
1.019 (0°C), 2.122 (25°C) [KRU93]

Density, g/cm³: 1.95 [CIC73]

Melting Point, °C: can explode; decomposed
by heating [MER06] [ALD93]

Reactions: transition from ortho-rhomb
to cub at 513 K [KIR79]

215

Compound: Ammonium permanganate

Formula: NH_4MnO_4

Molecular Formula: H_4MnNO_4

Molecular Weight: 136.975

CAS RN: 13446-10-1

Properties: dark purple; rhomb;
oxidizing agent [KIR78]

Solubility: g/100 g H_2O: 8 (15°C), 86 (25°C) [KIR78]

Density, g/cm³: 2.22 [KIR78]

Melting Point, °C: decomposes above 70 [KIR78]

216

Compound: Ammonium peroxydisulfate

Synonym: ammonium persulfate

Formula: $(NH_4)_2S_2O_8$

Molecular Formula: $H_8N_2O_8S_2$

Molecular Weight: 228.204

CAS RN: 7727-54-0

Properties: odorless, plate-like or prismatic cryst;
monocl; strong oxidizing agent [MER06]

Solubility: g/100 g soln, H_2O: 37.0 (0°C),
45.5 (25°C), 62.0 (80°C) [KRU93]

Density, g/cm³: 1.982 [ALF93]

Melting Point, °C: decomposes on heating
forming O_2 and $(NH_4)_2S_2O_7$ [MER06]

217

Compound: Ammonium perrhenate

Synonym: ammonium perrhenate(VII)

Formula: NH_4ReO_4

Molecular Formula: H_4NO_4Re

Molecular Weight: 268.244

CAS RN: 13598-65-7

Properties: colorless powd; weak oxidizing
agent [HAW93] [ALF93]

Solubility: sl s cold H_2O; s hot H_2O [HAW93]

Density, g/cm³: 3.97 [ALD93]

Melting Point, °C: decomposes at 365 [HAW93]

218

Compound: Ammonium phosphate dibasic

Synonym: diammonium hydrogen phosphate

Formula: $(NH_4)_2HPO_4$

Molecular Formula: $H_9N_2O_4P$

Molecular Weight: 132.07

CAS RN: 7883-28-0

Properties: odorless cryst or powd; used
to fireproof textiles, paper, wood, and
vegetable fibers, to impregnate lamp wicks,
and in soldering fluxes [MER06]

Solubility: 1 g/1.7 mL H_2O, 1 g in 0.5 mL
boiling H_2O [MER06]

Density, g/cm³: 1.619 [ALD94]

Melting Point, °C: decomposes at 155 [CRC10]

219

Compound: Ammonium phosphomolybdate

Synonym: ammonium molybdophosphate

Formula: $(NH_4)_3PO_4 \cdot 12MoO_3$

Molecular Formula: $H_{12}Mo_{12}N_3O_{40}P$

Molecular Weight: 1876.345

CAS RN: 54723-94-3

Properties: heavy, yellow, cryst powd [MER06]

Solubility: 0.2 g/L H_2O (20°C) [MER06]

Melting Point, °C: decomposes [ALF93]

220

Compound: Ammonium phosphotungstate dihydrate

Synonym: ammonium tungstophosphate

Formula: $(NH_4)_3PO_4 \cdot 12WO_3 \cdot 2H_2O$

Molecular Formula: $H_{16}N_3O_{42}PW_{12}$

Molecular Weight: 2967.176

CAS RN: 1311-90-6

Properties: microcryst powd [MER06]

Solubility: 0.15 g/L H_2O (20°C) [MER06]

221

Compound: Ammonium picrate

Synonym: ammonium carbazoate

Formula: $(NH_4)C_6H_2N_3O_7$

Molecular Formula: $C_6H_6N_4O_7$

Molecular Weight: 246.137

CAS RN: 131-74-8

Properties: bright yellow; ortho-rhomb; explodes
easily from heat or shock [MER06]

Solubility: 1 g/100 mL H_2O (20°C) [MER06]

Density, g/cm³: 1.72 [MER06]
Reactions: explodes at 423 [CRC10]

222
Compound: Ammonium polysulfide
Formula: $(NH_4)_2S_x$
Molecular Formula: $(NH_4)_2S_x$
CAS RN: 9080-17-5
Properties: exists only in solution; yellow unstable solution with an odor of H_2S; prepared by dissolving H_2S gas in 28% NH_4OH solution, then dissolving excess sulfur; used as an analytical chemistry reagent and as an insect spray [HAW93]
Reactions: decomposed by acids, evolving H_2S [HAW93]

223
Compound: Ammonium salicylate
Synonyms: salicylic acid, monoammonium salt
Formula: $(NH_4)C_7H_5O_3$
Molecular Formula: $C_7H_9NO_3$
Molecular Weight: 155.153
CAS RN: 528-94-9
Properties: odorless, lustrous cryst or white, cryst powd; discolors on exposure to light; loses some NH_3 on long exposure to air [MER06]
Solubility: 1 g/1 mL H_2O [MER06]
Boiling Point of 336.3°C: sublimes [CRC10]

224
Compound: Ammonium selenate
Formula: $(NH_4)_2SeO_4$
Molecular Formula: $H_8N_2O_4Se$
Molecular Weight: 179.035
CAS RN: 7783-21-3
Properties: white powd; colorless; monocl cryst; used to mothproof [HAW93] [MER06] [STR93]
Solubility: g/100 g soln, H_2O: 54.02 (25°C) [KRU93]; i alcohol [HAW93]
Density, g/cm³: 2.194 [MER06]
Melting Point, °C: decomposed by heating [MER06]

225
Compound: Ammonium selenite
Formula: $(NH_4)_2SeO_3$
Molecular Formula: $H_8N_2O_3Se$
Molecular Weight: 163.035
CAS RN: 7783-19-9
Properties: white or sl reddish cryst; deliq; used to color glass [HAW93] [MER06]
Solubility: g/100 g soln, H_2O: 49.21 (1°C), 54.70 (25°C), 69.08 (70°C); equilibrium solid phase, $(NH_4)_2SeO_3H_2O$ [KRU93]
Melting Point, °C: decomposes [MER06]

226
Compound: Ammonium sesquicarbonate
Synonym: hartshorn
Formula: $NH_2COONH_4 \cdot NH_4HCO_3$
Molecular Formula: $C_2H_{11}N_3O_5$
Molecular Weight: 157.126
CAS RN: 10361-29-2
Properties: mixture of ammonium bicarbonate and ammonium carbamate; colorless; hard translucent cryst with ammonia odor; obtained from a mixture of ammonium sulfate and calcium carbonate by sublimation; used in baking powd, as a mordant in dyeing, and as an expectorant; changes to bicarbonate in air [MER06] [HAW93]
Solubility: slowly s in 4 parts H_2O, decomposes in hot H_2O, evolving NH_3 and CO_2 [MER06] [HAW93]
Melting Point, °C: volatilizes at ~60 [MER06]

227
Compound: Ammonium stearate
Synonyms: octadecanoic acid, ammonium salt
Formula: $CH_3(CH_2)_{16}COONH_4$
Molecular Formula: $C_{18}H_{39}NO_2$
Molecular Weight: 301.514
CAS RN: 1002-89-7
Properties: yellowish white powd, softens at 1.7°C–4.4°C; used in vanishing creams, brushless shaving, and other cosmetics [MER06] [HAW93]
Solubility: s H_2O [MER06]; s hot toluene [HAW93]
Density, g/cm³: 0.89 [HAW93]
Melting Point, °C: 21–24 [MER06]

228
Compound: Ammonium sulfamate
Synonyms: sulfamic acid, monoammonium salt
Formula: $(NH_4)NH_2SO_3$
Molecular Formula: $H_6N_2O_3S$
Molecular Weight: 114.125
CAS RN: 7773-06-0
Properties: white, hygr cryst; made by reacting urea with fuming sulfuiric acid; used to flameproof textiles and in metal finishing [HAW93] [MER06]
Solubility: v s H_2O [MER06]
Melting Point, °C: 131 [MER06]
Boiling Point, °C: decomposes at 160 [HAW93]

229
Compound: Ammonium sulfate
Synonym: mascagnite
Formula: $(NH_4)_2SO_4$
Molecular Formula: $H_8N_2O_4S$

Molecular Weight: 132.141
CAS RN: 7783-20-2
Properties: brownish gray to white, odorless, ortho-
rhomb cryst; manufactured from NH_3 and sulfuric
acid; used as a nitrogen fertilizer, for water
treatment, and as a food additive [MER06] [HAW93]
Solubility: g/100 g soln, H_2O: 41.35 (0°C); 43.30 (25°C),
50.61 (100°C); equilibrium solid phase, $(NH_4)_2SO_4$
[KRU93]; 70.6 g/100 g H_2O (0°C), 103.8 g/100 g
H_2O (100°C); i alcohol, acetone [KIR78]
Density, g/cm³: 1.77 [MER06]
Melting Point, °C: decomposes at >280 [MER06]

230
Compound: Ammonium sulfide
Formula: $(NH_4)_2S$
Molecular Formula: H_8N_2S
Molecular Weight: 68.143
CAS RN: 12135-76-1
Properties: yellowish orange liq; cryst below −18°C;
stable only below −18°C, at higher temperatures
loses NH_3 to form NH_4HS and polysulfides; made
from NH_3 and H_2S; used in textile industry and
photography [STR93] [MER06] [KIR78] [HAW93]
Solubility: s H_2O, alcohol, alkalies [HAW93]
Density, g/cm³: 0.997 [ALD94]
Melting Point, °C: decomposes [HAW93]
Reactions: evolves NH_3 to form hydrosulfide
above −18°C [KIR78]

231
Compound: Ammonium sulfite
Formula: $(NH_4)_2SO_3$
Molecular Formula: $H_8N_2O_3S$
Molecular Weight: 116.140
CAS RN: 17026-44-7
Properties: white hygr cryst [CRC10]
Solubility: 64.2 g/100 g H_2O [CRC10]

232
Compound: Ammonium sulfite monohydrate
Synonyms: sulfurous acid, diammonium salt
Formula: $(NH_4)_2SO_3 \cdot H_2O$
Molecular Formula: $H_{10}N_2O_4S$
Molecular Weight: 134.156
CAS RN: 7783-11-1
Properties: colorless cryst; loses H_2O and gradually
oxidizes to $(NH_4)_2SO_4$ when heated in air;
hygr; used in medicine, metal lubricants, as a
chemical reducing agent [MER06] [HAW93]
Solubility: g anhydrous/100 g soln, H_2O: 32.40 (0°C),
39.29 (25°C), 60.44 (100°C); equilibrium phase:
monohydrate (0,25°C), anhydrous (100°C) [KRU93]

Density, g/cm³: 1.41 [HAW93]
Melting Point, °C: sublimes with
decomposition at 150 [HAW93]

233
Compound: Ammonium tartrate
Synonyms: tartaric acid, diammonium salt
Formula: $(NH_4)_2C_4H_4O_6$
Molecular Formula: $C_4H_{12}N_2O_6$
Molecular Weight: 184.148
CAS RN: 3164-29-2
Properties: white cryst; decomposes when heated;
used in medicine and in industry [HAW93]
Solubility: g/100 g H_2O: 45.0 (0°C), 55.0 (10°C),
63.0 (20°C), 70.5 (30°C), 76.5 (40°C), 86.9
(60°C) [LAN05]; s alcohol [HAW93]
Density, g/cm³: 1.601 [HAW93]
Melting Point, °C: decomposes [CRC10]

234
Compound: Ammonium tellurate
Synonym: ammonium tellurate(VI)
Formula: $(NH_4)_2TeO_4$
Molecular Formula: $H_8N_2O_4Te$
Molecular Weight: 227.675
CAS RN: 13453-06-0
Properties: −60 mesh with 99.5% purity;
white powd [ALF93] [STR93]
Density, g/cm³: 3.024 [STR93]
Melting Point, °C: decomposes [STR93]

235
Compound: Ammonium tetraborate tetrahydrate
Synonym: ammonium borate
Formula: $(NH_4)_2B_4O_7 \cdot 4H_2O$
Molecular Formula: $B_4H_{16}N_2O_{11}$
Molecular Weight: 263.377
CAS RN: 12228-87-4
Properties: tetr colorless cryst; unstable,
has an appreciable ammonia vapor
pressure; uses: fireproofing wood and
textiles [MER06] [STR93] [KIR78]
Solubility: % anhydrous by weight (H_2O): 3.75 (0°C),
9.00 (25°C), 21.2 (50°C), 52.7 (90°C) [KIR78]
Density, g/cm³: 1.58 [KIR78]
Melting Point, °C: decomposes [CRC10]

236
Compound: Ammonium tetrachloroaluminate
Synonym: aluminum ammonium chloride
Formula: NH_4AlCl_4
Molecular Formula: $AlCl_4H_4N$

Molecular Weight: 186.831
CAS RN: 7784-14-7
Properties: white cryst; preparation:
    from $AlCl_3$ and NH_4Cl; finds use in
    treating furs [MER06] [HAW93]
Solubility: s H_2O, ether [MER06]
Melting Point, °C: 304 [MER06]

237
Compound: Ammonium
    tetrachloroaurate(III) hydrate
Formula: $(NH_4)AuCl_4 \cdot xH_2O$
Molecular Formula: $AuCl_4H_4N$ (anhydrous)
Molecular Weight: 356.816 (anhydrous)
CAS RN: 13874-04-9
Properties: yellow cryst [STR93]
Melting Point, °C: 520 [ALD93]

238
Compound: Ammonium tetrachloropalladate(II)
Formula: $(NH_4)_2PdCl_4$
Molecular Formula: $Cl_4H_8N_2Pd$
Molecular Weight: 284.308
CAS RN: 13820-40-1
Properties: reddish brown powd; olive
    green cryst [STR93] [ALF93]
Density, g/cm³: 2.17 [ALD93]
Melting Point, °C: decomposes [CRC10]

239
Compound: Ammonium tetrachloroplatinate(II)
Synonym: ammonium platinous chloride
Formula: $(NH_4)_2PtCl_4$
Molecular Formula: $Cl_4H_8N_2Pt$
Molecular Weight: 372.968
CAS RN: 13820-41-2
Properties: dark ruby red cryst; used in
    photography [HAW93] [MER06]
Solubility: s H_2O [MER06]; i alcohol [HAW93]
Density, g/cm³: 2.936 [ALD93]
Melting Point, °C: decomposes at 140–150 [HAW93]

240
Compound: Ammonium tetrachlorozincate
Formula: $(NH_4)_2ZnCl_4$
Molecular Formula: $Cl_4H_8N_2Zn$
Molecular Weight: 243.298
CAS RN: 14639-97-5
Properties: white, ortho plates; hygr [CRC10]
Density, g/cm³: 1.879 [CRC10]
Melting Point, °C: decomposes at 150 [CRC10]

241
Compound: Ammonium tetrafluoroantimonate(III)
Formula: NH_4SbF_4
Molecular Formula: F_4H_4NSb
Molecular Weight: 215.789
CAS RN: 14972-90-8
Properties: white powd [STR93]

242
Compound: Ammonium tetrafluoroborate
Formula: NF_4BF_4
Molecular Formula: BF_8N
Molecular Weight: 104.844
CAS RN: 13826-83-0
Properties: white powd; ortho [CRC10]
Density, g/cm³: 1.871 [CRC10]
Melting Point, °C: decomposes at 487 [CRC10]

243
Compound: Ammonium
    tetranitrodiamminecobaltate(III)
Synonym: Erdmann's salt
Formula: $NH_4[Co(NH_3)_2(NO_2)_4]$
Molecular Formula: $CoH_{10}N_7O_8$
Molecular Weight: 295.054
CAS RN: 13600-89-0
Properties: reddish, pale brown
    rhomb [KIR79] [MER06]
Solubility: s H_2O [LID94]
Density, g/cm³: 1.876 [KIR79]

244
Compound: Ammonium tetrathiocyano-
    diammonochromate(III) monohydrate
Synonym: Reinecke salt
Formula: $NH_4[Cr(NH_3)_2(SCN)_4] \cdot H_2O$
Molecular Formula: $C_4H_{12}CrN_7OS_4$
Molecular Weight: 354.446
CAS RN: 13573-16-5
Properties: dark red cryst or red powd; can be
    produced by fusion of ammonium thiocyanate
    with ammonium dichromate; used to
    precipitate primary and secondary amines
    and as a reagent for mercury [MER06]
Solubility: sl s cold H_2O, s hot H_2O; can
    decompose in aq solutions [MER06]
Melting Point, °C: decomposes at 268–272 [ALD94]

245

Compound: Ammonium tetrathiomolybdate
Formula: $(NH_4)_2MoS_4$
Molecular Formula: $H_8MoN_2S_4$
Molecular Weight: 260.281
CAS RN: 15060-55-6
Properties: dark red cryst powd; preparation: by passing H_2S through a solution of ammonium molybdate [STR93] [ALF93] [KIR81]

246

Compound: Ammonium tetrathiotungstate
Formula: $(NH_4)_2WS_4$
Molecular Formula: $H_8N_2S_4W$
Molecular Weight: 348.181
CAS RN: 13862-78-7
Properties: orange cryst powd; H_2S odor; sensitive to heat; commonly made by adding NH_3 to a solution of tungstic acid, followed by saturation with H_2S; can be used as a source of WS_2 by decomposition in a nonoxidizing atm [KIR83] [HAW93]
Solubility: s H_2O, ammonia solutions [HAW93]
Density, g/cm³: 2.71 [ALD93]
Melting Point, °C: decomposes [HAW93]

247

Compound: Ammonium tetrathiovandate(IV)
Formula: $(NH_4)_3VS_4$
Molecular Formula: $H_{12}N_3S_4V$
Molecular Weight: 233.321
CAS RN: 14693-56-2
Properties: dark violet cryst [STR93]

248

Compound: Ammonium thiocyanate
Synonym: ammonium rhodanide
Formula: NH_4SCN
Molecular Formula: CH_4N_2S
Molecular Weight: 76.122
CAS RN: 1762-95-4
Properties: colorless; deliq cryst; formed from solution of NH_4CN with sulfur on boiling; used as fertilizer, in chemicals, and for dyeing fabrics [MER06] [HAW93]
Solubility: g/100 g soln, H_2O: 64.95 (26.33°C), 81.73 (71.53°C); equilibrium solid phase, NH_4SCN [KRU93]; s alcohol, acetone, ammonia solutions [HAW93]
Density, g/cm³: 1.3057 [HAW93]
Melting Point, °C: ~149 [MER06]
Boiling Point, °C: decomposes at 170 [HAW93]

249

Compound: Ammonium thiosulfate
Synonym: ammonium hyposulfite
Formula: $(NH_4)_2S_2O_3$
Molecular Formula: $H_8N_2O_3S_2$
Molecular Weight: 148.207
CAS RN: 7783-18-8
Properties: white cryst; used in photography, fungicides, silver plating, and hair wave preparations [MER06] [HAW93]
Solubility: 103.3 g/100 g H_2O at 100°C [CIC73]
Density, g/cm³: 1.679 [ALD93]
Melting Point, °C: decomposes at 150 [MER06]

250

Compound: Ammonium titanium oxalate monohydrate
Synonym: ammonium bis(oxalato)oxotitanate(IV)
Formula: $(VH_4)_2TiO(C_2O_4)_2 \cdot H_2O$
Molecular Formula: $C_4H_{10}N_2O_{10}Ti$
Molecular Weight: 293.997
CAS RN: 10580-03-7
Properties: hygr cryst; used as a mordant to dye leather and cellulosic fabrics [HAW93] [MER06] [ALD93]
Solubility: v s H_2O [MER06]

251

Compound: Ammonium tungstate(VI)
Formula: $(NH_4)_{10}W_{12}O_{41}$
Molecular Formula: $H_{40}N_{10}O_{41}W_{12}$
Molecular Weight: 3042.44
CAS RN: 11120-25-5
Properties: cryst powd [CRC10]
Solubility: s H_2O; i EtOH [CRC10]
Density, g/cm³: 2.3 [CRC10]

252

Compound: Ammonium tungstate pentahydrate
Synonym: ammonium tungstate(VI)
Formula: $(NH_4)_{10}W_{12}O_{41} \cdot 5H_2O$
Molecular Formula: $H_{50}N_{10}O_{46}W_{12}$
Molecular Weight: 3132.516
CAS RN: 1311-93-9
Properties: 99.999% pure plates or cryst powd; usually prepared by crystallization from a boiling solution [ALF93] [MER06]
Solubility: v s H_2O [MER06]
Density, g/cm³: 2.3 [ALF93]

253

Compound: Ammonium uranate(VI)
Synonym: ammonium diuranate

Formula: $(NH_4)_2U_2O_7$
Molecular Formula: $H_8N_2O_7U_2$
Molecular Weight: 624.131
CAS RN: 7783-22-4
Properties: −80 mesh with 99.5% purity; reddish yellow amorphous powd [MER06] [CER91]
Solubility: i H_2O [MER06]

254
Compound: Ammonium uranium fluoride
Formula: $UO_2(NH_4)_3F_5$
Molecular Formula: $F_5H_{12}N_3O_2U$
Molecular Weight: 419.135
CAS RN: 18433-40-4
Properties: green-yellow monocl cryst [CRC10]

255
Compound: Ammonium valerate
Synonyms: pentanoic acid, ammonium salt
Formula: $CH_3(CH_2)_3COONH_4$
Molecular Formula: $C_5H_{13}NO_2$
Molecular Weight: 119.164
CAS RN: 42739-38-8
Properties: very hygr cryst; used to flavor foods [HAW93] [MER06]
Solubility: v s H_2O [MER06]
Melting Point, °C: 108 [MER06]

256
Compound: Ammonium zirconyl carbonate dihydrate
Synonym: zirconium ammonium carbonate
Formula: $(NH_4)_3ZrOH(CO_3)_3 \cdot 2H_2O$
Molecular Formula: $C_3H_{17}N_3O_{12}Zr$
Molecular Weight: 378.404
CAS RN: 12616-24-9
Properties: large prisms from H_2O; unstable in air, gradually evolving CO_2 and NH_3; aq solution decomposes rapidly above 60°C; used in paper and textile water repellents [HAW93] [MER06]
Solubility: s H_2O [MER06]; decomposed by dil acid, alkalies [HAW93]
Density, g/cm³: aq solution: 1.238 [MER06]

257
Compound: Antimony
Synonym: stibium
Formula: Sb
Molecular Formula: Sb
Molecular Weight: 121.757
CAS RN: 7440-36-0

Properties: silver-white metal available with 99.999% purity; hex, a = 0.4307 nm, c = 1.1273 nm; hardness 3.0–3.5 Mohs; enthalpy of fusion 19.866 kJ/mol; enthalpy of vaporization 195.1 kJ/mol; electronegativity 1.82; electrical resistivity (0°C) 37 μohm · cm; uses: in alloys such as solder, type metal, and bearings, and as an evaporated semiconductor film [KIR78] [COT88] [CER91]
Solubility: oxidized by HNO_3 [HAW93]
Density, g/cm³: 6.697 [KIR78]
Melting Point, °C: 630.7 [KIR78]
Boiling Point, °C: 1587 [KIR78]; 1635, 1440 [MER06]
Reactions: forms $SbCl_3$ and $SbCl_5$ from reaction of Sb and Cl_2 [KIR78]
Thermal Conductivity, W/(m · K): 25.5 (0°C), 24.4 (25°C), 21.9 (100°C) [KIR78] [HO72]
Thermal Expansion Coefficient: coefficient of linear expansion at 20°C is $8–11 \times 10^{-6}$ m/(m · °C) [KIR78]

258
Compound: Antimony arsenide
Formula: Sb_3As
Molecular Formula: $AsSb_3$
Molecular Weight: 440.193
CAS RN: 12255-36-6
Properties: 6 mm pieces and smaller of 99.999% purity [CER91]

259
Compound: Antimony iodide sulfide
Formula: SbIS
Molecular Formula: ISSb
Molecular Weight: 280.727
CAS RN: 13816-38-1
Properties: dark red; −20 mesh [CRC10] [ALF95]
Solubility: i H_2O [CRC10]
Melting Point, °C: 392 [CRC10]
Boiling Point, °C: decomposes [CRC10]

260
Compound: Antimony phosphide
Formula: SbP
Molecular Formula: PSb
Molecular Weight: 152.731
CAS RN: 53120-23-3
Properties: black powd; −100 mesh with 99.5% purity [CER91] [AES93]

261
Compound: Antimony(III) acetate
Synonym: antimony triacetate

Formula: $Sb(CH_3COO)_3$
Molecular Formula: $C_6H_9O_6Sb$
Molecular Weight: 298.893
CAS RN: 3643-76-3
Properties: off-white powd; can be prepared
  by dissolution of Sb(III) salt in acetic acid,
  followed by crystallization [KIR78] [STR93]

262
Compound: Antimony(III) bromide
Synonym: antimony tribromide
Formula: $SbBr_3$
Molecular Formula: Br_3Sb
Molecular Weight: 361.472
CAS RN: 7789-61-9
Properties: yellow deliq cryst; enthalpy of vaporization
  59 kJ/mol; entropy of vaporization at 560°C is
  94.9 J/(mol·K); prepared by reacting Sb and Br_2;
  used as a mordant [HAW93] [KIR78] [CRC10]
Solubility: decomposed by H_2O; s dil
  HCl, HBr [HAW93] [MER06]
Density, g/cm³: 4.148 [HAW93]
Melting Point, °C: 96.0 [KIR78]
Boiling Point, °C: 280 [CRC10]

263
Compound: Antimony(III) chloride
Synonym: antimony trichloride
Formula: $SbCl_3$
Molecular Formula: Cl_3Sb
Molecular Weight: 228.118
CAS RN: 10025-91-9
Properties: colorless, ortho-rhomb cryst; very hygr;
  enthalpy of vaporization 45.19 kJ/mol; entropy
  of vaporization at 496°C is 93.3 J/(mol·K);
  enthalpy of fusion 12.70 kJ/mol; can be prepared
  from Sb or Sb_2O_3 and conc HCl; used as a
  catalyst to chlorinate olefins and to polymerize
  hydrocarbons [HAW93] [KIR78] [CRC10]
Solubility: g/100 g soln, H_2O: 601.6 (0°C),
  988.1 (25°C), infinite (72°C); solid phase at
  equilibrium: $SbCl_3$ [KRU93]; s $CHCl_3$ (22%),
  CCl_4 (13%), CS_2 and benzene [KIR78]
Density, g/cm³: 3.14 [MER06]
Melting Point, °C: 73.4 [KIR78]
Boiling Point, °C: 220.3 [CRC10]
Reactions: gradual hydrolysis to SbOCl in H_2O [MER06]

264
Compound: Antimony(III) fluoride
Synonym: antimony trifluoride
Formula: SbF_3
Molecular Formula: F_3Sb

Molecular Weight: 178.755
CAS RN: 7783-56-4
Properties: white to gray powd; hygr; ortho-rhomb;
  enthalpy of fusion 21.4 kJ/mol; entropy of fusion
  38.2 J/(mol·K); enthalpy of vaporization at
  298°C is 102.8 kJ/mol; entropy of vaporization
  at 25°C is 175.8 kJ/(mol·K); vapor pressure at
  mp is 26.34 kPa; slowly hydrolyzes in H_2O;
  can be prepared by dissolution of Sb_2O_3 in
  anhydrous HF; used as a fluorinating agent
  for organic compounds [HAW93] [KIR78]
Solubility: g/100 g H_2O: 384.7 (0°C), 492.4
  (25°C); 154 g/100 mL in methanol; i benzene,
  chlorobenzene, heptane [KRU93] [KIR78]
Density, g/cm³: 4.379 [MER06]
Melting Point, °C: 292 [COT88]
Boiling Point, °C: 319 [STR93]
Reactions: transforms from SbF_3 to $Sb_3O_2(OH)_2F_3$,
  then SbOF at 100°C [KIR78]

265
Compound: Antimony(III) hydride
Synonym: stibine
Formula: SbH_3
Molecular Formula: H_3Sb
Molecular Weight: 124.784
CAS RN: 7803-52-3
Properties: colorless gas; slowly decomposes at room
  temp, readily at 200°C; flammable material; enthalpy
  of vaporization 21.3 kJ/mol; distibine, Sb_2H_4, 14939-
  42-5, has been reported; preparation: by adding
  acid to a metal antimonide; used as η-type gas
  dopant in Si semiconductors [KIR78] [CRC10]
Solubility: sl s H_2O; s CS_2, ethanol [KIR78]
Density, g/cm³: 4.344 (air = 1.000), 15°C;
  liq at bp 2.204 [KIR78]
Melting Point, °C: −88 [KIR78]
Boiling Point, °C: −17 [KIR78]
Reactions: decomposes at 200°C to Sb + H_2 [KIR78]

266
Compound: Antimony(III) iodide
Synonym: antimony triiodide
Formula: SbI_3
Molecular Formula: I_3Sb
Molecular Weight: 502.473
CAS RN: 7790-44-5
Properties: ruby red trig cryst; volatile at high
  temperatures; enthalpy of fusion at 444°C
  is 22.7 kJ/mol; entropy of fusion at 444°C
  is 51.5 J/(mol·K); enthalpy of sublimation
  101.6 kJ/mol at 298°C; enthalpy of vaporization
  68.6 kJ/mol; can be obtained by reacting Sb and
  I_2 [MER06] [HAW93] [KIR78] [CRC10]

Solubility: decomposed in H_2O; s CS_2,
   HCl; i alcohol, $CHCl_3$ [HAW93]
Density, g/cm³: 4.921 [MER06]
Melting Point, °C: 170.5 [KIR78]
Boiling Point, °C: 401 [CRC10]
Reactions: decomposed by water and
   air to SbOI [MER06]

267
Compound: Antimony(III) iodide sulfide
Formula: SbIS
Molecular Formula: ISSb
Molecular Weight: 280.729
CAS RN: 13816-38-1
Properties: dark red prisms/needles [CRC10]
Melting point 400°C [CRC10]

268
Compound: Antimony(III) nitrate
Formula: $Sb(NO_3)_3$
Molecular Formula: N_3O_9Sb
Molecular Weight: 307.775
CAS RN: 20328-96-5
Properties: can be obtained by dissolution of
   Sb(III) salt in HNO_3 solution, followed
   by crystallization [KIR78]
Solubility: hydrolyzes [KIR78]

269
Compound: Antimony(III) oxide
Synonym: valentinite
Formula: Sb_2O_3
Molecular Formula: O_3Sb_2
Molecular Weight: 291.518
CAS RN: 1317-98-2
Properties: colorless; ortho-rhomb;
   stable above 570°C [KIR78]
Density, g/cm³: 5.67 [KIR78]
Melting Point, °C: 656 [KIR78]
Boiling Point, °C: 1425 [KIR78]

270
Compound: Antimony(III) oxide
Synonyms: antimony trioxide, antimony white
Formula: Sb_2O_3
Molecular Formula: O_3Sb_2
Molecular Weight: 291.518
CAS RN: 1309-64-4

Properties: white, odorless, cryst powd; obtained by
   igniting Sb in air; used to flameproof materials
   and in paints; evaporated material of 99.9%
   purity used in dielectric interference filter for
   ultraviolet radiation [HAW93] [CER91]
Solubility: sl s H_2O; i organic solvents [KIR78];
   s conc HCl, H_2SO_4, alkalies [HAW93]
Density, g/cm³: 5.67 [HAW93]
Melting Point, °C: 656 [KIR78]
Boiling Point, °C: 1425 [KIR78];
   sublimes at 1550 [STR93]

271
Compound: Antimony(III) oxide
Synonym: senarmontite
Formula: Sb_2O_3
Molecular Formula: O_3Sb_2
Molecular Weight: 291.518
CAS RN: 12412-52-1
Properties: colorless; cub; stable below 570°C; can
   be prepared by heating Sb in air; used as a flame
   retardant for fabrics and as a catalyst [KIR78]
Solubility: v sl s H_2O; i organic solvents [KIR78]
Density, g/cm³: 5.2 [KIR78]
Melting Point, °C: 656, in absence of O_2 [KIR78]
Boiling Point, °C: 1425, partial sublimation [KIR78]

272
Compound: Antimony(III) oxychloride
Formula: SbOCl
Molecular Formula: ClOSb
Molecular Weight: 173.212
CAS RN: 7791-08-4
Properties: white, monocl cryst [CRC10]
Solubility: reac H_2O; i EtOH, eth
Density, g/cm³: 5.7

273
Compound: Antimony(III) perchlorate trihydrate
Formula: $Sb(ClO_4)_3 \cdot 3H_2O$
Molecular Formula: $Cl_3H_6O_{15}Sb$
Molecular Weight: 474.157
CAS RN: 65277-48-7
Properties: can be prepared by dissolution of
   Sb(III) salt in perchloric acid, followed
   by crystallization [KIR78]
Solubility: hydrolyzes [KIR78]

274
Compound: Antimony(III) phosphate
Formula: $SbPO_4$

Molecular Formula: O_4PSb
Molecular Weight: 216.731
CAS RN: 12036-46-3
Properties: can be prepared by dissolving
Sb(III) compound in H_3PO_4, followed
by crystallization [KIR78]
Solubility: hydrolyzes [KIR78]

275
Compound: Antimony(III) potassium oxalate trihydrate
Formula: $K_3Sb(C_2O_4)_3 \cdot 3H_2O$
Molecular Formula: $C_6H_6K_3O_{15}Sb$
Molecular Weight: 557.158
CAS RN: 5965-33-3 (anhydrous compound)
Properties: cryst powd (CRC10)
Solubility: s H_2O [CRC10]

276
Compound: Antimony(III) selenide
Synonym: antimony triselenide
Formula: Sb_2Se_3
Molecular Formula: Sb_2Se_3
Molecular Weight: 480.400
CAS RN: 1315-05-5
Properties: gray powd; obtained by passing
H_2Se through a solution of potassium
antimonyl tartrate [MER06]
Solubility: v sl s H_2O [MER06]
Melting Point, °C: 611 [MER06]

277
Compound: Antimony(III) sulfate
Synonym: antimonous sulfate
Formula: $Sb_2(SO_4)_3$
Molecular Formula: $O_{12}S_3Sb_2$
Molecular Weight: 531.711
CAS RN: 7446-32-4
Properties: white, cryst powd; deliq; can be
obtained by dissolution of Sb(III) salt in H_2SO_4,
followed by crystallization; used in matches and
pyrotechnics [MER06] [KIR78] [HAW93]
Solubility: s H_2O, but can form insol basic salt [MER06]
Density, g/cm³: 3.62 [HAW93]
Melting Point, °C: decomposes [ALF93]

278
Compound: Antimony(III) sulfide
Synonym: antimony orange
Formula: Sb_2S_3
Molecular Formula: S_3Sb_2

Molecular Weight: 339.718
CAS RN: 1345-04-6
Properties: black cryst (stibnite) or amorphous
reddish orange powd; amorphous material
prepared by passing H_2S through $SbCl_3$ solution;
used in the form of 99.9% pure material as a
sputtering target to produce infrared filter with
high index in red part of visible spectrum,
used in pyrotechnics, certain matches, in
manufacturing ruby glass; the nonahydrate is a
lemon-yellow cryst [KIR78] [HAW93] [CER91]
Solubility: i H_2O, s conc HCl [MER06]
Density, g/cm³: 4.562 [HAW93]
Melting Point, °C: 550 [KIR78]
Boiling Point, °C: ~1150 [STR93]

279
Compound: Antimony(III) telluride
Synonym: antimony tritelluride
Formula: Sb_2Te_3
Molecular Formula: Sb_2Te_3
Molecular Weight: 626.314
CAS RN: 1327-50-0
Properties: gray; 3–12 mm fused pieces of
99.999% purity [CER91] [CRC10]
Density, g/cm³: 6.5 [ALD94]
Melting Point, °C: 629 [CRC10]

280
Compound: Antimony(IV) oxide
Synonym: antimony tetroxide
Formula: β-Sb_2O_4
Molecular Formula: O_4Sb_2
Molecular Weight: 307.518
CAS RN: 1332-81-6
Properties: monocl; formed by heating
valentinite in dry air at 1130°C; composed
of half Sb(III) and half Sb(V); used as an
oxidation catalyst [STR93] [KIR78]
Density, g/cm³: 5.82 [CRC10]
Melting Point, °C: vaporizes [KIR78]
Reactions: minus O at 930 [CRC10]

281
Compound: Antimony(IV) oxide
Synonym: cervantite
Formula: α-Sb_2O_4
Molecular Formula: O_4Sb_2
Molecular Weight: 307.518
CAS RN: 1332-81-6

Properties: colorless; ortho-rhomb; composition half Sb(III), half Sb(V); formed by heating valentinite in air at 460°C–540°C; used as an oxidation catalyst [STR93] [KIR78]
Density, g/cm³: 4.07 [KIR78]
Melting Point, °C: vaporizes [KIR78]

282
Compound: Antimony(V) chloride
Synonym: antimony pentachloride
Formula: SbCl₅
Molecular Formula: Cl₅Sb
Molecular Weight: 299.024
CAS RN: 7647-18-9
Properties: reddish yellow or colorless (if pure), hygr oily liq; fumes in air; enthalpy of vaporization at 449°C is 43.45 kJ/mol; entropy of vaporization at 449°C is 95.44 J/(mol·K); made by action of chlorine on molten SbCl₃; useful for providing chlorine for reactions such as formation of ICl from I₂; decomposes if distilled [HAW93] [KIR78] [MER06]
Solubility: hydrolyzes in H₂O; s HCl [MER06]
Density, g/cm³: 2.34 [HAW93]
Melting Point, °C: 3.2 [KIR78]
Boiling Point, °C: 68 (1.82 kPa), 176 (extrapolated) [KIR78]

283
Compound: Antimony(V) fluoride
Formula: SbF₅
Molecular Formula: F₅Sb
Molecular Weight: 216.752
CAS RN: 127386-54-3
Properties: hygr visc liq [CRC10]
Solubility: reac H₂O
Density, g/cm³: 3.10 [CRC10]
Melting Point, °C: 8.3 [CRC10]
Boiling Point, °C: 141 [CRC10]

284
Compound: Antimony(V) dichlorotrifluoride
Formula: SbCl₂F₃
Molecular Formula: Cl₂F₃Sb
Molecular Weight: 249.660
CAS RN: 7791-16-4
Properties: viscous liq; made by reacting SbF₃ and Cl₂; used as a catalyst in fluorocarbon manufacturing [MER06] [HAW93]

285
Compound: Antimony(V) fluoride
Synonym: antimony pentafluoride

Formula: SbF₅
Molecular Formula: F₅Sb
Molecular Weight: 216.752
CAS RN: 7783-70-2
Properties: colorless, hygr, viscous liq; viscosity 460 mPa·s at 20°C; tendency to polymerize, can be prevented by addition of 1% anhydrous HF; can be prepared by direct fluorination of SbF₃ or Sb powd; used as a catalyst in fluorinating reactions [HAW93] [KIR78]
Solubility: reacts vigorously with H₂O, becoming hydrolyzed [HAW93] [KIR78]
Density, g/cm³: 3.145 (15.5°C) [KIR78]
Melting Point, °C: 7 [KIR78]
Boiling Point, °C: 142.7 [KIR78]
Reactions: reacts with I₂, S, NO₂, graphite [KIR78]

286
Compound: Antimony(V) oxide
Synonym: antimony pentoxide
Formula: Sb₂O₅
Molecular Formula: O₅Sb₂
Molecular Weight: 323.517
CAS RN: 1314-60-9
Properties: yellowish powd; cub; always somewhat hydrated; prepared by reacting Sb or Sb₂O₃ with conc HNO₃; used as a flame retardant for textiles [HAW93] [MER06]
Solubility: sl s H₂O, i HNO₃; dissolves slowly in warm HCl, KOH [MER06]
Density, g/cm³: 3.78 [MER06]
Melting Point, °C: decomposes [MER06]
Reactions: loses oxygen at 300°C [MER06]

287
Compound: Antimony(V) oxide hydrate
Synonym: antimonio acid
Formula: Sb₂O₅·xH₂O
Molecular Formula: O₅Sb₂ (anhydrous)
Molecular Weight: 323.517 (anhydrous)
CAS RN: 12712-36-6
Properties: cub yellowish powd; material with approximate composition Sb₂O₅·3-1/2H₂O is prepared by hydrolysis of SbCl₅ [KIR78] [MER06]
Solubility: sl s H₂O; i HNO₃; s KOH [KIR78]
Density, g/cm³: 3.78 [MER06]
Reactions: forms cub white Sb₆O₁₃ ~700°C [KIR78]

288
Compound: Antimony(V) oxychloride
Synonym: basic antimony chloride
Formula: SbOCl

Molecular Formula: ClOSb
Molecular Weight: 173.212
CAS RN: 7791-08-4
Properties: white, monocl cryst or powd; can be prepared by adding $SbCl_3$ to water; used in flameproofing textiles [HAW93] [MER06] [KIR78]
Solubility: hydrolyzed by H_2O; s HCl [MER06]; i alcohol, ether [HAW93]
Melting Point, °C: decomposes at 170 [HAW93]
Reactions: heating to 250°C gives $Sb_2O_5Cl_2$, to >320°C gives Sb_2O_3 [MER06]

289
Compound: Antimony(V) sulfide
Synonym: golden sulfide of antimony
Formula: Sb_2S_5
Molecular Formula: S_5Sb_2
Molecular Weight: 403.850
CAS RN: 1315-04-4
Properties: yellow to orange to red solid; amorphous; odorless; can be formed by boiling Sb_2S_3 and sulfur in alkaline media, followed by precipitation with HCl; finds use as a red pigment and in the vulcanization of rubber [HAW93] [KIR78]
Solubility: i H_2O; s HCl to evolve H_2S [MER06]
Density, g/cm³: 4.12 [STR93]
Melting Point, °C: decomposes at 75 [STR93]

290
Compound: Argon
Formula: Ar
Molecular Formula: Ar
Molecular Weight: 39.948
CAS RN: 7440-37-1
Properties: colorless, odorless, tasteless inert gas; air contains 9.340 µL/L of argon; enthalpy of vaporization 6.469 kJ/mol; enthalpy of fusion 1.12 kJ/mol; sonic velocity (101.32 kPa, 0°C) 307.8 m/s; viscosity (101.32 kPa, 25°C) 22.64 Pa·s; critical temp −122.29°C; critical pressure 48.3 atm; crystallizes as fcc; triple point, −189.37°C; used as a carrier gas and for sputtering/VLSI [KIR78] [MER06] [CRC10]
Solubility: 33.6 mL/1000 g H_2O (20°C) [KIR78]; Henry's law constants, k × 10⁻⁴: 3.974 (25.0°C), 5.359 (65.1°C), 5.342 (91.1°C), 3.812 (222.7°C), 2.541 (267.3°C), 1.870 (287.9°C) [POT78]
Density, g/cm³: gas, 101.3 kPa, 0°C, 0.0017838 [K3R78]; solid, 1.623 at triple point [MER06]
Melting Point, °C: −189.35 [CRC10]
Boiling Point, °C: −185.87 [KIR78]
Thermal Conductivity, W/(m·K): gas (101.32 kPa, 0°C): 1.694 [KIR78]

291
Compound: Argon fluoride
Formula: ArF
Molecular Formula: ArF
Molecular Weight: 58.946
CAS RN: 56617-31-3
Properties: unstable gas; used as a light emitting source in lasers [KIR78]

292
Compound: Arsenic(α)
Formula: α-As
Molecular Formula: As
Molecular Weight: 74.92159
CAS RN: 7440-38-2
Properties: gray, shiny, brittle, metallic looking; rhomb, a = 0.376 nm, c = 1.0548 nm; oxidizes to As_2O_3 in air; hardness 3.5 Mohs; enthalpy of fusion 27.44 kJ/mol; enthalpy of sublimation 31.974 kJ/mol; specific heat (25°C) 24.6 J/(mol·K); electrical resistivity (0°C) 26 µohm·cm; electronegativity 2.20; used in semiconductors [KIR78] [MER06] [COT88] [CER91] [CRC10]
Solubility: i H_2O; s conc HNO_3 [KIR78]
Density, g/cm³: 5.778 [KIR78]
Melting Point, °C: 817.1, 28 atm [ALD94]
Boiling Point, °C: sublimes at 615 [KIR78]
Reactions: reacts with conc $HNO_3 \rightarrow H_3AsO_4$ [KIR78]
Thermal Conductivity, W/(m·K): 50.2 (25°C) [ALD94]
Thermal Expansion Coefficient: 20°C, linear coefficient of thermal expansion is 5.6 µm/(m·°C) [KIR78]

293
Compound: Arsenic(β)
Formula: β-As
Molecular Formula: As
Molecular Weight: 74.92159
CAS RN: 7440-38-2
Properties: dark gray amorphous solid; electrical resistivity 107 ohm·cm [KIR78]
Density, g/cm³: 4.700 [KIR78]
Melting Point, °C: sublimes [KIR78]
Reactions: transformation from amorphous to cryst at 280°C [KIR78]

294
Compound: Arsenic acid
Formula: H_3AsO_4
Molecular Formula: AsH_3O_4
Molecular Weight: 141.944
CAS RN: 7778-39-4
Properties: Exists only in solution [CRC10]

295
Compound: Arsenic acid hemihydrate
Formula: $H_3AsO_4 \cdot 0.5H_2O$
Molecular Formula: $AsH_4O_{4.5}$
Molecular Weight: 150.951
CAS RN: 7778-39-4
Properties: white, hygr cryst [CRC10]
Solubility: v s H_2O, EtOH [CRC10]
Density, g/cm³: 2.5 [CRC10]
Melting Point, °C: 36.1 [CRC10]

296
Compound: Arsenic disulfide
Synonym: realgar
Formula: As_4S_4
Molecular Formula: As_4S_4
Molecular Weight: 427.950
CAS RN: 12279-90-2
Properties: red or orange solid; naturally occurring
   mineral; can be manufactured by heating iron pyrites
   and arsenopyrite; used in pyrotechnics [KIR78]
Solubility: i H_2O, hot HCl; s warm alkali [KIR78]
Density, g/cm³: 3.5 [MER06]
Melting Point, °C: 307 [KER78]
Boiling Point, °C: 565 [KIR78]
Reactions: transforms to black allotropic
   modification at 267°C [KIR78]

297
Compound: Arsenic(III) ethoxide
Formula: $As(C_2H_5O)_3$
Molecular Formula: $C_6H_{15}AsO_6$
Molecular Weight: 210.103
CAS RN: 3141-12-6
Properties: liq
Density, g/cm³: 1.21 [CRC10]
Boiling Point, °C: 166 [CRC10]

298
Compound: Arsenic hemiselenide
Formula: As_2Se
Molecular Formula: As_2Se
Molecular Weight: 228.803
CAS RN: 1303-35-1
Properties: black cryst with metallic luster;
   formed by melting stoichiometric amounts
   of As and Se in nitrogen atm; used in
   glass manufacturing [MER06]
Solubility: i most solvents; decomposed by
   boiling alkali hydroxides [MER06]

299
Compound: Arsenic(II) iodide
Synonym: arsenic diiodide
Formula: AsI_2
Molecular Formula: AsI_2
Molecular Weight: 328.731
CAS RN: 13770-56-4
Properties: red solid; formula also given as As_2I_4 [KIR78]
Solubility: s organic solvents [KIR78]
Melting Point, °C: 130 [KIR78]
Reactions: with $H_2O \rightarrow AsI_3$ and As [KIR78]

300
Compound: Arsenic(II) sulfide
Synonym: realgar
Formula: As_2S_2
Molecular Formula: As_2S_2
Molecular Weight: 213.975
CAS RN: 1303-32-8
Properties: reddish brown monocl powd;
   α and β forms [CRC10] [ALF95]
Density, g/cm³: α: 3.506; β: 3.254 [CRC10]
Melting Point, °C: 360 [ALF95]
Boiling Point, °C: 565 [ALF95]
Reactions: α to β at 267°C [CRC10]

301
Compound: Arsenic(III) bromide
Synonym: arsenic tribromide
Formula: $AsBr_3$
Molecular Formula: $AsBr_3$
Molecular Weight: 314.634
CAS RN: 7784-33-0
Properties: colorless to yellow lumps; deliq;
   ortho-rhomb; fumes in moist air; enthalpy of
   vaporization 41.8 kJ/mol; enthalpy of fusion
   11.70 kJ/mol; dielectric constant 8.33 (35°C); can
   be formed from As and Br_2 dissolved in CS_2;
   used in analytical chemistry and in medicine
   [STR93] [KIR78] [MER06] [CRC10]
Solubility: decomposed in H_2O, forming HBr, As_2O_3;
   miscible with ether, benzene [MER06]
Density, g/cm³: 3.66 [KIR78]
Melting Point, °C: 31.1 [CRC10]
Boiling Point, °C: 221 [KIR78]

302
Compound: Arsenic(III) chloride
Synonym: arsenic trichloride
Formula: $AsCl_3$
Molecular Formula: $AsCl_3$

Molecular Weight: 181.280
CAS RN: 7784-34-1
Properties: colorless or pale yellow oily liq; fumes
in air; enthalpy of vaporization 35.01 kJ/mol;
enthalpy of fusion 10.10 kJ/mol; decomposed
by ultraviolet light; may be obtained by
reaction of As and Cl_2; used as an intermediate
in organic preparations and in ceramics
[MER06] [KIR78] [HAW93] [CRC10]
Solubility: decomposed by H_2O, giving $As(OH)_3$
and HCl products [MER06], s conc HCl
and most organic solvents [HAW93]
Density, g/cm³: 2.205 [KIR78]
Melting Point, °C: −16 [MER06]
Boiling Point, °C: 130.2 [MER06]

303
Compound: Arsenic(III) fluoride
Synonym: arsenic trifluoride
Formula: AsF_3
Molecular Formula: AsF_3
Molecular Weight: 131.917
CAS RN: 7784-35-2
Properties: colorless liq; fumes in air; enthalpy
of vaporization 29.7 kJ/mol; enthalpy of
fusion 10.40 kJ/mol; can be prepared by
fluorinating As_2O_3 with H_2SO_4 and CaF_2;
used as a fluorinating agent and to synthesize
AsF_5 [MER06] [KIR78] [CRC10]
Solubility: hydrolyzed by H_2O; s alcohol,
ether, benzene [MER06]
Density, g/cm³: 2.666 [KIR78]
Melting Point, °C: −5.9 [CRC10]
Boiling Point, °C: 57.8 [CRC10]

304
Compound: Arsenic(III) iodide
Synonym: arsenic triiodide
Formula: AsI_3
Molecular Formula: AsI_3
Molecular Weight: 455.635
CAS RN: 7784-45-4
Properties: red powd; enthalpy of formation
−58.2 kJ/mol; entropy 213.0 J/(mol · K);
enthalpy of vaporization 59.3 kJ/mol;
decomposes slowly in air at 100°C, rapidly
at 200°C, to give As, I_2, As_2O_3; made by
precipitation from a hot $AsCl_3$-HCl solution
by the addition of KI [KIR78] [STR93]
Solubility: 1 g/12 mL H_2O; not easily
hydrolyzed [MER06] [KIR78]
Density, g/cm³: 4.39 (15°C) [KIR78]

Melting Point, °C: 140 [COT88]
Boiling Point, °C: 424 [CRC10]

305
Compound: Arsenic(III) oxide
Synonym: arsenolite
Formula: As_2O_3
Molecular Formula: As_2O_3
Molecular Weight: 197.841
CAS RN: 1327-53-3
Properties: white, odorless and tasteless powd; cub;
may be obtained by strongly heating As in air or
by roasting arsenopyrite, FeAsS; used in pigments,
ceramic enamels, insecticide [HAW93] [KIR78]
Solubility: 1.7 g/100 g H_2O (25°C); s acids and
alkalies [KIR78]; s glycerol [HAW93]
Density, g/cm³: 3.865 [HAW93]
Melting Point, °C: 275 [KIR78]
Reactions: sublimes freely above 135°C [KIR78]

306
Compound: Arsenic(III) oxide
Synonym: claudetite
Formula: As_2O_3
Molecular Formula: As_2O_3
Molecular Weight: 197.841
CAS RN: 1327-53-3
Properties: white powd; monocl; thermodynamically
stable form; can be prepared by ignition of As in air;
used as a pigment in ceramics and as a decolorizing
agent in glass [HAW93] [STR93] [KIR78]
Solubility: g/100 g H_2O: 1.20 (0°C), 1.49 (10°C), 1.82
(20°C), 2.31 (30°C), 2.93 (40°C), 4.31 (60°C), 6.11
(80°C), 8.2 (100°C) [LAN05]; s dil HCl [MER06]
Density, g/cm³: 3.738 [STR93]
Melting Point, °C: 313 [MER06]
Boiling Point, °C: 465 [MER06]
Reactions: sublimes when slowly heated [MER06]

307
Compound: Arsenic(III) selenide
Synonym: arsenic triselenide
Formula: As_2Se_3
Molecular Formula: As_2Se_3
Molecular Weight: 386.723
CAS RN: 1303-36-2
Properties: black cryst; dark brown solid; preparation:
from melted As and Se; uses: vacuum
deposition [CER91] [STR93] [MER06]
Solubility: i H_2O; s HNO_3 [MER06]
Density, g/cm³: 4.75 [MER06]
Melting Point, °C: 260 [MER06]; ~360 [STR93]

308
Compound: Arsenic(III) sulfide
Synonyms: orpiment, arsenic trisulfide
Formula: As_2S_3
Molecular Formula: As_2S_3
Molecular Weight: 246.041
CAS RN: 1303-33-9
Properties: yellow or orange powd; forms when As_2O_3 is heated with sulfur; used as a pigment, reducing agent, and in the form of 99.9% or 99.99% material as a sputtering target to produce adherent, stable, nonhygr, antireflection films on germanium and silicon [HAW93] [KIR78] [MER06] [CER91]
Solubility: i H_2O; s alkalies, slowly s HCl; decomposes in HNO_3 [MER06]
Density, g/cm^3: 3.46 [MER06]
Melting Point, °C: 320 [KIR78]
Boiling Point, °C: 707 [KIR78]
Reactions: transition to red form at 170°C [HAW93]

309
Compound: Arsenic(III) telluride
Synonym: arsenic tritelluride
Formula: As_2Te_3
Molecular Formula: As_2Te_3
Molecular Weight: 532.643
CAS RN: 12044-54-1
Properties: black cryst; uses: vacuum deposition [CER91] [STR93]
Density, g/cm^3: 6.50 [STR93]
Melting Point, °C: 621 [STR93]

310
Compound: Arsenic(V) acid hemihydrate
Formula: $H_3AsO_4 \cdot 1/2H_2O$
Molecular Formula: $AsH_4O_{4.5}$
Molecular Weight: 150.951
CAS RN: 7778-39-4
Properties: white translucent; hygr cryst; acid, $K_1 = 5.6 \times 10^{-3}$, $K_2 = 1.7 \times 10^{-7}$, $K_3 = 3.0 \times 10^{-12}$; loses water above 300°C forming the anhydrous As_2O_3; can be obtained by treating As_2O_3 with conc HNO_3; used in glassmaking, wood treatment [HAW93] [KIR78] [MER06]
Solubility: v s H_2O, alcohol, glycerol [MER06]
Density, g/cm^3: 2–2.5 [HAW93]
Melting Point, °C: 35.5 [HAW93]
Reactions: minus H_2O forming H_4AsO_7 at 100°C; forms $HAsO_3$ >100°C [KIR78]

311
Compound: Arsenic(V) fluoride
Synonym: arsenic pentafluoride

Formula: AsF_5
Molecular Formula: AsF_5
Molecular Weight: 169.914
CAS RN: 7784-36-3
Properties: colorless gas; condenses to yellow liq; forms white clouds in moist air; enthalpy of vaporization 20.8 kJ/mol; dielectric constant, 12.8 (20°C); can be formed by reacting AsF_3 with fluorine; used as a doping agent for electroconductive polymers [HAW93] [KIR78] [MER06] [CRC10]
Solubility: hydrolyzed quickly in H_2O; s alcohol, benzene, ether [MER06]
Density, g/cm^3: liq: 2.33 at bp [KIR78]
Melting Point, °C: −88.7 [KIR78]
Boiling Point, °C: −53.2 [CRC10]

312
Compound: Arsenic(V) oxide
Synonym: arsenic pentoxide
Formula: As_2O_5
Molecular Formula: As_2O_5
Molecular Weight: 229.840
CAS RN: 1303-28-2
Properties: white, amorphous lumps or powd; uncertain structure; oxidizing agent, can liberate Cl_2 from HCl; deliq; obtained by reaction of As or As_2O_3 with O_2 under pressure; used as an insecticide, in the manufacture of colored glass, and for weed control [HAW93] [STR93] [KIR78] [MER06]
Solubility: g/100 g H_2O: 59.5 (0°C), 62.1 (10°C), 65.8 (20°C), 69.8 (30°C), 71.2 (40°C), 73.0 (60°C), 75.1 (80°C), 76.7 (100°C) [LAN05]; s alcohol [MER06]
Density, g/cm^3: 4.32 [STR93]
Melting Point, °C: 315 decomposes [STR93]

313
Compound: Arsenic(V) selenide
Synonym: arsenic pentaselenide
Formula: As_2Se_5
Molecular Formula: As_2Se_5
Molecular Weight: 544.643
CAS RN: 1303-37-3
Properties: black, brittle solid; metallic luster; obtained when stoichiometric amounts of As and Se are melted in a nitrogen atm [MER06]
Solubility: i H_2O, dil acids; s alkali hydroxides [MER06]
Melting Point, °C: decomposes on heating in air [MER]

314
Compound: Arsenic(V) sulfide
Synonym: arsenic pentasulfide

Formula: As_2S_5
Molecular Formula: As_2S_5
Molecular Weight: 310.173
CAS RN: 1303-34-0
Properties: yellow or orange powd; stable in air up to 95°C; can be prepared by fusing As and S or by passing H_2S through HCl solution of arsenic acid; used as a paint pigment, in light filters [HAW93] [KIR78]
Solubility: g/L soln, H_2O: 0.00136 (0°C) [KRU93]
Melting Point, °C: decomposes [KIR78]
Reactions: decomposes to As_2S_3 and S >95°C [KIR78]

315
Compound: Arsenious acid
Formula: H_3AsO_3
Molecular Formula: AsH_3O_3
Molecular Weight: 125.944
CAS RN: 13464-58-9
Properties: exists only in solution; is a weak acid, $K = 8 \times 10^{-16}$; structure: HO(OH)AsOH; preparation: 1 g As_2O_3, 5 mL dil HCl, dil to 100 mL with H_2O; uses: for skin and blood disorders in animals [MER06] [KIR78]

316
Compound: Arsine
Synonym: arsenic trihydride
Formula: AsH_3
Molecular Formula: AsH_3
Molecular Weight: 77.946
CAS RN: 7784-42-1
Properties: colorless gas; garlic-like odor; highly toxic; critical temp 105.4°C; critical pressure 6.60 MPa; enthalpy of vaporization 16.69 kJ/mol; decomposes at 230°C; formed by reaction of Zn, HCl, and As compound, and by hydride reduction, e.g., $NaBH_4$ in NaOH solution; used in organic synthesis, has some use in electronics industry [HAW93] [KIR78] [KIR80] [AIR87] [KOR91] [CRC10]
Solubility: mL/100 g H_2O (760 mm): 42 (0°C), 30 (10°C), 28 (20°C) [LAN05]
Density, g/cm³: liq: (−64.3°C) 1.640; gas: 2.695 g/L [KIR78] [KIR80]
Melting Point, °C: −116.3 [KIR78]
Boiling Point, °C: −62.4 [KIR78]
Reactions: becomes hydrated to $AsH_3 \cdot 6H_2O$ at −10°C [KIR78]

317
Compound: Astatine
Formula: At
Molecular Formula: At
Molecular Weight: 210
CAS RN: 7440-68-8
Properties: radioactive cryst halogen with 20 isotopes; heaviest of the halogens; ^{209}At, $t_{1/2} = 5.5$ h; ^{210}At, $t_{1/2} = 8.3$ h; more metallic than iodine; preparation: from Bi by α-particle bombardment; possible medical uses, concentrates in thyroid gland [HAW93] [MER06]
Solubility: s organic solvents [MER06]
Melting Point, °C: 302 [CRC10]
Boiling Point, °C: 337 (estimated) [CRC10]

318
Compound: Barium
Formula: Ba
Molecular Formula: Ba
Molecular Weight: 137.327
CAS RN: 7440-39-3
Properties: yellow-silver soft metal; bcc, a = 0.5025 nm; enthalpy of fusion 7.66 kJ/mol; enthalpy of vaporization 149.20 kJ/mol; vapor pressure, kPa: 0.00133 (629°C), 1.33 (1050°C), 101.3 (1640°C); easily air oxidized; Gruneisen parameter −0.2; electrical resistivity 29.4 μohm·cm for the pure element; electron work function 2.11 eV; Ba^{++} radius 0.143 nm; electronegativity 1.02 [CIC73] [KIR91] [MER06]
Solubility: s with H_2 evolution in cold H_2O and hot H_2O; sl s alcohol; i benzene [CRC10]
Density, g/cm³: 3.62 [CIC73]
Melting Point, °C: 729 [KNA91]
Boiling Point, °C: 1640 [KIR91]
Thermal Conductivity, W/(m·K): 18.4 (25°C) [CRC10]
Thermal Expansion Coefficient: coefficient of linear expansion 1.85×10^{-5} m/(m·°C) [KIR91]

319
Compound: Barium 2-ethylhexanoate
Formula: $[CH_3(CH_2)_3CHC_2H_5COO]_2Ba$
Molecular Formula: $C_{16}H_{30}BaO_4$
Molecular Weight: 423.739
CAS RN: 2457-01-4
Properties: precursor used in the preparation of thin-film superconductors [ALD94]
Melting Point, °C: >300 [ALD93]

320
Compound: Barium acetate
Synonyms: acetic acid, barium salt
Formula: $Ba(CH_3COO)_2$
Molecular Formula: $C_4H_6BaO_4$

Molecular Weight: 255.417
CAS RN: 543-80-6
Properties: white powd; crystallizes from H_2O as a trihydrate below 24.7°C, as a monohydrate from 24.7°C to 41°C, and as an anhydrous material above 41°C; can be prepared from acetic acid and either $BaCO_3$ or BaS, followed by crystallization and dehydration [KIR78] [STR93]
Solubility: g/100 g H_2O: 58.8 (0.3°C), 78.1 (24.1°C), 74.8 (99.2°C); solid equilibrium phase, $Ba(CH_3COO)_2 \cdot 3H_2O$ (0.3 and 24.1°C), $Ba(CH_3COO)_2$ (99.2°C) [KRU93]
Density, g/cm³: 2.47 [KIR78]

321
Compound: Barium acetate monohydrate
Synonyms: acetic acid, barium salt monohydrate
Formula: $Ba(CH_3COO)_2 \cdot H_2O$
Molecular Formula: $C_4H_8BaO_5$
Molecular Weight: 273.432
CAS RN: 5908-64-5
Properties: white cryst; made by the addition of acetic acid to barium sulfide solution, followed by evaporation and crystallization; used as a chemical reagent and as a textile mordant, used in paint and varnish driers [HAW93]
Solubility: 1 g/1.5 mL cold or boiling H_2O; 1 g/700 mL alcohol [MER06]
Density, g/cm³: 2.19 [KIR78]
Melting Point, °C: decomposes [HAW93]
Reactions: minus H_2O 110°C [MER06]

322
Compound: Barium acetylacetonate octahydrate
Synonyms: 2,4-pentanedione, barium derivative octahydrate
Formula: $Ba[CH_3COCH=C(O)CH_3]_2 \cdot 8H_2O$
Molecular Formula: $C_{10}H_{30}BaO_{12}$
Molecular Weight: 479.668
CAS RN: 12084-29-6
Properties: hygr powd [STR93] [ALD93]
Melting Point, °C: decomposes at 123 [ALD93]

323
Compound: Barium aluminate
Formula: $BaO \cdot Al_2O_3$
Molecular Formula: Al_2BaO_4
Molecular Weight: 255.288
CAS RN: 12004-04-5
Properties: nepheline-type structure; −100 mesh, 99.5% purity; a = 0.5224 nm, c = 0.8792 nm [TAY91b] [CER91] [CRC10]
Melting Point, °C: 1827 [KNA91]

Thermal Expansion Coefficient: from 25°C to 100°C (0.18), 200°C (0.45), 400°C (0.96), 600°C (1.50), 800°C (2.01), 1000°C (2.52), 1200°C (3.06) [TAY91b]

324
Compound: Barium aluminate
Formula: $3BaO \cdot Al_2O_3$
Molecular Formula: $Al_2Ba_3O_6$
Molecular Weight: 561.940
CAS RN: 11129-08-1
Properties: gray mass [HAW93]
Solubility: s H_2O, acid [HAW93]
Melting Point, °C: 1750 [KNA91]

325
Compound: Barium aluminide
Formula: $BaAl_4$
Molecular Formula: Al_4Ba
Molecular Weight: 245.253
CAS RN: 12672-79-6
Properties: 6 mm pieces and smaller [CER91]

326
Compound: Barium antimonide
Formula: Ba_3Sb_2
Molecular Formula: Ba_3Sb_2
Molecular Weight: 655.501
CAS RN: 55576-04-0
Properties: 6 mm pieces and smaller [CER91]

327
Compound: Barium arsenide
Formula: Ba_3As_2
Molecular Formula: As_2Ba_3
Molecular Weight: 561.824
CAS RN: 12255-50-4
Properties: brown; 6 mm pieces and smaller [CRC10] [CER91]
Density, g/cm³: 4.1 [CRC10]

328
Compound: Barium azide
Formula: $Ba(N_3)_2$
Molecular Formula: BaN_6
Molecular Weight: 221.367
CAS RN: 18810-58-7
Properties: cryst solid; monocl, a = 0.622 nm, b = 2.929 nm, c = 0.702 nm; M–N₃ bond length 0.2937 nm; unstable and can explode when heated or on impact; used in high explosives [CIC73] [HAW93] [CRC10]

Solubility: g/100 g H_2O: 12.5 (0°C), 16.1 (10°C), 17.4 (20°C) [LAN05]; alcohol 0.17 (16°C), i ether [CRC10]
Density, g/cm³: 2.936 [HAW93]
Boiling Point, °C: explodes [CRC10]
Reactions: evolves N_2 at 120°C [HAW93]

329
Compound: Barium bis(2,2,6,6-tetramethyl-3,5-heptanedionate) hydrate
Formula: $[(CH_3)_3CCOCH=C(O)C(CH_3)_3]_2Ba \cdot xH_2O$
Molecular Formula: $C_{22}H_{38}BaO_4$ (anhydrous)
Molecular Weight: 503.866 (anhydrous)
CAS RN: 17594-47-7
Properties: used in the preparation of superconducting thin films [ALD94]
Melting Point, °C: 175–180 [ALD94]

330
Compound: Barium bismuth oxide
Formula: $BaBi(III)_{0.5}Bi(V)_{0.5}O_3$
Molecular Formula: $BaBiO_3$
Molecular Weight: 394.305
CAS RN: 12785-50-1
Reactions: transitions: monocl to hex at 132°C; hex to cub at 440°C [TAY85]
Thermal Expansion Coefficient: from 25°C to: 100°C (0.15); 200°C (0.45); 400°C (1.26); 600°C (2.10); 800°C (3.96) [TAY85]

331
Compound: Barium bromate
Formula: $Ba(BrO_3)_2$
Molecular Formula: $BaBr_2O_6$
Molecular Weight: 393.131
CAS RN: 13967-90-3
Properties: can be prepared from potassium bromate and barium chloride [MER06]
Solubility: g/100 g soln, H_2O: 0.286 (0°C), 0.788 (25°C); 5.39 (99.65°C); equilibrium solid phase $Ba(BrO_3)_2$ [KRU93]

332
Compound: Barium bromate monohydrate
Formula: $Ba(BrO_3)_2 \cdot H_2O$
Molecular Formula: $BaBr_2H_2O_7$
Molecular Weight: 411.147
CAS RN: 10326-26-8

Properties: white, monocl cryst or powd; obtained by addition of bromine to hot barium hydroxide solution, followed by crystallization; used as an oxidizing agent and corrosion inhibitor [HAW93] [MER06]
Solubility: g/100 mL: 0.44 (10°C), 0.96 (30°C), 5.39 (100°C) [MER06]; i alcohol [HAW93]
Density, g/cm³: 3.99 [MER06]
Melting Point, °C: decomposes at 260 [MER06]

333
Compound: Barium bromide
Formula: $BaBr_2$
Molecular Formula: $BaBr_2$
Molecular Weight: 297.135
CAS RN: 10553-31-8
Properties: white powd; hygr; −20 mesh 99.9% and 99.995% purity; enthalpy of fusion 31.96 kJ/mol; made by reacting barium carbonate and hydrobromic acid [KIR78] [ALD94] [CIC73] [STR93] [CER91]
Solubility: g/100 g soln, H_2O: 47.5(0°C), 50.0 (25°C), 57.8 (100°C); equilibrium solid phase, $BaBr_2 \cdot 2H_2O$ [KRU93]
Density, g/cm³: 4.781 [KIR78]
Melting Point, °C: 857 [KNA91]
Boiling Point, °C: 1835 [KNA91]

334
Compound: Barium bromide dihydrate
Formula: $BaBr_2 \cdot 2H_2O$
Molecular Formula: $BaBr_2H_4O_2$
Molecular Weight: 333.166
CAS RN: 7791-28-8
Properties: white cryst; can be obtained from HBr and BaS solutions, followed by crystallization; used in manufacturing bromides [HAW93] [STR93]
Solubility: g/100 g H_2O: 98 (0°C), 101 (10°C), 104 (20°C), 109 (30°C), 114 (40°C), 123 (60°C), 135 (80°C), 149 (100°C) [LAN05]
Density, g/cm³: 3.58 [KIR78]
Melting Point, °C: see anhydrous $BaBr_2$
Boiling Point, °C: decomposes [CIC73]
Reactions: minus H_2O at 75°C; minus $2H_2O$ at 100°C [KIR78]

335
Compound: Barium calcium tungstate
Synonym: barium calcium tungsten oxide
Formula: Ba_2CaWO_6
Molecular Formula: Ba_2CaO_6W
Molecular Weight: 594.568
CAS RN: 15552-14-4

Properties: −325 mesh 99.9% purity [ALF93]
Melting Point, °C: decomposes at 1420 [ALF93]

336
Compound: Barium carbide
Formula: BaC_2
Molecular Formula: C_2Ba
Molecular Weight: 161.349
CAS RN: 50813-65-5
Properties: gray tetr; −8 mesh 99.5%
      purity [CER91] [CRC10]
Solubility: decomposes in H_2O yielding acetylene,
      HC≡CH; decomposed in acids [CRC10]
Density, g/cm³: 3.74 [CRC10]
Melting Point, °C: decomposes [KNA91]

337
Compound: Barium carbonate
Synonym: witherite
Formula: α-$BaCO_3$
Molecular Formula: $CBaO_3$
Molecular Weight: 197.336
CAS RN: 513-77-9
Properties: white, heavy powd; rhomb; witherite
      is naturally occurring mineral; hardness 3.0–
      3.75 Mohs; manufactured by precipitation from a
      solution of BaS by Na_2CO_3 at 60°C–70°C; used to
      remove sulfates from chlor-alkali cells, in brick-
      making, in oil well industry, to produce barium
      titanate, as a ceramic flux, and in radiation-resistant
      glass for color television [HAW93] [CIC73] [KIR78]
Solubility: g/1000 g H_2O: 0.0180 (25°C) [KRU93];
      s acid, NH_4Cl; i alcohol [CRC10]
Density, g/cm³: 4.2865 [MER06]
Melting Point, °C: 174 (90 atm), 811 (1 atm) [HAW93]
Reactions: ~ at 1300°C decomposes into BaO
      and CO_2 [MER06]

338
Compound: Barium chlorate
Formula: $Ba(ClO_3)_2$
Molecular Formula: $BaCl_2O_6$
Molecular Weight: 304.228
CAS RN: 13477-00-4
Properties: −80 mesh 99.9% purity; prepared by
      electrolysis of $BaCl_2$ solutions; monohydrate:
      colorless monocl [CER91] [CRC10] [MER06]
Solubility: 27.5 g/100 g soln (25°C);
      67 g/100 g soln (100°C) [CIC73]
Density, g/cm³: monohydrate: 3.18 [CRC10]
Reactions: monohydrate: minus H_2O at
      120°C, minus O at 250°C [CRC10]

339
Compound: Barium chlorate monohydrate
Formula: $Ba(ClO_3)_2 \cdot H_2O$
Molecular Formula: $BaCl_2H_2O_7$
Molecular Weight: 322.244
CAS RN: 10294-38-9
Properties: monocl, prismatic, white cryst;
      combustible, used in fireworks, explosives,
      and as a textile mordant [MER06]
Solubility: g/100 g H_2O: 20.3 (0°C), 26.9 (10°C), 33.9
      (20°C), 41.6 (30°C), 49.7 (40°C), 66.7 (60°C), 84.8
      (80°C), 105 (100°C); s HCl [LAN05] [MER06]
Density, g/cm³: 3.179 [MER06]
Melting Point, °C: 414 [MER06]
Reactions: minus H_2O at 120°C; evolves
      oxygen at 250°C [MER06]

340
Compound: Barium chloride
Formula: α-$BaCl_2$
Molecular Formula: $BaCl_2$
Molecular Weight: 208.232
CAS RN: 10361-37-2
Properties: white powd; two forms: monocl,
      cub; enthalpy of fusion 16.00 kJ/mol
      [KIR78] [STR93] [CIC73] [CRC10]
Solubility: g/100 g soln, H_2O: 23.8 (0°C), 27.1
      (25°C), 37.0 ± 0.3 (100°C); solid phase,
      $BaCl_2 \cdot 2H_2O$ (100°C) [KRU93]
Density, g/cm³: 3.856 [KIR78]
Melting Point, °C: 960 [KNA91]
Boiling Point, °C: 1560 [KNA91]
Reactions: transition α (monocl) to β (cub)
      at 925°C [SCH93] [KIR78]

341
Compound: Barium chloride dihydrate
Formula: $BaCl_2 \cdot 2H_2O$
Molecular Formula: $BaCl_2H_4O_2$
Molecular Weight: 244.263
CAS RN: 10326-27-9
Properties: white monocl; manufactured from
      BaS and HCl, followed by evaporation;
      used to make barium pigments and
      as a flux for Mg metal [KIR78]
Solubility: 31.7 g/100 g H_2O (0°C), 35.8 g/100 g
      H_2O (20°C), 58.7 g/100 g H_2O (100°C); i
      alcohol [LAN05] [KIR78] [HAW93]
Density, g/cm³: 3.097 [KIR78]
Reactions: minus $2H_2O$ at 113°C [KIR78]

342
Compound: Barium chloride fluoride
Formula: BaClF
Molecular Formula: BaClF
Molecular Weight: 191.788
CAS RN: 13718-55-3
Properties: white cryst [CRC10]

343
Compound: Barium chromate
Synonyms: lemon chrome, baryta yellow
Formula: $BaCrO_4$
Molecular Formula: $BaCrO_4$
Molecular Weight: 253.321
CAS RN: 10294-40-3
Properties: heavy, yellow powd; rhomb, monocl; prepared from $BaCl_2$ and Na_2CrO_4 solutions, followed by filtering resulting precipitate; used in safety matches, as a corrosion inhibitor [MER06] [HAW93]
Solubility: g/L soln, H_2O: 0.002 (0°C), 0.00291 (25°C) [KRU93]
Density, g/cm³: 4.50 [MER06]
Melting Point, °C: decomposes [KIR78]

344
Compound: Barium chromate(V)
Formula: $Ba_3(CrO_4)_2$
Molecular Formula: $Ba_3Cr_2O_8$
Molecular Weight: 643.968
CAS RN: 12345-14-1
Properties: greenish black cryst [KIR78]
Solubility: s, decomposing in H_2O; s dil acids [KIR78]

345
Compound: Barium citrate monohydrate
Formula: $Ba_3(C_6H_5O_7)_2 \cdot H_2O$
Molecular Formula: $C_{12}H_{12}Ba_3O_{15}$
Molecular Weight: 808.195
CAS RN: 512-25-4 (anhydrous parent compound)
Properties: gray white cryst [CRC10]
Solubility: Soluble in H_2O and acid

346
Compound: Barium copper yttrium oxide-I
Formula: $BaCuY_2O_5$
Molecular Formula: $BaCuO_6Y_2$
Molecular Weight: 485.682
CAS RN: 82642-06-6
Properties: green cryst; not superconducting [CRC10]

347
Compound: Barium copper yttrium oxide-II
Formula: $Ba_2Cu_3YO_7$
Molecular Formula: $Ba_2Cu_3O_7Y$
Molecular Weight: 666.194
CAS RN: 109064-29-1
Properties: black solid; high-temp superconductor [CRC10]

348
Compound: Barium copper yttrium oxide-III
Formula: $Ba_2Cu_4YO_8$
Molecular Formula: $Ba_2Cu_4O_8Y$
Molecular Weight: 745.739
CAS RN: 107539-20-8
Properties: high-temp superconductor [CRC10]

349
Compound: Barium copper yttrium oxide-IV
Formula: $Ba_4Cu_7Y_2O_{15}$
Molecular Formula: $Ba_4Cu_7O_{15}Y_2$
Molecular Weight: 1411.933
CAS RN: 124365-83-9
Properties: high-temp superconductor [CRC10]

350
Compound: Barium cyanide
Formula: $Ba(CN)_2$
Molecular Formula: C_2BaN_2
Molecular Weight: 189.362
CAS RN: 542-62-1
Properties: white, cryst powd; slowly decomposes in air; obtained by reaction of HCN and barium hydroxide, followed by crystallization; used in metallurgy and electroplating [HAW93] [MER06]
Solubility: 80.0 g/100 mL H_2O (14°C) [CRC10]; s alcohol [HAW93]

351
Compound: Barium dichromate dihydrate
Formula: $BaCr_2O_7 \cdot 2H_2O$
Molecular Formula: $BaCr_2H_4O_9$
Molecular Weight: 389.346
CAS RN: 10031-16-0
Properties: brownish red needles; used in ceramics [KIR78]
Solubility: decomposed in H_2O [KIR78]
Melting Point, °C: decomposes [CRC10]
Reactions: minus $2H_2O$ at 120°C [CRC10]

352
Compound: Barium diphenylamine-4-sulfonate
Synonyms: diphenylamine-4-sulfonic acid, barium salt
Formula: $(C_6H_5NHC_6H_4SO_3)_2Ba$
Molecular Formula: $C_{24}H_{20}BaO_6S_2$
Molecular Weight: 633.878
CAS RN: 6211-24-1
Properties: white, cryst leaflets; prepared by acetylation and subsequent sulfonation of diphenylamine; used as an oxidation/reduction indicator [ALD94] [HAW93] [MER06]
Solubility: sl s H_2O [MER06]

353
Compound: Barium disilicate
Formula: $BaSi_2O_5$
Molecular Formula: BaO_5Si_2
Molecular Weight: 273.495
CAS RN: 12650-28-1
Properties: white, ortho cryst [CRC10]
Density, g/cm³: 3.7 [CRC10]
Melting Point, °C: 1420 [CRC10]

354
Compound: Barium dithionate dihydrate
Synonym: barium hyposulfate dihydrate
Formula: $Ba(SO_3)_2 \cdot 2H_2O$
Molecular Formula: $BaH_4O_8S_2$
Molecular Weight: 333.486
CAS RN: 13845-17-5
Properties: colorless cryst; prepared from barium hydroxide and manganese dithionate [HAW93]
Solubility: s in four parts H_2O; sl s alcohol [MER06]
Density, g/cm³: 4.54 [MER06]
Reactions: minus SO_2 at >150°C forming $BaSO_4$ [MER06]

355
Compound: Barium ferrite
Formula: $BaFe_{12}O_{19}$
Molecular Formula: $BaFe_{12}O_{19}$
Molecular Weight: 1111.456
CAS RN: 11138-11-7
Properties: powd, −325 mesh 98% purity; used as a permanent magnet material [HAW93] [CER91]

356
Compound: Barium ferrocyanide hexahydrate
Synonym: barium hexacyanoferrate(II)
Formula: $Ba_2Fe(CN)_6 \cdot 6H_2O$
Molecular Formula: $C_6H_{12}Ba_2FeN_6O_6$
Molecular Weight: 594.696

CAS RN: 13821-06-2
Properties: yellow, becomes colorless with loss of H_2O; rectangular monocl [MER06]
Solubility: 0.17 g/100 mL (288 K) H_2O; i alcohol [CRC10]
Density, g/cm³: 2.666 [CRC10]
Reactions: minus H_2O (40°C); decomposes, losing HCN, at 80°C [MER06]

357
Compound: Barium fluoride
Formula: BaF_2
Molecular Formula: BaF_2
Molecular Weight: 175.324
CAS RN: 7787-32-8
Properties: white, cub cryst or 99.9% pure 3–6 mm melted pieces; enthalpy of fusion 23.36 kJ/mol; enthalpy of vaporization 347.3 kJ/mol; may be prepared by reacting barium carbonate with HF solution; used as a component in welding flux and to produce infrared transparent films [KIR78] [CIC73] [HAW93] [MER06] [CER91]
Solubility: g/L soln, H_2O: 1.586 (10°C), 1.617 ± 0.003 (25°C); s HCl, HNO_3 [KRU93] [MER06]
Density, g/cm³: 4.83 [MER06]
Melting Point, °C: 1354 [HAW93]
Boiling Point, °C: 2260 [CIC73]

358
Compound: Barium formate
Formula: $Ba(CHO_2)_2$
Molecular Formula: $C_2H_2BaO_4$
Molecular Weight: 227.362
CAS RN: 541-43-5
Properties: cryst
Solubility: s H_2O; i EtOH [CRC10]
Density, g/cm³: 3.21 [CRC10]

359
Compound: Barium hexaboride
Synonym: barium boride
Formula: BaB_6
Molecular Formula: B_6Ba
Molecular Weight: 202.193
CAS RN: 12046-08-1
Properties: metallic black cub; −100 mesh 99.5% purity [CER91] [CRC10]
Solubility: i H_2O; s HNO_3; i HCl [CRC10]
Density, g/cm³: 4.36 [CRC10]
Melting Point, °C: 2070 [KIR78]

360
Compound: Barium hexafluorogermanate
Formula: $BaGeF_6$

Molecular Formula: BaF_6Ge
Molecular Weight: 323.927
CAS RN: 60897-63-4
Properties: white, cryst solid [HAW93]
Density, g/cm³: 4.56 [HAW93]
Melting Point, °C: ~665 [HAW93]
Reactions: decomposes to BaF_2 and GeF_4 [HAW93]

361
Compound: Barium hexafluorosilicate
Formula: $BaSiF_6$
Molecular Formula: BaF_6Si
Molecular Weight: 279.403
CAS RN: 17125-80-3
Properties: white, ortho-rhomb needles; prolonged contact with water induces hydrolysis, which is accelerated by alkali; formed from $BaCl_2$ and H_2SiF_6; used in ceramics and insecticides [HAW93] [MER06]
Solubility: 0.015 g/100 mL H_2O (0°C), 0.0235 (25°C), 0.091 (100°C) [MER06]
Density, g/cm³: 4.29 [MER06]
Melting Point, °C: decomposes at 300 [MER06]

362
Compound: Barium hydride
Formula: BaH_2
Molecular Formula: BaH_2
Molecular Weight: 139.343
CAS RN: 13477-09-3
Properties: gray cryst; −60 mesh with 99.7% purity; sensitive to moisture; resembles CaH_2 in properties [KIR80] [CER91] [STR93]
Solubility: decomposes in water to $Ba(OH)_2 + H_2$; decomposes in acids [CRC10]
Density, g/cm³: 4.16 [KIR80]
Melting Point, °C: decomposes at 675 [STR93]
Boiling Point, °C: ~1673 [CRC10]

363
Compound: Barium hydrogen phosphate
Synonyms: barium phosphate, dibasic
Formula: $BaHPO_4$
Molecular Formula: $BaHO_4P$
Molecular Weight: 233.306
CAS RN: 10048-98-3
Properties: cryst, white powd; used as a flame retardant and in phosphors [STR93] [MER06] [HAW93]
Solubility: i H_2O; s dil HCl or HNO_3 [MER06]
Density, g/cm³: 4.16 [MER06]
Melting Point, °C: decomposes at 410 [CRC10]

364
Compound: Barium hydrosulfide
Formula: $Ba(HS)_2$
Molecular Formula: BaH_2S_2
Molecular Weight: 203.475
CAS RN: 25417-81-6
Properties: yellow cryst; hygr; preparation: reaction of H_2S with BaS, followed by precipitation of $Ba(HS)_2 \cdot 4H_2O$ by alcohol and dehydration [KIR78] [HAW93]
Solubility: g/100 g H_2O soln: 0°C (32.6); 20°C (32.8); 100°C (43.7) [KIR78]

365
Compound: Barium hydrosulfide tetrahydrate
Formula: $Ba(HS)_2 \cdot 4H_2O$
Molecular Formula: $BaH_{10}O_4S_2$
Molecular Weight: 275.536
CAS RN: 12230-74-9
Properties: yellow rhomb; obtained by passing H_2S through BaS solution, followed by addition of alcohol and subsequent crystallization [KIR78]
Solubility: g/100 g, H_2O: 32.6 (0°C), 32.8 (20°C), 43.7 (100°C) [KIR78]
Melting Point, °C: decomposes at 50 [KIR78]

366
Compound: Barium hydroxide
Synonyms: caustic baryta, anhydrous barium hydroxide
Formula: $Ba(OH)_2$
Molecular Formula: BaH_2O_2
Molecular Weight: 171.342
CAS RN: 17194-00-2
Properties: white powd, hygr; enthalpy of fusion 16.70 kJ/mol [CRC10] [STR93]
Solubility: g/100 g soln, H_2O: 1.67 (0°C), 4.68 (25°C), 101.4 (80°C); solid phase, $Ba(OH)_2 \cdot 8H_2O$ [KRU93]
Melting Point, °C: 408 [KNA91]
Boiling Point, °C: decomposes at 1032 (calculated) [KNA91]

367
Compound: Barium hydroxide monohydrate
Formula: $Ba(OH)_2 \cdot H_2O$
Molecular Formula: BaH_4O_3
Molecular Weight: 189.357
CAS RN: 22326-55-2
Properties: white powd; formed when barium hydroxide octahydrate is "boiled dry" in CO_2 free atm; used in manufacturing oil and grease additives, soaps, and in refining beet sugar [KIR78] [MER06] [HAW93]
Solubility: sl s H_2O; s acids [HAW93]

Density, g/cm³: 3.743 [MER06]
Reactions: minus H_2O <407°C, dehydrates
  to BaO at ~800°C [KIR78]

368
Compound: Barium hydroxide octahydrate
Formula: $Ba(OH)_2 \cdot 8H_2O$
Molecular Formula: $BaH_{18}O_{10}$
Molecular Weight: 315.464
CAS RN: 12230-71-6
Properties: white, monocl cryst; rapidly absorbs
  CO_2 from air; vapor pressure at mp 30.3 kPa;
  prepared by dissolution of BaO in hot water,
  followed by crystallization; used as plastic
  stabilizer, an additive in papermaking, a pigment
  dispersant, and to protect limestone materials
  from deterioration [KIR78] [STR93] [MER06]
Solubility: g $Ba(OH)_2$/100 g soln: 1.65 (0°C),
  3.76 (20°C), 48.5 (78°C) [KIR78]
Density, g/cm³: 2.18 [KIR78]
Melting Point, °C: 77.9, melts in waters
  of crystallization [KIR78]

369
Compound: Barium hypophosphite monohydrate
Formula: $Ba(H_2PO_2)_2 \cdot H_2O$
Molecular Formula: $BaH_6O_5P_2$
Molecular Weight: 285.320
CAS RN: 14871-79-5
Properties: monocl platelets; can be prepared
  by reacting white phosphorus and barium
  hydroxide; used in nickel plating; anhydrous
  material, $Ba(H_2PO_2)_2$, is a white, odorless,
  cryst powd [MER06] [HAW93]
Solubility: g/100 mL: 28.6 (17°C), 33.3
  (100°C); i alcohol [MER06]
Density, g/cm³: 2.90 [MER06]
Melting Point, °C: decomposes at 100–150 [CRC10]

370
Compound: Barium iodate
Formula: $Ba(IO_3)_2$
Molecular Formula: BaI_2O_6
Molecular Weight: 487.132
CAS RN: 10567-69-8
Properties: white, cryst powd [HAW93]
Solubility: g/L soln, H_2O: 0.395 (25°C)
  [KRU93]; g/100 g H_2O: 0.035 (20°C),
  0.046 (30°C), 0.057 (40°C) [LAN05]
Density, g/cm³: 5.23 [HAW93]
Melting Point, °C: decomposes at 476 [HAW93]

371
Compound: Barium iodate monohydrate
Formula: $Ba(IO_3)_2 \cdot H_2O$
Molecular Formula: $BaH_2I_2O_7$
Molecular Weight: 505.148
CAS RN: 7787-34-0
Properties: cryst [MER06]
Solubility: s in 3350 parts H_2O (25°C), 625 parts
  boiling H_2O; s HCl, HNO_3; i alcohol [MER06]
Density, g/cm³: 5.00 [MER06]
Reactions: minus H_2O at 130°C [MER06]

372
Compound: Barium iodide
Formula: BaI_2
Molecular Formula: BaI_2
Molecular Weight: 391.136
CAS RN: 13718-50-8
Properties: off-white powd; hygr; enthalpy of fusion
  26.53 kJ/mol [CIC73] [STR93] [CRC10]
Solubility: g/100 g soln, H_2O: 62.5 (0°C), 68.8 (25°C),
  73.35 (98.9°C); solid phase, $2BaI_2 \cdot 15H_2O$ (0°C,
  25°C), $BaI_2 \cdot 2H_2O + BaI_2 \cdot H_2O$ (98.9°C) [KRU93]
Density, g/cm³: 5.15 [KIR78]
Melting Point, °C: 711 [CIC73]
Boiling Point, °C: decomposes [CIC73]

373
Compound: Barium iodide dihydrate
Formula: $BaI_2 \cdot 2H_2O$
Molecular Formula: $BaH_4I_2O_2$
Molecular Weight: 427.167
CAS RN: 7787-33-9
Properties: colorless, odorless cryst; rapidly becomes
  reddish in air due to liberation of iodine; prepared
  from HI and barium hydroxide solutions with
  subsequent crystallization from hot solutions; used
  to prepare other iodides [MER06] [HAW93]
Solubility: g BaI_2/100 g soln: 169.4 (0°C), 271.0
  (100°C) [KIR78]; g/100 g H_2O: 182 (0°C),
  223 (20°C), 301 (100°C) [LAN05]
Density, g/cm³: 4.917 [KIR78]
Melting Point, °C: 740 [HAW93]
Reactions: minus $2H_2O$ at 150°C [KIR78]

374
Compound: Barium lead oxide
Formula: $BaPbO_3$
Molecular Formula: BaO_3Pb
Molecular Weight: 392.525
CAS RN: 12047-25-5

Properties: monocl [TAY85]
Reactions: transition from monocl to hex (127°C); from hex to cub (423°C) [TAY85]
Thermal Expansion Coefficient: from 25°C to: 500°C (1.29); 600°C (1.62); 800°C (2.22) [TAY85]

375

Compound: Barium manganate(VI)
Synonym: manganese green
Formula: $BaMnO_4$
Molecular Formula: $BaMnO_4$
Molecular Weight: 256.263
CAS RN: 7787-35-1
Properties: greenish gray cryst; sensitive to moisture; uses: oxidizes primary and secondary alcohols to carbonyl compounds, paint pigment [HAW93] [STR93] [ALD93]
Solubility: disproportionates in H_2O to $Ba(MnO_4)_2 + MnO_2$ [MER06]
Density, g/cm³: 4.85 [MER06]

376

Compound: Barium metaborate dihydrate
Synonym: barium borate dihydrate
Formula: $Ba(BO_2)_2 \cdot 2H_2O$
Molecular Formula: $B_2BaH_4O_6$
Molecular Weight: 258.977
CAS RN: 23436-05-7
Properties: prepared by precipitation by adding sodium metaborate solution to a solution of barium chloride at 90°C–95°C; at room temp the tetrahydrate precipitates; finds use as a fire retardant for paints, plastics, textiles, and paper products [KIR78]
Solubility: 12.5 g/L of $BaO \cdot B_2O_3 \cdot 4H_2O$ in H_2O (25°C) [KIR78]
Reactions: dehydrates above 140°C [KIR78]

377

Compound: Barium metaborate monohydrate
Synonym: barium borate monohydrate
Formula: $Ba(BO_2)_2 \cdot H_2O$
Molecular Formula: $B_2BaH_2O_5$
Molecular Weight: 240.962
CAS RN: 26124-86-7
Properties: white powd; manufactured from a solution of BaS and sodium tetraborate; used to add mold, corrosion, and fire resistance to paint [KIR78]
Solubility: 0.3% H_2O [KIR78]
Density, g/cm³: 3.25–3.35 [KIR78]
Melting Point, °C: >900 [KIR78]

378

Compound: Barium metaphosphate
Formula: $Ba(PO_3)_2$
Molecular Formula: BaO_6P_2
Molecular Weight: 295.271
CAS RN: 13466-20-1
Properties: white powd; used in glasses, porcelain, and enamel [HAW93]
Solubility: i H_2O; slowly dissolves in acids [HAW93]
Melting Point, °C: 1560 [ALF93]

379

Compound: Barium metasilicate
Synonym: monobarium silicate
Formula: $BaSiO_3$
Molecular Formula: BaO_3Si
Molecular Weight: 213.411
CAS RN: 13255-26-0
Properties: colorless, rhomb powd; can be formed by heating BaO, $BaCO_3$, or $BaSO_4$ to white heat with SiO_2, which also forms $3BaO \cdot SiO_2$. The tribarium silicate hydrolyzes to form $BaSiO_3$ and $Ba(OH)_2$, which is the basis for the Deguide process; used in ceramics [KIR78] [HAW93]
Solubility: i H_2O; s acids [HAW93]
Density, g/cm³: 4.4 [STR93]
Melting Point, °C: 1605 [KNA91]

380

Compound: Barium molybdate
Formula: $BaMoO_4$
Molecular Formula: $BaMoO_4$
Molecular Weight: 297.265
CAS RN: 7787-37-3
Properties: white powd; scheelite structure, c/a = 2.29; used in electronic and optical equipments and in paint pigments for protective coatings [HAW93] [MER06] [KIR81]
Solubility: 0.0055 g/100 g H_2O [KIR81]
Density, g/cm³: 4.975 [KIR81]
Melting Point, °C: 1450 [KNA91]

381

Compound: Barium niobate
Synonym: barium niobate(V)
Formula: $Ba(NbO_3)_2$
Molecular Formula: $BaNb_2O_6$
Molecular Weight: 419.136
CAS RN: 12009-14-2
Properties: yellow hex or ortho; −100 mesh, 99.9% purity [CER91] [LID94]

Density, g/cm³: 2.8 [LID94]
Melting Point, °C: 1450 [LID94]

382
Compound: Barium nitrate
Synonym: nitrobarite
Formula: Ba(NO₃)₂
Molecular Formula: BaN₂O₆
Molecular Weight: 261.336
CAS RN: 10022-31-8
Properties: white, cryst powd; prepared from BaCO₃ suspension and HNO₃, followed by crystallization; used in pyrotechnics, green flares, tracer bullets, and detonators [STR93] [MER06] [KIR78]
Solubility: g/100 g soln, H₂O: 4.72 (0°C), 9.27 (25°C), 25.6 (100°C); solid phase, Ba(NO₃)₂ [KRU93]
Density, g/cm³: 3.24 [KIR78]
Melting Point, °C: 592 [KIR78]
Reactions: decomposes above 590°C [MER06]

383
Compound: Barium nitride
Formula: Ba₃N₂
Molecular Formula: Ba₃N₂
Molecular Weight: 439.994
CAS RN: 12047-79-9
Properties: yellowish brown; −20 mesh, 99.7% purity [CER91] [CIC73]
Solubility: decomposed in H₂O [CRC10]
Density, g/cm³: 4.78 [ALF93]
Boiling Point, °C: 1000, vacuum [ALF93]

384
Compound: Barium nitrite
Formula: Ba(NO₂)₂
Molecular Formula: BaN₂O₄
Molecular Weight: 229.338
CAS RN: 13465-94-6
Properties: colorless, hex [CRC10]
Solubility: 67.5 g/100 mL (20°C) H₂O; sl s alcohol [CRC10]
Density, g/cm³: 3.234 [KIR78]
Melting Point, °C: 267 [KIR78]
Reactions: decomposes at 270°C to BaO, NO, and N₂ [KIR78]

385
Compound: Barium nitrite monohydrate
Formula: Ba(NO₂)₂·H₂O
Molecular Formula: BaH₂N₂O₅
Molecular Weight: 247.353
CAS RN: 7787-38-4

Properties: white to yellowish, hex, cryst powd; crystallized from a stoichiometric solution of BaCl₂ and NaNO₂; used as a corrosion inhibitor, in explosives, and for diazotization [HAW93] [KIR78]
Solubility: g Ba(NO₂)₂/100 g H₂O: 54.8 (0°C), 319 (100°C) [KIR78]; s alcohol [HAW93]
Density, g/cm³: 3.173 [KIR78]
Melting Point, °C: decomposes at 217 [HAW93]
Reactions: minus H₂O at 116°C [KIR78]

386
Compound: Barium oxalate
Synonyms: ethanedioic acid, barium salt
Formula: BaC₂O₄
Molecular Formula: C₂BaO₄
Molecular Weight: 225.347
CAS RN: 516-02-9
Properties: white powd, 99.999% purity [ALF93]
Solubility: g/1000 g soln, H₂O: 0.053 (0°C), 0.1087 (25°C), 0.285 (73°C); solid phase, BaC₂O₄·2H₂O [KRU93]
Density, g/cm³: 2.658 [STR93]
Melting Point, °C: decomposes at 400 [STR93]

387
Compound: Barium oxalate monohydrate
Synonyms: ethanedioic acid, barium salt monohydrate
Formula: BaC₂O₄·H₂O
Molecular Formula: C₂H₂BaO₅
Molecular Weight: 243.362
CAS RN: 13463-22-4
Properties: white, cryst powd; used in pyrotechnics, as an analytical reagent [HAW93]
Solubility: s in 10,000 parts cold H₂O, 5,000 parts boiling H₂O; s dil HCl, HNO₃ [MER06]
Density, g/cm³: 2.66 [MER06]

388
Compound: Barium oxide
Synonym: barium monoxide
Formula: BaO
Molecular Formula: BaO
Molecular Weight: 153.326
CAS RN: 1304-28-5
Properties: white to yellowish white powd; reacts with atm CO₂ and H₂O forming hydroxide and carbonate, with evolution of heat; two forms: cub, hex, a = 0.55391 nm; enthalpy of fusion 59.00 kJ/mol; made by heating BaCO₃ and carbon; uses: dehydrate solvents, additive in detergents, and lubricating oils [CIC73] [HAW93] [KIR78] [CRC10]
Solubility: 3.48 g/100 mL (0°C) H₂O; s dil acids, alcohol; i acetone, NH₃ [CRC10]

Density, g/cm³: 5.72 (cub), 5.32 (hex) [KIR78]
Melting Point, °C: 2013 [CRC10]
Boiling Point, °C: ~2000 [KIR78]
Reactions: BaO+O₂(g)=BaO₂ at 500°C [KIR78]
Thermal Expansion Coefficient: from 25°C to: 100°C
(0.33); 200°C (0.78); 400°C (1.77); 600°C (2.91);
800°C (4.08); 1000°C (5.25); 1200°C (6.51) [TAY84a]

389
Compound: Barium perchlorate
Formula: Ba(ClO₄)₂
Molecular Formula: BaCl₂O₈
Molecular Weight: 336.227
CAS RN: 13465-95-7
Properties: colorless, hex cryst; uses: efficient
desiccant [ALD93] [CRC10] [ALF93]
Solubility: g/100 g soln, H₂O: 67.3 (0°C), 74.3 (20°C),
84.9 (100°C); solid phase, Ba(ClO₄)₂·3H₂O
[KRU93]; s 125 g/100 g ethanol (25°C) [CIC73]
Density, g/cm³: 3.20 [ALF93]
Melting Point, °C: 505 [ALD93]

390
Compound: Barium perchlorate trihydrate
Formula: Ba(ClO₄)₂·3H₂O
Molecular Formula: BaCl₂H₆O₁₁
Molecular Weight: 390.273
CAS RN: 10294-39-0
Properties: colorless cryst; oxidizing agent; used
in the manufacture of explosives and in
rocket fuels [HAW93] [MER06] [STR93]
Solubility: g/100 g H₂O: 239 (0°C), 336 (20°C), 653
(100°C); s methanol [HAW93] [LAN05]
Density, g/cm³: 2.74 [HAW93]

391
Compound: Barium permanganate
Formula: Ba(MnO₄)₂
Molecular Formula: BaMn₂O₈
Molecular Weight: 375.198
CAS RN: 7787-36-2
Properties: brownish violet to black cryst;
oxidizing agent; used as a disinfectant,
in dry cell batteries [HAW93]
Solubility: 62.5 g/100 mL (10°C) H₂O; decomposed
by alcohol [MER06] [CRC10]
Density, g/cm³: 3.77 [MER06]
Melting Point, °C: decomposes at 200 [CRC10]

392
Compound: Barium peroxide
Synonym: barium dioxide

Formula: BaO₂
Molecular Formula: BaO₂
Molecular Weight: 169.326
CAS RN: 1304-29-6
Properties: white or grayish white, heavy powd,
−80 mesh 99% pure; decomposes slowly
in air; oxidizing agent; can be prepared by
heating BaO in oxygen or air at 500°C; used
to bleach materials and to decolorize glass
[HAW93] [MER06] [KIR78] [CER91]
Solubility: i H₂O, but slowly decomposed
by contact with H₂O [MER06]
Density, g/cm³: 4.96 [MER06]
Melting Point, °C: decomposes at 450 [STR93]
Reactions: decomposes at 700°C by reaction:
BaO₂ to BaO+O₂ [KIR78]

393
Compound: Barium potassium chromate
Synonym: Pigment E
Formula: BaK₂(CrO₄)₂
Molecular Formula: BaCr₂K₂O₈
Molecular Weight: 447.511
CAS RN: 27133-66-0
Properties: pale yellow solid; has lower chloride
and sulfate content than other chromate
pigments; prepared by reacting K₂Cr₂O₇ and
BaCO₃ at 500°C; used in paints to protect Fe
and steel from corrosion and to form strong
and elastic paint films [HAW93] [KIR78]
Solubility: g/100 g H₂O: 57.2 (0°C), 57.5
(30°C), 82.7 (100°C) [LAN05]
Density, g/cm³: 3.65 [KIR78]

394
Compound: Barium pyrophosphate
Formula: Ba₂P₂O₇
Molecular Formula: Ba₂O₇P₂
Molecular Weight: 448.597
CAS RN: 13466-21-2
Properties: white powd; rhomb [CRC10] [HAW93]
Solubility: 0.01 g/100 mL H₂O; s acids,
NH₄ salts [CRC10] [HAW93]
Density, g/cm³: 3.9 [CRC10]

395
Compound: Barium selenate
Formula: BaSeO₄
Molecular Formula: BaO₄Se
Molecular Weight: 280.285
CAS RN: 7787-41-9
Properties: ortho-rhomb cryst; preparation:
heating BaCO₃ and Se [MER06]

Solubility: g/L soln; H_2O; 0.081 (25°C); s
    HCl, i HNO_3 [MER06] [KRU93]
Density, g/cm³: 4.75 [MER06]
Reactions: heating causes decomposition [MER06]

396
Compound: Barium selenide
Formula: BaSe
Molecular Formula: BaSe
Molecular Weight: 216.287
CAS RN: 1304-39-8
Properties: cub microcryst powd, −20 mesh 99.5%
    purity; turns red in air; used in semiconductors
    and photocells [HAW93] [CER91] [MER06]
Solubility: decomposed by H_2O [MER06]
Density, g/cm³: 5.02 [MER06]

397
Compound: Barium selenite
Formula: $BaSeO_3$
Molecular Formula: BaO_3Se
Molecular Weight: 264.285
CAS RN: 13718-59-7
Properties: solid [ALD93]
Solubility: g/100 g soln, H_2O: 0.005 (0°C), 0.005
    (25°C); solid phase, $BaSeO_3$ [KRU93]

398
Compound: Barium silicate
Synonym: pentabarium octasilicate
Formula: $5BaO \cdot 8SiO_2$
Molecular Formula: $Ba_5O_{21}Si_8$
Molecular Weight: 1247.306
CAS RN: 12650-28-1
Properties: a=3.365 nm, b=0.4697 nm,
    c=1.3896 nm [TAY88a]
Melting Point, °C: 1445 [TAY88a]
Thermal Expansion Coefficient: from 25°C
    to: 100°C (0.21), 200°C (0.51), 400°C
    (1.29), 600°C (2.19), 800°C (3.24), 1000°C
    (4.47), 1200°C (5.85) [TAY88a]

399
Compound: Barium silicate
Synonym: sanbornite
Formula: BaO_2SiO_2
Molecular Formula: BaO_5Si_2
Molecular Weight: 273.495
CAS RN: 12650-28-1
Properties: monocl, a=2.3206 nm,
    b=0.4661 nm, c=1.3613 nm [TAY87]
Density, g/cm³: 3.70 [LID94]
Melting Point, °C: 1420 [TAY87]

Thermal Expansion Coefficient: from 25°C
    to: 100°C (0.27), 200°C (0.66), 400°C
    (1.44), 600°C (2.22), 800°C (3.06), 1000°C
    (3.90), 1200°C (4.74) [TAY87]

400
Compound: Barium silicate
Synonym: dibarium trisilicate
Formula: $2BaO \cdot 3SiO_2$
Molecular Formula: $Ba_2O_8Si_3$
Molecular Weight: 486.906
CAS RN: 14871-82-0
Properties: a=1.246 nm, b=0.4687 nm,
    c=1.3950 nm [TAY88a]
Melting Point, °C: 1446 [TAY88a]
Thermal Expansion Coefficient: from 25°C to:
    100°C (0.11), 200°C (0.25), 400°C (0.54), 600°C
    (0.84), 800°C (1.13), 1000°C (1.43) [TAY88a]

401
Compound: Barium silicide
Formula: $BaSi_2$
Molecular Formula: $BaSi_2$
Molecular Weight: 193.498
CAS RN: 1304-40-1
Properties: metallic gray lumps, 6 mm pieces and
    smaller, 98% pure; quite permanent in dry air, but
    decomposed by moisture to evolve H_2; metallurgic
    use to deoxidize steel [HAW93] [MER06] [CER91]
Melting Point, °C: 1180 [STR93]

402
Compound: Barium sodium niobium oxide
Formula: $Ba_2NaNb_5O_{15}$
Molecular Formula: $Ba_2NaNb_5O_{15}$
Molecular Weight: 1002.167
CAS RN: 12323-03-4
Properties: white powd of 99.999% purity; electro-
    optical cryst; used to produce coherent
    green light in lasers [HAW93] [ALF93]
Melting Point, °C: 1483 [ALD94]

403
Compound: Barium stannate
Formula: $BaSnO_3$
Molecular Formula: BaO_3Sn
Molecular Weight: 304.035
CAS RN: 12009-18-6
Properties: white cub cryst [CRC10]
Solubility: sl s H_2O [CRC10]
Density, g/cm³: 7.24 [CRC10]

404

Compound: Barium stannate trihydrate
Formula: $BaSnO_3 \cdot 3H_2O$
Molecular Formula: BaH_6O_6Sn
Molecular Weight: 358.081
CAS RN: 12009-18-6
Properties: anhydrous, 51404-76-3, −325 mesh 99% pure; cub, a=0.4117 nm; trihydrate is white, cryst powd; used in the production of special ceramic insulations requiring dielectric properties [HAW93] [TAY85] [CER91]
Solubility: sl s H_2O, s HCl [HAW93]
Thermal Expansion Coefficient: from 25°C to: 100°C (0.18), 200°C (0.45), 400°C (1.02), 600°C (1.62), 800°C (2.31), 1000°C (3.00), 1200°C (3.78) [TAY85]

405

Compound: Barium stearate
Formula: $Ba[CH_3(CH_2)_{16}COO]_2$
Molecular Formula: $C_{36}H_{70}BaO_4$
Molecular Weight: 704.277
CAS RN: 6865-35-6
Properties: white powd; used as a waterproofing agent, lubricant in metal working, and in wax compounding [HAW93] [STR93]
Solubility: i H_2O, alcohol [HAW93]
Density, g/cm³: 1.145 [HAW93]
Melting Point, °C: 160 [HAW93]

406

Compound: Barium strontium niobium oxide
Formula: $BaSr(NbO_3)_4$
Molecular Formula: $BaNb_4O_{12}Sr$
Molecular Weight: 788.566
CAS RN: 37185-09-4
Properties: −325 mesh white powd [ALF93]

407

Compound: Barium strontium tungsten oxide
Formula: Ba_2SrWO_6
Molecular Formula: Ba_2O_6SrW
Molecular Weight: 642.110
CAS RN: 14871-56-8
Properties: −325 mesh powd, 99.9% purity; sensitive to moisture [ALD94] [ALF93]
Melting Point, °C: 1400 [ALD94]

408

Compound: Barium sulfate
Synonym: barite

Formula: $BaSO_4$
Molecular Formula: BaO_4S
Molecular Weight: 233.391
CAS RN: 7727-43-7
Properties: white or yellowish, odorless and tasteless rhomb powd; hardness 3–3.5 Mohs; enthalpy of fusion 40.60 kJ/mol; obtained from mining; used in drilling muds in the form of an aq suspension, to lubricate and cool drill bits, and to plaster walls of drill holes [HAW93] [KIR78] [CRC10]
Solubility: g/L soln, H_2O: 0.00115 (0°C), 0.00223 (25°C), 0.0039 (100°C); s conc H_2SO_4 [KRU93] [KIR78]
Density, g/cm³: 4.50 [KIR78]
Melting Point, °C: 1350 [CRC10]
Boiling Point, °C: decomposes at 1580 [KIR78]
Reactions: transition from rhomb to monocl at 1150°C [KIR78]
Thermal Expansion Coefficient: (volume) 100°C (0.434), 200°C (1.023), 400°C (2.381) [CLA66]

409

Compound: Barium sulfide
Formula: BaS
Molecular Formula: BaS
Molecular Weight: 169.393
CAS RN: 21109-95-5
Properties: heavy, grayish-white or pale yellow powd; −100 mesh, 99% purity; used as depilatory, in luminous paints and in vulcanization of rubber [MER06] [CER91]
Solubility: g/100 g soln, H_2O: 2.88 (0°C), 8.95 (25°C), 60.29 (100°C) [KRU93]
Density, g/cm³: 4.36 [MER06]
Melting Point, °C: >2000 [MER06]
Reactions: slowly oxidizes in air [MER06]

410

Compound: Barium sulfide
Synonym: black ash
Formula: BaS
Molecular Formula: BaS
Molecular Weight: 169.393
CAS RN: 21109-95-5
Properties: black powd; colorless cub if pure; oxidizes in air; black ash is a commercial product produced by the reduction of $BaSO_4$ with carbon at 1000°C–1250°C; used as a precursor to produce $BaCO_3$, $BaCl_2$, and other Ba compounds, as a flame retardant, to dehair hides; other sulfides are: Ba_2S_3, 5311-28-7; BaS_2, 12230-99-8; BaS_3, 12231-01-5; $BaS_4 \cdot H_2O$, 12248-67-8; BaS_5; and $BaS \cdot 6H_2O$ [HAW93] [STR93] [MER06] [KIR78]
Density, g/cm³: 4.25 [KIR78]

Melting Point, °C: 1200 [STR93]
Reactions: in H_2O: $2BaS + 2H_2O = Ba(HS)_2 + Ba(OH)_2$ [KIR78]

411
Compound: Barium sulfite
Formula: $BaSO_3$
Molecular Formula: BaO_3S
Molecular Weight: 217.391
CAS RN: 7787-39-5
Properties: white powd, cub (hex) cryst; oxidizes gradually in air to $BaSO_4$; formed by reacting a soluble sulfite and soluble barium salt; used in paper manufacturing and in analysis [MER06] [KIR78] [HAW93]
Solubility: 0.0197 g/100 g H_2O (20°C); 0.0018 g/100 g (80°C); s dil HCl [KIR78] [HAW93]
Density, g/cm³: 4.44 [LID94]
Melting Point, °C: decomposed by heat [HAW93]

412
Compound: Barium tantalate
Formula: $Ba(TaO_3)_2$
Molecular Formula: BaO_6Ta_2
Molecular Weight: 595.219
CAS RN: 12047-34-6
Properties: −100 mesh 99% pure solid [CER91]

413
Compound: Barium tartrate
Formula: $BaC_4H_4O_6$
Molecular Formula: $C_4H_4BaO_6$
Molecular Weight: 285.399
CAS RN: 5908-81-6
Properties: white cryst; used in pyrotechnics [HAW93]
Solubility: s 3400 parts H_2O; i alcohol [MER06]
Density, g/cm³: 2.98 [HAW93]

414
Compound: Barium telluride
Formula: BaTe
Molecular Formula: BaTe
Molecular Weight: 264.927
CAS RN: 12009-36-8
Properties: −20 and −28 mesh yellow powd, 99.5% pure; cub [CER91] [ALF93] [CRC10]
Density, g/cm³: 5.13 [CRC10]

415
Compound: Barium tetracyanoplatinate(II) tetrahydrate
Formula: $BaPt(CN)_4 \cdot 4H_2O$

Molecular Formula: $C_4H_8BaN_4O_4Pt$
Molecular Weight: 508.540
CAS RN: 13755-32-3
Properties: yellow powd; large dichroic cryst; yellowish green by transmitted light, bluish violet by reflected light; used in x-ray screens [HAW93] [MER06] [STR93]
Solubility: s in about 35 parts H_2O, more in hot H_2O; i alcohol [HAW93] [MER06]
Density, g/cm³: 2.076 [STR93]; 3.05 [MER06]
Reactions: minus $2H_2O$ at 100°C [HAW93]

416
Compound: Barium tetraiodomercurate(II)
Synonym: mercuric barium iodide
Formula: $BaHgI_4$
Molecular Formula: $BaHgI_4$
Molecular Weight: 845.535
CAS RN: 10048-99-4
Properties: yellow or reddish, deliq cryst [MER06]
Solubility: v s H_2O, alcohol [MER06]

417
Compound: Barium thiocyanate
Formula: $Ba(SCN)_2$
Molecular Formula: $C_2BaN_2S_2$
Molecular Weight: 253.494
CAS RN: 2092-17-3
Properties: deliq cryst [MER06]
Solubility: v s H_2O, s acetone, methanol, ethanol [MER06]; g/100 g soln, H_2O: 62.63 (25°C); solid phase, $Ba(SCN)_2 \cdot 3H_2O$ [KRU93]

418
Compound: Barium thiocyanate trihydrate
Formula: $Ba(SCN)_2 \cdot 3H_2O$
Molecular Formula: $C_2H_6BaN_2O_3S_2$
Molecular Weight: 307.540
CAS RN: 68016-36-4
Properties: white cryst; needles from H_2O, deliq [STR93] [MER06]
Solubility: g/100 mL: 4.3 (20°C) H_2O; 35.0 (20°C) alcohol [CRC10]
Density, g/cm³: 2.286 [CRC10]
Reactions: loses H_2O at 160°C [CRC10]

419
Compound: Barium thiosulfate
Formula: BaS_2O_3
Molecular Formula: BaO_3S_2

Molecular Weight: 249.455
CAS RN: 35112-53-9
Properties: white, cryst powd [CRC10]
Solubility: i EtOH; 0.2^{20} g/100 g H_2O [CRC10]
Melting Point, °C: decomposes at 220 [CRC10]

420
Compound: Barium thiosulfate monohydrate
Synonym: barium hyposulfite monohydrate
Formula: $BaS_2O_3 \cdot H_2O$
Molecular Formula: $BaH_2O_4S_2$
Molecular Weight: 267.473
CAS RN: 7787-40-8
Properties: white, cryst powd; used in explosives, luminous paints, matches, varnishes, and photography [MER06] [HAW93]
Solubility: v sl s H_2O; i alcohol [HAW93] [MER06]
Density, g/cm³: 3.5 [HAW93]
Melting Point, °C: decomposed by heat [HAW93]

421
Compound: Barium titanate
Formula: $BaO \cdot 2TiO_2$
Molecular Formula: BaO_5Ti_2
Molecular Weight: 313.084
CAS RN: 12009-27-5
Thermal Expansion Coefficient: from 25°C to: 100°C (0.21), 200°C (0.48), 400°C (1.20), 600°C(1.95) [TOU77]

422
Compound: Barium titanate
Formula: $BaO \cdot 4TiO_2$
Molecular Formula: BaO_9Ti_4
Molecular Weight: 472.842
CAS RN: 12009-31-3
Properties: ortho-rhomb [WU88]
Density, g/cm³: 4.55 [WU88]
Thermal Expansion Coefficient: 100°C (0.18), 200°C (0.42), 400°C (0.96), 600°C (1.53) [TOU77]

423
Compound: Barium titanate
Synonym: barium metatitanate
Formula: $BaTiO_3$
Molecular Formula: BaO_3Ti
Molecular Weight: 233.192
CAS RN: 12047-27-7

Properties: white powd or sintered lumps; two forms: tetr, a=0.39932 nm, c=0.40347 nm; dielectric constant ~4000; preparation: by calcining and sintering barium carbonate and anatase powd at 1300°C–1450°C, by hydrothermal synthesis, and sol-gel process using Ti(IV) isopropylate; used in ferroelectric ceramic materials, as an evaporated ceramic at 99.995% purity in dielectric films and thin film capacitors [HAW93] [STR93] [KIR83] [CER91] [CHA90] [PHU89]
Solubility: i H_2O [CRC10]
Density, g/cm³: tetr, 6.017; hex, 5.806 [KNA91]
Melting Point, °C: ~1625 [KIR78]
Reactions: transition from hex to tetr (−5°C); from tetr to cub (120°C) [TAY85]
Thermal Expansion Coefficient: from 25°C to: 100°C (0.12), 200°C (0.36), 400°C (1.02), 600°C (1.80), 800°C (2.61), 1000°C (3.48), 1200°C (4.44) [TAY85]

424
Compound: Barium titanium silicate
Synonym: benitoite
Formula: $BaO \cdot TiO_2 \cdot 3SiO_2$
Molecular Formula: BaO_9Si_3Ti
Molecular Weight: 413.446
CAS RN: 15491-35-7
Properties: gemstone; a=0.6643 nm, c=0.9757 nm [TAY88b] [LAN05]
Solubility: s HF [LAN05]
Density, g/cm³: 3.6 [LAN05]
Thermal Expansion Coefficient: from 25°C to: 100°C (0.09), 200°C (0.21), 400°C (0.48), 600°C (0.72), 800°C (0.99), 1000°C (1.23) [TAY88b]

425
Compound: Barium tungstate
Synonym: barium white
Formula: $BaWO_4$
Molecular Formula: BaO_4W
Molecular Weight: 385.165
CAS RN: 7787-42-0
Properties: white powd, −200 mesh, 99.9% purity; tetr, a=0.5614 nm, c=1.2715 nm; used in x-ray photography and as a pigment [STR93] [HAW93] [TAY86] [CER91]
Solubility: i H_2O [HAW93]
Density, g/cm³: 5.04 [HAW93]
Thermal Expansion Coefficient: from 25°C to: 100°C (0.27), 200°C (0.63), 400°C (1.35), 600°C (2.04), 800°C (2.76), 1000°C (3.45), 1200°C(4.14) [TAY86]

426
Compound: Barium uranium oxide
Synonym: barium uranate

Formula: BaU_2O_7
Molecular Formula: BaO_7U_2
Molecular Weight: 725.381
CAS RN: 10380-31-1
Properties: orange or yellow powd; used in painting on porcelain [MER06]
Solubility: i H_2O, s acids [MER06]

427
Compound: Barium vanadate
Formula: $Ba_3(VO_4)_2$
Molecular Formula: $Ba_3O_8V_2$
Molecular Weight: 641.859
CAS RN: 39416-30-3
Properties: −200 mesh, 99.9% pure [CER91]
Melting Point, °C: 707 [KNA91]

428
Compound: Barium yttrium tungsten oxide
Formula: $Ba_3Y_3WO_9$
Molecular Formula: $Ba_3O_9WY_3$
Molecular Weight: 1006.534
CAS RN: 37265-86-4
Properties: −325 mesh, 99.9% purity [ALF93]
Melting Point, °C: decomposes at 1470 [ALF93]

429
Compound: Barium zirconate
Formula: $BaZrO_3$
Molecular Formula: BaO_3Zr
Molecular Weight: 276.549
CAS RN: 12009-21-1
Properties: light gray or white powd, −100, +200 mesh 99% purity; a = 0.4193 nm; used in the manufacture of white silicone rubber compounds [HAW93] [TAY85] [CER91]
Solubility: i H_2O, alkalies; sl s acids [HAW93]
Density, g/cm³: 5.52 [HAW93]
Melting Point, °C: 2500 [ALF93]
Thermal Expansion Coefficient: from 25°C to: 100°C (0.18), 200°C (0.42), 400°C (0.90), 600°C (1.38), 800°C (1.86), 1000°C (2.34) [TAY85]

430
Compound: Barium zirconium phosphate
Formula: $Ba_{0.5}Zr_2(PO_4)_3$
Molecular Formula: $Ba_{0.5}SO_{12}P_3Zr_2$
Molecular Weight: 536.026
CAS RN: 82045-86-1
Properties: NASICON structure, a = 0.8642 nm, c = 2.398 nm [TAY91a]

Thermal Expansion Coefficient: from 25°C to: 100°C (0.03), 200°C (0.06), 400°C (0.13), 600°C (0.20), 800°C (0.27), 1000°C (0.34) [TAY91a]

431
Compound: Barium zirconium silicate
Formula: $BaO \cdot ZrO_2 \cdot SiO_2$
Molecular Formula: BaO_5SiZr
Molecular Weight: 336.634β
CAS RN: 13708-68-4
Properties: white powd; uses: in producing electrical resistor ceramics and in glass opacifiers [HAW93]
Solubility: i H_2O, alkalies; sl s acids; s HF [HAW93]

432
Compound: Berkelium(α)
Formula: α-Bk
Molecular Formula: Bk
Molecular Weight: 247
CAS RN: 7440-40-6
Properties: metal; hex, a = 0.3416 nm, c = 1.1069 nm; in trivalent state, its properties are close to that of Ce^{+++}; ionic radius of Bk^{+++} is 0.096 nm, of Bk^{++++} is 0.0860 nm; enthalpy of vaporization 382 kJ/mol; enthalpy of fusion 7.92 kJ/mol; first discovered in 1949 [KIR78] [KAT86] [MER06]
Density, g/cm³: 14.78 (25°C) [KIR78]
Melting Point, °C: 1050 [KIR91]
Boiling Point, °C: ~2630 [KAT86]
Reactions: transfroms from hex to cub ~930°C [KAT86]

433
Compound: Berkelium(β)
Formula: β-Bk
Molecular Formula: Bk
Molecular Weight: 247
CAS RN: 7440-40-6
Properties: discovered in 1949; fcc, a = 0.4997 nm; stable <986°C [KIR78]
Density, g/cm³: 13.25 (25°C) [KIR78]
Melting Point, °C: 986 [MER06]
Boiling Point, °C: ~2630 [KAT86]

434
Compound: Beryllium
Synonym: glucinium
Formula: Be
Molecular Formula: Be
Molecular Weight: 9.012182
CAS RN: 7440-41-7

Properties: metal; two forms, α: gray; hex,
a=0.22856 nm, b=0.35832 nm, c/a=0.15677 nm;
β: bcc, a=0.2551 nm; enthalpy of fusion
7.90 kJ/mol; enthalpy of sublimation ~320 kJ/
mol; enthalpy of vaporization 230–310 kJ/
mol; electrical resistivity at 25°C is 4.266×10^{-8}
ohm·m; velocity of sound 12,600 m/s; reflectivity,
white light, 50%–55%; used in semiconductor
junctions [CIC73] [KIR78] [CER91] [CRC10]
Solubility: s acids except HNO_3; s alkalies [HAW93]
Density, g/cm³: 1.8477 [CIC73]
Melting Point, °C: 1278 [COT88]
Boiling Point, °C: 2970 [KIR78]
Reactions: transformation α to β at 1250°C [CIC73]
Thermal Conductivity, W/(m·K): 190
(25°C), 170 (100°C), 150 (200°C), 130
(400°C), 75 (800°C) [KIR78]
Thermal Expansion Coefficient: coefficient
of linear expansion, K^{-1}: 25°C–100°C,
11.5×10^{-6}; 25°C–200°C, 12.7×10^{-6};
25°C–400°C, 14.8×10^{-6} [KIR78]

435
Compound: Beryllium acetate
Synonyms: acetic acid, beryllium salt
Formula: $Be(CH_3COO)_2$
Molecular Formula: $C_4H_6BeO_4$
Molecular Weight: 127.101
CAS RN: 543-81-7
Properties: white cryst; preparation: can be crystallized
from hot acetic acid in pure form; uses: source
of pure beryllium salts; formula also given
as $Be_4O(CH_3COO)_6$ [MER06] [HAW93]
Solubility: s hot H_2O, with hydrolysis; i alcohol [MER06]
Reactions: decomposes at 60°C–100°C [MER06]

436
Compound: Beryllium acetylacetonate
Synonyms: 2,4-pentanedione, beryllium derivative
Formula: $Be[CH_3COCH=C(O)CH_3]_2$
Molecular Formula: $C_{10}H_{14}BeO_4$
Molecular Weight: 207.231
CAS RN: 10210-64-7
Properties: monocl cryst powd [MER06] [STR93]
Solubility: i H_2O, hydrolyzed in boiling H_2O
[MER06]; v s alcohol, ether [HAW93]
Density, g/cm³: 1.168 [MER06]
Melting Point, °C: 108 [MER06]
Boiling Point, °C: 270 [MER06]

437
Compound: Beryllium aluminate
Synonym: chrysoberyl

Formula: $BeAl_2O_4$
Molecular Formula: Al_2BeO_4
Molecular Weight: 126.973
CAS RN: 12004-06-7
Properties: ortho-rhomb; enthalpy of fusion
176 kJ/mol [JAN71] [CIC73]
Density, g/cm³: 3.76 [CRC10]
Melting Point, °C: 1870 [JAN71]

438
Compound: Beryllium aluminum silicate
Synonym: beryl
Formula: $3BeO \cdot Al_2O_3 \cdot 6SiO_2$
Molecular Formula: $Al_2Be_3O_{18}Si_6$
Molecular Weight: 537.502
CAS RN: 1302-52-9
Properties: gemstone; green or blue; chief
ore for beryllium; hex, a=0.9188 nm,
c=0.9189 nm [CIC73] [KIR78]
Density, g/cm³: 2.66 [CRC10]
Melting Point, °C: 1410 [CRC10]
Thermal Expansion Coefficient: (volume)
100°C (0.061), 200°C (0.079) [CLA66]

439
Compound: Beryllium basic acetate
Formula: $Be_4O(CH_3COO)_6$
Molecular Formula: $C_{12}H_{18}Be_4O_{13}$
Molecular Weight: 406.316
CAS RN: 1332-52-1
Properties: white cryst; can be crystallized in
very pure form from acetic acid; tetrahedra
when crystallized from chloroform; used as a
source of pure Be salts [HAW93] [MER06]
Solubility: i H_2O, hydrolyzed by hot H_2O and
dil acids; s glacial acetic acid, chloroform,
and other organic solvents except alcohol
and ether [MER06] [HAW93]
Density, g/cm³: 1.25 [MER06]; 1.360 [ALD93]
Melting Point, °C: 285–286 [MER06]
Boiling Point, °C: 330–331 [MER06]

440
Compound: Beryllium basic carbonate
Formula: $Be_3(OH)_2(CO_3)_2$
Molecular Formula: $C_2H_2Be_3O_8$
Molecular Weight: 181.069
CAS RN: 66104-24-3
Properties: white powd [CRC10]
Solubility: i H_2O; s acid, alk [CRC10]

441
Compound: Beryllium boride-I
Formula: Be_4B
Molecular Formula: BBe_4
Molecular Weight: 46.859
CAS RN: 12536-52-6
Properties: refractory materials; −80
    mesh with 98% purity
Melting Point, °C: 1160 [CRC10]

442
Compound: Beryllium boride-II
Formula: Be_2B
Molecular Formula: BBe_2
Molecular Weight: 28.835
CAS RN: 12536-51-5
Properties: pink cryst [CRC10]
Melting Point, °C: 1520 [CRC10]

443
Compound: Beryllium boride-III
Formula: BeB_2
Molecular Formula: B_2Be
Molecular Weight: 30.634
CAS RN: 12228-40-9
Properties: refrac solid [CRC10]
Melting Point, °C: 1970 [CRC10]

444
Compound: Beryllium boride-IV
Formula: BeB_6
Molecular Formula: B_6Be
Molecular Weight: 73.878
CAS RN: 12429-94-6
Properties: red solid [CRC10]
Solubility: Insoluble in H_2O and EtOH
Melting Point, °C: 2070 [CRC10]

445
Compound: Beryllium borohydride
Synonym: beryllium tetrahydroborate
Formula: $Be(BH_4)_2$
Molecular Formula: B_2BeH_8
Molecular Weight: 38.698
CAS RN: 17440-85-6
Properties: spontaneously flammable;
    obtained by reaction of diborane with
    dimethylberyllium [MER06]
Solubility: vigorous reaction with H_2O,
    HCl, evolving H_2 [MER06]

Melting Point, °C: sublimes at 91.3 [MER06]
Boiling Point, °C: decomposes at >123 [MER06]

446
Compound: Beryllium bromide
Formula: $BeBr_2$
Molecular Formula: $BeBr_2$
Molecular Weight: 168.820
CAS RN: 7787-46-4
Properties: ortho-rhomb; very hygr; −80 mesh, 99%
    purity; enthalpy of fusion 9.80 kJ/mol; made
    by reaction of Be and Br_2 at 500°C–700°C
    [KIR78] [CER91] [MER06] [CRC10]
Solubility: v s H_2O [MER06]
Density, g/cm³: 3.465 [MER06]
Melting Point, °C: 508 [CRC10]
Boiling Point, °C: 520 [MER06]
Reactions: sublimes at 473°C [MER06]

447
Compound: Beryllium carbide
Formula: Be_2C
Molecular Formula: CBe_2
Molecular Weight: 30.035
CAS RN: 506-66-1
Properties: brick red or yellowish red octahedra;
    hard, refractory material; −200 mesh, 98%
    purity; prepared by hot pressing mixture of
    Be and C to 900°C; used in nuclear-reactor
    cores [CER91] [HAW93] [KIR78] [MER06]
Solubility: decomposed very slowly by H_2O;
    hydrolysis yields methane and beryllium
    hydroxide [KIR78] [MER06]
Density, g/cm³: 1.90 [MER06]
Melting Point, °C: decomposes at >2100 [MER06]

448
Compound: Beryllium carbonate tetrahydrate
Formula: $BeCO_3 \cdot 4H_2O$
Molecular Formula: CH_8BeO_7
Molecular Weight: 141.083
CAS RN: 60883-64-9
Properties: unstable unless kept under CO_2
    atm; obtained by passing CO_2 through
    aq suspension of $Be(OH)_2$ [KIR78]

449
Compound: Beryllium chloride
Formula: $BeCl_2$
Molecular Formula: $BeCl_2$
Molecular Weight: 79.917

CAS RN: 7787-47-5
Properties: white to faint yellow powd, and sublimed
fibers and clumps of 99.5% purity; very deliq;
ortho-rhomb cryst; hydrolyzed by water vapor;
enthalpy of vaporization 105 kJ/mol; enthalpy of
fusion 8.66 kJ/mol; prepared by heating BeO, Cl_2,
and C at 600°C–800°C [KIR78] [MER06] [CRC10]
Solubility: v s H_2O with evolution of heat
[MER06]: g/100 g soln, H_2O: 40.35 (0°C), 41.72
(25°C); solid phase, $BeCl_2 \cdot 4H_2O$ [KRU93];
s alcohol, benzene, ether [HAW93]
Density, g/cm³: 1.90 [MER06]
Melting Point, °C: 405 [DOU83]
Boiling Point, °C: ~550 [DOU83]
Reactions: sublimes in vacuum at 300°C [MER06]

450
Compound: Beryllium fluoride
Formula: BeF_2
Molecular Formula: BeF_2
Molecular Weight: 47.009
CAS RN: 7787-49-7
Properties: 0.25 in. pieces and down; two forms;
glassy, hygr; tetr when heated above 230°C;
glassy form crystallizes spontaneously to
quartz modification; enthalpy of fusion 4.76 kJ/
mol; made by thermal decomposition of
$(NH_4)_2BeF_6$ [STR93] [KIR78] [MER06]
Solubility: v s H_2O [MER06]; sl s alcohol [HAW93]
Density, g/cm³: 1.986 [MER06]
Melting Point, °C: 552 [CRC10]
Boiling Point, °C: sublimes at 1036
under 1 mm Hg [MER06]
Reactions: transition from α to β form at 227°C [KIR78]

451
Compound: Beryllium formate
Synonyms: formic acid, beryllium salt
Formula: $Be(OOCH)_2$
Molecular Formula: $C_2H_2BeO_4$
Molecular Weight: 99.048
CAS RN: 1111-71-3
Properties: powd [MER06]
Solubility: very slowly hydrolyzed by H_2O [MER06]
Reactions: forms $Be_4O(HCOO)_6$ >250°C,
which sublimes at ~320°C [MER06]

452
Compound: Beryllium hydride
Formula: BeH_2
Molecular Formula: BeH_2
Molecular Weight: 11.028

CAS RN: 7787-52-2
Properties: white, amorphous solid; inert to laboratory
air; can be prepared by continuous thermal
decomposition of a di-t-butylberyllium ethyl
ether complex in a boiling hydrocarbon; has
found use as rocket fuel and as a moderator
for nuclear reactors [KIR80] [MER06]
Solubility: reacts slowly with H_2O, rapidly with
dil acids evolving $H_2(g)$ [MER06]
Density, g/cm³: 0.65 [LID94]
Reactions: rapid $H_2(g)$ evolution at 220°C [MER06]

453
Compound: Beryllium hydrogen phosphate
Formula: $BeHPO_4$
Molecular Formula: $BeHO_4P$
Molecular Weight: 104.991
CAS RN: 13598-15-7
Properties: cryst [CRC10]
Solubility: i H_2O

454
Compound: Beryllium hydroxide(α)
Formula: $Be(OH)_2$
Molecular Formula: BeH_2O_2
Molecular Weight: 43.027
CAS RN: 13327-32-7
Properties: amorphous powd or cryst; ortho-rhomb;
prepared by precipitation from beryllium acetate
solution with alkali [HAW93] [MER06] [KIR78]
Solubility: v sl s H_2O, dil alkali; s hot
NaOH, acids [MER06]
Density, g/cm³: 1.92 [MER06]
Reactions: minus H_2O at >950°C [KIR78]

455
Compound: Beryllium hydroxide(β)
Formula: $Be(OH)_2$
Molecular Formula: BeH_2O_2
Molecular Weight: 43.027
CAS RN: 13327-32-7
Properties: white powd; tetr; metastable; decomposes
to oxide at 138°C [KIR78] [HAW93]
Solubility: i H_2O; s acids, alkalies [HAW93]
Reactions: transition from β to α after
months of standing [KIR78]

456
Compound: Beryllium iodide
Formula: BeI_2
Molecular Formula: BeI_2

Molecular Weight: 262.821
CAS RN: 7787-53-3
Properties: −60 mesh, 99.5% pure and needles; very hygr; enthalpy of vaporization 70.5 kJ/mol; enthalpy of fusion 21.00 kJ/mol; obtained by reaction of Be and I_2 at 500°C–700°C [CER91] [KIR78] [MER06] [CRC10]
Solubility: reacts violently with H_2O giving off HI [MER06]
Density, g/cm³: 4.325 [CRC10]
Melting Point, °C: 510 [CRC10]
Boiling Point, °C: 590 [CRC10]

457

Compound: Beryllium nitrate trihydrate
Formula: $Be(NO_3)_2 \cdot 3H_2O$
Molecular Formula: $BeH_6N_2O_9$
Molecular Weight: 187.068
CAS RN: 13597-99-4
Properties: white to sl yellow; deliq; prepared from BeO and HNO_3 solution, followed by evaporation and crystallization; used as a reagent [HAW93] [MER06]
Solubility: g $Ba(NO_3)_2$/100 g H_2O: 97 (0°C), 108 (20°C), 178 (60°C); s alcohol [LAN05] [HAW93]
Density, g/cm³: 1.557 [CRC10]
Melting Point, °C: ~60 [MER06]
Boiling Point, °C: decomposes at 100–200 [HAW93]

458

Compound: Beryllium nitride
Formula: Be_3N_2
Molecular Formula: Be_3N_2
Molecular Weight: 55.050
CAS RN: 1304-54-7
Properties: hard white to grayish white; refractory; cub, a = 0.814 nm; obtained from a reaction of Be and NH_3 at 1100°C [KIR78] [HAW93] [CIC73]
Solubility: decomposes slowly in H_2O, quickly in acids and alkalies to evolve NH_3 [MER06]
Density, g/cm³: 2.71 [LID94]
Melting Point, °C: decomposes at 2200 [KIR78]
Boiling Point, °C: volatile [MER06]
Reactions: oxidized in air at 600°C [MER06]

459

Compound: Beryllium oxalate trihydrate
Formula: $BeC_2O_4 \cdot 3H_2O$
Molecular Formula: $C_2H_6BeO_7$
Molecular Weight: 151.078
CAS RN: 15771-43-4
Properties: rhomb; obtained by evaporating a solution of $Be(OH)_2$ in excess oxalic acid; used to prepare ultra pure BeO [CRC10] [KIR78]

Solubility: g/100 g soln, H_2O: 63.2 (25°C) [KRU93]
Melting Point, °C: 2200 [CRC10]
Boiling Point, °C: decomposes at 2240 [CRC10]
Reactions: decomposes to BeO above 320°C [KIR78]

460

Compound: Beryllium oxide
Synonym: beryllia
Formula: BeO
Molecular Formula: BeO
Molecular Weight: 25.011
CAS RN: 1304-56-9
Properties: light, amorphous, white powd; insulates electrically like a ceramic, conducts heat like a metal; electrical resistivity, $>1 \times 10^{+16}$ ohm · m; hardness 9 Mohs; tensile strength 150 MPa; compressive strength 1400 MPa; Poisson's ratio 0.164–0.380; modulus of rupture 250 MPa; modulus of elasticity 345 GPa; made from $Be(OH)_2$ and H_2SO_4; enthalpy of fusion 85.00 kJ/mol; used in electron tubes, resistor cores [HAW93] [MER06] [STR93] [KIR78] [CRC10]
Solubility: v sl s H_2O; slowly s conc acid, alkali [MER06]
Density, g/cm³: 3.01 [STR93]
Melting Point, °C: 2507 [CRC10]
Boiling Point, °C: 4300 [STR93]
Thermal Conductivity, W/(m·K): 25°C (290–330); 100°C (190–220); 500°C (65.4); 1000°C (20.3) [KIR80] [KIR78]
Thermal Expansion Coefficient: coefficient of thermal expansion at 100°C is 9.7×10^{-6}/K; at 500°C is 13.3×10^{-6}/K [KIR78]

461

Compound: Beryllium perchlorate tetrahydrate
Formula: $Be(ClO_4)_2 \cdot 4H_2O$
Molecular Formula: $BeCl_2H_8O_{12}$
Molecular Weight: 279.974
CAS RN: 7787-48-6
Properties: very hygr cryst; retains waters of crystallization tenaciously [MER06]
Solubility: g/100 g soln, H_2O: 59.5 (25°C); solid phase, $Be(ClO_4)_2 \cdot 4H_2O$ [KRU93]

462

Compound: Beryllium selenate tetrahydrate
Formula: $BeSeO_4 \cdot 4H_2O$
Molecular Formula: BeH_8SeO_8
Molecular Weight: 224.031
CAS RN: 10039-31-3
Properties: colorless, ortho-rhomb cryst [CRC10] [MER06]

Solubility: v s H_2O [MER06]
Density, g/cm³: 2.03 [MER06]
Reactions: minus $2H_2O$ at 100°C; minus
$4H_2O$ at 300°C [MER06]

463
Compound: Beryllium sulfate
Formula: $BeSO_4$
Molecular Formula: BeO_4S
Molecular Weight: 105.076
CAS RN: 13510-49-1
Properties: colorless cryst [HAW93]
Solubility: g/100 g soln, H_2O: 26.69 (0°C),
29.22 (25°C), 45.28 (100°C); solid phase,
$BeSO_4 \cdot 4H_2O$ [KRU93]; i alcohol [HAW93]
Density, g/cm³: 2.443 [CRC10]
Melting Point, °C: decomposes at 550–600 [CRC10]

464
Compound: Beryllium sulfate dihydrate
Formula: $BeSO_4 \cdot 2H_2O$
Molecular Formula: BeH_4O_6S
Molecular Weight: 141.106
CAS RN: 14215-00-0
Properties: forms when the tetrahydrate
is heated at 92°C [KIR78]
Reactions: minus $2H_2O$ at 400°C; decomposes
to BeO ~650°C [KIR78]

465
Compound: Beryllium sulfate tetrahydrate
Formula: $BeSO_4 \cdot 4H_2O$
Molecular Formula: BeH_8O_8S
Molecular Weight: 177.137
CAS RN: 7787-56-6
Properties: colorless cryst, lump 99.9% purity;
produced from $Be(OH)_2$ and H_2SO_4 solution,
followed by fractional crystallization; used to
produce BeO for ceramics [KIR78] [HAW93]
Solubility: v s H_2O [MER06]
Density, g/cm³: 1.71 [MER06]
Reactions: minus $2H_2O$ ~100°C [MER06]

466
Compound: Beryllium sulfide
Formula: BeS
Molecular Formula: BeS
Molecular Weight: 41.079
CAS RN: 13598-22-6
Properties: regular; −100 mesh with 99%
purity [CRC10] [CER91] [ALF93]

Solubility: decomposes in H_2O [CRC10]
Density, g/cm³: 2.36 [CRC10]

467
Compound: Bis(cyclopentadienyl)ruthenium
Synonym: ruthenocene
Formula: $(C_5H_5)_2Ru$
Molecular Formula: $C_{10}H_{10}Ru$
Molecular Weight: 231.259
CAS RN: 1287-13-4
Properties: light yellow cryst [STR93]
Melting Point, °C: 194–198 [STR93]

468
Compound: Bis(diethylamino)chlorophosphine
Formula: $[(C_2H_5)_2N]_2PCl$
Molecular Formula: $C_8H_{20}ClN_2P$
Molecular Weight: 210.687
CAS RN: 685-83-6
Properties: liq [ALF95]
Boiling Point, °C: 124–125 (15 mm Hg) [ALF95]

469
Compound: Bismuth
Formula: Bi
Molecular Formula: Bi
Molecular Weight: 208.980373
CAS RN: 7440-69-9
Properties: grayish white, soft, brittle metal, 99.999%
vacuum deposition grade; rhomb, a = 0.47457 nm;
enthalpy of fusion 11.30 kJ/mol; enthalpy of
vaporization 151 kJ/mol; Poisson's ratio 0.33;
electrical resistivity (20°C) 129 μohm · cm;
Brinell hardness 7; electronegativity 1.67;
used in ferromagnetic and resistive films, in
pharmaceuticals and medicine [KIR78] [MER06]
[CRC10] [COT88] [CER91] [ALD94]
Solubility: s dil HNO_3, conc HCl [MER06]
Density, g/cm³: 9.808 (25°C) [KIR78]
Melting Point, °C: 271.4 [KIR78]
Boiling Point, °C: 1564 [KIR78]
Thermal Conductivity, W/(m · K): 7.92 (25°C) [ALD94]
Thermal Expansion Coefficient: (volume)
100°C (0.37) [CLA66]

470
Compound: Bismuth acetate
Formula: $Bi(CH_3COO)_3$
Molecular Formula: $C_6H_9BiO_6$
Molecular Weight: 386.113
CAS RN: 22306-37-2

Properties: white cryst 99.999% pure; sensitive
to moisture [ALF93] [STR93] [ALD93]
Solubility: i H_2O [CRC10]
Melting Point, °C: decomposes [CRC10]

471
Compound: Bismuth antimonide
Formula: BiSb
Molecular Formula: BiSb
Molecular Weight: 330.740
CAS RN: 12323-19-2
Properties: 99.99% pure cryst; used as a
semiconductor material in the form of
a single cryst [HAW93] [ALF93]
Melting Point, °C: 475 [ALF93]

472
Compound: Bismuth arsenate
Formula: $BiAsO_4$
Molecular Formula: $AsBiO_4$
Molecular Weight: 347.900
CAS RN: 13702-38-0
Properties: white, monocl cryst [CRC10]
Solubility: i H_2O; sl conc HNO_3 [CRC10]
Density, g/cm³: 7.14 [CRC10]

473
Compound: Bismuth basic carbonate hemihydrate
Synonym: bismuth subcarbonate
Formula: $(BiO)_2CO_3 \cdot 1/2H_2O$
Molecular Formula: $CHBi_2O_{5.5}$
Molecular Weight: 518.976
CAS RN: 5892-10-4
Properties: odorless, tasteless, white powd; light
sensitive; used in a mixture with other compounds
in glazes for ceramics and to give a pearly surface
for plastics [ALD93] [ALF93] [CRC10] [MER06]
Solubility: i H_2O; s mineral acids,
glacial acetic acid [MER06]
Density, g/cm³: 6.86 [ALF93]
Boiling Point, °C: decomposes [CRC10]

474
Compound: Bismuth basic dichromate
Formula: $Bi_2O_3 \cdot 2CrO_3$
Molecular Formula: $Bi_2Cr_2O_9$
Molecular Weight: 665.948
CAS RN: 1304-75-2
Properties: reddish orange, amorphous powd;
preparation: reaction between $Bi(NO_3)_3$
and potassium chromate [HAW93]
Solubility: i H_2O; s acids, alkalies [HAW93]

475
Compound: Bismuth bromide
Synonym: bismuth tribromide
Formula: $BiBr_3$
Molecular Formula: $BiBr_3$
Molecular Weight: 448.692
CAS RN: 7787-58-8
Properties: yellowish cryst, −60 mesh 99.999% purity;
odor of HBr; hygr; enthalpy of fusion 21.8 kJ/mol;
enthalpy of sublimation 115 kJ/mol; enthalpy
of vaporization 75.4 kJ/mol; can be prepared by
dissolution of Bi_2O_3 in conc HBr solution, followed
by dewatering in gentle stream of N_2 and distillation
[CER91] [KIR78] [CRC10] [HAW93] [MER06]
Solubility: decomposed by H_2O forming
BiOBr; s dil HCl, acetone, ether; i alcohol
[HAW93] [KIR78] [MER06]
Density, g/cm³: solid: 5.72; liq 4.572
at 271.5°C [KIR78]
Melting Point, °C: 218 [MER06]
Boiling Point, °C: 453 [CRC10]

476
Compound: Bismuth chloride
Synonym: bismuth trichloride
Formula: $BiCl_3$
Molecular Formula: $BiCl_3$
Molecular Weight: 315.338
CAS RN: 7787-60-2
Properties: white to yellowish, very deliq cryst, −60
mesh 99.99% and 99.9% purity; HCl odor; enthalpy
of fusion 10.90 kJ/mol; enthalpy of sublimation
114 kJ/mol; enthalpy of vaporization at bp 72.61 kJ/
mol; a preparation is by chlorination of molten
metal; used as a catalyst for organic reactions
[KIR78] [HAW93] [MER06] [CER91] [CRC10]
Solubility: decomposed by H_2O forming BiOCl
[HAW93] [MER06]; s acids; i alcohol [HAW93]
Density, g/cm³: 4.75 [MER06]
Melting Point, °C: 230 [CRC10]
Boiling Point, °C: 447 [CRC10]
Reactions: sublimes at ~430°C [MER06]

477
Compound: Bismuth chloride monohydrate
Synonym: bismuth trichloride monohydrate
Formula: $BiCl_3 \cdot H_2O$
Molecular Formula: $BiCl_3H_2O$
Molecular Weight: 333.353
CAS RN: 39483-74-4
Properties: 99.99% pure cryst [ALF93]
Density, g/cm³: 4.75 [ALF93]

Melting Point, °C: 230–232 [ALF93]
Boiling Point, °C: 447 [ALF93]

478
Compound: Bismuth citrate
Formula: $BiC_6H_5O_7$
Molecular Formula: $C_6H_5BiO_7$
Molecular Weight: 398.082
CAS RN: 813-93-4
Properties: white powd; preparation: boiling bismuth subnitrate and citric acid solution; used in medicine [HAW93]
Solubility: i H_2O; s ammonia; sl s alcohol [HAW93]
Density, g/cm³: 3.458 [HAW93]
Melting Point, °C: decomposes [CRC10]

479
Compound: Bismuth fluoride
Synonym: bismuth trifluoride
Formula: BiF_3
Molecular Formula: BiF_3
Molecular Weight: 265.975
CAS RN: 7787-61-3
Properties: white to gray powd, −60 mesh of 99.9% and 99.999% purity; moisture sensitive; can be obtained by dissolving either bismuth oxide or oxyfluoride in HF, followed by careful evaporation; enthalpy of fusion 21.6 kJ/mol; enthalpy of sublimation 201 kJ/mol; used to prepare BiF_5 [KIR78] [MER06] [CER91] [ALD93] [CRC10]
Solubility: 0.00503 mol/L in H_2O [KIR78]
Density, g/cm³: 8.3 [MER06]
Melting Point, °C: 725 [COT88]
Boiling Point, °C: 900 [KIR78]
Reactions: volatilizes at >730°C, slowly, without decomposition [MER06]

480
Compound: Bismuth germanium oxide
Formula: $2Bi_2O_3 \cdot 3GeO_2$
Molecular Formula: $Bi_4Ge_3O_{12}$
Molecular Weight: 1245.744
CAS RN: 12233-56-6
Properties: white chips or powd, 99.99995% purity [ALF93]

481
Compound: Bismuth hexafluoroacetylacetonate
Synonym: bismuth hexafluoro-2,4-pentanedionate
Formula: $Bi(CF_3COCHCOCF_3)_3$
Molecular Formula: $C_{15}H_3BiF_{18}O_6$

Molecular Weight: 830.132
CAS RN: 142617-56-9
Properties: powd [CRC10]
Melting Point, °C: 96 [CRC10]

482
Compound: Bismuth hydride
Synonym: bismuthine
Formula: BiH_3
Molecular Formula: BiH_3
Molecular Weight: 212.004
CAS RN: 18288-22-7
Properties: colorless gas, unstable at room temp and rapidly decomposes to Bi and H_2; can be prepared by disproportionation of either methyl- or ethyl-bismuthine; enthalpy of vaporization about 25.15 kJ/mol; finds use in manufacturing Ge or Si semiconductors [MER06] [KIR78]
Density, g/cm³: 9.303 g/L [LID94]
Melting Point, °C: −67 [LID94]
Boiling Point, °C: ~16.8 [MER06]

483
Compound: Bismuth hydroxide
Formula: $Bi(OH)_3$
Molecular Formula: BiH_3O_3
Molecular Weight: 260.002
CAS RN: 10361-43-0
Properties: white to yellowish white, amorphous powd; used in plutonium separation, as an absorbent for rutin and quercetin [HAW93] [MER06]
Solubility: i H_2O [MER06]; s acids [HAW93]
Density, g/cm³: 4.962 [MER06]
Reactions: readily loses one H_2O to form metahydroxide [MER06]

484
Compound: Bismuth hydroxide nitrate oxide
Synonym: bismuth subnitrate
Formula: $4BiNO_3(OH)_2 \cdot BiO(OH)$
Molecular Formula: $Bi_5H_9N_4O_{22}$
Molecular Weight: 1461.987
CAS RN: 1304-85-4
Properties: odorless, tasteless, heavy, sl hygr, microcryst powd, 98+% [MER06] [ALF93]
Solubility: i H_2O; s dil HNO_3, HCl [MER06]
Reactions: decomposes at 260°C [ALF93]

485
Compound: Bismuth iodide
Synonym: bismuth triiodide

Formula: BiI_3
Molecular Formula: BiI_3
Molecular Weight: 589.693
CAS RN: 7787-64-6
Properties: black, minute, hex cryst or gray powd, −20 mesh and −40 mesh 99.999% and 99.9% purity; metallic sheen; enthalpy of fusion 39.1 kJ/mol; enthalpy of sublimation 134.3 kJ/mol; sensitive to moisture; prepared from Bi and I_2; used in analytical chemistry [HAW93] [MER06] [STR93] [ALF93]
Solubility: i H_2O, decomposed by hot H_2O, s alcohol and HI, KI solutions [HAW93]
Density, g/cm³: 5.778 [MER06]
Melting Point, °C: 408.6 [KIR78]
Boiling Point, °C: 542 [KIR78]
Reactions: sublimes at 439°C; decomposes at 500°C [MER06]

486
Compound: Bismuth iron molybdenum oxide
Formula: $Bi_3FeMo_2O_{12}$
Molecular Formula: $Bi_3FeMo_2O_{12}$
Molecular Weight: 1066.659
CAS RN: 59393-06-5
Properties: −325 mesh powd; used as an oxidation catalyst [ALF93]

487
Compound: Bismuth molybdate
Synonym: bismuth molybdenum oxide
Formula: $Bi_2(MoO_4)_3$
Molecular Formula: $Bi_2Mo_3O_{12}$
Molecular Weight: 897.774
CAS RN: 51898-99-8
Properties: colorless trig; hygr; −200 mesh and −325 mesh powd; used as an oxidation catalyst [CRC10] [CER91] [ALF93]
Density, g/cm³: 5.95 [LID94]
Melting Point, °C: decomposes at 30 [CRC10]

488
Compound: Bismuth molybdenum oxide
Formula: Bi_2MoO_6
Molecular Formula: Bi_2MoO_6
Molecular Weight: 609.897
CAS RN: 13565-96-3
Properties: −325 mesh powd; used as an oxidation catalyst [ALF93]
Density, g/cm³: 9.32 [KIR78]

489
Compound: Bismuth nitrate pentahydrate
Formula: $Bi(NO_3)_3 \cdot 5H_2O$

Molecular Formula: $BiH_{10}N_3O_{14}$
Molecular Weight: 485.071
CAS RN: 10035-06-0
Properties: lustrous, clear, colorless, hygr cryst, 99.999% purity; acid taste; odor of HNO_3; used to prepare other bismuth salts, in luminous paints and enamels [HAW93] [MER06] [STR93] [ALF93]
Solubility: decomposed by H_2O to subnitrate; s dil HNO_3 and glycerol, acetone [HAW93]
Density, g/cm³: 2.83 [MER06]
Boiling Point, °C: decomposes 75–80 [HAW93]

490
Compound: Bismuth oleate
Synonyms: oleic acid, bismuth salt
Formula: $[CH_3(CH_2)_7CH=CH(CH_2)_7COO]_3Bi$
Molecular Formula: $C_{54}H_{99}BiO_6$
Molecular Weight: 1053.356
CAS RN: 52951-38-9
Properties: yellowish brown, soft granular mass; obtained from Bi_2O_3, oleic acid, and acetic anhydride; used in catalysts for manufacturing aldehydes and alcohols by oxo process [MER06] [HAW93]
Solubility: i H_2O; s ether, s in about 1500 parts benzene [MER06] [HAW93]

491
Compound: Bismuth oxalate
Formula: $Bi_2(C_2O_4)_3$
Molecular Formula: $C_6Bi_2O_{12}$
Molecular Weight: 682.018
CAS RN: 6591-55-5
Properties: white powd [CRC10]
Solubility: i H_2O, EtOH; s dil acid [CRC10]

492
Compound: Bismuth oxide
Formula: Bi_2O_3
Molecular Formula: Bi_2O_3
Molecular Weight: 465.959
CAS RN: 1304-76-3
Properties: yellow, heavy, odorless powd or monocl cryst, of various grades: −30 mesh of 99.9999% purity, 3–12 mm pieces (sintered); used to enamel cast iron, in ceramic coloring, as an evaporated material and sputtering target for beam splitting, and as a base coating for gold films, which are used as transparent heating elements on glass [HAW93] [MER06] [CER91]
Solubility: i H_2O [MER06]; s acids [HAW93]
Density, g/cm³: 8.9 [STR93]

Melting Point, °C: 817 [STR93]
Boiling Point, °C: 1890 [STR93]

493
Compound: Bismuth oxybromide
Synonym: bismuth bromide oxide
Formula: BiOBr
Molecular Formula: BiBrO
Molecular Weight: 304.883
CAS RN: 7787-57-7
Properties: colorless cryst or amorphous powd; used in manufacture of dry cell cathodes [MER06] [CRC10]
Solubility: i H_2O, alcohol; s HCl, HBr, and HNO_3 [MER06]
Density, g/cm³: 8.082 (15°C) [CRC10]
Melting Point, °C: melts at red heat [MER06]

494
Compound: Bismuth oxychloride
Synonym: bismuth chloride oxide
Formula: BiOCl
Molecular Formula: BiClO
Molecular Weight: 260.432
CAS RN: 7787-59-9
Properties: white 99.999% fine powd or tetr cryst; used in face powd, to manufacture artificial pearls, in dry cell cathodes [MER06] [ALF93]
Solubility: i H_2O [MER06]
Density, g/cm³: 7.72 [MER06]
Melting Point, °C: melts at low red heat [MER06]

495
Compound: Bismuth oxyiodide
Synonym: bismuth iodide oxide
Formula: BiOI
Molecular Formula: BiIO
Molecular Weight: 351.883
CAS RN: 7787-63-5
Properties: brick red, heavy, odorless powd or copper-colored cryst; used in manufacturing dry cell cathodes and as an anti-infective [MER06]
Solubility: i H_2O, alcohol, chloroform; s HCl; decomposed by HNO_3 or alkali [MER06]
Density, g/cm³: 7.92 [MER06]
Melting Point, °C: fuses at red heat with partial decomposition [MER06]

496
Compound: Bismuth oxynitrate
Synonym: bismuth subnitrate
Formula: $BiONO_3$
Molecular Formula: $BiNO_4$

Molecular Weight: 286.985
CAS RN: 10361-46-3
Properties: heavy, white powd; sl hygr; uses: cosmetics, ceramic glasses, and enamel fluxes; there is a 99.99+% pure monohydrate, CAS RN 13595-83-0 [HAW93] [ALD94]
Solubility: i H_2O, alcohol; s acids [HAW93]
Density, g/cm³: 4.928 [HAW93]
Melting Point, °C: 260, decomposing [HAW93]

497
Compound: Bismuth oxyperchlorate monohydrate
Formula: $BiOClO_4 \cdot H_2O$
Molecular Formula: $BiClH_2O_6$
Molecular Weight: 342.445
CAS RN: 44584-78-3
Properties: white, cryst powd [STR93] [ALF93]

498
Compound: Bismuth pentafluoride
Synonym: bismuth(V) fluoride
Formula: BiF_5
Molecular Formula: BiF_5
Molecular Weight: 303.972
CAS RN: 7787-62-4
Properties: long, white needles; body-centered tetr cryst; very sensitive to moisture; discolors quickly in moist air; can be formed by fluorinating BiF_3 or Bi metal at 120°C; used as a fluorinating agent [KIR78] [MER06]
Solubility: violent reaction with H_2O to form BiF_3 and ozone [MER06]
Density, g/cm³: 5.55 [MER06]
Melting Point, °C: 151 [KIR78]
Boiling Point, °C: 230 [KIR78]

499
Compound: Bismuth phosphate
Formula: $BiPO_4$
Molecular Formula: BiO_4P
Molecular Weight: 303.951
CAS RN: 10049-01-1
Properties: odorless powd or monocl cryst; used as an antacid, to recover plutonium, and in optical glass [HAW93] [MER06]
Solubility: sl s H_2O; s conc HNO_3, HCl [MER06]
Density, g/cm³: 6.323 [MER06]
Melting Point, °C: does not melt on heating [MER06]

500
Compound: Bismuth potassium iodide
Formula: BiK_4I_7

Molecular Formula: BiI_7K_4
Molecular Weight: 1253.704
CAS RN: 41944-01-8
Properties: red cryst [CRC10]
Solubility: reac H_2O; s alk iodide soln [CRC10]

501
Compound: Bismuth selenide
Synonym: guanajuatite
Formula: Bi_2Se_3
Molecular Formula: Bi_2Se_3
Molecular Weight: 654.841
CAS RN: 12068-69-8
Properties: black cryst, 1–6 mm pieces (fused) of 99.999% purity; rhomb and hex; decomposes when heated in air, by conc HNO_3 and by aqua regia; used in semiconductors and in the form of a 99 or 99.999% pure material used as a sputtering target to produce multilayer thin films and magneto-resistant films [HAW93] [MER06] [CER91]
Solubility: i H_2O [MER06]
Density, g/cm³: 7.70 [MER06]
Melting Point, °C: 710 [MER06]
Thermal Conductivity, W/(m·K): 2.4 [CRC10]

502
Compound: Bismuth stannate
Formula: $Bi_2O_3 \cdot 2SnO_2$
Molecular Formula: $Bi_2O_7Sn_2$
Molecular Weight: 767.377
CAS RN: 12338-09-9
Properties: −200 mesh, 99.9% purity [CER91]
Reactions: pentahydrate: minus $5H_2O$ at ~140°C [HAW93]

503
Compound: Bismuth stannate pentahydrate
Formula: $Bi_2O_3 \cdot 3SnO_2 \cdot 5H_2O$
Molecular Formula: $Bi_2H_{10}O_{14}Sn_3$
Molecular Weight: 1008.162
CAS RN: 12777-45-6
Properties: light-colored cryst; used in ceramic capacitor such as barium titanate [HAW93]
Solubility: i H_2O [HAW93]
Reactions: minus $5H_2O$ at ~140°C [HAW93]

504
Compound: Bismuth strontium calcium copper oxide (1112)
Synonym: supercon 1112

Formula: $BiSrCaCu_2O_x$
Molecular Formula: $BiCaCuSr_2O_y$
CAS RN: 114901-61-0
Properties: 99.999% and 99.99% pure 20 μm powd; dry processed from 99.999% pure oxides and carbonates [STR93]

505
Compound: Bismuth strontium calcium copper oxide (2212)
Synonym: supercon 2212
Formula: $Bi_2Sr_2CaCu_2O_8$
Molecular Formula: $Bi_2CaCu_2O_8Sr_2$
Molecular Weight: 888.366
CAS RN: 114901-61-0
Properties: O_8 can be $O_{8.15}$ to $O_{8.20}$; superconductor; 99%–99.9% pure, 20 μm powd; dry processed from 99%–99.9% oxides and carbonates; intermediate precursor available as fine agglomerate and ballmilled powd; T_c 85 K [STR93] [ASM93] [ALF93]
Density, g/cm³: 6.40 [ALD93]

506
Compound: Bismuth strontium calcium copper oxide (2223)
Synonym: supercon 2223
Formula: $Bi_2Sr_2Ca_2Cu_3O_{10}$
Molecular Formula: $Bi_2Ca_2Cu_3O_{10}Sr_2$
Molecular Weight: 1023.989
CAS RN: 114901-61-0
Properties: superconductor; 99.999% and 99.9% pure 20 μm powd; dry processed from 99.999% and ACS grades, respectively, of oxides and carbonates; T_c 110 K; for $Bi_2Sr_2Ca_3Cu_4O_{12}$, T_c is <120 K [STR93] [ASM93]

507
Compound: Bismuth subacetate
Synonym: bismuth acetate oxide
Formula: $CH_3COOBiO$
Molecular Formula: $C_2H_3BiO_3$
Molecular Weight: 284.024
CAS RN: 5142-76-7
Properties: thin cryst plates; slight acetic acid odor [MER06]
Solubility: i H_2O; s glacial acetic acid [MER06]

508
Compound: Bismuth subnitrate
Formula: $Bi_5O(OH)_9(NO_3)_4$
Molecular Formula: $Bi_5H_9N_4O_{22}$
Molecular Weight: 1461.987
CAS RN: 1304-85-4

Properties: hygr cryst powd [CRC10]
Solubility: i H_2O, EtOH; s dil acid [CRC10]
Density, g/cm^3: 4.928 [CRC10]
Melting Point, °C: decomposes at 260 [CRC10]

509
Compound: Bismuth sulfate
Formula: $Bi_2(SO_4)_3$
Molecular Formula: $Bi_2O_{12}S_3$
Molecular Weight: 706.152
CAS RN: 7787-68-0
Properties: white needles or powd; used in the analysis of other metal sulfates [HAW93]
Solubility: i H_2O, alcohol; s dil HCl, HNO_3 [HAW93]
Density, g/cm^3: 5.08 [HAW93]
Melting Point, °C: decomposes at 405 [HAW93]

510
Compound: Bismuth sulfide
Synonyms: bismuth glance, stibnite
Formula: Bi_2S_3
Molecular Formula: Bi_2S_3
Molecular Weight: 514.159
CAS RN: 1345-07-9
Properties: blackish brown, −200 mesh 99.9% and 99.999% purity; ortho-rhomb bipyramidal cryst; hardness 2 Mohs [HAW93] [MER06] [CER91]
Solubility: i H_2O; s HNO_3, HCl [MER06]
Density, g/cm^3: 7.6–7.8 [HAW93]
Melting Point, °C: decomposes at 685 [STR93] [ALF93]

511
Compound: Bismuth telluride
Synonym: tetradymite
Formula: Bi_2Te_3
Molecular Formula: Bi_2Te_3
Molecular Weight: 800.761
CAS RN: 1304-82-1
Properties: gray hex platelets, various grades: 1–6 mm pieces (fused) −325 mesh powd, 99.999% purity; preparation: by heating stoichiometric quantities of Bi and Te at 475°C in a vacuum for several days; used as a semiconductor; resistivity 0.00033 ohm·cm; energy gap 0.15 eV [MER06] [CER91] [A1F93]
Solubility: i H_2O, alcohol [HAW93]
Density, g/cm^3: 7.642 [MER06]
Melting Point, °C: 585 [MER06]
Thermal Conductivity, W/(m·K): 3.0 [CRC10]

512
Compound: Bismuth tetroxide
Synonym: bismuth peroxide

Formula: Bi_2O_4
Molecular Formula: Bi_2O_4
Molecular Weight: 481.959
CAS RN: 12048-50-9
Properties: reddish orange to yellowish brown, heavy powd, −200 mesh as the dehydrate; used as a lubricant for metal-extrusion dies [HAW93] [MER06] [CER91]
Solubility: slowly decomposed by H_2O [MER06]
Density, g/cm^3: 5.6 [HAW93]
Melting Point, °C: 305 [HAW93]

513
Compound: Bismuth titanate
Formula: $Bi_2O_3 \cdot 4TiO_2$
Molecular Formula: $Bi_2O_{11}Ti_4$
Molecular Weight: 785.422
CAS RN: 12233-34-0
Properties: −325 mesh, 10 μm average or less or 99.9% purity; used as sputtering target for beam splitter and base coating for gold films to prepare heating elements on glass [CER91]

514
Compound: Bismuth titanate
Formula: $Bi_2O_3 \cdot 2TiO_2$
Molecular Formula: $Bi_2O_7Ti_2$
Molecular Weight: 625.723
CAS RN: 12048-51-0
Properties: white powd, −325 mesh 99.9% purity; used as sputtering target for beam splitter and base coating for gold films as heating elements on glass; there are two other compounds: $Bi_4Ti_3O_{12}$ (12010-77-4) and $Bi_2Ti_4O_{11}$ (12233-34-0) [CER91] [STR93]

515
Compound: Bismuth tungstate
Formula: $Bi_2(WO_4)_3$
Molecular Formula: $Bi_2O_{12}W_3$
Molecular Weight: 1161.479
CAS RN: 13595-87-4
Properties: off-white powd, −200 mesh of 99.9% purity [CER91] [STR93]

516
Compound: Bismuth vanadate
Synonym: pucherite
Formula: $Bi_2O_3 \cdot V_2O_5$ or $BiVO_4$
Molecular Formula: $Bi_2O_8V_2$
Molecular Weight: 647.839

CAS RN: 14059-33-7
Properties: reddish green; rhomb; −200 mesh,
    99.9% purity [CRC10] [CER91]
Density, g/cm³: 6.25 [CRC10]

517
Compound: Bismuth zirconate
Formula: $2Bi_2O_3 \cdot 3ZrO_2$
Molecular Formula: $Bi_4O_{12}Zr_3$
Molecular Weight: 1301.586
CAS RN: 37306-42-6
Properties: reacted powd, −325 mesh, 5 μm
    or less with 99% purity [CER91]

518
Compound: Borane carbonyl
Formula: BH_3CO
Molecular Formula: CBH_3O
Molecular Weight: 41.845
CAS RN: 13205-44-2
Properties: col gas [CRC10]
Density, g/L: 1.710 [CRC10]
Solubility: reac H_2O [CRC10]
Melting Point, °C: −137 [CRC10]
Boiling Point, °C: −64 [CRC10]

519
Compound: Borazole
Synonyms: borazine, s-triazoborzane
Formula: $B_3N_3H_6$
Molecular Formula: $B_3H_6N_3$
Molecular Weight: 80.501
CAS RN: 6569-51-3
Properties: inorganic analog of benzene; colorless
    liq; preparation: heating equimolar mixture
    of ammonia and BH_3 at 250°C–300°C
    for 30 min [MER06] [HAW93]
Solubility: hydrolyzes, evolving boron hydrides [HAW93]
Density, g/cm³: 0.824 (0°C) [HAW93]
Melting Point, °C: −58 [HAW93]
Boiling Point, °C: 53 [HAW93]

520
Compound: Orthoboric acid
Synonym: sassolite
Formula: $B(OH)_3$
Molecular Formula: BH_3O_3
Molecular Weight: 61.833
CAS RN: 10043-35-3
Properties: colorless, odorless cryst or white powd;
    tricl; used in borosilicate glass, as an ointment
    and eyewash [MER06] [HAW93] [KIR78]

Solubility: g/100 g soln in H_2O: 2.52 (0°C);
    4.72 (20°C); 27.53 (100°C) [KIR78]
Density, g/cm³: 1.5172 [KIR78]; 1.435 [STR93]
Melting Point, °C: 171 [JAN85]

521
Compound: Metaboric acid-α-Form
Formula: HBO_2
Molecular Formula: BHO_2
Molecular Weight: 43.818
CAS RN: 13460-50-9
Properties: col ortho cryst; hygr [CRC10]
Solubility: s H_2O
Density, g/cm³: 1.784 [CRC10]
Melting Point, °C: 176 [CRC10]

522
Compound: Metaboric acid-β-Form
Formula: HBO_2
Molecular Formula: BHO_2
Molecular Weight: 43.818
CAS RN: 13460-50-9
Properties: col monocl cryst; hygr [CRC10]
Solubility: s H_2O
Density, g/cm³: 2.045
Melting Point, °C: 201 [CRC10]

523
Compound: Metaboric acid-γ-Form
Formula: HBO_2
Molecular Formula: HBO_2
Molecular Weight: 43.818
CAS RN: 13460-50-9
Properties: col cub cryst [CRC10]
Density, g/cm³: 2.487 [CRC10]
Melting Point, °C: 236 [CRC10]

524
Compound: Boron
Formula: B
Molecular Formula: B
Molecular Weight: 10.811
CAS RN: 7440-42-8
Properties: black hard solid; polymorphic; α: rhomb;
    clear red cryst, almost as hard as diamond; β: rhomb;
    black; α′: tetr; black; opaque cryst; enthalpy of
    vaporization 480 kJ/mol; enthalpy of fusion 50.20 kJ/
    mol; Young's modulus of filamentary boron 3040–
    3330 MPa; tensile strength of filamentary boron
    3450–4830 MPa; amorphous: black or dark brown
    powd; hardness 11 Mohs [MER06] [KIR78] [CRC10]

Solubility: i H_2O [MER06]
Density, g/cm³: α: 2.46; β: 2.35; α': 2.31;
    amorphous: 2.350 [MER06]
Melting Point, °C: 2190 [KIR78]
Boiling Point, °C: 3660 [KIR78]
Thermal Conductivity, W/(m·K): 31.8 (0°C);
    27.4 (25°C); 18.8 (100°C) [HO72]

525
Compound: Boron arsenide
Formula: BAs
Molecular Formula: BAs
Molecular Weight: 85.733
CAS RN: 12005-69-5
Properties: brown cub cryst [CRC10]
Density, g/cm³: 5.22 [CRC10]
Melting Point, °C: decomposes at 1100 [CRC10]

526
Compound: Boron carbide
Synonym: norbide
Formula: B_4C
Molecular Formula: CB_4
Molecular Weight: 55.255
CAS RN: 12069-32-8
Properties: hard, black, shiny cryst, −325 mesh
    with 99.5% purity; rhomb; hardness 9.3 Mohs;
    less brittle than most ceramics; does not burn
    in oxygen flame; used as an abrasive; Knoop
    hardness ~27 GPa; produced by reducing
    B_2O_3 with carbon at 1400°C–2300°C;
    used in crucible form as a container for
    molten salts except molten caustic and as a
    99.5% pure sputtering target for producing
    semiconductor and wear-resistant films
    [KIR78] [HAW93] [MER06] [CER91]
Solubility: not attacked by hot HF, HNO_3,
    or chromic acid [MER06]
Density, g/cm³: 2.508–2.512 [MER06]
Melting Point, °C: 2350 [MER06]
Boiling Point, °C: >3500 [MER06]
Reactions: decomposed by molten
    alkalis at red heat [MER06]

527
Compound: Boron nitride
Formula: BN
Molecular Formula: BN
Molecular Weight: 24.818
CAS RN: 10043-11-5

Properties: white powd, 1 μm or less 99.5% pure; hex,
    most common form: a=0.2504 nm; c=0.6661 nm;
    fcc: a=0.3615 nm; hardness: hex like graphite, cub
    approaches that of diamond; band gap ~7.5 eV at
    300 K; dielectric 7.1; used in furnace insulation
    and in crucibles for melting aluminum, boron,
    iron, and silicon, also as sputtering target for
    dielectrics, diffusion masks, passivation layers
    [KIR81] [HAW93] [MER06] [CER91]
Density, g/cm³: hex: 2.34; fcc: 3.43 [CIC73]
Melting Point, °C: hex: 3000 under N_2 [KIR78]
Boiling Point, °C: sublimes sl below 3000 [MER06]
Reactions: decomposes in vacuum ~2700°C [MER06]
Thermal Conductivity, W/(m·K): hex: 15 [KIR81]
Thermal Expansion Coefficient:
    7.51×10^{-6}/°C [KIR81]

528
Compound: Boron oxide
Synonym: boron anhydride
Formula: B_2O_3
Molecular Formula: B_2O_3
Molecular Weight: 69.620
CAS RN: 1303-86-2
Properties: colorless, brittle, vitreous; 12 mm
    pieces and smaller (fused) of 99.9995%
    purity; hygr; heat of solution −75.9 kJ/
    mol [MER06] [KIR78] [CER91]
Solubility: %wt soln, H_2O: 4.72 (20°C),
    8.08 (40°C), 27.53 (100°C); in alcohol:
    94.4 g/L (25°C) [KIR78]
Density, g/cm³: amorphous: 1.8; cryst: 2.46 [MER06]
Melting Point, °C: cryst: 450 [MER06]
Boiling Point, °C: ~1860 [STR93]

529
Compound: Boron oxide glass
Synonym: vitreous boric oxide
Formula: B_2O_3
Molecular Formula: B_2O_3
Molecular Weight: 69.620
CAS RN: 1303-86-2
Properties: colorless, glassy solid; enthalpy of
    formation from elements, 25°C, −1252.2 kJ/mol;
    heat capacity (25°C) 62.969 J/(mol·K) [KIR78]
Density, g/cm³: 0°C, 1.8766;
    18°C–25°C, 1.844 [KIR78]
Boiling Point, °C: 2316, extrapolated [KIR78]

530
Compound: Boron phosphate
Synonym: borophosphoric acid

Formula: BPO_4
Molecular Formula: BO_4P
Molecular Weight: 105.782
CAS RN: 13308-51-5
Properties: white not hygr cryst; prepared by reacting boric acid and phosphoric acid up to 1200°C: $B(OH)_3 + H_3PO_4 = BPO_4 + 3H_2O$; used in special glasses [HAW93]
Solubility: s H_2O [HAW93]
Density, g/cm³: 1.873 [HAW93]
Melting Point, °C: vaporizes slowly without decomposition at >1200 [KIR78]

531
Compound: Boron phosphide
Formula: BP
Molecular Formula: BP
Molecular Weight: 41.785
CAS RN: 20205-91-8
Properties: maroon powd; −100 mesh at 97.5% purity; refractory; hardness 9.5 Mohs; band gap 2.0 eV (300 K); material with formula $B_{13}P_2$ [12008-82-1] is −325 mesh, 10 μm or less of 99% purity [HAW93] [CER91] [KIR82]
Solubility: reacts with H_2O and acids, releasing toxic fumes [HAW93]
Reactions: ignites at 200°C [CRC10]

532
Compound: Boron silicide
Formula: B_6Si
Molecular Formula: B_6Si
Molecular Weight: 92.952
CAS RN: 12008-29-6
Properties: black cryst; −200 mesh; also material with composition B_4Si [12007-81-7] and 98% purity [CRC10] [CER91] [ALF93]
Density, g/cm³: 2.47 [CRC10]

533
Compound: Boron tribromide
Formula: BBr_3
Molecular Formula: BBr_3
Molecular Weight: 250.523
CAS RN: 10294-33-4

Properties: 99.999% and 99.9+% purity doping grade; colorless fuming liq; sensitive to moisture; critical temp 300°C; enthalpy of vaporization 30.5 kJ/mol; used as a catalyst in the manufacture of diborane [HAW93] [MER06] [KIR78] [STR93] [CER91] [CRC10]
Solubility: decomposed by H_2O, alcohol [MER06]
Density, g/cm³: 2.698 [MER06]
Melting Point, °C: −46.0 [MER06]
Boiling Point, °C: 90 [ALD94]
Thermal Conductivity, W/(m·K): 0.112 (20°C) [KIR78]

534
Compound: Boron trichloride
Formula: BCl_3
Molecular Formula: BCl_3
Molecular Weight: 117.169
CAS RN: 10294-34-5
Properties: colorless, fuming liq or gas; critical temp 178.8°C; critical pressure 3901.0 kPa; enthalpy of vaporization 23.77 kJ/mol; enthalpy of fusion 2.10 kJ/mol; vapor pressure: 0.53 (−80°C), 8.9 (−40°C), 63.5 (0°C), 243 (40°C), 689 (80°C); used as catalyst in organic synthesis and as VLSI etchant in electronics industry [HAW93] [MER06] [KIR78] [AIR87] [CRC10]
Solubility: decomposed by H_2O, alcohol [MER06]
Density, g/cm³: 1.35 (12°C), 1.3728 (0°C) [MER06]; 1.434 (0°C) [KIR78]
Melting Point, °C: −107 [MER06]
Boiling Point, °C: 12.5 [MER06]

535
Compound: Boron trifluoride
Formula: BF_3
Molecular Formula: BF_3
Molecular Weight: 67.806
CAS RN: 7637-07-2
Properties: colorless gas, pungent suffocating odor; forms dense white fumes in moist air; triple point −128.4°C at 8.34 kPa; critical temp −12.25°C; critical pressure 4984 kPa; enthalpy of fusion 4.20 kJ/mol; enthalpy of vaporization 19.33 kJ/mol; can be produced by reacting borax, fluorspar, and sulfuric acid; used as a catalyst in organic synthesis and in electronics [HAW93] [KIR78] [MER06] [AIR87] [CRC10]
Solubility: 332 g/100 g H_2O (0°C) with some hydrolysis [MER06]
Density, g/cm³: gas: 3.07666 g/L (STP) [KIR78]
Melting Point, °C: −126.8 [CRC10]

Boiling Point, °C: −101 [CRC10]
Reactions: forms the solid complex
  $HNO_3 \cdot 2BF_3$ with HNO_3 [MER06]

536
Compound: Boron trifluoride etherate
Synonym: boron fluoride-ether
Formula: $BF_3(CH_3CH_2)_2O$
Molecular Formula: $C_4H_{10}BF_3O$
Molecular Weight: 141.929
CAS RN: 109-63-7
Properties: fuming liq; sensitive to moisture;
  prepared by vapor phase reaction between
  BF_3 and ethyl ether; used as catalyst in
  organic reactions [MER06] [ALF95]
Solubility: hydrolyzed by H_2O [MER06]
Density, g/cm³: 1.125 [MER06]
Melting Point, °C: −60.4 [MER06]
Boiling Point, °C: 125.7 [MER06]

537
Compound: Boron triiodide
Formula: BI_3
Molecular Formula: BI_3
Molecular Weight: 391.524
CAS RN: 13517-10-7
Properties: needles or cryst with 99.9%
  purity; unstable; enthalpy of vaporization
  40.5 kJ/mol [CRC10] [CER91]
Density, g/cm³: 3.35 [KIR78]
Melting Point, °C: 49.9 [COT88]
Boiling Point, °C: 210 [KIR78]

538
Compound: Boron trisulfide
Formula: B_2S_3
Molecular Formula: B_2S_3
Molecular Weight: 117.820
CAS RN: 12007-33-9
Properties: white; −200 mesh, 99.9%
  purity [CRC10] [CER91]
Density, g/cm³: 1.55 [CRC10]
Melting Point, °C: 310 [AES93]

539
Compound: Bromic acid
Formula: $HBrO_3$
Molecular Formula: $BrHO_3$
Molecular Weight: 128.910
CAS RN: 7789-31-3

Properties: colorless or sl yellowish liq; stable
  only in very dil aq solutions; oxidizing
  agent; used in dyes [HAW93]
Solubility: can exist only in aq media [HAW93]
Density, g/cm³: 3.28 [HAW93]
Melting Point, °C: decomposes at 100 [CRC10]

540
Compound: Bromine
Formula: Br_2
Molecular Formula: Br_2
Molecular Weight: 159.808 (atomic weight, 79.904)
CAS RN: 7726-95-6
Properties: dark reddish, volatile liq; suffocating odor;
  vaporizes rapidly at room temp; oxidant; viscosity
  (30°C) 0.288 mm²/s; surface tension 40.9 mN/m
  (25°C); electrical resistivity, $6.5 \times 10^{+10}$ ohm · cm
  at 25°C; dielectric constant $3.33 \times 10^{+5}$ Hz (25°C);
  enthalpy of vaporization 29.97 kJ/mol; enthalpy of
  fusion 10.57 kJ/mol; electronegativity 3.0; critical
  temp 311°C; critical pressure 10.3 atm; used in flame-
  retardant materials [MER06] [KIR78] [CRC10]
Solubility: 3.35 g/100 g soln, H_2O (25°C) [KIR78];
  forms HOBr with H_2O [MER06]
Density, g/cm³: 3.1055 (25°C); vapor
  density 7.139 g/L (0°C) [KIR78]
Melting Point, °C: −7.2 [CRC10]
Boiling Point, °C: 59.10 [CRC10]
Reactions: thermal dissociation at >600°C [KIR78]
Thermal Conductivity, W/(m · K):
  0.123 (25°C) [KIR78]
Thermal Expansion Coefficient:
  20°C–30°C, 0.0011/°C [KIR78]

541
Compound: Bromine azide
Formula: BrN_3
Molecular Formula: BrN_3
Molecular Weight: 121.924
CAS RN: 13973-87-0
Properties: cryst or red liq; oxidizing agent; used in
  detonators and other explosive devices [HAW93]
Melting Point, °C: ~45 [HAW93]
Boiling Point, °C: explodes [HAW93]

542
Compound: Bromine chloride
Formula: BrCl
Molecular Formula: BrCl
Molecular Weight: 115.357
CAS RN: 13863-41-7

Properties: reddish yellow liq; formed by reaction of bromide and chlorine, in the vapor or liq states; easily hydrolyzed by water; used as a disinfectant in industry, for wastewater treatment [HAW93] [KIR78]
Solubility: s H_2O with ready hydrolysis, s carbon disulfide, ether [HAW93]
Density, g/cm³: 5.062 g/L [LID94]
Melting Point, °C: −66 [HAW93]
Boiling Point, °C: decomposes evolving Cl_2 at 10 [HAW93]

543

Compound: Bromine dioxide
Formula: BrO_2 or Br_2O_4
Molecular Formula: BrO_2
Molecular Weight: 111.903
CAS RN: 21255-83-4
Properties: yellowish orange solid; obtained by ozonation of Br_2 in Freon 11 at −50°C, and subsequent evaporation [KIR78]
Melting Point, °C: decomposes at 0 [KIR78]

544

Compound: Bromine fluoride
Formula: BrF
Molecular Formula: BrF
Molecular Weight: 98.902
CAS RN: 13863-59-7
Properties: unstable red-brown gas [CRC10]
Density, g/cm³: 4.043 [CRC10]
Melting Point, °C: ~−33 [CRC10]
Boiling Point, °C: decomposes at ~20 [CRC10]

545

Compound: Bromine monofluoride
Formula: BrF
Molecular Formula: BrF
Molecular Weight: 98.902
CAS RN: 13863-59-7
Properties: red to brownish-red; very unstable, rapidly forming bromine and higher fluorides; prepared by reaction of Br_2 and F_2 [KIR78]
Density, g/cm³: 4.34 g/L [CRC10]
Melting Point, °C: ~−33 [KIR78]
Boiling Point, °C: ~20 [KIR78]

546

Compound: Bromine oxide
Synonym: bromine monoxide
Formula: Br_2O
Molecular Formula: Br_2O

Molecular Weight: 175.807
CAS RN: 21308-80-5
Properties: dark brown solid; stable below −40°C; formed by reaction of HgO and Br_2 in CCl_4 in the absence of light; used for bromination reactions [KIR78]
Melting Point, °C: decomposes at −17.5 [KIR78]

547

Compound: Bromine pentafluoride
Formula: BrF_5
Molecular Formula: BrF_5
Molecular Weight: 174.896
CAS RN: 7789-30-2
Properties: colorless, fuming liq; vapor pressure 7 psia (21.1°C); enthalpy of vaporization 30.6 kJ/mol; enthalpy of fusion 5.66 kJ/mol; specific conductivity (25°C) 9.1×10^{-8} ohm·cm; highly reactive, e.g., reacts with all known elements except the inert gases, nitrogen and oxygen; can be prepared by reacting Br_2 and F_2; used as a fluorinating agent in organic synthesis and as an oxidizing agent in liq rocket fuels [KIR78] [HAW93]
Solubility: explodes on contact with H_2O [MER06]
Density, g/cm³: 2.460 [MER06]
Melting Point, °C: −60.5 [MER06]
Boiling Point, °C: 40.76 [MER06]
Reactions: thermally stable up to 460°C [MER06]

548

Compound: Bromine trifluoride
Formula: BrF_3
Molecular Formula: BrF_3
Molecular Weight: 136.899
CAS RN: 7787-71-5
Properties: colorless liq, if pure; long prisms when solid; fumes in air; very reactive; vapor pressure 0.15 psia (21.1°C); formed by reaction of Br_2 and F_2; enthalpy of vaporization 47.57 kJ/mol; enthalpy of fusion 12.01 kJ/mol; used as a fluorinating agent and as a solvent for fluorides [HAW93] [MER06] [AIR87] [KIR78] [CRC10]
Solubility: violent reaction with H_2O [HAW93]
Density, g/cm³: 2.803 [MER06]
Melting Point, °C: 8.77 [MER06]
Boiling Point, °C: 125.75 [MER06]

549

Compound: Bromoauric(III) acid pentahydrate
Formula: $HAuBr_4 \cdot 5H_2O$
Molecular Formula: $AuBr_4H_{11}O_5$
Molecular Weight: 516.583
CAS RN: 17083-68-0

Properties: red-brown hygr cryst [CRC10]
Solubility: s H$_2$O, EtOH [CRC10]
Boiling Point, °C: 27 [CRC10]

550

Compound: Chloroauric(III) acid tetrahydrate
Formula: HAuCl$_4$·4H$_2$O
Molecular Formula: AuCl$_4$H$_9$O$_4$
Molecular Weight: 411.848
CAS RN: 16903-35-8
Properties: yellow monocl cryst; hygr [CRC10]
Density, g/cm^3: ~3.9 [CRC10]
Solubility: v sol H$_2$O, EtOH; s eth [CRC10]

551

Compound: Bromochloromethane
Formula: CH$_2$BrCl
Molecular Formula: CH$_2$BrCl
Molecular Weight: 129.383
CAS RN: 74-97-5
Properties: liq [ALF95]
Density, g/cm^3: 1.991 [ALF95]
Melting Point, °C: −88 [ALF95]
Boiling Point, °C: 68 [ALF95]

552

Compound: Bromogermane
Formula: GeH$_3$Br
Molecular Formula: BrH$_3$Ge
Molecular Weight: 155.57
CAS RN: 13569-42-2
Properties: col liq [CRC10]
Solubility: reac H$_2$O [CRC10]
Density, g/cm^3: 2.34 [CRC10]
Melting Point, °C: −32 [CRC10]
Boiling Point, °C: 52 [CRC10]

553

Compound: Bromosilane
Formula: SiH$_3$Br
Molecular Formula: BrH$_3$Si
Molecular Weight: 111.014
CAS RN: 13465-73-1
Properties: enthalpy of vaporization 24.4 kJ/mol; entropy
of vaporization 88.3 J/(mol·K) [CIC73] [CRC10]
Melting Point, °C: −94 [CIC73]
Boiling Point, °C: 1.9 [CIC73]

554

Compound: Dibromine trioxide
Formula: Br$_2$O$_3$

Molecular Formula: Br$_2$O$_3$
Molecular Weight: 207.806
CAS RN: 53809-75-9
Properties: orange needles [CRC10]
Melting Point, °C: decomposes at −40

555

Compound: Cacodylic acid
Synonyms: dimethylarsenic acid,
hydroxydimethylarsine oxide
Formula: (CH$_3$)$_2$As(O)OH
Molecular Formula: C$_2$H$_7$AsO$_2$
Molecular Weight: 137.998
CAS RN: 75-60-5
Properties: colorless, odorless cryst; hygr;
preparation: distillation of a mixture of As$_2$O$_3$
and CH$_3$COOK, followed by oxidation of
resulting product with HgO; uses: herbicide,
dermatology [MER06] [HAW93]
Solubility: s 0.5 parts H$_2$O; v s alcohol [MER06]
Melting Point, °C: 195–196 [ALD94]

556

Compound: Cadmium
Formula: Cd
Molecular Formula: Cd
Molecular Weight: 112.411
CAS RN: 7440-43-9
Properties: soft, silvery white with blue tinge metal;
distorted hex; easily cut with knife; slowly
oxidized by moist air to CdO; hardness 2.0 Mohs;
fusion enthalpy 6.19 kJ/mol; vaporization
enthalpy 99.87 kJ/mol; electrical resistivity
(22°C) 7.2 μohm·cm; Brinell hardness 16.23 kg/
mm^2; Poisson's ratio 0.33; used in easily fusible
alloys, solder for aluminum, photoelectric
cells, and as 99.999% pure sputtering target for
dielectric films [KIR78] [CER91] [CRC10]
Solubility: i H$_2$O; reacts with dil HNO$_3$,
slowly with hot HCl [MER06]
Density, g/cm^3: 8.65 [MER06]
Melting Point, °C: 321.07 [CRC10]
Boiling Point, °C: 767 [CRC10]
Thermal Conductivity, W/(m·K): 98 (0°C),
95 (100°C) 89 (300°C) [KIR78]
Thermal Expansion Coefficient: 20°C,
31.3 μm/(cm °C) [KIR78]

557

Compound: Cadmium acetate
Synonyms: acetic acid, cadmium salt
Formula: Cd(CH$_3$COO)$_2$

Molecular Formula: $C_4H_6CdO_4$
Molecular Weight: 230.501
CAS RN: 543-90-8
Properties: colorless cryst; used in ceramics
    and electroplating baths [HAW93]
Solubility: s H_2O, alcohol [HAW93]
Density, g/cm³: 2.34 [MER06]
Melting Point, °C: 255 [MER06]

558
Compound: Cadmium acetate dihydrate
Synonyms: acetic acid, cadmium salt dihydrate
Formula: $Cd(CH_3COO)_2 \cdot 2H_2O$
Molecular Formula: $C_4H_{10}CdO_6$
Molecular Weight: 266.530
CAS RN: 5743-04-4
Properties: white powd; cryst; slight acetic
    acid odor [MER06] [STR93]
Solubility: v s H_2O [MER06]
Density, g/cm³: 2.01 [HAW93]
Reactions: minus $2H_2O$ ~130°C [MER06]

559
Compound: Cadmium acetylacetonate
Synonyms: 2,4-pentanedione, cadmium(II) derivative
Formula: $Cd(CH_3COCH=C(O)CH_3)_2$
Molecular Formula: $C_{10}H_{14}CdO_4$
Molecular Weight: 310.630
CAS RN: 14689-45-3
Properties: white powd [STR93]
Melting Point, °C: decomposes [STR93]

560
Compound: Cadmium antimonide
Formula: CdSb
Molecular Formula: CdSb
Molecular Weight: 234.171
CAS RN: 12014-29-8
Properties: 6 mm pieces and smaller of 99.999%
    purity; ortho-rhomb, a=0.6471 nm, b=0.8253 nm;
    semiconductor; uses: in thermoelectric devices;
    enthalpy of fusion 32.05 kJ/mol; enthalpy of
    evaporation 138 kJ/mol; there is also Cd_3Sb_2,
    12014-29-8 [CER91] [KIR78] [HAW93]
Density, g/cm³: 6.92 [KIR78]
Melting Point, °C: 456 [KIR78]

561
Compound: Cadmium arsenide
Formula: Cd_3As_2
Molecular Formula: As_2Cd_3

Molecular Weight: 487.076
CAS RN: 12006-15-4
Properties: gray tetr cryst, a=0.8945 nm,
    c=1.265 nm [KIR78]
Density, g/cm³: 6.21 [KIR78]
Melting Point, °C: 721 [KIR78]

562
Compound: Cadmium azide
Formula: $Cd(N_3)_2$
Molecular Formula: CdN_6
Molecular Weight: 196.451
CAS RN: 14215-29-3
Properties: yellow-white ortho [LID94]
Density, g/cm³: 3.24 [LID94]

563
Compound: Cadmium borotungstate octadecahydrate
Formula: $Cd_5(BW_{12}O_{40}) \cdot 18H_2O$
Molecular Formula: $BCd_5H_{36}O_{58}W_{12}$
Molecular Weight: 3732.386
CAS RN: 1306-26-9
Properties: yellow, heavy cryst; used to
    separate minerals [HAW93]
Solubility: 1250 g/100 mL (19°C) H_2O, yielding a
    yellow to light brown solution [HAW93] [CRC10]
Melting Point, °C: 75 [CRC10]

564
Compound: Cadmium bromide
Formula: $CdBr_2$
Molecular Formula: Br_2Cd
Molecular Weight: 272.219
CAS RN: 7789-42-6
Properties: white powd, −80 mesh 99.9% pure; hygr;
    hex, a=0.395 nm, c=1.867 nm; enthalpy of fusion
    20.90 kJ/mol; enthalpy of vaporization 115 kJ/
    mol; pearly flakes; crystallizes as monohydrate
    below 36°C, as tetrahydrate above 36°C; finds
    use in lithography and photography [HAW93]
    [MER06] [STR93] [KIR78] [CER91] [CRC10]
Solubility: g/100 g soln, H_2O: 36.0 (0°C), 52.9
    (25°C), 61.65 (100°C); solid phase, $CdBr_2 \cdot 4H_2O$
    (0°C, 25°C), $CdBr_2$ (100°C) [KRU93];
    s acetone, alcohol, acids [HAW93]
Density, g/cm³: 5.192 [MER06]
Melting Point, °C: 568 [CRC10]
Boiling Point, °C: 1136 [COT88]

565
Compound: Cadmium bromide tetrahydrate
Formula: $CdBr_2 \cdot 4H_2O$

Molecular Formula: $Br_2CdH_8O_4$
Molecular Weight: 344.281
CAS RN: 13464-92-1
Properties: white to yellowish
efflorescent cryst [HAW93]
Solubility: 121 g/100 mL H_2O (10°C) [CRC10];
s acetone, alcohol, acids [HAW93]
Melting Point, °C: transition 36 [CRC10]

566
Compound: Cadmium carbonate
Synonym: otavite
Formula: $CdCO_3$
Molecular Formula: $CCdO_3$
Molecular Weight: 172.420
CAS RN: 513-78-0
Properties: −200 mesh with 99.999%, 99.9% and, 99%
purity; white, amorphous powd or rhomb leaflets;
a=0.61306 nm [HAW93] [MER06] [KIR78] [CER91]
Solubility: 2.8 μm/100 g H_2O; s dil
acids [KIR78] [MER06]
Density, g/cm³: 4.258 [HAW93]
Melting Point, °C: decomposes at 500 [HAW93]

567
Compound: Cadmium chlorate dihydrate
Formula: $Cd(ClO_3)_2 \cdot 2H_2O$
Molecular Formula: $CdCl_2H_4O_8$
Molecular Weight: 315.343
CAS RN: 7790-78-5
Properties: colorless cryst; hygr [HAW93]
Solubility: mol/100 mol soln, H_2O: 25.92 (0°C);
solid phase, $Cd(ClO_3)_2 \cdot 2H_2O$ [KRU93];
s alcohol, acetone [HAW93]; g/100 g H_2O: 299
(0°C), 322 (20°C), 455 (60°C) [LAN05]
Density, g/cm³: 2.28 (18°C) [HAW93]
Melting Point, °C: 80 [HAW93]

568
Compound: Cadmium chloride
Formula: $CdCl_2$
Molecular Formula: $CdCl_2$
Molecular Weight: 183.316
CAS RN: 10108-64-2
Properties: −80 mesh of 99.9% purity; white, odorless
hygr; hex, a=0.3854 nm, c=1.746 nm; enthalpy
of fusion 48.58 kJ/mol; enthalpy of vaporization
at bp 124.3 kJ/mol [MER06] [CRC10]
Solubility: g/100 g soln, H_2O: 47.3 (0°C),
54.65 (25°C), 59.55 (100°C); solid phase,
$CdCl_2 \cdot 2-1/2H_2O$ (0°C, 25°C), $CdCl_2 \cdot H_2O$(100°C)
[KRU93]; s acetone [HAW93]

Density, g/cm³: 4.05 [STR93]
Melting Point, °C: 564 [CRC10]
Boiling Point, °C: 960 [CRC10]

569
Compound: Cadmium chloride hemipentahydrate
Formula: $CdCl_2 \cdot 2-1/2H_2O$
Molecular Formula: $CdCl_2H_5O_{2.5}$
Molecular Weight: 228.354
CAS RN: 7790-78-5
Properties: white, efflorescent granules or
rhomb leaflets [MER06] [STR93]
Solubility: g/100 g H_2O: 90 (0°C), 113 (20°C),
132 (30°C) [LAN05]; s acetone [HAW93]
Density, g/cm³: 3.327 [HAW93]
Melting Point, °C: 960 [HAW93]

570
Compound: Cadmium chromate
Formula: $CdCrO_4$
Molecular Formula: $CdCrO_4$
Molecular Weight: 228.405
CAS RN: 14312-00-6
Properties: yellow solid; used in catalysts
and in pigments [KIR78]
Solubility: i H_2O [KIR78]
Density, g/cm³: 4.5 [LID94]

571
Compound: Cadmium cyanide
Formula: $Cd(CN)_2$
Molecular Formula: C_2CdN_2
Molecular Weight: 164.446
CAS RN: 542-83-6
Properties: cryst or white powd; turns brown
if heated in air; used to electroplate
cadmium [HAW93] [MER06]
Solubility: 1.71 g/100 mL H_2O (15°C) [MER06]
Density, g/cm³: 2.23 [MER06]
Melting Point, °C: decomposes at >200 [CRC10]

572
Compound: Cadmium dichromate monohydrate
Formula: $CdCr_2O_7 \cdot H_2O$
Molecular Formula: $CdCr_2H_2O_8$
Molecular Weight: 346.414
CAS RN: 69239-51-6
Properties: orange solid used in
metal finishing [KIR78]
Solubility: s H_2O [KIR78]

573
Compound: Cadmium 2-ethylhexanoate
Formula: $Cd(C_8H_{15}O_2)_2$
Molecular Formula: $C_{16}H_{30}CdO_4$
Molecular Weight: 398.818
CAS RN: 2420-98-6
Properties: powd [CRC10]

574
Compound: Cadmium fluoride
Formula: CdF_2
Molecular Formula: CdF_2
Molecular Weight: 150.408
CAS RN: 7790-79-6
Properties: 99.9% pure melted pieces, 3–6 mm; cub cryst, a = 0.53880 nm; enthalpy of fusion 22.60 kJ/mol; enthalpy of vaporization 214 kJ/mol; available as 99.89% pure cryst; used in electronic and optical materials, in the preparation of laser cryst, and as a sputtering target to produce multilayers [HAW93] [MER06] [CER91] [CRC10]
Solubility: mol/kg soln, H_2O: 0.4652 (0°C), 0.291 (25°C), 0.12 (100°C) [KRU93]; s acids, i alkalies [HAW93]
Density, g/cm³: 6.33 [MER06]; 6.64 [STR93]
Melting Point, °C: 1110 [CRC10]
Boiling Point, °C: 1748 [CRC10]

575
Compound: Cadmium hydroxide
Formula: $Cd(OH)_2$
Molecular Formula: CdH_2O_2
Molecular Weight: 146.426
CAS RN: 21041-95-2
Properties: white, amorphous powd or trig and hex cryst, a = 0.3475 nm, c = 0.467 nm; can absorb atm H_2O and CO_2 [HAW93] [MER06] [KIR78]
Solubility: i H_2O; s dil acids, NH_4OH [MER06]; mmol/L soln, H_2O: 0.038 (18°C) [KRU93]
Density, g/cm³: 4.79 [MER06]
Melting Point, °C: decomposes at 150 [KIR78]
Reactions: minus $2H_2O$ at 130°C–200°C [MER06]

576
Compound: Cadmium iodate
Formula: $Cd(IO_3)_2$
Molecular Formula: CdI_2O_6
Molecular Weight: 462.216
CAS RN: 7790-81-0
Properties: fine white powd; uses: oxidizing agent [HAW93]
Solubility: mol/L soln, H_2O: $(1.97 \pm 0.13) \times 10^{-3}$ (25°C) [KRU93]; s HNO_3, NH_4OH [HAW93]

Density, g/cm³: 6.48 [HAW93]
Melting Point, °C: decomposes [CRC10]

577
Compound: Cadmium iodide
Formula: CdI_2
Molecular Formula: CdI_2
Molecular Weight: 366.220
CAS RN: 7790-80-9
Properties: −40 mesh 99.5% pure; lustrous, flake-like cryst; odorless; two forms: α and β; α-form is hex, a = 0.424 nm, c = 0.684 nm, enthalpy of fusion 15.30 kJ/mol, enthalpy of vaporization 115 kJ/mol; becomes yellow on prolonged exposure to air and light; finds applications in photography and in lubricants [HAW93] [MER06] [CRC10]
Solubility: s H_2O, alcohol, ether, acetone [MER06]; g/100 g soln, H_2O: 44.05 (0°C), 46.30 (25°C), 55.55 (100°C); solid phase, CdI_2 [KRU93]
Density, g/cm³: α: 5.67; β: 5.30 [HAW93]
Melting Point, °C: α: 388; β: 404 [HAW93]
Boiling Point, °C: 742 [CRC10]

578
Compound: Cadmium metasilicate
Formula: $CdSiO_3$
Molecular Formula: CdO_3Si
Molecular Weight: 188.495
CAS RN: 13477-19-5
Properties: colorless; monocl, a = 1.504 nm, b = 0.710 nm, c = 0.696 nm [KIR78] [CRC10]
Solubility: v sl s H_2O [CRC10]
Density, g/cm³: 4.928 [KIR78]
Melting Point, °C: 1252 [KIR78]

579
Compound: Cadmium molybdate(VI)
Formula: $CdMoO_4$
Molecular Formula: $CdMoO_4$
Molecular Weight: 272.349
CAS RN: 13972-68-4
Properties: −200 mesh 99.9% pure; yellow cryst; scheelite structure, c/a = 2.174; applications in electronic and optical materials [HAW93] [KIR81]
Solubility: 0.0067 g/100 g H_2O, s acids [HAW93] [KIR81]
Density, g/cm³: 5.347 [HAW93]
Melting Point, °C: decomposes at ~900 [KIR81]

580
Compound: Cadmium niobate
Formula: $Cd_2Nb_2O_7$

Molecular Formula: $Cd_2Nb_2O_7$
Molecular Weight: 522.631
CAS RN: 12187-14-3
Properties: cub; −200 mesh 99.9% purity [LID94] [CER91]
Density, g/cm³: 6.2 [LID94]
Melting Point, °C: ~1450 [LID94]

581
Compound: Cadmium nitrate
Formula: $Cd(NO_3)_2$
Molecular Formula: CdN_2O_6
Molecular Weight: 236.427
CAS RN: 10325-94-7
Properties: white, amorphous pieces or hygr needles; cub, a=0.756 nm; used in coloring glass and in photographic emulsions [HAW93] [KIR78]
Solubility: g/100 g soln, H_2O: 55.1 (0.6°C), 61.3 (25°C), 87.2 (100°C); solid phase, $Cd(NO_3)_2 \cdot 4H_2O$ (0°C, 25°C), $Cd(NO_3)_2$ (100°C) [KRU93]; s alcohol, ammonia [HAW93]
Density, g/cm³: 3.6 [LID94]
Melting Point, °C: 350 [HAW93]

582
Compound: Cadmium nitrate tetrahydrate
Formula: $Cd(NO_3)_2 \cdot 4H_2O$
Molecular Formula: $CdH_8N_2O_{10}$
Molecular Weight: 308.482
CAS RN: 10022-68-1
Properties: colorless; hygr; ortho-rhomb cryst, a=0.583 nm, b=2.575 nm, c=1.099 nm; enthalpy of fusion 32.636 kJ/mol [MER06] [STR93] [KIR78]
Solubility: 132 g/100 g H_2O (0°C) [KIR78]
Density, g/cm³: 2.45 [STR93]
Melting Point, °C: 59.5 [MER06]
Boiling Point, °C: 132 [HAW93]

583
Compound: Cadmium oxalate
Synonyms: ethanedioic acid, cadmium salt
Formula: CdC_2O_4
Molecular Formula: C_2CdO_4
Molecular Weight: 200.431
CAS RN: 814-88-0
Properties: colorless [LAN05]
Solubility: mol/kg soln, H_2O: 0.00030 (25°C); equilibrium solid phase $CdC_2O_4 \cdot 3H_2O$ [KRU93]
Density, g/cm³: 3.32 [HAW93]
Melting Point, °C: decomposes at 330 [HAW93]

584
Compound: Cadmium oxalate trihydrate
Synonyms: ethanedioic acid, cadmium salt trihydrate
Formula: $CdC_2O_4 \cdot 3H_2O$
Molecular Formula: $C_2H_6CdO_7$
Molecular Weight: 254.477
CAS RN: 20712-42-9
Properties: white, amorphous powd [HAW93]
Solubility: i H_2O, alcohol; s dil acids, NH_4OH [HAW93]
Melting Point, °C: decomposes at 340 [HAW93]

585
Compound: Cadmium oxide
Formula: CdO
Molecular Formula: CdO
Molecular Weight: 128.410
CAS RN: 1306-19-0
Properties: −100 mesh 99.999% pure, −200 mesh 99.9% and 99.5% purity; two forms: (1) colorless amorphous powd and (2) brown or red cryst; cub cryst, a=0.46953 nm; enthalpy of fusion 243.5 kJ/mol; used in cadmium plating baths and ceramics [HAW93] [KIR78] [CER91]
Solubility: 0.00094 g/100 g H_2O; s dil acids [MER06] [KIR78]
Density, g/cm³: (1) 6.95; (2) 8.15 [HAW93]
Melting Point, °C: sublimes at 1540 [KIR78]
Thermal Conductivity, W/(m·K): 0.7 [CRC10]

586
Compound: Cadmium perchlorate
Formula: $Cd(ClO_4)_2$
Molecular Formula: $CdCl_2O_8$
Molecular Weight: 311.311
CAS RN: 79490-00-9
Solubility: g/100 g soln, H_2O: 58.7 (25°C), 66.9 (100°C); solid phase, $Cd(ClO_4)_2 \cdot 6H_2O$ [KRU93]

587
Compound: Cadmium perchlorate hexahydrate
Formula: $Cd(ClO_4)_2 \cdot 6H_2O$
Molecular Formula: $CdCl_2H_{12}O_{14}$
Molecular Weight: 419.403
CAS RN: 10326-28-0
Properties: white cryst [STR93]
Solubility: g/100 g H_2O: 180 (10°C), 188 (20°C), 272 (100°C) [LAN05]
Density, g/cm³: 2.37 [LID10]

588
Compound: Cadmium phosphate
Formula: $Cd_3(PO_4)_2$

Molecular Formula: CdO_8P_2
Molecular Weight: 527.176
CAS RN: 13477-17-3
Properties: powd [CRC10]
Solubility: i H_2O
Melting Point, °C: ~1500 [CRC10]

589
Compound: Cadmium phosphide
Formula: CdP_2
Molecular Formula: CdP_2
Molecular Weight: 174.359
CAS RN: 12133-44-7
Properties: −100 mesh of 99.9% purity; there is
a Cd_3P_2, 12014-28-7, melting point 700°C,
with same properties [CER91] [ALD94]
Density, g/cm³: Cd_3P_2: 5.60 [CRC10]
Melting Point, °C: Cd_3P_2: 700 [CRC10]

590
Compound: Cadmium selenate dihydrate
Formula: $CdSeO_4 \cdot 2H_2O$
Molecular Formula: CdH_4O_6Se
Molecular Weight: 291.399
CAS RN: 10060-09-0
Properties: ortho-rhomb cryst [MER06]
Solubility: g/100g H_2O: 72.44 (0°C), 61.23 (26°C), 22.01
(98.5°C); solid phase $CdSeO_4 \cdot 2H_2O$ [KRU93]
Density, g/cm³: 3.632 [MER06]
Melting Point, °C: decomposes at 100 [MER06]

591
Compound: Cadmium selenide
Formula: CdSe
Molecular Formula: CdSe
Molecular Weight: 191.371
CAS RN: 1306-24-7
Properties: 3–12 mm pieces (sintered), and −325
mesh 10 μm or less with 99.999% purity; white
to brown; cub or hex cryst; hex: a = 0.4309 nm,
c = 0.7021 nm, enthalpy of fusion 305.307 kJ/mol;
becomes red in sunlight; used as a red pigment,
in semiconductors, and as an evaporation material
and sputtering target to produce photoconductive
films and infrared filters; possible use in the
electrofabrication of microdiode arrays [HAW93]
[MER06] [CER91] [KLE93] [KRE91]
Solubility: i H_2O [MER06]
Density, g/cm³: α: 5.81 [KIR78]
Melting Point, °C: 1240 [CRC10]

592
Compound: Cadmium selenite
Formula: $CdSeO_3$
Molecular Formula: CdO_3Se
Molecular Weight: 239.369
CAS RN: 13814-59-0
Properties: −80 mesh 99.5% pure [CER91]

593
Compound: Cadmium stearate
Formula: $Cd(OOC_{18}H_{35})_2$
Molecular Formula: $C_{36}H_{70}CdO_4$
Molecular Weight: 679.361
CAS RN: 2223-93-0
Properties: white powd; uses: lubricant and
plastics stabilizer [HAW93] [STR93]
Density, g/cm³: 1.21 [KIR78]
Melting Point, °C: 104 [KIR78]

594
Compound: Cadmium succinate
Synonyms: succinic acid, cadmium salt
Formula: $Cd(OOCCH_2CH_2COO)$
Molecular Formula: $C_4H_4CdO_4$
Molecular Weight: 228.484
CAS RN: 141-00-4
Properties: white powd; needles or plates; preparation:
reaction of $CdCO_3$ with succinic acid; uses:
plant fungicide [MER06] [HAW93]
Solubility: 0.367 g/100mL (40°C) H_2O;
i alcohol [MER06] [HAW93]

595
Compound: Cadmium sulfate
Formula: $CdSO_4$
Molecular Formula: CdO_4S
Molecular Weight: 208.475
CAS RN: 10124-36-4
Properties: colorless, odorless cryst; ortho-
rhomb, a = 0.4717 nm, b = 0.6559 nm,
c = 0.4701 nm; enthalpy of fusion 20.084 kJ/
mol; used as a pigment [HAW93] [KIR78]
Solubility: g/100g H_2O: 75.6 ± 0.1 (0°C), 76.7 (25°C),
58.4 (99°C); solid phase, $CdSO_4 \cdot 8/3H_2O$ (0°C,
25°C), β-$CdSO_4 \cdot H_2O$ (99°C) [KRU93]
Density, g/cm³: 4.691 [KIR78]
Melting Point, °C: 1000 [HAW93]

596
Compound: Cadmium sulfate monohydrate
Formula: $CdSO_4 \cdot H_2O$

Molecular Formula: CdH_2O_5S
Molecular Weight: 226.490
CAS RN: 13477-20-8
Properties: monocl, a = 7.607 nm,
    b = 7.541 nm, c = 8.186 nm [KIR78]
Density, g/cm³: 3.79 [KIR78]
Melting Point, °C: 105 [KIR78]

597
Compound: Cadmium sulfate octahydrate
Formula: $3CdSO_4 \cdot 8H_2O$
Molecular Formula: $Cd_3H_{16}O_{20}S_3$
Molecular Weight: 769.546
CAS RN: 15244-35-6
Properties: colorless, odorless, monocl cryst;
    preparation: reaction of dil sulfuric acid with
    Cd or CdO; uses: pigment, electrolyte in
    Weston standard cell [MER06] [HAW93]
Solubility: v s H_2O [MER06]
Density, g/cm³: 3.08 [MER06]
Reactions: minus H_2O above 40°C, forms
    monohydrate by 80°C; does not become
    anhydrous if heated further [MER06]

598
Compound: Cadmium sulfide
Synonym: greenockite
Formula: CdS
Molecular Formula: CdS
Molecular Weight: 144.477
CAS RN: 1306-23-6
Properties: 3–6 mm pieces (highly dense, pressure
    sintered) and –325 mesh 99.999% pure; light
    yellow or orange colored cub or hex cryst; hex,
    a = 0.41348 nm, c = 0.6749 nm; enthalpy of fusion
    201.669 kJ/mol; used in pigments and inks and as
    evaporation material and pure sputtering target to
    produce photoconductive films, infrared filters, and
    solar cells [HAW93] [MER06] [KIR78] [CER91]
Solubility: 0.13 mg/100 g (18°C) [MER06]; mol/L
    in H_2O: 1.42×10^{-10} (25°C); 8.56×10^{-10} (100°C)
    [KRU93]; s acids, ammonia [HAW93]
Density, g/cm³: cub: 4.50; hex: 4.82 [MER06]
Melting Point, °C: 1750 [STR93];
    sublimes at 980 [MER06]
Reactions: decomposed by warm dil mineral
    acids to evolve H_2S [MER06]

599
Compound: Cadmium sulfite
Formula: $CdSO_3$
Molecular Formula: CdO_3S

Molecular Weight: 192.475
CAS RN: 13477-23-1
Properties: cryst [LAN05]
Solubility: mol/kg H_2O: 0.00221 (0°C),
    0.00207 (90°C) [KRU93]
Melting Point, °C: decomposes [LAN05]

600
Compound: Cadmium tantalate
Formula: $Cd_2Ta_2O_7$
Molecular Formula: $Cd_2O_7Ta_2$
Molecular Weight: 698.714
CAS RN: 12050-35-0
Properties: –200 mesh 99.9% pure [CER91]

601
Compound: Cadmium telluride
Formula: CdTe
Molecular Formula: CdTe
Molecular Weight: 240.011
CAS RN: 1306-25-8
Properties: 3–6 mm pieces (fused) of 99.999%
    purity; brownish black; cub cryst when
    prepared by sublimation in H_2 atm, also a
    hex form; hex, a = 0.457 nm, c = 0.747 nm;
    oxidizes eventually in moist air; used in
    phosphors, in semiconductor materials, and
    as an evaporation material and sputtering
    target to produce photoconductive films and
    infrared filters [HAW93] [MER06] [CER91]
Solubility: i H_2O, dil acids; s with
    decomposition in HNO_3 [MER06]
Density, g/cm³: 6.2 [MER06]
Melting Point, °C: 1041 [MER06]

602
Compound: Cadmium tellurite
Formula: $CdTeO_3$
Molecular Formula: CdO_3Te
Molecular Weight: 288.009
CAS RN: 15851-44-2
Properties: –80 mesh 99% pure [CER91]

603
Compound: Cadmium tetrafluoroborate
Formula: $Cd(BF_4)_2$
Molecular Formula: B_2CdF_8
Molecular Weight: 286.020
CAS RN: 14486-19-2
Properties: colorless liq [STR93]

604
Compound: Cadmium titanate
Formula: $CdTiO_3$
Molecular Formula: CdO_3Ti
Molecular Weight: 208.276
CAS RN: 12014-14-1
Properties: ortho-rhomb [KIR83]
Density, g/cm³: 6.5 [KIR83]

605
Compound: Cadmium tungstate(VI)
Formula: $CdWO_4$
Molecular Formula: CdO_4W
Molecular Weight: 360.249
CAS RN: 7790-85-4
Properties: −325 mesh, 10 μm or less of 99.9%
  purity; white or yellowish monocl cryst or
  powd; finds use in fluorescent paint, x-ray
  screens [HAW93] [MER06] [CER91]
Solubility: i H_2O [MER06]; s NH_4OH,
  alkali cyanides [HAW93]
Density, g/cm³: 8.0 [LID94]

606
Compound: Cadmium vanadate
Formula: CdV_2O_6
Molecular Formula: CdO_6V_2
Molecular Weight: 310.290
CAS RN: 12422-12-7
Properties: −200 mesh [CER91]

607
Compound: Cadmium zirconate
Formula: $CdZrO_3$
Molecular Formula: CdO_3Zr
Molecular Weight: 251.633
CAS RN: 12139-23-0
Properties: −200 mesh 99.5% purity [CER91]

608
Compound: Calcium
Formula: Ca
Molecular Formula: Ca
Molecular Weight: 40.078
CAS RN: 7440-70-2

Properties: lustrous, silver-white surface (freshly
  cut); much harder than sodium, softer than
  aluminum or magnesium; acquires bluish gray
  tarnish in moist air; enthalpy of fusion 8.54 kJ/
  mol; enthalpy of vaporization 161.5 kJ/mol;
  enthalpy of combustion 634.3 kJ/mol; electrical
  resistivity (20°C) 3.5 μohm·cm; Brinell
  hardness 17 [MER06] [CRC10] [KIR78]
Solubility: reacts with H_2O, alcohols, dil
  acids to evolve H_2 [MER06]
Density, g/cm³: 1.54 [MER06]
Melting Point, °C: 843 [COT88]
Boiling Point, °C: 1440 [MER06]
Reactions: burns in air; contact with alkali hydroxides
  or carbonates may cause explosion [MER06]
Thermal Conductivity, W/(m·K): 201 (25°C) [ALD94]
Thermal Expansion Coefficient: from 0 to 400°C
  coefficient is 22.3×10^{-6} m/(m·K) [KIR78]

609
Compound: Calcium acetate
Formula: $Ca(CH_3COO)_2$
Molecular Formula: $C_4H_6CaO_4$
Molecular Weight: 158.168
CAS RN: 62-54-4
Properties: very hygr; rod-shaped cryst [MER06]
Solubility: 26 g/100 g soln (25°C);
  23 g/100 g soln (100°C) [CIC73]
Density, g/cm³: 1.50 [MER06]
Melting Point, °C: decomposes [ALF93]
Reactions: decomposes at >160°C to
  acetone and $CaCO_3$ [MER06]

610
Compound: Calcium acetate dihydrate
Formula: $Ca(CH_3COO)_2 \cdot 2H_2O$
Molecular Formula: $C_4H_{10}CaO_6$
Molecular Weight: 194.197
CAS RN: 14977-17-4
Properties: colorless, long, transparent
  needles [CRC10] [MER06]
Solubility: g/100 g H_2O: 37.4 (0°C); 34.7
  (20°C); 29.7 (100°C) [LAN05]
Reactions: minus H_2O on standing in air
  to form monohydrate [MER06]

611
Compound: Calcium acetate monohydrate
Formula: $Ca(CH_3COO)_2 \cdot H_2O$
Molecular Formula: $C_4H_8CaO_5$
Molecular Weight: 176.183
CAS RN: 5743-26-0

Properties: brown, gray, or white powd; hygr; sl
   bitter taste; an odor of acetic acid; decomposes
   if heated; used in the manufacture of
   acetone, acetic acid; as a mordant in dyeing
   textiles [HAW93] [ALD94] [STR93]
Solubility: 48.6/100 mL (0°C), 34.3/100 mL (100°C)
   H_2O; sl s alcohol [CRC10] [HAW93]
Melting Point, °C: decomposes [CRC10]

612

Compound: Calcium acetylacetonate
Synonym: 2,4-pentanedionate, calcium derivative
Formula: $Ca(CH_3COCHCOCH_3)_2$
Molecular Formula: $C_{10}H_{14}CaO_4$
Molecular Weight: 238.294
CAS RN: 19372-44-2
Properties: cryst [CRC10]
Melting Point, °C: decomposes at 175 [CRC10]

613

Compound: Calcium acetylacetonate hydrate
Synonyms: 2,4-pentanedione, calcium derivative
Formula: $Ca(CH_3COCH=C(O)CH_3)_2 \cdot xH_2O$
Molecular Formula: $C_{10}H_{14}CaO_4$ (anhydrous)
Molecular Weight: 238.297 (anhydrous)
CAS RN: 19372-44-2
Properties: white powd; x ~ 0.5 [ALD94] [STR93]
Melting Point, °C: decomposes at 175 [STR93]

614

Compound: Calcium aluminate
Formula: $CaO \cdot Al_2O_3$
Molecular Formula: Al_2CaO_4
Molecular Weight: 158.039
CAS RN: 12042-68-1
Properties: white, monocl, tricl, or rhomb; −200
   mesh 99% powd [CER91] [CRC10]
Solubility: decomposed by H_2O [CRC10]
Density, g/cm³: 2.98 [ALF93]
Melting Point, °C: 1600 [ALF93]
Thermal Expansion Coefficient: (volume)
   100°C (0.12), 200°C (0.28), 400°C (0.63),
   800°C (1.49), 1200°C (2.37) [CLA66]

615

Compound: Calcium aluminate(β)
Formula: $3CaO \cdot Al_2O_3$
Molecular Formula: $Al_2Ca_3O_6$
Molecular Weight: 270.193
CAS RN: 12042-78-3
Properties: white cryst or powd; refractory
   material, important in cement [HAW93]

Solubility: s acids [HAW93]
Density, g/cm³: 3.038 [HAW93]
Melting Point, °C: decomposes at 1535 [HAW93]
Thermal Expansion Coefficient: (volume)
   100°C (0.18), 200°C (0.41), 400°C (0.97),
   800°C (2.23), 1200°C (3.66) [CLA66]

616

Compound: Calcium aluminum silicate
Synonym: gehlenite
Formula: $Ca_2Al_2SiO_7$
Molecular Formula: $Al_2Ca_2O_7Si$
Molecular Weight: 274.201
CAS RN: 1327-39-5
Properties: colorless, tetr; mineral; used in cement
   and refractories [MER06] [CRC10]
Density, g/cm³: 3.048 [CRC10]
Melting Point, °C: 1500 [CRC10]
Thermal Expansion Coefficient: (volume)
   100°C (0.20), 200°C (0.45), 400°C (0.93),
   800°C (1.97), 1200°C (3.07) [CLA66]

617

Compound: Calcium arsenate
Formula: $Ca_3(AsO_4)_2$
Molecular Formula: $As_2Ca_3O_8$
Molecular Weight: 398.072
CAS RN: 7778-44-1
Properties: −80 mesh 99% purity; white powd;
   decomposes if heated; preparation: from calcium
   chloride and sodium arsenate; used as an
   insecticide and germicide [HAW93] [CER91]
Solubility: 0.013 g/100 mL H_2O (25°C)
   [CRC10]; s dil acids [MER06]
Density, g/cm³: 3.62 [STR93]
Melting Point, °C: 1455 [CRC10]

618

Compound: Calcium arsenite
Formula: $CaHAsO_3$
Molecular Formula: $AsCaHO_3$
Molecular Weight: 163.206
CAS RN: 52740-16-6
Properties: white powd; uncertain composition;
   preparation: passing steam over mixture
   of CaO and As_2O_3; uses: insecticide,
   germicide [MER06] [HAW93]

619

Compound: Calcium
   bis(2,2,6,6-tetramethyl-3,5-heptanedionate)
Formula: $Ca[(CH_3)_3CCOCH=C(O)C(CH_3)_3]_2$

Molecular Formula: $C_{22}H_{38}CaO_4$
Molecular Weight: 406.619
CAS RN: 118448-18-3
Properties: uses: preparation of thin film semiconductors [ALD94]
Melting Point, °C: 221–224 [ALD94]

620
Compound: Calcium borate hexahydrate
Formula: $CaB_4O_7 \cdot 6H_2O$
Molecular Formula: $B_4CaH_{12}O_{13}$
Molecular Weight: 303.409
CAS RN: 13701-64-9
Properties: white powd; prepared by fusion of $CaCO_3$ and B_2O_3; used as a flux in metallurgy, in fire-retardant paint [MER06] [STR93]
Solubility: g/100 g H_2O: 2.32 (0°C), 2.72 (20°C), 8.70 (100°C) [LAN05]
Melting Point, °C: 986 (anhydrous) [CRC10]

621
Compound: Calcium boride
Synonym: calcium hexaboride
Formula: CaB_6
Molecular Formula: B_6Ca
Molecular Weight: 104.944
CAS RN: 12007-99-7
Properties: black cub; −200 mesh 99.5% pure; refractory material [KIR78] [CER91] [CRC10]
Density, g/cm³: 2.3 [ALD94]
Melting Point, °C: 2235 [KIR78]

622
Compound: Calcium bromate
Formula: $Ca(BrO_3)_2$
Molecular Formula: Br_2CaO_6
Molecular Weight: 295.882
CAS RN: 10102-75-7
Properties: 99% pure powd [ALF93]
Melting Point, °C: 180 [ALF93]

623
Compound: Calcium bromate monohydrate
Formula: $Ca(BrO_3)_2 \cdot H_2O$
Molecular Formula: $Br_2CaH_2O_7$
Molecular Weight: 313.898
CAS RN: 10102-75-7
Properties: white, monocl, cryst powd; oxidizing agent; used as a maturing agent, a dough conditioner [HAW93] [CRC10]

Solubility: v s H_2O [HAW93]
Density, g/cm³: 3.329 [HAW93]
Reactions: minus H_2O at 180°C [HAW93]

624
Compound: Calcium bromide
Formula: $CaBr_2$
Molecular Formula: Br_2Ca
Molecular Weight: 199.886
CAS RN: 7789-41-5
Properties: −80 mesh of 99.5% purity; white, odorless, deliq granules or rhomb cryst; becomes yellow on long exposure to air; sharp saline taste; enthalpy of fusion 29.08 kJ/mol; used in medicine, photography [HAW93] [MER06] [CER91] [STR93] [CRC10]
Solubility: v s H_2O; s alcohol, acetone [HAW93]
Density, g/cm³: 3.353 [MER06]
Melting Point, °C: 742 [CRC10]
Boiling Point, °C: 1815 [CRC10]
Reactions: when strongly heated in air, forms lime and bromine [MER06]

625
Compound: Calcium bromide dihydrate
Formula: $CaBr_2 \cdot 2H_2O$
Molecular Formula: $Br_2CaH_4O_2$
Molecular Weight: 235.917
CAS RN: 22208-73-7
Properties: white, cryst powd [ALF93] [STR93]

626
Compound: Calcium bromide hexahydrate
Formula: $CaBr_2 \cdot 6H_2O$
Molecular Formula: $Br_2CaH_{12}O_6$
Molecular Weight: 307.977
CAS RN: 13477-28-6
Properties: white, cryst or powd, odorless, sharp saline taste, very deliq; also $CaBr_2 \cdot xH_2O$ [HAW93] [STR93]
Solubility: g/100 g H_2O: 125 (0°C), 143 (20°C), 312 (105°C) [LAN05]; s alcohol, acetone [HAW93]
Density, g/cm³: 2.295 [STR93]
Melting Point, °C: 38.2 [STR93]
Boiling Point, °C: decomposes at 149 [HAW93]

627
Compound: Calcium carbide
Synonym: acetylenogen
Formula: CaC_2
Molecular Formula: C_2Ca

Molecular Weight: 64.100
CAS RN: 75-20-7
Properties: 9–40 mm grayish black, irregular lumps or ortho-rhomb cryst; garlic-like odor; can be prepared by reacting the metal or CaO with carbon in an electric furnace; used to generate acetylene gas, to reduce copper sulfide to metallic Cu [HAW93] [MER06] [ALF93] [COT88]
Solubility: decomposed by H_2O with the evolution of acetylene [MER06]
Density, g/cm³: 2.22 [MER06]
Melting Point, °C: 2300 [MER06]

628
Compound: Calcium carbonate
Synonym: aragonite
Formula: $CaCO_3$
Molecular Formula: $CCaO_3$
Molecular Weight: 100.087
CAS RN: 471-34-1
Properties: odorless, tasteless powd; ortho-rhomb; formed >30°C [MER06]
Solubility: mol/kg H_2O: 7.76×10^{-5} (0.7°C), 6.79×10^{-5} (25°C), 3.1×10^{-5} (90°C) [KRU93]; solubility data are also found in [PLU82]
Density, g/cm³: 2.83 [MER06]
Melting Point, °C: decomposes at 825 [MER06]
Reactions: transforms to calcite at ~400°C [KIR78]
Thermal Expansion Coefficient: (volume) 100°C (0.36), 200°C (1.00), 400°C (2.48) [CLA66]

629
Compound: Calcium carbonate
Synonym: calcite
Formula: $CaCO_3$
Molecular Formula: $CCaO_3$
Molecular Weight: 100.087
CAS RN: 471-34-1
Properties: −325 mesh 99.95% pure, 10 μm; white odorless, tasteless hex cryst or powd; formed below 30°C; occurs naturally as mineral calcite; enthalpy of fusion 53.10 kJ/mol; used as a source of lime [KIR78] [CER91] [HAW93] [CRC10] [MER06]
Solubility: mol/kg H_2O: 6.44×10^{-5} (0.2°C), 5.75×10^{-5} (25°C), 2.76×10^{-5} (89.7°C) [KRU93]; s acids, evolving CO_2 [HAW93]; solubility data are also found in [PLU82]
Density, g/cm³: 2.930 [STR93]
Reactions: decomposes to CaO at ~800°C [MER06]
Thermal Expansion Coefficient: (volume) 100°C (0.105), 200°C (0.285), 400°C (0.765), 600°C (1.395) [CLA66]

630
Compound: Calcium carbonate
Synonym: vaterite
Formula: $CaCO_3$
Molecular Formula: $CCaO_3$
Molecular Weight: 100.087
CAS RN: 471-34-1
Solubility: mol/kg H_2O: 1.34×10^{-4} (°C), 1.10×10^{-4} (25.1°C), 4.48×10^{-5} (90°C) [KRU93]; solubility data are also found in [PLU82]

631
Compound: Calcium chlorate dihydrate
Formula: $Ca(ClO_3)_2 \cdot 2H_2O$
Molecular Formula: $CaCl_2H_4O_8$
Molecular Weight: 243.010
CAS RN: 10035-05-9
Properties: white to yellowish cryst; prepared by reaction between chlorine and hot $Ca(OH)_2$ slurry; used in photography and as a powd to control poison ivy by dusting [HAW93]
Solubility: g/100 g soln, H_2O: 63.0 (0.5°C), 66.0 (25°C), 78.0 (93°C); solid phase, $Ca(ClO_3)_2 \cdot 2H_2O$ (25°C), $Ca(ClO_3)_2$ (93.0°C) [KRU93]
Density, g/cm³: 2.711 [HAW93]
Melting Point, °C: 340 [HAW93]

632
Compound: Calcium chloride
Synonym: hydrophilite
Formula: $CaCl_2$
Molecular Formula: $CaCl_2$
Molecular Weight: 110.983
CAS RN: 10043-52-4
Properties: 99.99% pure −325 mesh white powd; cub cryst, granules of fused masses; very hygr; enthalpy of fusion 28.54 kJ/mol; infinite enthalpy of solution −81.82 kJ/mol; used on roads to control ice and dust [HAW93] [MER06] [KIR78] [STR93] [CRC10]
Solubility: g/100 g soln, H_2O: 37.3 (0°C), 45.3 (25°C), 61.4 (100°C); solid phase, $CaCl_2 \cdot 6H_2O$ (0°C, 25°C), $CaCl_2 \cdot 2H_2O$ (100°C) [KRU93]
Density, g/cm³: 2.152 [MER06]
Melting Point, °C: 772 [MER06]
Boiling Point, °C: 1935 [KIR78]

633
Compound: Calcium chloride dihydrate
Formula: $CaCl_2 \cdot 2H_2O$
Molecular Formula: $CaCl_2H_4O_2$

Molecular Weight: 147.014
CAS RN: 10035-04-8
Properties: hygr granules, flakes or powd; enthalpy
    of fusion 88 J/g; enthalpy of infinite solution
    −44.78 kJ/mol [KIR78] [MER06] [ALF93]
Solubility: v s H_2O [MER06]
Density, g/cm³: 1.85 [KIR78]
Melting Point, °C: 176 [KIR78]

634
Compound: Calcium chloride hexahydrate
Synonym: antarcticite
Formula: $CaCl_2 \cdot 6H_2O$
Molecular Formula: $CaCl_2H_{12}O_6$
Molecular Weight: 219.074
CAS RN: 7774-34-7
Properties: white, deliq, trig cryst; enthalpy of
    fusion 209 J/g; enthalpy of infinite solution
    15.77 kJ/mol [KIR78] [MER06]
Solubility: g/100 g H_2O: 59.5 (0°C), 74.5
    (20°C), 159 (100°C) [LAN05]
Density, g/cm³: 1.68 [MER06]
Melting Point, °C: 29.9 [KIR78]
Reactions: minus $6H_2O$ at 200°C [MER06]

635
Compound: Calcium chloride monohydrate
Formula: $CaCl_2 \cdot H_2O$
Molecular Formula: $CaCl_2H_2O$
Molecular Weight: 128.998
CAS RN: 13477-29-7
Properties: white, deliq cryst, lumps, granules, flakes;
    enthalpy of fusion 17.28 kJ/mol; heat of infinite
    solution −52.24 kJ/mol [KIR78] [HAW93]
Solubility: s H_2O, alcohol [HAW93]
Density, g/cm³: 2.24 [KIR78]
Melting Point, °C: 187 [KIR78]

636
Compound: Calcium chloride tetrahydrate
Formula: $CaCl_2 \cdot 4H_2O$
Molecular Formula: $CaCl_2H_8O_4$
Molecular Weight: 183.045
CAS RN: 25094-02-4
Properties: enthalpy of fusion 29.86 kJ/mol; heat
    of infinite dilution −10.87 kJ/mol [KIR78]
Density, g/cm³: 1.83 [KIR78]

637
Compound: Calcium chlorite
Formula: $Ca(ClO_2)_2$

Molecular Formula: $CaCl_2O_4$
Molecular Weight: 174.981
CAS RN: 14674-72-7
Properties: cub white cryst; oxidizing
    agent [HAW93] [CRC10]
Solubility: decomposed by H_2O [HAW93]
Density, g/cm³: 2.71 [HAW93]

638
Compound: Calcium chromate
Synonym: calcium chrome yellow
Formula: $CaCrO_4$
Molecular Formula: $CaCrO_4$
Molecular Weight: 156.072
CAS RN: 10060-08-9
Properties: yellow cryst, monocl or rhomb; −20
    mesh 99.9% pure; used as a pigment and as
    a corrosion inhibitor [STR93] [CER91]
Solubility: sl s H_2O; s dil acids [MER06]

639
Compound: Calcium chromate dihydrate
Formula: $CaCrO_4 \cdot 2H_2O$
Molecular Formula: $CaCrH_4O_6$
Molecular Weight: 192.102
CAS RN: 13765-19-0
Properties: bright yellow powd; used as a pigment,
    corrosion inhibitor, and oxidizing agent [HAW93]
Solubility: g/100 g H_2O: 17.3 (0°C), 16.6
    (30°C), 16.1 (40°C) [LAN05]
Density, g/cm³: 2.5 [LID94]
Reactions: minus $2H_2O$ at 200°C [HAW93]

640
Compound: Calcium citrate tetrahydrate
Synonym: tricalcium dicitrate tetrahydrate
Formula: $[OOCCH_2C(OH)(COO)CH_2COO]_2Ca_3 \cdot 4H_2O$
Molecular Formula: $C_{12}H_{18}Ca_3O_{18}$
Molecular Weight: 570.498
CAS RN: 5785-44-4
Properties: white, odorless needles or powd;
    can be obtained from citrus fruits; used
    as a dietary supplement, sequesterant, and
    firming agent in foods, used to prepare
    citric acid [HAW93] [MER06] [KIR78]
Solubility: 0.088 g/100 mL H_2O at 18°C,
    0.096 g/100 mL at 23°C; 0.0065 g/100 mL
    alcohol at 18°C; i ether [KIR78]
Reactions: minus most of its water at 100°C,
    all H_2O at 120°C [MER06]

641

Compound: Calcium cyanamide
Synonym: calcium carbimide
Formula: N≡C–N=Ca
Molecular Formula: $CCaN_2$
Molecular Weight: 80.102
CAS RN: 156-62-7
Properties: −12 mesh; colorless cryst or powd; pure material glistens, hex cryst; used as a fertilizer and pesticide and in manufacturing iron and $Ca(CN)_2$ [HAW93] [MER06] [ALF93]
Solubility: i H_2O, but undergoes hydrolysis releasing acetylene and ammonia [HAW93] [MER06]
Density, g/cm³: 2.29 [MER06]; 1.083 [HAW93]
Melting Point, °C: ~1340 [MER06]
Reactions: sublimes at 1150°C–1200°C [MER06]

642

Compound: Calcium cyanide
Synonym: cyanogas
Formula: $Ca(CN)_2$
Molecular Formula: C_2CaN_2
Molecular Weight: 92.113
CAS RN: 592-01-8
Properties: colorless or white rhomb cryst or powd; decomposes in moist air liberating HCN; used as a rodenticide, as a fumigant in greenhouses, and in flour mills [HAW93] [MER06]
Solubility: s H_2O, gradually releasing HCN [MER06]
Melting Point, °C: 640 estimated [KIR78]

643

Compound: Calcium dichromate trihydrate
Formula: $CaCr_2O_7 \cdot 3H_2O$
Molecular Formula: $CaCr_2H_6O_{10}$
Molecular Weight: 310.112
CAS RN: 14307-33-6
Properties: bipyramidal, reddish orange cryst; not hygr, if pure; used as a catalyst in manufacturing $CrCl_3$ and CrO_3 and as a corrosion inhibitor [MER06]
Solubility: v s H_2O; i ether, CCl_4; reacts with alcohol [MER06]
Density, g/cm³: 2.37 [MER06]
Melting Point, °C: decomposes above 100 [MER06]
Reactions: decomposes when heated forming $CaCrO_4$ and CrO_3 [MER06]

644

Compound: Calcium dihydrogen phosphate monohydrate
Formula: $Ca(H_2PO_4)_2 \cdot H_2O$
Molecular Formula: $CaH_6O_9P_2$

Molecular Weight: 252.068
CAS RN: 10031-30-8
Properties: white, large, shining, tricl plates, cryst powd or granules; not hygr [MER06] [STR93]
Solubility: moderately s H_2O; s dil HCl, HNO_3, acetic acid [MER06]
Density, g/cm³: 2.22 [MER06]
Reactions: minus H_2O at 100°C; decomposes at 200°C [MER06]

645

Compound: Calcium 2-ethylhexanoate
Formula: $Ca(C_8H_{15}O_2)_2$
Molecular Formula: $C_{16}H_{30}CaO_4$
Molecular Weight: 326.485
CAS RN: 136-57-6
Properties: powd [CRC10]

646

Compound: Calcium ferrocyanide dodecahydrate
Formula: $Ca_2Fe(CN)_6 \cdot 12H_2O$
Molecular Formula: $C_6H_{24}Ca_2FeN_6O_{12}$
Molecular Weight: 508.291
CAS RN: 1327-39-5
Properties: yellow, tricl cryst; decomposes when heated; used to remove metallic impurities from citric, tartaric and, other acids [CRC10] [HAW93]
Solubility: 86.8 g/100 mL (25°C), 115 g/100 mL (65°C) H_2O; i alcohol [HAW93] [CRC10]
Density, g/cm³: 1.68 [HAW93]
Melting Point, °C: decomposes [CRC10]

647

Compound: Calcium fluoride
Synonyms: fluorspar, fluorite
Formula: CaF_2
Molecular Formula: CaF_2
Molecular Weight: 78.075
CAS RN: 7789-75-5
Properties: white powd or cub cryst; enthalpy of fusion 29.71 kJ/mol; enthalpy of vaporization 335 kJ/mol; hardness 4 Mohs; electrical conductivity 1.3×10^{-18} (ohm · cm)$^{-1}$ at 20°C, 6×10^{-5} at 650°C; becomes luminous on heating; pure material prepared from $CaCO_3$ and HF solution; mineral fluorspar is the main source of fluorine; 99.95% pure material used in spectroscopy, lasers, and electronics, for sputtering antireflection coating on glass [HAW93] [MER06] [CER91] [CRC10]
Solubility: g/L soln, H_2O: 0.013 (0°C), 0.016 (25°C) [KRU93]
Density, g/cm³: 3.18 [MER06]

Melting Point, °C: 1418 [CRC10]
Boiling Point, °C: 2500 [MER06]
Thermal Conductivity, W/(m·K): 10.96 [KIR78]
Thermal Expansion Coefficient: (volume)
    100°C (0.47), 200°C (1.12) [CLA66]

648
Compound: Calcium fluorophosphate
Synonym: fluoroapatite
Formula: $Ca_5(PO_4)_3F$
Molecular Formula: $Ca_5FO_{12}P_3$
Molecular Weight: 504.302
CAS RN: 1306-05-4
Properties: fluoroapatite is a mineral for fluorine;
    can be prepared from $CaCl_2$ and sodium
    monofluorophosphate; used as a laser cryst,
    may have the lowest energy threshold of any
    cryst at room temp [KIR78] [HAW93]

649
Compound: Calcium fluorophosphate dihydrate
Formula: $CaPO_3F \cdot 2H_2O$
Molecular Formula: $CaFH_4O_5P$
Molecular Weight: 174.079
CAS RN: 37809-19-1
Properties: monocl cryst powd; tendency to form twins;
    loses fluorine on heating [MER06] [STR93]
Solubility: 0.417 g/100 mL soln (27°C); i in most
    common organic solvents [MER06]

650
Compound: Calcium formate
Synonyms: formic acid, calcium salt
Formula: $Ca(HCOO)_2$
Molecular Formula: $C_2H_2CaO_4$
Molecular Weight: 130.114
CAS RN: 544-17-2
Properties: ortho-rhomb cryst or powd; slight odor
    similar to acetic acid; obtained from $Ca(OH)_2$ and
    CO reaction at high temperatures and pressures;
    used as food preservative, as binder for fine-ore
    briquettes, and in drilling fluids [MER06]
Solubility: g/100 g H_2O: 16.15 (0°C), 16.60 (20°C), 17.95
    (80°C), 18.40 (100°C) [LAN05]; i alcohol [MER06]
Density, g/cm³: 2.02 [MER06]
Melting Point, °C: 300 [HAW93]

651
Compound: Calcium hexaborate pentahydrate
Synonym: colemanite
Formula: $Ca_2B_6O_{11} \cdot 5H_2O$

Molecular Formula: $B_6Ca_2H_{10}O_{16}$
Molecular Weight: 411.091
CAS RN: 12291-65-5
Properties: monocl; forms when saturated solutions
    of inyoite or higher hydrates are heated [KIR78]
Solubility: about 1% in H_2O at 25°C [KIR78]
Density, g/cm³: 2.42 [KIR78]

652
Compound: Calcium hexafluoroacetylacetonate
    dihydrate
Synonyms: l,l,l,5,5,5-hexafluoro-2,4-
    pentanedione, calcium derivative
Formula: $Ca[CF_3COCH=C(O)CF_3]_2 \cdot 2H_2O$
Molecular Formula: $C_{10}H_6CaF_{12}O_6$
Molecular Weight: 490.211
CAS RN: 141572-90-9
Properties: off-white powd; precursor for metal oxide
    chemical vapor deposition [STR94] [STR93]
Melting Point, °C: 135–140 [STR93]
Boiling Point, °C: decomposes at 230–240 [STR93]
Reactions: sublimes at 180°C, 0.7 mm Hg [STR94]

653
Compound: Calcium hexafluorosilicate dihydrate
Formula: $CaSiF_6 \cdot 2H_2O$
Molecular Formula: $CaF_6H_4O_2Si$
Molecular Weight: 218.185
CAS RN: 16925-39-6
Properties: colorless, tetr; powd; obtained by reaction
    of a Ca salt and H_2SiF_6; used as a flotation agent
    and insecticide [HAW93] [MER06] [CRC10]
Solubility: i cold H_2O, partially decomposed
    in hot H_2O [MER06]
Density, g/cm³: 2.25 [MER06]

654
Compound: Calcium hydride
Formula: CaH_2
Molecular Formula: CaH_2
Molecular Weight: 42.094
CAS RN: 7789-78-8
Properties: −40 mesh, −20 mesh 98% pure; sensitive to
    moisture; ortho-rhomb cryst or powd; commercial
    product is gray; decomposes into Ca and H_2 at 990°C
    without melting; prepared by heating calcium metal
    to ~300°C under 1 atm of H_2; used to obtain some
    metals by reduction of their oxides and to dry gases
    and unreactive liq [KIR80] [MER06] [STR93]
Solubility: decomposed by H_2O, with
    the evolution of H_2 [HAW93]
Density, g/cm³: 1.7 [MER06]; 1.90 [KIR80]
Melting Point, °C: 816 under H_2 [STR93]

655

Compound: Calcium hydrogen phosphate
Synonyms: calcium phosphate, dibasic
Formula: $CaHPO_4$
Molecular Formula: $CaHO_4P$
Molecular Weight: 136.057
CAS RN: 7757-93-9
Properties: white, tasteless, tricl cryst; prepared by reacting phosphoric acid and milk of lime (calcium hydroxide suspended in water); used as food supplement, in medicine, constituent of a dentrifice and fertilizer [HAW93] [MER06]
Solubility: i H_2O, alcohol [MER06]
Reactions: at red heat, dehydrated to calcium pyrophosphate [MER06]

656

Compound: Calcium hydrogen phosphate dihydrate
Synonym: brushite
Formula: $CaHPO_4 \cdot 2H_2O$
Molecular Formula: CaH_5O_6P
Molecular Weight: 172.088
CAS RN: 7789-77-7
Properties: monocl [MER06]
Solubility: i H_2O, alcohol; s dil HCl, HNO_3 [MER06]
Density, g/cm³: 2.31 [MER06]
Reactions: minus H_2O <100°C; dehydrates at red heat to calcium pyrophosphate [MER06]

657

Compound: Calcium hydrogen sulfite
Formula: $Ca(HSO_3)_2$
Molecular Formula: $CaH_2O_6S_2$
Molecular Weight: 202.222
CAS RN: 13780-03-5
Properties: solution of calcium sulfite in an aq solution of sulfur dioxide; yellowish liq; used in bleaching textiles [HAW93]
Density, g/cm³: 1.06 [HAW93]

658

Compound: Calcium hydrosulfide hexahydrate
Formula: $Ca(HS)_2 \cdot 6H_2O$
Molecular Formula: $CaH_{14}O_6S_2$
Molecular Weight: 214.317
CAS RN: 12133-28-7
Properties: colorless, transparent cryst; used in the leather industry [HAW93]
Solubility: s H_2O, alcohol [HAW93]
Melting Point, °C: decomposes in air at 15–18 [HAW93]

659

Compound: Calcium hydroxide
Synonyms: portlandite, slaked lime
Formula: $Ca(OH)_2$
Molecular Formula: CaH_2O_2
Molecular Weight: 74.093
CAS RN: 1305-62-0
Properties: cryst or soft, odorless granules or powd; sl bitter, alkaline taste; readily absorbs CO_2 from air; manufactured from lime and water; used in mortar, plasters, cement [HAW93] [MER06]
Solubility: g/100 g H_2O: 0.189 (0°C), 0.173 (20°C), 0.076 (100°C) [LAN05]; s acids (caution) [MER06]; s glycerol; i alcohol [HAW93]
Density, g/cm³: 2.08–2.34 [MER06]
Reactions: minus H_2O at 580°C [HAW93]

660

Compound: Calcium hydroxide phosphate
Formula: $Ca_5(OH)(PO_4)_3$
Molecular Formula: $Ca_5HO_{13}P_3$
Molecular Weight: 502.311
CAS RN: 12167-74-7
Properties: col hex cryst [CRC10]
Solubility: i H_2O [CRC10]
Density, g/cm³: 3.155 [CRC10]
Melting Point, °C: decomposes at >900 [CRC10]

661

Compound: Calcium hypochlorite
Synonym: losantin
Formula: $Ca(OCl)_2$
Molecular Formula: $CaCl_2O_2$
Molecular Weight: 142.982
CAS RN: 7778-54-3
Properties: grayish white powd; oxidizing agent; used as an algicide, bactericide, deodorant, disinfectant, as a bleach, and to refine sugar; there is a dihydrate, 22464-76-2 [KIR78] [MER06] [HAW93] [STR93]
Solubility: decomposes in both H_2O and alcohol [HAW93]
Density, g/cm³: 2.35 [HAW93]
Melting Point, °C: decomposes at 100 [HAW93]

662

Compound: Calcium hypophosphite
Formula: $Ca(H_2PO_2)_2$
Molecular Formula: $CaH_4O_4P_2$
Molecular Weight: 170.055
CAS RN: 7789-79-9

Properties: white monocl, prismatic cryst or granular powd; used as a corrosion inhibitor and in medicine [MER06] [HAW93]
Solubility: s H_2O, sl s in glycerol; i alcohol [MER06]
Melting Point, °C: decomposes [ALF93]
Reactions: when heated at >300°C, evolves phosphine, which spontaneously ignites [MER06]

663
Compound: Calcium iodate
Synonym: lautarite
Formula: $Ca(IO_3)_2$
Molecular Formula: CaI_2O_6
Molecular Weight: 389.883
CAS RN: 7789-80-2
Properties: white powd; monocl prismatic cryst; the hexahydrate is ortho-rhomb; not hygr; obtained by passing Cl_2 into a hot solution of lime containing dissolved iodine; used as a deodorant, in mouthwashes, as a food additive, and dough conditioner [MER06] [STR93]
Solubility: s HNO_3; i alcohol [MER06]; g/100 g soln, H_2O: 0.0906 (0°C), 0.306 (25°C), 0.668 (90°C); solid phase, $Ca(IO_3)_2 \cdot 6H_2O$ (0°C, 25°C), $Ca(IO_3)_2$ (90°C) [KRU93]
Density, g/cm³: 4.519 [MER06]
Melting Point, °C: stable up to 540 [MER06]
Reactions: sensitive to reducing agents [MER06]

664
Compound: Calcium iodide
Formula: CaI_2
Molecular Formula: CaI_2
Molecular Weight: 293.887
CAS RN: 10102-68-8
Properties: −20 mesh 99.5% pure powd; very hygr; hex; becomes yellow and completely insoluble on exposure to air due to liberation of I_2 and absorption of CO_2; enthalpy of fusion 41.80 kJ/mol; finds use as an expectorant [MER06] [CRC10]
Solubility: v s H_2O, methanol, ethanol, acetone; i ether [MER06], g/100 g soln, H_2O: 64.6 (0°C), 68.3 (25°C), 81.0 (100°C) [KRU93]
Density, g/cm³: 4.0 [HAW93]
Melting Point, °C: 779 [CRC10]
Boiling Point, °C: 1100 [MER06]

665
Compound: Calcium iodide hexahydrate
Formula: $CaI_2 \cdot 6H_2O$
Molecular Formula: $CaH_{12}I_2O_6$

Molecular Weight: 401.978
CAS RN: 71626-98-7
Properties: white powd, hex, thick needles, plates or lumps; very hygr; becomes yellow in air; absorbs atm CO_2; used in photography and in medicine; formula also given as $CaI_2 \cdot xH_2O$ [HAW93] [MER06] [STR93]
Solubility: v s H_2O, alcohol [MER06]
Density, g/cm³: 2.55 [HAW93]
Melting Point, °C: 783 [HAW93]
Boiling Point, °C: ~1100 [HAW93]
Reactions: minus $6H_2O$ at 42°C [HAW93]

666
Compound: Calcium metaborate
Formula: $Ca(BO_2)_2$
Molecular Formula: B_2CaO_4
Molecular Weight: 125.698
CAS RN: 13701-64-9
Properties: powd [CRC10]
Solubility: g/100 g H_2O: 0.13²⁰ [CRC10]

667
Compound: Calcium metasilicate
Formula: $CaSiO_3$
Molecular Formula: CaO_3Si
Molecular Weight: 116.162
CAS RN: 1344-95-2
Properties: white, monocl cryst [CRC10]
Solubility: i H_2O
Density, g/cm³: 2.92 [CRC10]
Melting Point, °C: 1540 [CRC10]

668
Compound: Calcium molybdate
Synonym: powellite
Formula: $CaMoO_4$
Molecular Formula: $CaMoO_4$
Molecular Weight: 200.016
CAS RN: 7789-82-4
Properties: −200 mesh 99.9% pure; white, cryst powd; can be produced by reacting $CaSO_4$ with sodium molybdate; used in optical and electronic applications, as an alloying agent in iron and steel manufacturing [HAW93] [MER06] [CER91]
Solubility: 0.0050 g/100 g H_2O [KIR81]; s conc mineral acids [MER06]; mg/100 g soln, H_2O: 0.0022 (0°C), 0.0025 (22°C), 0.0085 (100°C) [KRU93]
Density, g/cm³: 4.38–4.53 [STR93]
Melting Point, °C: decomposes at 965 [KIR81]; ~1250 [HAW93]

669

Compound: Calcium nitrate
Formula: Ca(NO$_3$)$_2$
Molecular Formula: CaN$_2$O$_6$
Molecular Weight: 164.087
CAS RN: 10124-37-5
Properties: white, deliq granules; oxidizing agent; used in pyrotechnics, explosives, and fertilizers [HAW93] [MER06]
Solubility: v s H$_2$O, evolves heat; s methanol, ethanol, acetone [MER06]; g/100 g soln, H$_2$O: 50.50 (0°C), 57.98 (25°C), 78.43 (100°C); solid phase, Ca(NO$_3$)·4H$_2$O (0°C, 25°C), Ca(NO$_3$)$_2$ (100°C) [KRU93]
Density, g/cm^3: 2.36 [HAW93]
Melting Point, °C: 561 [HAW93]

670

Compound: Calcium nitrate tetrahydrate
Formula: Ca(NO$_3$)$_2$·4H$_2$O
Molecular Formula: CaH$_8$N$_2$O$_{10}$
Molecular Weight: 236.149
CAS RN: 13477-34-4
Properties: −4 mesh 99.999% pure; white cryst [STR93] [CER91]
Solubility: g/100 g H$_2$O: 102 (0°C), 129 (20°C), 363 (100°C) [LAN05]; s alcohol, acetone [HAW93]
Density, g/cm^3: 1.82 [STR93]
Melting Point, °C: 39.7 [STR93]

671

Compound: Calcium nitride
Formula: Ca$_3$N$_2$
Molecular Formula: Ca$_3$N$_2$
Molecular Weight: 148.247
CAS RN: 12013-82-0
Properties: 12 mm pieces and smaller, −200 mesh 99% pure; brown cryst; hex: a=0.3533 nm, c=0.411 nm; cub: a=1.138 nm [CIC73] [CER91]
Solubility: s H$_2$O, releasing ammonia; s dil acids; i absolute alcohol [HAW93]
Density, g/cm^3: hex: 2.62; cub: 2.54 [CIC73]
Melting Point, °C: 1195 [HAW93]

672

Compound: Calcium nitrite
Formula: Ca(NO$_2$)$_2$
Molecular Formula: CaN$_2$O$_4$
Molecular Weight: 132.089
CAS RN: 13780-06-8

Properties: white or yellowish deliq, hex cryst; prepared from nitric oxide and a mixture consisting of calcium ferrate and calcium nitrate; used to inhibit corrosion in lubricants and concrete [MER06]
Solubility: sl s alcohol [MER06]; g/100 g soln, H$_2$O: 38.3 (0°C), 43.0 (18.5°C), 71.2 (91°C); solid phase Ca(NO$_2$)$_2$·4H$_2$O (0°C, 18.5°C), Ca(NO$_2$)$_2$·2H$_2$O (91°C) [KRU93]
Density, g/cm^3: 2.23 [MER06]

673

Compound: Calcium nitrite monohydrate
Formula: Ca(NO$_2$)$_2$·H$_2$O
Molecular Formula: CaH$_2$N$_2$O$_5$
Molecular Weight: 150.104
CAS RN: 13780-06-8
Properties: colorless or yellowish cryst; hygr; used in lubricants as a corrosion inhibitor [HAW93]
Solubility: gCa(NO$_3$)$_2$·4H$_2$O/100 g H$_2$O: 63.9 (0°C), 104 (30°C), 178 (100°C) [LAN05]; sl s alcohol [HAW93]
Density, g/cm^3: 2.23 (anhydrous, 34°C) [HAW93]
Reactions: minus H$_2$O at 100°C [HAW93]

674

Compound: Calcium oleate
Synonyms: 9-octadecanoic acid, calcium salt
Formula: Ca(C$_{18}$H$_{33}$O$_2$)$_2$
Molecular Formula: C$_{36}$H$_{66}$CaO$_4$
Molecular Weight: 602.996
CAS RN: 142-17-6
Properties: pale yellow, transparent solid; slowly absorbs moisture from air to form monohydrate; used as a thickening agent for grease [HAW93] [MER06]
Solubility: 0.04 g/100 mL (25°C), 0.03 g/100 mL (50°C) H$_2$O; i alcohol, ether, acetone; s benzene, chloroform [MER06] [CRC10]
Melting Point, °C: decomposes at >140 [MER06]

675

Compound: Calcium oxalate
Synonyms: ethanedioic acid, calcium salt
Formula: CaC$_2$O$_4$
Molecular Formula: C$_2$CaO$_4$
Molecular Weight: 128.098
CAS RN: 563-72-4
Properties: white, cryst powd; obtained from calcium formate or calcium cyanamide; used to prepare oxalic acid, glazes, and to separate rare earths [HAW93] [MER06]
Solubility: g/L soln, H$_2$O: 0.0069 (25°C), 0.0142 (95°C); solid phase, CaC$_2$O$_4$·H$_2$O [KRU93]; s dil HCl, HNO$_3$ [HAW93]

Density, g/cm³: 2.2 [HAW93]
Melting Point, °C: decomposes [CRC10]

676
Compound: Calcium oxalate monohydrate
Synonyms: ethanedioic acid, calcium salt monohydrate
Formula: $CaC_2O_4 \cdot H_2O$
Molecular Formula: $C_2H_2CaO_5$
Molecular Weight: 146.113
CAS RN: 5794-28-5
Properties: colorless, cub cryst; hygr; uses: ceramic glazes [MER06] [CRC10]
Solubility: i H_2O, acetic acid; s dil HCl, HNO_3 [MER06]
Density, g/cm³: 2.2 [MER06]
Reactions: minus $2H_2O$ at 200°C; when ignited converts into $CaCO_3$ or CaO without appreciable charring [MER06]

677
Compound: Calcium oxide
Synonyms: lime, quicklime
Formula: CaO
Molecular Formula: CaO
Molecular Weight: 56.077
CAS RN: 1305-78-8
Properties: −325 mesh, 10 μm or less, 99.99% and 99.5% pure; cryst, white or grayish lumps or granular powd; readily absorbs CO_2 and H_2O from air; odorless; enthalpy of fusion 59.00 kJ/mol; produced from limestone; used in pulp and paper, dehairing of hides, in brick, mortar, and stucco [HAW93] [MER06] [CRC10]
Solubility: reacts with H_2O to form $Ca(OH)_2$; s acids [MER06]
Density, g/cm³: 3.32–3.35 [MER06]
Melting Point, °C: 2927 [CRC10]
Boiling Point, °C: 2850 [MER06]
Thermal Conductivity, W/(m·K): 8.0 (500°C), 7.8 (1000°C) [KIR80]
Thermal Expansion Coefficient: (volume) 100°C (0.225), 200°C (0.571), 400°C (1.402), 800°C (3.107), 1200°C (5.078) [CLA66]

678
Compound: Calcium oxide silicate
Formula: Ca_3OSiO_4
Molecular Formula: Ca_3O_5Si
Molecular Weight: 228.317
CAS RN: 12168-85-3
Properties: refrac solid [CRC10]
Melting Point, °C: 2150 [CRC10]

679
Compound: Calcium palmitate
Synonyms: hexadecanoic acid, calcium salt
Formula: $Ca(C_{16}H_{31}O_2)_2$
Molecular Formula: $C_{32}H_{62}CaO_4$
Molecular Weight: 550.920
CAS RN: 542-42-7
Properties: white or pale yellow powd; used for waterproofing, as a thickener for lubricating oils [HAW93] [MER06]
Solubility: i H_2O, alcohol, ether, acetone; sl s benzene [MER06]
Melting Point, °C: decomposes above 155 [MER06]

680
Compound: Calcium perborate heptahydrate
Formula: $Ca(BO_3)_2 \cdot 7H_2O$
Molecular Formula: $B_2CaH_{14}O_{13}$
Molecular Weight: 283.803
CAS RN: 12007-56-6
Properties: grayish white lumps or powd; uses: in medicine, as a bleach, and in tooth powd [HAW93]
Solubility: s H_2O, acids; evolves oxygen [HAW93]

681
Compound: Calcium perchlorate
Synonyms: perchloric acid, calcium salt
Formula: $Ca(ClO_4)_2$
Molecular Formula: $CaCl_2O_8$
Molecular Weight: 238.978
CAS RN: 13477-36-6
Properties: white cryst; oxidizing agent [HAW93]
Solubility: g/100 g soln, H_2O: 65.35 (25°C) [KRU93]; s alcohol [HAW93]
Density, g/cm³: 2.651 [HAW93]
Melting Point, °C: decomposes at 270 [HAW93]

682
Compound: Calcium perchlorate tetrahydrate
Formula: $Ca(ClO_4)_2 \cdot 4H_2O$
Molecular Formula: $CaCl_2H_8O_{12}$
Molecular Weight: 311.039
CAS RN: 15627-86-8
Properties: white cryst [STR93]

683
Compound: Calcium permanganate
Formula: $Ca(MnO_4)_2$
Molecular Formula: $CaMn_2O_8$
Molecular Weight: 277.949
CAS RN: 10118-76-0

Properties: violet or dark purple, deliq cryst;
made by reacting $KMnO_4$ and $CaCl_2$; used
in the textile industry, to sterilize water, and
as a deodorizer [HAW93] [MER06]
Solubility: v s H_2O; decomposed by alcohol [MER06]
Density, g/cm³: 2.4 [HAW93]

684
Compound: Calcium peroxide
Synonym: calcium dioxide
Formula: CaO_2
Molecular Formula: CaO_2
Molecular Weight: 72.077
CAS RN: 1305-79-9
Properties: tetr; white or yellowish, odorless,
almost tasteless powd; decomposes in moist
air; used to disinfect seeds, in dentrifices
[HAW93] [MER06] [CRC10]
Solubility: sl s H_2O; s in acids forming H_2O_2 [MER06]
Density, g/cm³: 2.92 [CRC10]
Melting Point, °C: decomposes at ~200 [HAW93]

685
Compound: Calcium phenoxide
Synonyms: calcium phenolate, calcium carbolate
Formula: $Ca(OC_6H_5)_2$
Molecular Formula: $C_{12}H_{10}CaO_2$
Molecular Weight: 226.288
CAS RN: 5793-84-0
Properties: reddish powd; decomposes in air;
used as a detergent for lubricating oil and
as an emulsifier [HAW93] [MER06]
Solubility: sl s H_2O, alcohol [MER06]

686
Compound: Calcium phosphate
Synonym: whitlockite
Formula: $Ca_3(PO_4)_2$
Molecular Formula: $Ca_3O_8P_2$
Molecular Weight: 310.177
CAS RN: 7758-87-4
Properties: amorphous, odorless, tasteless, white powd;
produced from phosphate rock; used in ceramics,
as a polishing powd [HAW93] [MER06] [STR93]
Solubility: i H_2O, alcohol, acetic acid;
s dil HCl, HNO_3 [MER06]
Density, g/cm³: 3.14 [MER06]
Melting Point, °C: 1670 [HAW93]

687
Compound: Calcium phosphate hydroxide
Synonyms: durapatite, hydroxylapatite

Formula: $3Ca_3(PO_4)_2 \cdot Ca(OH)_2$
Molecular Formula: $Ca_{10}H_{26}O_{26}P_6$
Molecular Weight: 748.143
CAS RN: 1306-06-5
Properties: occurs naturally as mineral; hex needles with
rosettes arrangement; preparation from $Ca(NO_3)_2$
and KH_2PO_4; uses: purification of DNA, Ca and
P supplement, prosthetic aid [ALD94] [MER06]
Solubility: i H_2O [MER06]
Melting Point, °C: decomposes at >1100 [MER06]

688
Compound: Calcium phosphide
Synonym: photophor
Formula: Ca_3P_2
Molecular Formula: Ca_3P_2
Molecular Weight: 182.182
CAS RN: 1305-99-3
Properties: 0.5 inch pieces and down; reddish
brown cryst powd or gray lumps; decomposed
by moist air, evolving flammable phosphine,
which can ignite spontaneously; obtained by
heating quicklime in phosphorus vapor; used
in signal fires, torpedoes, and pyrotechnics
[HAW93] [MER06] [STR93] [KIR82]
Solubility: decomposes in H_2O to form flammable
phosphine [MER06]; i alcohol, ether [HAW93]
Density, g/cm³: 2.51 (15°C) [HAW93]
Melting Point, °C: ~1600 [MER06]
Reactions: may be heated up to 1250°C in
H_2 without decomposition [KIR82]

689
Compound: Calcium phosphite monohydrate
Formula: $CaHPO_3 \cdot H_2O$
Molecular Formula: CaH_3O_4P
Molecular Weight: 138.073
CAS RN: 21056-98-4
Properties: cryst; used in fertilizers and
polymerization catalysts [MER06]
Solubility: sl s H_2O; i alcohol [MER06]
Reactions: minus H_2O at 200°C,
decomposes at >300°C [MER06]

690
Compound: Calcium phosphonate monohydrate
Formula: $CaHPO_3 \cdot H_2O$
Molecular Formula: CaH_3O_4P
Molecular Weight: 138.073
CAS RN: 25232-60-4
Properties: col monocl cryst [CRC10]

Solubility: sl H_2O; i EtOH [CRC10]
Melting Point, °C: decomposes at 150 [CRC10]

691
Compound: Calcium plumbate
Formula: Ca_2PbO_4
Molecular Formula: Ca_2O_4Pb
Molecular Weight: 351.354
CAS RN: 12013-69-3
Properties: orange to brown cryst powd; used as an oxidizing agent, in safety matches, and storage batteries [HAW93]
Solubility: i H_2O, decomposed by hot H_2O; s acids, with decomposition [HAW93]
Density, g/cm³: 5.71 [HAW93]
Melting Point, °C: decomposes [CRC10]

692
Compound: Calcium propionate
Synonyms: propionic acid, calcium salt
Formula: $(C_2H_5COO)_2Ca$
Molecular Formula: $C_6H_{10}CaO_4$
Molecular Weight: 186.221
CAS RN: 4075-81-4
Properties: white powd or monocl cryst; uses: mold-retardant additive for bread, tobacco, pharmaceuticals, antifungal agent [HAW93]
Solubility: s H_2O, sl s methanol, ethanol [MER06]

693
Compound: Calcium pyrophosphate
Synonym: calcium diphosphate
Formula: $Ca_2P_2O_7$
Molecular Formula: $Ca_2O_7P_2$
Molecular Weight: 254.099
CAS RN: 7790-76-3
Properties: 99.95% pure 6–8 μm; white, polymorphous cryst or powd; can be made by igniting $CaHPO_4$; used as a polishing agent in dentrifices, as a mild abrasive to polish metals [HAW93] [MER06] [STR93]
Solubility: i H_2O; s dil HCl, HNO_3 [MER06]
Density, g/cm³: 3.09 [MER06]
Melting Point, °C: 1230 [STR93]

694
Compound: Calcium selenate dihydrate
Formula: $CaSeO_4 \cdot 2H_2O$
Molecular Formula: CaH_4O_6Se
Molecular Weight: 219.066
CAS RN: 7790-74-1

Properties: colorless; monocl powd; used as a general pesticide [HAW93] [CRC10]
Solubility: g/100 g H_2O: 9.73 (0°C), 9.22 (20°C), 7.14 (40°C) [LAN05]
Density, g/cm³: 2.7 [HAW93]

695
Compound: Calcium selenide
Formula: CaSe
Molecular Formula: CaSe
Molecular Weight: 119.038
CAS RN: 1305-84-6
Properties: −20 mesh 99.5% pure; white powd; in air may turn red within a few minutes, and light brown in a few hours; obtained by reduction of $CaSeO_4$ with H_2 at 400°C–500°C; used in electron emitter devices [CER91] [MER06]
Solubility: decomposed by H_2O; releases H_2Se gas and forms red Se in HCl [MER06]
Density, g/cm³: 3.82 [MER06]

696
Compound: Calcium silicate
Synonym: wollastonite
Formula: β-$CaSiO_3$
Molecular Formula: CaO_3Si
Molecular Weight: 116.162
CAS RN: 1344-95-2
Properties: −200 mesh 99% white powd; used as an absorbent, antacid, as a filler paper, and for paper coatings [HAW93] [ALD93]
Solubility: i H_2O [HAW93]
Density, g/cm³: 2.9 [HAW93]
Reactions: transition to pseudowollastonite at 1200°C [ROB78]

697
Compound: Calcium silicide
Formula: CaSi
Molecular Formula: CaSi
Molecular Weight: 68.164
CAS RN: 12013-55-7
Properties: ortho cryst [CRC10]
Density, g/cm³: 2.39 [CRC10]
Boiling Point, °C: 1324 [CRC10]

698
Compound: Calcium silicide
Formula: $CaSi_2$
Molecular Formula: $CaSi_2$
Molecular Weight: 96.249

CAS RN: 12013-56-8
Properties: 3 mm pieces and smaller −140 mesh of 99.5% purity; powd [STR93] [CER91]
Solubility: i cold H_2O, decomposed by hot H_2O; s acids, alkalies [HAW93]
Density, g/cm³: 2.5 [HAW93]
Melting Point, °C: 1000 [ALF93]

699
Compound: Calcium stannate trihydrate
Formula: $CaSnO_3 \cdot 3H_2O$
Molecular Formula: CaH_6O_6Sn
Molecular Weight: 260.832
CAS RN: 12013-46-6
Properties: white, cryst powd; used as an additive for ceramic capacitors, in the production of ceramic colors; anhydrous is −325 mesh, 5 μm or less 99% pure [HAW93] [CER91]
Solubility: i H_2O [HAW93]
Reactions: minus $3H_2O$ ~350°C [HAW93]

700
Compound: Calcium stearate
Synonyms: octadecanoic acid, calcium salt
Formula: $Ca[CH_3(CH_2)_{16}COO]_2$
Molecular Formula: $C_{36}H_{70}CaO_4$
Molecular Weight: 607.028
CAS RN: 1592-23-0
Properties: white, granular, fatty powd; used as a water repellent; flattening agent in paints [HAW93] [MER06]
Solubility: i H_2O, ether, chloroform; sl s hot mineral oils [MER06]
Density, g/cm³: 1.12 [KIR78]
Melting Point, °C: 147–149 [MER06]

701
Compound: Calcium succinate trihydrate
Synonyms: butanedioic acid, calcium salt trihydrate
Formula: $Ca(OOCCH_2CH_2COO)_3 \cdot H_2O$
Molecular Formula: $C_4H_{10}CaO_7$
Molecular Weight: 210.197
CAS RN: 140-99-8
Properties: needles or granules [MER06]
Solubility: g/100 g H_2O: 1.127 (0°C), 1.28 (20°C), 0.66 (100°C) [LAN05]; i alcohol; s dil acids [MER06]

702
Compound: Calcium sulfate
Synonym: anhydrite
Formula: $CaSO_4$

Molecular Formula: CaO_4S
Molecular Weight: 136.142
CAS RN: 7778-18-9
Properties: ortho-rhomb; various colors; odorless; white with blue, gray, or red tinge; hardness 3–3.5 Mohs; enthalpy of fusion 28.03 kJ/mol; used in cement formulations and as a paper filler [MER06] [CRC10]
Solubility: g/100 g soln, H_2O: 0.63 (25°C), 0.151 (100°C); solid phase, $CaSO_4$ [KRU93]
Density, g/cm³: 2.96 [MER06]
Melting Point, °C: 1450 [CRC10]

703
Compound: Calcium sulfate dihydrate
Synonym: gypsum
Formula: $CaSO_4 \cdot 2H_2O$
Molecular Formula: CaH_4O_6S
Molecular Weight: 172.172
CAS RN: 10101-41-4
Properties: monocl; hardness 1.5–2.0 Mohs; lumps or white powd; used in manufacturing portland cement, plaster of paris and artificial marble [KIR78] [MER06] [STR93]
Solubility: g/100 mL soln, H_2O: 0.1759 (0°C), 0.2080 (25°C), 0.1619 (100°C) [KRU93]
Density, g/cm³: 2.32 [KIR79]
Reactions: minus 1.5 H_2O at 128°C, minus $2H_2O$ at 163°C [KIR78]
Thermal Expansion Coefficient: (volume) 100°C (0.58) [CLA66]

704
Compound: Calcium sulfate hemihydrate
Synonym: plaster of paris
Formula: $CaSO_4 \cdot 1/2H_2O$
Molecular Formula: $CaHO_{4.5}S$
Molecular Weight: 145.145
CAS RN: 10034-76-1
Properties: odorless and tasteless fine powd; hygr; uses: wall plasters, wallboards, and tiles [MER06] [ALD94]
Solubility: sets to hard mass when mixed with H_2O [MER06]; g/100 g soln, H_2O: 1.23 (0°C), 0.71 (25°C), 0.189 (100°C); solid phase, $CaSO_4 \cdot 1/2H_2O$ [KRU93]
Reactions: minus $1/2H_2O$ at 163°C [KIR78]

705
Compound: Calcium sulfide
Synonym: oldhamite
Formula: CaS
Molecular Formula: CaS
Molecular Weight: 72.144
CAS RN: 20548-54-3

Properties: −325 mesh, 10 μm or less, 99.99% and
99% pure; white powd, if pure; else may be
yellowish to a pale gray; odor of H_2S in moist
air; unpleasant alkaline taste; oxidizes in dry air
and decomposes in moist air; can be obtained by
reacting $CaCO_3$, H_2S, and H_2 at 1000°C [MER06]
Solubility: sl s cold H_2O, more s hot H_2O with
partial decomposition [MER06]
Density, g/cm³: 2.59 [MER06]
Melting Point, °C: >2000 [MER06]

706
Compound: Calcium sulfite dihydrate
Formula: $CaSO_3 \cdot 2H_2O$
Molecular Formula: CaH_4O_5S
Molecular Weight: 156.173
CAS RN: 10257-55-3
Properties: hex; white cryst or powd; slowly oxidizes in
air to $CaSO_4$; used in brewing, as a disinfectant in
sugar manufacturing [HAW93] [MER06] [CRC10]
Solubility: 0.0043 g/100 mL (18°C), 0.001 g/100 mL
(100°C) H_2O; sl s alcohol; s in acid solutions
with SO_2 evolution [MER06] [CRC10]
Reactions: minus $2H_2O$ at 100°C [HAW93]

707
Compound: Calcium tartrate tetrahydrate
Synonyms: 2,3-dihydroxybutanedioic acid,
calcium salt tetrahydrate
Formula: $CaC_4H_4O_6 \cdot 4H_2O$
Molecular Formula: $C_4H_{12}CaO_{10}$
Molecular Weight: 260.211
CAS RN: 3164-34-9
Properties: rhomb; white cryst; uses: food
preservative, antacid [HAW93] [CRC10]
Solubility: g/100 g H_2O: 0.026 (0°C), 0.034 (20°C),
0.130 (80°C) [LAN05]; s dil HCl, HNO_3 [MER06]
Melting Point, °C: decomposes [CRC10]

708
Compound: Calcium telluride
Formula: CaTe
Molecular Formula: CaTe
Molecular Weight: 167.678
CAS RN: 12013-57-9
Properties: cub; −20 mesh 99.5%
purity [CER91] [CRC10]
Density, g/cm³: 4.873 [CRC10]

709
Compound: Calcium tetrahydroaluminate
Formula: $Ca(AlH_4)_2$

Molecular Formula: Al_2CaH_8
Molecular Weight: 102.105
CAS RN: 16941-10-9
Properties: gray powd; flammable [CRC10]
Solubility: reac H_2O; s THF; i eth, bz [CRC10]

710
Compound: Calcium thiocyanate tetrahydrate
Formula: $Ca(SCN)_2 \cdot 4H_2O$
Molecular Formula: $C_2H_8CaN_2O_4S_2$
Molecular Weight: 228.307
CAS RN: 2092-16-2
Properties: hygr cryst or powd [MER06]
Solubility: v s H_2O; s methanol,
ethanol, acetone [MER06]
Reactions: decomposes if heated above 160°C [MER06]

711
Compound: Calcium thioglycollate trihydrate
Synonyms: mercaptoacetic acid, calcium salt trihydrate
Formula: $Ca(HSCH_2COO)_2 \cdot 3H_2O$
Molecular Formula: $C_4H_{12}CaO_7S_2$
Molecular Weight: 276.345
CAS RN: 814-71-1
Properties: white powd or prismatic rod cryst; odorless
or faint mercaptan odor; used in depilatories and
in hair wave preparations [HAW93] [MER06]
Solubility: s H_2O [MER06]
Reactions: slowly loses H_2O above 95°C;
darkens at 220°C; partially fuses with
decomposition at 280°C–290°C [MER06]

712
Compound: Calcium thiosulfate hexahydrate
Synonym: calcium hyposulfite hexahydrate
Formula: $CaS_2O_3 \cdot 6H_2O$
Molecular Formula: $CaH_{12}O_9S_2$
Molecular Weight: 260.300
CAS RN: 10124-41-1
Properties: tricl cryst; when dry, decomposes on standing
to form a yellow crust on the surface; spontaneously
decomposed at 43°C–49°C; used to treat dermatitis
and jaundice caused by arsphenamine [MER06]
Solubility: 100 g/100 mL H_2O (3°C), decomposed
by hot water; i alcohol [MER06] [CRC10]
Density, g/cm³: 1.87 [MER06]
Melting Point, °C: decomposes [CRC10]

713
Compound: Calcium titanate
Synonym: perovskite

Formula: $CaTiO_3$
Molecular Formula: CaO_3Ti
Molecular Weight: 135.956
CAS RN: 12049-50-2
Properties: -150, $+325$ mesh 99% pure; occurs naturally as the mineral perovskite; can be made by heating CaO and TiO_2 to 1350°C; used as an additive to $BaTiO_3$ [KIR83] [CER91]
Density, g/cm³: 4.10 [STR93]
Melting Point, °C: 1975 [STR93]

714
Compound: Calcium tungstate
Synonym: scheelite
Formula: $CaWO_4$
Molecular Formula: CaO_4W
Molecular Weight: 287.916
CAS RN: 7790-75-2
Properties: -325 mesh, 10 μm or less 99.9% pure; occurs in nature as mineral scheelite; white tetr powd, a$=0.524$ nm, c$=1.138$ nm; can be prepared by heating tungstic acid and CaO or $CaCO_3$; used in tumor treatment and in luminous paint [KIR83] [STR93] [MER06]
Solubility: i H_2O; decomposed by hot HCl, HNO_3 [MER06]
Density, g/cm³: 6.062 [STR93]
Melting Point, °C: 1620 [STR93]

715
Compound: Calcium vanadate
Formula: CaV_2O_6
Molecular Formula: CaO_6V_2
Molecular Weight: 237.957
CAS RN: 12135-52-3
Properties: -325 mesh, 10 μm or less, 99.9% pure [CER91]

716
Compound: Calcium zirconate
Formula: $CaZrO_3$
Molecular Formula: CaO_3Zr
Molecular Weight: 179.300
CAS RN: 12013-47-7
Properties: colorless, monocl; -100, $+200$ mesh, other sizes, 99% pure; white powd [STR93] [CER91] [CRC10]
Solubility: s HNO_3 [HAW93]
Density, g/cm³: 4.78 [STR93]
Melting Point, °C: 2550 [STR93]

717
Compound: Californium
Formula: Cf
Molecular Formula: Cf
Molecular Weight: 251
CAS RN: 7440-71-3
Properties: α-form: hex, a$=0.339$ nm, c$=1.101$ nm; β: fcc, a$=0.494$ nm; γ: fcc, a$=0.575$ nm; ionic radius of Cf^{+++} is 0.0934 nm, of Cf^{+++} is 0.0851 nm; discovered in 1950; ^{252}Cf is an intense neutron source, 1 g emits $2.4 \times 10^{+12}$ neutrons per sec; has application in neutron activation analysis and field use in mineral prospecting and oil-well logging, potential use in medical applications [KIR78]
Density, g/cm³: all at 25°C: α: 15.1; β: 13.7; γ: 8.70 [KIR78]
Melting Point, °C: 900 [KIR91]

718
Compound: Carbon
Synonym: fullerene
Formula: C_{60}
Molecular Formula: C_{60}
Molecular Weight: 720.642
CAS RN: 99685-96-8
Properties: yellow needles or plates [CRC10]
Solubility: s os [CRC10]
Melting Point, °C: >280 [CRC10]

719
Compound: Carbon
Synonym: graphite bromide
Formula: C_8Br
Molecular Formula: C_8Br
Molecular Weight: 175.992
CAS RN: 12079-58-2
Properties: -100 mesh 99.9% pure [CER91]

720
Compound: Carbon
Synonym: graphite oxide
Formula: $C_7O_2(H_2)$
Molecular Formula: $C_7H_2O_2$
Molecular Weight: 118.092
CAS RN: 1399-57-1
Properties: formula also given as $C_4O(OH)$; light yellow flakes or plates; preparation: oxidation of graphite with KNO_3 in nitric and sulfuric acids; uses: rocket propellant mixtures; membranes in the desalination of seawater by reverse osmosis [HAW93] [MER06]

721
Compound: Carbon
Synonym: graphite
Formula: C
Molecular Formula: C
Molecular Weight: 12.011
CAS RN: 7782-42-5
Properties: hex; soft, slippery feel; steel gray to black color with metallic sheen; electrical resistivity (20°C) 1375 μohm · cm; tensile strength 400–2000 psi; compressive strength ~2000–8000 psi; coefficient of friction 0.1 μ; enthalpy of fusion 104.6 kJ/mol; enthalpy of vaporization 711 kJ/mol; semiconductor: band gap 5.47 eV (300 K); mobility (300 K) cm^2/(V · s), 1800 electron, 1200 hole; effective mass: 0.2 electron, 0.25 hole [KIR78] [HAW93] [COT88] [ALD94] [CRC10]
Density, g/cm^3: 2.0–2.25 [HAW93]
Melting Point, °C: sublimes at 3650 [HAW93]
Thermal Conductivity, W/(m · K): 119–165 (25°C) [ALD94]; 13.4 (500°C), 9.9 (1000°C) [KIR80]
Thermal Expansion Coefficient: (linear) to 1000°C is 4.0×10^{-6}/°C [KIR80]

722
Compound: Carbon
Synonym: graphite fluoride
Formula: $(CF_x)_n$
Molecular Formula: x = 0.8–1.5
CAS RN: 11113-63-6
Properties: −200 mesh, 99.9% pure; polymer [ALD94] [CER91]

723
Compound: Carbon
Synonym: diamond
Formula: C
Molecular Formula: C
Molecular Weight: 12.011
CAS RN: 7782-40-3
Properties: cryst modification of carbon; fcc, a = 3.56683–3.56725 nm; specific heat 6.184 J/(mol · K); hardness 10 Mohs; resistivity, 20°C, >10^{+16} ohm · cm (Type I, most Type IIa), 10–10^{+3} (Type IIb); dielectric constant (27°C, 0–3 kHz) 5.58; Type I: diamonds containing 0.1%–0.2% nitrogen; Type IIa, free of nitrogen; Type IIb very pure, generally blue in color; obtained by mining; uses: jewelry, polishing, grinding [KIR78] [MER06]
Density, g/cm^3: 3.51524 [KIR78]
Melting Point, °C: >3550 [COT88]
Boiling Point, °C: 4827 [COT88]

Reactions: diamond to graphite transition >1500°C in absence of air [KIR78]
Thermal Conductivity, W/(m · K): 20°C: Type I, 900; Type IIa, 2600 [KIR78]
Thermal Expansion Coefficient: 20°C: 0.8×10^{-6}; −100°C: 0.4×10^{-6}; 100°C–900°C: $(1.5 \text{ to } 4.8) \times 10^{-6}$ [KIR78]

724
Compound: Carbon (amorphous)
Synonym: carbon black
Formula: C
Molecular Formula: C
Molecular Weight: 12.011
CAS RN: 7440-44-0
Properties: a quasi graphitic form of carbon of small particle size [MER06]

725
Compound: Carbon dioxide
Synonym: carbonic acid anhydride
Formula: CO_2
Molecular Formula: CO_2
Molecular Weight: 44.010
CAS RN: 124-38-9
Properties: colorless, odorless, noncombustible gas; faint acid taste; colorless, odorless volatile liq; white, snow-like flakes or cubes in the solid form; critical temp 31.3°C; critical pressure 7.38 MPa; enthalpy of vaporization 25.21 kJ/mol; enthalpy of fusion 9.02 kJ/mol [HAW93] [MER06] [AIR87] [CRC10]
Solubility: mL CO_2/100 mL H_2O at 760 mm: 171 (0°C), 88 (20°C), 36 (60°C) [MER06]
Density, g/cm^3: gas: 1.527 (air = 1) [MER06]
Melting Point, °C: −56.6 (5.2 atm) [MER06]
Boiling Point, °C: sublimes at −78.5 [MER06]]

726
Compound: Carbon diselenide
Formula: CSe_2
Molecular Formula: CSe_2
Molecular Weight: 169.931
CAS RN: 506-80-9
Properties: light sensitive, golden yellow, strongly refractive liq; odor of rotten radishes; turns brown to black on storage [MER06]
Solubility: i H_2O; miscible with CCl_4, CS_2, toluene [MER06]
Density, g/cm^3: 2.6824 [MER06]
Melting Point, °C: −40 to −45 [KIR82]
Boiling Point, °C: 125–126 [MER06]

727
Compound: Carbon disulfide
Formula: CS_2
Molecular Formula: CS_2
Molecular Weight: 76.143
CAS RN: 75-15-0
Properties: clear, colorless or faintly yellow liq; refractive, mobile, flammable; decomposes on standing for a long time; burns with blue flame to CO_2 and SO_2; enthalpy of fusion 4.39 kJ/ mol; enthalpy of vaporization at bp 26.74 kJ/ mol, 27.51 at 25°C; refractive index 1.6232; flash point −30°C; autoignition temp 100°C; used as a solvent, in the manufacture of viscose rayon, cellophane [HAW93] [CRC10] [MER06] [CIC73]
Solubility: g/100 g H_2O: 0.204 (0°C), 0.179 (20°C), 0.111 (40°C) [LAN05]; s alcohol, benzene, ether [HAW93]
Density, g/cm³: liq: 1.2632 (20°C); vapor: 2.67 [MER06]
Melting Point, °C: −111.6 [MER06]
Boiling Point, °C: 46 [COT88]

728
Compound: Carbon fluoride
Formula: C_4F
Molecular Formula: C_4F
Molecular Weight: 67.042
CAS RN: 12774-81-1
Properties: solid, nonconductor formed on carbon anodes when molten KF–HF mixtures are oxidized at carbon electrodes to generate fluorine [HAW93]
Melting Point, °C: decomposes at >60 [HAW93]

729
Compound: Carbon fullerenes
Synonym: (5,6) fullerene
Formula: C_{70}
Molecular Formula: C_{70}
Molecular Weight: 840.770
CAS RN: 115383-22-7
Properties: black powd; the C_{70} fullerene has five cryst structures, depending on the temp; fcc, T>67°C, a = 1.501 nm [DRE93] [STR93] [ALD94]
Solubility: s benzene, toluene [LID94]
Melting Point, °C: >280 [LID94]

730
Compound: Carbon monoxide
Formula: CO
Molecular Formula: CO
Molecular Weight: 28.010

CAS RN: 630-08-0
Properties: highly poisonous, odorless, colorless, tasteless gas; very flammable, burns with bright blue flame; autoignition temp 609°C; enthalpy of fusion 0.83 kJ/mol; enthalpy of vaporization 6.04 kJ/mol; critical temp −140.21°C; critical pressure 34.529 atm; critical density 0.3010 g/cm³; triple point 205.0°C at 115.4 mm Hg; produced by partial oxidation of hydrocarbon gases; used as a reducing agent in metallurgy, e.g., for Ni [CIC73] [CRC10] [MER06] [AIR87]
Solubility: mL/100 mL H_2O: 3.3 (0°C), 2.3 (20°C) [MER06]
Density, g/cm³: gas: 0.968 (air = 1.000) [MER06]
Melting Point, °C: −205.0 [MER06]
Boiling Point, °C: −191.5 [MER06]

731
Compound: Carbon oxyselenide
Synonym: carbonyl selenide
Formula: COSe
Molecular Formula: COSe
Molecular Weight: 106.970
CAS RN: 1603-84-5
Properties: colorless gas; light sensitive; unstable [KIR82] [CRC10]
Solubility: decomposed by H_2O [CRC10]
Density, g/cm³: 4.694 g/L [LID94]
Melting Point, °C: −124.4 [CRC10]
Boiling Point, °C: −21.7 [CRC10]

732
Compound: Carbon oxysulfide
Synonym: carbonyl sulfide
Formula: COS
Molecular Formula: COS
Molecular Weight: 60.076
CAS RN: 463-58-1
Properties: colorless gas with sulfide odor unless pure; flammable [HAW93] [COT88] [ALD94]
Solubility: mL/100 mL H_2O: 133.3 (0°C), 56.1 (20°C), 40.3 (30°C) [LAN05]; slowly decomposes in H_2O [COT88]; s alcohol [HAW93]
Density, g/cm³: 2.636 g/L [LID94]
Melting Point, °C: −138.8 [HAW93]
Boiling Point, °C: −50.2 [HAW93]

733
Compound: Carbon soot
Formula: Cx
Molecular Formula: C
CAS RN: 1333-86-4

Properties: black powd; contains 2%–20%
C_{60}/C_{70}; preparation: from resistive heating
of graphite, 5%–10% yield; uses: precursor
to the fullerenes [STR93] [ALD94]

734
Compound: Carbon suboxide
Synonym: l,2-propadiene-l,3-dione
Formula: C_3O_2
Molecular Formula: C_3O_2
Molecular Weight: 68.032
CAS RN: 504-64-3
Properties: colorless, highly refractive liq or
colorless gas, which burns with a blue sooty
flame; odor like acrolein or mustard oil;
structure: O=C=C=C=O [MER06]
Solubility: forms malonic acid with H_2O [MER06]
Density, g/cm³: 2.985 g/L [LID94]
Melting Point, °C: −111.3 [MER06]
Boiling Point, °C: 6.8 [MER06]

735
Compound: Carbon subsulfide
Formula: C_3S_2
Molecular Formula: C_3S_2
Molecular Weight: 100.162
CAS RN: 627-34-9
Properties: red liq [CRC10]
Solubility: reac H_2O [CRC10]
Density, g/cm³: 1.27 [CRC10]
Melting Point, °C: −1 [CRC10]
Boiling Point: decomposes at 90 [CRC10]

736
Compound: Carbon sulfide selenide
Formula: CSSe
Molecular Formula: CSSe
Molecular Weight: 123.037
CAS RN: 5951-19-9
Properties: yellow, oily liq; unstable,
sensitive to light [KIR82] [CRC10]
Solubility: i H_2O [CRC10]
Density, g/cm³: 1.9874 [CRC10]
Melting Point, °C: −85 [CRC10]
Boiling Point, °C: 84.5 [CRC10]

737
Compound: Carbon sulfide telluride
Synonym: carbon sulfotelluride
Formula: CSTe
Molecular Formula: CSTe
Molecular Weight: 171.677

CAS RN: 10340-06-4
Properties: reddish yellow liq; odor of garlic;
decomposed by light even at −50°C
forming CS_2 and Te [KIR83]
Density, g/cm³: 2.9 [CRC10]
Melting Point, °C: −54 [CRC10]
Boiling Point, °C: decomposes [CRC10]

738
Compound: Carbon tetrabromide
Synonym: tetrabromomethane
Formula: CBr_4
Molecular Formula: CBr_4
Molecular Weight: 331.627
CAS RN: 558-13-4
Properties: colorless cryst; slight decomposition
if boiled [HAW93] [COT88]
Solubility: i H_2O; s alcohol, ether, chloroform [HAW93]
Density, g/cm³: 3.42 [HAW93]
Melting Point, °C: 90.1 [HAW93]
Boiling Point, °C: 189.5 [HAW93]

739
Compound: Carbon tetrachloride
Synonym: tetrachloromethane
Formula: CCl_4
Molecular Formula: CCl_4
Molecular Weight: 153.822
CAS RN: 56-23-5
Properties: colorless, clear, nonflammable, heavy
liq; sweetish odor; refractive index 1.4607;
vapor pressure 91.3 mm Hg (20°C); enthalpy
of vaporization 29.82 kJ/mol; enthalpy of
fusion 3.28 kJ/mol [CRC10] [MER06]
Solubility: 1 mL dissolves in 2000 mL H_2O [MER06];
miscible with alcohol, ether [HAW93]
Density, g/cm³: 1.589 [MER06]
Melting Point, °C: −23 [MER06]
Boiling Point, °C: 76.8 [CRC10]

740
Compound: Carbon tetrafluoride
Synonyms: tetrafluoromethane, Freon 14
Formula: CF_4
Molecular Formula: CF_4
Molecular Weight: 88.003
CAS RN: 75-73-0
Properties: colorless, odorless gas; thermally stable;
chemically very inert; critical temp −45.7°C;
critical pressure 3.74 MPa; enthalpy of vaporization
11.98 kJ/mol; obtained by reaction of C or CO and
F_2; used as a gaseous insulator and in electronics
production [HAW93] [MER06] [AIR87]

Solubility: mL/100 mL in H_2O: 0.595 (10°C),
  0.490 (20°C), 0.366 (40°C) [LAN05]
Density, g/cm³: solid (−195°C), 1.98; liq
  (−183°C), 1.89 [MER06]
Melting Point, °C: −183.6 [MER06]
Boiling Point, °C: −127.8 [MER06]

741
Compound: Carbon tetraiodide
Synonym: tetraiodomethane
Formula: CI_4
Molecular Formula: CI_4
Molecular Weight: 519.629
CAS RN: 507-25-5
Properties: red cub cryst; odor of iodine;
  decomposes to iodine and tetraiodoethylene
  under influence of heat or light [MER06]
Solubility: sl s H_2O with hydrolysis; s
  benzene, chloroform [MER06]
Density, g/cm³: 4.32 [MER06]
Melting Point, °C: 171 [MER06]
Boiling Point, °C: decomposes [COT88]
Reactions: can be sublimed at low pressure [COT88]

742
Compound: Carbonyl bromide
Synonym: bromophosgene
Formula: $COBr_2$
Molecular Formula: CBr_2O
Molecular Weight: 187.818
CAS RN: 593-95-3
Properties: heavy, colorless liq with a strong odor;
  fumes in air; decomposed by light and heat;
  used in making cryst-violet type coloring
  agents and as a poison gas [HAW93]
Solubility: hydrolyzes in H_2O to form
  CO_2 and HBr [COT88]
Density, g/cm³: 2.5 (~15°C) [HAW93]
Melting Point, °C: 65 [COT88]

743
Compound: Carbonyl chloride
Synonyms: phosgene, carbon oxychloride
Formula: $COCl_2$
Molecular Formula: CCl_2O
Molecular Weight: 98.910
CAS RN: 75-44-5
Properties: colorless to light yellow gas; hay-like
  odor in small concentrations; enthalpy
  of fusion 5.74 kJ/mol; used in organic
  synsthesis for isocyanates, polyurethane, and
  polycarbonate resins [HAW93] [CRC10]

Solubility: sl s and hydrolyzed in H_2O;
  s benzene, toluene [HAW93]
Density, g/cm³: 4.34 g/L [LID94]
Melting Point, °C: −127.9 [CRC10]
Boiling Point, °C: 8.2 [HAW93]

744
Compound: Carbonyl fluoride
Synonym: fluorophosgene
Formula: COF_2
Molecular Formula: CF_2O
Molecular Weight: 66.007
CAS RN: 353-50-4
Properties: pungent, very hygr gas [MER06]
Solubility: instantly hydrolyzed by H_2O [MER06]
Density, g/cm³: solid: (−190°C), 1.388;
  liq: (−114°C), 1.139 [MER06]
Melting Point, °C: −114.0 [MER06]
Boiling Point, °C: −83.1 [MER06]

745
Compound: Ceric ammonium nitrate
Formula: $(NH_4)_2Ce(NO_3)_6$
Molecular Formula: $CeH_8N_8O_{18}$
Molecular Weight: 548.223
CAS RN: 16774-21-3
Properties: red-orange cryst [CRC10]
Solubility: v s H_2O

746
Compound: Ceric ammonium sulfate dihydrate
Formula: $(NH_4)_4Ce(SO_4)_4 \cdot 2H_2O$
Molecular Formula: $CeH_{20}N_4O_{18}S_4$
Molecular Weight: 632.551
CAS RN: 10378-47-9
Properties: cryst powd [CRC10]
Melting Point, °C: decomposes at 450 [CRC10]

747
Compound: Ceric basic nitrate trihydrate
Formula: $Ce(OH)(NO_3)_3 \cdot 3H_2O$
Molecular Formula: $CeH_7N_3O_{13}$
Molecular Weight: 397.183
Properties: long, red needle; prepartion:
  evaporating a solution of ceric hydroxide in
  nitric acid solution [KIR78] [CRC10]
Solubility: s H_2O [CRC10]

748
Compound: Ceric fluoride
Synonym: cerium(IV) fluoride

Formula: CeF_4
Molecular Formula: CeF_4
Molecular Weight: 216.109
CAS RN: 10060-10-3
Properties: white powd; hygr; minute cryst; thermally stable below 550°C; monocl, a = 1.2587 nm, c = 0.8041 nm; preparation: reaction of F_2 with CeF_3; uses: fluorinating agent [STR93] [MER06] [CRC10]
Solubility: i H_2O, very slowly hydrolyzed by cold H_2O [MER06]
Density, g/cm³: 4.77 [MER06]
Melting Point, °C: decomposes at 650 [STR93]

749
Compound: Ceric hydroxide
Synonym: cerium(IV) hydroxide
Formula: $Ce(OH)_4$
Molecular Formula: CeH_4O_4
Molecular Weight: 208.145
CAS RN: 12014-56-1
Properties: addition of NaOH or NH_4OH to a solution of Ce^{++++} results in a gelatinous precipitate of $CeO_2 \cdot xH_2O$ (x = 0.5–2); yellowish white powd when dried; granular $Ce(OH)_4$ obtained by boiling insoluble Ce^{++++} salt with NaOH [STR93] [KIR78]
Solubility: i H_2O, s conc acids [HAW93]

750
Compound: Ceric oxide
Synonym: cerianite
Formula: CeO_2
Molecular Formula: CeO_2
Molecular Weight: 172.114
CAS RN: 1306-38-3
Properties: brownish white powd or cub cryst, but usually pale yellow; refractory material; a = 0.54110 nm; used in ceramics, as an abrasive for polishing glass, as evaporated material of 99.9% purity in high index film for dielectric beam splitters, interference filters, and in multilayers as antireflection coating; can be prepared by calcining cerous oxalate or hydroxide [KIR78] [HAW93] [TAY85] [MER06] [CER91]
Solubility: i H_2O; s H_2SO_4, HNO_3; i dil acid [HAW93]
Density, g/cm³: 7.65 [HAW93]
Melting Point, °C: 2400 [KNA91]
Thermal Expansion Coefficient: from 25°C to: 100°C (0.24), 200°C (0.54), 400°C (1.20), 600°C (1.92), 800°C (2.70), 1000°C (3.51), 1200°C (4.38) [TAY85]

751
Compound: Ceric oxide hydrate
Synonym: cerium dioxide hydrate
Formula: $CeO_2 \cdot xH_2O$
Molecular Formula: CeO_2 (anhydrous)
Molecular Weight: 172.114 (anhydrous)
CAS RN: 12014-56-1
Properties: white powd; used to produce cerium salts, as an opacifier to impart a yellow color to glasses and enamels [HAW93]
Solubility: i H_2O; s conc mineral acids [HAW93]

752
Compound: Ceric sulfate tetrahydrate
Synonyms: sulfuric acid, cerium(IV) salt tetrahydrate
Formula: $Ce(SO_4)_2 \cdot 4H_2O$
Molecular Formula: $CeH_8O_{12}S_2$
Molecular Weight: 404.304
CAS RN: 10294-42-5
Properties: yellow to orange powd or ortho-rhomb cryst; oxidizer [MER06]
Solubility: s small quantity of H_2O, but decomposes in excess H_2O [MER06]; s dil H_2SO_4 [HAW93]
Density, g/cm³: 3.91 [HAW93]
Melting Point, °C: decomposes above 350 [MER06]
Reactions: minus $4H_2O$ at 180°C–200°C [MER06]

753
Compound: Ceric titanate
Synonym: cerium(IV) titanate
Formula: $CeTiO_4$
Molecular Formula: CeO_4Ti
Molecular Weight: 251.980
CAS RN: 52014-82-1
Properties: −325 mesh 10 µm or less at 99.9% purity [CER91]

754
Compound: Ceric vanadate
Synonym: cerium(IV) vanadate
Formula: $CeVO_4$
Molecular Formula: CeO_4V
Molecular Weight: 255.055
CAS RN: 13597-19-8
Properties: −200 mesh with 99.9% purity [CER91]

755
Compound: Ceric zirconate
Synonym: cerium(IV) zirconate
Formula: $CeZrO_4$

Molecular Formula: CeO$_4$Zr
Molecular Weight: 295.337
CAS RN: 53169-24-7
Properties: −325 mesh with 99.5% purity;
pyrochlore type structure [TAY88a] [CER91]
Thermal Expansion Coefficient: from 25°C
to: 100°C (0.24), 200°C (0.57), 400°C
(1.23), 600°C (1.92); 800°C (2.58); 1000°C
(3.24); 1200°C (3.90) [TAY88a]

756

Compound: Cerium
Formula: Ce
Molecular Formula: Ce
Molecular Weight: 140.115
CAS RN: 7440-45-1
Properties: gray metal; α-Ce, fcc; β-Ce, hex; γ-Ce, fcc;
δ-Ce, bcc; for γ: heat capacity 26.96 J/(mol·K);
compressibility 4.18×10^{-2} GPa; Young's modulus
30 GPa; shear modulus 12 GPa; Poisson's ratio,
0.248; Vicker's hardness 235 MPa; yield strength
91.2 MPa; enthalpy of fusion 5.46 kJ/mol; enthalpy of
sublimation 422.6 kJ/mol; electrical reistivity, 20°C,
73 μohm·cm; radius of atom 0.1824 nm; radius of
ion 0.1034 nm for Ce^{+++} [KIR82] [CRC10] [ALD94]
Solubility: s dil mineral acids [KIR78]
Density, g/cm^3: hex 6.689, cub 6.773 [CRC10], [KIR78]
Melting Point, °C: 798 [KIR78]
Boiling Point, °C: 3433 [KIR82]
Reactions: reacts vigorously with the
halogens >200°C [KIR78]
Thermal Conductivity, W/(m·K): 11.3, 25°C [ALD94]
Thermal Expansion Coefficient: 8.5×10^6/°C [KIR78]

757

Compound: Cerium carbide
Formula: CeC$_2$
Molecular Formula: C$_2$Ce
Molecular Weight: 164.137
CAS RN: 12012-32-7
Properties: red, hex; 6 mm pieces and smaller
of 99.5% purity [CER91] [CRC10]
Solubility: decomposes in H$_2$O; s acids [CRC10]
Density, g/cm^3: 5.23 [CRC10]
Melting Point, °C: 2420 [KNA91]

758

Compound: Cerium carbide
Formula: Ce$_2$C$_3$
Molecular Formula: C$_3$Ce$_2$
Molecular Weight: 316.264
CAS RN: 12115-63-8

Properties: yellow-brown cub cryst [CRC10]
Density, g/cm^3: 2.84 [CRC10]
Melting Point, °C: 1505 [CRC10]

759

Compound: Cerium dihydride
Formula: CeH$_2$
Molecular Formula: CeH$_2$
Molecular Weight: 142.131
CAS RN: 13569-50-1
Properties: black pyrophoric solid; ignites
spontaneously in air; can be prepared by reacting
cerium and hydrogen at 345°C; used to store
H$_2$ in the system CeMg$_2$ [KIR80] [KIR78]
Solubility: reacts with H$_2$O at 0°C [KIR80]
Density, g/cm^3: 5.45 [LID94]
Melting Point, °C: ignites [CRC10]

760

Compound: Cerium hexaboride
Formula: CeB$_6$
Molecular Formula: B$_6$Ce
Molecular Weight: 204.981
CAS RN: 12008-02-5
Properties: refractory material; blue cub;
−325 mesh, 10 μm or less at 99.9%
purity [CRC10] [KIR78] [CER91]
Solubility: i H$_2$O, HCl [CRC10]
Density, g/cm^3: 4.87 [LID94]
Melting Point, °C: 2190 [CRC10]

761

Compound: Cerium monosulfide
Formula: CeS
Molecular Formula: CeS
Molecular Weight: 172.181
CAS RN: 12014-82-3
Properties: yellow cub; 3 mm pieces and smaller
(fused) 99.9% [CER91] [LID94]
Density, g/cm^3: 5.9 [LID94]
Melting Point, °C: 2445 [KNA91]

762

Compound: Cerium nitride
Formula: CeN
Molecular Formula: CeN
Molecular Weight: 154.122
CAS RN: 25764-08-3
Properties: −60 mesh, 99.9% pure; NaCl cryst
system; a=0.501 nm [CIC73] [CER91]
Density, g/cm^3: 7.89 [LID94]
Melting Point, °C: 2557 [KNA89]

763
Compound: Cerium oxysulfide
Formula: Ce_2O_2S
Molecular Formula: Ce_2O_2S
Molecular Weight: 344.295
CAS RN: 12442-45-4
Properties: −200 mesh with 99.9% purity [CER91]

764
Compound: Cerium silicide
Formula: $CeSi_2$
Molecular Formula: $CeSi_2$
Molecular Weight: 196.286
CAS RN: 12014-85-6
Properties: 6.35 mm and down pieces, 99.9% purity [CER91] [ALF93]
Solubility: i H_2O [CRC10]
Density, g/cm³: 5.67 [CRC10]
Melting Point, °C: 1620 [LID94]

765
Compound: Cerium stannate
Formula: $CeO_2 \cdot SnO_2$
Molecular Formula: CeO_4Sn
Molecular Weight: 322.822
CAS RN: 53169-23-6
Properties: −325 mesh, 10 μm average reacted product of 99.9% purity; pyrochlore type structure [CER91] [TAY88a]
Thermal Expansion Coefficient: from 25°C to: 100°C (0.21), 200°C (0.48), 400°C (1.05), 600°C (1.62), 800°C (2.16), 1000°C (2.73), 1200°C (3.30) [TAY88a]

766
Compound: Cerium trihydride
Formula: CeH_3
Molecular Formula: CeH_3
Molecular Weight: 143.139
CAS RN: 13864-02-3
Properties: dark, black, amorphous powd [CRC10]
Solubility: decomposes in H_2O [CRC10]

767
Compound: Cerous acetate hemitrihydrate
Synonym: cerium(III) acetate hemitrihydrate
Formula: $Ce(CH_3COO)_3 \cdot 1\text{-}1/2H_2O$
Molecular Formula: $C_6H_{12}CeO_{7.5}$
Molecular Weight: 344.272

CAS RN: 537-00-8
Properties: white powd [STR93]
Solubility: g/100 mL H_2O: 26.5 (15°C), 16.2 (75°C) [CRC10]
Boiling Point, °C: decomposes [CRC10]
Reactions: minus 1-1/2 H_2O at 115°C [CRC10]

768
Compound: Cerous acetylacetonate hydrate
Synonyms: 2,4-pentanedione, cerium(III) derivative
Formula: $Ce(CH_3COCH=C(O)CH_3)_3 \cdot xH_2O$
Molecular Formula: $C_{15}H_{21}CeO_6$ (anhydrous)
Molecular Weight: 437.443 (anhydrous)
CAS RN: 15653-01-7
Properties: tan powd; hygr [STR93] [ALD94]
Melting Point, °C: 131–132 [CRC10]

769
Compound: Cerous ammonium nitrate tetrahydrate
Formula: $(NH_4)_2Ce(NO_3)_5 \cdot 4H_2O$
Molecular Formula: $CeH_8N_7O_{19}$
Molecular Weight: 558.279
CAS RN: 13083-04-0
Properties: col monocl cryst [CRC10]
Solubility: v s H_2O [CRC10]
Melting Point, °C: 74 [CRC10]

770
Compound: Cerous ammonium sulfate tetrahydrate
Formula: $NH_4Ce(SO_4)_2 \cdot 4H_2O$
Molecular Formula: $CeH_8NO_{12}S_2$
Molecular Weight: 422.341
CAS RN: 21995-38-0 (anhydrous compound)
Properties: monocl cryst [CRC10]
Solubility: s H_2O

771
Compound: Cerous bromide
Synonym: cerium(III) bromide
Formula: $CeBr_3$
Molecular Formula: Br_3Ce
Molecular Weight: 379.827
CAS RN: 14457-87-5
Properties: orange powd; hygr [STR93]
Melting Point, °C: 730 [KNA91]
Boiling Point, °C: 1457 [KNA91]

772
Compound: Cerous bromide heptahydrate
Formula: $CeBr_3 \cdot 7H_2O$
Molecular Formula: $Br_3CeH_{14}O_7$

Molecular Weight: 505.924
CAS RN: 14457-87-5
Properties: colorless, deliq needles; anhydrous $CeBr_3$
−20 mesh at 99.9% purity [MER06] [CER91]
Solubility: s H_2O, alcohol [MER06]
Melting Point, °C: 732 [MER06]

773
Compound: Cerous carbonate
Synonym: cerium(III) carbonate
Formula: $Ce_2(CO_3)_3$
Molecular Formula: $C_3Ce_2O_9$
Molecular Weight: 460.258
CAS RN: 537-01-9
Properties: white powd; if a solution of the
carbonate in water is boiling, the product
can be $Ce(OH)(CO_3)$ [KIR78] [MER06]
Solubility: i H_2O, s mineral acids [HAW93]
Reactions: decomposes at 500°C to CeO_2
with evolution of CO, CO_2 [KIR78]

774
Compound: Cerous carbonate pentahydrate
Synonym: cerium(III) carbonate pentahydrate
Formula: $Cc_2(CO_3)_3 \cdot 5H_2O$
Molecular Formula: $C_3H_{10}Ce_2O_{14}$
Molecular Weight: 550.334
CAS RN: 72520-94-6
Properties: white cryst; pentahydrate can be obtained
by adding an alkali bicarbonate solution to a
solution of Ce^{+++} [KIR78] [STR93] [MER06]
Solubility: i H_2O, s dil acid [MER06]

775
Compound: Cerous chloride
Synonym: cerium(III) chloride
Formula: $CeCl_3$
Molecular Formula: $CeCl_3$
Molecular Weight: 246.473
CAS RN: 7790-86-5
Properties: −20 mesh with 99.9% purity;
white very fine powd; can be prepared
by dissolving cerium carbonate in HCl
[CER91] [KIR78] [STR93] [MER06]
Solubility: s H_2O, alcohol (exothermic) [MER06]
Density, g/cm³: 3.97 [MER06]
Melting Point, °C: 807 [KNA91]
Boiling Point, °C: 1725 (estimated) [KNA91]

776
Compound: Cerous chloride heptahydrate
Synonym: cerium(III) chloride heptahydrate

Formula: $CeCl_3 \cdot 7H_2O$
Molecular Formula: $CeCl_3H_{14}O_7$
Molecular Weight: 372.580
CAS RN: 18618-55-8
Properties: colorless to yellow, deliq, ortho-rhomb
cryst; can be prepared by evaporating
$CeCl_3$ solution or by saturating conc $CeCl_3$
solution with HCl [KIR78] [MER06]
Solubility: v s H_2O, alcohol [MER06]
Density, g/cm³: 3.92 [ALD94]
Reactions: minus H_2O >90°C, becomes
anhydrous by 230°C [MER06]

777
Compound: Cerous chloride hydrate
Synonym: cerium(III) chloride hydrate
Formula: $CeCl_3 \cdot xH_2O$
Molecular Formula: $CeCl_3$ (anhydrous)
Molecular Weight: 246.473 (anhydrous)
CAS RN: 19423-76-8
Properties: −4 mesh with 99.9% purity; white
or off-white cryst [STR93] [CER91]

778
Compound: Cerous fluoride
Synonym: cerium(III) fluoride
Formula: CeF_3
Molecular Formula: CeF_3
Molecular Weight: 197.110
CAS RN: 7758-88-5
Properties: white powd or 99.9% pure melted pieces of
3–6 mm; hygr; hex, a=0.71306 nm, c=0.72805 nm;
melted pieces used as evaporation material
and sputtering target for multilayers and thin
film capacitors [STR93] [GME76] [CER91]
Solubility: i H_2O, but slowly hydrolyzed [MER06]
Density, g/cm³: 6.157 [MER06]
Melting Point, °C: 1437 [KNA91]
Boiling Point, °C: 2280 (estimated) [KNA91]

779
Compound: Cerous hydroxide
Synonym: cerium(III) hydroxide
Formula: $Ce(OH)_3$
Molecular Formula: CeH_3O_3
Molecular Weight: 191.137
CAS RN: 15785-09-8
Properties: white, gelatinous precipitate; however,
impurities impart yellow, brown, or pink coloration;
used to produce cerium salts to color glass [HAW93]
Solubility: i H_2O; s acids [HAW93]

780
Compound: Cerous iodide
Synonym: cerium(III) iodide
Formula: CeI_3
Molecular Formula: CeI_3
Molecular Weight: 520.828
CAS RN: 7790-87-6
Properties: −20 mesh with 99.9% purity; bright yellow, ortho-rhomb; decomposes in moist air [MER06] [CER91]
Solubility: s H_2O [MER06]
Melting Point, °C: 760 [KNA91]
Boiling Point, °C: 1500 (estimated) [KNA91]

781
Compound: Cerous iodide nonahydrate
Synonym: cerium(III) iodide nonahydrate
Formula: $CeI_3 \cdot 9H_2O$
Molecular Formula: $CeH_{18}I_3O_9$
Molecular Weight: 682.966
CAS RN: 7790-87-6
Properties: white or reddish white cryst [MER06]
Solubility: v s H_2O, solution decomposes with liberation of I_2; s alcohol [MER06]
Melting Point, °C: 752 [CRC10]
Boiling Point, °C: 1400 [CRC10]

782
Compound: Cerous nitrate hexahydrate
Synonym: cerium(III) nitrate hexahydrate
Formula: $Ce(NO_3)_3 \cdot 6H_2O$
Molecular Formula: $CeH_{12}N_3O_{15}$
Molecular Weight: 434.221
CAS RN: 10294-41-4
Properties: white cryst; deliq; oxidizing agent; used as a catalyst in the hydrolysis of phosphoric acid esters [HAW93]
Solubility: s H_2O, alcohol, acetone [HAW93]
Boiling Point, °C: decomposes at 200 [HAW93]
Reactions: minus $3H_2O$ at 150°C [HAW93]

783
Compound: Cerous oxalate nonahydrate
Synonym: cerium(III) oxalate nonahydrate
Formula: $Ce_2(C_2O_4)_3 \cdot 9H_2O$
Molecular Formula: $C_6H_{18}Ce_2O_{21}$
Molecular Weight: 706.426
CAS RN: 13266-83-6

Properties: white or sl pink, tasteless powd; odorless; decomposes upon heating; used in medicine and in the extraction of cerium metals; can be prepared by precipitation with oxalic acid from sl acidic cerium solutions [KIR78] [HAW93] [MER06]
Solubility: i H_2O; s H_2SO_4, HCl; i oxalic acid, alkali, ether, alcohol [KIR78]
Melting Point, °C: decomposes [STR93]

784
Compound: Cerous oxide
Synonym: cerium(III) oxide
Formula: Ce_2O_3
Molecular Formula: Ce_2O_3
Molecular Weight: 328.228
CAS RN: 1345-13-7
Properties: −100 mesh golden green with 99.9% purity; trig; can be prepared by heating powd carbon and CeO_2 at 1250°C in CO atm [KIR78] [CER91]
Solubility: i H_2O; s H_2SO_4; i HCl [CRC10]
Density, g/cm³: 6.86 [CRC10]
Melting Point, °C: 2177 [KNA91]

785
Compound: Cerous perchlorate hexahydrate
Synonym: cerium(III) perchlorate hexahydrate
Formula: $Ce(ClO_4)_3 \cdot 6H_2O$
Molecular Formula: $CeCl_3H_{12}O_{18}$
Molecular Weight: 546.557
CAS RN: 14017-47-1
Properties: white cryst [STR93]

786
Compound: Cerous phosphate hydrate
Synonym: monazite
Formula: $CePO_4 \cdot xH_2O$
Molecular Formula: $CePO_4$ (anhydrous)
Molecular Weight: 235.087 (anhydrous)
CAS RN: 13454-71-2
Properties: off-white powd [STR93]

787
Compound: Cerous selenate
Synonym: cerium(III) selenate
Formula: $Ce_2(SeO_4)_3$
Molecular Formula: $Ce_2O_{12}Se_3$
Molecular Weight: 709.103
CAS RN: 13454-73-4
Properties: rhomb [CRC10]

Solubility: g/100 g H_2O: 39.5 (0°C), 35.2
  (20°C), 2.1 (90°C) [LAN05]
Density, g/cm³: 4.456 [CRC10]

788
Compound: Cerous sulfate
Synonym: cerium(III) sulfate
Formula: $Ce_2(SO_4)_3$
Molecular Formula: $Ce_2O_{12}S_3$
Molecular Weight: 568.421
CAS RN: 13454-94-9
Properties: colorless to green, monocl or rhomb;
  prepared by heating hydrated salt at 350°C–400°C
  or by reducing a solution of ceric sulfate
  with H_2O_2 solution [KIR78] [CRC10]
Solubility: 10.1 g/100 mL H_2O (0°C),
  0.25 g/100 mL H_2O (100°C) [CRC10]
Density, g/cm³: 3.912 [CRC10]
Melting Point, °C: 630 [HAW93]; 920 [CRC10]

789
Compound: Cerous sulfate octahydrate
Synonym: cerium(III) sulfate octahydrate
Formula: $Ce_2(SO_4)_3 \cdot 8H_2O$
Molecular Formula: $Ce_2H_{16}O_{20}S_3$
Molecular Weight: 712.543
CAS RN: 10450-59-6
Properties: white; ortho-rhomb, octahedral
  cryst [MER06] [STR93]
Solubility: g/100 g H_2O: 9.43 (20°C), 5.70
  (40°C), 4.04 (60°C) [LAN05]
Density, g/cm³: 2.87 [MER06]
Melting Point, °C: 630 [ALD94]
Reactions: minus $8H_2O$ when 250°C is reached [MER06]

790
Compound: Cerous sulfide
Synonym: cerium(III) sulfide
Formula: Ce_2S_3
Molecular Formula: Ce_2S_3
Molecular Weight: 376.428
CAS RN: 12014-93-6
Properties: red cryst; dark brown powd or purple; −325
  mesh 10 μm or less 99.9% purity [CER91] [CRC10]
Solubility: i H_2O [CRC10]
Density, g/cm³: 5.02 [CRC10]
Melting Point, °C: 2100 [CRC10]

791
Compound: Cerous telluride
Synonym: cerium(III) telluride

Formula: Ce_2Te_3
Molecular Formula: Ce_2Te_3
Molecular Weight: 663.030
CAS RN: 12014-97-0
Properties: −20 mesh 99.9% purity [CER91]

792
Compound: Cerous tungstate
Synonym: cerium(III) tungstate
Formula: $Ce_2(WO_4)_3$
Molecular Formula: $Ce_2O_{12}W_3$
Molecular Weight: 1023.743
CAS RN: 13454-74-5
Properties: yellow tetr; −200 mesh at 99.9% purity;
  white, monocl powd, a = 1.151 nm, b = 1.172 nm,
  c = 0.782 nm [KIR83] [STR93] [CRC10]
Density, g/cm³: 6.77 [KIR83]
Melting Point, °C: 1089 [KIR83]

793
Compound: Cesium
Formula: Cs
Molecular Formula: Cs
Molecular Weight: 132.90543
CAS RN: 7440-46-2
Properties: bcc; atomic radius 0.274 nm; silvery white,
  ductile metal; oxidizes rapidly in moist air, can
  ignite spontaneously; hardness 0.2 Mohs; electrical
  resistivity 19 (0°C), 36.6 (30°C) μohm · cm; specific
  heat (20°C) 0.217 J/(g · K); enthalpy of fusion
  2.087 kJ/mol; enthalpy of vaporization 68.85 kJ/
  mol [KIR79] [HAW93] [MER06] [ALD94]
Solubility: reacts with H_2O to evolve
  H_2; s liq NH_3 [MER06]
Density, g/cm³: solid: (18°C) 1.892;
  liq: (40°C) 1.827 [KIR79]
Melting Point, °C: 28.44 [LID94]
Boiling Point, °C: 671 [LID94]
Thermal Conductivity, W/(m · K): 35.9 [ALD94];
  liq, at mp: 18.4; vapor at bp 0.0046 [KIR78]

794
Compound: Cesium acetate
Synonyms: acetic acid, cesium salt
Formula: CH_3COOCs
Molecular Formula: $C_2H_3CsO_2$
Molecular Weight: 191.950
CAS RN: 3396-11-0
Properties: lump; hygr [STR93]

Solubility: 945.1 g/100 mL (−2.5°C),
   1345.5 g/100 mL (88.5°C) [CRC10]
Melting Point, °C: 194 [STR93]

795
Compound: Cesium acetylacetonate
Synonyms: 2,4-pentanedione, cesium derivative
Formula: $Cs[CH_3COCH=C(O)CH_3]$
Molecular Formula: $C_5H_7CsO_2$
Molecular Weight: 232.015
CAS RN: 25937-78-4
Properties: hygr [ALD94]

796
Compound: Cesium aluminum sulfate dodecahydrate
Synonym: cesium alum
Formula: $CsAl(SO_4)_2 \cdot 12H_2O$
Molecular Formula: $AlCsH_{24}O_{20}S_2$
Molecular Weight: 568.198
CAS RN: 7784-17-0
Properties: colorless, cub cryst; used in
   mineral waters [HAW93] [CRC10]
Solubility: g anhydrous/100 g H_2O: 18.8 (0°C), 0.40
   (20°C), 22.7 (100°C) [LAN05]; i alcohol [HAW93]
Density, g/cm³: 2.0215 [HAW93]
Melting Point, °C: 117 [HAW93]

797
Compound: Cesium amide
Formula: $CsNH_2$
Molecular Formula: CsH_2N
Molecular Weight: 148.928
CAS RN: 22205-57-8
Properties: white needles; tetr [CRC10] [CIC73]
Solubility: decomposed by H_2O; v s
   liq NH_3 [CIC73] [CRC10]
Density, g/cm³: 3.44 [CRC10]
Melting Point, °C: 262 [CRC10]

798
Compound: Cesium azide
Formula: CsN_3
Molecular Formula: CsN_3
Molecular Weight: 174.925
CAS RN: 22750-57-8
Properties: colorless needles; hygr; tetr,
   a = 0.672 nm, c = 0.804 nm; Cs–N_3 bond
   length, 0.334 nm [CIC73] [CRC10]
Solubility: 224.2 g/100 mL H_2O (0°C) [CRC10]
Density, g/cm³: ~3.5 [LID94]
Melting Point, °C: 310 [CRC10]

799
Compound: Cesium bromate
Formula: $CsBrO_3$
Molecular Formula: $BrCsO_3$
Molecular Weight: 260.807
CAS RN: 13454-75-6
Properties: hex cryst [CRC10]
Solubility: g/100 g soln, H_2O: 3.66 (25°C); solid
   phase, $CsBrO_3$ [KRU93]; g/100 g H_2O:
   0.21 (0°C), 5.30 (35°C) [LAN05]
Density, g/cm³: 4.10 [LAN05]
Melting Point, °C: 420 [LAN05]

800
Compound: Cesium bromide
Formula: CsBr
Molecular Formula: BrCs
Molecular Weight: 212.809
CAS RN: 7787-69-1
Properties: white cryst; hygr; used in
   infrared spectroscopy, scintillation
   counters [HAW93] [STR93]
Solubility: s alcohol; i acetone [MER06];
   g/100 g H_2O: 55.24 (25°C) [KRU93]
Density, g/cm³: 4.44 [MER06]
Melting Point, °C: 636 [MER06]
Boiling Point, °C: ~1300 [MER06]

801
Compound: Chlorogermane
Formula: GeH_3Cl
Molecular Formula: $ClGe_5H_{12}$
Molecular Weight: 111.12
CAS RN: 13637-65-5
Properties: col liq [CRC10]
Solubility: reac H_2O [CRC10]
Density, g/cm³: 1.75 [CRC10]
Melting Point, °C: −52 [CRC10]
Boiling Point, °C: 28 [CRC10]

802
Compound: Cesium bromoiodide
Synonym: cesium dibromoiodide
Formula: $CsIBr_2$
Molecular Formula: Br_2CsI
Molecular Weight: 419.617
CAS RN: 18278-82-5
Properties: rhomb; −8 mesh with 99.9%
   purity [CRC10] [CER91]
Solubility: 4.61 g/100 mL (20°C) H_2O [CRC10]
Density, g/cm³: 4.25 [CRC10]

Melting Point, °C: 248 [CRC10]
Boiling Point, °C: decomposes at 330 [CRC10]

803

Compound: Cesium carbonate
Formula: Cs_2CO_3
Molecular Formula: CCs_2O_3
Molecular Weight: 325.820
CAS RN: 534-17-8
Properties: −20 mesh with 99.996% and 99.9% purity; white powd; very deliq cryst; preparation: addition of CO_2 to a solution of CsOH; uses: catalyst for ethylene oxide polymerization [KIR79] [STR93] [MER06] [CER91]
Solubility: 260.5 g/100 mL H_2O (15°C) [CRC10]; s alcohol, ether [MER06]
Density, g/cm³: 4.24 [LID94]
Melting Point, °C: decomposes at 610 [STR93]

804

Compound: Cesium chlorate
Formula: $CsClO_3$
Molecular Formula: $ClCsO_3$
Molecular Weight: 216.356
CAS RN: 13763-67-2
Properties: small cryst [CRC10]
Solubility: g/100 g H_2O: 2.46 (0°C), 7.6 (25°C), 79.0 (100°C) [KRU93]
Density, g/cm³: 3.57 [CRC10]

805

Compound: Cesium chloride
Formula: CsCl
Molecular Formula: ClCs
Molecular Weight: 168.358
CAS RN: 7647-17-8
Properties: −4 mesh with 99.999% and 99.9% purity; white, deliq, cub cryst; enthalpy of fusion 15.90 kJ/mol [CRC10] [MER06] [STR93] [CER91]
Solubility: v s H_2O; s alcohol [MER06]; g/100 g solution H_2O: 61.7 (0°C), 65.6 (25°C), 73.0 (100°C) [KRU93]; 11.382 ± 0.010 mol/(kg H_2O) at 25°C [RAR85b]
Density, g/cm³: 3.988 [STR93]
Melting Point, °C: 646 [MER06]
Boiling Point, °C: 1303 [MER06]

806

Compound: Cesium chromate
Formula: Cs_2CrO_4
Molecular Formula: $CrCs_2O_4$
Molecular Weight: 381.805
CAS RN: 56320-90-2

Properties: −20 mesh with 99.9% purity; yellow hex or ortho-rhomb; used in electronics [KIR78] [CER91]
Solubility: 71.4 g/100 mL H_2O (15°C), 95.95 g/100 mL (30°C) [CRC10]
Density, g/cm³: 4.23 [KIR78]

807

Compound: Cesium cyanide
Formula: CsCN
Molecular Formula: CCsN
Molecular Weight: 158.923
CAS RN: 21159-32-0
Properties: white cryst, has odor of HCN; very hygr [KIR78]
Solubility: v s H_2O [KIR78]
Density, g/cm³: 2.93 [CRC10]
Melting Point, °C: 350 [LID94]

808

Compound: Cesium fluoride
Formula: CsF
Molecular Formula: CsF
Molecular Weight: 151.903
CAS RN: 13400-13-0
Properties: −4 mesh of 99.9% purity; enthalpy of fusion 21.70 kJ/mol; hygr white powd; used in optics specialty glasses [HAW93] [STR93] [JAN85] [CER91]
Solubility: 366.5 parts CsF dissolves in 100 parts H_2O (18°C) [KIR79]; s methanol; i dioxane, pyridine [HAW93]
Density, g/cm³: 4.115 [HAW93]
Melting Point, °C: 703 [JAN71]
Boiling Point, °C: 1251 [HAW93]

809

Compound: Cesium fluoroborate
Formula: $CsBF_4$
Molecular Formula: $BCsF_4$
Molecular Weight: 219.710
CAS RN: 18909-69-8
Properties: white; ortho-rhomb below 140°C, a = 0.7647 nm, b = 0.9675 nm, c = 0.5885 nm; cub above 140°C [KIR78]
Solubility: 1.6 g/100 mL H_2O (17°C), 30 g/100 mL H_2O (100°C) [KIR78]
Density, g/cm³: 3.20 [KIR78]
Melting Point, °C: decomposes at 555 [KIR78]

810

Compound: Cesium formate
Formula: $CsCHO_2$

Molecular Formula: $CHCsO_2$
Molecular Weight: 177.923
CAS RN: 3495-36-1
Properties: white cryst [CRC10]
Solubility: v s H_2O [CRC10]
Density, g/cm^3: 1.017 [CRC10]

811

Compound: Cesium hexafluorogermanate
Formula: Cs_2GeF_6
Molecular Formula: Cs_2F_6Ge
Molecular Weight: 452.411
CAS RN: 16919-21-4
Properties: white, cryst solid [HAW93]
Solubility: sl s cold H_2O; v s hot H_2O; sl s acids [HAW93]
Density, g/cm^3: 4.10 [HAW93]
Melting Point, °C: ~675 [HAW93]

812

Compound: Cesium hydride
Formula: CsH
Molecular Formula: CsH
Molecular Weight: 133.913
CAS RN: 58724-12-2
Properties: white, cib cryst; flam [CRC10]
Solubility: reac H_2O
Density, g/cm^3: 3.42 [CRC10]
Melting Point, °C: 528 [CRC10]

813

Compound: Cesium hydrogen carbonate
Formula: $CsHCO_3$
Molecular Formula: $CHCsO_3$
Molecular Weight: 193.92
CAS RN: 29703-01-3
Properties: rhomb white powd [STR93] [CRC10]
Solubility: 209.3 g/100 mL (15°C) [CRC10]
Reactions: minus 1/2H_2O at 175°C [CRC10]

814

Compound: Cesium hydrogen fluoride
Formula: $CsHF_2$
Molecular Formula: CsF_2H
Molecular Weight: 171.910
CAS RN: 12280-52-3
Properties: tetr cryst [CRC10]
Density, g/cm^3: 3.86
Melting Point, °C: 170 [CRC10]

815

Compound: Cesium hydrogen sulfate
Formula: $CsHSO_4$

Molecular Formula: $CsHO_4S$
Molecular Weight: 229.976
CAS RN: 7789-16-4
Properties: col rhomb prisms [CRC10]
Solubility: s H_2O
Density, g/cm^3: 3.352 [CRC10]
Melting Point, °C: decomposes [CRC10]

816

Compound: Cesium hydroxide
Formula: CsOH
Molecular Formula: CsHO
Molecular Weight: 149.912
CAS RN: 21351-79-1
Properties: white or yellowish; fused; very deliq cryst mass; readily absorbs atm CO_2; strongest known base; preparation: by the electrolysis of Cs salts; uses: battery electrolyte, catalyst [KIR79] [MER06]
Solubility: s in about 0.25 parts H_2O, with evolution of heat [MER06]; g/100 g soln, H_2O: 75.18 (30°C) [KRU93]
Density, g/cm^3: 3.68 [MER06]
Melting Point, °C: 272 [MER06]

817

Compound: Cesium hydroxide monohydrate
Formula: $CsOH \cdot H_2O$
Molecular Formula: CsH_3O_2
Molecular Weight: 167.928
CAS RN: 35103-79-8
Properties: −4 mesh with 99.9% purity, contains up to 10% Cs_2CO_3; cryst; contains 15%–20% H_2O [STR93] [CER91]
Density, g/cm^3: 3.675 [STR93]
Melting Point, °C: 272 [ALD94]

818

Compound: Cesium iodate
Formula: $CsIO_3$
Molecular Formula: $CsIO_3$
Molecular Weight: 307.807
CAS RN: 13454-81-4
Properties: white, monocl; −4 mesh with 99.9% purity [CER91] [CRC10]
Solubility: 2.6 g/100 mL (24°C) [CRC10]
Density, g/cm^3: 4.85 [CRC10]

819

Compound: Cesium iodide
Formula: CsI
Molecular Formula: CsI

Molecular Weight: 259.809
CAS RN: 7789-17-5
Properties: −20 mesh with 99.9% purity; white; deliq cryst or cryst powd; used as an optical material in infrared spectrophotometers and in scintillation counters [MER06] [STR93] [CER91]
Solubility: s ethanol; sl s methanol; i acetone [MER06]; g/100 g soln, H_2O: 30.6 (0°C), 46.5 (25°C), 70.2 (102.8°C); solid phase, CsI [KRU93]
Density, g/cm³: 4.5 [MER06]
Melting Point, °C: 621 [MER06]
Boiling Point, °C: ~1280 [MER06]

820
Compound: Cesium metaborate
Formula: $CsBO_2$
Molecular Formula: $BCsO_2$
Molecular Weight: 175.715
CAS RN: 92141-86-1
Properties: cub cryst [CRC10]
Density, g/cm³: ~3.7 [CRC10]
Melting Point, °C: 732 [CRC10]

821
Compound: Cesium metavanadate
Formula: $CsVO_3$
Molecular Formula: CsO_3V
Molecular Weight: 231.845
CAS RN: 14644-55-4
Properties: −100 mesh with 99.9% purity [CER91]

822
Compound: Cesium molybdate
Formula: Cs_2MoO_4
Molecular Formula: Cs_2MoO_4
Molecular Weight: 425.749
CAS RN: 13597-64-3
Properties: −200 mesh with 99.9% purity; white [KIR81] [CER91]
Solubility: 67.07 g/100 g H_2O (18°C) [KIR81]
Melting Point, °C: 936 [KIR81]

823
Compound: Cesium niobate
Formula: $CsNbO_3$
Molecular Formula: $CsNbO_3$
Molecular Weight: 273.809
CAS RN: 12053-66-6
Properties: −200 mesh with 99.9% purity [CER91]

824
Compound: Cesium nitrate
Formula: $CsNO_3$
Molecular Formula: $CsNO_3$
Molecular Weight: 194.910
CAS RN: 7789-18-6
Properties: −4 mesh with 99.9% purity; white, lustrous hex or cub prisms; preparation: from pollucite (cesium aluminum silicate); uses: preparation of other cesium salts [MER06] [CER91] [HAW93]
Solubility: s acetone [MER06]; g/100 g soln, H_2O: 8.54 (0°C), 21.53 (25°C), 66.3 (100°C) [KRU93]
Density, g/cm³: 3.64–3.68 [MER06]
Melting Point, °C: 414 [MER06]
Boiling Point, °C: decomposes at >414 [MER06]

825
Compound: Cesium nitrite
Formula: $CsNO_2$
Molecular Formula: $CsNO_2$
Molecular Weight: 178.911
CAS RN: 13454-83-6
Properties: yellow cryst [CRC10]
Solubility: s H_2O
Melting Point, °C: 406

826
Compound: Cesium orthovanadate
Formula: Cs_3VO_4
Molecular Formula: Cs_3O_4V
Molecular Weight: 513.656
CAS RN: 34283-69-7
Properties: −100 mesh with 99.9% purity [CER91]

827
Compound: Cesium oxide
Formula: Cs_2O
Molecular Formula: Cs_2O
Molecular Weight: 281.810
CAS RN: 20281-00-9
Properties: 6 mm pieces and smaller with 99% purity; yellow-brown powd; lemon yellow at −80°C, reddish orange cryst at room temp, cherry red >180°C [KIR79] [HAW93] [STR93] [CER91]
Solubility: v s H_2O; s acids [HAW93]
Density, g/cm³: 4.25 [STR93]
Melting Point, °C: 490 [STR93]

828
Compound: Cesium perchlorate
Formula: $CsClO_4$

Molecular Formula: $ClCsO_4$
Molecular Weight: 232.356
CAS RN: 13454-84-7
Properties: −4 mesh with 99.9% purity; white cryst; hygr; oxidizing agent [HAW93] [STR93] [CER91]
Solubility: g/100 g H_2O: 0.8 (0°C), 2.0 (25°C), 30.0 (100°C) [KRU93]
Density, g/cm³: 3.327 [STR93]
Melting Point, °C: 250 [STR93]
Reactions: decomposes to CsCl at 575°C [KIR79]

829
Compound: Cesium periodate
Formula: $CsIO_4$
Molecular Formula: $CsIO_4$
Molecular Weight: 323.807
CAS RN: 13478-04-1
Properties: white, rhomb prisms [CRC10]
Solubility: g/100 g H_2O: 2.2^{15}
Density, g/cm³: 4.26 [CRC10]

830
Compound: Cesium pyrovanadate
Formula: $Cs_4V_2O_7$
Molecular Formula: $Cs_4O_7V_2$
Molecular Weight: 745.501
CAS RN: 55343-67-4
Properties: −100 mesh with 99.9% purity [CER91]

831
Compound: Cesium rubidium fullerene
Formula: Cs_2RbC_{60}
Molecular Formula: $C_{60}Cs_2Rb$
Molecular Weight: 1071.939
CAS RN: 141326-12-7
Properties: fcc, lattice parameter 1.4493 nm; superconductor, T_c 33 K [CEN92] [PRE93]

832
Compound: Cesium sulfate
Formula: Cs_2SO_4
Molecular Formula: Cs_2O_4S
Molecular Weight: 361.875
CAS RN: 10294-54-9
Properties: −20 mesh with 99.9% purity; white, hygr cryst; ortho-rhomb or hex prisms; enthalpy of fusion 35.70 kJ/mol [CRC10] [MER06] [STR93] [CER91]
Solubility: v s H_2O; i alcohol, acetone, pyridine [MER06]; g/100 g soln, H_2O: 62.6 (0°C), 64.5 (25°C), 68.8 (100°C) [KRU93]
Density, g/cm³: 4.24 [MER06]
Melting Point, °C: 1005 [CRC10]

833
Compound: Cesium sulfide
Formula: Cs_2S
Molecular Formula: Cs_2S
Molecular Weight: 297.877
CAS RN: 12214-16-3
Properties: −100 mesh with 99.9% purity; tetrahydrate: white cryst, hygr [CRC10] [CER91]
Solubility: tetrahydrate: v s H_2O [CRC10]

834
Compound: Cesium superoxide
Formula: CsO_2
Molecular Formula: CsO_2
Molecular Weight: 164.904
CAS RN: 12018-61-0
Properties: bright yellow cryst; oxidizing agent; preparation: by reaction of Cs metal with O_2 at 330°C [KIR79]
Density, g/cm³: 3.77 [LID94]
Melting Point, °C: 432 [LID94]
Reactions: forms Cs_2O_2 on heating at 280°C–360°C [KIR79]

835
Compound: Cesium tantalate
Formula: $CsTaO_3$
Molecular Formula: CsO_3Ta
Molecular Weight: 361.851
CAS RN: 12158-56-4
Properties: −200 mesh with 99.9% purity reacted product [CER91]

836
Compound: Cesium titanate
Formula: Cs_2TiO_3
Molecular Formula: Cs_2O_3Ti
Molecular Weight: 361.676
CAS RN: 51222-65-2
Properties: reacted product, −200 mesh with 99.9% purity [CER91]

837
Compound: Cesium trifluoroacetate
Synonyms: trifluoroacetic acid, cesium salt
Formula: CF_3COOCs
Molecular Formula: $C_2CsF_3O_2$
Molecular Weight: 245.921
CAS RN: 21907-50-6

Properties: hygr; uses: biochemical detection of DNA–DNA crosslinks and isolation of proteoglycans [ALD94]
Melting Point, °C: 114–116 [ALD94]

838
Compound: Cesium trioxide
Formula: Cs_2O_3
Molecular Formula: Cs_2O_3
Molecular Weight: 313.809
CAS RN: 12134-22-4
Properties: chocolate brown cryst [HAW93]
Solubility: decomposed by H_2O [HAW93]
Density, g/cm³: 4.25 [HAW93]
Melting Point, °C: 400 [HAW93]

839
Compound: Cesium tungstate
Formula: Cs_2WO_4
Molecular Formula: Cs_2O_4W
Molecular Weight: 513.649
CAS RN: 52350-17-1
Properties: −200 mesh with 99.9% purity [CER91]

840
Compound: Cesium zirconate
Formula: Cs_2ZrO_3
Molecular Formula: Cs_2O_3Zr
Molecular Weight: 405.033
CAS RN: 51222-66-3
Properties: reacted product, −200 mesh with 99.9% purity [CER91]

841
Compound: Chloric acid heptahydrate
Formula: $HClO_3 \cdot 7H_2O$
Molecular Formula: $ClH_{15}O_{10}$
Molecular Weight: 210.566
CAS RN: 7790-93-4
Properties: can occur only in an aq solution; oxidizing agent; preparation: reaction between H_2SO_4 and barium chlorate; used as a catalyst in the polymerization of acrylonitrile, as an oxidizing agent [MER06] [HAW93]
Solubility: v s H_2O [CRC10]
Density, g/cm³: 1.282 [CRC10]
Melting Point, °C: <−20 [CRC10]
Reactions: decomposes at 40°C [CRC10]

842
Compound: Chlorine
Formula: Cl_2
Molecular Formula: Cl_2
Molecular Weight: 70.906 (atomic weight 35.4527)
CAS RN: 7782-50-5
Properties: greenish yellow diatomic gas; suffocating odor; oxidizing agent; critical temp 144.0°C; critical pressure 78.525 atm; critical density 573 g/L; critical volume 1.763 L/kg; viscosity of gas 14.0 µPa·s at 20°C; specific volume 0.34 m³/kg (21.1°C); enthalpy of vaporization 20.41 kJ/mol; enthalpy of fusion 6.40 kJ/mol; strongly electronegative [HAW93] [AIR87] [MER06] [KIR78] [CRC10]
Solubility: 0.64 g/100 g H_2O [HAW93]; 0.062 mol/L H_2O (25°C) [MER06]
Density, g/cm³: 3.209 g/L (0°C) [KIR78]
Melting Point, °C: −101.5 [CRC10]
Boiling Point, °C: −34.05 [MER06]
Thermal Conductivity, W/(m·K): 0.0089 at 25°C [ALD94]

843
Compound: Chlorine dioxide
Formula: ClO_2
Molecular Formula: ClO_2
Molecular Weight: 67.452
CAS RN: 10049-04-4
Properties: strongly oxidizing, yellow to reddish yellow gas at room temp; unstable in light; reacts violently with organic material; vapor pressure at mp is 1.3 kPa; enthalpy of vaporization at bp 30 kJ/mol [MER06] [CRC10]
Solubility: 3.01 g/L H_2O (25°C) at 34.5 mm Hg [MER06]; g/100 g H_2O: 2.76 (0°C), 6.00 (10°C), 8.70 (15°C) [LAN05]
Density, g/cm³: 1.62 (11°C); liq: 1.765 (−59°C) [KIR78]
Melting Point, °C: −59 [KIR78]
Boiling Point, °C: 11 [MER06]

844
Compound: Chlorine fluoride
Formula: ClF
Molecular Formula: ClF
Molecular Weight: 54.451
CAS RN: 7790-89-8
Properties: col gas [CRC10]
Solubility: reacs H_2O [CRC10]
Density, g/L: 2.226 [CRC10]
Melting Point, °C: −155.6 [CRC10]
Boiling Point, °C: −101.1 [CRC10]

845
Compound: Chlorine heptoxide
Formula: Cl_2O_7
Molecular Formula: Cl_2O_7
Molecular Weight: 182.901
CAS RN: 10294-48-1
Properties: colorless, very volatile, oily liq;
explodes violently upon concussion or when
in contact with iodine or a flame; preparation:
dehydration of $HClO_4$ with phosphorus
pentoxide; uses: catalyst [KIR78] [MER06]
Solubility: slowly hydrolyzed in H_2O
to form $HClO_4$ [MER06]
Density, g/cm³: 1.86 [MER06]
Melting Point, °C: −91.5 [MER06]
Boiling Point, °C: 82 [MER06]
Reactions: decomposes with evolution of Cl_2
and O_2 at 0.2–10.7 kPa pressures and at
temperatures from 100°C–120°C [KIR78]

846
Compound: Chlorine monofluoride
Formula: ClF
Molecular Formula: ClF
Molecular Weight: 54.451
CAS RN: 7790-89-8
Properties: colorless gas; sl yellow when liq;
enthalpy of vaporization 24 kJ/mol; specific
conductivity 1.9×10^{-7} ohm·cm; destroys glass
instantly, attacks quartz readily in presence
of moisture; organic matter bursts into flame
instantly on contact [MER06] [CRC10]
Solubility: violent reaction with H_2O [MER06]
Density, g/cm³: gas: 2.389 g/L [LID94];
liq (−108°C), 1.67 [MER06]
Melting Point, °C: −155.6 [MER06]
Boiling Point, °C: −101.1 [CRC10]

847
Compound: Chlorine monoxide
Formula: Cl_2O
Molecular Formula: Cl_2O
Molecular Weight: 80.905
CAS RN: 7791-21-1
Properties: yellowish brown gas; disagreeable penetrating
odor; explodes on contact with organic matter;
decomposes at a moderate rate at room temp;
anhydride of hypochlorous acid; Henry's constant at
3.46°C is 14.23 kPa/(molarity); enthalpy of vaporization
25.9 kJ/mol; can be prepared by reacting Cl_2 with
HgO; used as an intermediate in manufacture of
calcium hypochlorite and in sterilization; reacts with
a variety of organic compounds [MER06] [KIR78]

Solubility: 1 volume of water dissolves more than
100 volumes (0°C); saturation solubility is
143.6 g/100 g H_2O (−9.4°C) [MER06] [KIR78]
Density, g/cm³: 3.813 g/L [LID94]
Melting Point, °C: −116 [COT88]
Boiling Point, °C: 2.2 [CRC10]
Reactions: forms HClO in water [MER06]

848
Compound: Chlorine pentafluoride
Formula: ClF_5
Molecular Formula: ClF_5
Molecular Weight: 130.445
CAS RN: 13637-63-3
Properties: colorless gas; critical temp 142.6°C;
enthalpy of vaporization 22.21 kJ/mol; specific
conductivity 1.25×10^{-9} ohm·cm [KIR78]
Density, g/cm³: 5.724 g/L [LID94]
Melting Point, °C: −103 [KIR78]
Boiling Point, °C: −13.1 [KIR78]

849
Compound: Chlorine perchlorate
Formula: $ClOClO_3$
Molecular Formula: Cl_2O_4
Molecular Weight: 134.904
CAS RN: 27218-16-2
Properties: unstable yellow liq [CRC10]
Density, g/cm³: 1.81 [CRC10]
Melting Point, °C: −117 [CRC10]
Boiling Point, °C: decomposes at ~45 [CRC10]

850
Compound: Chlorine trifluoride
Formula: ClF_3
Molecular Formula: ClF_3
Molecular Weight: 92.450
CAS RN: 7790-91-2
Properties: corrosive, colorless gas or pale yellow
liq; somewhat sweet, suffocating odor;
extremely reactive; critical temp 154.5°C;
enthalpy of vaporization 27.53 kJ/mol; specific
conductivity 4.9×10^{-9} ohm·cm; prepared by
reaction of F_2 and Cl_2; used as a fluorinating
agent for nuclear reactor fuels, rocket igniter,
and propellant and pyrolysis inhibitor for
fluoropolymers [KIR78] [MER06] [CRC10]
Solubility: violently hydrolyzed by H_2O [MER06]
Density, g/cm³: 1.825 (at bp) [KIR78]
Melting Point, °C: −76.34 [MER06]
Boiling Point, °C: 11.75 [MER06]

851
Compound: Chlorohydridotris
(triphenylphosphine) ruthenium(II)
Formula: [(C$_6$H$_5$)$_3$P]$_3$Ru(Cl)H
Molecular Formula: C$_{54}$H$_{46}$ClP$_3$Ru
Molecular Weight: 924.403
CAS RN: 55102-19-7
Properties: sensitive to moisture; used as a highly
active hydrogenation catalyst [ALD94]
Melting Point, °C: decomposes at 130 [ALD94]

852
Compound: Chloropentafluoroethane
Synonym: halocarbon-115
Formula: C$_2$Cl$_2$F$_5$
Molecular Formula: C$_2$Cl$_2$F$_5$
Molecular Weight: 189.919
CAS RN: 76-15-3
Properties: colorless, nonflammable gas with
ethereal odor; critical temp 80.0°C; critical
pressure 66.4 MPa; enthalpy of vaporization
23.59 kJ/mol; used in electronics [AIR87]
Melting Point, °C: −106.0 [AIR87]
Boiling Point, °C: −39.1 [AIR87]

853
Compound: Chlorosilane
Formula: SiH$_3$Cl
Molecular Formula: ClH$_3$Si
Molecular Weight: 66.563
CAS RN: 13465-78-6
Properties: colorless gas; enthalpy of vaporization
21 kJ/mol; entropy of vaporization
82.8 kJ/(mol · K) [CIC73] [CRC10]
Density, g/cm³: 3.033 g/L [CRC10]
Melting Point, °C: −118 [CIC73]
Boiling Point, °C: −30.4 [CIC73]

854
Compound: Chlorosulfonic acid
Formula: ClSO$_3$H
Molecular Formula: ClHO$_3$S
Molecular Weight: 116.525
CAS RN: 7790-94-5
Properties: colorless to light yellow, fuming liq;
pungent odor; used in synthetic detergents,
pharmaceuticals, and pesticides [HAW93]
Solubility: decomposed by H$_2$O to HCl and H$_2$SO$_4$,
decomposed by alcohol and acids [HAW93]
Density, g/cm³: 1.76−1.77 [HAW93]
Melting Point, °C: −80 [HAW93]
Boiling Point, °C: 158 [HAW93]

855
Compound: Chlorosyl trifluoride
Formula: ClOF$_3$
Molecular Formula: ClF$_3$O
Molecular Weight: 108.447
CAS RN: 3708-80-6
Properties: col liq [CRC10]
Solubility: reac H$_2$O
Melting Point, °C: −42 [CRC10]
Boiling Point, °C: 27 [CRC10]

856
Compound: Chloryl fluoride
Formula: ClO$_2$F
Molecular Formula: ClFO$_2$
Molecular Weight: 86.450
CAS RN: 13637-83-7
Properties: col gas [CRC10]
Solubility: reac H$_2$O [CRC10]
Density, g/L: 3.534 [CRC10]
Melting Point, °C: −115 [CRC10]
Boiling Point, °C: −6 [CRC10]

857
Compound: Chloryl trifluoride
Formula: ClO$_2$F$_3$
Molecular Formula: ClF$_3$O$_2$
Molecular Weight: 124.447
CAS RN: 38680-84-1
Properties: col gas [CRC10]
Solubility: reac H$_2$O [CRC10]
Density, g/L: 5.087 [CRC10]
Melting Point, °C: −81.2 [CRC10]
Boiling Point, °C: −21.6 [CRC10]

858
Compound: Chromium
Formula: Cr
Molecular Formula: Cr
Molecular Weight: 51.9961
CAS RN: 7440-47-3
Properties: bluish white, refractory metal; bcc,
a=0.2844−0.2848 nm; enthalpy of fusion
21.00 kJ/mol; enthalpy of vaporization (2680°C)
320.6 kJ/mol; specific heat 23.9 kJ/(mol · K);
electrical resistivity (20°C) 0.129 μohm · m;
elastic modulus 250 GPa [KIR78] [CRC10]
Solubility: reacts with dil HCl, H$_2$SO$_4$ [MER06]
Density, g/cm³: 7.19 (20°C) [KIR78]
Melting Point, °C: 1857 [COT88]
Boiling Point, °C: 2680 [COT88]

Thermal Conductivity, W/(m·K): 93.9 (25°C) [ALD94]
Thermal Expansion Coefficient: linear
    coefficient at 20°C is 6.2×10^{-6} [KIR78]

859
Compound: Chromic acid
Formula: H_2CrO_4
Molecular Formula: CrH_2O_4
Molecular Weight: 118.010
CAS RN: 7738-94-5
Properties: aq soln only [CRC10]
Solubility: s H_2O [CRC10]

860
Compound: Chromium(III) acetate
Formula: $Cr(C_2H_3O_2)_3$
Molecular Formula: $C_6H_9CrO_6$
Molecular Weight: 229.127
CAS RN: 1066-30-4
Properties: blue-green powd [CRC10]
Solubility: sl H_2O [CRC10]

861
Compound: Chromium(III) acetylacetonate
Synonym: chromium 2,4-pentanedionate
Formula: $Cr(CH_3COCHCOCH_3)_3$
Molecular Formula: $C_{12}H_{21}CrO_6$
Molecular Weight: 349.320
CAS RN: 21679-31-2
Properties: red, monocl cryst [CRC10]
Solubility: i H_2O; s bz [CRC10]
Density, g/cm³: 1.34 [CRC10]
Melting Point, °C: 208 [CRC10]
Boiling Point, °C: 345 [CRC10]

862
Compound: Chromium antimonide
Formula: CrSb
Molecular Formula: CrSb
Molecular Weight: 173.756
CAS RN: 12053-12-2
Properties: hex cryst; −100 mesh with
    99% purity [CER91]
Density, g/cm³: 7.11 [LID94]
Melting Point, °C: 1110 [LID94]

863
Compound: Chromium arsenide
Formula: Cr_2As
Molecular Formula: $AsCr_2$

Molecular Weight: 178.914
CAS RN: 12254-85-2
Properties: tetr cryst; −60 mesh with
    99% purity [LID94] [CER91]
Density, g/cm³: 7.04 [LID94]

864
Compound: Chromium boride
Formula: Cr_2B
Molecular Formula: BCr_2
Molecular Weight: 114.803
CAS RN: 12006-80-3
Properties: −325 mesh 10 μm or less with 99.5% purity;
    borides are generally used for wear-resistant films
    and to produce semiconductor films [CER91]

865
Compound: Chromium boride
Formula: Cr_5B_3
Molecular Formula: B_3Cr_5
Molecular Weight: 292.414
CAS RN: 12007-38-4
Properties: tetr cryst; −325 mesh 10 μm or less
    with 99.5% purity; borides are generally used
    to provide wear-resistant films and to produce
    semiconductor films [LID94] [CER91]
Density, g/cm³: 6.1 [LID94]
Melting Point, °C: 1900 [LID94]

866
Compound: Chromium carbide
Formula: Cr_3C_2
Molecular Formula: C_2Cr_3
Molecular Weight: 180.010
CAS RN: 12012-35-0
Properties: gray powd; ortho-rhomb cryst;
    microhardness 2700 kg/mm² Hg with 50 g load;
    used as a 99.5% pure sputtering target to produce
    wear-resistant films and semiconductor films; there
    are two other carbides: Cr_7C_3, 12075-40-0 and
    $Cr_{23}C_6$, 12105-81-6 [HAW93] [STR93] [CER91]
Density, g/cm³: 6.68 [STR93]
Melting Point, °C: 1890 [STR93]
Boiling Point, °C: 3800 [STR93]
Thermal Expansion Coefficient: 10.3×10^{-6}/K [KIR78]

867
Compound: Chromium carbonyl
Synonym: chromium hexacarbonyl
Formula: $Cr(CO)_6$
Molecular Formula: C_6CrO_6

Molecular Weight: 220.058
CAS RN: 13007-92-6
Properties: white cryst; ortho-rhomb; stable in air; preparation: reaction of Cr salt and CO gas in presence of Grignard reagent; uses: catalyst, gasoline additive [MER06] [KIR78] [DOU83]
Solubility: i H_2O, ether, ethanol, benzene; sl s CCl_4 [KIR78]
Density, g/cm³: 1.77 [KIR78]
Melting Point, °C: 154–155 [STR93]
Boiling Point, °C: decomposes at 130 [MER06]
Reactions: sinters at 90°C, explodes at 210°C [MER06]

868
Compound: Chromium diboride
Formula: CrB_2
Molecular Formula: B_2Cr
Molecular Weight: 73.618
CAS RN: 12007-16-8
Properties: −150, +325 mesh with 99.5% purity; refractory material; high melting point, very hard, very high corrosion resistance; used as metallurgical additive and as a sputtering target to produce films, which can be wear-resistant and semiconducting [KIR78] [HAW93] [CER91]
Density, g/cm³: 5.15 [HAW93]
Melting Point, °C: 1850 [HAW93]; 2130 [KIR78]

869
Compound: Chromium disilicide
Formula: $CrSi_2$
Molecular Formula: $CrSi_2$
Molecular Weight: 108.167
CAS RN: 12018-09-6
Properties: gray powd; used as 99.99% and 99.5% pure sputtering targets to fabricate wear-resistant interconnections and gate electrodes in IC devices [STR93] [CER91]
Density, g/cm³: 4.7 [STR93]
Melting Point, °C: 1490 [STR93]

870
Compound: Chromium monoboride
Formula: CrB
Molecular Formula: BCr
Molecular Weight: 62.807
CAS RN: 12006-79-0
Properties: silvery refractory material; powd; used as a sputtering target with 99.9% purity to produce wear-resistant films and semiconductor films [STR93] [CRC10] [KIR78] [CER91]
Solubility: i H_2O [CRC10]

Density, g/cm³: 6.17 [STR93]
Melting Point, °C: 2060 [KIR78]

871
Compound: Chromium nitride
Formula: Cr_2N
Molecular Formula: Cr_2N
Molecular Weight: 117.999
CAS RN: 12053-27-9
Properties: −325 mesh 15 μm or less with 99% purity; hex, a = 0.274 nm, c = 0.445 nm [CER91] [CIC73]
Density, g/cm³: 6.8 [LID94]

872
Compound: Chromium nitride
Formula: CrN
Molecular Formula: CrN
Molecular Weight: 66.003
CAS RN: 24094-93-7
Properties: gray; fcc, a = 0.4150 nm; electrical resistivity 640 μohm·cm; microhardness 1090; not superconductive; can be prepared by reacting NH_3 with chromium halide [KIR81]
Density, g/cm³: 6.14 [KIR81]
Melting Point, °C: decomposes at 1282 [COT88]
Thermal Conductivity, W/(m·K): 11.7 [KIR81]

873
Compound: Chromium(II,III) oxide
Synonym: Trichromium tetroxide
Formula: Cr_3O_4
Molecular Formula: Cr_3O_4
Molecular Weight: 219.986
CAS RN: 12018-34-7
Properties: cub cryst [CRC10]
Density, g/cm³: 6.1 [CRC10]

874
Compound: Chromium phosphide
Formula: CrP
Molecular Formula: CrP
Molecular Weight: 82.970
CAS RN: 26342-61-0
Properties: grayish-black cryst; −100 mesh with 99.5% purity [CER91] [CRC10]
Solubility: i H_2O [CRC10]
Density, g/cm³: 5.7 [CRC10]

875
Compound: Chromium selenide
Formula: CrSe

Molecular Formula: CrSe
Molecular Weight: 130.956
CAS RN: 12053-13-3
Properties: −325 mesh 10 μm or less
with 99.5% purity [CER91]
Density, g/cm³: 6.1 [LID94]

876
Compound: Chromium silicide
Formula: Cr₃Si
Molecular Formula: Cr₃Si
Molecular Weight: 184.074
CAS RN: 12018-36-9
Properties: cub cryst; −150, +325 mesh; in the form
of 99.99% and 99.5% pure material, used as a
sputtering target to produce resistant semiconductor
films and to fabricate interconnections and gate
electrodes in IC devices [CER91] [LID94]
Density, g/cm³: 6.4 [LID94]
Thermal Expansion Coefficient: (volume):
100°C (0.191), 200°C (0.468), 400°C (1.077),
800°C (2.516), 1000°C (3.368) [CLA66]

877
Compound: Chromium(II) acetate monohydrate
Synonym: chromous acetate monohydrate
Formula: Cr(CH₃COO)₂·H₂O
Molecular Formula: C₄H₈CrO₅
Molecular Weight: 188.101
CAS RN: 628-52-4
Properties: deep red powd or monocl cryst; easily
oxidized, especially when moist, to chromic
acetate; also listed as the dimer, CAS RN 14976-
80-8; uses: preparation of other Cr salts, to
absorb O₂ in gas analyses [MER06] [ALD94]
Solubility: sl s cold H₂O, readily s hot H₂O [MER06]
Density, g/cm³: 1.79 [MER06]
Reactions: minus H₂O when dried over
P₂O₅ at 100°C [MER06]

878
Compound: Chromium(II) bromide
Synonym: chromous bromide
Formula: CrBr₂
Molecular Formula: Br₂Cr
Molecular Weight: 211.804
CAS RN: 10049-25-9
Properties: white; monocl cryst, becomes
yellow when heated; stable in dry air,
oxidizes in moist air [MER06]

Solubility: s H₂O, exothermal, blue soln [MER06]
Density, g/cm³: 4.236 [MER06]
Melting Point, °C: 842 [MER06]

879
Compound: Chromium(II) chloride
Synonym: chromous chloride
Formula: CrCl₂
Molecular Formula: Cl₂Cr
Molecular Weight: 122.901
CAS RN: 10049-05-5
Properties: −80 mesh with 99.9% purity; white
cryst; tetr; has strongly reducing aq solution;
enthalpy of vaporization 197 kJ/mol; enthalpy
of fusion 32.20 kJ/mol; uses: to manufacture Cr
metal and Cr compounds, as a polymerization
catalyst [ALF95] [CRC10] [KIR78] [CER91]
Solubility: s H₂O, blue soln; absorbs O₂ [KIR78]
Density, g/cm³: 2.93 [KIR78]
Melting Point, °C: 815 [KIR78]
Boiling Point, °C: 1300 [CRC10]

880
Compound: Chromium(II) chloride tetrahydrate
Synonym: chromous chloride tetrahydrate
Formula: Cr(H₂O)₄Cl₂·4H₂O
Molecular Formula: Cl₂CrH₁₆O₈
Molecular Weight: 267.023
CAS RN: 13931-94-7
Properties: bright blue, hygr cryst; transforms to
isomeric green modification at >38°C [MER06]
Solubility: s H₂O, oxidized on standing
with H₂(g) liberated [MER06]
Reactions: minus H₂O to form
trihydrate at 51°C [MER06]

881
Compound: Chromium(II) fluoride
Synonym: chromous fluoride
Formula: CrF₂
Molecular Formula: CrF₂
Molecular Weight: 89.993
CAS RN: 10049-10-2
Properties: hygr, bluish-green, monocl cryst,
with iridescent sheen [MER06] [ALD94]
Solubility: sl s H₂O, s boiling HCl [MER06]
Density, g/cm³: 3.79 [MER06]
Melting Point, °C: 894 [MER06]
Boiling Point, °C: >1300 [CRC10]
Reactions: transforms to Cr₂O₃ when
heated in air [MER06]

882
Compound: Chromium(II) formate monohydrate
Synonym: chromous formate monohydrate
Formula: $Cr(HOOC)_2 \cdot H_2O$
Molecular Formula: $C_2H_4CrO_5$
Molecular Weight: 160.047
CAS RN: 4493-37-2
Properties: red needles; preparation: reaction between $CrCl_2$ and sodium formate; used in chromium electroplating solutions and as a catalyst for organic reactions [MER06] [HAW93]
Solubility: s H_2O to give blue soln [MER06]

883
Compound: Chromium(II) oxalate monohydrate
Synonym: chromous oxalate monohydrate
Formula: $CrC_2O_4 \cdot H_2O$
Molecular Formula: $C_2H_2CrO_5$
Molecular Weight: 158.031
CAS RN: 814-90-4
Properties: yellow to yellowish green, cryst powd; not appreciably oxidized by moist air [MER06]
Solubility: i cold H_2O, s hot H_2O [MER06]
Density, g/cm³: 2.468 [MER06]

884
Compound: Chromium(II) sulfate pentahydrate
Synonym: chromous sulfate pentahydrate
Formula: $CrSO_4 \cdot 5H_2O$
Molecular Formula: $CrH_{10}O_9S$
Molecular Weight: 238.136
CAS RN: 13825-86-0
Properties: blue cryst; stable in air if dry; water solutions rapidly oxidized by air [MER06]
Solubility: s H_2O; s dil H_2SO_4, decomposed by conc H_2SO_4 [MER06]

885
Compound: Chromium(III) acetate hexahydrate
Synonym: chromic acetate hexahydrate
Formula: $Cr(CH_3COO)_3 \cdot 6H_2O$
Molecular Formula: $C_6H_{21}CrO_{12}$
Molecular Weight: 337.222
CAS RN: 1066-30-4
Properties: bluish violet needles; solution in water is blue under incident light, red under transmitted light [MER06]
Solubility: readily s H_2O with partial hydrolysis [MER06]

886
Compound: Chromium(III) acetate hydroxide
Synonym: basic chromic acetate
Formula: $Cr(CH_3COO)_2(OH)$
Molecular Formula: $C_4H_7CrO_5$
Molecular Weight: 187.093
CAS RN: 39430-51-8
Properties: violet powd; commercial material used as a mordant in dyeing, in tanning, and as an oxidation catalyst; formula also written as $Cr_3(CH_3COO)_7(OH)_2$ [ALD94] [MER06]
Solubility: readily s H_2O [MER06]

887
Compound: Chromium(III) acetate monohydrate
Synonym: chromic acetate monohydrate
Formula: $Cr(CH_3COO)_3 \cdot H_2O$
Molecular Formula: $C_6H_{11}CrO_7$
Molecular Weight: 247.145
CAS RN: 1066-30-4
Properties: greenish gray powd or violet plates; used as a mordant for textiles and to harden emulsions [HAW93] [MER06]
Solubility: sl s H_2O; i alcohol [MER06]

888
Compound: Chromium(III) acetylacetonate
Synonyms: 2,4-pentanedione, chromium(III) derivative
Formula: $Cr(CH_3COCH=C(O)CH_3)_3$
Molecular Formula: $C_{15}H_{21}CrO_6$
Molecular Weight: 349.324
CAS RN: 21679-31-2
Properties: reddish violet cryst; monocl [KIR78]
Solubility: i H_2O; s benzene [KIR78]
Density, g/cm³: 1.34 [KIR78]
Melting Point, °C: 208 [KIR78]
Boiling Point, °C: 345 [KIR78]

889
Compound: Chromium(III) basic sulfate
Synonym: chromic basic sulfate
Formula: $Cr(OH)SO_4$
Molecular Formula: $CrHO_5S$
Molecular Weight: 165.067
CAS RN: 12336-95-7
Properties: prepared by reduction of sodium chromate in H_2SO_4 solution; used in tanning leather [KIR79]

890
Compound: Chromium(III) bromide
Synonym: chromic bromide

Formula: $CrBr_3$
Molecular Formula: Br_3Cr
Molecular Weight: 291.708
CAS RN: 10031-25-1
Properties: black cryst; used as a catalyst for
    polymerization of olefins [HAW93]
Solubility: i cold H_2O, s hot H_2O [HAW93]
Density, g/cm³: 4.25 [CRC10]
Melting Point, °C: 1130 [LID94]

891
Compound: Chromium(III) bromide hexahydrate
Synonym: chromic bromide hexahydrate
Formula: $CrBr_3 \cdot 6H_2O$
Molecular Formula: $Br_3CrH_{12}O_6$
Molecular Weight: 399.799
CAS RN: 13478-06-3
Properties: two isomeric forms:
    dibromotetraaquochromium bromide dihydrate:
    green, deliq cryst; hexaaquochromium
    tribromide: violet, deliq cryst [MER06]
Solubility: s H_2O; i alcohol, ether [MER06]
Density, g/cm³: 5.4 [ALF95]

892
Compound: Chromium(III) carbonate hydrate
Synonym: chromic carbonate hydrate
Formula: $Cr_2(CO_3)_3 \cdot xH_2O$
Molecular Formula: $C_3Cr_2O_9$ (anhydrous)
Molecular Weight: 284.019 (anhydrous)
CAS RN: 29689-14-3
Properties: bluish green, amorphous powd [MER06]
Solubility: i H_2O; s mineral acids [MER06]

893
Compound: Chromium(III) chloride
Synonym: chromic chloride
Formula: $CrCl_3$
Molecular Formula: Cl_3Cr
Molecular Weight: 158.354
CAS RN: 10025-73-7
Properties: bright purple plates; hex [KIR78]
Solubility: s H_2O extremely slowly [MER06]
Density, g/cm³: 2.87 [KIR78]
Melting Point, °C: 1152 [MER06]
Boiling Point, °C: dissociates above 1300 [MER06]

894
Compound: Chromium(III) chloride hexahydrate
Synonym: chromic chloride hexahydrate

Formula: $CrCl_3 \cdot 6H_2O$
Molecular Formula: $Cl_3CrH_{12}O_6$
Molecular Weight: 266.445
CAS RN: 10060-12-5
Properties: bright green cryst: tricl or
    monocl; violet cryst: rhomb; several
    known isomers [KIR78] [MER06]
Solubility: s H_2O, gives green or violet soln [KIR78]
Density, g/cm³: 1.835 [KIR78]
Melting Point, °C: tricl/monocl: 95; rhomb: 90 [KIR78]

895
Compound: Chromium(III) fluoride
Synonym: chromic fluoride
Formula: CrF_3
Molecular Formula: CrF_3
Molecular Weight: 108.991
CAS RN: 7788-97-8
Properties: dark green needles [MER06]
Solubility: i H_2O; s HCl, violet color [MER06]
Density, g/cm³: 3.8 [HAW93]
Melting Point, °C: 1400 [LID94]

896
Compound: Chromium(III) fluoride tetrahydrate
Synonym: chromic fluoride tetrahydrate
Formula: $CrF_3 \cdot 4H_2O$
Molecular Formula: $CrF_3H_8O_4$
Molecular Weight: 181.052
CAS RN: 123333-98-2
Properties: fine, green cryst [HAW93]
Solubility: i H_2O, alcohol; s HCl [HAW93]

897
Compound: Chromium(III) fluoride trihydrate
Synonym: chromic fluoride trihydrate
Formula: $CrF_3 \cdot 3H_2O$
Molecular Formula: $CrF_3H_6O_3$
Molecular Weight: 163.037
CAS RN: 16671-27-5
Properties: green cryst from solutions of Cr or
    $Cr(OH)_3$ in hydrofluoric acid [MER06]
Solubility: sl s H_2O [MER06]
Density, g/cm³: 2.2 [LID94]

898
Compound: Chromium(III) hydroxide trihydrate
Synonym: chromic hydroxide trihydrate
Formula: $Cr(OH)_3 \cdot 3H_2O$
Molecular Formula: CrH_9O_6
Molecular Weight: 157.063

CAS RN: 1308-14-1
Properties: bluish green powd [MER06]
Solubility: i H_2O; s dil mineral acids when freshly prepared [MER06]

899
Compound: Chromium(III) iodide
Synonym: chromic iodide
Formula: CrI_3
Molecular Formula: CrI_3
Molecular Weight: 432.709
CAS RN: 13569-75-0
Properties: black cryst; −60 mesh with 99.5% purity [CER91] [CRC10]
Density, g/cm³: 5.32 [LID94]
Melting Point, °C: decomposes at 500 [LID94]
Reactions: minus I_2 when heated in vacuum at 350°C [CRC10]

900
Compound: Chromium(III) nitrate
Synonym: chromic nitrate
Formula: $Cr(NO_3)_3$
Molecular Formula: CrN_3O_9
Molecular Weight: 238.011
CAS RN: 13548-38-4
Properties: pale green, extremely deliq powd [MER06]
Solubility: g/100 g soln, H_2O: 39.21 (5°C), 44.0 ± 0.8 (25°C); solid phase, $Cr(NO_3)_3 \cdot 9H_2O$ [KRU93]
Melting Point, °C: decomposes at >60 [MER06]

901
Compound: Chromium(III) nitrate nonahydrate
Synonym: chromic nitrate nonahydrate
Formula: $Cr(NO_3)_3 \cdot 9H_2O$
Molecular Formula: $CrH_{18}N_3O_{18}$
Molecular Weight: 400.148
CAS RN: 7789-02-8
Properties: greenish black, deep violet; rhomb, monocl cryst; hygr [MER06] [STR93]
Solubility: 74% H_2O (25°C) [KIR78]
Density, g/cm³: 1.80 [KIR78]
Melting Point, °C: 66.3 [KIR78]
Boiling Point, °C: decomposes above 100 [MER06]

902
Compound: Chromium(III) oxide
Synonyms: chromic oxide, eskolaite
Formula: Cr_2O_3
Molecular Formula: Cr_2O_3
Molecular Weight: 151.990

CAS RN: 1308-38-9
Properties: −325 mesh 10 µm or less, precipitated with 99.999% purity; green powd or hex cryst; cryst Cr_2O_3 is extremely hard and will scratch quartz, zircon, topaz; enthalpy of fusion 130.00 kJ/mol; evaporated material of 99.8% purity used for hard, durable light absorption films with a medium refractive index and as a sputtering target to produce an absorbent brown film with medium index [KIR78] [CRC10] [MER06] [CER91]
Solubility: i H_2O [KIR78]
Density, g/cm³: 5.22 [KIR78]
Melting Point, °C: 2330 [CRC10]
Boiling Point, °C: 4000 [HAW93]
Thermal Expansion Coefficient: (volume) 100°C (0.15), 200°C (0.36), 400°C (0.78), 800°C (1.68), 1200°C (2.61) [CLA66]

903
Compound: Chromium(III) perchlorate
Formula: $Cr(ClO_4)_3$
Molecular Formula: Cl_3CrO_{12}
Molecular Weight: 350.347
CAS RN: 55147-94-9
Properties: greenish blue cryst; there is also a hexahydrate [STR93] [ALD94]
Solubility: g/100 g soln, H_2O: 50.99 (0°C), 57.73 (25°C); solid phase, $Cr(ClO_4)_3 \cdot 9H_2O$ [KRU93]

904
Compound: Chromium(III) phosphate
Synonym: chromic phosphate
Formula: $CrPO_4$
Molecular Formula: CrO_4P
Molecular Weight: 146.967
CAS RN: 7789-04-0
Properties: grayish brown to black cryst or amorphous solid; partially oxidizes to CrO_3 on heating in air [MER06]
Solubility: i H_2O, acetic acid, HCl, aqua regia [MER06]
Density, g/cm³: 2.94 [MER06]
Melting Point, °C: >1800 [MER06]

905
Compound: Chromium(III) phosphate hemiheptahydrate
Synonyms: chromic phosphate hemiheptahydrate, Arnaudon's green, Plessy's green
Formula: $CrPO_4 \cdot 3\text{-}1/2H_2O$
Molecular Formula: $CrH_7O_{7.5}P$
Molecular Weight: 210.021
CAS RN: 7789-04-0

Properties: bluish green powd [MER06];
   formula probably also given as the
   tetrahydrate, see the reference [HAW93]
Solubility: i H_2O; s acids [MER06]
Density, g/cm³: 2.15 [MER06]

906
Compound: Chromium(III) phosphate hexahydrate
Synonym: chromic phosphate hexahydrate
Formula: $CrPO_4 \cdot 6H_2O$
Molecular Formula: $CrH_{12}O_{10}P$
Molecular Weight: 255.059
CAS RN: 84359-31-9
Properties: violet cryst; loses water gradually
   on heating, becomes anhydrous after 1 h
   at 800°C or 3–4 h at 500°C [MER06]
Solubility: i H_2O; sl s acetic acid solutions [MER06]
Density, g/cm³: 2.121 [MER06]
Melting Point, °C: decomposes at >500 [LID94]

907
Compound: Chromium(III) potassium
   oxalate trihydrate
Formula: $K_3Cr(C_2O_4)_3 \cdot 3H_2O$
Molecular Formula: $C_6H_6CrK_3O_{15}$
Molecular Weight: 487.394
CAS RN: 15275-09-9
Properties: bluish-green monocl cryst [CRC10]
Solubility: s H_2O

908
Compound: Chromium(III) potassium
   sulfate dodecahydrate
Synonym: chrome alum
Formula: $CrK(SO_4)_2 \cdot 12H_2O$
Molecular Formula: $CrH_{24}KO_{20}S_2$
Molecular Weight: 499.405
CAS RN: 7788-99-0
Properties: dark reddish violet cryst;
   efflorescent; used in tanning, as a textile
   mordant, in ceramics [HAW93]
Solubility: s H_2O [HAW93]
Density, g/cm³: 1.813 [HAW93]
Melting Point, °C: 89 [HAW93]
Reactions: minus all H_2O at 400°C [MER06]

909
Compound: Chromium(III) sulfate
Synonym: chromic sulfate
Formula: $Cr_2(SO_4)_3$
Molecular Formula: $Cr_2O_{12}S_3$

Molecular Weight: 392.183
CAS RN: 10101-53-8
Properties: peach-colored solid [MER06]
Solubility: i H_2O, acids [MER06]
Density, g/cm³: 3.012 [MER06]

910
Compound: Chromium(III) sulfate hydrate
Synonym: chromic sulfate hydrate
Formula: $Cr_2(SO_4)_3 \cdot xH_2O$
Molecular Formula: $Cr_2O_{12}S_3$ (anhydrous)
Molecular Weight: 392.183 (anhydrous)
CAS RN: 15244-38-9
Properties: greenish black powd;
   amorphous [STR93] [KIR78]

911
Compound: Chromium(III) sulfate octadecahydrate
Synonym: chromic sulfate octadecahydrate
Formula: $Cr_2(SO_4)_3 \cdot 18H_2O$
Molecular Formula: $Cr_2H_{36}O_{30}S_3$
Molecular Weight: 716.458
CAS RN: 10101-53-8
Properties: violet [KIR78]
Solubility: 120 g/100 mL H_2O (20°C) [CRC10]
Density, g/cm³: 1.7 [CRC10]
Reactions: minus $12H_2O$ on heating [CRC10]

912
Compound: Chromium(III) sulfide
Synonym: chromic sulfide
Formula: Cr_2S_3
Molecular Formula: Cr_2S_3
Molecular Weight: 200.190
CAS RN: 12018-22-3
Properties: −200 mesh with 99% purity;
   brownish black powd [CER91] [STR93]
Solubility: decomposed in H_2O [CRC10]
Density, g/cm³: 3.77 [STR93]
Reactions: minus sulfur at 1350°C [CRC10]

913
Compound: Chromium(III) telluride
Synonym: chromic telluride
Formula: Cr_2Te_3
Molecular Formula: Cr_2Te_3
Molecular Weight: 486.792
CAS RN: 12053-39-3
Properties: hex cryst; −325 mesh 10 μm or less;
   with 99.5% purity [CER91] [LID94]
Density, g/cm³: 7.0 [LID94]

914
Compound: Chromium(IV) chloride
Formula: $CrCl_4$
Molecular Formula: Cl_4Cr
Molecular Weight: 193.807
CAS RN: 15597-88-3
Properties: gas; stable only at high temp [LID94] [KIR78]
Boiling Point, °C: 830 [KIR78]

915
Compound: Chromium(IV) fluoride
Formula: CrF_4
Molecular Formula: CrF_4
Molecular Weight: 127.990
CAS RN: 10049-11-3
Properties: green cryst [CRC10]
Solubility: reac H_2O
Density, g/cm³: 2.89 [CRC10]
Melting Point, °C: 277 [CRC10]

916
Compound: Chromium(IV) oxide
Synonym: chromium dioxide
Formula: CrO_2
Molecular Formula: CrO_2
Molecular Weight: 83.995
CAS RN: 12018-01-8
Properties: dark brown or black powd; tetr; used in magnetic tapes [KIR78]; ferromagnetic; rutile structure [MER06]
Solubility: i H_2O, s acids giving Cr^{+++} and dichromate [KIR78]
Density, g/cm³: 4.98 (calc) [KIR78]
Boiling Point, °C: decomposes to Cr_2O_3 [KIR78]
Reactions: metastable in air; decomposes to Cr_2O_3 at many reported temperatures [MER06]

917
Compound: Chromium(V) fluoride
Formula: CrF_5
Molecular Formula: CrF_5
Molecular Weight: 146.988
CAS RN: 14884-42-5
Properties: red ortho cryst [CRC10]
Solubility: reac H_2O [CRC10]
Melting Point, °C: 34 [CRC10]
Boiling Point, °C: 117 [CRC10]

918
Compound: Chromium(V) oxide
Formula: Cr_2O_5

Molecular Formula: Cr_2O_5
Molecular Weight: 183.989
CAS RN: 12218-36-9
Properties: black needles [CRC10]
Melting Point, °C: decomposes at 200 [CRC10]

919
Compound: Chromium(VI) dichloride dioxide
Formula: CrO_2Cl_2
Molecular Formula: Cl_2CrO_2
Molecular Weight: 154.901
CAS RN: 14977-61-8
Properties: red liq [CRC10]
Density, g/cm³: 1.91 [CRC10]
Solubility: reac H_2O; s ctc, chl, bz [CRC10]
Melting Point, °C: −96.5 [CRC10]
Boiling Point, °C: 117 [CRC10]

920
Compound: Chromium(VI) difluoride dioxide
Formula: CrO_2F_2
Molecular Formula: CrF_2O_2
Molecular Weight: 121.922
CAS RN: 7788-96-7
Properties: red violet cryst [CRC10]
Solubility: reac H_2O

921
Compound: Chromium(VI) fluoride
Formula: CrF_6
Molecular Formula: CrF_6
Molecular Weight: 165.986
CAS RN: 13843-28-2
Properties: yellow solid; stable at low temp [CRC10]
Melting Point, °C: decomposes at −100 [CRC10]

922
Compound: Chromium(VI) morpholine
Synonym: morpholine chromate
Formula: $(OC_4H_8NH_2)_2CrO_4$
Molecular Formula: $C_8H_{20}CrN_2O_6$
Molecular Weight: 292.252
CAS RN: 36969-05-8
Properties: yellow, oily; corrosion inhibitor [KIR78]

923
Compound: Chromium(VI) oxide
Synonyms: chromium trioxide, chromic acid, chromic anhydride
Formula: CrO_3

Molecular Formula: CrO_3
Molecular Weight: 99.994
CAS RN: 1333-82-0
Properties: ruby red cryst; ortho-rhomb; oxidizing agent; deliq [HAW93] [KIR78]
Solubility: g/100 g H_2O: 164.8 (0°), 167.2 (20°C), 206.8 (100°C) [LAN05]; s acetic acid, H_2SO_4, ether [KIR78]
Density, g/cm³: 2.7 [KIR78]
Melting Point, °C: 197 [KIR78]
Boiling Point, °C: decomposes at 250 to Cr_2O_3 and O_2 [MER06]

924
Compound: Chromium(VI) tetrafluoride oxide
Formula: $CrOF_4$
Molecular Formula: CrF_4O
Molecular Weight: 143.989
CAS RN: 23276-90-6
Properties: dark red solid [CRC10]
Solubility: reac H_2O, ace, dmso [CRC10]
Melting Point, °C: 55 [CRC10]

925
Compound: Chromyl chloride
Synonym: chromium(VI) oxychloride
Formula: CrO_2Cl_2
Molecular Formula: Cl_2CrO_2
Molecular Weight: 154.900
CAS RN: 14977-61-8
Properties: cherry red liq; enthalpy of vaporization 35.1 kJ/mol [CRC10] [KIR78] [MER06]
Solubility: i H_2O, hydrolyzes; s CS_2, CCl_4 [KIR78]
Density, g/cm³: 1.91 [KIR78]
Melting Point, °C: −96.5 [KIR78]
Boiling Point, °C: 117 [ALD94]

926
Compound: Cobalt
Formula: Co
Molecular Formula: Co
Molecular Weight: 58.93320
CAS RN: 7440-48-4
Properties: gray, hard, magnetic, ductile, somewhat malleable metal; two allotropic forms, hex and cub, both can exist at room temp; transformation temp 417°C; enthalpy of transformation 14.79 kJ/mol; enthalpy of fusion 16.20 kJ/mol; enthalpy of vaporization 369.86 kJ/mol; electrical resistivity (20°C) 6.24 μohm·cm; Young's modulus 211 GPa; Poisson's ratio 0.32; Curie temp 1121°C; stable in air or water at ordinary temp [KIR79] [MER06] [CRC10]

Solubility: s dil HNO_3, very slowly attacked by cold H_2SO_4 or HCl [MER06]
Density, g/cm³: 8.92 [MER06]
Melting Point, °C: 1495 [CRC10]
Boiling Point, °C: 2927 [CRC10]
Thermal Conductivity, W/(m·K): 100 (25°C) [ALD94]
Thermal Expansion Coefficient: coefficient of thermal expansion 12.5×10^{-6}/°C at room temp (hex); 14.2×10^{-6}/°C (cub) at 417°C [KIR78]

927
Compound: Cobalt aluminate
Synonym: Thenard's blue
Formula: $CoAl_2O_4$
Molecular Formula: Al_2CoO_4
Molecular Weight: 176.894
CAS RN: 13820-62-7
Properties: blue cub solid; used in nickel and cobalt alloys [KIR79]
Solubility: i H_2O [KIR79]

928
Compound: Cobalt antimonide
Formula: CoSb
Molecular Formula: CoSb
Molecular Weight: 180.693
CAS RN: 12052-42-5
Properties: hex cryst; 6 mm pieces and smaller with 99.5% purity [CER91] [LID94]
Density, g/cm³: 8.8 [LID94]
Melting Point, °C: 1202 [LID94]

929
Compound: Cobalt arsenic sulfide
Synonym: cobaltite
Formula: CoAsS
Molecular Formula: AsCoS
Molecular Weight: 165.921
CAS RN: 12254-82-9
Properties: silvery white to gray mineral with metallic luster; 5.5 Mohs hardness; used as an important source of cobalt and in ceramics [HAW93]
Density, g/cm³: 6–6.3 [HAW93]
Melting Point, °C: decomposes [CRC10]

930
Compound: Cobalt arsenide
Formula: CoAs
Molecular Formula: AsCo
Molecular Weight: 133.855
CAS RN: 27016-73-5

Properties: ortho-rhomb cryst; −10 mesh
  with 99.5% purity [CER91] [LID94]
Density, g/cm³: 8.22 [LID94]
Melting Point, °C: 1180 [LID94]

931

Compound: Cobalt arsenide
Synonym: skutterudite
Formula: $CoAs_3$
Molecular Formula: As_3Co
Molecular Weight: 283.698
CAS RN: 12196-91-7
Properties: mineral, cub, hardness 6.0 Mohs;
  6mm pieces and smaller [CER91] [KIR79]
Density, g/cm³: 6.5 [KIR79]
Melting Point, °C: 942 [LID94]

932

Compound: Cobalt arsenide
Formula: $CoAs_2$
Molecular Formula: As_2Co
Molecular Weight: 208.776
CAS RN: 12044-42-7
Properties: cub mineral, also ortho-rhomb; hardness:
  cub 6.0 Mohs, ortho-rhomb 5.0 Mohs [KIR79]
Density, g/cm³: cub: 6.5; ortho-rhomb: 7.2 [KIR79]

933

Compound: Cobalt boride
Formula: CoB
Molecular Formula: BCo
Molecular Weight: 69.744
CAS RN: 12006-77-8
Properties: refractory material; cryst;
  used in ceramics [HAW93]
Solubility: decomposed by H_2O; s HNO_3 [HAW93]
Density, g/cm³: 7.25 [HAW93]
Melting Point, °C: 1460 [KIR78]

934

Compound: Cobalt boride
Formula: Co_2B
Molecular Formula: BCo_2
Molecular Weight: 128.677
CAS RN: 12045-01-1
Properties: refractory material; there is
  also a Co_3B, 12006-78-9 [KIR78]
Density, g/cm³: 8.1 [LID94]
Melting Point, °C: Co_2B: 1285; Co_3B:
  1125, decomposes [KIR78]

935

Compound: Cobalt carbonyl
Synonym: dicobalt octacarbonyl
Formula: $Co_2(CO)_8$
Molecular Formula: $C_8Co_2O_8$
Molecular Weight: 341.949
CAS RN: 10210-68-1
Properties: dark orange cryst; stabilized with 5%–10%
  hexane; sensitive to oxidation; freezing point
  of stabilized compound is −22.8°C; used as
  catalyst in Oxo process [HAW93] [STR93]
Solubility: i H_2O; s alcohol, ether,
  carbon disulfide [HAW93]
Density, g/cm³: 1.78 [HAW93]
Melting Point, °C: decomposes at 51–52 [STR93]

936

Compound: Cobalt disilicide
Formula: $CoSi_2$
Molecular Formula: $CoSi_2$
Molecular Weight: 115.104
CAS RN: 12017-12-8
Properties: gray rhomb powd [ALF93] [KIR79]
Solubility: s hot HCl [KIR79]
Density, g/cm³: 5.3 [KIR79]
Melting Point, °C: 1277 [ALF93]

937

Compound: Cobalt disulfide
Formula: CoS_2
Molecular Formula: CoS_2
Molecular Weight: 123.065
CAS RN: 12013-10-4
Properties: black cub; −200 mesh with
  99.5% purity [CRC10] [CER91]
Solubility: i H_2O; s HNO_3 [CRC10]
Density, g/cm³: 4.3 [LID94]
Melting Point, °C: decomposes at 269 [CRC10]

938

Compound: Cobalt dodecacarbonyl
Synonym: tetracobalt dodecacarbonyl
Formula: $Co_4(CO)_{12}$
Molecular Formula: $C_{12}Co_4O_{12}$
Molecular Weight: 571.858
CAS RN: 17786-31-1
Properties: black cryst; air sensitive [DOU83] [STR93]
Solubility: sl s H_2O [CRC10]
Density, g/cm³: 2.09 [STR93]
Melting Point, °C: decomposes at 60 [STR93]

939
Compound: Cobalt metaborate hydrate
Formula: $Co(BO_2)_2 \cdot xH_2O$ (x = 2,3)
Molecular Formula: B_2CoO_4 (x = 0)
Molecular Weight: 144.553 (x = 0)
CAS RN: 15293-77-3
Properties: prepartion: precipitation from an
    aq Co^{++} solution to which borax is added,
    the product is $CoO \cdot 3B_2O_3 \cdot 10H_2O$; stirring
    a Co^{++} solution with boric acid results in
    $CoB_4O_7 \cdot 4H_2O$; uses: as acid catalyst [KIR78]

940
Compound: Cobalt metatitanate
Formula: $CoTiO_3$
Molecular Formula: CoO_3Ti
Molecular Weight: 154.798
CAS RN: 12017-01-5
Properties: green; rhomb; −325 mesh 10 μm average or
    less, with 99.9% purity [CER91] [LID94] [KIR83]
Density, g/cm³: 5.0 [KIR83]

941
Compound: Cobalt molybdate
Formula: $CoMoO_4$
Molecular Formula: CoO_4Mo
Molecular Weight: 218.871
CAS RN: 13762-14-6
Properties: −325 mesh 10 μm or less with 99.9%
    purity; green powd; three forms: α, β, γ;
    obtained by reaction of the two oxides;
    used as a catalyst to desulfurize petroleum
    [KIR81] [HAW93] [STR93] [CER91]
Density, g/cm³: α: 3.6; β: 4.5; γ: 4.1 [KIR81]
Melting Point, °C: 1040 [KIR81]

942
Compound: Cobalt nitrosocarbonyl
Synonym: cobalt tricarbonyl nitrosyl
Formula: $Co(NO)(CO)_3$
Molecular Formula: C_3CoNO_4
Molecular Weight: 172.971
CAS RN: 14096-82-3
Properties: dark red liq; air sensitive [STR93]
Solubility: i H_2O [CRC10]
Melting Point, °C: −1.05 [CRC10]
Boiling Point, °C: 50 [STR93]

943
Compound: Cobalt nitrosodicarbonyl
Formula: $Co(NO)(CO)_2$
Molecular Formula: C_2CoNO_3

Molecular Weight: 144.960
CAS RN: 12021-68-0
Properties: cherry red liq [KIR79]
Solubility: i H_2O [KIR79]

944
Compound: Cobaltocene
Formula: $Co(C_5H_5)_2$
Molecular Formula: $C_{10}H_{10}Co$
Molecular Weight: 189.119
CAS RN: 1277-43-6
Properties: black purple cryst [CRC10]
Boiling Point, °C: 173 [CRC10]

945
Compound: Cobalt orthotitanate
Formula: Co_2TiO_4
Molecular Formula: Co_2O_4Ti
Molecular Weight: 229.731
CAS RN: 12017-38-8
Properties: greenish black, cub cryst [KIR79]
Solubility: s conc HCl [HAW93]
Density, g/cm³: 5.07–5.12 [HAW93]

946
Compound: Cobalt phosphide
Formula: Co_2P
Molecular Formula: Co_2P
Molecular Weight: 148.840
CAS RN: 12134-02-0
Properties: −100 mesh with 99% purity;
    gray needles [KIR79] [CER91]
Solubility: i H_2O; s HNO_3 [KIR79]
Density, g/cm³: 6.4 [KIR79]
Melting Point, °C: 1386 [KIR79]

947
Compound: Cobalt silicide
Formula: $CoSi_2$
Molecular Formula: $CoSi_2$
Molecular Weight: 115.104
CAS RN: 12017-12-8
Properties: gray cub cryst [CRC10]
Solubility: s hot HCl [CRC10]
Density, g/cm³: 4.9 [CRC10]
Melting Point, °C: 1326 [CRC10]

948
Compound: Cobalt stearate
Formula: $Co[CH_3(CH_2)_{16}COO]_2$

Molecular Formula: $C_{36}H_{70}CoO_4$
Molecular Weight: 625.883
CAS RN: 1002-88-6
Properties: purple pellets; used as drying
agent [KIR79] [STR93]
Density, g/cm³: 1.13 [KIR78]
Melting Point, °C: 140 [KIR78]

949
Compound: Cobalt zirconate
Formula: $CoZrO_3$
Molecular Formula: CoO_3Zr
Molecular Weight: 198.155
CAS RN: 39361-25-6
Properties: reacted product, −325 mesh 10 μm
or less with 99.5% purity [CER91]

950
Compound: Cobalt(II) acetate
Formula: $Co(C_2H_3O_2)_2$
Molecular Formula: $C_4H_6CoO_4$
Molecular Weight: 177.022
CAS RN: 71-48-7
Properties: pink cryst [CRC10]
Solubility: v s H_2O; s EtOH [CRC10]

951
Compound: Cobalt(II) acetate tetrahydrate
Synonym: cobaltous acetate tetrahydrate
Formula: $Co(C_2H_3O_2)_2 \cdot 4H_2O$
Molecular Formula: $C_4H_{14}CoO_8$
Molecular Weight: 249.083
CAS RN: 6147-53-1
Properties: pink; monocl, prismatic cryst; used as
a drying agent for lacquers and varnishes and
as a catalyst [KIR79] [MER06] [STR93]
Solubility: s H_2O, alcohol, dil acids [MER06]
Density, g/cm³: 1.705 [MER06]
Reactions: minus $4H_2O$ by 140°C [MER06]

952
Compound: Cobalt(II) acetylacetonate
Synonyms: 2,4-pentanedione, cobalt(II) derivative
Formula: $Co(CH_3COCH=C(O)CH_3)_2$
Molecular Formula: $C_{10}H_{14}CoO_4$
Molecular Weight: 257.152
CAS RN: 14024-48-7
Properties: black monocl; tetramer; used for
vapor deposition of chromium; hydrate is a
pink powd [KIR79] [STR93] [COT88]

Density, g/cm³: 1.43 [KIR79]
Melting Point, °C: 241 [KIR79]

953
Compound: Cobalt(II) aluminate
Formula: $CoAl_2O_4$
Molecular Formula: Al_2CoO_4
Molecular Weight: 176.894
CAS RN: 13820-62-7
Properties: blue cryst [CRC10]
Solubility: i H_2O [CRC10]
Density, g/cm³: 4.37 [CRC10]

954
Compound: Cobalt(II) arsenate octahydrate
Synonym: cobaltous arsenate octahydrate
Formula: $Co_3(AsO_4)_2 \cdot 8H_2O$
Molecular Formula: $As_2Co_3H_{16}O_{16}$
Molecular Weight: 598.760
CAS RN: 24719-19-5
Properties: pink to blood red; monocl,
fine needles [MER06]
Solubility: i H_2O; s dil mineral acids,
NH_4OH [MER06]
Density, g/cm³: 2.9–3.1 [MER06]
Melting Point, °C: decomposes by 1000
to $Co_4As_2O_{11}$ [MER06]
Reactions: minus $8H_2O$ by 400°C [MER06]

955
Compound: Cobalt(II) basic carbonate
Formula: $2CoCO_3 \cdot 3Co(OH)_2 \cdot H_2O$
Molecular Formula: $C_2H_8Co_5O_{13}$
Molecular Weight: 534.744
CAS RN: 7542-09-8
Properties: form of commercial cobalt carbonate;
reddish violet cryst; used in pigments
[HAW93] [MER06]
Solubility: i cold H_2O, decomposed by
hot H_2O; s acids [HAW93]
Melting Point, °C: decomposes [HAW93]

956
Compound: Cobalt(II) bromate hexahydrate
Formula: $Co(BrO_3)_2 \cdot 6H_2O$
Molecular Formula: $Br_2CoH_{12}O_{12}$
Molecular Weight: 422.829
CAS RN: 13476-01-2
Properties: violet cryst [STR93]; red octahedral [KIR79]
Solubility: 45.5 g/100 mL (17°C) H_2O; s NH_4OH [CRC10]
Density, g/cm³: ~2.462 [STR93]

957
Compound: Cobalt(II) bromide
Synonym: cobaltous bromide
Formula: CoBr$_2$
Molecular Formula: Br$_2$Co
Molecular Weight: 218.741
CAS RN: 7789-43-7
Properties: bright green solid or lustrous green cryst leaflets; hygr; forms hexahydrate in air [MER06]
Solubility: g/100 g H$_2$O: 91.9 (0°C), 112 (20°C), 257 (100°C) [LAN05]; s methanol, ethanol, acetone [MER06]
Density, g/cm^3: 4.91 [MER06]
Melting Point, °C: 678 (under HBr and N$_2$) [MER06]

958
Compound: Cobalt(II) bromide hexahydrate
Synonym: cobaltous bromide hexahydrate
Formula: CoBr$_2$ · 6H$_2$O
Molecular Formula: Br$_2$CoH$_{12}$O$_6$
Molecular Weight: 326.832
CAS RN: 13762-12-4
Properties: red to reddish purple; deliq, prismatic cryst [MER06] [ALD94]
Solubility: s H$_2$O to give red or blue solution, depending on temp [MER06]
Density, g/cm^3: 2.46 [MER06]
Melting Point, °C: 47–48 [MER06]
Reactions: minus 4H$_2$O at 100°C forming purple dihydrate [MER06]

959
Compound: Cobalt(II) carbonate
Synonym: spherocobaltite
Formula: CoCO$_3$
Molecular Formula: CCoO$_3$
Molecular Weight: 118.942
CAS RN: 513-79-1
Properties: −325 mesh 10 µm or less with 99.5% purity; pink powd or rhomb cryst; there is a monohydrate, CAS RN 137506-60-6 [MER06] [STR93] [CER91] [ALD94]
Solubility: i H$_2$O, alcohol, methyl acetate [MER06]
Density, g/cm^3: 4.13 [MER06]
Melting Point, °C: decomposes [KIR79]
Reactions: oxidized by air to cobalt(III) carbonate [MER06]

960
Compound: Cobalt(II) chlorate hexahydrate
Synonym: cobaltous chlorate hexahydrate

Formula: Co(ClO$_3$)$_2$ · 6H$_2$O
Molecular Formula: Cl$_2$CoH$_{12}$O$_{12}$
Molecular Weight: 333.926
CAS RN: 13478-33-6
Properties: red, cub, hygr [CRC10]
Solubility: mol/100 mol H$_2$O: 10.75 (0°C), 14.51 (21°C); solid phase, Co(ClO$_3$)$_2$ · 6H$_2$O (0°C), Co(ClO$_3$)$_2$ · 4H$_2$O (21°C) [KRU93]; g/100 g H$_2$O: 135 (0°C), 180 (20°C), 316 (60°C) [LAN05]
Density, g/cm^3: 1.92 [CRC10]
Melting Point, °C: 50 [CRC10]
Reactions: decomposes at 100°C [CRC10]

961
Compound: Copper(II) chloride hydroxide
Formula: Cu$_2$(OH)$_3$Cl
Molecular Formula: ClCu$_2$H$_3$O$_3$
Molecular Weight: 213.567
CAS RN: 1332-65-6
Properties: pale green cryst [CRC10]
Solubility: i H$_2$O; s acid [CRC10]

962
Compound: Cobalt(II) chloride
Synonym: cobaltous chloride
Formula: CoCl$_2$
Molecular Formula: Cl$_2$Co
Molecular Weight: 129.838
CAS RN: 7646-79-9
Properties: −80 mesh with 99.9% purity; pale blue powd; hygr leaflets; turns pink in moist air; enthalpy of fusion 45.00 kJ/mol [CRC10] [MER06] [CER91]
Solubility: s H$_2$O, alcohols, acetone, ether, glycerol, acetone [MER06]; g/100 g soln, H$_2$O: 30.3 (0°C), 36.0 (25°C), 51.5 (100°C); solid phase, CoCl$_2$ · 6H$_2$O (0°C, 25°C), CoCl$_2$ · 2H$_2$O (100°C) [KRU93]
Density, g/cm^3: 3.367 [MER06]
Melting Point, °C: 740 [CRC10]
Boiling Point, °C: 1049 [MER06]
Reactions: decomposes if subjected to lengthy heating in air at 400°C, sublimes at 500°C in HCl gas [MER06]

963
Compound: Cobalt(II) chloride dihydrate
Synonym: cobaltous chloride dihydrate
Formula: CoCl$_2$ · 2H$_2$O
Molecular Formula: Cl$_2$CoH$_4$O$_2$
Molecular Weight: 165.869
CAS RN: 16544-92-6
Properties: violet or blue cryst [MER06]

Solubility: s H$_2$O [CRC10]
Density, g/cm^3: 2.477 [MER06]

964

Compound: Cobalt(II) chloride hexahydrate
Synonym: cobaltous chloride hexahydrate
Formula: CoCl$_2 \cdot$ 6H$_2$O
Molecular Formula: Cl$_2$CoH$_{12}$O$_6$
Molecular Weight: 237.929
CAS RN: 7791-13-1
Properties: monocl cryst; pink to red, sl deliq [MER06]
Solubility: 76.7 g/100 mL cold H$_2$O, 190.7 g/100 mL hot H$_2$O; s alcohols, acetone, ether, glycerol [MER06]
Density, g/cm^3: 1.924 [MER06]
Melting Point, °C: 87 [MER06]
Reactions: minus 4H$_2$O at 52°C–56°C forming dihydrate; minus H$_2$O at 100°C giving monohydrate; loses remaining H$_2$O at 120°C–140°C [MER06]

965

Compound: Cobalt(II) chromate
Synonym: cobaltous chromate
Formula: CoCrO$_4$
Molecular Formula: CoCrO$_4$
Molecular Weight: 174.927
CAS RN: 24613-38-5
Properties: brown or yellowish brown powd, but pure material is gray-black; used in ceramics [HAW93]
Solubility: i H$_2$O; s mineral acids [HAW93]
Density, g/cm^3: ~4.0 [LID94]
Melting Point, °C: decomposes [CRC10]

966

Compound: Cobalt(II) chromite
Synonym: cobalt dichromium tetraoxide
Formula: CoCr$_2$O$_4$
Molecular Formula: CoCr$_2$O$_4$
Molecular Weight: 226.923
CAS RN: 12016-69-2
Properties: –200 mesh with 99.5% purity; brilliant greenish blue powd; cub spinel structure; used in pigments and catalysts [KIR79] [MER06] [CER91]
Solubility: almost i conc HCl, HNO$_3$ [MER06]

967

Compound: Cobalt(II) citrate dihydrate
Synonyms: citric acid, cobalt(II) salt dihydrate
Formula: Co$_3$(C$_6$H$_5$O$_7$)$_2 \cdot$ 2H$_2$O
Molecular Formula: C$_{12}$H$_{14}$Co$_3$O$_{16}$
Molecular Weight: 591.033

CAS RN: 18727-04-3
Properties: rose red [KIR79]
Solubility: 0.8 g/100 mL cold H$_2$O [KIR79]
Reactions: minus 2H$_2$O at 150°C [KIR79]

968

Compound: Cobalt(II) cyanide dihydrate
Synonym: cobaltous cyanide
Formula: Co(CN)$_2 \cdot$ 2H$_2$O
Molecular Formula: C$_2$H$_4$CoN$_2$O$_2$
Molecular Weight: 146.999
CAS RN: 20427-11-6
Properties: pink to reddish brown powd or needles [MER06]
Solubility: i H$_2$O, acids, methyl acetate; s alkali cyanides [MER06]
Density, g/cm^3: anhydrous: 1.872 [KIR79]
Melting Point, °C: decomposes at 300 [CRC10]
Reactions: minus 2H$_2$O at 280°C [KIR79]

969

Compound: Cobalt(II) cyanide trihydrate
Synonym: cobaltous cyanide trihydrate
Formula: Co(CN)$_2 \cdot$ 3H$_2$O
Molecular Formula: C$_2$H$_6$CoN$_2$O$_3$
Molecular Weight: 165.014
CAS RN: 20427-11-6
Properties: pink to reddish brown powd or needles [MER06]
Solubility: almost i H$_2$O, acids; s alkali cyanide solutions [MER06]
Reactions: minus 3H$_2$O at 250°C [CRC10]

970

Compound: Cobalt(II) diiron tetroxide
Synonym: cobaltous ferrite
Formula: CoFe$_2$O$_4$
Molecular Formula: CoFe$_2$O$_4$
Molecular Weight: 234.621
CAS RN: 12052-28-7
Properties: a magnetic ferrite [HAW93]

971

Compound: Cobalt(II) ferricyanide
Synonym: cobaltous ferricyanide
Formula: Co$_3$[Fe(CN)$_6$]$_2$
Molecular Formula: C$_{12}$Co$_3$Fe$_2$N$_{12}$
Molecular Weight: 600.703
CAS RN: 15415-49-3
Properties: red needles [KIR79]
Solubility: i H$_2$O, HCl; s NH$_4$OH [KIR79]

972
Compound: Cobalt(II) ferrocyanide hydrate
Synonym: cobaltous ferrocyanide
Formula: $Co_2[Fe(CN)_6] \cdot xH_2O$
Molecular Formula: $C_6Co_2FeN_6$ (anhydrous)
Molecular Weight: 329.819 (anhydrous)
CAS RN: 4049-81-1
Properties: gray green [KIR79]
Solubility: i H_2O, HCl; s KCN [KIR79]

973
Compound: Cobalt(II) fluoride
Synonym: cobaltous fluoride
Formula: CoF_2
Molecular Formula: CoF_2
Molecular Weight: 96.930
CAS RN: 10026-17-2
Properties: rosy red tetr cryst; pink powd; forms
    di-, tri-, and tetrahydrates, all soluble in
    H_2O; enthalpy of fusion 59.00 kJ/mol; can be
    prepared by reacting $CoCO_3$ with anhydrous
    HF; finds use in the manufacture of CoF_3
    [MER06] [STR93] [KIR78] [CRC10]
Solubility: sl s H_2O; readily s warm mineral
    acids [MER06]; g/100 ml soln, H_2O: 1.415
    (25°C); solid phase, $CoF_2 \cdot 4H_2O$ [KRU93]
Density, g/cm³: 4.46 [HAW93]
Melting Point, °C: 1127 [CRC10]
Boiling Point, °C: volatilizes ~1400 [MER06]

974
Compound: Cobalt(II) fluoride tetrahydrate
Synonym: cobaltous fluoride tetrahydrate
Formula: $CoF_2 \cdot 4H_2O$
Molecular Formula: $CoF_2H_8O_4$
Molecular Weight: 168.992
CAS RN: 13817-37-3
Properties: red ortho-rhomb; hygr; senstive
    to moisture [LID94] [ALD94]
Solubility: s H_2O [MER06]
Density, g/cm³: 2.19 [ALD94]
Boiling Point, °C: decomposes [MER06]

975
Compound: Cobalt(II) hexafluoroacetylacetonate
Synonym: 2,4-pentanedionate, cobalt(II) derivative
Formula: $Co(CF_3COCHCOCF_3)_2$
Molecular Formula: $C_8H_2CoF_{12}O_4$
Molecular Weight: 473.035

CAS RN: 19648-83-0
Properties: powd [CRC10]

976
Compound: Cobalt(II) hexafluorosilicate hexahydrate
Formula: $CoSiF_6 \cdot 6H_2O$
Molecular Formula: $CoF_6H_{12}O_6Si$
Molecular Weight: 309.100
CAS RN: 15415-49-3
Properties: pale red cryst; used in ceramics [HAW93]
Solubility: 118.1 g/100 mL cold H_2O [KIR79]
Density, g/cm³: 2.087 [HAW93]

977
Compound: Cobalt(II) hydroxide
Synonym: cobaltous hydroxide
Formula: $Co(OH)_2$
Molecular Formula: CoH_2O_2
Molecular Weight: 92.948
CAS RN: 21041-93-0
Properties: blue green cryst or rose-red powd
    or microscopic rhomb; red form most
    stable of two; easily oxidized by air to
    $Co(OH)_3$; amphoteric [MER06]
Solubility: v sl s H_2O; readily s in acids;
    almost i alkalies [MER06]; mol/L soln,
    H_2O: 2×10^{-5} (25°C) [KRU93]
Density, g/cm³: 3.597 [HAW93]
Melting Point, °C: decomposes [KIR79]
Reactions: minus H_2O on heating in vacuum [MER06]

978
Compound: Cobalt(II) iodate
Formula: $Co(IO_3)_2$
Molecular Formula: CoI_2O_6
Molecular Weight: 408.738
CAS RN: 13455-28-2
Properties: black violet needles [KIR79]
Solubility: mol/L soln, H_2O: 0.040 (30°C), 0.031 (100°C);
    solid phase, $Co(IO_3)_2$ [KRU93]; g/100 g H_2O: 1.02
    (20°C), 0.88 (40°C), 0.70 (100°C) [LAN05]
Density, g/cm³: 5.08 [KIR79]
Melting Point, °C: decomposes at 200 [KIR79]

979
Compound: Cobalt(II) iodide
Synonym: cobaltous iodide

Formula: CoI_2
Molecular Formula: CoI_2
Molecular Weight: 312.742
CAS RN: 15238-00-3
Properties: −60 mesh with 99.5% purity; two isomorphous forms: α-CoI_2 is black, graphite-like solid; very hygr; blackish green on exposure to air; β-CoI_2 is ochre yellow powd; blackens at 400°C; very hygr; deliq in moist air forming green droplets [MER06] [CER91]
Solubility: g/100 g soln in H_2O: 58.0 (0°C), 67.0 (25°C), 80.7 (100°C); solid phase: $CoI_2 \cdot H_2O$ (green) (0°C, 25°C), $CoI_2 \cdot H_2O$ (yellow) (100°C) [KRU93]
Density, g/cm³: α: 5.584; β: 5.45 [MER06]
Melting Point, °C: α: 515–520 (high vacuum); β: blackens at 400 [MER06]
Reactions: transition β to α at 400°C [MER06]

980
Compound: Cobalt(II) iodide dihydrate
Formula: $CoI_2 \cdot 2H_2O$
Molecular Formula: $CoH_4I_2O_2$
Molecular Weight: 348.773
CAS RN: 13455-29-3
Properties: green cryst; deliq [CRC10] [STR93]
Solubility: 376.2 g/100 mL H_2O (45°C) [CRC10]
Melting Point, °C: decomposes at 100 [STR93]

981
Compound: Cobalt(II) iodide hexahydrate
Synonym: cobaltous iodide hexahydrate
Formula: $CoI_2 \cdot 6H_2O$
Molecular Formula: $CoH_{12}I_2O_6$
Molecular Weight: 420.813
CAS RN: 13455-29-3
Properties: dark red; hex prisms; loses iodine when exposed to light and air [MER06]
Solubility: s H_2O to give red solution below 20°C, olive green color 20°C–40°C [MER06]
Density, g/cm³: 2.90 [MER06]
Reactions: minus $6H_2O$ by 130°C [MER06]

982
Compound: Cobalt(II) linoleate
Formula: $Co(C_{18}H_{31}O_2)_2$
Molecular Formula: $C_{36}H_{62}CoO_4$
Molecular Weight: 617.819
CAS RN: 14666-96-7

Properties: brown amorphous powd; used as a drier for paint and varnish, especially for enamels and white paints [HAW93]
Solubility: i H_2O; s alcohol, ether and acids [HAW93]

983
Compound: Cobalt(II) hydroxide monohydrate
Formula: $Co(OH)_2 \cdot H_2O$
Molecular Formula: CoH_4O_3
Molecular Weight: 110.963
CAS RN: 35340-84-2
Properties: blue solid [CRC10]
Melting Point, °C: decomposes at 136 [CRC10]

984
Compound: Cobalt(II) molybdate
Formula: $CoMoO_4$
Molecular Formula: $CoMoO_4$
Molecular Weight: 218.87
CAS RN: 13762-14-6
Properties: black monocl cryst [CRC10]
Density, g/cm³: 4.7 [CRC10]
Melting Point, °C: 1040 [CRC10]

985
Compound: Cobalt(II) molybdate monohydrate
Formula: $CoMoO_4 \cdot H_2O$
Molecular Formula: CoH_2MoO_5
Molecular Weight: 236.886
CAS RN: 18601-87-1
Properties: black powd [STR93]

986
Compound: Cobalt(II) nitrate
Synonym: cobaltous nitrate
Formula: $Co(NO_3)_2$
Molecular Formula: CoN_2O_6
Molecular Weight: 182.942
CAS RN: 10141-05-6
Properties: pale red powd [MER06]
Solubility: g/100 g soln, H_2O: 45.66 (0°), 50.5 ± 0.2 (25°C), 77.21 (91°C); solid phase, $Co(NO_3)_2 \cdot 6H_2O$ (0°C, 25°C), $Co(NO_3)_2 \cdot 3H_2O$ (91°C) [KRU93]
Density, g/cm³: 2.49 [MER06]
Melting Point, °C: decomposes at 100–105 [MER06]

987
Compound: Cobalt(II) nitrate hexahydrate
Synonym: cobaltous nitrate hexahydrate
Formula: $Co(NO_3)_2 \cdot 6H_2O$

Molecular Formula: $CoH_{12}N_2O_{12}$
Molecular Weight: 291.034
CAS RN: 10026-22-9
Properties: red monocl cryst; deliq [MER06]
Solubility: 133.8 g/100 mL H_2O (0°C),
0.217 g/100 mL H_2O (80°C) [CRC10]; s
alcohol, most organic solvents [MER06]
Density, g/cm³: 1.88 [MER06]
Melting Point, °C: 55–56 [STR93]
Reactions: decomposes to oxide at >74°C [MER06]

988
Compound: Cobalt(II) nitrite
Formula: $Co(NO_2)_2$
Molecular Formula: CoN_2O_4
Molecular Weight: 150.944
CAS RN: 18488-96-5
Solubility: g/100 g soln, H_2O: 0.076
(0°C), 0.49 (25°C) [KRU93]

989
Compound: Cobalt(II) oleate
Formula: $Co(C_{18}H_{33}O_2)_2$
Molecular Formula: $C_{36}H_{66}CoO_4$
Molecular Weight: 621.851
CAS RN: 14666-94-5
Properties: brown amorphous powd; used in
driers for paints and varnishes [HAW93]
Solubility: i H_2O; s alcohol, ether [HAW93]
Melting Point, °C: 235 [HAW93]

990
Compound: Cobalt(II) oxalate
Synonym: cobaltous oxalate
Formula: CoC_2O_4
Molecular Formula: C_2CoO_4
Molecular Weight: 146.953
CAS RN: 814-89-1
Properties: pink powd; readily absorbs moisture
from air to form hydrates [MER06] [STR93]
Solubility: i H_2O; s acids, NH_4OH [KIR79]
Density, g/cm³: 3.021 [MER06]
Melting Point, °C: decomposes at 250 [KIR79]

991
Compound: Cobalt(II) oxalate dihydrate
Synonym: cobaltous oxalate dihydrate
Formula: $CoC_2O_4 \cdot 2H_2O$
Molecular Formula: $C_2H_4CoO_6$
Molecular Weight: 182.984

CAS RN: 5965-38-8
Properties: light pink; microcryst powd or
needles; decomposes on heating with KOH
or Na_2CO_3 aq solutions [MER06]
Solubility: i H_2O; sl s acids; freely s aq ammonia
[MER06]; g/L soln, H_2O: 0.0346 (25°C) [KRU93]
Reactions: minus H_2O ~190°C [CRC10]

992
Compound: Cobalt(II) oxide
Synonym: cobaltous oxide
Formula: CoO
Molecular Formula: CoO
Molecular Weight: 74.932
CAS RN: 1307-96-6
Properties: −325 mesh 10 μm or less with 99.5%
purity; powd or cub or hex cryst; color varies from
olive green to red depending on particle size;
commercial material usually dark gray; readily
absorbs O_2, even at room temp [MER06] [CER91]
Solubility: i H_2O; s acids or alkalis [MER06]
Density, g/cm³: 5.7–6.7 [MER06]
Melting Point, °C: 1805 [JAN85]

993
Compound: Cobalt(II) perchlorate
Formula: $Co(ClO_4)_2$
Molecular Formula: Cl_2CoO_8
Molecular Weight: 257.833
CAS RN: 13455-31-7
Properties: red needles; used as a chemical
reagent, oxidizing agent [HAW93]
Solubility: g/100 g H_2O: 100.0 (0°C), 113.4
(26°C); solid phase, $Co(ClO_4)_2 \cdot 5H_2O$
[KRU93]; i alcohol, acetone [KIR79]
Density, g/cm³: 3.327 [HAW93]

994
Compound: Cobalt(II) perchlorate hexahydrate
Formula: $Co(ClO_4)_2 \cdot 6H_2O$
Molecular Formula: $Cl_2CoH_{12}O_{14}$
Molecular Weight: 365.925
CAS RN: 13478-33-6
Properties: red cryst; hygr [STR93]
Solubility: 225 g/100 mL H_2O (18°C) [CRC10]
Melting Point, °C: decomposes at 1534 [CRC10]

995
Compound: Cobalt(II) phosphate octahydrate
Synonym: cobaltous phosphate octahydrate
Formula: $Co_3(PO_4)_2 \cdot 8H_2O$

Molecular Formula: $Co_3H_{16}O_{16}P_2$
Molecular Weight: 510.865
CAS RN: 10294-50-5
Properties: pink to lavender amorphous powd [MER06]
Solubility: i H_2O; s mineral acids [MER06]
Density, g/cm^3: 2.77 [MER06]
Reactions: minus $8H_2O$ at 200°C [HAW93]

996
Compound: Cobalt(II) potassium sulfate hexahydrate
Formula: $CoK_2(SO_4)_2 \cdot 6H_2O$
Molecular Formula: $CoH_{12}K_2O_{14}S_2$
Molecular Weight: 437.347
CAS RN: 10026-20-7
Properties: red monocl cryst [CRC10]
Solubility: v sol H_2O [CRC10]
Density, g/cm^3: 2.22 [CRC10]
Melting Point, °C: decomposes at 75 [CRC10]

997
Compound: Cobalt phosphide
Formula: Co_2P
Molecular Formula: Co_2P
Molecular Weight: 148.840
CAS RN: 12134-02-0
Properties: gray needles [CRC10]
Density, g/m^3: 6.4 [CRC10]
Melting Point, °C: 1386 [CRC10]

998
Compound: Cobalt(II) selenate pentahydrate
Formula: $CoSeO_4 \cdot 5H_2O$
Molecular Formula: $CoH_{10}O_9Se$
Molecular Weight: 291.967
CAS RN: 14590-19-3
Properties: ruby red tricl [KIR79]
Solubility: v s H_2O [KIR79]
Density, g/cm^3: 2.512 [KIR79]
Melting Point, °C: decomposes [KIR79]

999
Compound: Cobalt(II) selenide
Formula: CoSe
Molecular Formula: CoSe
Molecular Weight: 137.893
CAS RN: 1307-99-9
Properties: 6 mm pieces and smaller with 99.5% purity; yellow hex [KIR79] [CER91]
Solubility: s HNO_3, aqua regia; i alkali [KIR79]
Density, g/cm^3: 7.65 [KIR79]
Melting Point, °C: 1055 [LID94]

1000
Compound: Cobalt(II) selenite dihydrate
Formula: $CoSeO_3 \cdot 2H_2O$
Molecular Formula: CoH_4O_5Se
Molecular Weight: 221.922
CAS RN: 19034-13-0
Properties: blue red powd [HAW93] [MER06]
Solubility: i H_2O [HAW93]

1001
Compound: Cobalt(II) silicate
Synonym: cobalt orthosilicate
Formula: Co_2SiO_4
Molecular Formula: Co_2O_4Si
Molecular Weight: 209.950
CAS RN: 12017-08-2
Properties: violet cryst [KIR79]
Solubility: i H_2O; s dil HCl [KIR79]
Density, g/cm^3: 4.63 [KIR79]
Melting Point, °C: 1345 [KIR79]

1002
Compound: Cobalt(II) stannate
Formula: Co_2SnO_4
Molecular Formula: Co_2O_4Sn
Molecular Weight: 300.574
CAS RN: 12139-93-4
Properties: greenish blue cub; used as paint and varnish drier [KIR79]
Solubility: s alkali [KIR79]
Density, g/cm^3: 6.30 [KIR79]

1003
Compound: Cobalt(II) stearate
Formula: $Co(C_{18}H_{35}O_2)_2$
Molecular Formula: $C_{36}H_{70}CoO_4$
Molecular Weight: 625.872
CAS RN: 1002-88-6
Properties: purple solid [CRC10]
Density, g/cm^3: 1.13 [CRC10]
Melting Point, °C: 74 [CRC10]

1004
Compound: Cobalt(II) sulfate
Synonym: cobaltous sulfate
Formula: $CoSO_4$
Molecular Formula: CoO_4S
Molecular Weight: 154.997
CAS RN: 10124-43-3

Properties: red to lavender dimorphic, ortho-rhomb cryst; used in ceramics [MER06] [KIR79]
Solubility: dissolves slowly in boiling H_2O [MER06]; g/100 g soln, H_2O: 19.7 ± 0.1 (0°C), 27.2 ± 0.1 (25°C), 27.8 (100°C); solid phase, $CoSO_4 \cdot 7H_2O$ (0°C, 25°C), $CoSO_4 \cdot H_2O$ (100°C) [KRU93]
Density, g/cm³: 3.71 [MER06]
Melting Point, °C: decomposes at 1140 [JAN85]

1005
Compound: Cobalt(II) sulfate heptahydrate
Synonym: bieberite
Formula: $CoSO_4 \cdot 7H_2O$
Molecular Formula: $CoH_{14}O_{11}S$
Molecular Weight: 281.103
CAS RN: 10026-24-1
Properties: pink to red monocl, prismatic cryst [MER06]
Solubility: 60.4 g/100 mL cold H_2O, 67 g/100 mL hot H_2O; sl s methanol, ethanol [MER06] [KIR78]
Density, g/cm³: 1.948 [STR93]
Melting Point, °C: 96.8 [STR93]
Reactions: minus H_2O at 41.5°C; minus $6H_2O$ at 71°C [MER06]

1006
Compound: Cobalt(II) sulfate monohydrate
Formula: $CoSO_4 \cdot H_2O$
Molecular Formula: CoH_2O_5S
Molecular Weight: 173.012
CAS RN: 13455-34-0
Properties: red cryst; used as a pigment for porcelain and glazes [KIR78]
Solubility: s H_2O [KIR79]
Density, g/cm³: 3.075 [KIR79]
Melting Point, °C: decomposes [KIR79]

1007
Compound: Cobalt(II) sulfide
Synonyms: sycoporite, cobaltous sulfide
Formula: CoS
Molecular Formula: CoS
Molecular Weight: 90.999
CAS RN: 1317-42-6
Properties: −200 mesh with 99.5% purity; exists in two forms; α-CoS: black, amorphous powd; forms Co(OH)S in air; β-CoS: gray powd or reddish-silver octahedral cryst [MER06] [CER91]
Solubility: 0.00038 g/100 mL cold H_2O; s acids [MER06] [KIR79]
Density, g/cm³: 5.45 [MER06]
Melting Point, °C: 1182 [LID94]

1008
Compound: Cobalt(II) telluride
Formula: CoTe
Molecular Formula: CoTe
Molecular Weight: 186.533
CAS RN: 12017-13-9
Properties: hex cryst; 6 mm pieces and smaller [LID94] [CER91]
Density, g/cm³: ~8.8 [LID94]

1009
Compound: Cobalt(II) thiocyanate
Synonym: cobaltous thiocyanate
Formula: $Co(SCN)_2$
Molecular Formula: $C_2CoN_2S_2$
Molecular Weight: 175.100
CAS RN: 3017-60-5
Properties: yellowish brown powd [MER06]
Solubility: s H_2O, giving rose-colored soln [MER06]; g/100 g soln, H_2O: 50.7 (25°C) [KRU93]

1010
Compound: Cobalt(II) thiocyanate trihydrate
Synonym: cobaltous thiocyanate trihydrate
Formula: $Co(SCN)_2 \cdot 3H_2O$
Molecular Formula: $C_2H_6CoN_2O_3S_2$
Molecular Weight: 229.146
CAS RN: 97126-35-7
Properties: violet to brownish, violet rhomb cryst; red in transmitted light; used as an indicator of relative humidity [KIR79] [MER06]
Solubility: s H_2O, resulting in blue soln [MER06]
Reactions: minus $3H_2O$ at 105°C [KIR79]

1011
Compound: Cobalt(II) titanate
Formula: $CoTiO_3$
Molecular Formula: CoO_3Ti
Molecular Weight: 154.798
CAS RN: 12017-01-5
Properties: green rhomb cryst [CRC10]
Density, g/cm³: 5.0 [CRC10]

1012
Compound: Cobalt(II) tungstate
Formula: $CoWO_4$
Molecular Formula: CoO_4W
Molecular Weight: 306.771
CAS RN: 12640-47-0

Properties: −325 mesh 10μm or less with 99.9%
purity; reddish orange powd; used as a pigment
and antiknock agent [HAW93] [CER91]
Solubility: i H_2O; s hot conc acids [HAW93]
Density, g/cm³: 8.42 [HAW93]

1013
Compound: Cobalt(II,III) oxide
Synonym: cobaltic-cobaltous oxide
Formula: Co_3O_4
Molecular Formula: Co_3O_4
Molecular Weight: 240.798
CAS RN: 1308-06-1
Properties: −325 mesh 10μm or less with 99.5% purity;
black or gray octahedral cryst; nonstoichiometric;
commercial material is black [MER06] [CER91]
Solubility: i H_2O; s acids, alkalies [MER06]
Density, g/cm³: 6.11 [MER06]
Melting Point, °C: decomposes at 947 [JAN85]
Reactions: minus O_2 >900°C forming CoO [MER06]

1014
Compound: Cobalt(III) acetate
Synonym: cobaltic acetate
Formula: $Co(CH_3COO)_3$
Molecular Formula: $C_6H_9CoO_6$
Molecular Weight: 236.066
CAS RN: 917-69-1
Properties: dark green, very hygr powd or green
octahedral cryst; used as a catalyst for
cumene hydroperoxide [KIR79] [MER06]
Solubility: s H_2O with slow hydrolysis, becoming
rapid at 60°C–70°C [MER06]
Melting Point, °C: decomposes at 100 [KIR79]
Reactions: becomes black and decomposes
on heating to 100°C [MER06]

1015
Compound: Cobalt(III) acetylacetonate
Synonyms: 2,4-pentanedione, cobalt(III) derivative
Formula: $Co(CH_3COCH=C(O)CH_3)_3$
Molecular Formula: $C_{15}H_{21}CoO_6$
Molecular Weight: 356.261
CAS RN: 21679-46-9
Properties: sensitive to moisture; green
cryst [STR93] [ALD94]
Melting Point, °C: 216 [STR93]
Boiling Point, °C: 340 [STR93]

1016
Compound: Cobalt(III) fluoride
Synonyms: cobaltic fluoride, cobalt trifluoride

Formula: CoF_3
Molecular Formula: CoF_3
Molecular Weight: 115.928
CAS RN: 10026-18-3
Properties: light brown; hex cryst; strong oxidizing
agent; hygr and discolors rapidly in moist air; can
be obtained by reacting F_2 with $CoCl_2$ or CoF_2
at 150°C–180°C; used as a fluorinating agent
to replace hydrogen by fluorine in halocarbons
and hydrocarbons [KIR78] [MER06]
Solubility: reacts with H_2O to evolve O_2 [MER06]
Density, g/cm³: 3.88 [MER06]
Melting Point, °C: 927 [LID94]
Reactions: comparatively stable to heat [MER06]

1017
Compound: Cobalt(III) fluoride dihydrate
Formula: $Co_2F_6 \cdot 2H_2O$
Molecular Formula: $Co_2F_6H_4O_2$
Molecular Weight: 267.887
CAS RN: 54496-71-8
Properties: two forms: α is red, rhomb;
β is rose powd [KIR79]
Solubility: s H_2O; i alcohol [KIR79]
Density, g/cm³: 2.192 [KIR79]

1018
Compound: Cobalt(III) hydroxide
Formula: $Co(OH)_3$
Molecular Formula: CoH_3O_3
Molecular Weight: 109.955
CAS RN: 1307-86-4
Properties: brown powd [CRC10]
Solubility: i H_2O; s acid [CRC10]
Density, g/cm³: ~4 [CRC10]
Melting Point, °C: decomposes [CRC10]

1019
Compound: Cobalt(III) hydroxide trihydrate
Formula: $Co(OH)_3 \cdot 3H_2O$
Molecular Formula: CoH_9O_6
Molecular Weight: 164.001
CAS RN: 1307-86-4
Properties: dark brown or pink powd; uses:
catalyst [STR93] [HAW93]
Solubility: i H_2O, alcohol; s cold
conc acids [HAW93]
Density, g/cm³: 4.46 [HAW93]
Melting Point, °C: decomposes [STR93]
Reactions: minus water at 100°C [HAW93]

1020
Compound: Cobalt(III) nitrate
Formula: Co(NO$_3$)$_3$
Molecular Formula: CoN$_3$O$_9$
Molecular Weight: 244.948
CAS RN: 15520-84-0
Properties: green hygr, reacts vigorously
with organic solvents [KIR79]
Solubility: s H$_2$O [LID94]
Density, g/cm^3: ~3.0 [LID94]

1021
Compound: Cobalt(III) oxide
Synonyms: cobaltic oxide; cobalt black
Formula: Co$_2$O$_3$
Molecular Formula: Co$_2$O$_3$
Molecular Weight: 165.864
CAS RN: 1308-04-9
Properties: steel gray or black powd; used as a pigment
to color enamels and in glazing pottery [HAW93]
Solubility: i H$_2$O; s conc acids [HAW93]
Density, g/cm^3: 4.81–5.60 [HAW93]
Melting Point, °C: decomposes at 895 [HAW93]

1022
Compound: Cobalt(III) oxide hydroxide
Synonym: cobaltic oxide monohydrate
Formula: CoO(OH)
Molecular Formula: CoHO$_2$
Molecular Weight: 91.940
CAS RN: 12016-80-7
Properties: dark brown to black powd; hex; formula
also written as Co$_2$O$_3$·H$_2$O [MER06]
Solubility: i H$_2$O; s HCl, evolving Cl$_2$;
s HNO$_3$, H$_2$SO$_4$ [MER06]
Reactions: transforms to Co$_3$O$_4$ if heated
to 148°C–150°C [MER06]

1023
Compound: Cobalt(III) oxide monohydrate
Formula: Co$_2$O$_3$·H$_2$O
Molecular Formula: Co$_2$H$_2$O$_4$
Molecular Weight: 109.955
CAS RN: 12016-80-7
Properties: brown-black hex cryst [CRC10]
Solubility: i H$_2$O; s acid [CRC10]
Melting Point, °C: decomposes at 150 [CRC10]

1024
Compound: Cobalt(III) potassium nitrite sesquihydrate
Formula: CoK$_3$(NO$_2$)$_6$·1.5H$_2$O

Molecular Formula: CoH$_3$K$_3$N$_6$O$_{13.5}$
CAS RN: 13782-01-9 [For anhydrous compound]
Properties: yellow cub cryst [CRC10]
Solubility: sl H$_2$O; reac acid; i EtOH [CRC10]
Density, g/cm^3: 2.6 [CRC10]

1025
Compound: Cobalt(III) sepulchrate trichloride
Synonym: 1,3,6,8,10,13,16,19-octaazabicyclo(6,6,6)
eicosanecobalt trichloride
Molecular Weight: 451.72
Molecular Formula: C$_{12}$H$_{30}$Cl$_3$CoN$_8$
CAS RN: 71963-57-0
Properties: promising sensitizer for water-
splitting systems [ALD94] [HOU82]
Melting Point, °C: decomposes at 262 [ALD94]

1026
Compound: Cobalt(III) sulfide
Formula: Co$_2$S$_3$
Molecular Formula: Co$_2$S$_3$
Molecular Weight: 214.064
CAS RN: 1332-71-4
Properties: black cryst [KIR79]
Solubility: decomposes in acid, aqua regia [KIR79]
Density, g/cm^3: 4.8 [KIR79]

1027
Compound: Cobalt(III) titanate
Formula: Co$_2$TiO$_4$
Molecular Formula: Co$_2$O$_4$Ti
Molecular Weight: 229.731
CAS RN: 1207-38-8
Properties: green-black cub cryst [CRC10]
Solubility: s conc HCl [CRC10]
Density, gm/cm^3: 5.1 [CRC10]

1028
Compound: Cobaltocene
Synonym: bis(cyclopentadienyl)cobalt(II)
Formula: (C$_5$H$_5$)$_2$Co
Molecular Formula: C$_{10}$H$_{10}$Co
Molecular Weight: 189.122
CAS RN: 1277-43-6
Properties: purplish-black cryst; sensitive to
light, air, and heat; forms intercalation
compound SnS$_2$[Co(C$_5$H$_5$)$_2$] by reaction
with SnS$_2$ [CEN94] [STR93] [ALD94]
Melting Point, °C: decomposes at 176–180 [ALD94]

1029
Compound: Cobaltocenium hexafluorophosphate
Synonym: bis(cyclopentadienyl) cobalt(II) hexafluorophosphate
Formula: $(C_5H_5)_2CoPF_6$
Molecular Formula: $C_{10}H_{10}CoF_6P$
Molecular Weight: 334.086
CAS RN: 12427-42-8
Properties: yellow cryst; moisture sensitive [STR93] [ALD94]

1030
Compound: Copper
Formula: Cu
Molecular Formula: Cu
Molecular Weight: 63.546
CAS RN: 7440-50-8
Properties: reddish, lustrous, ductile metal; fcc; excellent conductor of electricity; becomes dull when exposed to air, coated with green basic carbonate in moist air; hardness 3.0 Mohs; electrical resistivity (20°C) 1.673 μohm·cm; electronegativity 2.43; enthalpy of fusion 13.26 kJ/mol; enthalpy of vaporization 300.4 kJ/mol; ionic radius of Cu^{++} 0.096 nm [KIR78] [CRC10] [MER06] [ALD94]
Solubility: very slowly attacked by cold HCl and dil H_2SO_4, readily by HNO_3 [MER06]
Density, g/cm³: 8.94 [MER06]
Melting Point, °C: 1084.62 [CRC10]
Boiling Point, °C: 2562 [CRC10]
Thermal Conductivity, W/(m·K): 394 (25°C) [ALD94]
Thermal Expansion Coefficient: linear coefficient of expansion at 20°C is 16.5×10^{-6}/°C [KIR78]

1031
Compound: Copper arsenide
Formula: Cu_3As
Molecular Formula: $AsCu_3$
Molecular Weight: 265.560
CAS RN: 12005-75-3
Properties: 6 mm pieces and smaller with 99% purity [CER91]

1032
Compound: Copper citrate hemipentahydrate
Formula: $Cu_2C_6H_4O_7 \cdot 2\text{-}1/2H_2O$
Molecular Formula: $C_6H_9Cu_2O_{9.5}$
Molecular Weight: 360.219
CAS RN: 10402-15-0
Properties: green or bluish green; odorless; cryst powd; can be prepared by reacting hot citric acid with copper sulfate solutions; used as an antiseptic and astringent [MER06]

Solubility: sl s H_2O; s ammonia, dil acids [MER06]
Reactions: minus 2.5 H_2O at 100°C [MER06]

1033
Compound: Copper nitride
Formula: Cu_3N
Molecular Formula: Cu_3N
Molecular Weight: 204.645
CAS RN: 1308-80-1
Properties: dark green powd; −200 mesh with 99.5% purity; cub, a = 0.380 nm [CER91] [CIC73] [CRC10]
Solubility: decomposes in H_2O [CRC10]
Density, g/cm³: 5.84 [CIC73]
Melting Point, °C: 300, decomposes [ALF93]

1034
Compound: Copper phosphide
Formula: Cu_3P
Molecular Formula: Cu_3P
Molecular Weight: 221.612
CAS RN: 12019-57-7
Properties: grayish-black; −100 mesh of 99.5% purity; other phosphides are Cu_3P_2, 12134-35-9, Cu_2P, 12324-28-6, CuP_2, 12019-11-3 [KIR82] [CER91]
Solubility: i H_2O [CRC10]
Density, g/cm³: 6.4−6.8 [CRC10]
Melting Point, °C: decomposes [CRC10]

1035
Compound: Copper silicide
Formula: Cu_5Si
Molecular Formula: Cu_5Si
Molecular Weight: 345.816
CAS RN: 12159-07-8
Properties: 6.35 mm and down pieces [ALF93]
Melting Point, °C: 825 [ALF93]

1036
Compound: Copper vanadate
Formula: CuV_2O_6
Molecular Formula: CuO_6V_2
Molecular Weight: 261.425
CAS RN: 12789-09-2
Properties: −200 mesh with 99.5% purity [CER91]

1037
Compound: Copper zirconate
Formula: $CuZrO_3$
Molecular Formula: CuO_3Zr
Molecular Weight: 202.768

CAS RN: 70714-64-6
Properties: reacted product of CuO + ZrO$_2$, −200 mesh with 99.5% purity [CER91]

1038
Compound: Copper(I) acetate
Synonyms: acetic acid, copper(I) salt
Formula: Cu(CH$_3$COO)
Molecular Formula: C$_2$H$_3$CuO$_2$
Molecular Weight: 122.591
CAS RN: 598-54-9
Properties: tan powd; stable when dry, decomposes slowly in H$_2$O [STR93] [KIR78]
Solubility: rapidly hydrolyzed by H$_2$O to form yellow Cu$_2$O [MER06]
Melting Point, °C: volatilizes by heating [MER06]
Boiling Point, °C: decomposes if strongly heated [MER06]

1039
Compound: Copper(I) acetylide
Synonym: cuprous acetylide
Formula: CuC≡CCu
Molecular Formula: C$_2$Cu$_2$
Molecular Weight: 151.114
CAS RN: 1117-94-8
Properties: amorphous red powd; unstable; oxidizes in air to Cu$_2$O, carbon, and water, structure CuC≡CCu; can explode when subjected to shock or heated; obtained by reacting acetylene, HC≡CH, with soluble cuprous salt in water; used in detonators and other explosives [HAW93] [KIR78]
Solubility: v sl s H$_2$O [CRC10]
Melting Point, °C: explodes [CRC10]

1040
Compound: Copper(I) azide
Formula: CuN$_3$
Molecular Formula: CuN$_3$
Molecular Weight: 105.566
CAS RN: 14336-80-2
Properties: colorless tetr, a = 0.865 nm, c = 0.559 nm; M−N$_3$ bond length, 0.223 nm; highly explosive [CRC10] [CIC73]
Solubility: 0.00075 g/100 mL H$_2$O (20°C) [CRC10]
Density, g/cm^3: 3.26 [CRC10]

1041
Compound: Copper(I) bromide
Synonym: cuprous bromide
Formula: CuBr

Molecular Formula: BrCu
Molecular Weight: 143.450
CAS RN: 7787-70-4
Properties: −80 mesh with 99.999% and 99% purity; white powd; cub cryst; hygr; turns green to dark blue in sunlight; used as a catalyst for organic reactions [HAW93] [MER06] [STR93] [CER91]
Solubility: mol/dm^3 CuBr (mol/dm^3 KBr), 24.8°C: 0.00064 (0.0989), 0.00633 (0.3955), 0.01539 (0.5933), 0.0496 (0.9888), 0.334 (1.978), 1.005 (2.966), 1.860 (3.955) [FRI87]
Density, g/cm^3: 4.72 [ALD94]
Melting Point, °C: 497 [CRC10]
Boiling Point, °C: 1345 [KIR78]
Reactions: oxidizes slowly in air becoming green [KIR78]
Thermal Conductivity, W/(m · K): 1.25 [CRC10]
Thermal Expansion Coefficient: 15.4 × 10^{-6}/K [CRC10]

1042
Compound: Copper(I) chloride
Synonyms: nantokite, cuprous chloride
Formula: CuCl
Molecular Formula: ClCu
Molecular Weight: 98.999
CAS RN: 7758-89-6
Properties: −80 mesh with 99.999% and 99% purity; white cub cryst; stable to air and light, if dry; turns green in presence of moisture, sensitive to light, becoming brown; enthalpy of fusion 10.20 kJ/mol; used as a catalyst, preservative, and fungicide [HAW93] [MER06] [CER91] [CRC10]
Solubility: sl s H$_2$O; s conc HCl, conc NH$_4$OH with complex formation [MER06]
Density, g/cm^3: 4.14 [MER06]
Melting Point, °C: 430 [CRC10]
Boiling Point, °C: 1490 [STR93]; 1366 [KIR78]
Thermal Conductivity, W/(m · K): 0.84 [CRC10]
Thermal Expansion Coefficient: 12.1 × 10^{-6}/K (300 K) [CRC10]

1043
Compound: Copper(I) cyanide
Synonym: cuprous cyanide
Formula: CuCN
Molecular Formula: CCuN
Molecular Weight: 89.564
CAS RN: 544-92-3
Properties: white to cream colored powd; colorless or dark green ortho-rhomb cryst; used in plating copper, antifouling paints, and insecticide [HAW93] [MER06]

Solubility: i H_2O, alcohol, cold dil acids; s NH_4OH, sodium and potassium cyanide solutions [HAW93] [MER06]
Density, g/cm³: 1.9 [HAW93]; 2.92 [STR93]
Melting Point, °C: 474 [MER06]
Boiling Point, °C: decomposes [STR93]
Reactions: decomposed by HNO_3 or boiling HCl [MER06]

1044
Compound: Copper(I) fluoride
Formula: CuF
Molecular Formula: CuF
Molecular Weight: 82.544
CAS RN: 13478-41-6
Properties: cub cryst [CRC10]
Density, g/cm³: 7.1 [CRC10]

1045
Compound: Copper(I) hydride
Formula: CuH
Molecular Formula: CuH
Molecular Weight: 64.554
CAS RN: 13517-00-5
Properties: light reddish brown; composition can vary from $Cu_{0.6}$ to CuH; can be prepared by precipitation with hypophosphorous acid at 65°C; used as a reducing agent in some organic reactions [KIR78]
Solubility: decomposes in H_2O at 65°C [CRC10]
Density, g/cm³: 6.38 [CRC10]
Melting Point, °C: decomposes at >60 [KIR78]

1046
Compound: Copper(I) iodide
Synonyms: marshite, cuprous iodide
Formula: CuI
Molecular Formula: CuI
Molecular Weight: 190.450
CAS RN: 7681-65-4
Properties: −80 mesh with 99.999% purity and −60 mesh with 99% purity; pure white or brownish powd; light sensitive; can be prepared by reacting solution of Cu^{++} with I^- to precipitate CuI; has been used in the manufacture of photographic emulsions, catalyst, and in cloud seeding to assist rainfall [KIR78] [CER91] [ALD94]
Solubility: i H_2O, dil acids; s aq NH_3 solutions [MER06]
Density, g/cm³: 5.63 [MER06]
Melting Point, °C: 605 [CRC10]
Boiling Point, °C: ~1290 [MER06]
Reactions: decomposed by conc H_2SO_4 and HNO_3 [MER06]

Thermal Conductivity, W/(m·K): 1.68 (25°C) [CRC10]
Thermal Expansion Coefficient: 19.2×10^{-6}/K [CRC10]

1047
Compound: Copper(I) mercury iodide
Synonym: cuprous mercuric iodide
Formula: Cu_2HgI_4
Molecular Formula: Cu_2HgI_4
Molecular Weight: 835.300
CAS RN: 13876-85-2
Properties: −60 mesh with 99.5% purity; α-form tetra, deep red; β-form cub, brownish; thermochromic; used for detecting overheating of machine bearings; red color changes to brownish-black at 60°C–70°C, then back to red when cooled; obtained by precipitation from a solution of K_2HgI_4 and cuprous chloride [KIR81] [MER06] [CER91] [CRC10]
Solubility: i H_2O [KIR81]
Density, g/cm³: α: 6.116; β: 6.102 [CRC10]

1048
Compound: Copper(I) oxide
Synonyms: cuprite, cuprous oxide
Formula: Cu_2O
Molecular Formula: Cu_2O
Molecular Weight: 143.091
CAS RN: 1317-39-1
Properties: −200 mesh with 99% purity; cub cryst; color may be yellow, red or brown, depending on method of preparation; brownish red mineral; enthalpy of fusion 64.8 kJ/mol; stable at high temp, can be made by thermal decomposition of CuO above 1030°C; used as a catalyst, fungicide, in purification of helium, and as an antioxidant in lubricants [MER06] [KIR78] [CER91] [JAN85]
Solubility: i H_2O; s NH_4OH, HCl; reacts with dil H_2SO_4, HNO_3 [MER06]
Density, g/cm³: 6.0 [MER06]
Melting Point, °C: 1235 [KIR78]
Boiling Point, °C: decomposes at >1800 [KIR78]

1049
Compound: Copper(I) selenide
Synonym: cuprous selenide
Formula: Cu_2Se
Molecular Formula: Cu_2Se
Molecular Weight: 206.052
CAS RN: 20405-64-5
Properties: 6 mm pieces and smaller with 99.5% purity; bluish black to black; tetr or cub cryst; metallic luster; semiconductor [MER06] [CER91]
Solubility: i H_2O; s HCl, evolves H_2Se [MER06]

Density, g/cm³: 6.84 [MER06]
Melting Point, °C: 1113 [MER06]

1050

Compound: Copper(I) sulfide
Synonyms: chalcocite, cuprous sulfide
Formula: Cu_2S
Molecular Formula: Cu_2S
Molecular Weight: 159.158
CAS RN: 22205-45-4
Properties: −200 mesh with 99.999% purity;
blue to grayish black lustrous powd; ortho-
rhomb cryst; mineral chalcocite has hardness
2.5–3 Mohs [KIR78] [MER06] [CER91]
Solubility: i H_2O, acetic acid; sl s HCl [MER06]
Density, g/cm³: 5.6 [MER06]
Melting Point, °C: ~1100 [MER06]
Reactions: decomposed by HNO_3 and
conc H_2SO_4 [MER06]

1051

Compound: Copper(I) sulfite hemihydrate
Synonyms: Etard's salt, cuprous sulfite hemihydrate
Formula: $Cu_2SO_3 \cdot 1/2H_2O$
Molecular Formula: $Cu_2HO_{3.5}S$
Molecular Weight: 216.164
CAS RN: 13982-53-1
Properties: white to pale yellow; hex
cryst; fungicide [MER06]
Solubility: sl s H_2O; s HCl, NH_4OH,
alkali solutions [MER06]

1052

Compound: Copper(I) sulfite monohydrate
Synonyms: Rogojski's salt, cuprous sulfite monohydrate
Formula: $Cu_2SO_3 \cdot H_2O$
Molecular Formula: $Cu_2H_2O_4S$
Molecular Weight: 225.172
CAS RN: 13982-53-1
Properties: white cryst powd or brick red solid;
shown to have an equimolar composition of
metallic Cu and Chevreul's salt: $Cu_3O_6S_2$, cupro-
cupric sulfate; used as a catalyst, fungicide,
and in dyeing textiles [HAW93] [MER06]
Solubility: i H_2O; s NH_4OH, decomposes
in HCl [HAW93]
Density, g/cm³: 3.83 [HAW93]
Melting Point, °C: decomposes [CRC10]

1053

Compound: Copper(I) telluride
Formula: Cu_2Te

Molecular Formula: Cu_2Te
Molecular Weight: 254.692
CAS RN: 12019-52-2
Properties: black hex; 3 mm pieces and smaller
with 99.5% purity [CER91] [LID94]
Density, g/cm³: 4.6 [STR93]

1054

Compound: Copper(I) thiocyanate
Synonym: cuprous thiocyanate
Formula: CuSCN
Molecular Formula: CCuNS
Molecular Weight: 121.630
CAS RN: 1111-67-7
Properties: white to yellow, amorphous powd; used
in manufacturing organic chemicals, antifouling
paints, and in printing textiles [HAW93] [MER06]
Solubility: i H_2O, dil acids, alcohol, acetone;
s NH_4OH, ether [MER06]
Density, g/cm³: 2.843 [HAW93]
Melting Point, °C: 1084 [HAW93]
Reactions: decomposed by conc mineral acids [MER06]

1055

Compound: Copper(I,II) sulfite dihydrate
Synonym: Chevreul's salt
Formula: $Cu_2SO_3 \cdot CuSO_3 \cdot 2H_2O$
Molecular Formula: $Cu_3H_4O_8S_2$
Molecular Weight: 386.797
CAS RN: 13814-81-8
Properties: red; microcryst powd or
prismatic cryst [MER06]
Solubility: i H_2O, alcohol; s NH_4OH, HCl [MER06]
Density, g/cm³: 3.57 [CRC10]
Melting Point, °C: decomposes at 200 [CRC10]

1056

Compound: Copper(II) acetate
Synonyms: acetic acid, Cu(II) salt
Formula: $Cu(CH_3COO)_2$
Molecular Formula: $C_4H_6CuO_4$
Molecular Weight: 181.636
CAS RN: 142-71-2
Properties: bluish green powd; hygr [STR93]

1057

Compound: Copper(II) acetate metaarsenite
Synonyms: Paris green, cupric acetoarsenite
Formula: $Cu(CH_3COO)_2 \cdot 3Cu(AsO_2)_2$
Molecular Formula: $C_4H_6As_6Cu_4O_{16}$

Molecular Weight: 1013.796
CAS RN: 12002-03-8
Properties: emerald green cryst powd; stable to air and light; used as an insecticide, wood preservative, and paint pigment [KIR78] [MER06]
Solubility: i H_2O, decomposed by prolonged heating; unstable in acids [MER06]

1058

Compound: Copper(II) acetate monohydrate
Synonyms: neutral verdigris, cupric acetate monohydrate
Formula: $Cu(CH_3COO)_2 \cdot H_2O$
Molecular Formula: $C_4H_8CuO_5$
Molecular Weight: 199.651
CAS RN: 6046-93-1
Properties: dark green; monocl cryst; forms dimeric units; effloresces in dry air [KIR78] [MER06]
Solubility: s H_2O, alcohol; sl s ether, glycerol [MER06]
Density, g/cm³: 1.882 [MER06]
Melting Point, °C: 115 [MER06]
Boiling Point, °C: decomposes at 240 [MER06]

1059

Compound: Copper(II) acetylacetonate
Synonyms: 2,4-pentanedione, copper(II) derivative
Formula: $Cu(CH_3COCH=C(O)CH_3)_2$
Molecular Formula: $C_{10}H_{14}CuO_4$
Molecular Weight: 261.765
CAS RN: 13395-16-9
Properties: cryst blue powd; does not hydrolyze readily [HAW93] [STR93]
Solubility: sl s H_2O, alcohol; s chloroform [HAW93]
Melting Point, °C: decomposes at 284–285 [STR93]
Boiling Point, °C: sublimes 78 at 0.05 mm Hg pressure [STR93]

1060

Compound: Copper(II) acetylide
Formula: CuC_2
Molecular Formula: C_2Cu
CAS RN: 12540-13-5
Molecular Weight: 87.567
Properties: brown-black solid; explodes [CRC10]
Melting Point, °C: explodes at 100 [CRC10]

1061

Compound: Copper(II) arsenate
Formula: $Cu_3(AsO_4)_2$
Molecular Formula: $As_2Cu_3O_8$
Molecular Weight: 468.476
CAS RN: 10103-61-4

Properties: light blue, blue, or greenish blue; can have a variable composition; used as an insecticide and fungicide [HAW93]
Solubility: i H_2O, alcohol; s dil acids, NH_4OH [HAW93]

1062

Compound: Copper(II) arsenite
Synonym: Scheele's green
Formula: $CuHAsO_3$ (also, $Cu(AsO_2)_2$)
Molecular Formula: $AsCuHO_3$
Molecular Weight: 187.474
CAS RN: 10290-12-7
Properties: yellowish green powd; used as an insecticide, fungicide, paint pigment, and as a wood preservative [HAW93] [MER06]
Solubility: i H_2O, alcohol; s acids, ammonia [MER06]
Melting Point, °C: decomposes [HAW93]

1063

Compound: Copper(II) azide
Formula: $Cu(N_3)_2$
Molecular Formula: CuN_6
Molecular Weight: 147.586
CAS RN: 14215-30-6
Properties: brownish-red; ortho-rhomb, a = 0.923 nm, b = 1.323 nm, c = 0.307 nm [CRC10] [CIC73]
Solubility: 0.008 g/100 mL H_2O (20°C) [CRC10]
Density, g/cm³: 2.604 [CRC10]
Reactions: explodes at 215°C [CRC10]

1064

Compound: Copper(II) basic acetate
Synonym: blue verdigris
Formula: $Cu(CH_3COO)_2 \cdot CuO \cdot 6H_2O$
Molecular Formula: $C_4H_{18}Cu_2O_{11}$
Molecular Weight: 369.272
CAS RN: 52503-64-7
Properties: formula is approximate; compounds are in the form of blue cryst or blue to green powd [MER06] [CRC10]
Solubility: sl s H_2O, alcohol; s dil acids, ammonia [MER06]

1065

Compound: Copper(II) basic chromate
Synonym: basic cupric chromate
Formula: $CuCrO_4 \cdot 2Cu(OH)_2$
Molecular Formula: $CrCu_3H_4O_8$
Molecular Weight: 340.646
CAS RN: 12433-14-6

Properties: light chocolate brown powd; used as
   a mordant in dyeing, as a wood preservative,
   and to treat seeds for fungus [HAW93]
Solubility: i H_2O; s HNO_3 [HAW93]
Reactions: minus water at 260°C [HAW93]

1066
Compound: Copper(II) basic nitrite
Formula: $Cu(NO_2)_2 \cdot 3Cu(OH)_2$
Molecular Formula: $Cu_4H_6N_2O_{10}$
Molecular Weight: 448.239
CAS RN: 14984-71-5
Properties: green powd [HAW93]
Solubility: sl s H_2O; s NH_4OH; decomposes
   in dil acids [HAW93]
Melting Point, °C: decomposes at 120 [HAW93]
Reactions: decomposed by hot H_2O [CRC10]

1067
Compound: Copper(II) benzoate dihydrate
Formula: $Cu(C_6H_5COO)_2 \cdot 2H_2O$
Molecular Formula: $C_{14}H_{14}CuO_6$
Molecular Weight: 341.807
CAS RN: 6046-97-5
Properties: blue, cryst, odorless powd [HAW93]
Solubility: sl s cold H_2O, alcohol, and acids [HAW93]
Reactions: minus $2H_2O$ at 100°C [HAW93]

1068
Compound: Copper(II) borate
Synonym: copper(II) metaborate
Formula: $Cu(BO_2)_2$
Molecular Formula: B_2CuO_4
Molecular Weight: 149.166
CAS RN: 39290-85-2
Properties: amorphous or bluish green cryst
   powd; can be prepared by adding borax to an
   aq Cu^{++} sulfate or chloride solution, which
   precipitates the borate; used as an oil pigment,
   a dehydrogenation catalyst, wood preservative,
   and fire retardant additive [HAW93] [KIR78]
Solubility: i H_2O; s acids [HAW93]
Density, g/cm³: 3.859 [HAW93]

1069
Compound: Copper(II) bromide
Synonym: cupric bromide
Formula: $CuBr_2$
Molecular Formula: Br_2Cu
Molecular Weight: 223.354
CAS RN: 7789-45-9
Properties: almost black, deliq, monocl cryst
   or gray powd [MER06] [STR93]

Solubility: s alcohol, acetone, ammonia; i benzene,
   ether [MER06]; g/100 g soln, H_2O: 55.7 (0°);
   solid phase, $CuBr_2$ [KRU93]; g/100 g H_2O: 107
   (0°C), 126 (20°C), 131 (50°C) [LAN05]
Density, g/cm³: 4.710 [MER06]
Melting Point, °C: 498 [MER06]
Boiling Point, °C: 900 [MER06]

1070
Compound: Copper(II) butanoate monohydrate
Formula: $Cu(C_4H_7O_2)_2 \cdot H_2O$
Molecular Formula: $C_8H_{16}CuO_5$
Molecular Weight: 255.756
CAS RN: 540-16-9
Properties: green monocl plates
Solubility: s H_2O, dioxane, benzene; sl EtOH [CRC10]

1071
Compound: Copper(II) butyrate monohydrate
Synonym: cupric butyrate monohydrate
Formula: $Cu(CH_3CH_2CH_2COO)_2 \cdot H_2O$
Molecular Formula: $C_8H_{16}CuO_5$
Molecular Weight: 255.758
CAS RN: 540-16-9
Properties: large, dark green; monocl, hex
   plates; becomes dull and disintegrates after
   several days of exposure to air [MER06]
Solubility: s H_2O, dioxane, benzene; sl s
   alcohol, chloroform [MER06]

1072
Compound: Copper(II) carbonate
Formula: $CuCO_3$
Molecular Formula: $CCuO_3$
Molecular Weight: 123.555
CAS RN: 1184-64-1
Properties: cryst [CRC10]
Solubility: i H_2O

1073
Compound: Copper(II) carbonate hydroxide
Synonyms: malachite, Bremen green
Formula: $CuCO_3 \cdot Cu(OH)_2$
Molecular Formula: $CH_2Cu_2O_5$
Molecular Weight: 221.116
CAS RN: 12069-69-1
Properties: green to blue amorphous powd or
   dark green monocl cryst; malacite hardness
   3.5–4 Mohs [KIR78] [MER06]
Solubility: i H_2O, alcohol; s dil acids,
   ammonia [MER06]

Density, g/cm³: 4.0 [STR93]
Melting Point, °C: decomposes at 200 [STR93]

1074
Compound: Copper(II) chlorate hexahydrate
Synonym: cupric chlorate hexahydrate
Formula: $Cu(ClO_3)_2 \cdot 6H_2O$
Molecular Formula: $Cl_2CuH_{12}O_{12}$
Molecular Weight: 338.539
CAS RN: 14721-21-2
Properties: blue to green; deliq;
   octahedral cryst [MER06]
Solubility: mol/100 mol H_2O: 11.02 (0.8°C),
   equilibrium solid phase $Cu(ClO_3)_2 \cdot 4H_2O$
   [KRU93]; v s alcohol [MER06]
Melting Point, °C: 65 [MER06]
Boiling Point, °C: decomposes at 100 [MER06]

1075
Compound: Copper(II) chloride
Synonym: cupric chloride
Formula: $CuCl_2$
Molecular Formula: Cl_2Cu
Molecular Weight: 134.451
CAS RN: 7447-39-4
Properties: −80 mesh with 99% purity; yellow to
   brown; deliq; microcryst powd; usually exists as
   blue-green dihydrate; enthalpy of fusion 20.40 kJ/
   mol [CRC10] [MER06] [KIR78] [CER91]
Solubility: s H_2O, alcohol, acetone [MER06]; g/100 g
   soln, H_2O: 40.7 ± 10.2 (0°C), 43.8 ± 0.6 (25°C),
   54.6 (100°C); solid phase, $CuCl_2 \cdot 2H_2O$ [KRU93]
Density, g/cm³: 3.386 [STR93]
Melting Point, °C: 620 [ALD94]
Boiling Point, °C: decomposes to CuCl at 993 [KIR78]

1076
Compound: Copper(II) chloride dihydrate
Synonyms: eriochalcite, cupric chloride dihydrate
Formula: $CuCl_2 \cdot 2H_2O$
Molecular Formula: $Cl_2CuH_4O_2$
Molecular Weight: 170.482
CAS RN: 10125-13-0
Properties: green to blue powd or ortho-rhomb
   cryst; deliq in moist air, effloresces in dry
   air; used as catalyst in organic synthesis, e.g.,
   chlorination; used as pigment [MER06] [KIR78]
Solubility: 1 g dihydrate/1 mL H_2O (25°C)
   [KIR78]; s methanol, ethanol [MER06]
Density, g/cm³: 2.51 [MER06]
Melting Point, °C: ~100 [MER06]
Reactions: minus $2H_2O$ if heated at 120°C
   in a stream of HCl [KIR78]

1077
Compound: Copper(II) chromate
Formula: $CuCrO_4$
Molecular Formula: $CrCuO_4$
Molecular Weight: 179.540
CAS RN: 13548-42-0
Properties: reddish brown; used as a fungicide,
   to weatherproof textiles, in epoxy adhesives,
   and to preserve wood [KIR78]
Solubility: mol/L soln, H_2O: 0.0020 (25°C);
   solid phase, $CuCrO_4 \cdot H_2O$ [KRU93]
Melting Point, °C: decomposes at 400 [KIR78]
Reactions: minus O_2 to form blue-black
   $CuOCr_2O_3$ at 400°C [KIR78]

1078
Compound: Copper(II) chromite
Synonym: cupric chromite
Formula: $CuCr_2O_4$
Molecular Formula: Cr_2CuO_4
Molecular Weight: 231.536
CAS RN: 12018-10-9
Properties: grayish black to black; tetr cryst;
   used in automobile exhaust catalysts; there
   is also $2CuO \cdot Cr_2O_3$, CAS RN 12053-
   18-8 [KIR78] [MER06] [ALD94]
Solubility: i H_2O, dil acids, conc HCl [MER06]
Density, g/cm³: 5.4 [LID94]

1079
Compound: Copper(II) citrate hemipentahydrate
Formula: $Cu_2C_6H_4O_7 \cdot 2.5H_2O$
Molecular Formula: $C_6H_9O_{9.5}$
Molecular Weight: 360.221
CAS RN: 10402-15-0
Properties: blue-green cryst [CRC10]
Solubility: sl H_2O; s dil acid
Melting Point, °C: decomposes at 100 [CRC10]

1080
Compound: Copper(II) cyanide
Synonym: cupric cyanide
Formula: $Cu(CN)_2$
Molecular Formula: C_2CuN_2
Molecular Weight: 115.581
CAS RN: 14763-77-0
Properties: green powd; used to electroplate
   copper onto iron [HAW93]
Solubility: i H_2O; s acids and alkalies [HAW93]
Melting Point, °C: decomposes [CRC10]

1081
Compound: Copper(II) cyclohexanebutanoate
Formula: $Cu(C_{10}H_{17}O_2)_2$
Molecular Formula: $C_{20}H_{34}CuO_4$
Molecular Weight: 402.028
CAS RN: 2218-80-6
Properties: powd [CRC10]
Melting Point, °C: decomposes at 126 [CRC10]

1082
Compound: Copper(II) dichromate dihydrate
Formula: $CuCr_2O_7 \cdot 2H_2O$
Molecular Formula: $Cr_2CuH_4O_9$
Molecular Weight: 315.565
CAS RN: 13675-47-3
Properties: reddish brown; tricl; used in catalysts and as a wood preservative [KIR78]
Solubility: v s H_2O [KIR78]
Density, g/cm³: 2.286 [KIR78]
Reactions: minus $2H_2O$ at 100°C [CRC10]

1083
Compound: Copper(II) ethanolate
Formula: $Cu(C_2H_5O)_2$
Molecular Formula: $C_4H_{10}CuO_2$
Molecular Weight: 153.667
CAS RN: 2850-65-9
Properties: blue hygr solid [CRC10]
Solubility: i organic solvents [CRC10]
Melting Point, °C: decomposes at 120 [CRC10]

1084
Compound: Copper(II) ethylacetoacetate
Formula: $Cu(C_2H_5CO_2CHCOCH_3)_2$
Molecular Formula: $C_{12}H_{18}CuO_6$
Molecular Weight: 321.813
CAS RN: 14284-06-1
Properties: green cryst [CRC10]
Solubility: s EtOH, chloroform [CRC10]
Melting Point, °C: 192 [CRC10]

1085
Compound: Copper(II) 2-ethylhexanoate
Formula: $Cu(C_8H_{15}O_2)_2$
Molecular Formula: $C_{16}H_{30}CuO_4$
Molecular Weight: 349.953
CAS RN: 149-11-1
Properties: powd [CRC10]
Melting Point, °C: decomposes at 252 [CRC10]

1086
Compound: Copper(II) ferrate
Synonym: copper diiron tetroxide
Formula: $CuFe_2O_4$
Molecular Formula: $CuFe_2O_4$
Molecular Weight: 239.234
CAS RN: 12018-79-0
Properties: black cryst; spinel structure; magnetic properties; formed by reaction of CuO and Fe_2O_3 at ~900°C or by coprecipitation and sintering at 900°C; has been used to manufacture piezomagnets and magnetic tapes; also used as a catalyst, e.g., exhaust control [KIR78] [STR93]

1087
Compound: Copper(II) ferrocyanide
Synonyms: Hatchett's brown, cupric ferrocyanide
Formula: $Cu_2Fe(CN)_6$
Molecular Formula: $C_6Cu_2FeN_6$
Molecular Weight: 339.043
CAS RN: 13601-13-3
Properties: reddish brown powd or cub cryst; precipitates as gel or colloid [MER06]
Solubility: i H_2O, dil acids; s NH_4OH [MER06]

1088
Compound: Copper(II) ferrous sulfide
Synonym: chalcopyrite
Formula: $CuFeS_2$
Molecular Formula: $CuFeS_2$
Molecular Weight: 183.523
CAS RN: 1308-56-1
Properties: yellow brass colored cryst; tetr; hardness 3.5–4.0; chalcopyrite is an ore; can be synthesized by reacting KFeS with ammonia copper solution; used in semiconductor research [MER06]
Solubility: s HNO_3, aqua regia; i HCl [MER06]
Density, g/cm³: 4.1–4.3 [MER06]
Melting Point, °C: 950 [MER06]
Thermal Expansion Coefficient: (volume) 100°C (0.42) [CLA66]

1089
Compound: Copper(II) fluoride
Synonym: cupric fluoride
Formula: CuF_2
Molecular Formula: CuF_2
Molecular Weight: 101.543
CAS RN: 7789-19-7

Properties: −100 mesh with 99.5% purity; white powd; monocl cryst; hygr, turns blue in moist air, forming dihydrate; enthalpy of fusion 55.23 kJ/mol; prepared by reaction of copper oxide or basic carbonate with anhydrous HF, followed by dehydration in stream of anhydrous HF at high temp; finds use as a catalyst in organic reactions, as a fluorinating agent, and in high-density batteries [KIR78] [MER06] [STR93] [CER91] [CRC10] [JAN82]

Solubility: 4.7 g/100 mL H_2O (20°C); hydrolyzed to CuFOH in hot H_2O [MER06]; g/100 ml soln, H_2O: 0.075 (25°C); solid phase, $CuF_2 \cdot 2H_2O$ [KRU93]

Density, g/cm³: 4.23 [STR93]

Melting Point, °C: 836 [CRC10]

Boiling Point, °C: 1678 [JAN85]

1090

Compound: Copper(II) fluoride dihydrate

Synonym: cupric fluoride dihydrate

Formula: $CuF_2 \cdot 2H_2O$

Molecular Formula: $CuF_2H_4O_2$

Molecular Weight: 137.574

CAS RN: 13454-88-1

Properties: blue; monocl cryst; a trihydrate is also listed with density 2.93 [MER06] [ALD94]

Solubility: sl s cold H_2O, hydrolyzed to CuFOH in hot H_2O [MER06]

Density, g/cm³: 2.934 [MER06]

Melting Point, °C: decomposes at >130 [MER06]

1091

Compound: Copper(II) formate

Synonym: cupric formate

Formula: $Cu(HCOO)_2$

Molecular Formula: $C_2H_2CuO_4$

Molecular Weight: 153.582

CAS RN: 544-19-4

Properties: three forms of anhydrous formate exist: powd blue, turquoise or royal blue cryst [MER06]

Solubility: 12.5 g/100 mL H_2O [CRC10]; i most organic solvents [MER06]

Density, g/cm³: 1.831 [STR93]

1092

Compound: Copper(II) formate tetrahydrate

Synonym: cupric formate tetrahydrate

Formula: $Cu(HCOO)_2 \cdot 4H_2O$

Molecular Formula: $C_2H_{10}CuO_8$

Molecular Weight: 225.642

CAS RN: 5893-61-8

Properties: large, light blue; monocl; powd blue modification formed by dehydration over $CaCl_2$, under reduced pressure [MER06]

Solubility: 6.2 g/100 mL H_2O [CRC10]; v sl s alcohol [MER06]

Density, g/cm³: 1.81 [CRC10]

Reactions: minus H_2O at 130°C [CRC10]

1093

Compound: Copper(II) gluconate

Formula: $Cu(CH_2OH(CHOH)_4COO)_2$

Molecular Formula: $C_{12}H_{22}CuO_{14}$

Molecular Weight: 453.845

CAS RN: 527-09-3

Properties: forms light blue to bluish green water soluble cryst; used as a feed additive, dietary supplement, mouth deodorant, and to treat arthritis [HAW93] [KIR78] [STR93]

Solubility: s H_2O; i alcohol, ether, acetone [HAW93]

1094

Compound: Copper(II) glycinate monohydrate

Synonym: cupric glycinate hydrate

Formula: $Cu(H_2NCH_2COO)_2 \cdot H_2O$

Molecular Formula: $C_4H_{10}CuN_2O_5$

Molecular Weight: 229.679

CAS RN: 13479-54-4

Properties: deep blue; long rhomb needles [MER06]

Solubility: s H_2O, sl s alcohol [MER06]

Melting Point, °C: chars 213 [MER06]

Boiling Point, °C: decomposes with evolution of gas at 228 [MER06]

1095

Compound: Copper(II) hexafluoroacetylacetonate

Synonyms: 1,1,1,5,5,5-hexafluoro-2,4-pentanedione, Cu(II) derivative

Formula: $Cu(CF_3COCHCOCF_3)_2$

Molecular Formula: $C_{10}H_2CuF_{12}O_4$

Molecular Weight: 477.650

CAS RN: 14781-45-4

Properties: blue-green cryst; hygr [STR94]

Melting Point, °C: 85–89 [STR93]

Boiling Point, °C: sublimes at 70 (0.05 mm Hg) [STR93]

Reactions: decomposes at 220°C [STR93]

1096

Compound: Copper(II) hexafluorosilicate tetrahydrate

Synonym: cupric hexafluorosilicate tetrahydrate

Formula: $CuSiF_6 \cdot 4H_2O$

Molecular Formula: $CuF_6H_8O_4Si$
Molecular Weight: 277.684
CAS RN: 12062-24-7
Properties: blue, monocl, efflorescent cryst; decomposed by heating [HAW93] [MER06]
Solubility: g anhydrous/100 g H_2O: 73.5 (0°C), 81.6 (20°C), 93.2 (75°C) [LAN05]; sl s alcohol [HAW93] [MER06]
Density, g/cm³: 2.56 [MER06]

1097
Compound: Copper(II) hexafluoroacetylacetonate hydrate
Synonyms: 1,1,1,5,5,5-hexafluoro-2,4-pentanedione, Cu(II) derivative hydrate
Formula: $Cu[CF_3COCH=C(O)CF_3]_2 \cdot xH_2O$
Molecular Formula: $C_{10}H_2CuF_{12}O_4$ (anhydrous)
Molecular Weight: 477.650 (anhydrous)
CAS RN: 14781-45-4
Properties: bluish-green cryst; precursor for metal oxide chemical vapor deposition [STR94] [STR93] [ALD94]
Melting Point, °C: 130–134 [ALD94]

1098
Compound: Copper(II) hydroxide
Synonym: cupric hydroxide
Formula: $Cu(OH)_2$
Molecular Formula: CuH_2O_2
Molecular Weight: 97.561
CAS RN: 20427-59-2
Properties: blue to bluish green gel or light blue cryst powd, which decomposes to CuO by sl warming; can be prepared by anodic dissolution of a copper anode in sodium sulfate or trisodium phosphate solution; used as a fungicide and as a constituent of antifouling marine paints [MER06]
Solubility: i H_2O; when freshly precipitated, s conc alkali; s acids, NH_4OH [MER06]; mol/L soln, H_2O: 3×10^{-5} (25°C) [KRU93]
Density, g/cm³: 3.37 [MER06]
Reactions: minus H_2O on heating [CRC10]

1099
Compound: Copper(II) hydroxy chloride
Formula: $CuCl_2 \cdot 3Cu(OH)_2$
Molecular Formula: $Cl_2Cu_4H_6O_6$
Molecular Weight: 427.133
CAS RN: 16004-08-3

Properties: this basic chloride as well as other compositions can be synthesized by precipitation at controlled pH; other compositions include Cu_2OCl_2, 12167-76-6, and $Cu_4O_3Cl_2$, 12356-86-4; compounds have been used in crop protection, electronics, metallurgy, and as catalysts [KIR78]
Solubility: i H_2O [CRC10]
Density, g/cm³: 3.75 [CRC10]
Reactions: minus H_2O at 250°C [CRC10]

1100
Compound: Copper(II) iodate
Formula: $Cu(IO_3)_2$
Molecular Formula: CuI_2O_6
Molecular Weight: 413.351
CAS RN: 13454-89-2
Properties: green monocl cryst [CRC10]
Density, g/cm³: 5.241 [CRC10]
Solubility: g/100 g H_2O: 0.15^{20}; s dil acid [CRC10]
Melting Point, °C: decomposes [CRC10]

1101
Compound: Copper(II) iodate monohydrate
Formula: $Cu(IO_3)_2 \cdot H_2O$
Molecular Formula: $CuH_2I_2O_7$
Molecular Weight: 431.367
CAS RN: 13454-90-5
Properties: blue tricl cryst [CRC10]
Solubility: g/100 g H_2O: 0.15^{20}; s dil H_2SO_4 [CRC10]
Melting Point, °C: decomposes at 248 [CRC10]
Density, g/cm³: 4.872 [CRC10]

1102
Compound: Copper(II) lactate dihydrate
Formula: $Cu(C_3H_5O_3)_2 \cdot 2H_2O$
Molecular Formula: $C_6H_{14}CuO_8$
Molecular Weight: 277.718
CAS RN: 16039-52-4
Properties: greenish blue cryst or granules; fungicide [HAW93]
Solubility: 16.7 g/100 mL H_2O [CRC10]; NH_4OH [HAW93]

1103
Compound: Copper(II) molybdate
Formula: $CuMoO_4$
Molecular Formula: $CuMoO_4$
Molecular Weight: 223.484
CAS RN: 13767-34-5

Properties: light green cryst; used as a paint pigment, corrosion inhibitor, and in protective coatings [HAW93] [KIR81] [MOI86]
Solubility: 0.038 g/100 g H_2O [KIR81]
Density, g/cm³: 3.4 [HAW93]
Melting Point, °C: decomposes at 820 [KIR81]

1104
Compound: Copper(II) nitrate
Synonym: cupric nitrate
Formula: $Cu(NO_3)_2$
Molecular Formula: CuN_2O_6
Molecular Weight: 187.555
CAS RN: 3251-23-8
Properties: large, bluish green; deliq; ortho-rhomb cryst; can be obtained by sublimation of $Cu(NO_3)_2$ under vacuum from a well mixed mixture of $AgNO_3$ and $CuBr_2$ at 200°C; used as ceramic color, mordant in dyeing, and a catalyst [MER06]
Solubility: s H_2O, ethyl acetate, dioxane; reacts with ether [MER06]; g/100 g soln, H_2O: 45.5 (0°C), 60.1 (25°C), 71.2 (100°C); solid phase, $Cu(NO_3)_2 \cdot 6H_2O$ (0°C, 25°C), $Cu(NO_3)_2 \cdot 2\text{-}1/2H_2O$ (100°C) [KRU93]
Melting Point, °C: 255–256 [MER06]

1105
Compound: Copper(II) nitrate hexahydrate
Synonym: cupric nitrate hexahydrate
Formula: $Cu(NO_3)_2 \cdot 6H_2O$
Molecular Formula: $CuH_{12}N_2O_{12}$
Molecular Weight: 295.647
CAS RN: 13478-38-1
Properties: blue; deliq; prismatic cryst [MER06]
Solubility: 243.7 g/100 mL H_2O (0°C) [CRC10]; s alcohol [MER06]
Density, g/cm³: 2.07 [MER06]
Reactions: minus $3H_2O$ at 26.4°C [HAW93]

1106
Compound: Copper(II) nitrate trihydrate
Synonyms: cupric nitrate trihydrate, gerhardite
Formula: $Cu(NO_3)_2 \cdot 3H_2O$
Molecular Formula: $CuH_6N_2O_9$
Molecular Weight: 241.602
CAS RN: 10031-43-3
Properties: blue; deliq; rhomb plates; there is a hemipentahydrate with CAS RN 19004-19-4 with same density and melting point as the trihydrate [MER06] [ALD94]

Solubility: 137.8 g/100 mL H_2O (0°C), 1270 g/100 mL H_2O (100°C) [CRC10]; v s alcohol; i ethyl acetate [MER06]
Density, g/cm³: 2.32 [HAW93]
Melting Point, °C: 114.5 [MER06]
Boiling Point, °C: decomposes at 170 [HAW93]

1107
Compound: Copper(II) oleate
Synonyms: 9-octadecanoic acid, Cu(II) salt
Formula: $Cu(C_{17}H_{33}COO)_2$
Molecular Formula: $C_{36}H_{66}CuO_4$
Molecular Weight: 626.464
CAS RN: 1120-44-1
Properties: blue to green solid; can be prepared by reacting the acid with CuO or basic carbonate of Cu; coalesces mercury drops; improves oil combustion by reducing smoke and fumes of burning oil; used as textile fungicide, used in antifouling marine paints [MER06]
Solubility: i H_2O; sl s alcohol; s ether [MER06]

1108
Compound: Copper(II) oxalate
Synonyms: ethanedioic acid, Cu(II) salt
Formula: CuC_2O_4
Molecular Formula: C_2CuO_4
Molecular Weight: 151.566
CAS RN: 814-91-5
Properties: bluish white powd [MER06]
Solubility: i H_2O, alcohol, ether, acetic acid; s NH_4OH [MER06]
Melting Point, °C: decomposes at 310 [MER06]

1109
Compound: Copper(II) oxalate hemihydrate
Synonym: cupric oxalate hemihydrate
Formula: $CuC_2O_4 \cdot 1/2H_2O$
Molecular Formula: $C_2HCuO_{4.5}$
Molecular Weight: 160.573
CAS RN: 814-91-5
Properties: bluish white powd; loses any hydrated water by 200°C [MER06]
Solubility: i H_2O, alcohol, ether, acetic acid; s NH_4OH [MER06]
Melting Point, °C: anhydrous decomposes at ~300 to copper oxide [HAW93]

1110
Compound: Copper(II) oxide
Synonyms: tenorite [1317-92-6], cupric oxide

Formula: CuO
Molecular Formula: CuO
Molecular Weight: 79.545
CAS RN: 1317-38-0
Properties: −20 mesh with 99.999% and −200 mesh with 99.9% purity; black to brownish black amorphous or cryst powd or granules; enthalpy of fusion 11.80 kJ/mol; can be prepared by oxidation of Cu turnings at 800°C; used as a fungicide, herbicide, and in cloud seeding, as a heat collector surface in solar energy; reduces tar in tobacco smoke [MER06] [KIR78] [CER91] [CRC10]
Solubility: 0.000320 moles/kg alkaline phosphate: 0.203×10^{-6} mol/kg H_2O (26.1°C), 0.675×10^{-6} mol/kg H_2O (94.4°C), 5.241×10^{-6} moles/kg H_2O (190.0°C), 20.43×10^{-6} m/kg H_2O (260.0°C) [ZIE92a]
Density, g/cm³: 6.315 [MER06]
Melting Point, °C: 1446 [CRC10]

1111

Compound: Copper(II) oxychloride
Synonym: Brunswick green
Formula: $CuCl_2 \cdot 3CuO \cdot 3\text{-}1/2H_2O$
Molecular Formula: $Cl_2Cu_4H_7O_{6.5}$
Molecular Weight: 436.141
CAS RN: 1332-40-7
Properties: bluish green powd; used as a pigment, pesticide, and to control fungus growth in grapevines [HAW93]
Solubility: i H_2O; s acids, ammonia [HAW93]
Reactions: minus $3H_2O$ at 140°C [CRC10]

1112

Compound: Copper(II) perchlorate
Synonym: cupric perchlorate
Formula: $Cu(ClO_4)_2$
Molecular Formula: Cl_2CuO_8
Molecular Weight: 262.446
CAS RN: 13770-18-8
Properties: very pale green; hygr cryst; volatilized by heating; thermally stable up to 130°C [MER06]
Solubility: s H_2O, ether, dioxane, ethyl acetate; i benzene, CCl_4 [MER06]; g/100 g soln, H_2O: 54.3 (0°C), 59.3 (30°C); solid phase, $Cu(ClO_4)_2 \cdot 6H_2O$ [KRU93]
Melting Point, °C: ~230–240 [MER06]
Reactions: decomposes to basic perchlorate >130°C [MER06]

1113

Compound: Copper(II) perchlorate hexahydrate
Synonym: cupric perchlorate hexahydrate

Formula: $Cu(ClO_4)_2 \cdot 6H_2O$
Molecular Formula: $Cl_2CuH_{12}O_{14}$
Molecular Weight: 370.538
CAS RN: 10294-46-9
Properties: deep blue; monocl cryst [MER06]
Solubility: v H_2O, methanol, ethanol, acetic acid [MER06]
Density, g/cm³: 2.225 [STR93]
Melting Point, °C: 82 [STR93]
Boiling Point, °C: decomposes at 120 [STR93]

1114

Compound: Copper(II) phosphate
Formula: $Cu_3(PO_4)_2$
Molecular Formula: $Cu_3O_8P_2$
Molecular Weight: 380.581
CAS RN: 7798-23-4
Properties: blue-green tricl cryst [CRC10]
Solubility: i H_2O; s acid, NH_4OH [CRC10]

1115

Compound: Copper(II) phosphate trihydrate
Synonym: cupric phosphate trihydrate
Formula: $Cu_3(PO_4)_2 \cdot 3H_2O$
Molecular Formula: $Cu_3H_6O_{11}P_2$
Molecular Weight: 434.627
CAS RN: 10031-48-8
Properties: blue or olive green; ortho-rhomb cryst; used in chemical analysis, as a fungicide, catalyst, and to inhibit oxidation of metals [HAW93] [MER06]
Solubility: i cold H_2O, sl s hot H_2O; s acids, NH_4OH [MER06]
Reactions: decomposes if heated [MER06]

1116

Compound: Copper(II) phthalocyanine
Synonym: Pigment blue 15C
Molecular Formula: $C_{32}H_{16}CuN_8$
Molecular Weight: 576.079
CAS RN: 147-14-8
Properties: bright blue cryst; both α and β forms, β more stable; used in inks and paints [MER06] [ALD94]
Solubility: s 98% H_2SO_4; i H_2O [MER06]
Melting Point, °C: sublimes at ~580 at low pressure of N_2 [MER06]
Reactions: decomposed by hot HNO_3 [MER06]

1117

Compound: Copper(II) pyrophosphate hydrate
Synonym: Unichrome
Formula: $Cu_2P_2O_7 \cdot xH_2O$

Molecular Formula: $Cu_2O_7P_2$ (anhydrous)
Molecular Weight: 301.035 (anhydrous)
CAS RN: 10102-90-6
Properties: used in electroplating copper [HAW93]

1118
Compound: Copper(II) selenate pentahydrate
Synonym: cupric selenate pentahydrate
Formula: $CuSeO_4 \cdot 5H_2O$
Molecular Formula: $CuH_{10}O_9Se$
Molecular Weight: 296.580
CAS RN: 10031-45-5
Properties: light blue; tricl cryst; used as a black
colorant for copper [HAW93] [MER06]
Solubility: g/100 g soln H_2O: 10.6 ± 0.2 (0°C),
16.0 ± 0.1 (25°C), solid phase $CuSeO_4 \cdot 5H_2O$
[KRU93]; s acids, NH_4OH; v sl s acetone;
i alcohol [HAW93] [MER06]
Density, g/cm³: 2.56 [MER06]
Reactions: forms monohydrate at 150°C–200°C,
anhydrous by 265°C; decomposes to
selenite and basic selenite at ~480°C,
then to CuO at ~700°C [MER06]

1119
Compound: Copper(II) selenide
Synonym: cupric selenide
Formula: CuSe
Molecular Formula: CuSe
Molecular Weight: 142.506
CAS RN: 1317-41-5
Properties: 6 mm pieces and smaller with 99.5%
purity; bluish black to greenish black;
prismatic needles or hex plates; decomposes
at dull red heat [MER06] [CER91]
Solubility: s HCl, evolves H_2Se; s H_2SO_4,
evolves SO_2 [MER06]
Density, g/cm³: 5.99 [STR93]
Melting Point, °C: decomposes at 550 [STR93]
Reactions: oxidized to $CuSeO_3$ by HNO_3 [MER06]

1120
Compound: Copper(II) selenite dihydrate
Synonym: cupric selenite dihydrate
Formula: $CuSeO_3 \cdot 2H_2O$
Molecular Formula: CuH_4O_5Se
Molecular Weight: 226.535
CAS RN: 15168-20-4
Properties: blue; ortho-rhomb or monocl cryst [MER06]
Solubility: i H_2O, H_2SeO_3; s acids, NH_4OH [MER06]
Density, g/cm³: 3.31 [MER06]

Reactions: minus $2H_2O$ by 265°C; decomposes
to $CuSeO_3 \cdot CuO$ at >460°C; decomposes
to CuO at >660°C [MER06]

1121
Compound: Copper(II) silicate dihydrate
Synonym: chrysocolla
Formula: $CuSiO_3 \cdot 2H_2O$
Molecular Formula: CuH_4O_5Si
Molecular Weight: 175.661
CAS RN: 26318-99-0
Properties: green to blue ortho-rhomb;
hardness 2.4 Mohs [KIR78]
Density, g/cm³: 2–2.24 [CRC10]

1122
Compound: Copper(II) stannate
Formula: $CuSnO_3$
Molecular Formula: CuO_3Sn
Molecular Weight: 230.254
CAS RN: 12019-07-7
Properties: blue powd [STR93]

1123
Compound: Copper(II) stearate
Synonyms: octadecanoic acid, Cu(II) salt
Formula: $Cu[CH_3(CH_2)_{16}COO]_2$
Molecular Formula: $C_{36}H_{70}CuO_4$
Molecular Weight: 630.496
CAS RN: 660-60-6
Properties: pale blue to bluish green; amorphous
powd; used in coatings for xerographic plates and
in heat sensitive coatings [MER06] [KIR78]
Solubility: i H_2O, ethanol, ether; s pyridine,
dioxane [MER06]
Density, g/cm³: 1.10 [KIR78]
Melting Point, °C: 112 [KIR78]

1124
Compound: Copper(II) sulfate
Synonym: chalcocyanite
Formula: $CuSO_4$
Molecular Formula: CuO_4S
Molecular Weight: 159.610
CAS RN: 7758-98-7
Properties: grayish white to greenish white;
rhomb cryst or amorphous powd [MER06]
Solubility: i alcohol [MER06]; g/100 g soln,
H_2O: 12.7 ± 0.3 (0°C), 18.4 ± 0.2 (25°C),
42.9 ± 0.5 (100°C); solid phase, $CuSO_4 \cdot 5H_2O$
(0°C, 25°C), $CuSO_4 \cdot 3H_2O$ (100°C) [KRU93]

Density, g/cm³: 3.603 [STR93]
Melting Point, °C: decomposes at 560 [LID94]

1125
Compound: Copper(II) sulfate, basic
Formula: $Cu_3(OH)_4SO_4$
Molecular Formula: $Cu_3H_4O_8S$
Molecular Weight: 354.730
CAS RN: 1332-14-5
Properties: green rhomb cryst [CRC10]
Solubility: i H_2O [CRC10]
Density, g/cm³: 3.88 [CRC10]

1126
Compound: Copper(II) sulfate pentahydrate
Synonym: blue vitriol
Formula: $CuSO_4 \cdot 5H_2O$
Molecular Formula: $CuH_{10}O_9S$
Molecular Weight: 249.686
CAS RN: 7758-99-8
Properties: blue; tricl cryst; can be made by
    reaction of copper with hot, conc H_2SO_4; used
    as fungicide, source of Cu in animal nutrition;
    slowly efflorescent in air [KIR78] [MER06]
Solubility: v s H_2O; s methanol; sl s ethanol [MER06]
Density, g/cm³: 2.286 [MER06]
Melting Point, °C: 110, decomposes [STR93]
Reactions: minus $2H_2O$ at 30°C; minus $2H_2O$ at
    110°C; becomes anhydrous by 250°C [MER06]

1127
Compound: Copper(II) sulfide
Synonyms: covellite, cupric sulfide
Formula: CuS
Molecular Formula: CuS
Molecular Weight: 95.612
CAS RN: 1317-40-4
Properties: −100 mesh with 99.999% and −200
    mesh with 99.5% purity; black powd; stable
    in dry air, oxidized to $CuSO_4$ in moist air;
    used in antifouling paints, catalysts; covellite
    mineral is indigo-blue or darker; hex; hardness
    1.5–2 Mohs [KIR78] [MER06] [CER91]
Solubility: i H_2O, alcohol, dil acids and
    alkalies; s KCN, hot HNO_3 [MER06]
Density, g/cm³: 4.6 [STR93]
Melting Point, °C: decomposes at 220 [HAW93]

1128
Compound: Copper(II) tartrate trihydrate
Synonym: cupric tartrate trihydrate

Formula: $CuC_4H_4O_6 \cdot 3H_2O$
Molecular Formula: $C_4H_{10}CuO_9$
Molecular Weight: 265.664
CAS RN: 815-82-7
Properties: bluish green cryst [STR93]
Solubility: 0.02 g/100 mL H_2O (15°C), 0.14 g/100 mL H_2O
    (85°C) [CRC10]; s acids, alkali solutions [MER06]
Melting Point, °C: decomposes [CRC10]

1129
Compound: Copper(II) telluride
Formula: CuTe
Molecular Formula: CuTe
Molecular Weight: 191.146
CAS RN: 12019-23-7
Properties: −60 mesh with 99.5% purity;
    formula is generally $Cu_{1.4}Te$ [CER91]
Density, g/cm³: 7.1 [LID94]

1130
Compound: Copper(II) tellurite
Formula: $CuTeO_3$
Molecular Formula: CuO_3Te
Molecular Weight: 239.144
CAS RN: 13812-58-3
Properties: black glassy; −60 mesh with
    99% purity [CER91] [CRC10]
Solubility: i H_2O [CRC10]

1131
Compound: Copper(II) tetraammine sulfate monohydrate
Synonym: tetraaminecopper(II) sulfate monohydrate
Formula: $Cu(NH_3)_4SO_4 \cdot H_2O$
Molecular Formula: $CuH_{14}N_4O_5S$
Molecular Weight: 245.747
CAS RN: 10380-29-7
Properties: dark blue cryst; odor of ammonia;
    decomposes in air; made by dissolving
    $CuSO_4$ in water containing ammonia, then
    by precipitating with ethanol; used in textile
    printing and as a fungicide [MER06]
Solubility: 18.5 g/100 mL H_2O (21.5°C) [MER06]
Density, g/cm³: 1.810 [ALD94]
Reactions: minus H_2O and $2NH_3$ at 120°C,
    additional $2NH_3$ at 160°C [MER06]

1132
Compound: Copper(II) tetrafluoroborate
Synonym: copper(II) fluoroborate
Formula: $Cu(BF_4)_2$
Molecular Formula: B_2CuF_8

Molecular Weight: 237.155
CAS RN: 14735-84-3
Properties: prepared by neutralizing HBF with cupric hydroxide or cupric carbonate, then crystallizing; usually a hydrate; used in electroplating bath formulation [KIR78]

1133
Compound: Copper(II) titanate
Formula: $CuTiO_3$
Molecular Formula: CuO_3Ti
Molecular Weight: 159.411
CAS RN: 12019-08-8
Properties: reacted product −325 mesh 10 μm with 99.5% purity; gray powd [STR93] [CER91]

1134
Compound: Copper(II) trifluoroacetylacetonate
Synonyms: 1,1,1-trifluoro-2,4-pentanedione, Cu(II) derivative
Formula: $Cu(CF_3COCHCOCH_3)_2$
Molecular Formula: $C_{10}H_8CuF_6O_4$
Molecular Weight: 369.707
CAS RN: 14324-82-4
Properties: purple powd [STR93]
Melting Point, °C: 194–196 [STR94]
Boiling Point, °C: sublimes at 140 (0.1 mm Hg) [STR93]
Reactions: decomposes at 260°C [STR93]

1135
Compound: Copper(II) tungstate
Synonym: cupric tungstate
Formula: $CuWO_4$
Molecular Formula: CuO_4W
Molecular Weight: 311.384
CAS RN: 13587-35-4
Properties: −200 mesh with 99.5% purity; brown powd [STR93] [CER91]
Density, g/cm³: 7.5 [LID94]

1136
Compound: Copper(II) tungstate dihydrate
Synonym: cupric tungstate dihydrate
Formula: $CuWO_4 \cdot 2H_2O$
Molecular Formula: CuH_4O_6W
Molecular Weight: 347.414
CAS RN: 13587-35-4
Properties: light green powd; turns brown to grayish yellow by heating, with loss of H_2O; used in semiconductors, nuclear reactors, catalyst for polyester formation [MER06]

Solubility: i H_2O; sl s acetic acid; s NH_4OH, H_3PO_4 [MER06]
Reactions: decomposed by mineral acids [MER06]

1137
Compound: Copper(II) vanadate
Formula: $Cu(VO_3)_2$
Molecular Formula: CuO_6V_2
Molecular Weight: 261.425
CAS RN: 12789-09-2
Properties: powd

1138
Compound: Curium(α)
Formula: α-Cm
Molecular Formula: Cm
Molecular Weight: 247
CAS RN: 7440-51-9
Properties: silvery white, hard brittle metal; chemistry of trivalent state similar to that of trivalent lanthanides; α-emitter; hex, a = 0.3496 nm, c = 1.1331 nm; enthalpy of vaporization 1340 kJ/mol; ionic radius of Cm^{+++} is 0.0970 nm; discovered in 1944; used in generating thermoelectric power for remote locations and in space; β-Cm is fcc, which is stable at <1340°C [HAW93] [MER06] [KIR78]
Density, g/cm³: 13.51 (25°C) [KIR78]
Melting Point, °C: 1345 [KIR91]
Boiling Point, °C: 3110 [KIR91]

1139
Compound: Curium(β)
Formula: β-Cm
Molecular Formula: Cm
Molecular Weight: 247
CAS RN: 7440-51-9
Properties: fcc, a = 0.5039 nm; silvery, hard, brittle metal; oxidizes rapidly in the presence of traces of oxygen; chemistry of trivalent state similar to that of trivalent lanthanides; discovered in 1944; stable at <1340°C [KIR78] [MER06]
Density, g/cm³: 12.66 (25°C) [KIR78]
Melting Point, °C: 1350 [MER06]

1140
Compound: Cyanogen
Synonym: dicyan
Formula: N≡C–C≡N
Molecular Formula: C_2N_2
Molecular Weight: 52.035
CAS RN: 460-19-5

Properties: colorless, highly poisonous gas; almond-like odor; burns with pink flame with a bluish border; critical pressure 59.6 atm; critical temp 123.3°C; used in organic synthesis, welding, and as a rocket propellant [HAW93] [MER06]
Solubility: 1.1–1.3 g/100 g H_2O; 26 g/100 g alcohol; 5 g/100 g ether [CIC73]
Density, g/cm³: liq at b.p: 0.9537 [MER06]; gas: 2.321 g/L [CIC73]
Melting Point, °C: −28 [COT88]
Boiling Point, °C: −21.17 [MER06]
Reactions: slowly hydrolyzed to oxalic acid and ammonia [MER06]

1141

Compound: Cyanogen azide
Synonym: carbon pernitride
Formula: N=N=N–C≡N
Molecular Formula: CN_4
Molecular Weight: 68.038
CAS RN: 764-05-6
Properties: clear, colorless, oily liq; can detonate on thermal, electrical, or mechanical shock; decomposes in acetonitrile solvent; made by suspending NaN_3 in dry acetonitrile, followed by distillation of cyanogen chloride into the cooled suspension; used in organic synthesis for example reacts with alkanes to produce primary alkyl cyanamides [MER06]
Thermal Expansion Coefficient: from 25°C to: 100°C (0.18), 200°C (0.48), 400°C (1.11), 600°C (1.77) [TOU77]

1142

Compound: Cyanogen bromide
Synonym: bromocyanide
Formula: BrC≡N
Molecular Formula: CBrN
Molecular Weight: 105.922
CAS RN: 506-68-3
Properties: white cryst; sensitive to moisture; corrodes most metals; prepared from Br_2 and KCN; used in organic synthesis, as a fumigant, and in gold extraction [HAW93] [STR93]
Solubility: decomposed slowly by cold H_2O [HAW93]
Density, g/cm³: 2.015 [STR93]
Melting Point, °C: 52 [STR93]
Boiling Point, °C: 61.4 [STR93]

1143

Compound: Cyanogen chloride
Synonym: chlorocyanide

Formula: ClC≡N
Molecular Formula: CClN
Molecular Weight: 61.470
CAS RN: 506-77-4
Properties: colorless gas or liq with irritating vapor; prepared from Cl_2 and HCN; used in chemical syntheses [HAW93] [MER06]
Solubility: s H_2O, ether, alcohol [MER06]
Density, g/cm³: 2.697 g/L [LID94]
Melting Point, °C: −6 [MER06]
Boiling Point, °C: 13.8 [MER06]

1144

Compound: Cyanogen fluoride
Synonym: fluorocyanide
Formula: FC≡N
Molecular Formula: CFN
Molecular Weight: 45.016
CAS RN: 1495-50-7
Properties: colorless gas; made by reaction of AgF and cyanogen iodide; used in tear gas and in organic synthesis [HAW93]
Solubility: i H_2O [HAW93]
Density, g/cm³: 1.975 g/L [LID94]
Melting Point, °C: −82 [LID94]
Boiling Point, °C: −46 [LID94]

1145

Compound: Cyanogen iodide
Formula: ICN
Molecular Formula: CIN
Molecular Weight: 152.922
CAS RN: 506-78-5
Properties: col needles [CRC10]
Solubility: s H_2O, EtOH, eth [CRC10]
Density, g/cm³: 2.84 [CRC10]
Melting Point, °C: 146.7 [CRC10]

1146

Compound: Cyclohexadiene iron tricarbonyl
Formula: $C_6H_8Fe(CO)_3$
Molecular Formula: $C_{11}H_8FeO_3$
Molecular Weight: 220.008
CAS RN: 12152-72-6
Properties: orange-yellow liq; air sensitive [STR93]
Melting Point, °C: 8 [STR93]

1147

Compound: Cyclooctatetraene iron tricarbonyl
Formula: $C_8H_8Fe(CO)_3$

Molecular Formula: $C_{11}H_8FeO_3$
Molecular Weight: 244.029
CAS RN: 12093-05-9
Properties: red-brown cryst; sensitive to air [STR93]
Melting Point, °C: 93–95 [STR93]

1148
Compound: Cyclopentadienylindium(I)
Formula: C_5H_5In
Molecular Formula: C_5H_5In
Molecular Weight: 179.915
CAS RN: 34822-89-4
Properties: off-white cryst; sensitive to air, light, and heat [STR93]
Melting Point, °C: sublimes at 50 (0.01 mm Hg) [STR93]

1149
Compound: Cyclopentadienyliron dicarbonyl dimer
Formula: $[C_5H_5Fe(CO)_2]_2$
Molecular Formula: $C_{14}H_{10}Fe_2O_4$
Molecular Weight: 353.925
CAS RN: 12154-95-9
Properties: purple-red cryst; air sensitive [STR93]
Melting Point, °C: decomposes at 194 [STR93]

1150
Compound: Cyclopentadienylniobium tetrachloride
Formula: $C_5H_5NbCl_4$
Molecular Formula: $C_5H_5Cl_4Nb$
Molecular Weight: 229.812
CAS RN: 33114-15-7
Properties: red-brown powd; sensitive to moisture [STR93]
Melting Point, °C: decomposes at 180 [STR93]

1151
Compound: Decaborane(14)
Formula: $B_{10}H_{14}$
Molecular Formula: $B_{10}H_{14}$
Molecular Weight: 122.221
CAS RN: 17702-41-9
Properties: white cryst [STR93]
Solubility: sl s cold H_2O [MER06]
Density, g/cm³: 0.94 [STR93]
Melting Point, °C: 100 [STR93]
Boiling Point, °C: 213 (extrapolated) [COT88]
Reactions: hydrolyzed by hot H_2O [MER06]

1152
Compound: Decaborane(16)
Formula: $B_{10}H_{16}$
Molecular Formula: $B_{10}H_{16}$
Molecular Weight: 124.237
CAS RN: 71595-75-0
Properties: col cryst [CRC10]
Density, g/cm³: sublimes [CRC10]
Melting Point, °C: ~81
Boiling Point, °C: decomposes at 170

1153
Compound: Deuterium
Synonym: heavy water
Formula: D_2
Molecular Formula: D_2
Molecular Weight: 4.032
CAS RN: 7782-39-0
Properties: colorless, odorless gas; flammable; stable, not radioactive; specific volume 6.00 m³/kg at 21.1°C and 101.3 kPa; critical temp −234.75°C; critical pressure 16.432 atm; enthalpy of vaporization 1.23 kJ/mol; enthalpy of fusion 197 J/mol; used extensively at trace levels in measuring rates of chemical reactions [MER06] [AIR87] [KIR78]
Density, g/cm³: gas: 0.00018; liq: 0.169 at −252.7°C [MER06] [ALF93]
Melting Point, °C: −254.6 [ALF93]
Boiling Point, °C: −249.7 [ALF93]
Thermal Conductivity, W/(m·K): 0.126 (−252.7°C) [KIR78]

1154
Compound: Deuterium bromide
Formula: DBr
Molecular Formula: BrD
Molecular Weight: 81.919
CAS RN: 13536-59-9
Properties: corrosive [ALD94]
Density, g/cm³: 1.537 [ALD93]
Boiling Point, °C: 126 [ALD93]

1155
Compound: Deuterium chloride
Formula: DCl
Molecular Formula: ClD
Molecular Weight: 37.468
CAS RN: 7698-05-7
Properties: gas [CRC10]
Solubility: 11.9 cm³/100 mL H_2O (25°C), 8.4 cm³/100 mL H_2O (40°C) [CRC10]

Melting Point, °C: −254.6 [CRC10]
Boiling Point, °C: −249.7 [CRC10]

1156
Compound: Deuterium iodide
Formula: DI
Molecular Formula: DI
Molecular Weight: 128.919
CAS RN: 14104-45-1
Properties: hygr [ALD94]

1157
Compound: Deuterium oxide
Synonym: water-d$_2$
Formula: D$_2$O
Molecular Formula: D$_2$O
Molecular Weight: 20.028
CAS RN: 7789-20-0
Properties: ordinary water contains about one part of D$_2$O to 6500 parts of H$_2$O; triple point 3.82°C; critical temp 371.5°C; enthalpy of fusion 6.280 kJ/mol; enthalpy of vaporization 41.493 kJ/mol; dielectric constant (25°C) 78.06; finds use in the study of rates and mechanisms of chemical reactions [MER06] [HAW93]
Density, g/cm³: 1.1056 [HAW93]
Melting Point, °C: 3.81 [MER06]
Boiling Point, °C: 101.42 [MER06]

1158
Compound: Deuterosulfuric acid
Formula: D$_2$SO$_4$
Molecular Formula: D$_2$O$_4$S
Molecular Weight: 100.094
CAS RN: 13813-19-9
Properties: liq; 96%–98%, in D$_2$O [ALF93]
Density, g/cm³: 1.878 [ALD94]

1159
Compound: Diborane(6)
Synonym: boroethane
Formula: B$_2$H$_6$
Molecular Formula: B$_2$H$_6$
Molecular Weight: 27.670
CAS RN: 19287-45-7
Properties: gas; spontaneously flammable in air; critical temp 16.7°C; critical pressure 4.00 MPa; enthalpy of vaporization 14.28 kJ/mol; can be prepared from NaBH$_4$ and H$_3$PO$_4$; used as a catalyst and as a reducing agent [AIR87] [COT88] [CRC10] [MER06]

Solubility: hydrolyzes quickly [COT88]
Density, g/cm³: 1.214 g/L [LID94]
Melting Point, °C: −164.9 [AIR87]
Boiling Point, °C: −87.55 [CRC10]
Reactions: decomposes at red heat to B + H$_2$ [MER06]

1160
Compound: Dibromogermane
Formula: GeH$_2$Br$_2$
Molecular Formula: Br$_2$GeH$_2$
Molecular Weight: 234.46
CAS RN: 13769-36-3
Properties: col liq [CRC10]
Solubility: reac H$_2$O [CRC10]
Density, g/cm³: 2.80 [CRC10]
Melting Point, °C: −15 [CRC10]
Boiling Point, °C: 89 [CRC10]

1161
Compound: Dibromine pentoxide
Formula: Br$_2$O$_5$
Molecular Formula: Br$_2$O$_5$
Molecular Weight: 239.805
CAS RN: 58572-43-3
Properties: col cryst (low temp) [CRC10]
Melting Point, °C: decomposes at −20 [CRC10]

1162
Compound: Dibromosilane
Formula: SiH$_2$Br$_2$
Molecular Formula: Br$_2$H$_2$Si
Molecular Weight: 189.910
CAS RN: 13768-94-0
Properties: colorless liq; flammable; enthalpy of vaporization 31 kJ/mol; entropy of vaporization 91.21 kJ/(mol·K) [CRC10] [CIC73]
Solubility: decomposed by H$_2$O [CRC10]
Density, g/cm³: 2.17 (0°C) [CRC10]
Melting Point, °C: −70.1 [CIC73]
Boiling Point, °C: 66 [CIC73]

1163
Compound: Dicarbonylacetylacetonate iridium(I)
Formula: Ir(CO)$_2$(CH$_3$COCHCOCH$_3$)
Molecular Formula: C$_7$H$_7$IrO$_4$
Molecular Weight: 347.350
CAS RN: 14023-80-4
Properties: copper-brown cryst [STR93]

1164
Compound: Dichlorine heptoxide
Formula: Cl_2O_7
Molecular Formula: Cl_2O_7
Molecular Weight: 182.902
CAS RN: 10294-48-1
Properties: col oily liq; explosive [CRC10]
Density, g/cm³: 1.9 [CRC10]
Melting Point, °C: −91.5 [CRC10]
Boiling Point, °C: 82 [CRC10]

1165
Compound: Dichlorine hexoxide
Formula: Cl_2O_6
Molecular Formula: Cl_2O_6
Molecular Weight: 166.902
CAS RN: 12442-63-6
Properties: red liq [CRC10]
Solubility: reac H_2O [CRC10]
Melting Point, °C: 3.5 [CRC10]
Boiling Point, °C: ~200 [CRC10]

1166
Compound: Dichlorine trioxide
Formula: Cl_2O_3
Molecular Formula: Cl_2O_3
Molecular Weight: 118.904
CAS RN: 17496-59-2
Properties: dark brown solid [CRC10]
Melting Point, °C: explodes at <25 [CRC10]

1167
Compound: Dichlorodiamminepalladium(II)
Formula: $Pd(NH_3)_2Cl_2$
Molecular Formula: $Cl_2H_6N_2Pd$
Molecular Weight: 211.386
CAS RN: *cis*: 15684-18-1; *trans*: 13782-33-7
Properties: yellow powd [STR93]
Density, g/cm³: trans: 2.50 [ALD94]

1168
Compound: Dichlorodiammineplatinum(II) (cis)
Synonym: cisplatin
Formula: $Pt(NH_3)_2Cl_2$
Molecular Formula: $Cl_2H_6N_2Pt$
Molecular Weight: 300.046
CAS RN: 15663-27-1
Properties: deep yellow solid; uses: antineoplastic, has antitumor activity [MER06] [ALD94]

Solubility: 0.253 g/L H_2O (25°C) [MER06]
Melting Point, °C: decomposes at 270 [MER06]
Reactions: *cis* → *trans* change slowly in aq media [MER06]

1169
Compound: Dichlorodiammineplatinum(II)-trans
Formula: $Pt(NH_3)_2Cl_2$
Molecular Formula: $Cl_2H_6N_2Pt$
Molecular Weight: 300.046
CAS RN: 14913-33-8
Melting Point, °C: decomposes at 340 [ALD94]

1170
Compound: Dichlorodifluorogermane
Formula: GeF_2Cl_2
Molecular Formula: Cl_2F_2Ge
Molecular Weight: 181.54
CAS RN: 24422-21-7
Properties: col gas [CRC10]
Density, g/L: 7.149 [CRC10]
Melting Point, °C: −51.8 [CRC10]
Boiling Point, °C: −2.8 [CRC10]

1171
Compound: Dichlorodimethylgermane
Formula: $Ge(CH_3)_2Cl_2$
Molecular Formula: $C_2H_6Cl_2Ge$
Molecular Weight: 173.62
CAS RN: 1528-48-2
Properties: liq
Density, g/cm³: 1.49 [CRC10]
Melting Point, °C: −22 [CRC10]
Boiling Point, °C: 124 [CRC10]

1172
Compound: Dichlorodifluoromethane
Synonym: halocarbon-12
Formula: CCl_2F_2
Molecular Formula: CCl_2F_2
Molecular Weight: 120.913
CAS RN: 75-71-8
Properties: colorless gas; nonflammable; critical temp 112.04°C; critical pressure 4.14 MPa; enthalpy of vaporization 19.99 kJ/mol; used in electronics industry [AIR87]
Melting Point, °C: −157.8 [AIR87]
Boiling Point, °C: −29.8 [AIR87]

1173
Compound: Dichlorogermane
Formula: GeH_2Cl_2
Molecular Formula: Cl_2GeH_2
Molecular Weight: 145.56
CAS RN: 15230-48-5
Properties: col liq [CRC10]
Solubility: reac H_2O [CRC10]
Density, g/cm³: 1.90 [CRC10]
Melting Point, °C: −68 [CRC10]
Boiling Point, °C: 69.5 [CRC10]

1174
Compound: Dichlorosilane
Formula: SiH_2Cl_2
Molecular Formula: Cl_2H_2Si
Molecular Weight: 101.007
CAS RN: 4109-96-0
Properties: colorless, flammable gas; sharp
    pungent odor; autoignition temp 100°C; critical
    temp 176.0°C; critical pressure 4.68 MPa;
    enthalpy of vaporization 25 kJ/mol; entropy
    of vaporization 89.5 kJ/(mol·K); used in
    electronics industry [AIR87] [CIC73] [CRC10]
Solubility: hydrolyzes in H_2O [AIR87]
Density, g/cm³: 3.47 (air = 1) [AIR87]
Melting Point, °C: −122 [CIC73]
Boiling Point, °C: 8.3 [CIC73]

1175
Compound: Diethylaluminum chloride
Synonym: DEAC
Formula: $(C_2H_5)_2AlCl$
Molecular Formula: $C_4H_{10}AlCl$
Molecular Weight: 120.558
CAS RN: 96-10-6
Properties: colorless, volatile liq; flammable; sensitive to
    moisture; prepared from ethyl halide and Al; uses:
    aldol condensation reagent [ALD94] [MER06]
Density, g/cm³: 0.961 [ALD94]
Melting Point, °C: −50 [ALD94]
Boiling Point, °C: 125–126 at 50 mm Hg [ALD94]
Reactions: can be cleaved by H_2O [MER06]

1176
Compound: Diethylzinc
Synonym: zinc diethyl
Formula: $(C_2H_5)_2Zn$
Molecular Formula: $C_4H_{10}Zn$
Molecular Weight: 123.513
CAS RN: 557-20-0

Properties: liq; can ignite in air; preparation: reaction
    of Zn and diethyl iodide; uses: organic synthesis,
    preservation of archival papers [MER06]
Solubility: miscible with ether, petroleum
    ether, benzene [MER06]
Density, g/cm³: 1.2065 [MER06]
Melting Point, °C: −28 [ALD94]
Boiling Point, °C: 117 [ALD94]

1177
Compound: Difluorophosphoric acid
Formula: HPO_2F_2
Molecular Formula: F_2HO_2P
Molecular Weight: 101.978
CAS RN: 13779-41-4
Properties: mobile, colorless liq; fumes in air [KIR78]
Density, g/cm³: 1.583 [KIR78]
Melting Point, °C: −96.5 or −91.3 [KIR78]
Boiling Point, °C: decomposes at 107–111 [KIR78]

1178
Compound: Difluorosilane
Formula: SiH_2F_2
Molecular Formula: F_2H_2Si
Molecular Weight: 68.099
CAS RN: 13824-36-7
Properties: enthalpy of vaporization 16.3 kJ/
    mol; entropy of vaporization 84 kJ/
    (mol·K) [CIC73] [CRC10]
Melting Point, °C: −122 [CIC73]
Boiling Point, °C: −77.8 [CRC10]

1179
Compound: Digermane
Formula: Ge_2H_6
Molecular Formula: Ge_2H_6
Molecular Weight: 151.33
CAS RN: 13818-89-8
Properties: col liq, flammable [CRC10]
Density, g/cm³: 1.98 [at melting point] [CRC10]
Melting Point, °C: −109 [CRC10]
Boiling Point, °C: 29 [CRC10]

1180
Compound: Dihydrazine sulfate
Formula: $(N_2H_4)_2 \cdot H_2SO_4$
Molecular Formula: $H_{10}N_4O_4S$
Molecular Weight: 162.170
CAS RN: 13464-80-7
Properties: white cryst; deliq [LAN05] [MER06]

Solubility: g/100 g H_2O: 221 (30°C), 300
(40°C), 554 (60°C) [LAN05]
Melting Point, °C: ~104 [MER06]
Boiling Point, °C: decomposes [MER06]

1181
Compound: Diiodosilane
Formula: SiH_2I_2
Molecular Formula: H_2I_2Si
Molecular Weight: 283.911
CAS RN: 13760-02-6
Properties: enthalpy of vaporization 36.8 kJ/mol;
entropy of vaporization 87.0 kJ/(mol·K) [CIC73]
Melting Point, °C: −1 [CIC73]
Boiling Point, °C: 150 [CIC73]

1182
Compound: Diisobutylaluminum chloride
Formula: $[(CH_3)_2CHCH_2]_2AlCl$
Molecular Formula: $C_8H_{18}AlCl$
Molecular Weight: 176.665
CAS RN: 1779-25-5
Properties: liq; pyrophoric; sensitive to air;
uses: transmetallating reagent [ALD94]
Density, g/cm³: 0.905 [ALD94]
Melting Point, °C: −40 [ALD94]
Boiling Point, °C: 152 (10 mm Hg) [ALD94]

1183
Compound: Dimethylaminotrimethyltin
Synonym: pentamethylstannanamine
Formula: $(CH_3)_3SnN(CH_3)_2$
Molecular Formula: $C_5H_{15}NSn$
Molecular Weight: 207.891
CAS RN: 993-50-0
Properties: liq; uses: a dehydrochlorinating
agent [ALD94]
Density, g/cm³: 1.274 [ALD94]
Melting Point, °C: 1 [ALD94]
Boiling Point, °C: 126 [ALD94]

1184
Compound: Dimethylgermanium dichloride
Formula: $(CH_3)_2GeCl_2$
Molecular Formula: $C_2H_6Cl_2Ge$
Molecular Weight: 173.585
CAS RN: 1529-48-2
Properties: liq; flammable [ALD94]
Density, g/cm³: 1.505 [ALD94]
Melting Point, °C: −22 [ALD94]
Boiling Point, °C: 123 [ALD94]

1185
Compound: Dimethylmercury
Synonym: methylmercury
Formula: $(CH_3)_2Hg$
Molecular Formula: C_2H_6Hg
Molecular Weight: 230.659
CAS RN: 593-74-8
Properties: colorless, volatile liq; flammable;
uses: inorganic reagent [MER06]
Solubility: i H_2O; s ether, alcohol [MER06]
Density, g/cm³: 2.961 [ALD94]
Melting Point, °C: −43 [ALD94]
Boiling Point, °C: 93–94 [ALD94]

1186
Compound: Disilane
Formula: Si_2H_6
Molecular Formula: H_6Si_2
Molecular Weight: 62.219
CAS RN: 1590-87-0
Properties: colorless gas; extremely reactive, ignites in
air spontaneously; critical temp 150.9°C; critical
pressure 5.15 MPa; enthalpy of vaporization
21.21 kJ/mol; critical temp 109°C; made by
photolysis of SiH_4 and H_2 mixture; used in
electronics industry [AIR87] [CIC73] [COT88]
Solubility: s CS_2, ethanol, benzene [MER06]
Density, g/cm³: liq, at bp: 0.69 [CIC73]
Melting Point, °C: −132.5 [CIC73]
Boiling Point, °C: −14.5 [CIC73]
Reactions: decomposes at 300°C; KOH
causes evolution of H_2 [MER06]

1187
Compound: Disulfur decafluoride
Synonym: sulfur pentafluoride
Formula: S_2F_{10}
Molecular Formula: $F_{10}S_2$
Molecular Weight: 254.116
CAS RN: 5714-22-7
Properties: colorless gas; odor of SO_2; vapor
pressure is 561 torr at 20°C [HAW93]
Solubility: i H_2O [HAW93]
Density, g/cm³: 2.08 [HAW93]
Melting Point, °C: −92 [HAW93]
Boiling Point, °C: 29 [HAW93]

1188
Compound: Dodecaborane(16)
Formula: $B_{10}H_{16}$
Molecular Formula: $B_{10}H_{16}$

Molecular Weight: 124.237
CAS RN: 71595-75-0
Properties: col cryst [CRC10]
Density, g/cm³: sublimes [CRC10]
Melting Point, °C: ~81 [CRC10]
Boiling Point, °C: decomposes at 170 [CRC10]

1189
Compound: Dysprosium
Formula: Dy
Molecular Formula: Dy
Molecular Weight: 162.50
CAS RN: 7429-91-6
Properties: silver metal; tarnishes in moist air; hex close-packed; forms greenish yellow salts; enthalpy of fusion 10.782 kJ/mol; enthalpy of sublimation 290.4 kJ/mol; radius of atom 0.17743 nm; radius of ion 0.0908 nm for Dy^{+++}; light yellow solutions; electrical reisitivity at 20°C: 89 µohm · cm [KIR82] [MER06] [ALD94]
Solubility: reacts slowly with H_2O; s dil acids [HAW93]
Density, g/cm³: 8.55 [KIR82]
Melting Point, °C: 1412 [KIR82]
Boiling Point, °C: 2567 [KIR82]
Thermal Conductivity, W/(m·K): 10.7 (25°C) [ALD94]
Thermal Expansion Coefficient: 9.9×10^{-6}/K [CRC10]

1190
Compound: Dysprosium acetate tetrahydrate
Formula: $Dy(CH_3COO)_3 \cdot 4H_2O$
Molecular Formula: $C_6H_{17}DyO_{10}$
Molecular Weight: 411.695
CAS RN: 15280-55-4
Properties: yellow cryst [ALF93]
Solubility: s H_2O [CRC10]
Melting Point, °C: decomposes at 120 [ALF93]

1191
Compound: Dysprosium acetylacetonate
Synonyms: 2,4-pentanedione, dysprosium(III) derivative
Formula: $Dy(CH_3COCH=C(O)CH_3)_3$
Molecular Formula: $C_{15}H_{21}DyO_6$
Molecular Weight: 459.828
CAS RN: 14637-88-8
Properties: powd [STR93]

1192
Compound: Dysprosium boride
Formula: DyB_4
Molecular Formula: B_4Dy

Molecular Weight: 205.744
CAS RN: 12310-43-9
Properties: −60 mesh with 99.9% purity; there is a DyB_6 material, −60 mesh and 99.9% purity, 12008-04-7 [CER91]
Density, g/cm³: 6.98 [LID94]
Melting Point, °C: 2500 [LID94]

1193
Compound: Dysprosium bromide
Formula: $DyBr_3$
Molecular Formula: Br_3Dy
Molecular Weight: 402.212
CAS RN: 14456-48-5
Properties: colorless cryst; −20 mesh with 99.9% purity [CER91] [CRC10]
Solubility: s H_2O [CRC10]
Melting Point, °C: 881 [AES93]
Boiling Point, °C: 1480 [CRC10]

1194
Compound: Dysprosium carbonate tetrahydrate
Formula: $Dy_2(CO_3)_3 \cdot 4H_2O$
Molecular Formula: $C_3H_8Dy_2O_{13}$
Molecular Weight: 577.089
CAS RN: 38245-35-1
Properties: white powd; cryst [ALF93]
Solubility: i H_2O [CRC10]
Reactions: minus $3H_2O$ at 15°C [CRC10]

1195
Compound: Dysprosium chloride
Formula: $DyCl_3$
Molecular Formula: Cl_3Dy
Molecular Weight: 268.858
CAS RN: 10025-74-8
Properties: −20 mesh with 99.9% purity; white powd [STR93]; yellow, shining cryst [MER06] [CER91]
Density, g/cm³: 3.67 [MER06]
Melting Point, °C: 680 [MER06]; 718 [STR93]

1196
Compound: Dysprosium chloride hexahydrate
Formula: $DyCl_3 \cdot 6H_2O$
Molecular Formula: $Cl_3DyH_{12}O_6$
Molecular Weight: 376.949
CAS RN: 15059-52-6
Properties: −4 mesh with 99.9% purity; light yellow cryst; hygr [STR93] [CER91]
Melting Point, °C: 718 [ALF93]

1197
Compound: Dysprosium fluoride
Formula: DyF_3
Molecular Formula: DyF_3
Molecular Weight: 219.495
CAS RN: 13569-80-7
Properties: white powd or 99.9% pure melted pieces of 3–12 mm; hygr; melted pieces used as an evaporation material for possible application to multilayers [STR93] [CER91]
Melting Point, °C: 1154 [LID94]
Boiling Point, °C: >2200 [STR93]

1198
Compound: Dysprosium hydride
Formula: DyH_3
Molecular Formula: DyH_3
Molecular Weight: 165.524
CAS RN: 13537-09-2
Properties: hex; −40 mesh with 99.9% purity; lump [ALF93] [CER91] [LID94]
Density, g/cm³: 7.1 [LID94]

1199
Compound: Dysprosium hydroxide
Formula: $Dy(OH)_3$
Molecular Formula: DyH_3O_3
Molecular Weight: 213.522
CAS RN: 1308-85-6
Properties: gelatinous precipitate; forms a blue colloidal solution [MER06]

1200
Compound: Dysprosium iodide
Formula: DyI_3
Molecular Formula: DyI_3
Molecular Weight: 544.213
CAS RN: 15474-63-2
Properties: greenish yellow cryst; −20 mesh with 99.9% purity; yellow powd [STR93] [CER91] [CRC10]
Solubility: s H_2O [CRC10]
Melting Point, °C: 955 [STR93]
Boiling Point, °C: 1320 [STR93]

1201
Compound: Dysprosium nitrate pentahydrate
Formula: $Dy(NO_3)_3 \cdot 5H_2O$
Molecular Formula: $DyH_{10}N_3O_{14}$
Molecular Weight: 438.591
CAS RN: 10031-49-9
Properties: yellow cryst; hygr [STR93]
Solubility: s H_2O [MER06]
Melting Point, °C: at 88.6, melts in waters of hydration [MER06]

1202
Compound: Dysprosium nitride
Formula: DyN
Molecular Formula: DyN
Molecular Weight: 176.507
CAS RN: 12019-88-4
Properties: −60 mesh with 99.9% purity [CER91]
Density, g/cm³: 9.93 [LID94]

1203
Compound: Dysprosium oxalate decahydrate
Formula: $Dy_2(C_2O_4)_3 \cdot 10H_2O$
Molecular Formula: $C_6H_{20}Dy_2O_{22}$
Molecular Weight: 769.210
CAS RN: 24670-07-3
Properties: white cryst [STR93]
Solubility: i H_2O [CRC10]
Melting Point, °C: 40 [ALF93]
Reactions: minus H_2O at 40°C [CRC10]

1204
Compound: Dysprosium oxide
Synonym: dysprosia
Formula: Dy_2O_3
Molecular Formula: Dy_2O_3
Molecular Weight: 372.998
CAS RN: 1308-87-8
Properties: white powd; can be prepared by heating the oxalate or sulfate; more magnetic than ferric oxide; sl hygr; absorbs atm H_2O and CO_2; used with nickel in cermets and as an evaporated film of 99.9% purity it is reactive to radio frequencies [HAW93] [MER06] [CER91]
Solubility: s acids [HAW93]
Density, g/cm³: 7.81 [MER06]
Melting Point, °C: 2330–2350 [STR93]

1205
Compound: Dysprosium perchlorate hydrate
Formula: $Dy(ClO_4)_3 \cdot xH_2O$
Molecular Formula: Cl_3DyO_{12} (anhydrous)
Molecular Weight: 460.851 (anhydrous)
CAS RN: 14692-17-2
Properties: white cryst; hygr [STR93]

1206
Compound: Dysprosium silicide
Formula: $DySi_2$
Molecular Formula: $DySi_2$
Molecular Weight: 218.671
CAS RN: 12133-07-2
Properties: ortho-rhomb; 10 mm and down lump, 6 mm pieces and smaller with 99.9% purity [ALF93] [CER91] [LID94]
Density, g/cm³: 5.2 [LID94]

1207
Compound: Dysprosium sulfate octahydrate
Formula: $Dy_2(SO_4)_3 \cdot 8H_2O$
Molecular Formula: $Dy_2H_{16}O_{20}S_3$
Molecular Weight: 757.313
CAS RN: 10031-50-2
Properties: yellow cryst; can be prepared by dissolving the oxide in H_2SO_4, then precipitating with alcohol; stable in air at 110°C [MER06]
Solubility: s H_2O [HAW93]
Reactions: minus $8H_2O$ at 360°C [MER06]

1208
Compound: Dysprosium sulfide
Formula: Dy_2S_3
Molecular Formula: Dy_2S_3
Molecular Weight: 421.198
CAS RN: 12133-10-7
Properties: reddish brown monocl; −200 mesh with 99.9% purity [CER91] [LID94]
Density, g/cm³: 6.08 [LID94]

1209
Compound: Dysprosium telluride
Formula: Dy_2Te_3
Molecular Formula: Dy_2Te_3
Molecular Weight: 707.800
CAS RN: 12159-43-2
Properties: −20 mesh with 99.9% purity [CER91]

1210
Compound: Einsteinium
Formula: Es
Molecular Formula: Es
Molecular Weight: 252
CAS RN: 7429-92-7

Properties: man-made radioisotope; identified by Ghiorso and colleagues at Berkeley in December 1952, as part of debris from first large thermonuclear explosion; chemical properties similar to those of holmium; ionic radius of Es^{+++} is 0.0925 nm; has lowest enthalpy of vaporization of any of the divalent elements; cub, a = 0.575 nm; discovered in 1952; $t_{1/2}$ of ^{253}Es is 20.5 days, $t_{1/2}$ of ^{254}Es is 276 days, $t_{1/2}$ of ^{255}Es is 40 days [HAW93] [KIR78]
Melting Point, °C: 860 [KIR91]

1211
Compound: Erbium
Formula: Er
Molecular Formula: Er
Molecular Weight: 167.26
CAS RN: 7440-52-0
Properties: soft, malleable, dark, gray, metallic solid; hex close-pack cryst; similar to other rare earths; used in nuclear controls, room temp laser; enthalpy of fusion 19.90 kJ/mol; enthalpy of sublimation 317.10 kJ/mol; electrical resistivity at 20°C 86 μohm · cm; radius of atom 0.17566 nm; radius of ion 0.0881 nm, Er^{+++}, pink-colored solutions [MER06] [HAW93] [KIR82] [ALD94]
Solubility: i H_2O; s acids [HAW93]
Density, g/cm³: 9.066 [KIR82]
Melting Point, °C: 1529 [KIR82]
Boiling Point, °C: 2868 [KIR82]
Thermal Conductivity, W/(m · K): 14.5 (25°C) [CRC10]
Thermal Expansion Coefficient: 12.2×10^{-6}/K [CRC10]

1212
Compound: Erbium acetate tetrahydrate
Formula: $Er(CH_3COO)_3 \cdot 4H_2O$
Molecular Formula: $C_6H_{17}ErO_{10}$
Molecular Weight: 416.455
CAS RN: 15280-57-6
Properties: pink cryst; tricl [STR93] [CRC10]
Density, g/cm³: 2.114 [STR93]

1213
Compound: Erbium acetylacetonate hydrate
Synonyms: 2,4-pentanedione, erbium(III) derivative
Formula: $Er(CH_3COCH=C(O)CH_3)_3 \cdot xH_2O$
Molecular Formula: $C_{15}H_{21}ErO_6$ (anhydrous)
Molecular Weight: 464.588 (anhydrous)
CAS RN: 14553-08-3
Properties: off-white powd [STR93]

1214
Compound: Erbium barium copper oxide
Formula: $ErBa_2Cu_3O_x$
Molecular Formula: $Ba_2Cu_3ErO_x$
CAS RN: 109457-23-0
Properties: 99.9% and 99.999%, $0.2\,\mu m$ and $20\,\mu m$ powd, high T_c superconductor [ALF93]

1215
Compound: Erbium boride
Formula: ErB_4
Molecular Formula: B_4Er
Molecular Weight: 210.504
CAS RN: 12310-44-0
Properties: tetr; −100 mesh with 99.9% purity [CER91] [LID94]
Density, g/cm³: 7.0 [LID94]
Melting Point, °C: 2450 [LID94]

1216
Compound: Erbium bromide
Formula: $ErBr_3$
Molecular Formula: Br_3Er
Molecular Weight: 406.972
CAS RN: 13536-73-7
Properties: −20 mesh with 99.9% purity; powd [ALF93] [CER91]
Melting Point, °C: 923 (under argon) [ALF95]

1217
Compound: Erbium bromide hexahydrate
Formula: $ErBr_3 \cdot 6H_2O$
Molecular Formula: $Br_3ErH_{12}O_6$
Molecular Weight: 515.062
CAS RN: 14890-44-9
Properties: pink cryst [CRC10]
Solubility: s H_2O

1218
Compound: Erbium bromide nonahydrate
Formula: $ErBr_3 \cdot 9H_2O$
Molecular Formula: $Br_3ErH_{18}O_9$
Molecular Weight: 569.110
CAS RN: 13536-73-7
Properties: rose cryst; deliq [MER06]
Solubility: s H_2O [CRC10]
Boiling Point, °C: 1460 [CRC10]

1219
Compound: Erbium carbonate hydrate
Formula: $Er_2(CO_3)_3 \cdot xH_2O$

Molecular Formula: $C_3Er_2O_9$ (anhydrous)
Molecular Weight: 514.548 (anhydrous)
CAS RN: 22992-83-2
Properties: pink powd [STR93]

1220
Compound: Erbium chloride
Formula: $ErCl_3$
Molecular Formula: Cl_3Er
Molecular Weight: 273.618
CAS RN: 10138-41-7
Properties: −20 mesh with 99.9% purity; pinkish powd [MER06] [CER91]
Density, g/cm³: 4.1 [MER06]
Melting Point, °C: 774 [STR93]
Boiling Point, °C: 1500 [STR93]

1221
Compound: Erbium chloride hexahydrate
Formula: $ErCl_3 \cdot 6H_2O$
Molecular Formula: $Cl_3ErH_{12}O_6$
Molecular Weight: 381.709
CAS RN: 10025-75-9
Properties: pink; hygr cryst [STR93]
Solubility: s H_2O; sl s alcohol [MER06]
Melting Point, °C: 1500 [ALF93]

1222
Compound: Erbium fluoride
Formula: ErF_3
Molecular Formula: ErF_3
Molecular Weight: 224.255
CAS RN: 13760-83-3
Properties: rose powd or 99.9% pure melted pieces of 3–6 mm; hygr; melted pieces used as evaporation material for possible application to multilayers [STR93] [CER91]
Density, g/cm³: 7.814 [STR93]
Melting Point, °C: 1350 [STR93]
Boiling Point, °C: 2200 [STR93]

1223
Compound: Erbium hydride
Formula: ErH_3
Molecular Formula: ErH_3
Molecular Weight: 170.284
CAS RN: 13550-53-3
Properties: hex; −60 mesh with 99.9% purity; lump [ALF93] [CER91] [LID94]
Density, g/cm³: ~7.6 [LID94]

1224
Compound: Erbium hydroxide
Formula: Er(OH)₃
Molecular Formula: ErH₃O₃
Molecular Weight: 218.282
CAS RN: 14646-16-3
Properties: pale pink, gelatinous precipitate [MER06]

1225
Compound: Erbium iodide
Formula: ErI₃
Molecular Formula: ErI₃
Molecular Weight: 547.973
CAS RN: 13813-42-8
Properties: −20 mesh with 99.9% purity;
  red hygr powd [STR93] CER91]
Density, g/cm³: ~5.5 [LID94]
Melting Point, °C: 1020 [STR93]
Boiling Point, °C: 1280 [STR93]

1226
Compound: Erbium nitrate pentahydrate
Formula: Er(NO₃)₃·5H₂O
Molecular Formula: ErH₁₀N₃O₁₄
Molecular Weight: 443.351
CAS RN: 10031-51-3
Properties: reddish cryst; deliq [MER06]
Solubility: 5.5223 ± 0.0053 mol/(kg·H₂O) at 25°C;
  reference is uncertain about the number
  of hydrated waters, and pentahydrate is
  assumed for this solubility [RAR85b]
Reactions: minus 4H₂O at 130°C [MER06]

1227
Compound: Erbium nitride
Formula: ErN
Molecular Formula: ErN
Molecular Weight: 181.267
CAS RN: 12020-21-2
Properties: cub; −60 mesh with 99.9%
  purity [CER91] [LID94]
Density, g/cm³: 10.6 [LID94]

1228
Compound: Erbium oxalate decahydrate
Formula: Er₂(C₂O₄)₃·10H₂O
Molecular Formula: C₆H₂₀Er₂O₂₂
Molecular Weight: 778.732
CAS RN: 30618-31-6
Properties: pink cryst [STR93]

Solubility: s H₂O, dil acids [HAW93]
Density, g/cm³: 2.64 [HAW93]
Melting Point, °C: decomposes at 575 [HAW93]

1229
Compound: Erbium oxide
Synonym: erbia
Formula: Er₂O₃
Molecular Formula: Er₂O₃
Molecular Weight: 382.518
CAS RN: 12061-16-4
Properties: pinkish powd; changes into cub
  cryst at 1300°C; readily absorbs atm H₂O
  and CO₂; used as a phosphor activator, in
  infrared absorbing glass, and as an evaporated
  material of 99.9% purity it is reactive to radio
  frequencies [HAW93] [MER06] [CER91]
Solubility: 1.28×10⁻⁵ g mol/L H₂O
  (29°C); v s a [MER06]
Density, g/cm³: 8.64 [MER06]
Melting Point, °C: 2400 [STR93]

1230
Compound: Erbium perchlorate hydrate
Formula: Er(ClO₄)₃·xH₂O
Molecular Formula: Cl₃ErO₁₂ (anhydrous)
Molecular Weight: 465.611 (anhydrous)
CAS RN: 61565-07-9
Properties: pink cryst; hygr [STR93]

1231
Compound: Erbium silicide
Formula: ErSi₂
Molecular Formula: ErSi₂
Molecular Weight: 223.431
CAS RN: 12020-28-9
Properties: ortho-rhomb; 10 mm and
  down lump [ALF93] [LID94]
Density, g/cm³: 7.26 [LID94]

1232
Compound: Erbium sulfate
Formula: Er₂(SO₄)₃
Molecular Formula: Er₂O₁₂S₃
Molecular Weight: 622.711
CAS RN: 13478-49-4
Properties: powd; hygr; dissociates on heating
  in H₂O with evolution of heat [MER06]
Density, g/cm³: 3.678 [MER06]
Melting Point, °C: decomposes [LID94]

1233
Compound: Erbium sulfate octahydrate
Formula: $Er_2(SO_4)_3 \cdot 8H_2O$
Molecular Formula: $Er_2H_{16}O_{20}S_3$
Molecular Weight: 766.833
CAS RN: 10031-52-4
Properties: pink; monocl cryst [MER06]
Solubility: parts/100 parts H_2O: 16
   (20°C), 6.53 (40°C) [MER06]
Density, g/cm³: 3.217 [STR93]
Melting Point, °C: decomposes [STR93]
Reactions: minus $8H_2O$ at 400°C [HAW93]

1234
Compound: Erbium sulfide
Formula: Er_2S_3
Molecular Formula: Er_2S_3
Molecular Weight: 430.718
CAS RN: 12159-66-9
Properties: reddish brown monocl; −200 mesh
   with 99.9% purity [CER91] [LID94]
Density, g/cm³: 6.07 [LID94]
Melting Point, °C: 1730 [LID94]

1235
Compound: Erbium telluride
Formula: Er_2Te_3
Molecular Formula: Er_2Te_3
Molecular Weight: 717.320
CAS RN: 12020-39-2
Properties: ortho-rhomb; −20 mesh with
   99.9% purity [CER91] [LID94]
Density, g/cm³: 7.11 [LID94]
Melting Point, °C: 1213 [LID94]

1236
Compound: Ethylenediaminetetraacetic
   acid dihydrate disodium salt
Synonym: Edetate disodium
Formula: see under Properties
Molecular Formula: $C_{10}H_{14}N_2Na_2O_8$
Molecular Weight: 336.209
CAS RN: 6381-92-6
Properties: Used as a sequestering chelating
   agent for metals; forms strong chelates
   with most metals [MER06]
Solubility: g/100 g H_2O: 10.6 (0°C), 11.1 (20°C),
   27.0 (98°C) [LAN05]; pH ~5.3 [MER06]
Melting Point, °C: decomposes at 252 [MER06]

1237
Compound: Ethylenediaminetetraacetic acid
Synonym: Edetic acid, EDTA
Formula: $(HOOCCH_2)_2NCHCH_2CH_2N(CH_2COOH)_2$
Molecular Formula: $C_{10}H_{16}N_2O_8$
Molecular Weight: 292.246
CAS RN: 60-00-4
Properties: colorless cryst; prepared by addition
   of NaCN and formaldehyde to basic solution
   of ethylenediamine, forming the tetrasodium
   salt; used as food antioxidant and as chelating
   agent for pharmaceutics, added to detergents,
   shampoos, liq soaps [HAW93] [MER06]
Solubility: 0.5 g/L H_2O (25°C) [MER06]
Melting Point, °C: decomposes at 220 [MER06]
Reactions: decarboxylates when heated
   to 150°C [MER06]

1238
Compound: Europium
Formula: Eu
Molecular Formula: Eu
Molecular Weight: 151.965
CAS RN: 7440-53-1
Properties: soft, silvery metal; enthalpy of fusion
   9.1 kJ/mol; radius of atom 0.20418 nm; radius of
   Eu^{+++} is 0.0950 nm; bcc, a = 0.4582 nm; enthalpy of
   vaporization 176 kJ/mol; can be made by reduction
   of Eu_2O_3 with La under vacuum, followed by
   distillation; used as a neutron absorber, in color
   TV phosphors, and in phosphors for postage-stamp
   glues for electronic identification of first-class mail
   [ALD94] [HAW93] [MER06] [RAR85a] [KIR82]
Solubility: reacts with H_2O to evolve hydrogen gas;
   s liq ammonia [HAW93] [MER06]
Density, g/cm³: 5.234 [KIR82]
Melting Point, °C: 822 [ALD94]
Boiling Point, °C: 1527 [ALD94]
Thermal Conductivity, W/(m·K): 13.9
   (25°C) [ALD94]
Thermal Expansion Coefficient: 35×10^{-6}/K [CRC10]

1239
Compound: Europium boride
Formula: EuB_6
Molecular Formula: B_6Eu
Molecular Weight: 216.831
CAS RN: 12008-05-8
Properties: cub; −60 mesh with 99.9% purity
   [LID94] [CER91]
Density, g/cm³: 4.91 [LID94]
Melting Point, °C: ~2600 [LID94]

1240
Compound: Europium hydride
Formula: EuH$_{2-3}$
Molecular Formula: EuH$_2$; EuH$_3$
Molecular Weight: EuH$_2$: 153.981; EuH$_3$: 154.989
CAS RN: 70446-10-5
Properties: −60 mesh with 99.9% purity [CER91]

1241
Compound: Europium nitride
Formula: EuN
Molecular Formula: EuN
Molecular Weight: 165.972
CAS RN: 12020-58-5
Properties: −60 mesh with 99.9% purity [CER91]

1242
Compound: Europium silicide
Formula: EuSi$_2$
Molecular Formula: EuSi$_2$
Molecular Weight: 208.136
CAS RN: 12434-24-1
Properties: tetr; 6 mm pieces and smaller with
    99.9% purity, and 10 mm and down lump
    [ALF93] [LID94] [CER91]
Density, g/cm^3: 5.46 [LID94]
Melting Point, °C: 1500 [LID94]

1243
Compound: Europium(II) chloride
Synonym: europous chloride
Formula: EuCl$_2$
Molecular Formula: Cl$_2$Eu
Molecular Weight: 222.870
CAS RN: 13769-20-5
Properties: white, ortho-rhomb cryst;
    amorphous powd [LID94] [MER06]
Solubility: s H$_2$O [MER06]
Density, g/cm^3: 4.9 [LID94]
Melting Point, °C: 731 [LID94]

1244
Compound: Europium(II) fluoride
Synonym: europous fluoride
Formula: EuF$_2$
Molecular Formula: EuF$_2$
Molecular Weight: 189.962
CAS RN: 14077-37-5
Properties: yellow, cub; −200 mesh (precipitated)
    with 99.9% purity [CER91] [STR93] [LID94]

Solubility: i H$_2$O [CRC10]
Density, g/cm^3: 6.495 [CRC10]
Melting Point, °C: 1380 [CRC10]
Boiling Point, °C: >2400 [CRC10]

1245
Compound: Europium(II) iodide
Synonym: europous iodide
Formula: EuI$_2$
Molecular Formula: EuI$_2$
Molecular Weight: 405.774
CAS RN: 22015-35-6
Properties: olive green cryst; −20 mesh with
    99.9% purity [CER91] [CRC10]
Solubility: s H$_2$O [CRC10]
Density, g/cm^3: 5.50 [CRC10]
Melting Point, °C: 580 [LID94]
Boiling Point, °C: 1580 [CRC10]

1246
Compound: Europium(II) selenide
Formula: EuSe
Molecular Formula: EuSe
Molecular Weight: 230.925
CAS RN: 12020-66-5
Properties: brown cub; −100 mesh with
    99.9% purity [LID94] [CER91]
Density, g/cm^3: 6.45 [LID94]
Melting Point, °C: 2027 [CRC10]
Thermal Conductivity, W/(m·K): 0.24 [CRC10]

1247
Compound: Europium(II) sulfate
Synonym: europous sulfate
Formula: EuSO$_4$
Molecular Formula: EuO$_4$S
Molecular Weight: 248.029
CAS RN: 10031-54-6
Properties: ortho-rhomb; colorless cryst
    [MER06] [CRC10]
Solubility: i H$_2$O, dil acids [MER06]
Density, g/cm^3: 4.989 [CRC10]

1248
Compound: Europium(II) sulfide
Formula: EuS
Molecular Formula: EuS
Molecular Weight: 184.031
CAS RN: 12020-65-4
Properties: cub; −200 mesh with 99.9% purity
    [LID94] [CER91]
Density, g/cm^3: 5.7 [LID94]

1249
Compound: Europium(II) telluride
Formula: EuTe
Molecular Formula: EuTe
Molecular Weight: 279.565
CAS RN: 12020-69-8
Properties: black, cub; −20 mesh with 99.9% purity [LID94] [CER91]
Density, g/cm³: 6.48 [LID94]
Melting Point, °C: 1526 [LID94]

1250
Compound: Europium(III) acetylacetonate
Synonyms: 2,4-pentanedione, Eu(III) derivative
Formula: $Eu(CH_3COCH=C(O)CH_3)_3 \cdot xH_2O$
Molecular Formula: $C_{15}H_{21}EuO_6$ (anhydrous)
Molecular Weight: 449.293 (anhydrous)
CAS RN: 14284-86-7
Properties: hygr [ALD94]
Melting Point, °C: decomposes at 140 [ALD94]

1251
Compound: Europium(III) bromide
Synonym: europic bromide
Formula: $EuBr_3$
Molecular Formula: Br_3Eu
Molecular Weight: 391.677
CAS RN: 13759-88-1
Properties: gray cryst; −20 mesh with 99.9% purity [LID94] [CER91]
Solubility: s H_2O [CRC10]
Melting Point, °C: 702 [CRC10]
Boiling Point, °C: decomposes [CRC10]

1252
Compound: Europium(III) carbonate hydrate
Formula: $Eu_2(CO_3)_3 \cdot xH_2O$
Molecular Formula: $C_3Eu_2O_9$ (anhydrous)
Molecular Weight: 483.958 (anhydrous)
CAS RN: 86546-99-8
Properties: white powd; hygr [STR93] [ALD94]

1253
Compound: Europium(III) chloride
Synonym: europic chloride
Formula: $EuCl_3$
Molecular Formula: Cl_3Eu
Molecular Weight: 258.323
CAS RN: 10025-76-0

Properties: −20 mesh with 99.9% purity; yellowish-white powd; hygr [STR93] [CER91]
Density, g/cm³: 4.89 [STR93]
Melting Point, °C: 623 [LID94]
Reactions: yields $EuCl_2$ by reduction with H_2 at 600°C [MER06]

1254
Compound: Europium(III) chloride hexahydrate
Formula: $EuCl_3 \cdot 6H_2O$
Molecular Formula: $Cl_3EuH_{12}O_6$
Molecular Weight: 366.414
CAS RN: 13759-92-7
Properties: yellow needles; white cryst; hygr [HAW93] [STR93]
Solubility: s H_2O [HAW93]
Density, g/cm³: 4.89 (20°C) [HAW93]
Melting Point, °C: 850 [HAW93]

1255
Compound: Europium(III) fluoride
Formula: EuF_3
Molecular Formula: EuF_3
Molecular Weight: 208.960
CAS RN: 13765-25-8
Properties: white powd or 99.9% pure melted pieces of 3–6mm; hygr; pieces used as evaporation material for possible applications in multilayers [STR93] [CER91]
Melting Point, °C: 1390 [STR93]
Boiling Point, °C: 2280 [STR93]

1256
Compound: Europium(III) nitrate hexahydrate
Synonym: europic nitrate hexahydrate
Formula: $Eu(NO_3)_3 \cdot 6H_2O$
Molecular Formula: $EuH_{12}N_3O_{15}$
Molecular Weight: 446.071
CAS RN: 10031-53-5
Properties: white to pale pink cryst; hygr [HAW93] [STR93]
Solubility: 4.2732 ± 0.0061 mol/kg in H_2O (25°C) [RAR84]
Melting Point, °C: 85 [HAW93]

1257
Compound: Europium(III) nitrate pentahydrate
Formula: $Eu(NO_3)_3 \cdot 5H_2O$
Molecular Formula: $EuH_{10}N_3O_{14}$
Molecular Weight: 428.071
CAS RN: 63026-01-7
Properties: white cryst [STR93]

1323
Compound: Ferrous oxalate dihydrate
Synonym: iron(II) oxalate dihydrate
Formula: $FeC_2O_4 \cdot 2H_2O$
Molecular Formula: $C_2H_4FeO_6$
Molecular Weight: 179.895
CAS RN: 6047-25-2
Properties: pale yellow, odorless, cryst powd; decomposes at 160°C, evolving carbon monoxide; used as a photographic developer, a pigment for glass, in paints [HAW93] [MER06]
Solubility: sl s H_2O; s mineral acids [MER06]
Density, g/cm³: 2.28 [MER06]
Melting Point, °C: decomposes at 150 [LID94]

1324
Compound: Ferrous oxide
Synonym: wustite
Formula: FeO
Molecular Formula: FeO
Molecular Weight: 71.844
CAS RN: 1345-25-1
Properties: jet black powd; easily oxidized in air; strong base; readily absorbs CO_2; enthalpy of fusion 24.00 kJ/mol; used as a catalyst, glass colorant [HAW93] [MER06] [CRC10]
Solubility: i H_2O, alkalies; s acid [HAW93] [MER06]
Density, g/cm³: 5.7 [HAW93]
Melting Point, °C: 1377 [CRC10]
Thermal Expansion Coefficient: (volume) 100°C (0.30), 200°C (0.63), 400°C (1.32), 600°C (2.10) [CLA66]

1325
Compound: Ferrous perchlorate hexahydrate
Synonym: iron(II) perchlorate hexahydrate
Formula: $Fe(ClO_4)_2 \cdot 6H_2O$
Molecular Formula: $Cl_2FeH_{12}O_{14}$
Molecular Weight: 362.839
CAS RN: 13520-69-9
Properties: green cryst [AES93]
Solubility: g/100 g soln, H_2O: 63.39 (0°C), 67.76 (25°C); solid phase, $Fe(ClO_4)_2 \cdot 6H_2O$ [KRU93]
Melting Point, °C: decomposes at >100 [CRC10]

1326
Compound: Ferrous phosphate octahydrate
Synonym: vivianite
Formula: $Fe_3(PO_4)_2 \cdot 8H_2O$
Molecular Formula: $Fe_3H_{16}O_{16}P_2$

Molecular Weight: 501.600
CAS RN: 14940-41-1
Properties: grayish blue powd or monocl cryst; hygr; used in ceramics and as a catalyst [HAW93] [MER06]
Solubility: i H_2O; s mineral acids [MER06]
Density, g/cm³: 2.58 [MER06]

1327
Compound: Ferrous phosphide
Formula: Fe_2P
Molecular Formula: Fe_2P
Molecular Weight: 142.664
CAS RN: 1310-43-6
Properties: gray; hex needles or bluish gray powd; ferromagnetic; used in the manufacture of iron and steel; there is also an FeP, 26508-33-8, and Fe_3P, 12023-53-9; can be formed by heating phosphorus rock, silica, and coke [KIR82] [HAW93] [MER06] [CER91]
Solubility: i H_2O, dil acid, dil alkali; reacts with hot mineral acids [MER06]
Density, g/cm³: 6.85 [MER06]
Melting Point, °C: 1290 [HAW93]

1328
Compound: Ferrous selenide
Synonym: iron(II) selenide
Formula: FeSe
Molecular Formula: FeSe
Molecular Weight: 134.805
CAS RN: 1310-32-3
Properties: black mass with metallic luster; stable in air; decomposes on heating in O_2; used in semiconductor technology [MER06] [CER91] [HAW93]
Solubility: i H_2O; s HCl evolving H_2Se [MER06]
Density, g/cm³: 6.78 [MER06]

1329
Compound: Ferrous sulfate
Synonym: iron(II) sulfate
Formula: $FeSO_4$
Molecular Formula: FeO_4S
Molecular Weight: 151.909
CAS RN: 7720-78-7
Properties: white, ortho-rhomb, hygr [LID94]
Solubility: g/100 g soln, H_2O: 13.6 (0°C), 22.8 (25°C), 24.0 (100°C); solid phase, $FeSO_4 \cdot 7H_2O$ (0°C, 25°C), $FeSO_4 \cdot H_2O$ (100°C) [KRU93]

Density, g/cm³: 3.65 [LID94]
Melting Point, °C: decomposes at 671 [JAN85]

1330
Compound: Ferrous sulfate heptahydrate
Synonym: melanterite
Formula: FeSO₄·7H₂O
Molecular Formula: FeH₁₄O₁₁S
Molecular Weight: 278.015
CAS RN: 7782-63-0
Properties: off-white powd; bluish green; monocl cryst or granules; efflorescent in dry air; oxidizes in moist air; odorless with saline taste; used as a pigment, in water and sewage treatment, process engraving; minus 7H₂O by 300°C [HAW93] [STR93] [MER06]
Solubility: g/100 g H₂O: 28.8 (0°C), 48.0 (20°C), 57.8 (100°C) [LAN05]
Density, g/cm³: 1.897 [MER06]
Melting Point, °C: decomposes at ~60 [LID94]
Reactions: minus 3H₂O at 56.6°C; minus 6H₂O at 65°C [MER06]

1331
Compound: Ferrous sulfate monohydrate
Synonym: szomolnokite
Formula: FeSO₄·H₂O
Molecular Formula: FeH₂O₅S
Molecular Weight: 169.924
CAS RN: 17375-41-6
Properties: white to yellow cryst powd [MER06]
Solubility: s H₂O [MER06]
Density, g/cm³: 2.970 [CRC10]
Reactions: minus H₂O at ~300°C; decomposes at higher temp [MER06]

1332
Compound: Ferrous sulfide
Synonyms: troilite, iron(II) sulfide
Formula: FeS
Molecular Formula: FeS
Molecular Weight: 87.911
CAS RN: 1317-37-9
Properties: when pure: colorless, hex cryst; usually gray to brownish black lumps; oxidized by moist air to S and Fe₂O₃; enthalpy of fusion 31.50 kJ/mol [MER06] [CRC10]
Solubility: i H₂O; s acids, evolving H₂S [MER06]
Density, g/cm³: 4.84 [MER06]
Melting Point, °C: 1194 [MER06]
Boiling Point, °C: decomposes [HAW93]

1333
Compound: Ferrous tantalate
Formula: Fe(TaO₃)₂
Molecular Formula: FeO₆Ta₂
Molecular Weight: 513.737
CAS RN: 12140-41-9
Properties: brown tetr cryst [CRC10]
Density, g/cm³: 7.33 [CRC10]

1334
Compound: Iron(II) tartrate
Formula: FeC₄H₄O₆
Molecular Formula: C₄H₄FeO₆
Molecular Weight: 203.916
CAS RN: 2044-65-2
Properties: white cryst [CRC10]
Solubility: g/100 g H₂O: 0.88; v s acid; s NH₄OH [CRC10]

1335
Compound: Ferrous thiocyanate trihydrate
Synonym: iron(II) thiocyanate trihydrate
Formula: Fe(SCN)₂·3H₂O
Molecular Formula: C₂H₆FeN₂O₃S₂
Molecular Weight: 226.058
CAS RN: 6010-09-9
Properties: pale green, monocl prisms; rapidly oxidized when exposed to air [MER06]
Solubility: s H₂O, alcohol, ether [MER06]
Reactions: decomposed by heat [MER06]

1336
Compound: Ferrous titanate
Synonym: ilmenite
Formula: FeTiO₃
Molecular Formula: FeO₃Ti
Molecular Weight: 151.710
CAS RN: 12168-52-4
Properties: opaque black with almost metallic luster; rhomb; occurs naturally as the mineral ilmenite; finds extensive use in manufacturing Ti paint pigments; there is also Fe₂TiO₅, 12789-64-9, −100 mesh with 99.9% purity [KIR83] [CER91]
Density, g/cm³: 4.72 [KIR83]
Melting Point, °C: ~1470 [KIR83]

1337
Compound: Fluorine
Formula: F₂
Molecular Formula: F₂

Molecular Weight: 18.9984032 (atomic wt)
CAS RN: 7782-41-4
Properties: pale yellow diatomic gas, condenses
  to yellowish orange liq at −188°C, solidifies
  to yellow solid at −220°C; critical temp
  −129°C; critical pressure 55 atm; most
  reactive nonmetal; enthalpy of fusion
  0.51 kJ/mol; enthalpy of vaporization 6.62 kJ/
  mol; enthalpy of dissociation 157.7 kJ/mol; reacts
  vigorously with most oxidizable substances;
  produced by electrolysis of dil solution of KF in
  anhydrous HF [KIR78] [MER06] [CRC10]
Density, g/cm³: gas: 1.695 g/L [KIR78];
  liq: 1.5127 at bp [MER06]
Melting Point, °C: −219.66 [CRC10]
Boiling Point, °C: −188.11 [CRC10]
Thermal Conductivity, W/(m·K): 24.77×10^{-7}
  at 0°C [KIR78]; 0.0277 at 25°C [ALD94]

1338
Compound: Fluorine dioxide
Synonym: dioxygen difluoride
Formula: F_2O_2
Molecular Formula: F_2O_2
Molecular Weight: 69.996
CAS RN: 7783-44-0
Properties: thermally unstable gas at room temp;
  pale yellow solid or yellow liq; enthalpy of
  vaporization 19.1 kJ/mol; produced by reacting
  O_2 and F_2 at cryogenic temperatures in an
  electrical discharge [KIR78] [MER06] [CRC10]
Density, g/cm³: 3.071 g/L [LID94]
Melting Point, °C: −154 [MER06]
Boiling Point, °C: −57 [CRC10]
Reactions: decomposes to F_2 and
  O_2 at −100°C [MER06]

1339
Compound: Fluorine monoxide
Formula: F_2O
Molecular Formula: F_2O
Molecular Weight: 53.996
CAS RN: 7783-41-7
Properties: colorless gas; yellowish brown when
  liq; gas may be kept over water unchanged for a
  month; does not attack glass in the cold; enthalpy
  of vaporization 11.09 kJ/mol [CRC10] [MER06]
Solubility: 6.8 mL gas/100 mL H_2O (0°C) [MER06]
Density, g/cm³: 2.369 g/L [LID94]; liq:
  1.90 (−224°C) [MER06]
Melting Point, °C: −223.8 [MER06]
Boiling Point, °C: −144.75 [CRC10]

1340
Compound: Fluorine nitrate
Formula: FNO_3
Molecular Formula: FNO_3
Molecular Weight: 81.003
CAS RN: 7789-26-6
Properties: colorless gas; moldy, acrid odor; liq
  explodes on slight percussion; hydrolyzed
  by water to OF_2, O_2, HF, and HNO_3; burns
  with alcohol, ether, aniline [MER06]
Solubility: s acetone [MER06]
Density, g/cm³: 3.554 g/L [LID94]
Melting Point, °C: −175 [LID94]
Boiling Point, °C: −46 [LID94]

1341
Compound: Fluorine perchlorate
Synonym: chlorine tetroxyfluoride
Formula: $FOClO_3$
Molecular Formula: $ClFO_4$
Molecular Weight: 118.449
CAS RN: 10049-03-3
Properties: colorless gas; pungent, acrid
  odor; readily explodes on contact with
  solids or on heating [MER06]
Solubility: reacts with H_2O [LID94]
Density, g/cm³: 5.197 g/L [LID94]
Melting Point, °C: −167.3 [MER06]
Boiling Point, °C: −15.9 (755 mm) [MER06]

1342
Compound: Fluorine tetroxide
Formula: F_2O_4
Molecular Formula: F_2O_4
Molecular Weight: 101.995
CAS RN: 107782-11-6
Properties: red-brown solid [CRC10]
Melting Point, °C: −191 [CRC10]
Boiling Point, °C: decomposes at 185 [CRC10]

1343
Compound: Fluoroantimonic acid
Formula: $HF \cdot SbF_5$
Molecular Formula: F_6HSb
Molecular Weight: 236.758
CAS RN: 16950-06-4
Properties: superacid; moisture sensitive
  [KIR78] [ALD94]

1344
Compound: Fluoroboric acid
Formula: HBF_4

Molecular Formula: BF$_4$H
Molecular Weight: 87.813
CAS RN: 16872-11-0
Properties: does not exist as a free pure material; colorless, strongly acid liq; stable in conc solutions; produced by reacting 70% HF with boric acid, H$_3$BO$_3$; used to produce fluoroborates, in electrolytic brightening of aluminum [HAW93] [KIR78]
Solubility: miscible with H$_2$O, alcohol [HAW93]
Density, g/cm^3: ~1.84 [HAW93]
Boiling Point, °C: decomposes at 130 [HAW93]

1345
Compound: Fluorogermane
Formula: GeH$_3$F
Molecular Formula: FgeH$_3$
Molecular Weight: 94.66
CAS RN: 13537-30-9
Properties: col gas [CRC10]
Solubility: reac H$_2$O [CRC10]
Density, g/L: 3.868 [CRC10]

1346
Compound: Fluorosilane
Formula: SiH$_3$F
Molecular Formula: FH$_3$Si
Molecular Weight: 50.108
CAS RN: 13537-33-2
Properties: enthalpy of vaporization 18.8 kJ/mol; entropy of vaporization 107.9 kJ/(m·K) [CIC73]
Boiling Point, °C: −98.6 [CIC73]

1347
Compound: Fluorosulfonic acid
Formula: HSO$_3$F
Molecular Formula: FHO$_3$S
Molecular Weight: 100.070
CAS RN: 7789-21-1
Properties: colorless liq; fumes in moist air; stable up to 900°C; considerably more acidic than 100% H$_2$SO$_4$; does not attack glass when anhydrous and pure; viscosity 1.56 MPa·s; dielectric constant ~120; specific conductance 1.08×10^{-6} (ohm·m)$^{-1}$; used as a catalyst in organic synthesis, in electropolishing, and as a fluorinating agent [HAW93] [MER06] [KIR78]
Solubility: hydrolyzes violently in H$_2$O; reddish-brown color in acetone [MER06]
Density, g/cm^3: 1.726 [MER06]
Melting Point, °C: freezing point −89 [MER06]
Boiling Point, °C: 163 [MER06]

1348
Compound: Fluorotrimethylsilane
Synonym: trimethylsilyl fluoride
Formula: (CH$_3$)$_3$SiF
Molecular Formula: C$_3$H$_9$FSi
Molecular Weight: 92.158
CAS RN: 420-56-4
Properties: gas; flammable; sensitive to moisture [ALD94]
Density, g/cm^3: 0.793 [ALD94]
Melting Point, °C: −74 [ALD94]
Boiling Point, °C: 16 [ALD94]

1349
Compound: Francium
Formula: Fr
Molecular Formula: Fr
Molecular Weight: 223
CAS RN: 7440-73-5
Properties: heaviest of the alkali metal family; may exist only as radioactive isotopes; only natural isotope is ^{223}Fr with a half-life of 21 min; discovered in 1939 by Mll. M. Perey, Curie Inst., Paris; formed from α-decay of actinium [CRC10] [HAW93]
Melting Point, °C: 27 [CRC10]
Boiling Point, °C: 677 [CRC10]

1350
Compound: Fullerene
Synonym: carbon fullerenes
Formula: C$_{60}$
Molecular Formula: C$_{60}$
Molecular Weight: 720.660
CAS RN: 99685-96-8
Properties: fcc, lattice constant 1.417 nm; mean ball diameter 0.683 nm; compressibility 6.9×10^{-12} cm^2/dyne; bulk modulus 14 GPa; binding energy per atom 7.40 eV; structural phase transitions −18°C, −108°C; sound velocity vt 2.1×10^5 cm/s, vl 3.6×10^5 cm/s; Debye temp −88°C; static dielectric constant 4.0–4.5; synthesized by ac discharge of graphite electrodes under He at 200 torr [DRE93]
Solubility: s organic solvents [LID94]
Density, g/cm^3: 1.72 [DRE93]
Melting Point, °C: >280 [LID94]
Reactions: bromination gives C$_{60}$Br$_6$, C$_{60}$Br$_8$ [IUP93]
Thermal Conductivity, W/(m·K): 0.4 (7°C) [DRE93]
Thermal Expansion Coefficient: volume thermal expansion 6.2×10^{-5}/K [DRE93]

1351
Compound: Fullerenes
Formula: C$_{60}$/C$_{70}$

Molecular Formula: C_{60}/C_{70}
Molecular Weight: C_{60}: 720.660; C_{70}: 840.777
CAS RN: 131159-39-2
Properties: black powd; contains
    10%–15% C_{70} [STR93]
Density, g/cm³: 1.6 [STR93]

1352
Compound: Fullerene fluoride
Formula: $C_{60}F_{60}$
Molecular Formula: $C_{60}F_{60}$
Molecular Weight: 1860.546
CAS RN: 134929-59-2
Properties: col plates [CRC10]
Solubility: v s ace; s THF; i chl [CRC10]
Melting Point, °C: 287 [CRC10]

1353
Compound: Gadolinium
Formula: Gd
Molecular Formula: Gd
Molecular Weight: 157.25
CAS RN: 7440-54-2
Properties: colorless or faintly yellowish metal;
    tarnishes in moist air; hex close-packed; magnetic,
    especially at low temperatures; enthalpy of fusion
    9.81 kJ/mol; enthalpy of sublimation 397.5 kJ/
    mol; electrical resistivity (20°C) 126 µohm·cm;
    radius of atom 0.10813 nm; radius of Gd^{+++}
    ion 0.0938 nm; solutions are colorless; used in
    neutron shielding, garnets for microwave filter
    [KIR82] [MER06] [HAW93] [CRC10] [ALD94]
Solubility: reacts slowly with H_2O; s dil acid [HAW93]
Density, g/cm³: 7.9004 [KIR82]
Melting Point, °C: 1312 [MER06]
Boiling Point, °C: 3273 [KIR82]
Thermal Conductivity, W/(m·K): 10.5 (25°C) [ALD94]
Thermal Expansion Coefficient: 9×10^{-6} K [CRC10]

1354
Compound: Gadolinium acetate tetrahydrate
Synonyms: acetic acid, Gd(III) salt
Formula: $Gd(CH_3COO)_3 \cdot 4H_2O$
Molecular Formula: $C_6H_{17}GdO_{10}$
Molecular Weight: 406.445
CAS RN: 15280-53-2
Properties: tricl white cryst [STR93] [CRC10]
Solubility: 11.6 g/100 mL H_2O (25°C) [CRC10]
Density, g/cm³: 1.611 [STR93]
Boiling Point, °C: decomposes [ALF95]

1355
Compound: Gadolinium acetylacetonate dihydrate
Synonyms: 2,4-pentanedione, gadolinium(III) derivative
Formula: $Gd(CH_3COCH=C(O)CH_3)_3 \cdot 2H_2O$
Molecular Formula: $C_{15}H_{25}GdO_8$
Molecular Weight: 490.609
CAS RN: 14284-87-8
Properties: off-white powd [STR93]
Melting Point, °C: decomposes at 143 [ALD94]

1356
Compound: Gadolinium boride
Formula: GdB_6
Molecular Formula: B_6Gd
Molecular Weight: 222.116
CAS RN: 12008-06-9
Properties: brownish-black cub; –325 mesh
    10 µm or less with 99.9% purity; refractory
    material [LID94] [KIR78] [CER91]
Density, g/cm³: 5.31 [LID94]
Melting Point, °C: 2100 [KIR78]

1357
Compound: Gadolinium bromide
Formula: $GdBr_3$
Molecular Formula: Br_3Gd
Molecular Weight: 396.962
CAS RN: 13818-75-2
Properties: white, hygr cryst; –20 mesh with
    99.9% purity [LID94] [CER91]
Melting Point, °C: 770 [LID94]

1358
Compound: Gadolinium chloride
Formula: $GdCl_3$
Molecular Formula: Cl_3Gd
Molecular Weight: 263.608
CAS RN: 10138-52-0
Properties: –20 mesh with 99.9% purity; white;
    monocl cryst; hygr [STR93] [MER06] [CER91]
Solubility: s H_2O [MER06]
Density, g/cm³: 4.52 (0°C) [MER06]
Melting Point, °C: ~609 [MER06]

1359
Compound: Gadolinium chloride hexahydrate
Formula: $GdCl_3 \cdot 6H_2O$
Molecular Formula: $Cl_3GdH_{12}O_6$
Molecular Weight: 371.654
CAS RN: 13450-84-5

Properties: −4 mesh with 99.9% purity; colorless deliq cryst, obtained from aq solutions; used as a source of Gd metal [MER06] [CER91] [HAW93]
Solubility: s H_2O [HAW93]
Density, g/cm³: 2.424 [MER06]

1360
Compound: Gadolinium fluoride
Formula: GdF_3
Molecular Formula: F_3Gd
Molecular Weight: 214.245
CAS RN: 13765-26-9
Properties: white powd or 99.9% pure melted pieces of 3–6 mm; pieces used as evaporation material for possible application to multilayers [STR93] [CER91]
Melting Point, °C: 1231 [LID94]

1361
Compound: Gadolinium gallium garnet
Formula: $Gd_3Ga_5O_{12}$
Molecular Formula: $Ga_5Gd_3O_{12}$
Molecular Weight: 1012.358
CAS RN: 12024-36-1
Properties: lump [ALF95]
Density, g/cm³: 7.09 [ALD94]

1362
Compound: Gadolinium hydride
Formula: GdH_{2-3}
Molecular Formula: GdH_2; GdH_3
Molecular Weight: GdH_2: 159.266; GdH_3: 160.274
CAS RN: 13572-97-9
Properties: −60 mesh with 99.9% purity [CER91]

1363
Compound: Gadolinium iodide
Formula: GdI_3
Molecular Formula: GdI_3
Molecular Weight: 537.963
CAS RN: 13572-98-0
Properties: yellow; −20 mesh with 99.9% purity [CER91] [CRC10]
Solubility: s H_2O [CRC10]
Melting Point, °C: 926 [AES93]
Boiling Point, °C: 1340 [CRC10]

1364
Compound: Gadolinium nitrate hexahydrate
Formula: $Gd(NO_3)_3 \cdot 6H_2O$
Molecular Formula: $GdH_{12}N_3O_{15}$

Molecular Weight: 451.356
CAS RN: 19598-90-4
Properties: deliq; tricl cryst [MER06]
Solubility: s H_2O, alcohol [MER06]
Density, g/cm³: 2.332 [MER06]
Melting Point, °C: 91 [MER06]

1365
Compound: Gadolinium nitrate pentahydrate
Formula: $Gd(NO_3)_3 \cdot 5H_2O$
Molecular Formula: $GdH_{10}N_3O_{14}$
Molecular Weight: 433.341
CAS RN: 52788-53-1
Properties: hygr white cryst [STR93]
Solubility: i H_2O [MER06]
Density, g/cm³: 2.406 [MER06]
Melting Point, °C: 92 [MER06]

1366
Compound: Gadolinium nitride
Formula: GdN
Molecular Formula: GdN
Molecular Weight: 171.257
CAS RN: 25764-15-2
Properties: −60 mesh with 99.9% purity; NaCl cryst system, a = 0.499 nm [CIC73] [CER91]
Density, g/cm³: 9.10 [LID94]

1367
Compound: Gadolinium oxalate decahydrate
Formula: $Gd_2(C_2O_4)_3 \cdot 10H_2O$
Molecular Formula: $C_6H_{20}Gd_2O_{22}$
Molecular Weight: 758.712
CAS RN: 22992-15-0
Properties: monocl white powd [STR93] [CRC10]
Solubility: i H_2O; sl s acids [HAW93]
Reactions: minus $6H_2O$ at 110°C [HAW93]

1368
Compound: Gadolinium oxide
Synonym: gadolinia
Formula: Gd_2O_3
Molecular Formula: Gd_2O_3
Molecular Weight: 362.498
CAS RN: 12064-62-9
Properties: −325 mesh 5 μm or less with 99.999% purity; white to cream-colored powd; hygr; absorbs CO_2 from air; used in neutron shields, in special glasses, and as an evaporated material of 99.9% purity, it is reactive to radio frequencies [HAW93] [MER06] [CER91]

Solubility: i H_2O; s in acids [HAW93]
Density, g/cm³: 7.407 [MER06]
Melting Point, °C: 2310 [STR93]

1369
Compound: Gadolinium perchlorate hydrate
Formula: $Gd(ClO_4)_3 \cdot xH_2O$
Molecular Formula: Cl_3GdO_{12} (anhydrous)
Molecular Weight: 455.601 (anhydrous)
CAS RN: 14017-52-8
Properties: white cryst; hygr; x = 6 [ALF95] [STR93]

1370
Compound: Gadolinium(II) selenide
Formula: GdSe
Molecular Formula: GdSe
Molecular Weight: 236.21
CAS RN: 12024-81-6
Properties: cub cryst
Density, g/cm³: 8.1 [CRC10]
Melting Point, °C: 2170 [CRC10]

1371
Compound: Gadolinium silicide
Formula: $GdSi_2$
Molecular Formula: $GdSi_2$
Molecular Weight: 213.421
CAS RN: 12134-75-7
Properties: 10 mm and down lump; 6 mm pieces and smaller with 99.9% purity [ALF93] [CER91]
Density, g/cm³: 5.9 [LID94]

1372
Compound: Gadolinium sulfate
Formula: $Gd_2(SO_4)_3$
Molecular Formula: $Gd_2O_{12}S_3$
Molecular Weight: 602.691
CAS RN: 13450-87-8
Properties: colorless [CRC10]
Solubility: g/100 g H_2O: 3.98 (0°C), 2.60 (20°C), 2.32 (40°C) [LAN05]
Density, g/cm³: 4.139 [LAN05]
Melting Point, °C: decomposes at 500 [CRC10]

1373
Compound: Gadolinium sulfate octahydrate
Formula: $Gd_2(SO_4)_3 \cdot 8H_2O$
Molecular Formula: $[Gd_2H]_6O_{20}S_3$
Molecular Weight: 746.813
CAS RN: 13450-87-8

Properties: colorless, monocl cryst; used in cryogenic research [MER06] [HAW93]
Solubility: 3.28 g/100 mL H_2O (20°C), 2.54 g/100 mL H_2O (40°C) [CRC10]
Density, g/cm³: 3.010 [STR93]
Reactions: minus $8H_2O$ at 400°C [MER06]

1374
Compound: Gadolinium sulfide
Formula: Gd_2S_3
Molecular Formula: Gd_2S_3
Molecular Weight: 410.698
CAS RN: 12134-77-9
Properties: yellow, hygr; −200 mesh, 99.9% purity [CER91]
Solubility: decomposed by H_2O [CRC10]
Density, g/cm³: 6.1 [LID94]

1375
Compound: Gadolinium telluride
Formula: Gd_2Te_3
Molecular Formula: Gd_2Te_3
Molecular Weight: 697.300
CAS RN: 12160-99-5
Properties: ortho-rhomb; −20 mesh with 99.9% purity [LID94] [CER91]
Density, g/cm³: 7.7 [LID94]
Melting Point, °C: 1255 [LID94]

1376
Compound: Gadolinium titanate
Formula: $Gd_2Ti_2O_7$
Molecular Formula: $Gd_2O_7Ti_2$
Molecular Weight: 522.230
CAS RN: 12024-89-4
Properties: −100 mesh with 99.9% purity [CER91]

1377
Compound: Gadopentetic acid
Synonyms: diethylenetriaminepentaacetic acid, gadolinium(III) salt
Molecular Formula: $C_{14}H_{20}GdN_3O_{10}$
Molecular Weight: 547.577
CAS RN: 80529-93-7
Properties: uses: metal ion complex for MRI, diagnosis of cerebral tumors [ALD94] [MER06]
Melting Point, °C: decomposes at 129 [ALD94]

1378
Compound: Gallium
Formula: Ga

Molecular Formula: Ga
Molecular Weight: 69.723
CAS RN: 7440-55-3
Properties: silvery white liq or grayish metal; has tendency to remain in supercooled state; contracts on melting; ortho-rhomb, a = 0.45198 nm, b = 0.76602 nm, c = 0.45258 nm; electron affinity, 0.18 eV; enthalpy of fusion 5.59 kJ/mol; enthalpy of vaporization 254 kJ/mol; electrical resistivity 15.05 μohm·cm (20°C) for polycrystalline form, 25.79 μohm·cm for liq (30°C); radius of atom 0.138 nm, radius of Ga^{+++} 0.133 nm [CRC10] [KIR78] [MER87] [CIC73]
Solubility: reacts with alkalies to evolve H_2 [MER06]
Density, g/cm³: solid: 5.907; liq: 6.095 [KIR78] [CIC73]
Melting Point, °C: 29.78 [ALD94]
Boiling Point, °C: 2403 [ALD94]
Reactions: attacked by halogens and cold conc HCl [MER06]
Thermal Conductivity, W/(m·K): a-axis: 88.4, b-axis: 16.0, c-axis: 40.8 (20°C); liq 28.7 (77°C) [KIR78]
Thermal Expansion Coefficient: cup coefficient/°C: 0°C–20°C, 5.98×10^{-5} (solid); liq 1.2×10^{-4} (103°C), 1.03×10^{-4} (600°C) [KIR78]

1379
Compound: Galliumn acetylacetonate
Synonyms: 2,4-pentanedione, gallium(III) derivative
Formula: $Ga(CH_3COCH=C(O)CH_3)_3$
Molecular Formula: $C_{15}H_{21}GaO_6$
Molecular weight: 367.051
CAS RN: 14405-43-7
Properties: monocl white powd [STR93] [CRC10]
Density, g/cm³: 1.42 [STR93]
Melting Point, °C: 192–194 [STR93]
Boiling Point, °C: 140 (10 mm Hg) sublimes [STR93]

1380
Compound: Gallium antimonide
Formula: GaSb
Molecular Formula: GaSb
Molecular Weight: 191.483
CAS RN: 12064-03-8
Properties: cub; 6 mm pieces and smaller with 99.99% purity; band gap, eV, 0.81 (0 K), 0.72 (300 K); mobility (300 K), cm²/(V·s), 5000 for electrons, 850 for holes; effective mass 0.042 for electrons, 0.40 for holes; dielectric constant 15.7; enthalpy of fusion 25.10 kJ/mol; used in semiconducting devices; obtained by direct reaction of Ga and Sb at high temp [HAW93] [KIR82] [CER91] [CRC10]
Density, g/cm³: 6.096 [KIR78]
Melting Point, °C: 703 [CRC10]

Thermal Conductivity, W/(m·K): 27 [CRC10]
Thermal Expansion Coefficient: 6.1×10^{-6}/K [CRC10]

1381
Compound: Gallium arsenide
Formula: GaAs
Molecular Formula: AsGa
Molecular Weight: 144.645
CAS RN: 1303-00-0
Properties: cub cryst; 3–12 mm pieces of 99.999% purity, 25 mm and down polycrystalline pieces; dark gray with metallic sheen; hardness 4.5; dielectric constant 13.1; band gap, eV, 1.52 (0 K), 1.42 (300 K); mobility (300 K), cm²/(V·s), 8500 electrons and 400 holes; electroluminescent in infrared light; obtained by direct reaction of Ga and As at high temp; used as a semiconductor in light-emitting diodes for telephone dials [HAW93] [MER06] [STR93] [KIR82]
Density, g/cm³: 5.3176 [LID94]
Melting Poing, °C: 1238 [MER06]
Thermal Conductivity, W/(m·K): 0.52 [MER06]
Thermal Expansion Coefficient: 5.9×10^{-6}/°C [MER06]

1382
Compound: Gallium azide
Synonym: gallium(III) azide
Formula: $Ga(N_3)_3$
Molecular Formula: GaN_9
Molecular Weight: 195.784
CAS RN: 73157-11-6
Properties: prepared by decomposition of $GaF_3 \cdot NH_3$, 73157-06-9, at ~250°C [KIR78]

1383
Compound: Gallium nitride
Formula: GaN
Molecular Formula: GaN
Molecular Weight: 83.730
CAS RN: 25617-97-4
Properties: gray powd; −100 mesh with 99.9% purity; hex, a = 0.319 nm, c = 0.518 nm; has both semiconductor and electroluminescence properties; band gap, eV, 3.50 (0 K) and 3.36 (300 K); electron mobility (300 K) 380 cm²/(V·s); effective mass 0.19 electrons and 0.60 holes; dielectric 12.2; can be prepared by reaction of Ga with ammonia at ~1000°C [KIR81] [CIC73] [KIR82] [CER91] [CRC10]
Density, g/cm³: 6.1 [LID94]
Melting Point, °C: 600 (vacuum) [CIC73]
Boiling Point, °C: decomposes at >600 [KIR78]
Thermal Conductivity, W/(m·K): 6.56 [CRC10]

1384
Compound: Gallium phosphide
Formula: GaP
Molecular Formula: GaP
Molecular Weight: 100.697
CAS RN: 12063-98-8
Properties: amber; cub; 6 mm pieces and smaller with
99.999% purity; translucent, amber-colored cryst;
dielectric constant 11.1; band gap, eV, 2.34 (0 K)
and 2.26 (300 K); mobility (300 K), cm²/(V · s), 75
holes and 110 electrons; effective mass 0.82 electrons
and 0.60 holes; obtained by direct reaction of Ga
and P at high temperatures; used in semiconductor
devices; electroluminescent in visible light
[HAW93] [MER06] [KIR82] [CER91] [STR93]
Density, g/cm³: 4.138 [LID94]
Melting Point, °C: 1457 [LID94]
Thermal Conductivity, W/(m · K): 75.2 (25°C) [CRC10]
Thermal Expansion Coefficient: 5.3×10^{-6}/K [CRC10]

1385
Compound: Gallium suboxide
Formula: Ga_2O
Molecular Formula: Ga_2O
Molecular Weight: 155.445
CAS RN: 12024-20-3
Properties: brown powd; obtained by heating Ga_2O_3
and Ga at 700°C; stable in dry air [MER06]
Solubility: i H_2O [CRC10]
Density, g/cm³: 4.77 [KIR78]
Melting Point, °C: decomposes above 800 [MER06]
Reactions: oxidized to the trivalent state
by HNO_3 or Br_2 [MER06]

1386
Compound: Gallium(II) chloride
Synonym: gallium dichloride
Formula: $GaCl_2$
Molecular Formula: Cl_2Ga
Molecular Weight: 140.628
CAS RN: 24597-12-4
Properties: white; deliq; cryst; can be prepared by
heating $GaCl_3$ with Ga [MER06] [CRC10]
Solubility: decomposed by H_2O [CRC10]
Density, g/cm³: 2.74 [LID94]
Melting Point, °C: 172.4 [MER06]; 164 [STR93]
Boiling Point, °C: 535 [STR93]

1387
Compound: Gallium(II) selenide
Formula: GaSe
Molecular Formula: GaSe

Molecular Weight: 148.683
CAS RN: 12024-11-2
Properties: dark red; hex; 6 mm pieces and smaller
with 99.999% purity [CER91] [KIR78] [CRC10]
Density, g/cm³: 5.01 [KIR78]
Melting Point, °C: 960–965 [KIR78]

1388
Compound: Gallium(II) sulfide
Formula: GaS
Molecular Formula: GaS
Molecular Weight: 101.789
CAS RN: 12024-10-1
Properties: 6 mm pieces and smaller with
99.999% purity; hex, lamellar structure; air
sensitive [KIR78] [STR93] [CER91]
Density, g/cm³: 3.86 [STR93]
Melting Point, °C: ~965 [STR93]

1389
Compound: Gallium(II) telluride
Formula: GaTe
Molecular Formula: GaTe
Molecular Weight: 197.323
CAS RN: 12024-14-5
Properties: 6 mm pieces and smaller with 99.999%
purity; monocl or hex [KIR78] [CER91]
Density, g/cm³: 5.44 [KIR78]
Melting Point, °C: 825 [KIR78]

1390
Compound: Gallium(III) bromide
Formula: $GaBr_3$
Molecular Formula: Br_3Ga
Molecular Weight: 309.435
CAS RN: 13450-88-9
Properties: −8 mesh with 99.999% purity; ortho-
rhomb white cryst; enthalpy of vaporization
38.9 kJ/mol; enthalpy of fusion 11.70 kJ/mol;
formed by reacting Br_2 vapor with Ga in N_2 atm;
has seven hydrates with 1, 2, 2.5, 3, 4, 6, and
15 H_2O [CER91] [STR93] [KIR78] [CRC10]
Density, g/cm³: 3.69 [STR93]
Melting Point, °C: 121.5 [CRC10]
Boiling Point, °C: 279 [CRC10]

1391
Compound: Gallium(III) chloride
Formula: $GaCl_3$
Molecular Formula: Cl_3Ga
Molecular Weight: 176.081
CAS RN: 13450-90-3

Properties: solid ingot in glass with 99.999% purity; tricl, colorless needles; enthalpy of vaporization 23.9 kJ/mol; enthalpy of fusion 10.90 kJ/mol; prepared by reacting Ga metal with Cl_2 or HCl in nitrogen atm at 200°C; a trihydrate, 23306-52-7, is known [MER06] [KIR78] [CER91] [CRC10]
Solubility: >800 g$GaCl_3$/L H_2O [KIR78]
Density, g/cm³: 2.47 [STR93]
Melting Point, °C: 78 [ALD94]
Boiling Point, °C: 201 [CRC10]

1392
Compound: Gallium(III) fluoride
Synonym: gallium trifluoride
Formula: GaF_3
Molecular Formula: F_3Ga
Molecular Weight: 126.718
CAS RN: 7783-51-9
Properties: trig; −60 mesh with 99.95% purity; white powd, colorless needles; formed by thermal decomposition of ammonium hexafluorogallate in Ar atm [MER06] [KIR78] [CER91]
Solubility: 0.0024 g/100 mL H_2O (25°C) [MER06]
Density, g/cm³: 4.47 (heated in F_2 atm, 630°C) [MER06]
Melting Point, °C: >1000 [MER06]
Reactions: can be sublimed in N_2 atm at 800°C without decomposition [MER06]

1393
Compound: Gallium(III) fluoride trihydrate
Formula: $GaF_3 \cdot 3H_2O$
Molecular Formula: $F_3GaH_6O_3$
Molecular Weight: 180.764
CAS RN: 22886-66-4
Properties: −60 mesh with 99.5% purity; white cryst; obtained by dissolution of Ga or $Ga(OH)_3$ in HF [KIR78] [STR93] [CER91]
Solubility: more soluble than anhydrous GaF_3 in H_2O [MER06]
Melting Point, °C: >140 [MER06]
Reactions: thermally decomposed to $16[Ga(OH,F)_3] \cdot 6H_2O$ at 200°C [KIR78]

1394
Compound: Gallium(III) hydride
Synonym: gallane
Formula: GaH_3
Molecular Formula: GaH_3
Molecular Weight: 72.747
CAS RN: 13572-93-5

Properties: viscous liq; can be prepared by reacting $(CH_3)_3N \cdot GaH_3$, 19528-13-3, with BF_3 at −20°C [LID94] [KIR78]
Melting Point, °C: decomposes above −15 [KIR78]

1395
Compound: Gallium(III) hydroxide
Formula: $Ga(OH)_3$
Molecular Formula: GaH_3O_3
Molecular Weight: 120.745
CAS RN: 12023-99-3
Properties: white; unstable gelatinous precipitate; obtained by adding ammonia to a solution of Ga(III) salt [MER06] [KIR78] [CRC10]
Solubility: i H_2O [CRC10]
Melting Point, °C: decomposes at 440 [CRC10]

1396
Compound: Gallium(III) iodide
Formula: GaI_3
Molecular Formula: GaI_3
Molecular Weight: 450.436
CAS RN: 13450-91-4
Properties: −20 mesh with 99.999% purity; monocl; enthalpy of vaporization 56.5 kJ/mol; enthalpy of fusion 16.30 kJ/mol; can be prepared by direct reaction of Ga and I_2 [CRC10] [CER91] [KIR78]
Density, g/cm³: 4.15 [KIR78]
Melting Point, °C: 212 [KIR78]
Boiling Point, °C: sublimes at 340 [ALD94]

1397
Compound: Gallium(III) nitrate
Formula: $Ga(NO_3)_3$
Molecular Formula: GaN_3O_9
Molecular Weight: 255.738
CAS RN: 13494-90-1
Properties: white; cryst powd [MER06]
Solubility: s warm and cold H_2O, absolute alcohol, ether [MER06]
Melting Point, °C: decomposes at 110 [AES93]

1398
Compound: Gallium(III) nitrate hydrate
Formula: $Ga(NO_3)_3 \cdot xH_2O$
Molecular Formula: GaN_3O_9 (anhydrous)
Molecular Weight: 255.738 (anhydrous)
CAS RN: 69365-72-6
Properties: obtained by dissolving Ga metal or the oxide in conc HNO_3 [MER06]
Solubility: s H_2O [CRC10]

Melting Point, °C: decomposes 110 [AES93]
Reactions: decomposes to Ga_2O_3 at 200°C [CRC10]

1399
Compound: Gallium(III) oxide
Formula: Ga_2O_3
Molecular Formula: Ga_2O_3
Molecular Weight: 187.444
CAS RN: 12024-21-4
Properties: α, β, γ, δ, ε forms, β is the most stable;
white cryst; α and β obtained by thermal
decomposition of salts; γ formed by rapid
dehydration of $Ge(OH)_3$ gels at ~400°C, δ prepared
by decomposing $Ge(NO_3)_3$ at ~250°C, ε formed
by briefly heating δ form at ~550°C; used in
spectroscopic analysis and as an evaporated
material and sputtering target of 99.999% purity
in dielectric films [HAW93] [MER06] [CER91]
Solubility: s hot acid [HAW93]
Density, g/cm³: α: 6.44; β: 5.88 [HAW93]
Melting Point, °C: 1725 [LID94]
Reactions: reacts violently with Mg to give Ga [MER06]

1400
Compound: Gallium(III) oxide hydroxide
Formula: GaOOH
Molecular Formula: $GaHO_2$
Molecular Weight: 102.730
CAS RN: 20665-52-5
Properties: ortho-rhomb; prepared by oxidation of Ga
with H_2O at ~200°C under pressure [KIR78]
Density, g/cm³: 5.23 [LID94]
Reactions: GaOOH → α-Ga_2O_3 at
300°C–500°C [KIR78]

1401
Compound: Gallium(III) perchlorate hexahydrate
Formula: $Ga(ClO_4)_3 \cdot 6H_2O$
Molecular Formula: $Cl_3GaH_{12}O_{18}$
Molecular Weight: 476.166
CAS RN: 17835-81-3
Properties: cryst [ALF95]
Melting Point, °C: decomposes at 175 [ALF95]

1402
Compound: Gallium(III) selenide
Formula: Ga_2Se_3
Molecular Formula: Ga_2Se_3
Molecular Weight: 376.326
CAS RN: 12024-24-7
Properties: monocl; 6 mm pieces and smaller
with 99.999% purity [CER91] [KIR78]

Density, g/cm³: 4.95 [KIR78]
Melting Point, °C: 1005–1010 [KIR78]
Thermal Conductivity, W/(m·K): 5 [CRC10]
Thermal Expansion Coefficient: 8.9×10^{-6}/K [CRC10]

1403
Compound: Gallium(III) sulfate
Formula: $Ga_2(SO_4)_3$
Molecular Formula: $Ga_2O_{12}S_3$
Molecular Weight: 427.637
CAS RN: 13494-94-2
Properties: white powd; prepared by evaporation of a
solution of GaOOH, 20665-52-5, in 50% sulfuric
acid, followed by drying at 360°C; crystallizes from
aq solution as the octadecahydrate [KIR78] [CRC10]
Solubility: v s H_2O [CRC10]

1404
Compound: Gallium(III) sulfate octadecahydrate
Formula: $Ga_2(SO_4)_3 \cdot 18H_2O$
Molecular Formula: $Ga_2H_{36}O_{30}S_3$
Molecular Weight: 751.912
CAS RN: 13780-42-2
Properties: octahedral cryst; formed by dissolving
Ga_2O_3 or $Ga(OH)_3$ in sulfuric acid and
precipitating with ether or alcohol [MER06]
Solubility: s H_2O, 60% alcohol [MER06]
Density, g/cm³: 3.86 [STR93]

1405
Compound: Gallium(III) sulfide
Formula: Ga_2S_3
Molecular Formula: Ga_2S_3
Molecular Weight: 235.644
CAS RN: 12024-22-5
Properties: −100 mesh with 99.95% purity;
monocl [CER91] [KIR78]
Density, g/cm³: 3.77 [KIR78]
Melting Point, °C: 1090 [KIR78]

1406
Compound: Gallium(III) telluride
Formula: Ga_2Te_3
Molecular Formula: Ga_2Te_3
Molecular Weight: 522.246
CAS RN: 12024-27-0
Properties: 6 mm pieces and smaller with 99.999%
purity; two forms: cub and tetra; tetra, 73623-48-0,
is stable only at 400°C–495°C; the pentavalent
telluride Ga_2Te_5, 73623-48-0, is stable only in
the range 400°C–495°C [KIR78] [CER91]

Density, g/cm³: cub: 5.57; tetr: 5.85;
Ga₂Te₅: 5.85 [KIR78]
Melting Point, °C: 792 [KIR78]
Thermal Conductivity, W/(m·K): 4.7 [CRC10]

1407
Compound: Germanium
Formula: Ge
Molecular Formula: Ge
Molecular Weight: 72.61
CAS RN: 7440-56-4
Properties: grayish white, brittle metalloid; stable to oxidation in air up to 400°C; cub, a = 0.56574 nm; enthalpy of fusion 36.94 kJ/mol; enthalpy of vaporization 334 kJ/mol; Poisson's ratio 0.278; hardness 6 Mohs; resistivity 53,000 µohm·cm (25°C); electronegativity 1.8–1.9; band gap, eV, 0.74 (0 K), 0.66 (300 K); mobility (300 K), cm²/(V·s), 3900 electron and 1900 holes; used in transistors and semiconductor applications [MER06] [KIR78] [COT88] [CRC10]
Solubility: i H₂O, HCl, dil alkali hydroxides; attacked by aqua regia [MER06]
Density, g/cm³: 5.323 [MER06]
Melting Point, °C: 938.25 [LID94]
Boiling Point, °C: 2830 [KIR78]
Thermal Conductivity, W/(m·K): 60.2 (25°C) [ALD94]
Thermal Expansion Coefficient: 6.1 × 10/°C [MER06]

1408
Compound: Germanium nitride
Formula: Ge₃N₄
Molecular Formula: Ge₃N₄
Molecular Weight: 273.857
CAS RN: 12065-36-0
Properties: brownish-white powd; −200 mesh with 99.999% purity; prepared by reacting Ge powd and ammonia at 700°C–850°C; ortho-rhomb; a = 1.384 nm, b = 0.406 nm, c = 0.818 nm [CIC73] [CER91] [CRC10]
Solubility: i H₂O; does not react with most mineral acids, aqua regia or caustic solutions [KIR78] [CRC10]
Density, g/cm³: 5.25 [CRC10]
Melting Point, °C: decomposes at 900–1000 [CIC73]

1409
Compound: Germanium tetrahydride
Synonym: germane
Formula: GeH₄
Molecular Formula: GeH₄
Molecular Weight: 76.642

CAS RN: 7782-65-2
Properties: colorless gas; spontaneously flammable in air; enthalpy of vaporization 14.06 kJ/mol; can be prepared in small quantity by reaction: GeCl₄ + 4NaB₄ + 12H₂O = GeH₄(gas) + 4NaCl + 4B(OH)₃ + 12H₂(gas); used to produce highly pure electronic grade germanium by thermal decomposition at ~350°C [KIR80] [CRC10]
Solubility: i H₂O; s liq ammonia, sl s hot HCl [HAW93]
Density, g/cm³: liq: 1.523 at −142°C; gas: 3.43 g/L (0°C) [KIR80]
Melting Point, °C: −165 [HAW93]
Boiling Point, °C: −90 [MER06]

1410
Compound: Germanium(II) bromide
Formula: GeBr₂
Molecular Formula: Br₂Ge
Molecular Weight: 232.45
CAS RN: 24415-00-7
Properties: yellow monocl cryst [CRC10]
Solubility: reac H₂O [CRC10]
Melting Point, °C: 122 [CRC10]
Boiling Point, °C: decomposes at 150 [CRC10]

1411
Compound: Germanium(II) chloride
Formula: GeCl₂
Molecular Formula: Cl₂Ge
Molecular Weight: 143.515
CAS RN: 10060-11-4
Properties: white powd; unstable; decomposes into polymer subchloride at low temp [MER06] [HAW93]
Solubility: decomposes in H₂O; s ether, benzene; i alcohol and chloroform [HAW93] [MER06]
Melting Point, °C: decomposes [HAW93]

1412
Compound: Germanium(II) fluoride
Formula: GeF₂
Molecular Formula: F₂Ge
Molecular Weight: 110.607
CAS RN: 13940-63-1
Properties: white solid; decomposes above 130°C to form GeF₄(gas), Ge and GeF(gas); deliq in moist air forming Ge(II) hydroxide; can be formed by reduction of GeF₄ with metallic Ge [KIR78]
Solubility: s HF solutions [KIR78]
Melting Point, °C: 110 [LID94]
Boiling Point, °C: decomposes at 130 [LID94]

1413
Compound: Germanium(II) iodide
Formula: GeI$_2$
Molecular Formula: GeI$_2$
Molecular Weight: 326.419
CAS RN: 13573-08-5
Properties: −10 mesh with 99.999% purity;
    yellow powd [STR93] [CER91]
Solubility: s H$_2$O [CRC10]
Density, g/cm^3: 5.73 [STR93]
Melting Point, °C: decomposes at 550 [LID94]

1414
Compound: Germanium(II) oxide
Synonym: germanium monoxide
Formula: GeO
Molecular Formula: GeO
Molecular Weight: 88.609
CAS RN: 20619-16-3
Properties: black solid; stable at room temp; best
    prepared in a pure form by heating Ge and
    GeO$_2$ in oxygen free atm, GeO sublimes above
    710°C and is condensed [KIR78] [HAW93]
Solubility: s in about 250 parts cold H$_2$O, 100
    parts boiling H$_2$O; s acids [MER06]
Melting Point, °C: sublimes at 710 [HAW93]

1415
Compound: Germanium(II) selenide
Formula: GeSe
Molecular Formula: GeSe
Molecular Weight: 151.570
CAS RN: 12065-10-0
Properties: gray ortho-rhomb or brown
    powd; 6 mm pieces and smaller with
    99.999% purity [CER91] [LID94]
Density, g/cm^3: 5.6 [LID94]
Melting Point, °C: 667 [LID94]

1416
Compound: Germanium(II) sulfide
Formula: GeS
Molecular Formula: GeS
Molecular Weight: 104.676
CAS RN: 12025-32-0
Properties: reddish-yellow, amorphous or rhomb cryst;
    −20 mesh with 99.95% purity [CER91] [CRC10]
Solubility: 0.24 g/100 mL H$_2$O [CRC10]
Density, g/cm^3: amorphous: 3.31; rhomb: 4.01 [CRC10]
Melting Point, °C: 615 [LID94]
Reactions: sublimes at 430°C [CRC10]

1417
Compound: Germanium(II) telluride
Formula: GeTe
Molecular Formula: GeTe
Molecular Weight: 200.210
CAS RN: 12025-39-7
Properties: 6 mm pieces and smaller with 99.999%
    purity; good semiconductor [HAW93] [CER91]
Density, g/cm^3: 6.14 [CRC10]
Melting Point, °C: 725 [HAW93]

1418
Compound: Germanium(IV) bromide
Synonym: germanium tetrabromide
Formula: GeBr$_4$
Molecular Formula: Br$_4$Ge
Molecular Weight: 392.226
CAS RN: 13450-92-5
Properties: solid ingot in glass with 99.999% purity;
    white cryst; enthalpy of vaporization 41.4 kJ/
    mol; can be prepared readily by reacting Ge
    with Br$_2$ or with GeO$_2$ and HBr solutions
    [KIR78] [STR93] [CER91] [CRC10]
Density, g/cm^3: 3.132 [STR93]
Melting Point, °C: 26.1 [STR93]
Boiling Point, °C: 186.5 [ALD94]

1419
Compound: Germanium(IV) chloride
Synonym: germanium tetrachloride
Formula: GeCl$_4$
Molecular Formula: Cl$_4$Ge
Molecular Weight: 214.421
CAS RN: 10038-98-9
Properties: 99.9999% purity; colorless liq; fumes in
    air; appreciably volatile at room temp; refractive
    index 1.464; enthalpy of vaporization 27.9 kJ/mol;
    vapor pressure, Pa: 100 (−48°C), 1,000 (−20°C),
    10,000 (21°C), 10^5 (83°C), 10^6 (190°C); prepared by
    reacting germanium oxides or germanates with HCl
    [KIR78] [HAW93] [MER06] [CER91] [CRC10]
Solubility: hydrolyzed in H$_2$O; s benzene, ether,
    carbon disulfide, alcohol, chloroform [HAW93]
Density, g/cm^3: 1.874 [HAW93]
Melting Point, °C: −49.5 [HAW93]
Boiling Point, °C: 86.55 [CRC10]

1420
Compound: Germanium(IV) ethoxide
Formula: Ge(OC$_2$H$_5$)$_4$
Molecular Formula: C$_8$H$_{20}$GeO$_4$
Molecular Weight: 252.84

CAS RN: 14165-55-0
Properties: liq [ALF95]
Solubility: s alcohol, benzene [ALF95]
Density, g/cm³: 1.140 [ALF95]
Melting Point, °C: −72 [ALF95]
Boiling Point, °C: 185.5 [ALF95]
Reactions: hydrolyzed by H_2O [ALF95]

1421
Compound: Germanium(IV) fluoride
Synonym: germanium tetrafluoride
Formula: GeF_4
Molecular Formula: F_4Ge
Molecular Weight: 148.604
CAS RN: 7783-58-6
Properties: 99.99% purity; colorless gas; fumes strongly in air; odor of garlic; thermally stable up to ~1000°C; triple point reported as −50°C and 404.1 kPa; vapor pressure ~100 kPa at −36.5°C; pure GeF_4 usually prepared by decomposing $BaGeF_6$ at ~700°C [KIR78] [MER06] [CER91]
Solubility: hydrolyzes to GeO_2 and H_2GeF_6 in H_2O [MER06]
Density, g/cm³: liq: 2.162; solid (−195°C): 3.148 [MER06]
Melting Point, °C: −15 (3032 mm pressure) [MER06]
Boiling Point, °C: sublimes at −36.5 [MER06]
Reactions: corrodes Hg and grease [MER06]

1422
Compound: Germanium(IV) fluoride trihydrate
Synonym: germanium tetrafluoride trihydrate
Formula: $GeF_4 \cdot 3H_2O$
Molecular Formula: $F_4GeH_6O_3$
Molecular Weight: 202.650
CAS RN: 7783-58-6
Properties: white cryst; deliq; obtained by slow evaporation of GeO_2 in 20% HF [MER06] [CRC10]
Solubility: s H_2O [CRC10]
Melting Point, °C: decomposes [CRC10]

1423
Compound: Germanium(IV) iodide
Synonym: germanium tetraiodide
Formula: GeI_4
Molecular Formula: GeI_4
Molecular Weight: 580.228
CAS RN: 13450-95-8
Properties: −10 mesh with 99.999% purity; reddish orange cryst; a method of preparation is to react Ge with I_2 or GeO_2 with HI solutions [KIR78] [STR93] [CER91]
Density, g/cm³: 4.416 [STR93]

Melting Point, °C: 146 [STR93]
Boiling Point, °C: 350 [STR93]

1424
Compound: Germanium(IV) oxide
Synonym: germanium dioxide
Formula: GeO_2
Molecular Formula: GeO_2
Molecular Weight: 104.609
CAS RN: 1310-53-8
Properties: white powd; two forms: hex, tetr amorphous (vitreous); hex can be produced by the hydrolysis of $GeCl_4$ in H_2O or by igniting Ge sulfides; tetr form is insoluble, it can be produced by heating the hex form at 300°C–900°C; amorphous material formed when the hex or tetr forms are melted and cooled; used in phosphors, transistors, and diodes, in infrared transmitting glass; hex form of 99.999% purity is a sputtering target for dielectric film preparation [HAW93] [CER91]
Solubility: (probably hex form) g/100 g H_2O: 0.49 (10°C), 0.43 (20°C), 0.61 (40°C) [LAN05]; hex: s HCl, HF, NaOH solutions [KIR78]
Density, g/cm³: hex: 4.228; tetr: 6.239; amorphous: 3.637 [KIR78]
Melting Point, °C: hex: 1116; tetr: 1086 [KIR78]

1425
Compound: Germanium(IV) selenide
Synonym: germanium diselenide
Formula: $GeSe_2$
Molecular Formula: $GeSe_2$
Molecular Weight: 230.530
CAS RN: 12065-11-1
Properties: 6 mm pieces and smaller with 99.999% purity; orange cryst [STR93] [CER91]
Solubility: i H_2O [CRC10]
Density, g/cm³: 4.56 [STR93]
Melting Point, °C: 707 [STR93]
Boiling Point, °C: decomposes [CRC10]

1426
Compound: Germanium(IV) sulfide
Synonym: germanium disulfide
Formula: GeS_2
Molecular Formula: GeS_2
Molecular Weight: 136.742
CAS RN: 12025-34-2
Properties: black cryst; can be prepared by reacting GeO_2 and sulfur [KIR78] [STR93]
Solubility: decomposed by H_2O [CRC10]
Density, g/cm³: 2.94 [CRC10]
Melting Point, °C: 530 [STR93]

1427
Compound: Germanium(IV) telluride
Synonym: germanium ditelluride
Formula: GeTe$_2$
Molecular Formula: GeTe$_2$
Molecular Weight: 327.810
CAS RN: 12260-55-8
Properties: reacted product, 6 mm pieces and smaller with 99.999% purity [CER91]

1428
Compound: Gold
Formula: Au
Molecular Formula: Au
Molecular Weight: 196.96654
CAS RN: 7440-57-5
Properties: yellow; soft metal; cub, a=0.407 nm; electronegativity, 2.88; electrical resistivity (20°C) 2.35 μohm·cm; temp coefficient of electrical resistivity (0°C–100°C) 0.004; enthalpy of fusion 12.55 kJ/mol; enthalpy of vaporization 324 kJ/mol; specific heat (18°C) 131 J/(kg·°C); hardness 2.5–3.0 Mohs; forms AuTe$_2$ (calaverite), 12006-61-0, by reaction with Te at ~475°C [KIR78] [MER06] [CRC10]
Solubility: s aqua regia, alkali cyanide solutions [MER06]
Density, g/cm³: 19.32 (20°C) [KIR78]
Melting Point, °C: 1064.43 [ALD94]
Boiling Point, °C: 2808 [ALD94]
Reactions: extremely stable: not attacked by acids or air [MER06]
Thermal Conductivity, W/(m·K): 318 (25°C) [ALD94]
Thermal Expansion Coefficient: 100°C: 14.16×10⁻⁶/K [KIR78]

1429
Compound: Gold(I) bromide
Synonym: aurous bromide
Formula: AuBr
Molecular Formula: AuBr
Molecular Weight: 276.871
CAS RN: 10294-27-6
Properties: yellowish gray mass [HAW93]
Solubility: i H$_2$O [HAW93]
Density, g/cm³: 7.9 [CRC10]
Melting Point, °C: decomposes at 165 [HAW93]

1430
Compound: Gold(I) carbonyl chloride
Formula: Au(CO)Cl
Molecular Formula: CAuClO

Molecular Weight: 260.430
CAS RN: 50960-82-2
Properties: off-white powd; sensitive to atm oxygen [STR93]

1431
Compound: Gold(I) chloride
Synonym: aurous chloride
Formula: AuCl
Molecular Formula: AuCl
Molecular Weight: 232.420
CAS RN: 10294-29-8
Properties: yellowish powd [MER06]
Solubility: i H$_2$O with slow decomposition [MER06]
Density, g/cm³: 7.57 [MER06]
Melting Point, °C: decomposes at ~289 [MER06]
Reactions: decomposes to Au and Cl$_2$ at 289°C [MER06]

1432
Compound: Gold(I) cyanide
Synonym: aurous cyanide
Formula: AuCN
Molecular Formula: CAuN
Molecular Weight: 222.985
CAS RN: 506-65-0
Properties: yellow powd; hex; odorless; iridescent in sunlight; slowly decomposes in presence of moisture [MER06]
Solubility: i H$_2$O, alcohol; dil acid; s NH$_3$, NaCN soln [MER06]
Density, g/cm³: 7.14 [MER06]
Melting Point, °C: decomposes to Au and CN when ignited [MER06]
Reactions: evolves HCN gas if warmed with HCl [MER06]

1433
Compound: Gold(I) iodide
Synonym: aurous iodide
Formula: AuI
Molecular Formula: AuI
Molecular Weight: 323.871
CAS RN: 10294-31-2
Properties: yellowish to greenish yellow powd; decomposes slowly at ordinary temp, rapidly at elevated temp [MER06]
Solubility: i H$_2$O; s alkali iodide or cyanide [MER06]
Density, g/cm³: 8.25 [MER06]
Melting Point, °C: decomposes at 120 [STR93]
Reactions: decomposed by warm acids [MER06]

1434
Compound: Gold(I) sulfide
Synonym: aurous sulfide
Formula: Au_2S
Molecular Formula: Au_2S
Molecular Weight: 425.999
CAS RN: 1303-60-2
Properties: brownish black powd; forms colloid in H_2O when freshly prepared by treatment of acidified $KAu(CN)_2$ solutions with H_2S [MER06] [KIR78]
Solubility: i H_2O, dil single acids; s aqua regia, alkali cyanide solutions [MER06] [KIR78]
Density, g/cm³: ~11 [LID94]
Melting Point, °C: decomposes at 240 [CRC10]

1435
Compound: Gold(III) bromide
Synonym: auric bromide
Formula: $AuBr_3$
Molecular Formula: $AuBr_3$
Molecular Weight: 436.679
CAS RN: 10294-28-7
Properties: brownish orange powd; used to test alkaloids and for testing spermatic fluid, medicinal uses [HAW93] [STR93]
Solubility: s H_2O, alcohol, glycerol [MER06]
Melting Point, °C: decomposes at ~160 [MER06]
Reactions: slowly decomposed by alcohol and glycerol [MER06]

1436
Compound: Gold(III) chloride
Synonym: auric chloride
Formula: $AuCl_3$
Molecular Formula: $AuCl_3$
Molecular Weight: 303.325
CAS RN: 13453-07-1
Properties: −8 mesh with 99% purity; yellow to red cryst; can be made by heating Au and Cl_2 at 200°C; exists as dimer Au_2Cl_3 in both solid and gas phases under Cl_2 below 254°C [HAW93] [CER91]
Solubility: s H_2O, alcohol, ether [HAW93]
Density, g/cm³: 4.7 [LID94]
Melting Point, °C: decomposes at >160 [LID94]
Reactions: decomposes to AuCl, then Au, in Cl_2 atm at >254°C [KIR78]

1437
Compound: Gold(III) cyanide trihydrate
Synonym: cyanoauric acid
Formula: $Au(CN)_3 \cdot 3H_2O$

Molecular Formula: $C_3H_6AuN_3O_3$
Molecular Weight: 329.066
CAS RN: 535-37-5
Properties: colorless; deliq cryst; used as an electrolyte for plating gold [HAW93] [MER06]
Solubility: v s H_2O; sl s alcohol, ether [HAW93]
Melting Point, °C: decomposes at 50 [MER06]

1438
Compound: Gold(III) fluoride
Formula: AuF_3
Molecular Formula: AuF_3
Molecular Weight: 253.962
CAS RN: 14720-21-9
Properties: orange-yellow hex cryst [CRC10]
Density, g/cm³: 6.75 [CRC10]
Melting Point, °C: 300 [CRC10]
Boiling Point, °C: sublimes [CRC10]

1439
Compound: Gold(III) hydroxide
Synonym: auric hydroxide
Formula: $Au(OH)_3$
Molecular Formula: AuH_3O_3
Molecular Weight: 247.989
CAS RN: 1303-52-2
Properties: brown powd; decomposed by sunlight to Au metal; also decomposes on standing; decomposes to Au and O_2 at >160°C; precipitates from $AuCl_4^-$ solutions following addition of alkali hydroxides; used in gilding liq, in decorating porcelain [HAW93] [MER06] [KIR78]
Solubility: i H_2O; s in NaCN soln, HCl, conc HNO_3 [MER06]
Melting Point, °C: decomposes at ~100 [LID94]
Reactions: $Au(OH)_3 + NH_3 \rightarrow$ gold fulminate (explosive) [MER06]

1440
Compound: Gold(III) iodide
Synonym: auric iodide
Formula: AuI_3
Molecular Formula: AuI_3
Molecular Weight: 577.680
CAS RN: 13453-24-2
Properties: green powd, unstable, converts to AuI, 10294-31-2 [KIR78] [STR93]
Solubility: i cold H_2O, decomposed by hot H_2O [CRC10]

1441
Compound: Gold(III) oxide
Synonyms: auric oxide, gold trioxide

Formula: Au_2O_3
Molecular Formula: Au_2O_3
Molecular Weight: 441.931
CAS RN: 1303-58-8
Properties: −100 mesh with 99.9% purity; brownish orange powd; slowly decomposed by sunlight; forms when $Au(OH)_3$ is heated at 140°C; used in gold plating [MER06] [HAW93] [CER91] [KIR78]
Solubility: i H_2O; s HCl, conc HNO_3, NaCN solns [MER06]
Melting Point, °C: decomposes at ~150 [LID94]
Reactions: evolves O_2 at 110°C; decomposed to metal at 250°C [MER06]

1442
Compound: Gold(III) selenate
Synonym: auric selenate
Formula: $Au_2(SeO_4)_3$
Molecular Formula: $Au_2O_{12}Se_3$
Molecular Weight: 822.806
CAS RN: 10294-32-3
Properties: small, yellow cryst; decomposes in light [MER06]
Solubility: i H_2O; s H_2SO_4, HNO_3 [MER06]

1443
Compound: Gold(III) selenide
Synonym: auric selenide
Formula: Au_2Se_3
Molecular Formula: Au_2Se_3
Molecular Weight: 630.813
CAS RN: 1303-62-4
Properties: black amorphous solid [MER06]
Solubility: s aqua regia, alkali cyanides [MER06]
Density, g/cm³: 4.65 [MER06]
Melting Point, °C: decomposed by heat [MER06]

1444
Compound: Gold(III) sulfide
Synonyms: auric sulfide, gold trisulfide
Formula: Au_2S_3
Molecular Formula: Au_2S_3
Molecular Weight: 490.131
CAS RN: 1303-61-3
Properties: black powd [MER06]; not very stable, can be prepared by adding H_2S to ether solution of $AuCl_3$ [KIR78]
Solubility: i H_2O [CRC10]
Density, g/cm³: 8.75 [CRC10]
Reactions: decomposes to metal and sulfur at 200°C [MER06]

1445
Compound: Hafnium
Formula: Hf
Molecular Formula: Hf
Molecular Weight: 178.49
CAS RN: 7440-58-6
Properties: gray, highly lustrous, hard ductile metal; two forms; thermal neutron cross section 115 barns; good corrosion resistance and high strength; electrical resistivity 3.57×10^{-7} ohm·m (0°C), 6.24×10^{-7} (200°C); enthalpy of vaporization 571 kJ/mol; enthalpy of fusion 27.20 kJ/mol; entropy of fusion 54.4 J/(kg·K); used in controls for nuclear reactors, in light bulb filaments, electrodes, and special glasses [HAW93] [MER06] [KIR80] [CRC10] [ALD94]
Solubility: s HF; slowly reacts with conc H_2SO_4, aqua regia [KIR80]
Density, g/cm³: 13.28 [KIR80]
Melting Point, °C: 2227 [ALD94]
Boiling Point, °C: 4602 [ALD94]
Reactions: transition α to β at 1777°C [KIR80]
Thermal Conductivity, W/(m·K): 23.0 (25°C) [ALD94]; 22.3 (50°C), 20.7 (400°C) [KIR80]
Thermal Expansion Coefficient: 5.9×10^{-6}/K [KIR80]

1446
Compound: Hafnium acetylacetonate
Synonyms: 2,4-pentanedione, hafnium(IV) derivative
Formula: $Hf(CH_3COCH=C(O)CH_3)_4$
Molecular Formula: $C_{20}H_{28}HfO_8$
Molecular Weight: 574.927
CAS RN: 17475-67-1
Properties: powd [STR93]

1447
Compound: Hafnium boride
Formula: HfB_2
Molecular Formula: B_2Hf
Molecular Weight: 200.112
CAS RN: 12007-23-7
Properties: gray, hex, cryst solid, a = 0.3141 nm, c = 0.3470 nm; can be prepared by heating $HfO_2 + C + B_2O_3$; hardness 2900 kgf/mm²; resistivity 8.8 μohm·cm; used as a refractory material and as a sputtering target with 99.5% purity to produce films, which may be wear-resistant and semiconducting [HAW93] [KIR80] [CER91]
Density, g/cm³: 10.5 (theoretical 11.2) [KIR80]
Melting Point, °C: 3250 [KIR78]
Reactions: attacked by HF, else highly resistant [KIR80]
Thermal Expansion Coefficient: 5.7×10^{-6} [KIR80]

1448
Compound: Hafnium(II) bromide
Formula: $HfBr_2$
Molecular Formula: Br_2Hf
Molecular Weight: 338.30
CAS RN: 13782-95-1
Properties: blue black cryst [CRC10]
Melting Point, °C: decomposes at 400 [CRC10]

1449
Compound: Hafnium bromide
Synonym: hafnium tetrabromide
Formula: $HfBr_4$
Molecular Formula: Br_4Hf
Molecular Weight: 498.106
CAS RN: 13777-22-5
Properties: −20 mesh with 99.7% purity;
    hygr; white cub, a=0.095 nm; or tan
    powd [KIR80] [STR93] [CER91]
Density, g/cm³: 4.90 (5.09 theoretical) [KIR80]
Melting Point, °C: 424 (3.34 MPa) [KIR80]
Boiling Point, °C: sublimation point 322 [KIR80]

1450
Compound: Hafnium carbide
Formula: HfC
Molecular Formula: CHf
Molecular Weight: 190.501
CAS RN: 12069-85-1
Properties: dark, gray, brittle solid; fcc, a=0.4640 nm;
    high cross section for absorption of thermal
    neutrons; resistivity 8.8 μohm · cm; most
    refractory binary material known; hardness
    2300 kgf/mm²; used in control rods of nuclear
    reactors; can be prepared by heating HfO_2 with
    lampblack under H_2 at 1900°C–2300°C; used
    in crucible form for melting hafnium oxide,
    other oxides [KIR80] [HAW93] [CER91]
Density, g/cm³: 12.2 (theoretical 12.7) [KIR80]
Melting Point, °C: 3950 [KIR80] [CIC73]
Thermal Expansion Coefficient:
    6.59×10^{-6}/K [KIR80]

1451
Compound: Hafnium(II) chloride
Formula: $HfCl_2$
Molecular Formula: Cl_2Hf
Molecular Weight: 249.40
CAS RN: 13782-92-8
Properties: black solid [CRC10]
Melting Point, °C: decomposes at 400 [CRC10]

1452
Compound: Hafnium chloride
Synonym: hafnium tetrachloride
Formula: $HfCl_4$
Molecular Formula: Cl_4Hf
Molecular Weight: 320.301
CAS RN: 13499-05-3
Properties: −80 mesh with 99.99% and 99% purity;
    white, cryst monocl, a=0.631 nm, b=0.7407 nm,
    c=0.6256 nm; can be obtained by heating the oxide in
    Cl_2 with carbon at >317°C [MER06] [CER91] [KIR80]
Solubility: hydrolyzed by H_2O to $HfOCl_2$ [MER06]
Melting Point, °C: 432 [KIR80]
Boiling Point, °C: sublimation point 317 [KIR80]

1453
Compound: Hafnium fluoride
Synonym: hafnium tetrafluoride
Formula: HfF_4
Molecular Formula: F_4Hf
Molecular Weight: 254.484
CAS RN: 13709-52-9
Properties: white or highly dense pressure sintered
    pieces of 3–6 mm; monocl, a=0.957 nm,
    b=0.993 nm, c=0.7730 nm; can be formed by
    careful thermal decomposition of ammonium
    fluorohafnate, 16925-24-9; sintered pieces used as
    evaporation material for possible application to low-
    index material as a replacement for ThF_4, 99.95%
    pure material used as a sputtering target to produce
    antireflection coatings on glass [KIR80] [CER91]
Density, g/cm³: 7.1 [LID94]
Melting Point, °C: >968 [KIR80]
Boiling Point, °C: sublimes at 968 [KIR80]

1454
Compound: Hafnium hydride
Formula: HfH_2
Molecular Formula: H_2Hf
Molecular Weight: 180.506
CAS RN: 12770-26-2
Properties: −325 mesh 10 μm or less with 99.8%
    purity; brittle solid, fcc; prepared by reacting
    Hf and H_2 above 250°C [KIR80] [CER91]
Density, g/cm³: 11.4 [LID94]

1455
Compound: Hafnium iodide
Formula: HfI_3
Molecular Formula: HfI_3
Molecular Weight: 559.20

CAS RN: 13779-73-2
Properties: black cryst [CRC10]
Melting Point, °C: decomposes [CRC10]

1456

Compound: Hafnium iodide
Synonym: hafnium tetraiodide
Formula: HfI_4
Molecular Formula: HfI_4
Molecular Weight: 686.108
CAS RN: 13777-23-6
Properties: yellowish orange cub, a = 1.176 nm, or red
    powd; sensitive to moisture [STR93] [KIR80]
Density, g/cm³: 5.6 [LID94]
Melting Point, °C: 449 (3.34 MPa) [KIR80]
Boiling Point, °C: sublimation point 393 [KIR80]

1457

Compound: Hafnium nitride
Formula: HfN
Molecular Formula: HfN
Molecular Weight: 192.497
CAS RN: 25817-87-2
Properties: yellowish brown cryst; fcc, a = 0.4518 nm;
    most refractory of all known metal nitrides; hardness
    1640 kgf/mm²; electrical resistivity 33 μohm·cm;
    can be prepared by heating Hf in N_2 or NH_3 atm
    at 1000°C–1500°C; as a 99.5% pure material,
    used as a sputtering target to increase electrical
    stability of diodes, transistors, and integrated
    circuits [KIR80] [KIR81] [HAW93] [CER91]
Density, g/cm³: 13.84 (theoretical) [KIR80]
Melting Point, °C: 3305 [HAW93]
Thermal Conductivity, W/(m·K): 11.1 [KIR81]
Thermal Expansion Coefficient: 6.9×10^{-6} [KIR80]

1458

Compound: Hafnium oxide
Synonym: hafnia
Formula: HfO_2
Molecular Formula: HfO_2
Molecular Weight: 210.489
CAS RN: 12055-23-1
Properties: white solid; cub, a = 0.51156 nm,
    b = 0.51722 nm, c = 0.52948 nm; obtained by
    ignition of the hydroxide, oxalate, or sulfate;
    hardness 1050 kgf/mm²; resistivity >10^{+8}
    μohm·cm; used as a refractory metal oxide, an
    evaporated material of 99.9% purity for dielectric
    coatings, to coat wires for emitters, and as a
    99.95% pure sputtering target to provide very
    hard, adherent film; stabilized with 10%–15%
    CaO [HAW93] [MER06] [KIR80] [CER91]

Solubility: i H_2O [HAW93]
Density, g/cm³: 9.68 [MER06]
Melting Point, °C: 2900 [KIR80]
Reactions: transformation tetr to cub
    above 2700°C [KIR80]
Thermal Expansion Coefficient: (volume)
    100°C (0.144), 200°C (0.319), 400°C (0.696),
    800°C (1.499), 1000°C (2.085) [CLA66]

1459

Compound: Hafnium oxychloride octahydrate
Formula: $HfOCl_2 \cdot 8H_2O$
Molecular Formula: $Cl_2H_{16}HfO_9$
Molecular Weight: 409.517
CAS RN: 14456-34-9
Properties: −6 mesh with 99.99% purity; white powd
    or tetr cryst; produced by addition of $HfCl_4$ to
    H_2O or by dissolution of HfO_2 in HCl; when
    heated, first dissolves in water of crystallization,
    then decomposes [KIR80] [STR93] [CER91]
Solubility: s H_2O [KIR80]
Melting Point, °C: decomposes [KIR80]

1460

Compound: Hafnium phosphide
Formula: HfP
Molecular Formula: HfP
Molecular Weight: 209.464
CAS RN: 12325-59-6
Properties: −100 mesh with 99% purity; hex,
    a = 0.365 nm, c = 1.237 nm [KIR80] [CER91]
Density, g/cm³: 9.78 (theoretical) [KIR80]

1461

Compound: Hafnium selenide
Synonym: hafnium diselenide
Formula: $HfSe_2$
Molecular Formula: $HfSe_2$
Molecular Weight: 336.410
CAS RN: 12162-21-9
Properties: −325 mesh 10 μm or less with 99.5%
    purity; dark brown; hex, a = 0.375 nm, b = 0.616 nm;
    resistivity 20 μohm·cm [CER91] [KIR80]
Density, g/cm³: 7.46 [KIR80]

1462

Compound: Hafnium silicate
Formula: $HfSiO_4$
Molecular Formula: HfO_4Si
Molecular Weight: 270.574
CAS RN: 37248-04-7

Properties: zircon cryst structure; a = 0.65725 nm, c = 0.59632 nm [SUB90]
Thermal Expansion Coefficient: 1020°C is 3.1 × 10^{-6}/°C [SUB90]

1463
Compound: Hafnium silicide
Formula: $HfSi_2$
Molecular Formula: $HfSi_2$
Molecular Weight: 234.661
CAS RN: 12401-56-8
Properties: gray powd; rhomb, a = 0.3677 nm, b = 1.455 nm, c = 0.3649 nm; hardness 930 kgf/mm^2; as 99.5% pure material, used as sputtering target to produce wear-resistant films and semiconducting films for use in integrated circuits [KIR80] [STR93] [CER91]
Density, g/cm^3: 7.2 (8.03 theoretical) [KIR80]
Melting Point, °C: 1680 [STR93]; 1750 [KIR80]

1464
Compound: Hafnium sulfate
Formula: $Hf(SO_4)_2$
Molecular Formula: HfO_8S_2
Molecular Weight: 370.617
CAS RN: 15823-43-5
Properties: can be prepared by reacting fuming sulfuric acid with $HfCl_4$ [MER06]
Melting Point, °C: decomposes at >500 [MER06]

1465
Compound: Hafnium sulfide
Formula: HfS_2
Molecular Formula: HfS_2
Molecular Weight: 242.622
CAS RN: 18855-94-2
Properties: −200 mesh with 99.9% purity; purple brown; hex, a = 0.364 nm, b = 0.584 nm; resistivity at room temp 1 µohm·cm; can be prepared by reacting the elements at 500°C; used as a solid lubricant [HAW93] [CER91]
Density, g/cm^3: 6.03 [KIR80]

1466
Compound: Hafnium telluride
Formula: $HfTe_2$
Molecular Formula: $HfTe_2$
Molecular Weight: 433.690
CAS RN: 39082-23-0
Properties: −325 mesh 10 µm or less with 99.5% purity [CER91]

1467
Compound: Hafnium titanate
Formula: $HfTiO_4$
Molecular Formula: HfO_4Ti
Molecular Weight: 290.355
CAS RN: 12055-24-2
Properties: −325 mesh 10 µm or less with 99.5% purity; off-white powd [STR93] [CER91]

1468
Compound: Hafnocene dichloride
Synonym: bis(cyclopentadienyl)hafnium dichloride
Molecular Weight: 379.59
Molecular Formula: $C_{10}H_{10}Cl_2Hf$
CAS RN: 12116-66-4
Properties: moisture sensitive; uses: synthesis of many organometallic compounds and early-transition-metal complexes [ALD94]
Melting Point, °C: 230–233 [ALD94]

1469
Compound: Helium
Formula: He
Molecular Formula: He
Molecular Weight: 4.002602
CAS RN: 7440-59-7
Properties: inert gas; nonflammable, colorless, odorless, tasteless; critical temp −267.9°C; critical pressure 227.5 kPa; enthalpy of vaporization 81.70 J/mol; enthalpy of fusion 0.0138 kJ/mol; heat capacity (101.32 kPa, 25°C) 20.78 J/(mol·K); sonic velocity (101.32 kPa, 0°C) 973 m/s; viscosity (101.32 kPa, 0°C) 19.86 Pa·s, specific volume (21.1°C, 101.3 kPa) 6.04 m^3/kg [MER06] [KIR78] [AIR87] [ALD94]
Solubility: 8.61 mL/1000 g H_2O (101.32 kPa, 0°C) [KIR78]; Henry's law constants, k × 10^{-4}: 9.856 (104°C), 6.739 (149.4°C), 2.524 (250.6°C), 1.796 (275.1°C) [POT78]
Density, g/cm^3: gas (101.3 kPa, 0°C) 0.00017850 [KIR78]
Melting Point, °C: −272.2 (25 atm) [HAW93]
Boiling Point, °C: −268.93 [KIR78]
Thermal Conductivity, W/(m·K): 0.14184 (gas) (101.32 kPa, 0°C) [KIR78]

1470
Compound: Helium-3
Formula: ^3He
Molecular Formula: He
Molecular Weight: 3.0160
CAS RN: 14762-55-7

Properties: gas; enthalpy of vaporization
0.0829 kJ/mol; heat capacity (101.32 kPa,
25°C) 20.78 J/(mol·K); viscosity (101.32 kPa,
25°C) ~17.2 Pa·s [KIR78] [CRC10]
Density, g/cm³: 0.0001347 (gas, 101.3 kPa, 0°C) [KIR78]
Boiling Point, °C: −268.93 [CRC10]
Thermal Conductivity, W/(m·K): gas
(101.32 kPa, 0°C) ~0.1636 [KIR78]

1471
Compound: Hexaamminecobalt(III) chloride
Formula: Co(NH₃)₆Cl₃
Molecular Formula: Cl₃CoH₁₈N₆
Molecular Weight: 267.474
CAS RN: 10534-89-1
Properties: wine red monocl [KIR79]
Solubility: 5.99 g/100 mL cold H₂O; s conc
HCl; i alcohol, NH₄OH [KIR79]
Density, g/cm³: 1.71 [KIR79]
Reactions: evolves NH₃ at 215°C [KIR79]

1472
Compound: Hexaammineruthenium(III) chloride
Formula: Ru(NH₃)₆Cl₃
Molecular Formula: Cl₃H₁₈N₆Ru
Molecular Weight: 309.612
CAS RN: 14282-91-8
Properties: off-white powd; there is also a
hexaamineruthenium(II) chloride, CAS
RN 15305-72-3 [STR93] [ALD94]

1473
Compound: Hexaborane(10)
Formula: B₆H₁₀
Molecular Formula: B₆H₁₀
Molecular Weight: 74.945
CAS RN: 23777-80-2
Properties: colorless liq; slowly
decomposes at 25°C [COT88]
Solubility: hydrolyzed in H₂O if heated [COT88]
Density, g/cm³: 0.67 [LID94]
Melting Point, °C: −62.3 [KIR78]
Boiling Point, °C: 108 [KIR78]

1474
Compound: Hexaborane(12)
Formula: B₆H₁₂
Molecular Formula: B₆H₁₂
Molecular Weight: 76.961
CAS RN: 12008-19-4
Properties: col liq [CRC10]

Solubility: reac H₂O [CRC10]
Melting Point, °C: −82.3 [CRC10]
Boiling Point, °C: ~85 [CRC10]

1475
Compound: Hexachlorodisilane
Formula: Cl₃SiSiCl₃
Molecular Formula: Cl₆Si₂
Molecular Weight: 268.89
CAS RN: 13465-77-5
Properties: colorless liq; precursor to substituted
disilanes [ALD94] [CRC10]
Density, g/cm³: 1.562 [ALD94]
Melting Point, °C: −1 [CRC10]
Boiling Point, °C: 144–145.5 [ALD94]

1476
Compound: Hexadecaborane(20)
Formula: B₁₆H₂₀
Molecular Formula: B₁₆H₂₀
Molecular Weight: 193.135
CAS RN: 28265-11-4
Properties: col cryst [CRC10]
Solubility: s cyhex, thf [CRC10]
Melting Point, °C: ~110 [CRC10]

1477
Compound: Hexafluorophosphoric acid
Formula: HPF₆
Molecular Formula: F₆HP
Molecular Weight: 145.972
CAS RN: 16940-81-1
Properties: clear liq; fumes due to evolving HF;
forms a hexahydrate [KIR78] [STR93]
Density, g/cm³: 1.651 [ALD94]

1478
Compound: Holmium
Formula: Ho
Molecular Formula: Ho
Molecular Weight: 164.93032
CAS RN: 7440-60-0
Properties: cryst solid, has metallic luster; hex; forms
yellow-green salts; sensitive to air and moisture;
electrical resistivity (20°C) 94 µohm·cm; enthalpy
of fusion 11.76 kJ/mol; enthalpy of sublimation
300.8 kJ/mol; atom radius 0.17661 nm; radius of
Ho⁺⁺⁺ ion 0.0894 nm, has yellow colored solutions;
used as a getter in vacuum tubes, applications to
electrochemical and spectrochemical research
[HAW93] [MER06] [KIR82] [CRC10] [ALD94]

Solubility: reacts slowly with H_2O, s dil acids [HAW93]
Density, g/cm³: 8.7947 [KIR82]
Melting Point, °C: 1470 [ALD94]
Boiling Point, °C: 2700 [ALD94]
Thermal Conductivity, W/(m·K): 16.2 (25°C) [CRC10]
Thermal Expansion Coefficient: 11.2×10^{-6}/K [CRC10]

1479
Compound: Holmium acetate monohydrate
Formula: $Ho(CH_3COO)_3 \cdot H_2O$
Molecular Formula: $C_6H_{11}HoO_7$
Molecular Weight: 360.079
CAS RN: 25519-09-9
Properties: peach powd [STR93]

1480
Compound: Holmium bromide
Formula: $HoBr_3$
Molecular Formula: Br_3Ho
Molecular Weight: 404.642
CAS RN: 13825-76-8
Properties: −20 mesh with 99.9% purity; off-white powd; hygr [CER91] [STR93]
Melting Point, °C: 914 [MER06]
Boiling Point, °C: 1470 [STR93]

1481
Compound: Holmium carbonate hydrate
Formula: $Ho_2(CO_3)_3 \cdot xH_2O$
Molecular Formula: $C_3Ho_2O_9$ (anhydrous)
Molecular Weight: 509.888 (anhydrous)
CAS RN: 38245-34-0
Properties: white powd; hygr [STR93] [ALD94]

1482
Compound: Holmium chloride
Formula: $HoCl_3$
Molecular Formula: Cl_3Ho
Molecular Weight: 271.288
CAS RN: 10138-62-2
Properties: −20 mesh with 99.9% purity; off-white powd; bright yellow solid; hygr [HAW93] [CER91] [MER06] [STR93]
Solubility: s H_2O [HAW93]
Melting Point, °C: 718 [MER06]
Boiling Point, °C: 1500 [HAW93]

1483
Compound: Holmium chloride hexahydrate
Formula: $HoCl_3 \cdot 6H_2O$

Molecular Formula: $Cl_3H_{12}HoO_6$
Molecular Weight: 379.380
CAS RN: 14914-84-2
Properties: −4 mesh with 99.9% purity; off-white cryst [CER91] [STR93]
Melting Point, °C: 718 [AES93]

1484
Compound: Holmium fluoride
Formula: HoF_3
Molecular Formula: F_3Ho
Molecular Weight: 221.925
CAS RN: 13760-78-6
Properties: bright yellow solid or 99.9% pure melted pieces of 3–6 mm; hygr; melted pieces used as evaporation material for possible application in multilayers [STR93] [HAW93] [CER91]
Solubility: s H_2O [HAW93]
Density, g/cm³: 7.644 [STR93]
Melting Point, °C: 1143 [STR93]
Boiling Point, °C: >2200 [STR93]

1485
Compound: Holmium hydride
Formula: HoH_{2-3}
Molecular Formula: H_2Ho; H_3Ho
Molecular Weight: H_2Ho: 166.946; H_3Ho: 167.954
CAS RN: 13598-41-9
Properties: −60 mesh with 99.9% purity [CER91]

1486
Compound: Holmium iodide
Formula: HoI_3
Molecular Formula: HoI_3
Molecular Weight: 545.643
CAS RN: 13813-41-7
Properties: −20 mesh with 99.9% purity; light yellow solid [CER91] [MER06]
Melting Point, °C: 1010 [MER06]
Boiling Point, °C: 1300 [CRC10]

1487
Compound: Holmium nitrate pentahydrate
Formula: $Ho(NO_3)_3 \cdot 5H_2O$
Molecular Formula: $H_{10}HoN_3O_{14}$
Molecular Weight: 441.022
CAS RN: 14483-18-2
Properties: orange cryst; hygr [STR93]

1488
Compound: Holmium nitride
Formula: HoN

Molecular Formula: HoN
Molecular Weight: 178.937
CAS RN: 12029-81-1
Properties: cub; −60 mesh with 99.9%
    purity [LID94] [CER91]
Density, g/cm³: 10.6 [LID94]

1489
Compound: Holmium oxalate decahydrate
Formula: $Ho_2(C_2O_4)_3 \cdot 10H_2O$
Molecular Formula: $C_6H_{20}Ho_2O_{22}$
Molecular Weight: 774.072
CAS RN: 28965-57-3
Properties: off-white powd [STR93]
Reactions: minus H_2O, 40°C [AES93]

1490
Compound: Holmium oxide
Synonym: holmia
Formula: Ho_2O_3
Molecular Formula: Ho_2O_3
Molecular Weight: 377.859
CAS RN: 12055-62-8
Properties: light yellow solid; sl hygr; used in
    refractories and as a special catalyst, and as an
    evaporated film of 99.9% purity, it is reactive
    to radio frequencies [HAW93] [CER91]
Solubility: s inorganic acids [HAW93]
Density, g/cm³: 8.36 [STR93]
Melting Point, °C: 2415 [LID94]

1491
Compound: Holmium perchlorate hexahydrate
Formula: $Ho(ClO_4)_3 \cdot 6H_2O$
Molecular Formula: $Cl_3H_{12}HoO_{18}$
Molecular Weight: 571.373
CAS RN: 14017-54-0
Properties: cryst; hygr [STR93]

1492
Compound: Holmium silicide
Formula: $HoSi_2$
Molecular Formula: $HoSi_2$
Molecular Weight: 221.101
CAS RN: 12136-24-2
Properties: hex; 10 mm and down lump,
    6 mm pieces and smaller with 99.9%
    purity [ALF93] [LID94] [CER91]
Density, g/cm³: 7.1 [LID94]

1493
Compound: Holmium sulfate octahydrate
Formula: $Ho_2(SO_4)_3 \cdot 8H_2O$
Molecular Formula: $H_{16}Ho_2O_{20}S_3$
Molecular Weight: 762.174
CAS RN: 13473-57-9
Properties: hygr cryst [AES93] [ALD94]
Solubility: g/100 g H_2O: 8.18 (20°C), 6.71
    (25°C), 4.52 (40°C) [LAN05]

1494
Compound: Holmium sulfide
Formula: Ho_2S_3
Molecular Formula: Ho_2S_3
Molecular Weight: 426.059
CAS RN: 12162-59-3
Properties: monocl; −200 mesh with
    99.9% purity [LID94] [CER91]
Density, g/cm³: 5.92 [LID94]

1495
Compound: Holmium telluride
Formula: Ho_2Te_3
Molecular Formula: Ho_2Te_3
Molecular Weight: 712.661
CAS RN: 12162-61-7
Properties: −20 mesh with 99.9% purity [CER91]

1496
Compound: Holmium boride
Synonym: holmium tetraboride
Formula: HoB_4
Molecular Formula: B_4Ho
Molecular Weight: 208.174
CAS RN: 12045-77-1
Properties: −100 mesh of 99.9% purity [CER91]

1497
Compound: Hydrazine
Formula: H_2NNH_2
Molecular Formula: H_4N_2
Molecular Weight: 32.045
CAS RN: 302-01-2
Properties: colorless, oily liq, fuming in air; burns
    with violet flame; N–N distance 0.146 nm; vapor
    pressure 14.38 mm (25°C); viscosity 0.009 dyne
    s/cm² (25°C); surface tension 66.67 dyne/cm
    (25°C); dielectric constant 58.5 (0°C); enthalpy
    of vaporization 41.8 kJ/mol; enthalpy of fusion
    12.60 kJ/mol [CRC10] [MER06] [CIC73]

Solubility: miscible with H_2O and the following
 alcohols: methyl, ethyl, propyl, isobutyl [MER06]
Density, g/cm³: liq: 1.00; solid: 1.146 (−5°C) [CIC73]
Melting Point, °C: 1.4 [CRC10]
Boiling Point, °C: 113.55 [CRC10]
Reactions: good reducing agent [MER06]

1498
Compound: Hydrazine acetate
Formula: $H_2NNH_2 \cdot (CH_3COOH)$
Molecular Formula: $C_2H_8N_2O_2$
Molecular Weight: 92.098
CAS RN: 13255-48-6
Melting Point, °C: 100–102 [ALD94]

1499
Compound: Hydrazine azide
Formula: $N_2H_4 \cdot HN_3$
Molecular Formula: H_5N_5
Molecular Weight: 75.074
CAS RN: 14662-04-5
Properties: white; deliq prism [CRC10]
Solubility: v s H_2O [CRC10]
Melting Point, °C: explodes at 75.4 [CIC73]

1500
Compound: Hydrazine dihydrochloride
Formula: $N_2H_4 \cdot 2HCl$
Molecular Formula: $Cl_2H_6N_2$
Molecular Weight: 104.966
CAS RN: 5341-61-7
Properties: white, cryst powd; ortho-rhomb, a = 1.249 nm,
 b = 2.185 nm, c = 0.441 nm [CIC73] [MER06]
Solubility: 27.2 g/100 g H_2O (32°C)
 [CIC73]; sl s alcohol [HAW93]
Density, g/cm³: 1.42 [CIC73]
Melting Point, °C: 198 (minus HCl) [CIC73]
Boiling Point, °C: decomposes at 200 [CIC73]

1501
Compound: Hydrazine dinitrate
Formula: $N_2H_4 \cdot 2HNO_3$
Molecular Formula: $H_6N_4O_6$
Molecular Weight: 158.071
CAS RN: 13464-98-7
Properties: needles [LAN05]
Solubility: 20.2 g/100 g H_2O (35°C) [CIC73]
Melting Point, °C: 104 [LAN05]
Boiling Point, °C: decomposes [LAN05]
Reactions: decomposes quickly at 104°C,
 slowly at 80°C [CIC73]

1502
Compound: Hydrazine hydrate
Formula: $H_2NNH_2 \cdot xH_2O$
Molecular Formula: H_4N_2 (anhydrous)
Molecular Weight: 32.048 (anhydrous)
CAS RN: 10217-52-4
Properties: x~1.5 [ALD94]
Density, g/cm³: 1.029 [ALD94]
Melting Point, °C: 95 [ALD94]

1503
Compound: Hydrazine monohydrate
Formula: $N_2H_4 \cdot H_2O$
Molecular Formula: H_6N_2O
Molecular Weight: 50.060
CAS RN: 7803-57-8
Properties: fuming refractive liq; trig, a = 0.4873 nm,
 c = 1.094 nm [CIC73] [MER06]
Solubility: miscible with H_2O, alcohol; i
 chloroform, ether [MER06]
Density, g/cm³: 1.0305 [CIC73]
Melting Point, °C: −51.7 [CIC73]
Boiling Point, °C: 118.5 [CIC73]
Reactions: reducing agent [MER06]

1504
Compound: Hydrazine monohydrobromide
Formula: $N_2H_4 \cdot HBr$
Molecular Formula: BrH_5N_2
Molecular Weight: 112.957
CAS RN: 13775-80-9
Properties: white, cryst flakes; used in soldering
 flux; monocl, a = 1.285 nm, b = 0.454 nm,
 c = 1.194 nm [CIC73] [HAW93]
Solubility: s H_2O, lower alcohols; i most
 organic solvents [HAW93]
Density, g/cm³: 2.3 [LID94]
Melting Point, °C: 81–87 [HAW93]
Boiling Point, °C: decomposes at ~190 [HAW93]

1505
Compound: Hydrazine monohydrochloride
Formula: $N_2H_4 \cdot HCl$
Molecular Formula: ClH_5N_2
Molecular Weight: 68.506
CAS RN: 2644-70-4
Properties: white, cryst flakes; ortho-rhomb,
 a = 1.249 nm, b = 2.185 nm, c = 0.441 nm [CIC73]
Solubility: 37 g/100 g H_2O (20°C); i most
 organic solvents [HAW93]
Density, g/cm³: 1.5 [LID94]

Melting Point, °C: 89 [CIC73]
Boiling Point, °C: decomposes at 240 [CIC73]

1506
Compound: Hydrazine monohydroiodide
Formula: $N_2H_4 \cdot HI$
Molecular Formula: H_5IN_2
Molecular Weight: 159.957
CAS RN: 10039-55-1
Properties: colorless prism [CRC10]
Solubility: s H_2O [CIC73]
Melting Point, °C: 125 [CIC73]

1507
Compound: Hydrazine mononitrate
Formula: $N_2H_4 \cdot HNO_3$
Molecular Formula: $H_5N_3O_3$
Molecular Weight: 95.058
CAS RN: 37836-27-4
Properties: colorless needles; explosive; α, β
   forms; monocl, a = 1.123 nm, b = 1.173 nm,
   c = 0.517 nm [HAW93] [CIC73] [CRC10]
Solubility: 327 g/100 g H_2O (25°C) [CIC73]; g/100 g H_2O:
   175 (10°C), 266 (20°C), 2127 (60°C) [LAN05]
Melting Point, °C: α: 70.71; β: 62.09 [CIC73]
Boiling Point, °C: sublimes at 140 [CRC10]

1508
Compound: Hydrazine monooxalate
Formula: $2N_2H_4 \cdot H_2C_2O_4$
Molecular Formula: $C_2H_{10}N_4O_4$
Molecular Weight: 154.13
CAS RN: 108249-27-0
Properties: monocl, a = 0.3580 nm,
   b = 0.3321 nm, c = 0.5097 nm [CIC73]
Solubility: 200 g/100 mL H_2O (15°C) [CRC10]
Melting Point, °C: 148 [CRC10]

1509
Compound: Hydrazine perchlorate hemihydrate
Formula: $N_2H_4 \cdot HClO_4 \cdot 1/2H_2O$
Molecular Formula: $ClH_6N_2O_{4.5}$
Molecular Weight: 141.511
CAS RN: 13762-65-7
Properties: solid; used as a rocket propellant [HAW93]
Solubility: decomposes in H_2O; s alcohol; i ether,
   benzene, chloroform, carbon disulfide [HAW93]
Density, g/cm³: 1.939 [CIC73]
Melting Point, °C: 137 [HAW93]
Boiling Point, °C: 145 [HAW93]
Reactions: can explode [CRC10]

1510
Compound: Hydrazine sulfate
Formula: $N_2H_4 \cdot H_2SO_4$
Molecular Formula: $H_6N_2O_4S$
Molecular Weight: 130.125
CAS RN: 10034-93-2
Properties: glass-like plates or prisms; ortho-
   rhomb cryst, a = 0.8251 nm, b = 0.9159 nm,
   c = 0.5532 nm [MER06] [CIC73]
Solubility: 3.415 g/100 g H_2O (25°C)
   [CIC73]; g/100 g H_2O: 2.87 (20°C), 4.15
   (40°C), 14.39 (80°C) [LAN05]
Density, g/cm³: 1.378 [MER06]
Melting Point, °C: 254 [CIC73]

1511
Compound: Hydrazoic acid
Synonym: hydrogen azide
Formula: HN_3
Molecular Formula: HN_3
Molecular Weight: 43.028
CAS RN: 7782-79-8
Properties: colorless, volatile liq; intolerable pungent
   odor; highly explosive; enthalpy of vaporization
   30.5 kJ/mol [CRC10] [MER06] [HAW93]
Solubility: v s H_2O [HAW93]
Density, g/cm³: 1.092 [CRC10]
Melting Point, °C: −80 [MER06]
Boiling Point, °C: 35.7 [CRC10]

1512
Compound: Hydrofluoric acid, 70%
Formula: HF
Molecular Formula: HF
Molecular Weight: 20.006
CAS RN: 7664-39-3
Properties: colorless, fuming, mobile liq; solid phase
   is $HF \cdot H_2O$ at freezing point; specific conductivity
   0.79 (ohm·cm)$^{-1}$ at 0°C; attacks glass, silica;
   manufactured by reacting fluorspar with sulfuric
   acid to form calcium sulfate and HF(gas); used
   in aluminum production, acidizing oil wells,
   gasoline production [HAW93] [KIR78]
Density, g/cm³: 1.22 [KIR78]
Melting Point, °C: −69 [KIR78]
Boiling Point, °C: 66.4 [KIR78]

1513
Compound: Hydrogen
Formula: H_2
Molecular Formula: H_2

Molecular Weight: 2.016 (at wt 1.00794)
CAS RN: 1333-74-0
Properties: colorless, odorless, tasteless, diatomic gas; flammable or explosive when mixed with air, oxygen, chlorine; critical temp −239.96°C; critical pressure 1315 kPa; critical volume 66.949 cm³/mol; enthalpy of vaporization 0.898 kJ/mol; enthalpy of fusion 0.12 kJ/mol; velocity of sound (0°C) 1246 m/s; viscosity (0°C) 0.00843 mPa·s; dielectric constant at bp 1.231, 1.000271 at 0°C [MER06] [CRC10]
Solubility: s in about 50 vols H_2O (0°C) [MER06]
Density, g/cm³: 0.088 g/L [LID94]
Melting Point, °C: −259.34 [CRC10]
Boiling Point, °C: −252.87 [CRC10]
Thermal Conductivity, W/(m·K): liq: at bp: 0.10; at triple point: 0.074; gas: 0.1739 [KIR80]

1514

Compound: Hydrogen-d_2
Formula: D_2
Molecular Formula: D_2
Molecular Weight: 4.028
CAS RN: 7782-39-0
Properties: col gas [CRC10]
Density, g/L: 0.164 [CRC10]
Melting Point, °C: −254.42 [CRC10]
Boiling Point, °C: −249.48 [CRC10]

1515

Compound: Hydrogen-t_2
Formula: T_2
Molecular Formula: T_3
Molecular Weight: 6.032
CAS RN: 10028-17-8
Properties: col gas [CRC10]
Density, g/L: 0.246 [CRC10]
Melting Point, °C: −252.53 [CRC10]
Boiling Point, °C: −248.11 [CRC10]

1516

Compound: Hydrogen-d_1
Formula: HD
Molecular Formula: DH
Molecular Weight: 3.022
CAS RN: 13983-20-5
Properties: col gas [CRC10]
Density, g/L: 0.123 [CRC10]
Melting Point, °C: −256.55 [CRC10]
Boiling Point, °C: −251.02

1517

Compound: Hydrogen-t_1
Formula: HT
Molecular Formula: HT
Molecular Weight: 4.024
CAS RN: 14885-60-0
Properties: col gas [CRC10]
Melting Point, °C: −254.7 [CRC10]
Boiling Point, °C: −249.6 [CRC10]

1518

Compound: Hydrogen-d_1,t_1
Formula: DT
Molecular Formula: DT
Molecular Weight: 5.030
CAS RN: 14885-61-1
Properties: col gas [CRC10]
Melting Point, °C: −252.5 [CRC10]
Boiling Point, °C: −238.9 [CRC10]

1519

Compound: Hydrogen bromide-d
Formula: DBr
Molecular Formula: BrD
Molecular Weight: 81.918
CAS RN: 13536-59-9
Properties: col gas [CRC10]
Solubility: s H_2O [CRC10]
Melting Point, °C: −87.54 [CRC10]
Boiling Point, °C: −66.9 [CRC10]

1520

Compound: Hydrogen bromide
Formula: HBr
Molecular Formula: BrH
Molecular Weight: 80.912
CAS RN: 10035-10-6
Properties: colorless, corrosive, nonflammable gas; fumes in moist air; specific conductance 1.4×10^{-10} (ohm·cm)$^{-1}$ at −84°C; dielectric constant 7.33 at −84°C; enthalpy of vaporization (25°C) 12.69 kJ/mol; enthalpy of fusion 2.41 kJ/mol [CRC10] [MER06] [COT88]
Solubility: g/100 g H_2O: 221.2 (0°C), 204.0 (15°C), 171.5 (50°C), 130.0 (100°C) [LAN05]; s alcohol [HAW93]
Density, g/cm³: 3.55 g/L [LID94]
Melting Point, °C: −86.81 [CRC10]
Boiling Point, °C: −66.38 [CRC10]

1521
Compound: Hydrogen chloride
Formula: HCl
Molecular Formula: ClH
Molecular Weight: 36.461
CAS RN: 7647-01-0
Properties: colorless gas; fumes in air; suffocating odor; enthalpy of fusion 2.00 kJ/mol; enthalpy of vaporization 16.15 kJ/mol; triple point −114.25°C; critical temp 54.4°C; critical pressure 8.316 MPa; critical volume 0.069 L/mol; critical density 424 g/L; dielectric constant liq: 14.2 (10°C), gas: 1.0046 (25°C); specific conductance 3.5×10^{-9} (−85°C) [KIR80] [MER06] [COT88] [CRC10]
Solubility: g/100 g H_2O: 82.3 (0°C); 67.3 (30°C); 56.1 (60°C) [MER06]
Density, g/cm³: 1.268 (air = 1.000) [MER06]
Melting Point, °C: −114.18 [CRC10]
Boiling Point, °C: −85 [CRC10]
Thermal Conductivity, W/(m·K): liq: (−154.99°C) 3.35; vapor: (0°C) 1.34 [KIR80]

1522
Compound: Hydrogen chloride-d
Formula: DCl
Molecular Formula: ClD
Molecular Weight: 37.467
CAS RN: 7698-05-7
Properties: col gas [CRC10]
Solubility: s H_2O [CRC10]
Melting Point, °C: −114.72 [CRC10]
Boiling Point, °C: −84.4 [CRC10]

1523
Compound: Hydrogen chloride dihydrate
Formula: HCl·2H₂O
Molecular Formula: ClH_5O_2
Molecular Weight: 72.492
CAS RN: 13465-05-9
Properties: col liq [CRC10]
Density, gm/cm³: 1.46 [CRC10]
Melting Point, °C: −17.7 [CRC10]

1524
Compound: Hydrogen cyanide
Synonyms: hydrocyanic acid, prussic acid
Formula: HCN
Molecular Formula: CHN
Molecular Weight: 27.026
CAS RN: 74-90-8

Properties: colorless gas or liq; burns in air with blue flame; weakly acidic solutions; critical pressure 55 atm; critical temp 183.5°C; viscosity at 20°C 0.00201 dyne s/cm²; dielectric constant at 15.6°C is 123; enthalpy of fusion 8.41 kJ/mol [CIC73] [MER06] [CRC10]
Solubility: miscible with H_2O, alcohol; sl s ether [MER06]
Density, g/cm³: gas: 0.941 (air = 1); liq: 0.687 [MER06]
Melting Point, °C: −13.4 [MER06]
Boiling Point, °C: 25.6 [MER06]

1525
Compound: Hydrogen disulfide
Formula: H_2S_2
Molecular Formula: H_2S_2
Molecular Weight: 66.146
CAS RN: 13465-07-1
Properties: col liq [CRC10]
Density, g/cm³: 1.334 [CRC10]
Boiling Point, °C: 70.7 [CRC10]

1526
Compound: Hydrogen fluoride
Formula: HF
Molecular Formula: FH
Molecular Weight: 20.006
CAS RN: 7664-39-3
Properties: colorless liq or gas; fumes in air; irritating, corrosive, and poisonous; vapor pressure 122.9 MPa at 25°C; enthalpy of vaporization 7493 J/mol; enthalpy of fusion 4.58 kJ/mol; critical temp 188°C; critical pressure 6.480 MPa; critical density 0.29 g/cm³; viscosity 0.25 mPa·s at 0°C; dielectric constant 83.6 at 0°C; specific conductance 1.6×10^{-6} (ohm·cm)$^{-1}$ at 0°C [KIR78] [MER06] [COT88] [CRC10]
Solubility: v s H_2O, alcohol; sl s ether, other organic solvents [MER06]
Density, g/cm³: liq: 0.9576 (25°C) [KIR78]
Melting Point, °C: −83.36 [CRC10]
Boiling Point, °C: 19.51 [MER06]

1527
Compound: Hydrogen hexachloroiridate(IV) hydrate
Synonym: chloroiridic acid
Formula: H₂IrCl₆·xH₂O
Molecular Formula: Cl_6H_2Ir (anhydrous)
Molecular Weight: 406.952 (anhydrous)
CAS RN: 110802-84-1
Properties: black cryst; hygr [STR93]

1528
Compound: Hydrogen hexachloroplatinate(IV)
Synonym: platinic acid
Formula: H_2PtCl_6
Molecular Formula: Cl_6H_2Pt
Molecular Weight: 409.812
CAS RN: 16941-12-1
Properties: color red to brown; liq [KIR82] [ALF95]
Solubility: v s H_2O, alcohol [KIR82]

1529
Compound: Hydrogen hexachloroplatinate(IV)
   hexahydrate
Synonym: platinic acid hexahydrate
Formula: $H_2PtCl_6 \cdot 6H_2O$
Molecular Formula: $Cl_6H_{14}O_6Pt$
Molecular Weight: 517.903
CAS RN: 16941-12-1
Properties: brownish yellow; very deliq
   cryst; sensitive to light [MER06]
Solubility: v s H_2O, alcohol [MER06]
Density, g/cm³: 2.431 [MER06]
Melting Point, °C: 60 [MER06]

1530
Compound: Hydrogen hexafluorosilicic acid
Formula: H_2SiF_6
Molecular Formula: F_6H_2Si
Molecular Weight: 144.092
CAS RN: 16961-83-4
Properties: aq solution: colorless fuming liq; attacks
   glass and stoneware; produced as a byproduct
   of the reaction between H_2SO_4 and phosphate
   rocks, which contain fluorides and silica; used
   to fluoridate water, to increase the hardness of
   ceramics, and in electroplating [HAW93]
Density, g/cm³: 1.22 [ALD94]

1531
Compound: Hydrogen hexahydroxyplatinate(IV)
Formula: $H_2Pt(OH)_6$
Molecular Formula: H_8O_6Pt
Molecular Weight: 299.15
CAS RN: 51850-20-5
Properties: yellow needles; hygr [ALD94]
Reactions: minus $2H_2O$ at 100; minus
   $3H_2O$ at 120°C [CRC10]

1532
Compound: Hydrogen iodide
Formula: HI

Molecular Formula: HI
Molecular Weight: 127.912
CAS RN: 10034-85-2
Properties: pale yellow or colorless, nonflammable
   gas; fumes in moist air; decomposed by light;
   critical temp 150°C; critical pressure 8.3 MPa;
   dielectric constant 3.57 at −45°C; specific
   conductance 8.5×10^{-10} $(ohm \cdot cm)^{-1}$ at −45°C;
   enthalpy of vaporization 19.76 kJ/mol; enthalpy
   of fusion 2.87 kJ/mol; produced by reaction
   of I_2 and hydrazine in the presence of H_2O
   [KIR81] [MER06] [CRC10] [COT88]
Solubility: g/100 g H_2O: 234 (10°C), 900
   (0°C); s organic solvents [MER06]
Density, g/cm³: gas: 5.23 g/L (25°C); liq:
   2.85 g/cm³ (−4.7°C) [KIR81]
Melting Point, °C: −50.77 [CRC10]
Boiling Point, °C: −35.55 [CRC10]

1533
Compound: Hydrogen iodide-d
Formula: DI
Molecular Formula: DI
Molecular Weight: 128.918
CAS RN: 14104-45-1
Properties: col gas [CRC10]
Solubility: s H_2O [CRC10]
Melting Point, °C: −51.93 [CRC10]
Boiling Point, °C: −36.2 [CRC10]

1534
Compound: Hydrogen peroxide
Formula: H_2O_2
Molecular Formula: H_2O_2
Molecular Weight: 34.015
CAS RN: 7722-84-1
Properties: clear, colorless liq; weakly acidic,
   dissociation constant 1.78×10^{-12}; viscosity
   1.245 mPa·s (20°C); surface tension, 80.4 mN/m
   (20°C); enthalpy of dissociation 34.3 kJ/mol;
   enthalpy of vaporization 51.6 kJ/g at 25°C; enthalpy
   of fusion 12.50 kJ/mol; specific conductance
   (25°C) 4×10^{-7} ohm·cm [CRC10] [KIR81]
Solubility: miscible with H_2O [KIR81]
Density, g/cm³: 1.443 [KIR81]
Melting Point, °C: −0.43 [CRC10]
Boiling Point, °C: 150.2 [KIR81]

1535
Compound: Hydrogen selenide
Formula: H_2Se
Molecular Formula: H_2Se

Molecular Weight: 80.976
CAS RN: 7783-07-5
Properties: colorless gas with disagreeable odor; flammable; toxic; liquefies at 0°C under 6.6 atm pressure; thermodynamically unstable at room temp, but rate of decomposition is slow; reducing agent; enthalpy of vaporization 19.7 kJ/mol; can be prepared by adding HCl to ferrous selenide [KIR82] [MER06] [CRC10]
Solubility: mL/100 mL H_2O: 377 (4°C), 270 (22.5°C) [MER06]; mL/100 g H_2O at standard temp, pressure: 386 (0°C), 351 (10°C), 289 (20°C) [LAN05]
Density, g/cm³: 3.553 g/L [LID94]; (−42°C) 2.12 [MER06]
Melting Point, °C: −65.73 [MER06]
Boiling Point, °C: −41.3 [KIR82]
Reactions: reacts directly with most metals to form highly insoluble selenides [MER06]

1536
Compound: Hydrogen sulfide
Formula: H_2S
Molecular Formula: H_2S
Molecular Weight: 34.082
CAS RN: 7783-06-4
Properties: colorless, flammable gas with characteristic rotten egg odor; burns in air with blue flame; critical temp 100.5°C; critical pressure 9.02 MPa; enthalpy of vaporization 18.67 kJ/mol; enthalpy of fusion 23.80 kJ/mol [CRC10] [AIR87] [HAW93] [MER06]
Solubility: 1 g dissolves in H_2O: 187 mL (10°C), 242 mL (20°C), 314 (30°C) [MER06]
Density, g/cm³: 1.19 (air = 1) [MER06]
Melting Point, °C: −85.49 [MER06]
Boiling Point, °C: −59.55 [CRC10]

1537
Compound: Hydrogen telluride
Formula: H_2Te
Molecular Formula: H_2Te
Molecular Weight: 129.616
CAS RN: 7783-09-7
Properties: colorless gas with offensive odor, like garlic; 1 L weighs 6.234 g; liq H_2Te readily decomposed by light; dry gas stable to light, but decomposes in presence of dust; enthalpy of vaporization 19.2 kJ/mol; can be prepared by adding $AlTe_3$ to water in the absence of air; easily oxidized [KIR83] [MER06] [CRC10]
Solubility: s H_2O, unstable solution; s alcohol and alkalies [HAW93]

Density, g/cm³: 5.687 g/L [LID94]; (−12°C) 2.68 [MER06]
Melting Point, °C: −49 [MER06]
Boiling Point, °C: −2 [MER06]

1538
Compound: Hydrogen tetrabromoaurate(III) pentahydrate
Formula: $HAuBr_4 \cdot 5H_2O$
Molecular Formula: $AuBr_4H_{11}O_5$
Molecular Weight: 607.667
CAS RN: 17083-68-0
Properties: dark reddish brown, needle-shaped cryst or granular masses; odorless; acidic taste [HAW93]
Solubility: s H_2O, alcohol [HAW93]
Melting Point, °C: 27 [HAW93]

1539
Compound: Hydrogen tetracarbonylferrate(II)
Formula: $H_2Fe(CO)_4$
Molecular Formula: $C_4H_2FeO_4$
Molecular Weight: 169.903
CAS RN: 12002-28-7
Properties: colorless cryst [MER06]
Solubility: s alkalies [MER06]
Melting Point, °C: −70 [MER06]

1540
Compound: Hydrogen tetrachloroaurate(III) hydrate
Formula: $HAuCl_4 \cdot xH_2O$
Molecular Formula: $AuCl_4H$ (anhydrous)
Molecular Weight: 339.785 (anhydrous)
CAS RN: 27988-77-8
Properties: yellowish orange cryst; hygr; sensitive to light [STR93]
Melting Point, °C: decomposes [STR93]

1541
Compound: Hydrogen tetrachloroaurate(III) tetrahydrate
Formula: $HAuCl_4 \cdot 4H_2O$
Molecular Formula: $AuCl_4H_9O_4$
Molecular Weight: 411.847
CAS RN: 16903-35-8
Properties: golden yellow to reddish yellow; very hygr; deliq; monocl cryst; readily affected by sunlight; there is a trihydrate CAS RN 16961-25-4 [MER06] [ALD94]
Solubility: v s H_2O, alcohol; s ether [MER06]

Density, g/cm³: ~3.9 [MER06]
Reactions: decomposes if strongly heated
forming Cl_2, HCl, Au [MER06]

1542
Compound: Hydroxylamine
Formula: H_2NOH
Molecular Formula: H_3NO
Molecular Weight: 33.030
CAS RN: 7803-49-8
Properties: unstable, large, white flakes or needles;
ortho-rhomb, a=0.729 nm, b=0.439 nm,
c=0.488 nm; very hygr; vapor pressure
5.3 mm (32°C), 400 mm (99.2°C); dielectric
constant 77.63–77.85; used as a reducing
agent in photography [CIC73] [MER06]
Solubility: v s H_2O, methanol; decomposed
by hot H_2O [MER06]
Density, g/cm³: liq: 1.204 (33°C) [CIC73]
Melting Point, °C: 32.05 [CIC73]
Boiling Point, °C: 56–57 (22 mm) [CIC73]

1543
Compound: Hydroxylamine hydrobromide
Formula: $H_2NOH \cdot HBr$
Molecular Formula: BrH_4NO
Molecular Weight: 113.942
CAS RN: 41591-55-3
Properties: monocl, a=0.729 nm,
b=0.613 nm, c=0.804 nm [CIC73]
Density, g/cm³: 2.3514 [CIC73]

1544
Compound: Hydroxylamine hydrochloride
Formula: $H_2NOH \cdot HCl$
Molecular Formula: ClH_4NO
Molecular Weight: 69.491
CAS RN: 5470-11-1
Properties: colorless; monocl, a=0.695 nm,
b=0.595 nm, c=0.770 nm [CIC73] [CRC10]
Solubility: 94.4 g/100 g H_2O (25°C) [CIC73]
Density, g/cm³: 1.680 [CIC73]
Melting Point, °C: decomposes at 152 [CIC73]

1545
Compound: Hydroxylamine perchlorate
Formula: $H_2NOH \cdot HClO_4$
Molecular Formula: ClH_4NO_5
Molecular Weight: 133.489
CAS RN: 15598-62-2
Properties: ortho-rhomb, a=0.752 nm,
b=0.714 nm, c= 1.599 nm [CIC73]

Melting Point, °C: 88–89 [CIC73]
Boiling Point, °C: decomposes at 120 [CIC73]

1546
Compound: Hydroxylamine sulfate
Formula: $(H_2NOH)_2 \cdot H_2SO_4$
Molecular Formula: $H_8N_2O_6S$
Molecular Weight: 164.139
CAS RN: 10039-54-0
Properties: colorless, monocl cryst [CRC10]
Solubility: 32.9 g/100 mL H_2O (0°C),
68.5 g/100 mL (20°C) [CRC10]
Melting Point, °C: decomposes at 170 [CIC73]

1547
Compound: Hypobromous acid
Formula: HOBr
Molecular Formula: BrHO
Molecular Weight: 96.911
CAS RN: 13517-11-8
Properties: stable only in solution, produced by
the hydrolysis of bromine chloride; used as a
bactericide and a disinfectant [HAW93]
Solubility: s H_2O, decomposed by hot H_2O [CRC10]

1548
Compound: Hypochlorous acid
Formula: HOCl
Molecular Formula: ClHO
Molecular Weight: 52.460
CAS RN: 7790-92-3
Properties: greenish yellow; can exist only in aq solution;
very unstable weak acid, which decomposes to HCl
and oxygen; used in bleaching textiles and fibers, in
water purification, and as an antiseptic [HAW93]

1549
Compound: Hypophosphoric acid
Formula: $H_4O_6P_2$
Molecular Formula: $H_4O_6P_2$
Molecular Weight: 161.976
CAS RN: 7803-60-3
Properties: plate-like cryst; available commercially
as a water solution; readily forms hydrates;
conc solution decomposes; used in baking powd
in the form of the sodium salt [HAW93]
Melting Point, °C: decomposes at 73 [LID94]

1550
Compound: Hypophosphorous acid
Synonym: phosphinic acid

Formula: H_3PO_2
Molecular Formula: H_3O_2P
Molecular Weight: 65.997
CAS RN: 6303-21-5
Properties: colorless, oily liq or deliq cryst; sour odor; reducing agent; enthalpy of fusion 9.70 kJ/mol; strong monobasic acid; sold in the form of a solution; can be prepared by heating baryta water with white phosphorus, followed by treatment with H_2SO_4 and filtering; used in electroplating baths [HAW93] [CRC10]
Density, g/cm³: 1.439 [HAW93]
Melting Point, °C: 26.5 [HAW93]
Boiling Point, °C: decomposes at 130 [CRC10]

1551

Compound: Indium
Formula: In
Molecular Formula: In
Molecular Weight: 114.818
CAS RN: 7440-74-6
Properties: soft, white metal with bluish tinge; ductile; quite stable in air; hardness 1.2 Mohs; tetr, a = 0.5979 nm, c = 0.49467 nm; enthalpy of fusion 3.28 kJ/mol; enthalpy of vaporization 55.57 kJ/mol; electrical resistivity (22°C) 8.8 μohm·cm; tensile strength 2.645 MPa; elongation 22%; modulus of elasticity 10.8 GPa; superconductor at 3.38 K; uses: production of bearings and in solder [MER06] [KIR81] [CRC10]
Solubility: i H_2O; attacked by mineral acids; not attacked by alkalies [MER06]
Density, g/cm³: 7.31 [CIC73]
Melting Point, °C: 156.6 [KIR81]
Boiling Point, °C: 2080 [KIR81]
Thermal Conductivity, W/(m·K): 81.8 (25°C) [ALD94]
Thermal Expansion Coefficient: linear expansion is 25×10^{-6}/°C from 0°C–100°C [KIR81]

1552

Compound: Indium acetate
Formula: $In(CH_3COO)_3$
Molecular Formula: $C_6H_9InO_6$
Molecular Weight: 291.951
CAS RN: 25114-58-3
Properties: white, hygr cryst [STR93]

1553

Compound: Indium acetylacetonate
Synonyms: 2,4-pentanedione, indium(III) derivative
Formula: $In(CH_3COCH=C(O)CH_3)_3$
Molecular Formula: $C_{15}H_{21}InO_6$

Molecular Weight: 412.146
CAS RN: 14405-45-9
Properties: off-white powd [STR93]
Melting Point, °C: 180–185 [STR93]

1554

Compound: Indium antimonide
Formula: InSb
Molecular Formula: InSb
Molecular Weight: 236.578
CAS RN: 1312-41-0
Properties: 6 mm pieces and smaller (fused) with 99.999% purity; black cryst; semiconductor; band gap, eV, 0.23 (0 K) and 0.17 (300 K); electron mobility 80,000 cm²/(V·s) and hole mobility 1250 cm²/(V·s); dielectric constant 17.7; effective mass 0.0145 for electrons and 0.40 for holes; enthalpy of fusion 25.50 kJ/mol [KIR82] [MER06] [STR93] [CER91] [CRC10]
Density, g/cm³: 5.7747 [LID94]; liq: 6.48 [MER06]
Melting Point, °C: 525 [CRC10]
Thermal Conductivity, W/(m·K): 16 [CRC10]
Thermal Expansion Coefficient: 4.7×10^{-6}/K [CRC10]

1555

Compound: Indium arsenide
Formula: InAs
Molecular Formula: AsIn
Molecular Weight: 189.740
CAS RN: 1303-11-3
Properties: 6 mm pieces and smaller (fused) with 99.999% purity; semiconductor; metallic appearance; band gap, 0.42 (0 K) and 0.36 (300 K); mobility (300 K), cm²/(V·s), 33,000 electrons and 460 holes; dielectric constant 14.6; effective mass 0.023 for electrons and 0.40 for holes [KIR82] [MER06] [CER91]
Solubility: i acids [HAW93]
Density, g/cm³: 5.67 [LID94]
Melting Point, °C: 943 [MER06]

1556

Compound: Indium nitride
Formula: InN
Molecular Formula: InN
Molecular Weight: 128.825
CAS RN: 25617-98-5
Properties: −100 mesh with 99.999% purity; wurtzite system, a = 0.353 nm, c = 0.570 nm; can be prepared by reacting In_2O_3 with ammonia at high temp; has semiconductor and electroluminescence properties [KIR81] [CIC73] [CER91]

Density, g/cm³: 6.89 [CIC73]
Melting Point, °C: decomposes above 600 [KIR81]
Thermal Conductivity, W/(m·K): 55.6 [CRC10]

1557
Compound: Indium phosphide
Formula: InP
Molecular Formula: InP
Molecular Weight: 145.792
CAS RN: 22398-80-7
Properties: black cryst; 6 mm pieces and smaller with
99.999% purity; semiconductor; band gap, eV,
1.42 (0 K) and 1.35 (300 K); mobility (300 K), cm²/
(V·s), 4600 electrons and 150 holes; dielectric
constant 12.4; effective mass 0.077 electrons
and 0.64 holes [KIR82] [CER91] [STR93]
Density, g/cm³: 4.81 [LID94]
Melting Point, °C: 1070 [AES93]

1558
Compound: Indium(I) bromide
Formula: InBr
Molecular Formula: BrIn
Molecular Weight: 194.722
CAS RN: 14280-53-6
Properties: can be prepared by reacting In
with InBr₃ vapor; enthalpy of vaporization
92 kJ/mol [KIR81] [CRC10]
Solubility: reacts in H₂O to form In and InBr₃ [KIR81]
Density, g/cm³: 4.96 [CRC10]
Melting Point, °C: 220 [AES93]
Boiling Point, °C: 656 [CRC10]

1559
Compound: Indium(I) chloride
Formula: InCl
Molecular Formula: ClIn
Molecular Weight: 150.271
CAS RN: 13465-10-6
Properties: golden yellow powd; sensitive to atm
oxygen and moisture; enthalpy of fusion 17.20 kJ/
mol; can be prepared by passing InCl₃ vapor
over In metal [KIR81] [STR93] [CRC10]
Solubility: decomposes in H₂O to In and InCl₃ [KIR81]
Density, g/cm³: 4.19 [STR93]
Melting Point, °C: 225 [CRC10]
Boiling Point, °C: 608 [STR93]

1560
Compound: Indium(I) iodide
Formula: InI
Molecular Formula: IIn

Molecular Weight: 241.722
CAS RN: 13966-94-4
Properties: reddish solid; −8 mesh with
99.999% purity; enthalpy of vaporization
90.8 kJ/mol [CRC10] [CER91]
Density, g/cm³: 5.31 [CRC10]
Melting Point, °C: 351 [AES93]
Boiling Point, °C: 712 [CRC10]

1561
Compound: Indium(II) bromide
Formula: InBr₂
Molecular Formula: Br₂In
Molecular Weight: 274.626
CAS RN: 21264-43-7
Properties: pale yellow solid; preparation: by reduction
of InBr₃ with H₂/HBr mixture [KIR81] [CRC10]
Solubility: reacts in H₂O to form In and InBr₃ [KIR81]
Density, g/cm³: 4.22 [CRC10]
Melting Point, °C: 235 [CRC10]
Boiling Point, °C: sublimes at 632 [CRC10]

1562
Compound: Indium(II) chloride
Formula: InCl₂
Molecular Formula: Cl₂In
Molecular Weight: 185.723
CAS RN: 13465-11-7
Properties: colorless, rhomb cryst; can be prepared by
heating indium to 200°C in HCl or by reducing InCl₃
in H₂/HCl mixture below 600°C [KIR81] [CRC10]
Solubility: decomposes in H₂O to In and InCl₃ [KIR81]
Density, g/cm³: 3.655 [CRC10]
Melting Point, °C: 235 [CRC10]
Boiling Point, °C: 550–570 [CRC10]

1563
Compound: Indium(II) sulfide
Formula: InS
Molecular Formula: InS
Molecular Weight: 146.884
CAS RN: 12030-14-7
Properties: red-brown solid, −100 mesh with
99.999% purity [CER91] [KIR81]
Density, g/cm³: 5.18 [CRC10]
Melting Point, °C: 692 [CRC10]
Boiling Point, °C: sublimes at 850 in vacuum [CRC10]

1564
Compound: Indium(III) bromide
Formula: InBr₃

Molecular Formula: Br₃In
Molecular Weight: 354.530
CAS RN: 13465-09-3
Properties: −60 mesh with 99.999% purity; white to light yellow powd; hygr [STR93] [CER91]
Density, g/cm³: 4.74 [STR93]
Melting Point, °C: ~436 [STR93]

1565
Compound: Indium(III) chloride
Synonym: indium trichloride
Formula: InCl₃
Molecular Formula: Cl₃In
Molecular Weight: 221.176
CAS RN: 10025-82-8
Properties: sublimed flakes with 99.999% purity; tan to yellowish, deliq cryst; can be made by heating indium in presence of excess Cl₂; used in electroplating baths; has high vapor pressure [KIR81] [HAW93] [MER06] [CER91]
Solubility: v s H₂O [MER06]; s alcohol [HAW93]
Density, g/cm³: 4.0 [MER06]; 3.46 [HAW93]
Melting Point, °C: 586 [MER06]
Boiling Point, °C: sublimes at 600 [CRC10]

1566
Compound: Indium(III) chloride tetrahydrate
Formula: InCl₃·4H₂O
Molecular Formula: Cl₃H₈InO₄
Molecular Weight: 293.238
CAS RN: 22519-64-8
Properties: white cryst [STR93]

1567
Compound: Indium(III) fluoride
Synonym: indium trifluoride
Formula: InF₃
Molecular Formula: F₃In
Molecular Weight: 171.813
CAS RN: 7783-52-0
Properties: −40 mesh with 99.999% purity; white powd; hygr; stable in hot and cold H₂O [STR93] [MER06] [CER91]
Solubility: 0.04 g/200 mL H₂O (25°C); v s dil acids [MER06]
Density, g/cm³: 4.39 [MER06]
Melting Point, °C: 1170 [MER06]
Boiling Point, °C: >1200 [MER06]

1568
Compound: Indium(III) fluoride trihydrate
Synonym: indium trifluoride trihydrate

Formula: InF₃·3H₂O
Molecular Formula: F₃H₆InO₃
Molecular Weight: 225.859
CAS RN: 14166-78-0
Properties: white cryst [STR93]
Solubility: 8.49 g/100 mL H₂O (22°C) [MER06]
Melting Point, °C: decomposes at 100 [STR93]

1569
Compound: Indium(III) hydroxide
Formula: In(OH)₃
Molecular Formula: H₃InO₃
Molecular Weight: 165.840
CAS RN: 20661-21-6
Properties: yellow-white powd; −100 mesh with 99.999% purity [AES93] [CER91]
Density, g/cm³: 4.4 [LID94]
Reactions: minus H₂O at 150°C [AES93]

1570
Compound: Indium(III) iodide
Formula: InI₃
Molecular Formula: I₃In
Molecular Weight: 495.531
CAS RN: 13510-35-5
Properties: −40 mesh with 99.999% purity; yellow-red solid; hygr [STR93] [CER91]
Density, g/cm³: 4.69 [STR93]
Melting Point, °C: 210 [STR93]

1571
Compound: Indium(III) nitrate trihydrate
Formula: In(NO₃)₃·3H₂O
Molecular Formula: H₆InN₃O₁₂
Molecular Weight: 354.879
CAS RN: 13770-61-1
Properties: −80 mesh with 99.999% purity; can be prepared by dissolving In or the oxide in nitric acid [CER91] [KIR81]
Solubility: v s [KIR81]
Melting Point, °C: decomposes [AES93]
Reactions: minus 2H₂O at 100°C [KIR81]

1572
Compound: Indium(III) oxide
Formula: In₂O₃
Molecular Formula: In₂O₃
Molecular Weight: 277.634
CAS RN: 1312-43-2

Properties: white to pale yellow powd in both amorphous and cryst forms; turns red brown if heated; volatilizes at 850°C; can be prepared by heating indium in air or by calcining indium carbonate; forms In_2O, 12030-22-7 and InO, 12136-26-4, if carefully reduced; used to impart yellow color to glass, as an evaporated film of 99.999% purity provides protective coating for metal minors, and as a sputtering target for transparent conductive films in electro-optical displays [KIR82] [MER06]
Solubility: i H_2O; s hot mineral acids [MER06]
Density, g/cm³: 7.179 [HAW93]
Melting Point, °C: ~2000 [LID94]

1573
Compound: Indium(III) perchlorate octahydrate
Formula: $In(ClO_4)_3 \cdot 8H_2O$
Molecular Formula: $Cl_3H_{16}InO_{20}$
Molecular Weight: 557.291
CAS RN: 13465-15-1
Properties: white cryst [STR93]
Melting Point, °C: ~80 [STR93]
Boiling Point, °C: decomposes at 200 [STR93]

1574
Compound: Indium(III) phosphate
Formula: $InPO_4$
Molecular Formula: InO_4P
Molecular Weight: 209.789
CAS RN: 14693-82-4
Properties: prepared by adding phosphate solution to an indium solution [KIR81]
Solubility: i H_2O [KIR81]
Density, g/cm³: 4.9 [LID94]

1575
Compound: Indium(III) selenide
Formula: In_2Se_3
Molecular Formula: In_2Se_3
Molecular Weight: 466.516
CAS RN: 12056-07-4
Properties: black cryst; ortho-rhomb; used in the form of 99.999% pure material as a sputtering target for production of semiconductors [MER06] [CER91]
Density, g/cm³: 5.67 [ALD94]
Melting Point, °C: 660 [MER06]

1576
Compound: Indium(III) sulfate
Formula: $In_2(SO_4)_3$

Molecular Formula: $In_2O_{12}S_3$
Molecular Weight: 517.827
CAS RN: 13464-82-9
Properties: −80 mesh with 99.999% purity; grayish deliq powd; decomposed by heat; formed by dissolution of In or In oxides in warm sulfuric acid [KIR81] [HAW93] [CER91]
Solubility: s H_2O [MER06]
Density, g/cm³: 3.438 [HAW93]
Melting Point, °C: decomposes at 600 [AES93]

1577
Compound: Indium(III) sulfide
Formula: In_2S_3
Molecular Formula: In_2S_3
Molecular Weight: 325.834
CAS RN: 12030-24-9
Properties: −200 mesh with 99.999% purity; orange powd; preparation is by heating In and S or by precipitating with H_2S from weakly acidic solutions [KIR81] [STR93] [CER91]
Density, g/cm³: 4.45 [STR93]
Melting Point, °C: 1050 [STR93]

1578
Compound: Indium(III) telluride
Formula: In_2Te_3
Molecular Formula: In_2Te_3
Molecular Weight: 612.436
CAS RN: 1312-45-4
Properties: 6 mm pieces and smaller with 99.999% purity; black or gray cryst; used in semiconductor technology [HAW93] [STR93] [CER91]
Density, g/cm³: (x-ray) 5.798 [MER06]
Melting Point, °C: 667 [MER06]

1579
Compound: Iodic acid
Formula: HIO_3
Molecular Formula: HIO_3
Molecular Weight: 175.910
CAS RN: 7782-68-5
Properties: colorless, rhomb cryst or white, cryst powd; darkens on exposure to light; it is a moderately strong acid; used in analytical chemistry and in medicine [HAW93] [MER06]
Solubility: g/mL H_2O: 269 (20°C), 295 (40°C); i alcohol, ether [MER06]; 310 g/100 g H_2O [KIR81]
Density, g/cm³: 4.629 [MER06]
Melting Point, °C: 110, with decomposition [MER06]
Reactions: decomposes to I_2O_5 at 220°C [MER06]

1580
Compound: Iodine
Formula: I_2
Molecular Formula: I_2
Molecular Weight: 253.809 (atomic wt 126.90447)
CAS RN: 7553-56-2
Properties: heavy, bluish black scales or plates; rhomb; metallic luster; a=0.47761 nm, b=0.72501 nm, c=0.97711 nm; readily sublimes to a violet, corrosive vapor; dielectric constant (23°C) 10.3; enthalpy of fusion 15.52 kJ/mol; enthalpy of sublimation (113.6°C) 238.24 J/g; enthalpy of vaporization 41.57 kJ/mol; electrical resistivity (25°C) $5.85 \times 10^{+6}$ ohm·cm; used in aniline dyes, antiseptics, feed and food additives [HAW93] [MER06] [KIR81] [CRC10]
Solubility: g/100g H_2O: 0.014 (0°C), 0.029 (20°C), 0.445 (100°C) [LAN05]; s benzene, CCl_4, alcohol, glycerol, ether, CS_2 and alkaline iodide solutions [HAW93]
Density, g/cm³: solid 4.940; liq (120°C) 3.960 [KIR81]
Melting Point, °C: 113.60 [MER06]
Boiling Point, °C: 185.25 [CRC10]
Thermal Conductivity, W/(m·K): 0.421 (24.4°C) [KIR81]
Thermal Expansion Coefficient: cub coefficient is 2.81×10^{-4}/°C (0°C–113.6°C) [KIR81]

1581
Compound: Iodine bromide
Formula: IBr
Molecular Formula: BrI
Molecular Weight: 206.808
CAS RN: 7789-33-5
Properties: black ortho cryst [CRC10]
Solubility: s H_2O, EtOH, eth [CRC10]
Density, g/cm³: 4.3 [CRC10]
Melting Point, °C: 40 [CRC10]
Boiling Point, °C: decomposes at 116 [CRC10]

1582
Compound: Iodine chloride
Formula: ICl
Molecular Formula: ClI
Molecular Weight: 162.357
CAS RN: 7790-99-0
Properties: red cryst or oily liq [CRC10]
Density, g/cm³: 3.24 [CRC10]
Melting Point, °C: 27.38 [CRC10]
Boiling Point, °C: decomposes at 94.4 [CRC10]

1583
Compound: Iodine cyanide
Synonym: cyanogen iodide

Formula: ICN
Molecular Formula: CIN
Molecular Weight: 152.922
CAS RN: 506-78-5
Properties: colorless needles; very pungent odor and acrid taste; used in taxidermists' preservatives [HAW93]
Solubility: s H_2O, alcohol, ether [HAW93]
Density, g/cm³: 1.84 [HAW93]
Melting Point, °C: 146.5 [HAW93]

1584
Compound: Iodine dioxide
Formula: IO_2
Molecular Formula: IO_2
Molecular Weight: 158.903
CAS RN: 13494-92-3
Properties: lemon yellow cryst [KIR82]
Density, g/cm³: 4.2 [KIR81]
Reactions: decomposes at 85°C [KIR81]

1585
Compound: Iodine fluoride
Formula: IF
Molecular Formula: FI
Molecular Weight: 145.902
CAS RN: 13873-84-2
Properties: white powd (−78°C) [CRC10]
Melting Point, °C: −14 [CRC10]

1586
Compound: Iodine heptafluoride
Formula: IF_7
Molecular Formula: F_7I
Molecular Weight: 259.893
CAS RN: 16921-96-3
Properties: colorless liq; enthalpy of vaporization 24.7 kJ/mol; specific conductivity 10^{-9} ohm·cm [KIR78] [MER06]
Solubility: s H_2O with some decomposition [MER06]
Density, g/cm³: 2.669 [KIR78]
Melting Point, °C: 6.45 [MER06]
Boiling Point, °C: sublimes at 4.77 [MER06]

1587
Compound: Iodine hexoxide
Formula: I_2O_6
Molecular Formula: I_2O_6
Molecular Weight: 349.805
CAS RN: 65355-99-9

Properties: yellow solid [CRC10]
Solubility: reac H_2O [CRC10]
Melting Point, °C: decomposes at 150 [CRC10]

1588

Compound: Iodine monobromide
Formula: IBr
Molecular Formula: BrI
Molecular Weight: 206.808
CAS RN: 7789-33-5
Properties: brownish black cryst or very hard solid;
  uses: organic synthesis [HAW93] [MER06]
Solubility: s H_2O, alcohol, ether, CS_2,
  glacial acetic acid [MER06]
Density, g/cm³: 4.416 [MER06]
Melting Point, °C: 40 [MER06]
Boiling Point, °C: 116, decomposes [MER06]

1589

Compound: Iodine monochloride
Formula: ICl
Molecular Formula: ClI
Molecular Weight: 162.357
CAS RN: 7790-99-0
Properties: reddish brown oily liq; viscosity at 35°C is
  1.21 cSt; electrical conductivity 4.60×10^{-3} at 35°C;
  enthalpy of vaporization 256.4 kJ/kg; enthalpy of
  fusion 11.60 kJ/mol; readily supercooled; polar
  solvent; best prepared by direct reaction between
  I_2 and liq Cl_2; used in analytical chemistry and in
  organic synthesis [KIR81] [CRC10] [HAW93]
Solubility: s in H_2O (decomposes),
  alcohol, dil HCl [HAW93]
Density, g/cm³: 3.24 (liq at 34°C) [HAW93]
Melting Point, °C: 27.38 [CRC10]
Boiling Point, °C: decomposes at 101 [HAW93]

1590

Compound: Iodine nonoxide
Formula: I_4O_9
Molecular Formula: I_4O_9
Molecular Weight: 651.613
CAS RN: 73560-00-6
Properties: yellow hygr powd [KIR81]
Reactions: decomposes at 75°C [KIR81]

1591

Compound: Iodine pentafluoride
Formula: IF_5
Molecular Formula: F_5I
Molecular Weight: 221.896
CAS RN: 7783-66-6
Properties: fuming straw-colored liq; critical temp
  300.7°C; critical pressure 5.16 MPa; enthalpy of
  vaporization 41.3 kJ/mol; enthalpy of fusion 11.21 kJ/
  mol; specific conductivity 5.4×10^{-6} ohm·cm;
  attacks glass; used as a fluorinating and incendiary
  agent [KIR78] [HAW93] [AIR87] [CRC10]
Solubility: reacts violently with H_2O [HAW93]
Density, g/cm³: 3.19 [MER06]
Melting Point, °C: 9.43 [MER06]
Boiling Point, °C: 100.5 [MER06]

1592

Compound: Iodine pentoxide
Synonym: iodine(V) oxide
Formula: I_2O_5
Molecular Formula: I_2O_5
Molecular Weight: 333.806
CAS RN: 12029-98-0
Properties: −80 mesh with 99.9% purity; white,
  cryst powd; oxidizing agent, used in organic
  synthesis; oxidizes CO to CO_2 at ordinary
  temperatures; can be prepared by dehydration
  of iodic acid [KIR81] [HAW93] [CER91]
Solubility: s H_2O, HNO_3; i absolute
  alcohol, ether, CS_2 [HAW93]
Density, g/cm³: 4.980 [KIR81]
Melting Point, °C: decomposes at 300–350 [STR93]

1593

Compound: Iodine tetroxide
Formula: I_2O_4
Molecular Formula: I_2O_4
Molecular Weight: 317.807
CAS RN: 12399-08-5
Properties: yellow cryst [CRC10]
Solubility: sl H_2O
Density, g/cm³: 4.2 [CRC10]
Melting Point, °C: 130 [CRC10]
Boiling Point, °C: decomposes at >85 [CRC10]

1594

Compound: Iodine trichloride
Formula: ICl_3
Molecular Formula: Cl_3I
Molecular Weight: 233.262
CAS RN: 865-44-1
Properties: yellowish orange; deliq cryst powd;
  pungent irritating odor; electrical conductivity
  8.60×10^{-3} at 102°C; can be formed by adding
  iodine to liq chlorine; used to introduce iodine
  and chlorine in organic synthesis, used as
  a topical antiseptic [HAW93] [KIR81]

Solubility: decomposes in H_2O; s
    alcohol, benzene [HAW93]
Density, g/cm³: 3.111 [KIR81]
Melting Point, °C: 33 [HAW93]
Reactions: decomposes at 65°C [KIR81]

1595
Compound: Iodine trifluoride
Formula: IF_3
Molecular Formula: F_3I
Molecular Weight: 183.899
CAS RN: 22520-96-3
Properties: yellow solid; stable at low temp [CRC10]
Melting Point, °C: −28 [CRC10]

1596
Compound: Iodosyl pentafluoride
Formula: IOF_5
Molecular Formula: F_5IO
Molecular Weight: 237.895
CAS RN: 16056-61-4
Properties: col liq [CRC10]
Melting Point, °C: 4.5 [CRC10]

1597
Compound: Iodosyl trifluoride
Formula: IOF_3
Molecular Formula: F_3IO
Molecular Weight: 199.898
CAS RN: 19058-78-7
Properties: hygr col needles [CRC10]
Solubility: reac H_2O [CRC10]
Melting Point, °C: decomposes at >110 [CRC10]

1598
Compound: Iodogermane
Formula: GeH_3I
Molecular Formula: GeH_3I
Molecular Weight: 202.57
CAS RN: 13573-02-9
Properties: liq [CRC10]
Solubility: reac H_2O [CRC10]
Melting Point, °C: ~90 [CRC10]

1599
Compound: Iodyl trifluoride
Formula: IO_2F_3
Molecular Formula: F_3IO_2
Molecular Weight: 215.898
CAS RN: 25402-50-0

Properties: yellow solid [CRC10]
Melting Point, °C: 41 [CRC10]
Boiling Point, °C: sublimes [CRC10]

1600
Compound: Iridium
Formula: Ir
Molecular Formula: Ir
Molecular Weight: 192.217
CAS RN: 7439-88-5
Properties: silver white metal; most corrosion resistant
    of the elements; cub, a=0.384 nm; hardness 6.5
    Mohs; Poisson's ratio 0.26; not attacked by acids,
    including aqua regia; attacked by fluorine, chlorine
    at red heat; modulus of elasticity is 75,000,000
    psi, one of the highest; enthalpy of fusion 41.12 kJ/
    mol; enthalpy of vaporization, 231.84 kJ/mol;
    electrical resistivity at 20°C, 4.71 μohm·cm
    [HAW93] [MER06] [KIR82] [CRC10] [ALD94]
Solubility: s slowly in aqua regia and
    in fused alkalies [HAW93]
Density, g/cm³: 22.65, highest of any element [MER06]
Melting Point, °C: 2410 [ALD94]
Boiling Point, °C: 4130 [ALD94]
Thermal Conductivity, W/(m·K): 147 (25°C) [ALD94]
Thermal Expansion Coefficient: 6.4×10^{-6}/K [CRC10]

1601
Compound: Iridium carbonyl
Synonym: tetrairidium dodecacarbonyl
Formula: $Ir_4(CO)_{12}$
Molecular Formula: $C_{12}Ir_4O_{12}$
Molecular Weight: 1104.993
CAS RN: 11065-24-0
Properties: yellow powd; stable in air [DOU83] [STR93]
Melting Point, °C: decomposes at 230 [STR93]

1602
Compound: Iridium hexafluoride
Formula: IrF_6
Molecular Formula: F_6Ir
Molecular Weight: 306.207
CAS RN: 7783-75-7
Properties: golden yellow; cub; very hygr;
    enthalpy of vaporization 36 kJ/mol; enthalpy
    of fusion 8.40 kJ/mol [MER06] [CRC10]
Solubility: decomposes in H_2O [MER06]
Density, g/cm³: 4.8 [LID94]
Melting Point, °C: 44 [CRC10]
Boiling Point, °C: 53 [MER06]
Reactions: volatilized by slow heating; reduced
    to IrF_4 by halogens [MER06]

1603
Compound: Iridium pentafluoride
Formula: $(IrF_5)_4$
Molecular Formula: $F_{20}Ir_4$
Molecular Weight: 1148.836
CAS RN: 14568-19-5
Properties: tetramer [KIR82]

1604
Compound: Iridium(I) chlorotricarbonyl
Synonym: chlorotricarbonyliridium(I)
Formula: $[IrCl(CO)_3]n$
Molecular Formula: C_3ClIrO_3 (n = 1)
Molecular Weight: 311.701 (n = 1)
CAS RN: 32594-40-4
Properties: brown powd [STR93]
Melting Point, °C: decomposes at 235 [STR93]

1605
Compound: Iridium(III) acetylacetonate
Synonyms: 2,4-pentanedione, Ir(III) derivative
Formula: $Ir[CH_3COCH=C(O)CH_3]_3$
Molecular Formula: $C_{15}H_{21}IrO_6$
Molecular Weight: 489.550
CAS RN: 15635-87-7
Properties: orange-yellow cryst [CRC10]
Melting Point, °C: 269–271 [ALD94]

1606
Compound: Ir(III) bromide
Formula: $IrBr_3$
Molecular Formula: Br_3Ir
Molecular Weight: 431.929
CAS RN: 10049-24-8
Properties: red-brown monocl [CRC10]
Solubility: i H_2O, acid, alk [CRC10]
Density, g/cm³: 6.82 [CRC10]

1607
Compound: Iridium(III) bromide tetrahydrate
Synonym: iridium tribromide tetrahydrate
Formula: $IrBr_3 \cdot 4H_2O$
Molecular Formula: $Br_3H_8IrO_4$
Molecular Weight: 503.991
CAS RN: 10049-24-8
Properties: olive green, brown or
black cryst [HAW93]
Solubility: s H_2O; i alcohol [HAW93]
Reactions: minus $3H_2O$ at 100°C [AES93]

1608
Compound: Iridium(III) chloride
Synonym: iridium trichloride
Formula: $IrCl_3$
Molecular Formula: Cl_3Ir
Molecular Weight: 298.575
CAS RN: 10025-83-9
Properties: −8 mesh with 99.5% purity; α-$IrCl_3$:
brown, monocl cryst; β-$IrCl_3$: red, ortho-
rhomb cryst [MER06] [CER91]
Solubility: i H_2O, acids, alkalies [MER06]
Density, g/cm³: 5.30 [STR93]
Melting Point, °C: decomposes at 763 [MER06]

1609
Compound: Iridium(III) chloride hydrate
Formula: $IrCl_3 \cdot xH_2O$
Molecular Formula: Cl_3Ir (anhydrous)
Molecular Weight: 298.575 (anhydrous)
CAS RN: 14996-61-3
Properties: black cryst [STR93]

1610
Compound: Iridium(III) fluoride
Formula: IrF_3
Molecular Formula: F_3Ir
Molecular Weight: 249.212
CAS RN: 23370-59-4
Properties: black hex cryst [CRC10]
Solubility: i H_2O, dil acid [CRC10]
Density, g/cm³: ~8.0 [CRC10]
Melting Point, °C: decomposes at 250 [CRC10]

1611
Compound: Iridium(III) iodide
Formula: IrI_3
Molecular Formula: I_3Ir
Molecular Weight: 572.930
CAS RN: 7790-41-2
Properties: dark brown monocl cryst [CRC10]
Solubility: i H_2O, acid, bz, chl; s alk [CRC10]
Density, g/cm³: ~7.4 [CRC10]

1612
Compound: Iridium(III) oxide
Synonym: iridium trioxide
Formula: Ir_2O_3
Molecular Formula: Ir_2O_3
Molecular Weight: 432.432
CAS RN: 1312-46-5

Properties: bluish black powd [MER06]
Solubility: i H_2O; slowly dissolves in
   boiling HCl [MER06]
Reactions: minus O at 400°C; oxidized to
   IrO_2 by HNO_3 [MER06] [CRC10]

1613
Compound: Iridium(III) sulfide
Formula: Ir_2S_3
Molecular Formula: Ir_2S_3
Molecular Weight: 480.629
CAS RN: 12136-42-4
Properties: ortho cryst [CRC10]
Density, g/cm³: 10.2 [CRC10]

1614
Compound: Iridium(IV) oxide
Formula: IrO_2
Molecular Formula: IrO_2
Molecular Weight: 224.216
CAS RN: 12030-49-8
Properties: −325 mesh 10 μm or less with 99.9%
   purity; brown powd [STR93] [CER91]
Density, g/cm³: 11.66 [STR93]
Melting Point, °C: decomposes at 1100 [STR93]

1615
Compound: Iron
Formula: Fe
Molecular Formula: Fe
Molecular Weight: 55.845
CAS RN: 7439-89-6
Properties: cub; silver-white or gray; soft, ductile;
   somewhat magnetic; tensile strength 30,000 psi;
   Brinell hardness 82–100; electrical resistivity
   (20°C) 9.71 μohm·cm; magnetic permeability
   88,400 gauss (25°C); tensile strength 245–280 MPa;
   yield strength 70–140 MPa; enthalpy of fusion
   13.81 kJ/mol; enthalpy of vaporization 340 kJ/
   mol; reducing agent; oxidizes readily in moist
   air; elongation in 5 cm at 20°C is 40%–60%
   [MER06] [KIR81] [CRC10] [ALD94]
Solubility: s HCl, H_2SO_4, dil HNO_3 [HAW93]
Density, g/cm³: 7.86 [MER06]
Melting Point, °C: 1535 [MER06]
Boiling Point, °C: 2861 [LID94]
Reactions: reacts with steam to produce
   hydrogen gas [HAW93]
Thermal Conductivity, W/(m·K): 80.4 (25°C) [ALD94]
Thermal Expansion Coefficient: (volume)
   100°C (0.326), 200°C (0.747), 400°C (1.645),
   800°C (3.523), 1200°C (2.986) [CLA66]

1616
Compound: Iron antimonide
Formula: Fe_3Sb_2
Molecular Formula: Fe_3Sb_2
Molecular Weight: 411.055
CAS RN: 39356-80-4
Properties: 6 mm pieces and smaller
   with 99.5% purity [CER91]

1617
Compound: Iron arsenide
Formula: FeAs
Molecular Formula: AsFe
Molecular Weight: 130.767
CAS RN: 12044-16-5
Properties: white; 6 mm pieces and smaller with 99.5%
   purity; there are two other arsenides: Fe_2As, 12005-
   88-8, and $FeAs_2$, 12006-21-2 [CER91] [CRC10]
Solubility: v sl s H_2O [CRC10]
Density, g/cm³: 7.83 [CRC10]
Melting Point, °C: 1020 [CRC10]

1618
Compound: Iron boride
Formula: FeB
Molecular Formula: BFe
Molecular Weight: 66.656
CAS RN: 12006-84-7
Properties: gray cryst; −35 mesh with 99% purity;
   refractory material [KIR78] [CER91] [CRC10]
Density, g/cm³: 7.15 [ALF93]
Melting Point, °C: ~1550 [KIR78]

1619
Compound: Iron boride
Formula: Fe_2B
Molecular Formula: BFe_2
Molecular Weight: 122.501
CAS RN: 12006-85-8
Properties: −35 mesh with 99% purity;
   refractory material [KIR78] [CER91]

1620
Compound: Iron carbide
Formula: Fe_3C
Molecular Formula: CFe_3
Molecular Weight: 179.546
CAS RN: 12011-67-5
Properties: gray cub cryst [CRC10]
Density, g/cm³: 7.694 [CRC10]
Melting Point, °C: 1227 [CRC10]

1731
Compound: Lead sulfide
Synonym: galena
Formula: PbS
Molecular Formula: PbS
Molecular Weight: 239.266
CAS RN: 1314-87-0
Properties: black powd or silvery cub cryst; enthalpy of fusion 19.00 kJ/mol; photoconductive; can be made by heating Pb in sulfur vapor; used in ceramics, infrared radiation detection, and semiconductors, and in the form of 99.9% or 99.999% pure material as a sputtering target for metallic high reflecting films; galena is natural lead sulfide ore, it has a lead gray color with a metallic luster, a 2.5 Mohs hardness [KIR78] [HAW93] [CER91] [CRC10]
Solubility: 0.01244 g/100 mL H_2O (20°C); s HNO_3, hot dil HCl [KIR78] [MER06]
Density, g/cm³: 7.61 [CRC10]
Melting Point, °C: 1113 [CRC10]
Thermal Conductivity, W/(m·K): 2.3 [CRC10]
Thermal Expansion Coefficient: (volume) 100°C (0.490), 200°C (1.099), 400°C (2.402), 600°C (3.878) [CLA66]

1732
Compound: Lead sulfite
Formula: $PbSO_3$
Molecular Formula: O_3PbS
Molecular Weight: 287.264
CAS RN: 7446-10-8
Properties: white powd [HAW93]
Solubility: i H_2O; s HNO_3 [HAW93]
Melting Point, °C: decomposes [HAW93]

1733
Compound: Lead tantalate
Formula: $PbTa_2O_6$
Molecular Formula: O_6PbTa_2
Molecular Weight: 665.092
CAS RN: 12065-68-8
Properties: ortho; −200 mesh with 99.9% purity [LID94] [CER91]
Density, g/cm³: 7.9 [LID94]

1734
Compound: Lead telluride
Synonym: altaite
Formula: PbTe
Molecular Formula: PbTe
Molecular Weight: 334.800
CAS RN: 1314-91-6
Properties: silver gray cub cryst, 99.999% pure melted pieces of 3–12 mm and 1–3 mm; semiconductor, photoconductor; hardness 3 Mohs; made by melting mixture of Pb and Te; used as an evaporation material and sputtering target for high-index film in infrared filters and infrared detectors [KIR78] [MER06] [CER91]
Solubility: not attacked by HCl, HF, HClO [MER06]
Density, g/cm³: 8.164 [STR93]
Melting Point, °C: 905 [MER06]
Thermal Conductivity, W/(m·K): 2.3 [CRC10]

1735
Compound: Lead tellurite
Formula: $PbTeO_3$
Molecular Formula: O_3PbTe
Molecular Weight: 382.798
CAS RN: 15851-47-5
Properties: −100 mesh with 99.9% purity [CER91]

1736
Compound: Lead tetraacetate
Synonym: lead(IV) acetate
Formula: $Pb(CH_3COO)_4$
Molecular Formula: $C_8H_{12}O_8Pb$
Molecular Weight: 443.378
CAS RN: 546-67-8
Properties: white to light brown; monocl prisms from glacial acetic acid; readily turns pink; unstable in air, moisture sensitive; hydrolyzes in H_2O to form brown PbO_2 and acetic acid; oxidizing agent; obtained by adding warm glacial water free acetic acid to Pb_3O_4, then cooling; used as a laboratory reagent, as an oxidant in organic synthesis [KIR78] [HAW93] [MER06] [STR93]
Solubility: decomposes in cold H_2O and alcohol; s hot glacial acetic acid, benzene, chloroform [MER06] [KIR78]
Density, g/cm³: 2.228 [MER06]
Melting Point, °C: 175–180 [MER06]
Reactions: reacts with conc halogen acids, HX, to form haloplumbic acids, H_2PbX [MER06]

1737
Compound: Lead tetrachloride
Formula: $PbCl_4$
Molecular Formula: Cl_4Pb
Molecular Weight: 349.0
CAS RN: 13463-30-4
Properties: yellow oily liq [CRC10]
Melting Point: −15 [CRC10]
Boiling Point, °C: decomposes at ~50 [CRC10]

1738
Compound: Lead tetrafluoride
Formula: PbF_4
Molecular Formula: F_4Pb
Molecular Weight: 283.194
CAS RN: 7783-59-7
Properties: white tetr cryst; readily hydrolyzes and forms PbO_2 in the presence of moisture; can be produced from a reaction of F_2 with PbF_2; used as a very effective fluorinating agent for olefins [KIR78] [MER06]
Density, g/cm³: 6.7 [MER06]
Melting Point, °C: ~600 decomposes [MER06]

1739
Compound: Lead thiocyanate
Formula: $Pb(SCN)_2$
Molecular Formula: $C_2N_2PbS_2$
Molecular Weight: 323.367
CAS RN: 592-87-0
Properties: white or light yellow odorless powd; used as an ingredient of priming mixture for small arms cartridges, safety matches, and used in dyeing [HAW93] [MER06]
Solubility: s ~200 parts cold, 50 parts boiling H_2O [MER06]; g/L H_2O: 0.0137 (18°) [KRU93]
Density, g/cm³: 3.82 [MER06]
Melting Point, °C: decomposes at 190 [ALD94]

1740
Compound: Lead thiosulfate
Formula: PbS_2O_3
Molecular Formula: O_3PbS_2
Molecular Weight: 319.330
CAS RN: 13478-50-7
Properties: white cryst [HAW93]
Solubility: i H_2O; s acids and sodium thiosulfate solutions [HAW93]
Density, g/cm³: 5.18 [HAW93]
Melting Point, °C: decomposes [HAW93]

1741
Compound: Lead titanate
Synonym: lead metatitanate
Formula: $PbTiO_3$
Molecular Formula: O_3PbTi
Molecular Weight: 303.065
CAS RN: 12060-00-3

Properties: yellow tetr cryst <490°C, cub >490°C; can be made by calcination of stoichiometric amounts of PbO and TiO_2 at 400°C; also prepared by precipitation from aq solution of lead nitrate, titanium tetrachloride, and ammonium hydroxide, followed by calcining at 900°C; has been used as a paint pigment; and in 99.9% purity as a sputtering target for thin film capacitors [KIR78] [KIR83] [STR93] [CER91] [SAF87]
Solubility: i H_2O; decomposed in HCl solution to $PbCl_2$ and TiO_2 [KIR78] [HAW93]
Density, g/cm³: 7.52 [HAW93]

1742
Compound: Lead tungstate
Synonym: raspite
Formula: $PbWO_4$
Molecular Formula: O_4PbW
Molecular Weight: 455.038
CAS RN: 7759-01-5
Properties: −200 mesh with 99.9% purity; colorless monocl [CER91] [KIR83]
Solubility: 0.03 g/100 mL H_2O [CRC10]
Density, g/cm³: 8.46 [KIR83]
Melting Point, °C: 1123 [KIR83]
Reactions: transforms to stolzite ~400°C [LID94]

1743
Compound: Lead tungstate
Synonym: stolzite
Formula: $PbWO_4$
Molecular Formula: O_4PbW
Molecular Weight: 455.038
CAS RN: 7759-01-5
Properties: white powd; used as a pigment [HAW93]
Solubility: i H_2O, cold HNO_3; s in fixed alkali hydroxides [MER06]
Density, g/cm³: 8.24 [HAW93]
Melting Point, °C: 1130 [HAW93]

1744
Compound: Lead vanadate
Synonym: lead metavanadate
Formula: $Pb(VO_3)_2$
Molecular Formula: O_6PbV_2
Molecular Weight: 405.079
CAS RN: 10099-79-3
Properties: yellow powd; used to prepare other vanadium compounds and as a pigment [HAW93]
Solubility: i H_2O; decomposed by HNO_3 [MER06]

1745
Compound: Lead zirconate
Formula: $PbZrO_3$
Molecular Formula: O_3PbZr
Molecular Weight: 346.422
CAS RN: 12060-01-4
Properties: colorless cub perovskite >230°, pseudotetr or ortho-rhomb <230°C; can be obtained by heating stoichiometric amounts of lead and zirconium oxides; has high piezoelectric properties; used in high power acoustic transducer, hydrophones, and as a 99.7% pure sputtering target for thin film capacitors [STR93] [CER91] [KIR78]
Solubility: i H_2O, alkalies; s mineral acids [KIR78]
Density, g/cm³: 7.0 [STR93]

1746
Compound: Lead(II) butanoate
Formula: $Pb(C_4H_7O_2)_2$
Molecular Formula: $C_8H_{14}PbO_4$
Molecular Weight: 381.4
CAS RN: 819-73-8
Properties: col solid [CRC10]
Solubility: i H_2O; s dil HNO_3 [CRC10]
Melting Point, °C: ~90 [CRC10]

1747
Compound: Lead(II) carbonate, basic
Formula: $Pb(OH)_2 \cdot 2PbCO_3$
Molecular Formula: $C_2H_2O_8Pb_3$
Molecular Weight: 775.6
CAS RN: 1319-46-6
Properties: white, hex cryst [CRC10]
Solubility: i H_2O, EtOH; s acid [CRC10]
Density, g/cm³: ~6.5 [CRC10]
Melting Point, °C: decomposes at 400 [CRC10]

1748
Compound: Lead(II) chloride fluoride
Formula: $PbClF$
Molecular Formula: $ClFPb$
Molecular Weight: 261.7
CAS RN: 13847-57-9
Properties: tetr cryst [CRC10]
Solubility: g/100 g H_2O: 0.035²⁰ [CRC10]
Density, g/cm³: 7.05 [CRC10]

1749
Compound: Lead(II) chromate(VI) oxide
Formula: $PbCrO_4 \cdot PbO$

Molecular Formula: CrO_5Pb_2
Molecular Weight: 546.4
CAS RN: 18454-12-1
Properties: red powd [CRC10]
Solubility: i H_2O [CRC10]

1750
Compound: Lead(II) 2-ethylhexanoate
Formula: $Pb(C_7H_{15}CO_2)_2$
Molecular Formula: $C_{16}H_{30}O_4Pb$
Molecular Weight: 493.6
CAS RN: 301-08-6
Properties: visc liq [CRC10]
Density, g/cm³: 1.56 [CRC10]

1751
Compound: Lead(II) formate
Formula: $Pb(CHO_2)_2$
Molecular Formula: $C_2H_2O_4Pb$
Molecular Weight: 297.2
CAS RN: 811-54-1
Properties: white prisms or needles [CRC10]
Solubility: g/100 g H_2O: 1.6¹⁶; i EtOH [CRC10]
Melting Point, °C: decomposes at 190 [CRC10]
Density, g/cm³: 4.63 [CRC10]

1752
Compound: Lead(II) hydrogen arsenate
Formula: $PbHAsO_4$
Molecular Formula: $AsHO_4Pb$
Molecular Weight: 347.1
CAS RN: 7784-40-9
Properties: white, monocl cryst [CRC10]
Solubility: i H_2O; s HNO_3, alk [CRC10]
Melting Point, °C: decomposes at 280 [CRC10]
Density, g/cm³: 5.943 [CRC10]

1753
Compound: Lead(II) lactate
Formula: $Pb(C_3H_5O_3)_2$
Molecular Formula: $C_6H_{10}O_6Pb$
Molecular Weight: 385.3
CAS RN: 18917-82-3
Properties: white, cryst powd [CRC10]
Solubility: s H_2O, hot EtOH [CRC10]

1754
Compound: Lead(II) oleate
Formula: $Pb(C_{18}H_{33}O_2)_2$

Molecular Formula: $C_{36}H_{66}O_4Pb$
Molecular Weight: 770.1
CAS RN: 1120-46-3
Properties: wax-like solid [CRC10]

1755
Compound: Lead(II) oxide hydrate
Formula: $3PbO \cdot H_2O$
Molecular Formula: $H_2O_4Pb_3$
Molecular Weight: 687.6
CAS RN: 1311-11-1
Properties: white powd [CRC10]
Solubility: i H_2O; s dil acid [CRC10]
Density, g/cm³: 7.41 [CRC10]

1756
Compound: Lead(IV) bromide
Formula: $PbBr_4$
Molecular Formula: Br_4Pb
Molecular Weight: 526.8
CAS RN: 13702-91-2
Properties: unstable liq [CRC10]

1757
Compound: Lead(IV) chloride
Formula: $PbCl_4$
Molecular Formula: Cl_4Pb
Molecular Weight: 349.0
CAS RN: 13463-30-4
Properties: yellow oily liq [CRC10]

1758
Compound: Lead(IV) fluoride
Formula: PbF_4
Molecular Formula: F_4Pb
Molecular Weight: 283.2
CAS RN: 7783-59-7
Properties: white, tetr cryst; hygr [CRC10]
Density: 6.7 [CRC10]
Melting Point, °C: ~600 [CRC10]

1759
Compound: Lead(II,III) oxide
Synonyms: minium, red lead
Formula: Pb_3O_4
Molecular Formula: O_4Pb_3
Molecular Weight: 685.598
CAS RN: 1314-41-6

Properties: bright red, heavy powd; spinel structure; evolves Cl in contact with hot HCl; oxidizing agent; manufactured by heating PbO in air at 450°C–500°C; used in storage batteries, in the purification of alcohol, as a pigment in corrosion-protective paints [HAW93] [MER06] [KIR78]
Solubility: i H_2O, alcohol; s in excess glacial acetic acid, hot HCl [MER06]
Density, g/cm³: 9.1 [MER06]
Melting Point, °C: 830 (under O_2) [KIR78]
Boiling Point, °C: decomposes at 500 [KIR78]

1760
Compound: Lithium
Formula: Li
Molecular Formula: Li
Molecular Weight: 6.941
CAS RN: 7439-93-2
Properties: very soft silver-white metal; bcc from ~–195°C–180°C, a=0.350 nm; hardness 0.6 Mohs; specific heat 3.55 J/g; enthalpy of fusion 3.00 kJ/mol; enthalpy of vaporization ~147.8 kJ/mol; electrical resistivity 9.446 µohm·cm; ionic radius 0.060 nm; vapor pressure, kPa: 0.065 (702°C), 0.376 (802°C), 1.61 (902°C), 5.47 (1002°C), 9.4 (1052°C), 12.13 (1077°C); manufactured by electrolysis; used to prepare the amide, nitride, and hydride [KIR81] [CRC10]
Solubility: reacts vigorously with H_2O, dil HCl, H_2SO_4, HNO_3 evolving H_2 [MER06]
Density, g/cm³: 0.531 (20°C) [KIR81]
Melting Point, °C: 180.5 [KIR81]
Boiling Point, °C: 1336 [KIR81]
Reactions: transformation to fcc at –133°C; bcc to hex at –199°C [KIR78]
Thermal Conductivity, W/(m·K): 84.7 [CRC10]
Thermal Expansion Coefficient: 46×10^{-6}/K [CRC10]

1761
Compound: Lithium acetate
Formula: CH_3COOLi
Molecular Formula: $C_2H_3LiO_2$
Molecular Weight: 65.986
CAS RN: 546-89-4
Properties: used as an alcoholysis catalyst in the manufacture of alkyl resins [KIR81]
Solubility: g/100 g soln, H_2O: 23.76 (0°C), 31.28 (25.8°C), 66.73 (102.8°C); solid phase, $CH_3COOLi \cdot 2H_2O$ (0°C, 25°C), CH_3COOLi (100°C) [KRU93]
Melting Point, °C: 291 [KIR81]

1762

Compound: Lithium acetate dihydrate
Synonyms: acetic acid, lithium salt
Formula: $LiCH_3COO \cdot 2H_2O$
Molecular Formula: $C_2H_7LiO_4$
Molecular Weight: 102.017
CAS RN: 6108-17-4
Properties: white powd; rhomb cryst; obtained readily from reaction of acetic acid and Li_2CO_3 or LiOH; used as catalyst in production of polyester, as an anticorrosion agent [KIR81] [STR93] [MER06] [FMC93]
Solubility: $gLiC_2H_3O_2/100\,g\ H_2O$: 31 (25°C), 66 g/100 g H_2O (100°C) [KIR81]
Density, g/cm³: 1.3 [STR93]
Melting Point, °C: 57.8; anhydrous, 291 [KIR81]

1763

Compound: Lithium acetylacetonate
Synonyms: 2,4-pentanedione, lithium derivative
Formula: $Li[CH_3COCH=C(O)CH_3]$
Molecular Formula: $C_5H_7LiO_2$
Molecular Weight: 106.051
CAS RN: 18115-70-3
Properties: white powd; hygr [STR93]
Melting Point, °C: decomposes at 250 [ALD94]

1764

Compound: Lithium aluminum deuteride
Formula: $LiAlD_4$
Molecular Formula: AlD_4Li
Molecular Weight: 41.983
CAS RN: 14128-54-2
Properties: white to gray cryst; sensitive to moisture; decomposes at >140°C liberating deuterium; used to introduce deuterium atoms into molecules [HAW93]
Solubility: s ether, tetrahydrofuran [HAW93]
Density, g/cm³: 1.02 [HAW93]
Melting Point, °C: decomposes at 175 [ALD94]

1765

Compound: Lithium aluminum hydride
Synonym: lithium tetrahydridoaluminate
Formula: $LiAlH_4$
Molecular Formula: AlH_4Li
Molecular Weight: 37.955
CAS RN: 16853-85-3

Properties: cryst or gray powd; monocl; stable in dry air, decomposes in moist air; prepared by reaction of LiH with $AlCl_3$; used as a reducing agent for organics to convert esters, aldehydes, and ketones to alcohols, used in perfumes, and in pharmaceuticals [HAW93] [MER06]
Solubility: reacts rapidly with H_2O; 30 parts/100 parts ether [MER06]
Density, g/cm³: 0.917 [STR93]
Melting Point, °C: decomposes at >125 [MER06]
Reactions: slowly evolves H_2 at 120°C [MER06]

1766

Compound: Lithium aluminum silicate
Synonym: α-spodumene
Formula: $\alpha\text{-}LiAlSi_2O_6$
Molecular Formula: $AlLiO_6Si_2$
Molecular Weight: 186.089
CAS RN: 12068-40-5
Properties: white powd; used in ceramic flux formulations, as a solid electrolyte, and in heat sinks for solar and nuclear applications [FMC93] [STR93]

1767

Compound: Lithium amide
Formula: $LiNH_2$
Molecular Formula: H_2LiN
Molecular Weight: 22.964
CAS RN: 7782-89-0
Properties: gray-to-white powd; tetr; manufactured by reacting LiH and NH_3 gas; used in the pharmaceutical industry to make antihistamines and analgesics [KIR81] [MER06] [FMC93]
Solubility: decomposed by H_2O [MER06]
Density, g/cm³: 1.178 [ALD94]
Melting Point, °C: 375 [KIR81]
Reactions: transforms to imide at >400°C [KIR81]

1768

Compound: Lithium arsenate
Formula: Li_3AsO_4
Molecular Formula: $AsLi_3O_4$
Molecular Weight: 159.743
CAS RN: 13478-14-3
Properties: white powd [HAW93]
Solubility: sl s H_2O; s dil acetic acid [HAW93]
Density, g/cm³: 3.07 (15°C) [HAW93]

1769
Compound: Lithium azide
Formula: LiN$_3$
Molecular Formula: LiN$_3$
Molecular Weight: 48.961
CAS RN: 19597-69-4
Properties: colorless; hygr; body-center
    rhomb [CRC10] [CIC73]
Solubility: g/100 g H$_2$O: 61.3 (0°C), 67.2
    (20°C), 86.6 (60°C) [LAN05]
Density, g/cm^3: 1.83 [LID94]
Melting Point, °C: decomposes at 115–298 [CRC10]

1770
Compound: Lithium borate
Synonym: lithium metaborate
Formula: LiBO$_2$
Molecular Formula: BLiO$_2$
Molecular Weight: 49.751
CAS RN: 13453-69-5
Properties: −80 mesh with 99.9% purity; used
    in special glass and enamel formulations
    as a flux, as an electrolyte component
    for lithium batteries; there is a dihydrate,
    LiBO$_2$·2H$_2$O [KIR81] [CER91] [FMC93]
Solubility: g/100 g H$_2$O: 0.9 (0°C), 2.7
    (20°C), 5.78 (30°C) [LAN05]
Density, g/cm^3: 2.18 [LID94]
Melting Point, °C: 849 [KIR81]

1771
Compound: Lithium borohydride
Formula: LiBH$_4$
Molecular Formula: BH$_4$Li
Molecular Weight: 21.784
CAS RN: 16949-15-8
Properties: white to gray cryst powd; ortho-rhomb;
    decomposes in moist air; used as a source
    of hydrogen; reducing agent for aldehydes,
    ketones, and esters [MER06] [HAW93]
Solubility: s H$_2$O above pH 7, ether, tetrahydrofuran,
    aliphatic amines [MER06]
Density, g/cm^3: 0.66 [MER06]
Melting Point, °C: 268 [MER06]
Boiling Point, °C: decomposes at 380 [MER06]
Reactions: reacts with HCl to form hydrogen,
    diborane, and LiCl [MER06]

1772
Compound: Lithium bromate
Formula: LiBrO$_3$

Molecular Formula: BrLiO$_3$
Molecular Weight: 134.843
CAS RN: 13550-28-2
Solubility: g/100 g soln, H$_2$O: 61.2 (0°C), 65.4
    (25°C), 78.0 (100°C); solid phase, LiBrO$_3$·H$_2$O
    (0°C, 25°C), LiBrO$_3$ (100°C) [KRU93]

1773
Compound: Lithium bromide
Formula: LiBr
Molecular Formula: BrLi
Molecular Weight: 86.845
CAS RN: 7550-35-8
Properties: white cub; very deliq; bitter taste;
    enthalpy of fusion 17.60 kJ/mol manufactured by
    neutralizing HBr with LiOH or Li$_2$CO$_3$; used in
    pharmaceuticals, air conditioning, in batteries,
    LiBr solution is used as a component of refrigrant
    in absorption air conditioning and as a swelling
    agent for proteins [CRC10] [HAW93] [FMC93]
Solubility: s alcohol, glycol, ether, amyl alcohol
    [KIR81] [MER06]; g/100 g soln, H$_2$O: 58.4 ±
    0.4 (0°C), 63.3 ± 1.8 (25°C), 72.7 (100°C);
    solid phase, LiBr·3H$_2$O (0°C), LiBr·2H$_2$O
    (25°C), LiBr·H$_2$O (100°C) [KRU93]
Density, g/cm^3: 3.464 [LID94]
Melting Point, °C: 552 [CRC10]
Boiling Point, °C: 1310 [KIR81]; 1265 [STR93]

1774
Compound: Lithium bromide monohydrate
Formula: LiBr·H$_2$O
Molecular Formula: BrH$_2$LiO
Molecular Weight: 104.860
CAS RN: 13453-70-8
Properties: white powd; obtained by
    crystallization from a hot solution of HBr
    and either LiOH or Li$_2$CO$_3$; can be dried to
    the anhydrous salt [KIR81] [STR93]

1775
Compound: Lithium carbide
Formula: Li$_2$C$_2$
Molecular Formula: C$_2$Li$_2$
Molecular Weight: 37.904
CAS RN: 1070-75-3
Properties: cryst white powd; decomposes in water,
    evolves acetylene when dissolved in acid [HAW93]
Density, g/cm^3: 1.65 (18°C) [HAW93]

1776

Compound: Lithium carbonate
Formula: Li_2CO_3
Molecular Formula: CLi_2O_3
Molecular Weight: 73.891
CAS RN: 554-13-2
Properties: white, light alkaline powd; enthalpy of fusion 41.00 kJ/mol; produced by reacting hot conc soda ash with LiCl, Li_2SO_4; used in ceramics and porcelain glazes; Li_2CO_3 slurry can be dissolved by passing CO_2 through the slurry, carbonate reprecipitates if heated [HAW93] [MER06] [CRC10]
Solubility: g/100 g H_2O: 1.52 (0°C), 1.31 (20°C), 0.71 (100°C); i alcohol [KIR81]; g/100 g H_2O: 1.54 (0°C), 1.29 (25°C), 0.72 (100°C) [KRU93]
Density, g/cm³: 2.11 [MER06]
Melting Point, °C: 726 [KIR81]
Boiling Point, °C: decomposes at 1310 [STR93]

1777

Compound: Lithium chlorate
Formula: $LiClO_3$
Molecular Formula: $ClLiO_3$
Molecular Weight: 90.392
CAS RN: 13453-71-9
Properties: needle-like cryst; deliq; oxidizing agent; decomposes at 270°C; used in air conditioning, in propellants [HAW93]
Solubility: g/100 g soln, H_2O: 71.1 (1.5°C), 81.7 (22.1°C), 94.9 (99°C); solid phase, $LiClO_3 \cdot 3H_2O + LiClO_3 \cdot H_2O$ (1.5°C), $4LiClO_3 \cdot H_2O$ (22.1°C), α-$LiClO_3$ + β-$LiClO_3$ (100°C) [KRU93]
Density, g/cm³: 1.119 [HAW93]
Melting Point, °C: 128 [HAW93]
Boiling Point, °C: decomposes [HAW93]

1778

Compound: Lithium chloride
Formula: LiCl
Molecular Formula: ClLi
Molecular Weight: 42.394
CAS RN: 7447-41-8
Properties: white powd; cub; very hygr; sharp saline taste; enthalpy of fusion 19.90 kJ/mol; obtained from reaction of HCl and LiOH or Li_2CO_3 at >95°C; used in air conditioning, welding, and in soldering flux; component of dry batteries, catalyst for some oxidation reactions, chlorinating agent for steroid substrates [KIR81] [MER06] [STR93] [FMC93]

Solubility: s water, alcohol, acetone, amyl alcohol, pyridine [KIR81] [MER06]; g/100 g soln, H_2O: 40.9 (0°C), 45.8 (25°C), 56.2 (100°C); solid phase, $LiCl \cdot 2H_2O$ (0°C), $LiCl \cdot H_2O$ (25°C), LiCl (100°C) [KRU93]
Density, g/cm³: 2.068 [HAW93]
Melting Point, °C: 610 [CRC10]
Boiling Point, °C: 1360 [HAW93]

1779

Compound: Lithium chloride monohydrate
Formula: $LiCl \cdot H_2O$
Molecular Formula: ClH_2LiO
Molecular Weight: 60.409
CAS RN: 16712-20-2
Properties: white cryst; prepared by reacting HCl with LiOH or Li_2CO_3 and crystallizing below 95°C; solution is used for deicing, in fire extinguishers, catalysts and for dehumidifying [FMC93] [KIR81] [STR93]
Solubility: 45.9 g/100 g saturated solution in water (25°C) [KIR81]
Density, g/cm³: 1.78 [CRC10]
Reactions: minus H_2O at >98°C [CRC10]

1780

Compound: Lithium chromate
Formula: Li_2CrO_4
Molecular Formula: $CrLi_2O_4$
Molecular Weight: 129.876
CAS RN: 14307-35-8
Properties: yellow cryst; clear yellow solution is used as a corrosion inhibitor and as an additive to industrial batteries [STR93] [FMC93]
Solubility: g/100 g soln, H_2O: 47.27 (0.7°C), 48.60 (20°C), 56.82 (100°C); solid phase, $Li_2CrO_4 \cdot 2H_2O$ (0.7°C, 20°C), Li_2CrO_4 (100°C) [KRU93]
Melting Point, °C: 495 [AES93]

1781

Compound: Lithium chromate dihydrate
Formula: $Li_2CrO_4 \cdot 2H_2O$
Molecular Formula: $CrH_4Li_2O_6$
Molecular Weight: 165.906
CAS RN: 7789-01-7
Properties: yellow cryst; ortho-rhomb; deliq powd; oxidizing agent; eutectic in aq solutions is at −60°C; used as a corrosion inhibitor [HAW93] [MER06]
Solubility: 49.6% H_2O (30°C); s methanol, ethanol [KIR78] [MER06]

Density, g/cm³: 2.15 [KIR78]
Reactions: minus $2H_2O$ (74.6°C) to
    become anhydrous [KIR78]

1782

Compound: Lithium citrate tetrahydrate
Synonyms: citric acid, trilithium salt
Formula: $LiOOCCH_2C(OH)(COOLi)CH_2COOLi \cdot 4H_2O$
Molecular Formula: $C_6H_{13}Li_3O_{11}$
Molecular Weight: 281.985
CAS RN: 6680-58-6
Properties: white granules or cryst powd; deliq
    in moist air; barely perceptive alkaline taste;
    used in beverages and pharmaceuticals, as a
    clay dispersant, in electroplating solutions,
    and as a buffer for ion chromatography
    [FMC93] [HAW93] [MER06]
Solubility: 74.5 g/100 mL H_2O (25°C), 66.7 g/100 mL
    (100°C) H_2O [CRC10]; sl s alcohol [MER06]
Melting Point, °C: 209.92 [ALD94]
Reactions: minus $4H_2O$ at 105°C
    becoming anhydrous [MER06]

1783

Compound: Lithium cobaltite
Formula: $LiCoO_2$
Molecular Formula: $CoLiO_2$
Molecular Weight: 97.873
CAS RN: 12190-79-3
Properties: dark gray powd; a = 0.2817 nm, c = 1.4059 nm;
    can be prepared by reacting Li_2CO_3 or $LiOH \cdot H_2O$
    with $CoCO_3$ at 850°C–900°C for 24 h in air; used
    in ceramics and as an insertion electrode in Li
    battery systems; has fluxing properties of lithium
    oxide, and enhances adhesion similar to cobalt
    oxide [HAW93] [STR93] [DAH90] [GUM92]
Solubility: i H_2O [HAW93]

1784

Compound: Lithium cyanide
Formula: LiCN
Molecular Formula: CLiN
Molecular Weight: 32.959
CAS RN: 2408-36-8
Properties: colorless to light yellow liq; sensitive
    to moisture; freezing point 58°C [STR93]
Density, g/cm³: 1.075 (fused) [KIR78]
Melting Point, °C: 160 [KIR78]
Reactions: decomposes to cyanamide and
    carbon below ~600°C [KIR78]

1785

Compound: Lithium cyclopentadienide
Formula: LiC_5H_5
Molecular Formula: C_5H_5Li
Molecular Weight: 72.036
CAS RN: 16733-97-4
Properties: off-white powd; air and moisture
    sensitive [STR93] [ALF95]

1786

Compound: Lithium deuteride
Formula: LiD
Molecular Formula: DLi
Molecular Weight: 8.956
CAS RN: 13587-16-1
Properties: off-white powd; sensitive to air and
    moisture; thermally stable up to its melting point;
    used in thermonuclear fusion [HAW93] [STR93]
Density, g/cm³: 0.820 [STR93]
Melting Point, °C: ~680 [STR93]

1787

Compound: Lithium dichromate dihydrate
Formula: $Li_2Cr_2O_7 \cdot 2H_2O$
Molecular Formula: $Cr_2H_4Li_2O_9$
Molecular Weight: 265.901
CAS RN: 10022-48-7
Properties: reddish orange cryst powd; deliq; eutectic
    in aq solutions is at −70°C; used to dehumidify
    and in refrigeration [HAW93] [KIR78]
Solubility: g/100 g soln, H_2O: 62.36 (0.8°C),
    65.25 (30°C), 73.55 (100°C); solid
    phase $Li_2Cr_2O_7 \cdot 2H_2O$ [KRU93]
Density, g/cm³: 2.34 (30°C) [KIR78]
Melting Point, °C: 130 [HAW93]
Reactions: minus $2H_2O$ at 110°C [CRC10]

1788

Compound: Lithium dihydrogen phosphate
Formula: LiH_2PO_4
Molecular Formula: H_2LiO_4P
Molecular Weight: 103.928
CAS RN: 13453-80-0
Properties: white powd; used as a constituent in
    low-expansion porcelain enamel glazes and
    in some laser glasses [STR93] [FMC93]
Solubility: g/100 g soln, H_2O: 55.8 (0°C);
    solid phase, LiH_2PO_4 [KRU93]
Density, g/cm³: 2.461 [STR93]
Melting Point, °C: >100 [AES93]

1789
Compound: Lithium diisopropylamide
Synonym: LDA
Formula: [(CH$_3$)$_2$CH]$_2$NLi
Molecular Formula: C$_6$H$_{14}$LiN
Molecular Weight: 107.125
CAS RN: 4111-54-0
Properties: pyrophoric powd; hindered, non-nucleophilic strong base; sensitive to air and moisture; uses: generation of carbanions [ALD94] [MER06]
Melting Point, °C: decomposes [MER06]

1790
Compound: Lithium iron silicide
Formula: LiFeSi
Molecular Formula: FeLiSi
Molecular Weight: 90.872
CAS RN: 64082-35-5
Properties: dark brittle cryst
Solubility: reac H$_2$O

1791
Compound: Lithium fluoride
Formula: LiF
Molecular Formula: FLi
Molecular Weight: 25.939
CAS RN: 7789-24-4
Properties: cub cryst or white fluffy powd; does not form a hydrate; enthalpy of fusion 27.09 kJ/ mol; enthalpy of vaporization 147 kJ/mol; index of refraction 1.3915; manufactured by reacting lithium carbonate or lithium hydroxide with dil HF; used as a welding and soldering flux, used in ceramics to reduce firing temperatures and to improve thermal shock resistance; as a 99.9% pure sputtering target for low-index, antireflection film [HAW93] [MER06] [KIR78] [CER91] [CRC10]
Solubility: 0.133 g/100 soln in H$_2$O (25°C) [KIR81]; g/100 g soln, H$_2$O: 0.120 (0°C), 0.134 ± 0.008 (25°C) [KRU93]; s acids; i alcohol [HAW93]
Density, g/cm^3: 2.640 [MER06]
Melting Point, °C: 848 [KIR81]
Boiling Point, °C: 1673 [CRC10]
Reactions: volatilizes at 1100°C–1200°C [MER06]
Thermal Expansion Coefficient: (volume) 100°C (0.912), 200°C (2.086), 400°C (4.759) [CLA66]

1792
Compound: Lithium formate monohydrate
Formula: Li(CHO$_2$)·H$_2$O

Molecular Formula: CHLiO$_3$
Molecular Weight: 69.974
CAS RN: 6108-23-2
Properties: col-white cryst [CRC10]
Solubility: sol H$_2$O [CRC10]
Density, g/cm^3: 1.46 [CRC10]

1793
Compound: Lithium hexafluoroantimonate
Formula: LiSbF$_6$
Molecular Formula: F$_6$LiSb
Molecular Weight: 242.691
CAS RN: 18424-17-4
Properties: powd; hygr [STR93]
Melting Point, °C: decomposes [ALF95]

1794
Compound: Lithium hexafluoroarsenate
Synonym: lithium hexafluorarsenate(V)
Formula: LiAsF$_6$
Molecular Formula: AsF$_6$Li
Molecular Weight: 195.853
CAS RN: 29935-35-1
Properties: white, hygr powd; has been used as an electrolyte for organic solvent lithium batteries [KIR78]

1795
Compound: Lithium hexafluorophosphate
Formula: LiPF$_6$
Molecular Formula: F$_6$LiP
Molecular Weight: 151.905
CAS RN: 21324-40-3
Properties: white to off-white powd; hygr [STR93]

1796
Compound: Lithium hexafluorosilicate
Formula: Li$_2$SiF$_6$
Molecular Formula: F$_6$Li$_2$Si
Molecular Weight: 155.958
CAS RN: 17347-95-4
Properties: white powd [STR93]

1797
Compound: Lithium hexafluorostannate(IV)
Formula: Li$_2$SnF$_6$
Molecular Formula: F$_6$Li$_2$Sn

Molecular Weight: 246.582
CAS RN: 17029-16-2
Properties: white powd [STR93]

1798
Compound: Lithium hydride
Formula: LiH
Molecular Formula: HLi
Molecular Weight: 7.949
CAS RN: 7580-67-8
Properties: gray powd; sensitive to moisture; cub; darkens rapidly on exposure to light; very stable thermally, melts without decomposition; enthalpy of fusion 22.59 kJ/mol; can be prepared by adding H_2O to molten lithium at 680°C–900°C under ~1 atm H_2 pressure; used as a source of hydrogen in military applications and buoyancy devices and in organic synthesis [KIR80] [CRC10] [KIR81] [MER06]
Solubility: reacts vigorously with H_2O [KIR81]; s ether, i benzene, toluene [HAW93]
Density, g/cm³: 0.78 [KIR81]
Melting Point, °C: 688.7 [CRC10]
Reactions: forms LiOH and H_2 in H_2O; reacts with lower alcohols, carboxylic acids, chlorine, and ammonia with evolution of H_2 [MER06]

1799
Compound: Lithium hydrogen carbonate
Synonym: lithia water
Formula: LiHCO₃
Molecular Formula: CHLiO₃
Molecular Weight: 67.958
CAS RN: 10377-37-4
Properties: white; prepared by dissolving lithium carbonate in water that contains excess dissolved carbon dioxide; used in medicine and in the preparation of mineral water [CRC10] [HAW93]
Solubility: 5.5 g/100 mL H_2O (13°C) [CRC10]

1800
Compound: Lithium hydroxide
Formula: LiOH
Molecular Formula: HLiO
Molecular Weight: 23.948
CAS RN: 1310-65-2
Properties: colorless; granular, free-flowing powd; tetr; acrid; readily absorbs CO_2 and H_2O from atm; enthlapy of vaporization 188 kJ/mol; enthalpy of fusion 20.88 kJ/mol; can be prepared from Li_2CO_3 and $Ca(OH)_2$; used to manufacture lithium stearate, in storage battery electrolytes, and to absorb CO_2 in space vehicles [HAW93] [CRC10] [MER06] [KIR81]

Solubility: g/100 g H_2O: 10.7 (0°C), 11.3 (40°C), 14.8 (100°C); sl s alcohol [KIR81] [MER06]; g/100 g soln, H_2O: 12.7 (0°C), 12.9 (25°C), 17.5 (100°C); solid phase, LiOH·H_2O [KRU93]
Density, g/cm³: 1.45 [LID94]
Melting Point, °C: 471.2 [CRC10]
Boiling Point, °C: 1626 [CRC10]

1801
Compound: Lithium hydroxide monohydrate
Formula: LiOH·H_2O
Molecular Formula: H₃LiO₂
Molecular Weight: 41.964
CAS RN: 1310-66-3
Properties: white powd; hygr; monocl; solid phase in equilibrium with dissolved LiOH from 0°C–100°C; used in manufacturing lithium-based greases, as an additive in alkaline battery electrolyte, dye solubilizer for textiles, and as a heat sink in nuclear reactors [KIR81] [STR93] [KIR81] [FMC93]
Solubility: w/w solubility, H_2O: 10.7% (0°C), 10.9% (20°C), 14.8% (100°C); sl s in alcohol [MER06]
Density, g/cm³: 1.51 [MER06]
Melting Point, °C: 680 [AES93]
Reactions: minus H_2O at >100°C [KIR81]

1802
Compound: Lithium hypochlorite
Formula: LiOCl
Molecular Formula: ClLiO
Molecular Weight: 58.393
CAS RN: 13840-33-0
Properties: white granules; oxidant; can ignite organic materials; can be prepared by action of chlorine on a solution of LiOH; used as a bleach and an oxidizing agent, sanitizer for swimming pools, cooling water treatment [HAW93] [KIR81] [FMC93]

1803
Compound: Lithium iodate
Formula: LiIO₃
Molecular Formula: ILiO₃
Molecular Weight: 181.843
CAS RN: 13765-03-2
Properties: −80 mesh with 99.9% purity; has two forms, α and β; white powd; oxidizing agent [HAW93] [CER91]
Solubility: s H_2O; i alcohol [HAW93]
Density, g/cm³: 4.487 [HAW93]
Melting Point, °C: 50–60 [HAW93]
Reactions: transition from α to β at 50°–60° [HAW93]

1804

Compound: Lithium iodide
Formula: LiI
Molecular Formula: ILi
Molecular Weight: 133.845
CAS RN: 10377-51-2
Properties: white powd; hygr; enthalpy of fusion 14.60 kJ/mol; formed by neutralizing HI solutions with LiOH or Li_2CO_3 to obtain the trihydrate followed by careful dehydration in vacuum [KIR81] [STR93] [CRC10]
Solubility: g/100 g soln, H_2O: 62.6 (25°C), 71.1 (75°C) [KIR81]; g/100 g H_2O: 149 (0°C), 163 ± 3 (25°C), 476 (99°C); solid phase, LiI·$3H_2O$ (0°C, 25°C), LiI·H_2O (99°C) [KRU93]
Density, g/cm³: 3.494 [STR93]
Melting Point, °C: 469 [CRC10]
Boiling Point, °C: 1142 [KIR81]; 1180 [STR93]
Reactions: minus iodine when heated in air [KIR81]

1805

Compound: Lithium iodide trihydrate
Formula: LiI·$3H_2O$
Molecular Formula: H_6ILiO_3
Molecular Weight: 187.891
CAS RN: 7790-22-9
Properties: white cryst; extremely hygr; granules of fused masses; becomes yellow due to liberation of I_2 when exposed to atm; used in air conditioning, as a catalyst in acetal formation; there are two other hydrates: LiI·$2H_2O$, 17023-25-5, and LiI·H_2O, 17023-24-4 [KIR81] [HAW93] [MER06]
Solubility: s in about 0.5 parts H_2O or alcohol; v s in amyl alcohol or acetone [MER06]
Density, g/cm³: 3.48 [HAW93]
Melting Point, °C: 73 [MER06]
Boiling Point, °C: 1171 [HAW93]
Reactions: minus $3H_2O$ at 450°C [HAW93]

1806

Compound: Lithium manganate
Formula: $LiMn_2O_3$
Molecular Formula: $LiMn_2O_3$
Molecular Weight: 164.815
CAS RN: 12057-17-9
Properties: monocl, a=0.4921 nm, b=0.8526 nm, c=0.9606 nm; prepared by reacting LiOH and γ-MnO_2 in air at 400°C for several days or at 700°C for 24 h; used in battery research [ROS91] [RIO92]
Density, g/cm³: 3.90 [ROS91] [RIO92]

1807

Compound: Lithium manganite
Formula: Li_2MnO_3
Molecular Formula: Li_2MnO_3
Molecular Weight: 116.818
CAS RN: 12163-00-7
Properties: reddish brown powd; very highly stable; used as a smelter addition in the manufacture of frit and as a cathode material for lithium batteries [HAW93] [FMC93]
Solubility: i H_2O [HAW93]

1808

Compound: Lithium metaaluminate
Formula: $LiAlO_2$
Molecular Formula: $AlLiO_2$
Molecular Weight: 65.922
CAS RN: 12003-67-7
Properties: white powd [STR93]
Density, g/cm³: 2.55 [STR93]
Melting Point, °C: >1625 [STR93]

1809

Compound: Lithium metaborate
Formula: $LiBO_2$
Molecular Formula: $BLiO_2$
Molecular Weight: 49.751
CAS RN: 13453-69-5
Properties: white, monocl cryst; hygr [CRC10]
Solubility: g/100 g H_2O: 2.6^{20}; sl H_2O; sol EtOH [CRC10]
Density, g/cm³: 2.18 [CRC10]
Melting Point, °C: 844 [CRC10]

1810

Compound: Lithium metaborate dihydrate
Formula: $LiBO_2$·$2H_2O$
Molecular Formula: BH_4LiO_4
Molecular Weight: 85.782
CAS RN: 15293-74-0
Properties: white, cryst powd; used in ceramics as a flux, in welding and brazing [HAW93]
Solubility: s H_2O [HAW93]
Density, g/cm³: 1.8 [STR93]

1811

Compound: Lithium metaphosphate
Formula: $LiPO_3$

Molecular Formula: LiO$_3$P
Molecular Weight: 85.913
CAS RN: 13762-75-9
Properties: white cryst or glassy transparent particles; used as a consituent in low-expansion procelain enamel and in selected laser glasses [FMC93] [AES93]
Solubility: g/100 g H$_2$O: 0.101 (0°C), 0.058 (25°C), 0.048 (40°C) [LAN05]
Melting Point, °C: 656 [AES93]

1812

Compound: Lithium metasilicate
Formula: Li$_2$SiO$_3$
Molecular Formula: Li$_2$O$_3$Si
Molecular Weight: 89.966
CAS RN: 10102-24-6
Properties: white powd; ortho-rhomb needles; enthalpy of fusion 28.00 kJ/mol; obtained by fusing lithium carbonate and SiO$_2$; used as a flux in glazes and ceramic enamels [HAW93] [MER06] [FMC93]
Solubility: i cold H$_2$O, decomposes in boiling H$_2$O [MER06]
Density, g/cm^3: 2.52 [MER06]
Melting Point, °C: 1201 [MER06]

1813

Compound: Lithium molybdate
Synonym: lithium molybdate(VI)
Formula: Li$_2$MoO$_4$
Molecular Formula: Li$_2$MoO$_4$
Molecular Weight: 173.820
CAS RN: 13568-40-6
Properties: white cryst; phenacite structure, c/a = 1.153; used in steel coating and in petroleum cracking catalysts [HAW93] [KIR81]
Solubility: g/100 g soln, H$_2$O: 45.24 (0°C), 44.81 (25°C), 42.50 (98°C); solid phase, 4Li$_2$MoO$_4$ · 3H$_2$O [KRU93]
Density, g/cm^3: 2.66 [STR93]
Melting Point, °C: 705 [STR93]

1814

Compound: Lithium niobate
Synonym: lithium niobate(V)
Formula: LiNbO$_3$
Molecular Formula: LiNbO$_3$
Molecular Weight: 147.845
CAS RN: 12031-63-9

Properties: white powd, also single cryst; can be prepared by hydrolysis of equimolar amounts of lithium ethoxide and niobium ethoxide in absolute alcohol by refluxing at 78.5°C for 24 h, then crystallizing the resulting precipitate by heating 2 h in a stream of oxygen gas at 250°C–350°C; ferroelectric; used in infrared detectors, in transducers for lasers, and as a sputtering target of 99.9% purity for piezoelectric applications [HAW93] [STR93] [HIR87] [CER91]
Melting Point, °C: 1240 [LID94]

1815

Compound: Lithium nitrate
Formula: LiNO$_3$
Molecular Formula: LiNO$_3$
Molecular Weight: 68.946
CAS RN: 7790-69-4
Properties: white, cryst powd; very hygr; enthalpy of fusion 24.90 kJ/mol; can be prepared by reaction of HNO$_3$ with LiOH or Li$_2$CO$_3$, followed by evaporation to dryness and then heating at ~200°C in vacuum; used in ceramics, pyrotechnics, molten salt baths, rocket propellants, refrigerators [HAW93] [CRC10] [MER06] [KIR81] [FMC93]
Solubility: 43 g/100 g soln, H$_2$O (20°C); s alcohol [KIR81] [MER06]; g/100 g soln, H$_2$O: 34.6 (0°C), 45.8 (25°C), 68.0 (90°C); solid phase, LiNO$_3$ · 3H$_2$O (0°C, 25°C), LiNO$_3$ (90°C) [KRU93]
Density, g/cm^3: 2.38 [MER06]
Melting Point, °C: 253 [CRC10]

1816

Compound: Lithium nitride
Formula: Li$_3$N
Molecular Formula: Li$_3$N
Molecular Weight: 34.830
CAS RN: 26134-62-3
Properties: reddish brown cryst or freely flowing powd; slowly decomposed by atm moisture; ruby red; hex, a = 0.3658 nm, c = 0.3882 nm; conductivity, 227°C, 0.04 (ohm · cm)$^{-1}$; one of most effective solid ionic conductors; can be prepared by direct reaction of Li and nitrogen; used as a nitriding agent in metallurgy [HAW93] [STR93] [CIC73] [KIR81]
Solubility: reacts with H$_2$O, yielding LiOH and ammonia; i polyethers [HAW93]
Density, g/cm^3: 1.27 [LID94]
Melting Point, °C: 813 [KIR81]; 845 [HAW93]

1817

Compound: Lithium nitrite
Formula: LiNO$_2$

Molecular Formula: $LiNO_2$
Molecular Weight: 52.947
CAS RN: 13568-33-7
Properties: white, hygr cryst [CRC10]
Solubility: v s H_2O [CRC10]
Melting Point, °C: 222 [CRC10]

1818
Compound: Lithium nitrite monohydrate
Formula: $LiNO_2 \cdot H_2O$
Molecular Formula: H_2LiNO_3
Molecular Weight: 70.962
CAS RN: 13568-33-7
Properties: colorless needles [CRC10]
Solubility: g/100 g soln, H_2O: 41.5 (0°C), 50.9 (25°C), 76.4 (99°C); solid phase, $LiNO_2 \cdot H_2O$ (0°C, 25°C), $LiNO_2$ (99°C) [KRU93]
Density, g/cm³: 1.615 [CRC10]
Melting Point, °C: >100 [CRC10]
Boiling Point, °C: decomposes [CRC10]

1819
Compound: Lithium orthosilicate
Formula: Li_4SiO_4
Molecular Formula: Li_4O_4Si
Molecular Weight: 119.848
CAS RN: 13453-84-4
Properties: rhomb; white powd; −100 mesh with 99.9% purity; used as a flux in ceramic formulations [FMC93] [CER91] [CRC10]
Density, g/cm³: 2.39 [CRC10]
Melting Point, °C: 1256 [CRC10]

1820
Compound: Lithium oxalate
Formula: $Li_2C_2O_4$
Molecular Formula: $C_2Li_2O_4$
Molecular Weight: 101.902
CAS RN: 30903-87-8
Properties: white, cryst powd; used as an anticoagulant in blood analysis [FMC93] [STR93]
Solubility: s in 15 part H_2O [MER06]; g/100 g soln, H_2O: 5.87 (25°C) [KRU93]
Density, g/cm³: 2.12 [MER06]
Melting Point, °C: decomposes [STR93]

1821
Compound: Lithium oxide
Synonym: lithia
Formula: Li_2O
Molecular Formula: Li_2O

Molecular Weight: 29.881
CAS RN: 12057-24-8
Properties: finely divided white powd or crusty material; readily absorbs CO_2 and H_2O from the atm; made by heating LiOH to ~800°C in a vacuum or by thermal decomposition of lithium peroxide; used in ceramics and special glass formulations and in lithium thermal batteries [HAW93] [MER06] [KIR81] [FMC93]
Density, g/cm³: 2.013 [MER06]
Melting Point, °C: 1570 [MER06]
Reactions: attacks glass, silica, many metals at elevated temperatures [MER06]

1822
Compound: Lithium perchlorate
Formula: $LiClO_4$
Molecular Formula: $ClLiO_4$
Molecular Weight: 106.392
CAS RN: 7791-03-9
Properties: white powd or ortho-rhomb cryst; hygr; oxidizing agent; enthalpy of fusion 29.00 kJ/mol; prepared from a saturated solution, which forms the trihydrate, followed by drying at 300°C; used in solid rocket propellants, as an electrolyte constituent for lithium batteries, and as a catalyst and oxidizing agent [HAW93] [STR93] [KIR79] [FMC93]
Solubility: v s alcohol, acetone, ether, ethyl acetate [MER06]; g/100 g soln, H_2O: 29.90 (0°C), 37.48 (25°C), 71.4 (100°C); solid phase, $LiClO_4 \cdot 3H_2O$ (0°C, 25°C), $LiClO_4 \cdot H_2O$ (100°C) [KRU93]
Density, g/cm³: 2.428 [STR93]
Melting Point, °C: 236 [CRC10]
Boiling Point, °C: 430 [STR93]
Reactions: decomposes rapidly at 450°C to LiCl and O_2 [KIR81]

1823
Compound: Lithium perchlorate trihydrate
Formula: $LiClO_4 \cdot 3H_2O$
Molecular Formula: ClH_6LiO_7
Molecular Weight: 160.438
CAS RN: 13453-78-6
Properties: white powd; hygr; can be prepared by crystallization from a saturated solution of lithium perchlorate [KIR81] [STR93]
Solubility: 37.5 g/100 g saturated solution in water (25°C) [KIR81]
Density, g/cm³: 1.841 [STR93]
Melting Point, °C: 95 [KER79]
Boiling Point, °C: decomposes at 470 [KIR79]
Reactions: minus $3H_2O$ at 300°C [KIR81]

1824
Compound: Lithium peroxide
Formula: Li_2O_2
Molecular Formula: Li_2O_2
Molecular Weight: 45.881
CAS RN: 12031-80-0
Properties: light yellow to tan powd; can be made from H_2O_2 and LiOH solution in boiling ethanol; used in fuel cells and as an oxidizing agent [HAW93] [FMC93]
Solubility: solubility in H_2O is 8% (20°C); solubility in acetic acid is 5.6% (20°C); i absolute alcohol (20°C) [HAW93]
Density, g/cm³: 2.14 (20°C) [HAW93]

1825
Compound: Lithium phosphate
Synonyms: lithium phosphate, tribasic
Formula: Li_3PO_4
Molecular Formula: Li_3O_4P
Molecular Weight: 115.794
CAS RN: 10377-52-3
Properties: white powd [STR93]
Solubility: g/100 g soln, H_2O: 0.022 (0°C); 0.0297 g/100 mL soln, H_2O (25°C); solid phase, Li_3PO_4 [KRU93]
Density, g/cm³: 2.46 [LID94]
Melting Point, °C: 1205 [STR93]

1826
Compound: Lithium selenate monohydrate
Formula: $Li_2SeO_4 \cdot H_2O$
Molecular Formula: $H_2Li_2O_5Se$
Molecular Weight: 174.855
CAS RN: 15593-52-9
Properties: monocl [MER06]
Solubility: v s H_2O [MER06]
Density, g/cm³: 2.565 [MER06]

1827
Compound: Lithium selenite monohydrate
Formula: $Li_2SeO_3 \cdot H_2O$
Molecular Formula: $H_2Li_2O_4Se$
Molecular Weight: 158.856
CAS RN: 15593-51-8
Properties: cryst; hygr [MER06]
Solubility: g/100 g H_2O: 25.0 (0°C), 21.5 (20°C), 9.9 (100°C) [LAN05]

1828
Compound: Lithium stearate
Synonyms: stearic acid, lithium salt
Formula: $CH_3(CH_2)_{16}COOLi$
Molecular Formula: $C_{18}H_{35}LiO_2$
Molecular Weight: 290.416
CAS RN: 4485-12-5
Properties: white cryst; forms gels in mineral oils; used in cosmetics, plastics, waxes, greases [HAW93]
Solubility: i H_2O, alcohol, ethyl acetate [HAW93]
Density, g/cm³: 1.025 [HAW93]
Melting Point, °C: 220 [HAW93]

1829
Compound: Lithium sulfate
Formula: Li_2SO_4
Molecular Formula: Li_2O_4S
Molecular Weight: 109.946
CAS RN: 10377-48-7
Properties: white, hygr cryst; enthalpy of fusion 7.50 kJ/mol; can be obtained from sulfuric acid and LiOH or Li_2CO_3 solutions to form the monohydrate, followed by heating to obtain the anhydrous salt; used in ceramic compositions, as a solubilizer in photographic developers, and as an additive to specialty Portland cements [KIR81] [FMC93]
Solubility: g/100 g soln, H_2O: 25.7 (25°C), 23.6 (100°C) [KIR81]; g/100 g soln, H_2O: 25.9 ± 0.5 (0°C), 25.7 ± 0.1 (25°C), 23.5 (100.1°C); solid phase, $Li_2SO_4 \cdot H_2O$ [KRU93]
Melting Point, °C: 860 [KIR81]

1830
Compound: Lithium sulfate monohydrate
Formula: $Li_2SO_4 \cdot H_2O$
Molecular Formula: $H_2Li_2O_5S$
Molecular Weight: 127.961
CAS RN: 10102-25-7
Properties: colorless cryst; does not form alums; obtained by reacting solution of H_2SO_4 with LiOH or Li_2CO_3; used in ceramics and in pharmaceutical products [HAW93] [KIR81] [STR93]
Solubility: soluble in 2.6 parts H_2O; sl s alcohol [MER06]
Density, g/cm³: 2.06 [MER06]
Melting Point, °C: 130 [HAW93]
Reactions: minus H_2O at >100°C [KIR81]

1831
Compound: Lithium sulfide
Formula: Li_2S

Molecular Formula: Li_2S
Molecular Weight: 45.948
CAS RN: 12136-58-2
Properties: off-white powd; sensitive
to moisture [STR93]
Density, g/cm³: 1.64 [LID94]
Melting Point, °C: 1372 [LID94]

1832
Compound: Lithium tantalate
Formula: $LiTaO_3$
Molecular Formula: LiO_3Ta
Molecular Weight: 235.887
CAS RN: 12031-66-2
Properties: white powd [STR93]

1833
Compound: Lithium tellurite
Formula: Li_2TeO_3
Molecular Formula: Li_2O_3Te
Molecular Weight: 189.480
CAS RN: 14929-69-2
Properties: −100 mesh with 99.5% purity [CER91]

1834
Compound: Lithium tetraborate
Formula: $Li_2B_4O_7$
Molecular Formula: $B_4Li_2O_7$
Molecular Weight: 169.122
CAS RN: 12007-60-2
Properties: white powd; −100 mesh with 99.9% purity;
used as a flux in x-ray fluorescence analysis, in
grease formulations, and as an electrolyte component
in lithium batteries [FMC93] [CER91] [STR93]
Solubility: 2.89 g/100 mL H_2O (20°C),
5.45 g/100 mL H_2O (100°C) [CRC10]
Melting Point, °C: 917 [STR93]

1835
Compound: Lithium tetraborate pentahydrate
Formula: $Li_2B_4O_7 \cdot 5H_2O$
Molecular Formula: $B_4H_{10}Li_2O_{12}$
Molecular Weight: 211.200
CAS RN: 1303-94-2
Properties: white, cryst powd; used in
ceramics, in vacuum spectroscopy, in metal
refining and degassing [HAW93]
Solubility: v s H_2O; i alcohol [HAW93]
Reactions: minus $5H_2O$ at 200°C [HAW93]

1836
Compound: Lithium thiocyanate
Formula: LiSCN
Molecular Formula: CLiNS
Molecular Weight: 65.024
CAS RN: 556-65-0
Properties: white, hygr cryst [CRC10]
Solubility: g/100 g H_2O: 120^{25} [CRC10]

1837
Compound: Lithium tetrachlorocuprate
Formula: Li_2CuCl_4
Molecular Formula: Cl_4CuLi_2
Molecular Weight: 219.239
CAS RN: 15489-27-7
Properties: 0.1 M in THF; freezing point −17°C;
orange liq; sensitive to moisture [STR93]

1838
Compound: Lithium tetracyanoplatinate(II) pentahydrate
Formula: $Li_2Pt(CN)_4 \cdot 5H_2O$
Molecular Formula: $C_4H_{10}Li_2N_4O_5Pt$
Molecular Weight: 403.109
CAS RN: 14402-73-4
Properties: greenish yellow cryst [MER06]
Solubility: sl s H_2O [MER06]

1839
Compound: Lithium tetrafluoroborate
Formula: $LiBF_4$
Molecular Formula: BF_4Li
Molecular Weight: 93.746
CAS RN: 14283-07-9
Properties: white, hygr powd; hygr; can
be prepared by reacting LiOH with
fluoroboric acid [KIR78] [STR93]
Solubility: v s H_2O [KIR78]
Melting Point, °C: decomposes [STR93]

1840
Compound: Lithium thiocyanate hydrate
Formula: $LiSCN \cdot xH_2O$
Molecular Formula: CLiNS (anhydrous)
Molecular Weight: 65.025 (anhydrous)
CAS RN: 123333-85-7
Properties: white, deliq cryst [CRC10]
Solubility: g/100 g soln, H_2O: 54.5 (25°C);
solid phase, $LiSCN \cdot 2H_2O$ [KRU93]

1841
Compound: Lithium titanate
Formula: Li_2TiO_3
Molecular Formula: Li_2O_3Ti
Molecular Weight: 109.747
CAS RN: 12031-82-2
Properties: white powd; has strong fluxing
    properties at low concentrations for use in
    titanium-bearing enamels [HAW93]
Solubility: i H_2O [HAW93]
Melting Point, °C: 1520–1564 [STR93]

1842
Compound: Lithium tungstate
Formula: Li_2WO_4
Molecular Formula: Li_2O_4W
Molecular Weight: 261.720
CAS RN: 13568-45-1
Properties: trig; white powd [CRC10] [STR93]
Solubility: s H_2O [HAW93]
Density, g/cm³: 3.71 [STR93]
Melting Point, °C: 742 [STR93]

1843
Compound: Lithium vanadate
Formula: $LiVO_3$
Molecular Formula: LiO_3V
Molecular Weight: 105.881
CAS RN: 15060-59-0
Properties: −100 mesh with 99.9% purity; hydrate
    is yellowish powd [HAW93] [CER91]

1844
Compound: Lithium zirconate
Formula: Li_2ZrO_3
Molecular Formula: Li_2O_3Zr
Molecular Weight: 153.104
CAS RN: 12031-83-3
Properties: white powd; used as a flux in glasses, which
    contain zirconium dioxide [HAW93] [FMC93]

1845
Compound: Lutetium
Synonym: cassiopeium
Formula: Lu
Molecular Formula: Lu
Molecular Weight: 174.967
CAS RN: 7439-94-3

Properties: silvery white metal; hex; soft and ductile;
    electrical resistivity (20°C) 54 µohm · cm;
    enthalpy of fusion 78.03 kJ/mol; enthalpy
    of sublimation 427.6 kJ/mol; atom radius is
    0.17349 nm; ion radius is 0.0850 nm for Lu^{+++},
    colorless solutions [HAW93] [KIR82] [ALD94]
Solubility: reacts slowly with H_2O;
    s in dil acids [HAW93]
Density, g/cm³: 9.840 [KIR82]
Melting Point, °C: 1663 [KIR82]
Boiling Point, °C: 3402 [KIR82]
Thermal Conductivity, W/(m · K): 16.4 at 25°C [CRC10]
Thermal Expansion Coefficient: 9.9×10^{-6}/K [CRC10]

1846
Compound: Lutetium acetate hydrate
Formula: $Lu(CH_3COO)_3 \cdot xH_2O$
Molecular Formula: $C_6H_9LuO_6$ (anhydrous)
Molecular Weight: 352.101 (anhydrous)
CAS RN: 18779-08-3
Properties: hygr white cryst [STR93] [ALD94]

1847
Compound: Lutetium boride
Formula: LuB_4
Molecular Formula: B_4Lu
Molecular Weight: 218.211
CAS RN: 12688-52-7
Properties: tetr; −100 mesh of 99.9%
    purity [LID94] [CER91]
Density, g/cm³: ~7.0 [LID94]
Melting Point, °C: 2600 [LID94]

1848
Compound: Lutetium bromide
Formula: $LuBr_3$
Molecular Formula: Br_3Lu
Molecular Weight: 414.679
CAS RN: 14456-53-2
Properties: white, hygr cryst; −20 mesh with
    99.9% purity [CER91] [LID94]
Solubility: s H_2O [CRC10]
Density, g/cm³: 1.025 [ALF95]
Melting Point, °C: 1400 [CRC10]

1849
Compound: Lutetium chloride
Formula: $LuCl_3$
Molecular Formula: Cl_3Lu
Molecular Weight: 281.325
CAS RN: 10099-66-8

Properties: colorless cryst [MER06]
Solubility: s H$_2$O [MER06]
Density, g/cm^3: 3.98 [STR93]
Melting Point, °C: 905 [HAW93]
Reactions: sublimes at >750°C [MER06]

1850
Compound: Lutetium chloride hexahydrate
Formula: LuCl$_3$·6H$_2$O
Molecular Formula: Cl$_3$H$_{12}$LuO$_6$
Molecular Weight: 389.416
CAS RN: 15230-79-2
Properties: −4 mesh with 99.9% purity;
    white cryst [CER91] [STR93]
Melting Point, °C: 892 [ALF95]

1851
Compound: Lutetium fluoride
Formula: LuF$_3$
Molecular Formula: F$_3$Lu
Molecular Weight: 231.962
CAS RN: 13760-81-1
Properties: ortho; 99.9% pure melted pieces of 3–6 mm;
    used as an evaporation material for possible
    application to multilayers [LID94] [CER91]
Solubility: i H$_2$O [HAW93]
Density, g/cm^3: 8.3 [LID94]
Melting Point, °C: 1182 [HAW93]
Boiling Point, °C: 2200 [HAW93]

1852
Compound: Lutetium hydride
Formula: LuH$_{2-3}$
Molecular Formula: H$_2$Lu; H$_3$Lu
Molecular Weight: H$_2$Lu: 176.983; H$_3$Lu: 177.991
CAS RN: 13598-44-2
Properties: −60 mesh with 99.9% purity;
    lumps, under argon [ALF95] [CER91]

1853
Compound: Lutetium iodide
Formula: LuI$_3$
Molecular Formula: I$_3$Lu
Molecular Weight: 555.680
CAS RN: 13813-45-1
Properties: powd, under argon; −20 mesh
    with 99.9% purity [CER91] [ALF95]
Density, g/cm^3: ~5.6 [LID94]
Melting Point, °C: 1050 [AES93]

1854
Compound: Lutetium iron oxide
Synonym: lutetium garnet
Formula: Lu$_3$Fe$_5$O$_{12}$
Molecular Formula: Fe$_5$Lu$_3$O$_{12}$
Molecular Weight: 996.119
CAS RN: 12023-71-1
Properties: used in 99.9% purity as a sputtering target in
    the preparation of bubble memory devices [CER91]

1855
Compound: Lutetium nitrate
Formula: Lu(NO$_3$)$_3$
Molecular Formula: LuN$_3$O$_9$
Molecular Weight: 360.982
CAS RN: 10099-67-9
Properties: hygr col solid [CRC10]
Solubility: s H$_2$O, EtOH [CRC10]

1856
Compound: Lutetium nitrate hydrate
Formula: Lu(NO$_3$)$_3$·xH$_2$O
Molecular Formula: LuN$_3$O$_9$ (anhydrous)
Molecular Weight: 360.982 (anhydrous)
CAS RN: 10099-67-9
Properties: white cryst [STR93]

1857
Compound: Lutetium nitride
Formula: LuN
Molecular Formula: LuN
Molecular Weight: 188.974
CAS RN: 12125-25-6
Properties: −60 mesh with 99.9% purity [CER91]
Density, g/cm^3: 11.6 [LID94]

1858
Compound: Lutetium oxalate hexahydrate
Formula: Lu$_2$(C$_2$O$_4$)$_3$·6H$_2$O
Molecular Formula: C$_6$H$_{12}$Lu$_2$O$_{18}$
Molecular Weight: 722.084
CAS RN: 26677-69-0
Properties: white cryst [ALF95]

1859
Compound: Lutetium oxide
Formula: Lu$_2$O$_3$
Molecular Formula: Lu$_2$O$_3$
Molecular Weight: 397.932

CAS RN: 12032-20-1
Properties: white powd; cub cryst; absorbs
H_2O and CO_2; as an evaporated material of
99.9% purity is reactive to radio frequencies
[HAW93] [CER91] [STR93] [MER06]
Density, g/cm³: 9.41 [STR93]
Melting Point, °C: 2487 [STR93]

1860
Compound: Lutetium perchlorate hexahydrate
Formula: $Lu(ClO_4)_3 \cdot 6H_2O$
Molecular Formula: $Cl_3H_{12}LuO_{18}$
Molecular Weight: 581.409
CAS RN: 14646-29-8
Properties: white cryst; hygr [STR93]

1861
Compound: Lutetium silicide
Formula: $LuSi_2$
Molecular Formula: $LuSi_2$
Molecular Weight: 231.138
CAS RN: 12032-13-2
Properties: 6 mm pieces and smaller
with 99.9% purity [CER91]

1862
Compound: Lutetium sulfate
Formula: $Lu_2(SO_4)_3$
Molecular Formula: $Lu_2O_{12}S_3$
Molecular Weight: 638.125
CAS RN: 14986-89-1
Properties: white powd [STR93]
Solubility: 0.6260 ± 0.0017 mol/kg
in H_2O (25°C) [RAR88]

1863
Compound: Lutetium sulfate octahydrate
Formula: $Lu_2(SO_4)_3 \cdot 8H_2O$
Molecular Formula: $H_{16}Lu_2O_{20}S_3$
Molecular Weight: 782.247
CAS RN: 13473-77-3
Properties: white cryst [STR93]
Solubility: g/100 g H_2O: 42.27 (20°C),
16.93 (40°C) [MER06]

1864
Compound: Lutetium sulfide
Formula: Lu_2S_3
Molecular Formula: Lu_2S_3
Molecular Weight: 446.132
CAS RN: 12163-20-1

Properties: gray rhomb cryst; −200 mesh
with 99.9% purity [CER91] [LID94]
Density, g/cm³: 6.26 [LID94]
Melting Point, °C: decomposes at 1750 [LID94]

1865
Compound: Lutetium telluride
Formula: Lu_2Te_3
Molecular Formula: Lu_2Te_3
Molecular Weight: 732.734
CAS RN: 12163-22-3
Properties: ortho cryst; −20 mesh with
99.9% purity [LID94] [CER91]
Density, g/cm³: 7.8 [LID94]

1866
Compound: Magnesium
Formula: Mg
Molecular Formula: Mg
Molecular Weight: 24.3050
CAS RN: 7439-95-4
Properties: silver-white metal; hex, a = 0.3203 nm,
c = 0.5199 nm; slowly oxidizes in moist air; electrical
resistivity (20°C) 4.46 µohm cm; enthalpy of fusion
8.48 kJ/mol; enthalpy of sublimation 150 kJ/mol;
enthalpy of combustion 606 kJ/mol; thermal
diffusivity (20°C) 0.87 cm²/s; Poisson's ratio
0.35; electronegativity 1.56; many uses including
ferromagnetic films; preparation by diffusion with
bismuth [DOU83] [MER06] [KIR81] [CER91]
Solubility: reacts slowly with H_2O; evolves
H_2 with dil acids [MER06]
Density, g/cm³: 1.738 (20°C) [CIC73]
Melting Point, °C: 649 [CIC73]
Boiling Point, °C: 1105 [CIC73]
Thermal Conductivity, W/(m · K): 155 at 20°C [KIR81]
Thermal Expansion Coefficient: 24.8×10^{-6}/K [CRC10]

1867
Compound: Magnesium acetate
Synonym: cromosan
Formula: $Mg(CH_3COO)_2$
Molecular Formula: $C_4H_6MgO_4$
Molecular Weight: 142.395
CAS RN: 142-72-3
Properties: white; two forms: α ortho-rhomb,
a = 1.127 nm, b = 1.501 nm, c = 1.100 nm,
obtained by reacting MgO with 13%–33%
acetic acid in boiling ethyl acetate, and β tricl,
a = 1.034 nm, b = 1.29 nm, c = 7.726 nm, obtained
from 5%–6% acetic acid; odor of acetic acid;
used as a dye fixative in textile printing, as a
deodorant, and disinfectant [KIR81] [HAW93]

Solubility: g/100 g soln, H_2O: 36.2 (0.1°C), 39.6 (24.9°C); solid phase, $Mg(CH_3COO)_2 \cdot 4H_2O$ [KRU93]; s dil alcohol [HAW93]
Density, g/cm³: α: 1.507, β: 1.502 [KIR81]
Melting Point, °C: decomposes at 323 [KIR81]

1868
Compound: Magnesium acetate monohydrate
Formula: $Mg(CH_3COO)_2 \cdot H_2O$
Molecular Formula: $C_4H_8MgO_5$
Molecular Weight: 160.410
CAS RN: 60582-92-5
Properties: ortho-rhomb, a = 1.175 nm, b = 1.753 nm, c = 0.6662 nm; can be obtained by reacting MgO and acetic acid in isobutyl alcohol, which contains some H_2O [KIR81]
Density, g/cm³: 1.553 [KIR81]

1869
Compound: Magnesium acetate tetrahydrate
Synonyms: acetic acid, magnesium salt
Formula: $Mg(CH_3COO)_2 \cdot 4H_2O$
Molecular Formula: $C_4H_{14}MgO_8$
Molecular Weight: 214.455
CAS RN: 16674-78-5
Properties: colorless or white; monocl, a = 0.8550 nm, b = 1.1995 nm, c = 0.4807 nm; deliq cryst; crystallizes from aq solution as only stable phase below 68°C; there is a β phase [KIR81] [MER06]
Solubility: v s H_2O, alcohol [MER06]
Density, g/cm³: 1.45 [LID94]
Melting Point, °C: decomposes at 80 [LID94]

1870
Compound: Magnesium acetylacetonate dihydrate
Synonyms: 2,4-pentanedione, magnesium derivative dihydrate
Formula: $Mg(CH_3COCH=C(O)CH_3)_2 \cdot 2H_2O$
Molecular Formula: $C_{10}H_{18}MgO_6$
Molecular Weight: 258.554
CAS RN: 68488-07-3
Properties: white powd [STR93]
Melting Point, °C: decomposes at 265 [ALD94]

1871
Compound: Magnesium aluminum oxide
Synonym: spinel
Formula: $Mg(AlO_2)_2$
Molecular Formula: Al_2MgO_4
Molecular Weight: 142.266
CAS RN: 12068-51-8

Properties: 3–12 mm pieces (fused); used as an evaporated ceramic of 99.9% purity to form a high temp dielectric [CER91] [MIT72] [YAM87]
Density, g/cm³: 3.58 [KIR80]
Melting Point, °C: 2135 [KIR80]
Thermal Conductivity, W/(m·K): 9.1 (500°C), 5.8 (1000°C) [KIR80]
Thermal Expansion Coefficient: (volume) 100°C (0.18), 200°C (0.39), 400°C (0.90), 800°C (2.01), 1200°C (3.24) [CLA66]

1872
Compound: Magnesium aluminum silicate
Synonym: cordierite
Formula: $Mg_2Al_3(AlSi_5O_{18})$
Molecular Formula: $Al_4Mg_2O_{18}Si_5$
Molecular Weight: 584.953
CAS RN: 61027-88-1
Properties: dielectric constant (26°C–300°C) 4.00–4.42, hydrothermally prepared material [MOY86]; sol-gel synthesis in [MAE90] and [KAZ90]
Density, g/cm³: when sintered at 1200°C–1400°C: 2.63–2.35 [KAZ90]
Thermal Expansion Coefficient: 50°C–650°C: 0.3–2.5×10^{-6}/°C [MOY86]

1873
Compound: Magnesium aluminum zirconate
Synonym: zirconium spinel
Formula: $MgO \cdot Al_2O_3 \cdot ZrO_2$
Molecular Formula: Al_2MgO_6Zr
Molecular Weight: 265.488
CAS RN: 53169-11-2
Properties: 3–12 sintered pieces powd; used for vacuum deposition [ALF95] [CER91]

1874
Compound: Magnesium amide
Formula: $Mg(NH_2)_2$
Molecular Formula: H_4MgN_2
Molecular Weight: 56.350
CAS RN: 7803-54-5
Properties: white powd or cryst; flammable in air; used as a catalyst for polymerization [HAW93] [MER06]
Solubility: reacts violently with H_2O releasing NH_3 [MER06]
Density, g/cm³: 1.39 [MER06]
Melting Point, °C: decomposes when heated [HAW93]

1875
Compound: Magnesium ammonium phosphate hexahydrate
Synonym: guanite

Formula: $MgNH_4PO_4 \cdot 6H_2O$
Molecular Formula: $H_{16}MgNO_{10}P$
Molecular Weight: 245.407
CAS RN: 13478-16-5
Properties: white powd; used as a fire retardant
 for fabrics and in fertilizer [HAW93]
Solubility: i H_2O, alcohol; s in acids [HAW93]
Density, g/cm³: 1.71 [HAW93]
Melting Point, °C: decomposes to $Mg_2P_2O_7$ [HAW93]

1876
Compound: Magnesium antimonide
Formula: Mg_3Sb_2
Molecular Formula: Mg_3Sb_2
Molecular Weight: 316.435
CAS RN: 12057-75-9
Properties: hex; 6 mm pieces and smaller
 with 99.5% purity [CER91] [CRC10]
Density, g/cm³: 4.088 [CRC10]
Melting Point, °C: 1245 [LID94]

1877
Compound: Magnesium arsenate hydrate
Formula: $Mg_3(AsO_4)_2 \cdot xH_2O$
Molecular Formula: $As_2Mg_3O_8$ (anhydrous)
Molecular Weight: 350.753 (anhydrous)
CAS RN: 10103-50-1
Properties: white powd; used as an insecticide [HAW93]
Solubility: i H_2O [HAW93]

1878
Compound: Magnesium arsenide
Formula: Mg_3As_2
Molecular Formula: As_2Mg_3
Molecular Weight: 222.758
CAS RN: 12044-49-4
Properties: 6 mm pieces and smaller
 of 99.5% purity [CER91]
Density, g/cm³: 3.148 [ALF95]
Melting Point, °C: 800 [ALF95]

1879
Compound: Magnesium basic carbonate pentahydrate
Formula: $4MgCO_3 \cdot Mg(OH)_2 \cdot 5H_2O$
Molecular Formula: $C_4H_{12}Mg_5O_{19}$
Molecular Weight: 485.653
CAS RN: 56378-72-4
Properties: white, colorless, bulky powd [MER06]
Solubility: s in ~3300 parts H_2O; more
 soluble if H_2O contains dissolved CO_2;
 s dil acids; i alcohol [MER06]
Reactions: converts to MgO at ~700°C [MER06]

1880
Compound: Magnesium
 bis(pentamethylcyclopentadienyl)
Synonym: bis(pentamethylcyclopentadienyl)magnesium
Formula: $[(CH_3)_5C_5]_2Mg$
Molecular Formula: $C_{20}H_{30}Mg$
Molecular Weight: 294.763
CAS RN: 74507-64-5
Properties: cryst [ALF95]

1881
Compound: Magnesium borate octahydrate
Formula: $Mg(BO_2)_2 \cdot 8H_2O$
Molecular Formula: $B_2H_{16}MgO_{12}$
Molecular Weight: 254.047
CAS RN: 13703-82-7
Properties: white powd [MER06]
Solubility: sl s H_2O [MER06]
Density, g/cm³: 2.30 [CRC10]

1882
Compound: Magnesium boride
Formula: MgB_2
Molecular Formula: B_2Mg
Molecular Weight: 45.927
CAS RN: 12007-25-9
Properties: hex cryst; −100 mesh with 99% purity;
 refractory material [CER91] [KIR78] [LID94]
Density, g/cm³: 2.57 [LID94]
Melting Point, °C: decomposes at 800 [KIR78]

1883
Compound: Magnesium boride
Formula: MgB_6
Molecular Formula: B_6Mg
Molecular Weight: 89.171
CAS RN: 12008-22-9
Properties: refractory material [KIR78]
Melting Point, °C: decomposes at 1100 [KIR78]

1884
Compound: Magnesium bromate hexahydrate
Formula: $Mg(BrO_3)_2 \cdot 6H_2O$
Molecular Formula: $Br_2H_{12}MgO_{12}$
Molecular Weight: 388.201
CAS RN: 7789-36-8
Properties: colorless or white cryst; used as an
 oxidizing agent [HAW93] [MER06]
Solubility: 42 g/100 mL H_2O (18°C) [CRC10]

Density, g/cm³: 2.29 [HAW93]
Reactions: minus 6H$_2$O at ~200°C; decomposes
at higher temp [MER06]

1885
Compound: Magnesium bromide
Formula: MgBr$_2$
Molecular Formula: Br$_2$Mg
Molecular Weight: 184.113
CAS RN: 7789-48-2
Properties: hex, a=0.3822 nm, c=0.6269 nm; off-
white powd; hygr; enthalpy of fusion 39.30 kJ/
mol; occurs in seawater, brines, the Dead Sea;
used in medicine as a sedative and in some dry cell
electrolytes for batteries [STR93] [KIR81] [CRC10]
Solubility: g/100 g H$_2$O: 100.6 (25°C), 125.4
(100°C); solid phase, MgBr$_2$·6H$_2$O [KRU93]
Density, g/cm³: 3.72 [STR93]
Melting Point, °C: 711 [CRC10]

1886
Compound: Magnesium bromide hexahydrate
Formula: MgBr$_2$·6H$_2$O
Molecular Formula: Br$_2$H$_{12}$MgO$_6$
Molecular Weight: 292.204
CAS RN: 13446-53-2
Properties: colorless monocl, a=1.0286 nm,
b=0.7331 nm, c=0.6211 nm; very deliq cryst or
white granules; bitter taste; used as a sedative and
in organic synthesis [HAW93] [KIR81] [MER06]
Solubility: s 0.3 parts H$_2$O; s alcohol [MER06]
Density, g/cm³: 2.00 [STR93]
Melting Point, °C: ~165 with decomposition [MER06]

1887
Compound: Magnesium carbonate
Synonym: magnesite
Formula: MgCO$_3$
Molecular Formula: CMgO$_3$
Molecular Weight: 84.314
CAS RN: 546-93-0
Properties: light, bulky white powd; trig, a=0.46332 nm,
c=1.5015 nm; magnesite mineral, 13717-00-5,
hardness is 3.5–4.5 Mohs; can be prepared in
aq systems under high CO$_2$ pressure; used in
heat insulation and inks [HAW93] [KIR81]
Solubility: g MgCO$_3$/100 g soln at CO$_2$ pressure, kPa,
18°C: 3.5 (203), 4.28 (405), 5.90 (1010), 7.49
(1820), 7.49 (5670); at 0°C 8.58 (3445), at 60°C
5.56 (3445); s acids; i alcohol [HAW93] [KIR81]
Density, g/cm³: 3.009 (calculated) [KIR81]

Melting Point, °C: decomposes at 350 [HAW93]
Reactions: minus CO$_2$ at 900°C [CRC10]

1888
Compound: Magnesium carbonate dihydrate
Synonym: barringtonite
Formula: MgCO$_3$·2H$_2$O
Molecular Formula: CH$_4$MgO$_5$
Molecular Weight: 120.345
CAS RN: 5145-48-2
Properties: colorless; tricl, a=0.9115 nm,
b=0.6202 nm, c=0.6092 nm [KIR81]
Density, g/cm³: 2.825 (calculated) [KIR81]

1889
Compound: Magnesium carbonate
hydroxide tetrahydrate
Synonym: hydromagnesite
Formula: 4MgCO$_3$·Mg(OH)$_2$·4H$_2$O
Molecular Formula: C$_4$H$_{10}$Mg$_5$O$_{18}$
Molecular Weight: 467.637
CAS RN: 39409-82-0
Properties: white; monocl, a=1.011 nm, b=0.315 nm,
c=0.622 nm; there is a pentahydrate (dypingite)
56378-72-6, and an octahydrate, 75300-49-1 [KIR81]
Density, g/cm³: 2.254 [KIR81]

1890
Compound: Magnesium carbonate hydroxide trihydrate
Synonym: artinite
Formula: MgCO$_3$·Mg(OH)$_2$·3H$_2$O
Molecular Formula: CH$_8$Mg$_2$O$_8$
Molecular Weight: 198.680
CAS RN: 12143-96-3
Properties: white; monocl, a=1.656 nm,
b=0.315 nm, c=0.622 nm [KIR81]
Density, g/cm³: 2.039 [KIR81]

1891
Compound: Magnesium carbonate pentahydrate
Synonym: lansfordite
Formula: MgCO$_3$·5H$_2$O
Molecular Formula: CH$_{10}$MgO$_8$
Molecular Weight: 174.390
CAS RN: 61042-72-6
Properties: white monocl [KIR81]
Solubility: 0.176 g/100 mL H$_2$O (7°C),
0.375 g/100 mL H$_2$O (20°C) [CRC10]
Density, g/cm³: 1.73 (calculated) [KIR81]
Melting Point, °C: decomposes [CRC10]

1892
Compound: Magnesium carbonate trihydrate
Synonym: nesquehonite
Formula: $MgCO_3 \cdot 3H_2O$
Molecular Formula: CH_6MgO_6
Molecular Weight: 138.360
CAS RN: 14457-83-1
Properties: colorless to white; monocl, a = 1.2112 nm,
b = 0.5365 nm, c = 0.7697 nm [KIR81]
Solubility: 0.179 g/100 mL H_2O (16°C) [CRC10]
Density, g/cm³: 1.837 (calculated) [KIR81]
Reactions: minus $3H_2O$ at 100°C [CRC10]

1893
Compound: Magnesium chlorate hexahydrate
Formula: $Mg(ClO_3)_2 \cdot 6H_2O$
Molecular Formula: $Cl_2H_{12}MgO_{12}$
Molecular Weight: 299.298
CAS RN: 10326-21-3
Properties: white; very deliq cryst or cryst powd;
bitter taste; used as a defoliant and dessicant;
oxidizing agent [HAW93] [MER06]
Solubility: mol/100 mol H_2O: 10.73 (0°C), 13.52 (25°C),
26.38 (93°C); solid phase $Mg(ClO_3)_2 \cdot 6H_2O$ (0°C,
25°C), $Mg(ClO_3)_2 \cdot 2H_2O$ (93°C) [KRU93]
Density, g/cm³: 1.80 [MER06]
Melting Point, °C: ~35 [MER06]

1894
Compound: Magnesium chloride
Synonym: magnogene
Formula: $MgCl_2$
Molecular Formula: Cl_2Mg
Molecular Weight: 95.210
CAS RN: 7786-30-3
Properties: white lustrous, soft highly deliq leaflets;
hex, a = 0.3632 nm, c = 1.7795 nm; can be distilled
in H_2; attacks fused silica when melted; evolves
heat when dissolved in H_2O; can be obtained by
dissolution of MgO, $MgCO_3$ or $Mg(OH)_2$ in HCl,
followed by cooling and dehydration; enthalpy
of fusion 43.10 kJ/mol; used in disinfectants, in
fire extinguishers, for fireproofing wood, and in
ceramics [HAW93] [CRC10] [MER06] [KIR81]
Solubility: g/100 g soln, H_2O: 34.6 (0°C), 35.5 (25°C),
42.3 (100°C); solid phase, $MgCl_2 \cdot 6H_2O$ [KRU93];
g/100 g alcohol: 3.61 (0°C), 15.89 (60°C) [KIR81]
Density, g/cm³: 2.325 [KIR81]
Melting Point, °C: 714 [CRC10]
Boiling Point, °C: 1412 [CIC73]
Reactions: slow heating releases Cl_2 at 300°C [MER06]

1895
Compound: Magnesium chloride hexahydrate
Synonym: bischofite
Formula: $MgCl_2 \cdot 6H_2O$
Molecular Formula: $Cl_2H_{12}MgO_6$
Molecular Weight: 203.301
CAS RN: 7791-18-6
Properties: colorless or white, highly deliq, monocl
cryst, a = 0.9871 nm, b = 0.7113 nm, c = 0.6097 nm;
only stable hydrate from 0°C–100°C;
obtained from a solution of MgO, $MgCO_3$,
or $Mg(OH)_2$ in HCl [KIR81] [HAW93]
Solubility: 5.8101 ± 0.0017 mol/(kg · H_2O) at 25°C
[RAR85b]; 1 g/2 mL alcohol [MER06]
Density, g/cm³: 1.56 [MER06]
Melting Point, °C: decomposes at ~118 [MER06]
Reactions: minus $2H_2O$ at 95°C–115°C, minus
$4H_2O$ at 135°C–180°C, minus $5H_2O$ at
>230°C and decomposes [KIR81]

1896
Compound: Magnesium chromate pentahydrate
Formula: $MgCrO_4 \cdot 5H_2O$
Molecular Formula: $CrH_{10}MgO_9$
Molecular Weight: 230.375
CAS RN: 16569-85-0
Properties: small yellow cryst; tricl; used as a
corrosion inhibitor in the water coolant of
gas turbine engines [KIR78] [HAW93]
Solubility: 35.39% H_2O (25°C) [KIR78]
Density, g/cm³: 1.954 [KIR78]
Reactions: transforms to $7H_2O$ (17.2°C) [KIR78]

1897
Compound: Magnesium chromite
Formula: $MgCr_2O_4$
Molecular Formula: Cr_2MgO_4
Molecular Weight: 192.295
CAS RN: 12053-26-8
Properties: brown cub spinel; used
as a refractory [KIR78]
Density, g/cm³: 4.415 [KIR78]

1898
Compound: Magnesium citrate
Formula: $Mg_3(C_6H_5O_7)_2$
Molecular Formula: $C_{12}H_{10}Mg_3O_{14}$
Molecular Weight: 451.114
CAS RN: 3344-18-1
Properties: white cryst [CRC10]
Solubility: sl H_2O [CRC10]

1899

Compound: Magnesium citrate pentahydrate
Synonym: magnesium dibasic citrate
Formula: $MgC_6H_6O_7 \cdot 5H_2O$
Molecular Formula: $C_6H_{16}MgO_{12}$
Molecular Weight: 304.491
CAS RN: 7779-25-1
Properties: white or sl yellow granules
or powd; odorless [MER06]
Solubility: 20 g/100 mL H_2O (20°C) [CRC10]

1900

Compound: Magnesium citrate tetradecahydrate
Synonyms: citric acid, magnesium salt tetradecahydrate
Formula: $Mg_3(C_6H_5O_7)_2 \cdot 14H_2O$
Molecular Formula: $C_{12}H_{38}Mg_3O_{28}$
Molecular Weight: 703.332
CAS RN: 144-23-0
Properties: white, odorless, cryst powd or
granules; used as a cathartic [MER06]
Solubility: sl s H_2O; s dil acids [MER06]

1901

Compound: Magnesium diboride
Formula: MgB_2
Molecular Formula: B_2Mg
Molecular Weight: 45.927
CAS RN: 12007-25-9
Properties: hex cryst [CRC10]
Density, g/cm³: 2.57 [CRC10]
Melting Point, °C: decomposes at 800 [CRC10]

1902

Compound: Magnesium dichromate hexahydrate
Formula: $MgCr_2O_7 \cdot 6H_2O$
Molecular Formula: $Cr_2H_{12}MgO_{13}$
Molecular Weight: 348.384
CAS RN: 16569-85-0
Properties: reddish orange ortho-rhomb; deliq [HAW93]
Solubility: 58.52% H_2O (30°C) [KIR78]
Density, g/cm³: 2.002 [KIR78]
Reactions: minus H_2O at 48.5°C [KIR78]

1903

Compound: Magnesium dititanate
Synonym: magnesium pyrotitanate
Formula: $MgTi_2O_5$
Molecular Formula: MgO_5Ti_2
Molecular Weight: 176.461
CAS RN: 12032-35-8

Properties: ortho-rhomb; −325 mesh with
99.9% purity [CER91] [KIR83]
Melting Point, °C: 1645 [KIR83]

1904

Compound: Magnesium dodecaboride
Formula: MgB_{12}
Molecular Formula: $B_{12}Mg$
Molecular Weight: 154.037
CAS RN: 12230-32-9
Properties: refrac solid
Density, g/cm³: [CRC10]
Melting Point, 1300°C: [CRC10]
Boiling Point, °C: [CRC10]

1905

Compound: Magnesium fluoride
Synonym: sellaite
Formula: MgF_2
Molecular Formula: F_2Mg
Molecular Weight: 62.302
CAS RN: 7783-40-6
Properties: white powd or cryst or 99.9% pure melted
pieces of 3–6 mm or 0.8–3 mm; enthalpy of fusion
58.2 kJ/mol; enthalpy of vaporization 264 kJ/
mol; manufactured by reacting hydrofluoric acid
with MgO or $MgCO_3$; hardness 6 Mohs; used
in ceramics, melted pieces used as evaporation
material and sputtering material for widely
used antireflection films, low-index film used in
multilayers [HAW93] [MER06] [KIR78] [CER91]
Solubility: g/L soln, H_2O: 0.130 (25°C) [KRU93]
Density, g/cm³: 3.148 [MER06]
Melting Point, °C: 1263 [CIC73]
Boiling Point, °C: 2227 [CIC73]

1906

Compound: Magnesium formate dihydrate
Formula: $Mg(CHO_2)_2 \cdot 2H_2O$
Molecular Formula: $C_2H_6MgO_6$
Molecular Weight: 150.370
CAS RN: 6150-82-9
Properties: white cryst [CRC10]
Solubility: s H_2O; i EtOH [CRC10]
Melting Point, °C: decomposes [CRC10]

1907

Compound: Magnesium germanate
Formula: Mg_2GeO_4
Molecular Formula: $GeMg_2O_4$
Molecular Weight: 185.218

CAS RN: 12025-13-7
Properties: white precipitate; −325 mesh 10 μm or
　　less with 99.9% purity [CER91] [CRC10]
Solubility: 0.0016 g/100 mL H_2O (25°C) [CRC10]

1908
Compound: Magnesium germanide
Formula: Mg_2Ge
Molecular Formula: $GeMg_2$
Molecular Weight: 121.220
CAS RN: 1310-52-7
Properties: cub cryst; used in semiconductor
　　research [LID94] [MER06]
Density, g/cm³: 3.09 [LID94]
Melting Point, °C: 1115 [MER06]

1909
Compound: Magnesium
　　hexafluoroacetylacetonate dihydrate
Synonyms: 1,1,1,5,5,5-hexafluoro-2,4-
　　pentanedione, magnesium derivative
Formula: $Mg(CF_3COCH=C(O)CF_3)_2 \cdot H_2O$
Molecular Formula: $C_{10}H_6F_{12}MgO_6$
Molecular Weight: 474.440
CAS RN: 19648-85-2
Properties: white powd [STR93] [ALF95]

1910
Compound: Magnesium hexafluorosilicate hexahydrate
Formula: $MgSiF_6 \cdot 6H_2O$
Molecular Formula: $F_6H_{12}MgO_6Si$
Molecular Weight: 274.472
CAS RN: 60950-56-3
Properties: white, efflorescent, odorless cryst; used
　　to mothproof textile fabrics [MER06] [STR93]
Solubility: anhydrous, g/100 g H_2O: 26.3 (0°C),
　　30.8 (20°C), 44.4 (80°C) [LAN05]
Density, g/cm³: 1.788 [MER06]
Reactions: minus SiF_4 at ~120°C [MER06]

1911
Compound: Magnesium hydride
Formula: MgH_2
Molecular Formula: H_2Mg
Molecular Weight: 26.321
CAS RN: 60616-74-2
Properties: white; nonvolatile mass or tetr cryst;
　　strong reducing agent; readily oxidized;
　　reactivity depends on method of preparation,
　　e.g., if prepared from diethylmagnesium it
　　is very reactive; spontaneously ignites in air
　　forming MgO and H_2O [MER06] [KIR80]

Solubility: reacts violently with H_2O,
　　evolving H_2 [MER06]
Density, g/cm³: 1.45 [MER06]
Melting Point, °C: decomposes 280
　　in high vacuum [MER06]

1912
Compound: Magnesium hydrogen phosphate trihydrate
Synonym: newberyite
Formula: $MgHPO_4 \cdot 3H_2O$
Molecular Formula: H_7MgO_7P
Molecular Weight: 174.331
CAS RN: 7757-86-0
Properties: white, cryst powd; decomposes to
　　$Mg_2P_2O_7$ when heated; used to fireproof wood
　　and as a stabilizer for plastics [HAW93]
Solubility: sl s H_2O; s dil acids [MER06]
Density, g/cm³: 2.13 [MER06]
Melting Point, °C: decomposes at 550–650 [HAW93]

1913
Compound: Magnesium hydroxide
Synonym: brucite
Formula: $Mg(OH)_2$
Molecular Formula: H_2MgO_2
Molecular Weight: 58.320
CAS RN: 1309-42-8
Properties: white powd; absorbs CO_2 when H_2O
　　is present; hex, a=0.3147 nm, c=0.4769 nm;
　　hardness 2.5 Mohs; produced from seawater and
　　brines by precipitation of soluble magnesium
　　with $Ca(OH)_2$; used in sugar refining, as an
　　antacid [HAW93] [KIR81] [MER06]
Solubility: mg/L, H_2O: 11.7 (25°C), 4.08 (100°C); s dil
　　acid [KIR81]; mol/L soln, H_2O: 0.5×10^{-4} (0°C),
　　$(2.6 \pm 1.5) \times 10^{-4}$ (25°C), 7.2×10^{-5} (100°C) [KRU93]
Density, g/cm³: 2.37 [KIR81]
Melting Point, °C: 350 [KIR81]

1914
Compound: Magnesium iodate tetrahydrate
Formula: $Mg(IO_3)_2 \cdot 4H_2O$
Molecular Formula: $H_8I_2MgO_{10}$
Molecular Weight: 446.171
CAS RN: 7790-32-1
Properties: white, monocl [CRC10]
Solubility: g/100 g soln, H_2O: 8.55 (25°C),
　　13.5 (90°C); solid phase, $Mg(IO_3)_2 \cdot 4H_2O$
　　(25°C), $Mg(IO_3)_2$ (90°C) [KRU93]
Density, g/cm³: 3.3 [CRC10]
Melting Point, °C: decomposes at 210 [CRC10]
Reactions: minus $4H_2O$ at 210°C [CRC10]

1915
Compound: Magnesium iodide
Formula: MgI_2
Molecular Formula: I_2Mg
Molecular Weight: 278.114
CAS RN: 10377-58-9
Properties: white; hex, a = 0.4148 nm, c = 0.6894 nm; very hygr; decomposes in air evolving I_2; enthalpy of fusion 29.00 kJ/mol; can be obtained by heating the hexahydrate in stream of dry H_2 [KIR81] [CRC10]
Solubility: g/100 g soln, H_2O: 54.7 (0°C), 59.6 (25°C), 65.2 (100°C); solid phase, $MgI_2 \cdot 8H_2O$ (0°C, 25°C), $MgI_2 \cdot 6H_2O$ (100°C) [KRU93]
Density, g/cm³: 4.43 [KIR81]
Melting Point, °C: decomposes at 637 [KIR81]

1916
Compound: Magnesium iodide hexahydrate
Formula: $MgI_2 \cdot 6H_2O$
Molecular Formula: $H_{12}I_2MgO_6$
Molecular Weight: 386.2005
CAS RN: 75535-11-4
Properties: white; monocl, a = 1.1159 nm, b = 0.7740 nm, c = 0.6323 nm [KIR81]
Density, g/cm³: 2.353 [KIR81]

1917
Compound: Magnesium iodide octahydrate
Formula: $MgI_2 \cdot 8H_2O$
Molecular Formula: $H_{16}I_2MgO_8$
Molecular Weight: 422.236
CAS RN: 7790-31-0
Properties: white; ortho-rhomb, a = 0.9948 nm, b = 1.5652 nm, c = 0.8585 nm; deliq powd; discolors readily in air and light [MER06] [KIR81]
Solubility: 81 g/100 mL H_2O (20°C), 90.3 g/100 mL H_2O (80°C) [CRC10]
Density, g/cm³: 2.098 [KIR81]
Melting Point, °C: decomposes at 41 [KIR81]

1918
Compound: Magnesium metaborate octahydrate
Formula: $Mg(BO_2)_2 \cdot 8H_2O$
Molecular Formula: $B_2H_{16}MgO_{12}$
Molecular Weight: 254.047
CAS RN: 13703-82-7 [For anhydrous compound]
Properties: white powd [CRC10]
Solubility: sl H_2O [CRC10]
Melting Point, °C: 988 (anhydrous parent compound) [CRC10]

1919
Compound: Magnesium metasilicate
Formula: $FeSi_2$
Molecular Formula: $FeSi_2$
Molecular Weight: 112.016
CAS RN: 12022-99-0
Properties: gray tetr cryst [CRC10]
Density, g/cm³: 4.74
Melting Point, °C: 1220

1920
Compound: Magnesium metatitanate
Formula: $MgTiO_3$
Molecular Formula: MgO_3Ti
Molecular Weight: 120.183
CAS RN: 12032-30-3
Properties: rhombohedral white powd; used mainly as an additive for ceramic dielectric materials, also as a gemstone and a pigment in ultraviolet cured systems [KIR83] [STR93]
Density, g/cm³: 3.36 [STR93]
Melting Point, °C: 1610 [STR93]

1921
Compound: Magnesium molybdate
Synonym: magnesium molybdate(VI)
Formula: $MgMoO_4$
Molecular Formula: $MgMoO_4$
Molecular Weight: 184.243
CAS RN: 12013-21-7
Properties: −200 mesh with 99.9% purity; white powd; used in electronic and optical applications [CER91] [STR93] [HAW93]
Solubility: g/100 g soln, H_2O: 15.90 (25°C), 9.38 (95°C); solid phase, $MgMoO_4 \cdot 5H_2O$ (25°C), $MgMoO_4 \cdot 2H_2O$ (95°C) [KRU93]
Density, g/cm³: 2.208 [STR93]
Melting Point, °C: ~1060 [HAW93]

1922
Compound: Magnesium niobate
Formula: $MgNb_2O_6$
Molecular Formula: $MgNb_2O_6$
Molecular Weight: 306.114
CAS RN: 12163-26-7
Properties: −200 mesh with 99.9% purity [CER91]

1923
Compound: Magnesium nitrate
Formula: $Mg(NO_3)_2$
Molecular Formula: MgN_2O_6

Molecular Weight: 148.314
CAS RN: 10377-60-3
Properties: white, cub cryst; difficult to isolate in anhydrous form; can be prepared at room temp by dissolution of MgO, Mg(OH)$_2$, or MgCO$_3$ in HNO$_3$, followed by solvent evaporation and crystallization; finds use as a fertilizer and in the manufacture of ammonium nitrate [KIR81] [LID94]
Solubility: g/100 g soln, H$_2$O: 38.5 (1.0°C), 42.1 (25°C), 71.7 (100°C); solid phase, Mg(NO$_3$)$_2$·6H$_2$O (1.0°C, 25°C), Mg(NO$_3$)$_2$·2H$_2$O (100°C) [KRU93]
Density, g/cm^3: ~2.3 [LID94]

1924
Compound: Magnesium nitrate dihydrate
Formula: Mg(NO$_3$)$_2$·2H$_2$O
Molecular Formula: H$_4$MgN$_2$O$_8$
Molecular Weight: 184.345
CAS RN: 15750-45-5
Properties: white cryst; deliq; used in pyrotechnics [HAW93]
Solubility: s H$_2$O, alcohol [HAW93]
Density, g/cm^3: 1.45 [HAW93]
Melting Point, °C: 95–100 [HAW93]
Boiling Point, °C: decomposes at 330 [HAW93]

1925
Compound: Magnesium nitrate hexahydrate
Formula: Mg(NO$_3$)$_2$·6H$_2$O
Molecular Formula: H$_{12}$MgN$_2$O$_{12}$
Molecular Weight: 256.406
CAS RN: 13446-18-9
Properties: colorless, clear deliq cryst; monocl [KIR81] [MER06]
Solubility: s in 0.8 parts H$_2$O; v sl alcohol [MER06]
Density, g/cm^3: 1.464 [MER06]
Melting Point, °C: ~95 [MER06]

1926
Compound: Magnesium nitride
Formula: Mg$_3$N$_2$
Molecular Formula: Mg$_3$N$_2$
Molecular Weight: 100.928
CAS RN: 12057-71-5
Properties: −325 mesh 10 μm or less of 99.6% purity; bcc, a = 0.993 nm [CER91] [CIC73]
Density, g/cm^3: 2.71 [ALD94]
Melting Point, °C: decomposes at 271 [CIC73]

1927
Compound: Magnesium nitrite trihydrate
Formula: Mg(NO$_2$)$_2$·3H$_2$O
Molecular Formula: H$_6$MgN$_2$O$_7$

Molecular Weight: 170.362
CAS RN: 15070-34-5
Properties: white prism; hygr [CRC10]
Solubility: g/100 g soln, H$_2$O: 47.0 (25.65°C); solid phase, Mg(NO$_2$)$_2$·6H$_2$O [KRU93]
Melting Point, °C: decomposes at 100 [CRC10]

1928
Compound: Magnesium orthosilicate
Formula: Mg$_2$SiO$_4$
Molecular Formula: Mg$_2$O$_4$Si
Molecular Weight: 140.694
CAS RN: 26686-77-1
Properties: white, ortho cryst [CRC10]
Density, g/cm^3: 3.21 [CRC10]
Melting Point, °C: 1897

1929
Compound: Magnesium orthotitanate
Formula: Mg$_2$TiO$_4$
Molecular Formula: Mg$_2$O$_4$Ti
Molecular Weight: 160.475
CAS RN: 12032-52-9
Properties: cub [KIR83]
Density, g/cm^3: 3.53 [KIR83]
Melting Point, °C: 1840 [KIR83]

1930
Compound: Magnesium oxalate
Formula: MgC$_2$O$_4$
Molecular Formula: C$_2$MgO$_4$
Molecular Weight: 112.324
CAS RN: 547-66-0
Properties: white powd [CRC10]
Solubility: g/100 g H$_2$O: 0.038^{25} [CRC10]

1931
Compound: Magnesium oxalate dihydrate
Formula: MgC$_2$O$_4$·2H$_2$O
Molecular Formula: C$_2$H$_4$MgO$_6$
Molecular Weight: 148.335
CAS RN: 6150-88-5
Properties: white powd [MER06]
Solubility: g/L solution in H$_2$O: 0.38 ± 0.04 (25°C), 0.4 (92°C); solid phase MgC$_2$O$_4$·2H$_2$O [KRU93]
Density, g/cm^3: 2.45 [CRC10]
Melting Point, °C: decomposes at 150 [AES93]

1932
Compound: Magnesium oxide
Synonyms: periclase, magnesia

Formula: MgO
Molecular Formula: MgO
Molecular Weight: 40.304
CAS RN: 1309-48-4
Properties: white powd; highly reflective in visible and near ultraviolet regions; cub, a = 0.4213 nm; hardness 5.5 Mohs; resistivity $1.3 \times 10^{+15}$ ohm·cm (27°C); enthalpy of fusion 78 kJ/mol; used as a refractory material, particularly for steel furnace linings, as a sputtering target of 99.95% and 99.9% purity to prepare high temp dielectrics, and in crucible form to contain melting corrosive salts such as fluorides [MER06] [HAW93] [KIR81] [CER91] [CRC10]
Solubility: v sl s pure H_2O; s dil acids; i alcohol [MER06]
Density, g/cm³: 3.581 [KIR81]
Melting Point, °C: 2852 [KIR81]
Boiling Point, °C: 3600 [STR93]
Reactions: absorbs CO_2 and H_2O from atm [MER06]
Thermal Conductivity, W/(m·K): 60.0 (27°C), 43.1 (127°C); 13.9 (550°C), 7.0 (1000°C) [KIR80] [KIR81]
Thermal Expansion Coefficient: (volume) 100°C (0.219), 200°C (0.588), 400°C (1.386), 800°C (3.150), 1000°C (4.050) [CLA66]

1933
Compound: Magnesium perborate heptahydrate
Formula: $Mg(BO_3)_2 \cdot 7H_2O$
Molecular Formula: $B_2H_{14}MgO_{13}$
Molecular Weight: 268.030
CAS RN: 14635-87-1
Properties: white powd; decomposes, evolving oxygen; used in driers, in bleaching, and as an antiseptic for tooth powd [HAW93]
Solubility: sl s H_2O [HAW93]

1934
Compound: Magnesium perchlorate
Synonym: dehydrite
Formula: $Mg(ClO_4)_2$
Molecular Formula: Cl_2MgO_8
Molecular Weight: 223.205
CAS RN: 10034-81-8
Properties: white, very hygr powd; oxidant; evolves heat when dissolving in H_2O; crystallizes from H_2O as the hexahydrate; used as a regenerable drying agent for gases [HAW93] [MER06]
Solubility: mol/100 mol H_2O: 0.410 (0°C), 0.448 ± 0.001 (25°C); solid phase, $Mg(ClO_4)_2 \cdot 6H_2O$ [KRU93]
Density, g/cm³: 2.21 [STR93]
Melting Point, °C: decomposes at 251 [STR93]

1935
Compound: Magnesium perchlorate hexahydrate
Formula: $Mg(ClO_4)_2 \cdot 6H_2O$
Molecular Formula: $Cl_2H_{12}MgO_{14}$
Molecular Weight: 331.297
CAS RN: 13446-19-0
Properties: white cryst; hygr [HAW93]
Solubility: v s H_2O, alcohol [HAW93]
Density, g/cm³: 1.98 [HAW93]
Melting Point, °C: 185–190 [HAW93]

1936
Compound: Magnesium permanganate hexahydrate
Formula: $Mg(MnO_4)_2 \cdot 6H_2O$
Molecular Formula: $H_{12}MgMn_2O_{14}$
Molecular Weight: 370.268
CAS RN: 10377-62-5
Properties: bluish black cryst; deliq; used as an antiseptic and as a polymerization catalyst [HAW93]
Solubility: s H_2O [HAW93]
Density, g/cm³: 2.18 [HAW93]
Melting Point, °C: decomposes [HAW93]

1937
Compound: Magnesium peroxide
Synonym: magnesium dioxide
Formula: MgO_2
Molecular Formula: MgO_2
Molecular Weight: 56.304
CAS RN: 1335-26-8
Properties: white, tasteless, odorless powd; used as a bleaching and oxidizing agent, as an antacid [HAW93] [MER06]
Solubility: i H_2O; s dil acids forming H_2O_2 [MER06]
Melting Point, °C: decomposes at >100 [HAW93]
Reactions: gradually decomposed by H_2O evolving O_2 [MER06]

1938
Compound: Magnesium phosphate octahydrate
Synonym: bobierrite
Formula: $Mg_3(PO_4)_2 \cdot 8H_2O$
Molecular Formula: $H_{16}Mg_3O_{16}P_2$
Molecular Weight: 406.980
CAS RN: 13446-23-6
Properties: soft, bulky, white powd; odorless and tasteless; used as a dentifrice polishing agent, as an antacid [HAW93] [ALF95]
Solubility: i H_2O; s acids [HAW93]
Density, g/cm³: 2.195 [CRC10]
Reactions: minus $5H_2O$ at 150°C [CRC10], minus $8H_2O$ at 400°C [HAW93]

1939
Compound: Magnesium phosphate pentahydrate
Formula: $Mg_3(PO_4)_2 \cdot 5H_2O$
Molecular Formula: $H_{10}Mg_3O_{13}P_2$
Molecular Weight: 352.934
CAS RN: 7757-87-1
Properties: white, cryst powd [MER06]
Solubility: i H_2O; s dil mineral acids [MER06]
Reactions: minus last H_2O at ~400°C [MER06]

1940
Compound: Magnesium phosphide
Formula: Mg_3P_2
Molecular Formula: Mg_3P_2
Molecular Weight: 134.863
CAS RN: 12057-74-8
Properties: shiny grayish yellow; stable in dry air; decomposed by moisture; can be prepared directly from magnesium and phosphorus; used with igniting agent such as 1% HNO_3 in sea flares [KIR82]
Solubility: decomposed by H_2O [KIR82]
Density, g/cm³: 2.055 [CRC10]

1941
Compound: Magnesium pyrophosphate
Formula: $Mg_2P_2O_7$
Molecular Formula: $Mg_2O_7P_2$
Molecular Weight: 222.553
CAS RN: 13446-24-7
Properties: colorless; monocl [CRC10]
Density, g/cm³: 2.559 [CRC10]
Melting Point, °C: 1383 [HAW93]

1942
Compound: Magnesium pyrophosphate trihydrate
Formula: $Mg_2P_2O_7 \cdot 3H_2O$
Molecular Formula: $H_6Mg_2O_{10}P_2$
Molecular Weight: 276.600
CAS RN: 10102-34-8
Properties: white powd [MER06]
Solubility: i H_2O; s mineral acids [MER06]
Density, g/cm³: 2.56 [MER06]
Reactions: minus $3H_2O$ at 100°C [MER06]

1943
Compound: Magnesium salicylate tetrahydrate
Formula: $Mg(C_7H_5O_3)_2 \cdot 4H_2O$
Molecular Formula: $C_{14}H_{18}MgO_{10}$
Molecular Weight: 370.596
CAS RN: 18917-95-8
Properties: white, odorless, efflorescent, cryst powd; used as an anti-infective in medicine [HAW93] [MER06]
Solubility: s 13 parts H_2O; s alcohol [MER06]

1944
Compound: Magnesium selenate hexahydrate
Formula: $MgSeO_4 \cdot 6H_2O$
Molecular Formula: $H_{12}MgO_{10}Se$
Molecular Weight: 275.354
CAS RN: 14986-91-5
Properties: monocl cryst [MER06]
Solubility: g/100g soln in H_2O: 31.41 (0°C), 35.70 (25°C), 46.50 (99.5°C); solid phase, $MgSeO_4 \cdot 7H_2O$ (0°C), $MgSeO_4 \cdot 6H_2O$ (25°C), $MgSeO_4 \cdot 4\text{-}1/2H_2O$ (99.5°C) [KRU93]
Density, g/cm³: 1.928 [MER06]

1945
Compound: Magnesium selenide
Formula: MgSe
Molecular Formula: MgSe
Molecular Weight: 103.265
CAS RN: 1313-04-8
Properties: light brown powd; unstable in air [MER06]
Solubility: decomposes in H_2O [MER06]
Density, g/cm³: 4.21 [MER06]

1946
Compound: Magnesium selenite hexahydrate
Formula: $MgSeO_3 \cdot 6H_2O$
Molecular Formula: $H_{12}MgO_9Se$
Molecular Weight: 259.355
CAS RN: 15593-61-0
Properties: colorless ortho-rhomb cryst [LID94] [MER06]
Solubility: i H_2O; s dil acids [MER06]
Density, g/cm³: 2.09 [LID94]
Reactions: loses $5H_2O$ to form monohydrate at 100°C [MER06]

1947
Compound: Magnesium silicate
Synonym: clinoenstatite
Formula: $MgSiO_3$
Molecular Formula: MgO_3Si
Molecular Weight: 100.389
CAS RN: 1343-88-0
Properties: white, monocl cryst [MER06]
Solubility: i H_2O [MER06]
Density, g/cm³: 3.192 [MER06]

Melting Point, °C: decomposes at 1557 [MER06]
Thermal Expansion Coefficient: (volume)
100°C (0.19), 200°C (0.42), 400°C (0.96),
800°C (2.28), 1200°C (3.68) [CLA66]

1948

Compound: Magnesium silicate
Synonym: forsterite
Formula: Mg_2SiO_4
Molecular Formula: Mg_2O_4Si
Molecular Weight: 140.694
CAS RN: 26686-77-1
Properties: white; ortho-rhomb; enthalpy
of fusion 71.00 kJ/mol [CRC10]
Density, g/cm³: 3.22 [KIR80]
Melting Point, °C: 1898 [KIR80]
Thermal Conductivity, W/(m·K): 3.1 (500°C),
2.4 (1000°C) [KIR80]
Thermal Expansion Coefficient: linear expansion
to 1000°C, 9.5×10^{-6}/°C [KIR80]

1949

Compound: Magnesium silicide
Formula: Mg_2Si
Molecular Formula: Mg_2Si
Molecular Weight: 76.696
CAS RN: 22831-39-6
Properties: −20 mesh with 99.5% purity; gray powd;
slate blue, cub cryst; used in semiconductor
technology [CER91] [HAW93] [STR93] [MER06]
Solubility: decomposed by H_2O, HCl [MER06]
Density, g/cm³: 1.94 [STR93]
Melting Point, °C: 1085 [MER06]; 1102 [ALF93]

1950

Compound: Magnesium stannate trihydrate
Formula: $MgSnO_3 \cdot 3H_2O$
Molecular Formula: H_6MgO_6Sn
Molecular Weight: 245.059
CAS RN: 12032-29-0
Properties: white, cryst powd; decomposes at ~340°C;
used as an additive in ceramic capacitors;
anhydrous form −325 mesh 10 μm or less with
99% and 99.9% purity [CER91] [HAW93]
Solubility: s H_2O [HAW93]

1951

Compound: Magnesium stannide
Formula: Mg_2Sn
Molecular Formula: Mg_2Sn
Molecular Weight: 167.320

CAS RN: 1313-08-2
Properties: bluish white, metallic compound; resistivity
(25°C) 42,000 μohm·cm; used in semiconductors,
thermoelectric research [HAW93] [MER06]
Solubility: s H_2O, dil HCl [MER06]
Density, g/cm³: 3.60 [LID94]
Melting Point, °C: 775 [HAW93]

1952

Compound: Magnesium stearate
Formula: $Mg[CH_3(CH_2)_{16}COO]_2$
Molecular Formula: $C_{36}H_{70}MgO_4$
Molecular Weight: 591.255
CAS RN: 557-04-0
Properties: soft, light white powd; odorless, tasteless;
nonflammable used in baby dusting powd and
as tablet lubricant [HAW93] [MER06]
Solubility: i H_2O; decomposed by dil acids [MER06]
Density, g/cm³: 1.028 [HAW93]
Melting Point, °C: 88.5 [HAW93]

1953

Compound: Magnesium sulfate
Formula: $MgSO_4$
Molecular Formula: MgO_4S
Molecular Weight: 120.369
CAS RN: 7487-88-9
Properties: colorless; ortho-rhomb, a = 0.5182 nm,
b = 0.7893 nm, c = 0.6506 nm; occurs widely in
minerals; enthalpy of fusion 14.60 kJ/mol; saline
bitter taste; prepared by dehydration of its hydrates;
used in fireproofing, for warp sizing, and loading
textile goods [HAW93] [KIR81] [CRC10]
Solubility: g/100 g soln, H_2O: 20.5 (0°C), 27.6 (25°C),
42.9 (100°C); solid phase, $MgSO_4 \cdot 6H_2O$ [KRU93]
Density, g/cm³: 2.66 [STR93]
Melting Point, °C: decomposes at 1124 [HAW93]

1954

Compound: Magnesium sulfate heptahydrate
Synonym: epsomite
Formula: $MgSO_4 \cdot 7H_2O$
Molecular Formula: $H_{14}MgO_{11}S$
Molecular Weight: 246.475
CAS RN: 10034-99-8
Properties: ortho-rhomb, a = 1.186 nm, b = 1.199 nm,
c = 0.6858 nm; colorless efflorescent cryst or
powd; bitter, saline cooling taste; stable from
~−5°C–48.2°C; there is a hexahydrate, 17830-18-1,
stable from 48.2°C to 67.5°C [MER06] [KIR81]
Solubility: g/100 mL H_2O: 71 (20°C), 91 (40°C);
sl s alcohol [MER06]

Density, g/cm^3: 1.67 [MER06]
Melting Point, °C: decomposes at ~150 [KIR81]
Reactions: loses 6H$_2$O at 150°C, minus
  H$_2$O at 200°C [HAW93]

1955
Compound: Magnesium sulfate monohydrate
Synonym: kieserite
Formula: MgSO$_4$·H$_2$O
Molecular Formula: H$_2$MgO$_5$S
Molecular Weight: 138.384
CAS RN: 14168-73-1
Properties: colorless cryst; monocl, a=0.690 nm,
  b=0.771 nm, c=0.754 nm [KIR81]
Solubility: H$_2$O: 37.1% (67.5°C), 8%
  (170°C), 0.5% (240°C) [KIR81]
Density, g/cm^3: 2.571 [KIR81]
Melting Point, °C: decomposes at ~150 [KIR81]

1956
Compound: Magnesium sulfide
Formula: MgS
Molecular Formula: MgS
Molecular Weight: 56.371
CAS RN: 12032-36-9
Properties: powd; sensitive to moisture;
  reddish brown cryst; used as a source of
  hydrogen sulfide [STR93] [HAW93]
Solubility: decomposes in H$_2$O [HAW93]
Density, g/cm^3: 2.68 [LID94]
Melting Point, °C: decomposes at >2000 [STR93]

1957
Compound: Magnesium sulfite
Formula: MgSO$_3$
Molecular Formula: MgO$_3$S
Molecular Weight: 104.369
CAS RN: 7757-88-2
Properties: used in systems for flue gas
  desulfurization in which Mg(OH)$_2$ is
  the alkaline scrubber [KIR81]
Solubility: g/100 g soln, H$_2$O: 0.338 (0°C), 0.646
  (25°C), 0.615 (98°C); solid phase, MgSO$_3$·6H$_2$O
  (0°C, 25°C), MgSO$_3$·3H$_2$O (98°C) [KRU93]

1958
Compound: Magnesium sulfite hexahydrate
Formula: MgSO$_3$·6H$_2$O
Molecular Formula: H$_{12}$MgO$_9$S
Molecular Weight: 212.461

CAS RN: 13446-29-2
Properties: white; hex, a=0.88385 nm,
  c=0.9080 nm; gradually oxidizes to sulfate
  in air; used in the manufacture of paper
  pulp [HAW93] [MER06] [KIR81]
Solubility: s in ~150 parts H$_2$O; sl more soluble
  in hot H$_2$O [MER06]; i alcohol [HAW93]
Density, g/cm^3: 1.725 [HAW93]
Melting Point, °C: decomposes at 200 [KIR81]
Reactions: minus all H$_2$O at 200°C [MER06]

1959
Compound: Magnesium sulfite trihydrate
Formula: MgSO$_3$·3H$_2$O
Molecular Formula: H$_6$MgO$_6$S
Molecular Weight: 158.415
CAS RN: 19086-20-5
Properties: colorless; ortho-rhomb, a=0.939 nm,
  b=0.9584 nm, c=0.5523 nm; used in
  flue gas desulfurization [KIR81]
Density, g/cm^3: 2.117 [KIR81]

1960
Compound: Magnesium tantalate
Formula: MgTa$_2$O$_6$
Molecular Formula: MgO$_6$Ta$_2$
Molecular Weight: 482.197
CAS RN: 12293-61-7
Properties: −200 mesh with 99.9% purity [CER91]

1961
Compound: Magnesium tetrahydrogen
  phosphate dihydrate
Formula: MgH$_4$(PO$_4$)$_2$·2H$_2$O
Molecular Formula: H$_8$MgO$_{10}$P$_2$
Molecular Weight: 254.311
CAS RN: 13092-66-5
Properties: white, cryst powd; hygr; decomposes
  when heated to metaphosphate; preparation:
  reaction between phosphoric acid and
  magnesium hydroxide; used to fireproof wood
  and as a stabilizer for plastics [HAW93]
Solubility: s H$_2$O, acids; i alcohol [HAW93]

1962
Compound: Magnesium thiocyanate tetrahydrate
Formula: Mg(SCN)$_2$·4H$_2$O
Molecular Formula: C$_2$H$_8$MgN$_2$O$_4$S$_2$
Molecular Weight: 212.534
CAS RN: 306-61-6
Properties: colorless or white deliq cryst [MER06]
Solubility: v s H$_2$O, alcohol [MER06]

1963

Compound: Magnesium thiosulfate hexahydrate
Synonym: magnesium hyposulfite hexahydrate
Formula: $MgS_2O_3 \cdot 6H_2O$
Molecular Formula: $H_{12}MgO_9S_2$
Molecular Weight: 244.527
CAS RN: 10124-53-5
Properties: colorless or white cryst [MER06]
Solubility: g/100 g soln, H_2O: 30.69 (0°C), 34.51 (28°C); solid phase, $MgS_2O_3 \cdot 6H_2O$ [KRU93]; i alcohol [MER06]
Density, g/cm³: 1.82 [MER06]
Melting Point, °C: decomposes at 1700 [AES93]
Reactions: minus $3H_2O$ at 170°C [MER06]

1964

Compound: Magnesium trifluoroacetylacetonate dihydrate
Synonyms: 1,1,1-trifluoro-2,4-pentandione, magnesium derivative
Formula: $Mg(CF_3COCH=C(O)CH_3)_2 \cdot 2H_2O$
Molecular Formula: $C_{10}H_{12}F_6MgO_6$
Molecular Weight: 366.497
CAS RN: 53633-79-7
Properties: white powd [STR93]

1965

Compound: Magnesium trisilicate
Formula: $Mg_2Si_3O_8$
Molecular Formula: $Mg_2O_8Si_3$
Molecular Weight: 260.862
CAS RN: 14987-04-3
Properties: white powd [CRC10]
Solubility: i H_2O, EtOH [CRC10]

1966

Compound: Magnesium tungstate
Synonym: magnesium tungstate(VI)
Formula: $MgWO_4$
Molecular Formula: MgO_4W
Molecular Weight: 272.143
CAS RN: 13573-11-0
Properties: −325 mesh with 99.9% purity, 10 μm or less; white, cryst powd; used in fluorescent x-ray screens and in luminescent paint [CER91] [HAW93] [MER06]
Solubility: i H_2O and alcohol; s in acids [HAW93]
Density, g/cm³: 5.66 [HAW93]

1967

Compound: Magnesium vanadate
Formula: $2MgO \cdot V_2O_5$
Molecular Formula: $Mg_2O_7V_2$
Molecular Weight: 262.489
CAS RN: 13568-63-3
Properties: tricl, a = 1.3767 nm, b = 0.5414 nm, c = 0.4912 nm; other vanadates are MgV_2O_6, 13573-13-2, MgV_3O_8, 12181-49-6, $Mg_{1.9}V_3O_8$ and $Mg_3V_2O_8$, 13568-68-8; formed by addition of MgO dispersed in oil into the flame zone of utility boilers in order to remove vanadium compounds and thereby to reduce corrosion caused by the presence of vanadium [KIR81] [CER91]
Density, g/cm³: 3.1 [KIR81]

1968

Compound: Magnesium zirconate
Formula: $MgZrO_3$
Molecular Formula: MgO_3Zr
Molecular Weight: 163.527
CAS RN: 12032-31-4
Properties: reacted product, −100 and +200 mesh of 99% purity; powd; used in electronics [CER91] [HAW93]
Density, g/cm³: 4.23 [HAW93]
Melting Point, °C: 2060 [HAW93]

1969

Compound: Magnesium zirconium silicate
Formula: $MgO \cdot ZrO_2 \cdot SiO_2$
Molecular Formula: MgO_5SiZr
Molecular Weight: 223.612
CAS RN: 52110-05-1
Properties: white solid; used in electrical resistor, ceramics, as an opacifier for glazes [HAW93]
Solubility: i H_2O, alkalies; sl s in acids [HAW93]

1970

Compound: Manganese
Formula: Mn
Molecular Formula: Mn
Molecular Weight: 54.93805
CAS RN: 7439-96-5
Properties: steel gray, lustrous, hard, brittle metal; has four allotropes: α-Mn, cub, a = 0.89 nm, stable <710°C; β-Mn, cub, a = 0.63 nm, stable 710°C–1079°C; γ-Mn, fcc, a = 0.387 nm, stable 1079°C–1143°C; δ-Mn, bcc, a = 0.309 nm, stable 1143°C to melting; hardness 5.0 Mohs; enthalpy of fusion 12.91 kJ/mol; enthalpy of vaporization 220.9 kJ/mol; electrical resistivity at 20°C for α is 160 μohm·cm [MER06] [KIR81] [CRC10]
Solubility: decomposes slowly in cold H_2O, rapidly in hot H_2O; evolves hydrogen with dil mineral acids, dissolving as Mn++ [MER06]

Density, g/cm³: α: 7.47; β: 7.26; γ: 6.37; δ: 6.28 [MER06]
Melting Point, °C: 1244 [MER06]
Boiling Point, °C: 2095 [MER06]
Thermal Conductivity, W/(m·K): 7.82 (25°C) [CRC10]
Thermal Expansion Coefficient: 100°C:
    $25.2 \times 10^{-6}/°C$ (α), $43.0 \times 10^{-6}/°C$ (β),
    $45.2 \times 10^{-6}/°C$ (γ), $41.6 \times 10^{-6}/°C$ (δ) [KIR81]

1971
Compound: Manganese aluminide
Formula: $MnAl_3$
Molecular Formula: Al_3Mn
Molecular Weight: 135.883
CAS RN: 12253-13-3
Properties: 6 mm pieces and smaller
    with 99.5% purity [CER91]

1972
Compound: Manganese ammonium sulfate hexahydrate
Formula: $MnSO_4 \cdot (NH_4)_2SO_4 \cdot 6H_2O$
Molecular Formula: $H_{20}MnN_2O_{14}S_2$
Molecular Weight: 391.235
CAS RN: 7785-19-5
Properties: light red cryst [HAW93]
Solubility: s H_2O [HAW93]
Density, g/cm³: 1.83 [HAW93]

1973
Compound: Manganese antimonide
Formula: MnSb
Molecular Formula: MnSb
Molecular Weight: 176.698
CAS RN: 12032-82-5
Properties: hex cryst; 6 mm pieces and smaller with
    99.5% purity; there is also an antimonide with the
    formula Mn_2Sb, 12032-97-2 [CER91] [LID94]
Density, g/cm³: 6.9 [LID94]
Melting Point, °C: 840 [LID94]

1974
Compound: Manganese antimonide
Formula: Mn_2Sb
Molecular Formula: Mn_2Sb
Molecular Weight: 231.636
CAS RN: 12032-97-2
Properties: tetr cryst [CRC10]
Density, g/cm³: 7.0 [CRC10]
Melting Point, °C: 948 [CRC10]

1975
Compound: Manganese bis(cyclopentadienyl)
Synonym: bis(cyclopentadienyl)manganese

Formula: $(C_5H_5)_2Mn$
Molecular Formula: $C_{10}H_{10}Mn$
Molecular Weight: 185.127
CAS RN: 1271-27-8
Properties: sublimed powd, in ampoules [ALF95]
Melting Point, °C: 292 [ALF95]

1976
Compound: Manganese boride
Formula: MnB
Molecular Formula: BMn
Molecular Weight: 65.749
CAS RN: 12045-15-7
Properties: powd; −80 mesh with 99% purity,
    there is also MnB_2, 12228-50-1, −200 mesh;
    refractory material [CER91] [KIR81] [CRC91]
Density, g/cm³: 6.2 [CRC10]
Melting Point, °C: 1890 [KIR81]

1977
Compound: Manganese boride
Formula: Mn_2B
Molecular Formula: BMn_2
Molecular Weight: 120.687
CAS RN: 12045-16-8
Properties: red-brown tetr cryst [CRC10]
Density, g/cm³: 7.20 [CRC10]
Melting Point, °C: 1580 [CRC10]

1978
Compound: Manganese carbide
Formula: Mn_3C
Molecular Formula: CMn_3
Molecular Weight: 176.825
CAS RN: 12266-65-8
Properties: tetr; mixture of Mn_5C_2 and possibly other
    Mn-C phases, 6 mm pieces and smaller with
    99.5% purity; there is also a material with
    formula $Mn_{23}C_6$, 72266-65-8, −80 mesh
    with 99.5% purity [CER91] [CRC10]
Density, g/cm³: 6.89 [CRC10]
Melting Point, °C: 1520 [LID94]

1979
Compound: Manganese carbonyl
Synonym: dimanganese decacarbonyl
Formula: $Mn_2(CO)_{10}$
Molecular Formula: $C_{10}Mn_2O_{10}$
Molecular Weight: 389.980
CAS RN: 10170-69-1

Properties: golden yellow, monocl cryst; stable
   under CO gas, less stable to air, heat and
   light in solution; used as an antiknock
   agent in gasoline [HAW93] [MER06]
Solubility: i H_2O; s organic solvents [MER06]
Density, g/cm^3: 1.75 [HAW93]
Melting Point, °C: 154 [HAW93]
Boiling Point, °C: 80 [ALF95]

1980
Compound: Manganese diboride
Formula: MnB_2
Molecular Formula: B_2Mn
Molecular Weight: 76.560
CAS RN: 12228-50-1
Properties: gray-violet cryst; refractory
   material [KIR81] [CRC10]
Solubility: decomposed by H_2O [CRC10]
Density, g/cm^3: 5.3 [LID94]
Melting Point, °C: 1827 [LID94]

1981
Compound: Manganese niobate
Formula: $MnNb_2O_6$
Molecular Formula: $MnNb_2O_6$
Molecular Weight: 336.747
CAS RN: 12032-69-8
Properties: –200 mesh with 99.9% purity [CER91]

1982
Compound: Manganese nitride
Formula: MnN
Molecular Formula: MnN
Molecular Weight: 68.945
CAS RN: 36678-21-4
Properties: formed by reaction of Mn and N_2 above
   740°C; other nitrides are Mn_6N_5 (64886-63-1),
   Mn_3N_2 (12033-03-3), Mn_2N (12163-53-0), and
   Mn_4N (12033-07-7); used in steelmaking as
   nitrogen containing intermediate alloys [KIR81]

1983
Compound: Manganese pentacarbonyl bromide
Formula: $Mn(CO)_5Br$
Molecular Formula: C_5BrMnO_5
Molecular Weight: 274.894
CAS RN: 14516-54-2
Properties: yellowish orange cryst [STR93]

1984
Compound: Manganese phosphide
Formula: MnP
Molecular Formula: MnP
Molecular Weight: 85.912
CAS RN: 12032-78-9
Properties: ortho cryst [CRC10]
Density, g/cm^3: 5.49 [CRC10]
Melting Point, °C: 1147 [CRC10]

1985
Compound: Manganese phosphide
Formula: Mn_2P
Molecular Formula: Mn_2P
Molecular Weight: 140.850
CAS RN: 12333-54-9
Density, g/cm^3: 6.0 [CRC10]
Melting Point, °C: 1327 [CRC10]

1986
Compound: Manganese phosphide
Formula: Mn_3P_2
Molecular Formula: Mn_3P_2
Molecular Weight: 226.762
CAS RN: 12397-32-9
Properties: dark gray; mixture of MnP and Mn_2P;
   –100 mesh with 99% purity [CER91] [CRC10]
Density, g/cm^3: 5.12 [CRC10]
Melting Point, °C: 1095 [AES93]

1987
Compound: Manganese selenide
Formula: MnSe
Molecular Formula: MnSe
Molecular Weight: 133.90
CAS RN: 1313-22-0
Properties: gray cub cryst [CRC10]
Solubility: i H_2O [CRC10]
Density, g/cm^3: 5.45 [CRC10]
Melting Point, °C: 1460 [CRC10]

1988
Compound: Manganese silicate
Synonyms: rhodonite, manganjustite, tephroite
Formula: $MnSiO_3$
Molecular Formula: MnO_3Si
Molecular Weight: 131.022
CAS RN: 7759-00-4

Properties: red cryst or yellowish red powd; prepared from manganous salts and sodium silicate; used to color glass and pottery [MER06] [HAW93]
Solubility: i H_2O [MER06]
Density, g/cm³: 3.48 [MER06]
Melting Point, °C: 1323 [HAW93]

1989
Compound: Manganese silicide
Synonym: manganese disilicide
Formula: $MnSi_2$
Molecular Formula: $MnSi_2$
Molecular Weight: 111.109
CAS RN: 12032-86-9
Properties: gray; possibly $Mn_{15}Si_{26}$; −325 mesh 10 μm or less with 99.5% purity [CER91] [CRC10]
Density, g/cm³: 5.24 [CRC10]

1990
Compound: Manganese vanadate
Formula: MnV_2O_6
Molecular Formula: MnO_6V_2
Molecular Weight: 252.817
CAS RN: 14986-94-8
Properties: −200 mesh with 99.9% purity [CER91]

1991
Compound: Manganocene
Formula: $Mn(C_5H_5)_2$
Molecular Formula: $C_{10}H_{10}Mn$
Molecular Weight: 185.124
CAS RN: 1271-27-8
Properties: yellow-brown cryst [CRC10]
Solubility: s py, thf; sl bz [CRC10]
Melting Point, °C: 173 [CRC10]

1992
Compound: Manganese(II) acetate tetrahydrate
Formula: $Mn(CH_3COO)_2 \cdot 4H_2O$
Molecular Formula: $C_4H_{14}MnO_8$
Molecular Weight: 245.088
CAS RN: 6156-78-1
Properties: pale red cryst; monocl; used in textile dyeing, as an oxidation catalyst; anhydrous manganese(II) acetate has CAS RN 638-38-0 [ALF93] [HAW93]
Solubility: s H_2O, alcohol [HAW93]
Density, g/cm³: 1.589 [KIR81]
Melting Point, °C: 80 [HAW93]; 180 [AES93]

1993
Compound: Manganese(II) acetylacetonate
Synonyms: 2,4-pentanedione, manganese(II) derivative
Formula: $Mn(CH_3COCH=C(O)CH_3)_2$
Molecular Formula: $C_{10}H_{14}MnO_4$
Molecular Weight: 253.157
CAS RN: 14024-58-9
Properties: tan powd; trimer; hygr [STR93] [COT88]
Melting Point, °C: decomposes at 216 [ALD94]

1994
Compound: Manganese(II) borate octahydrate
Formula: $MnB_4O_7 \cdot 8H_2O$
Molecular Formula: $B_4H_{16}MnO_{15}$
Molecular Weight: 354.300
CAS RN: 12228-91-0
Properties: brownish-white powd; preparation: by addition of borax to an aq manganese(II) sulfate solution; used in drying varnishes and oils and in the leather industry [MER06] [KIR81]
Solubility: i H_2O, alcohol; s dil acids [MER06]
Reactions: decomposes on standing in H_2O [MER06]

1995
Compound: Manganese(II) bromide
Formula: $MnBr_2$
Molecular Formula: Br_2Mn
Molecular Weight: 214.746
CAS RN: 13446-03-2
Properties: −80 mesh with 99.5% purity; pink powd; hygr [STR93] [CER91]
Solubility: g/100 g soln, H_2O: 56.0 (0°C), 60.2 (25°C), 69.5 (100°C); solid phase, $MnBr_2 \cdot 4H_2O$ (0°C, 25°C), $MnBr_2 \cdot 2H_2O$ (100°C) [KRU93]
Density, g/cm³: 4.385 [STR93]
Melting Point, °C: 698 [LID94]

1996
Compound: Manganese(II) bromide tetrahydrate
Formula: $MnBr_2 \cdot 4H_2O$
Molecular Formula: $Br_2H_8MnO_4$
Molecular Weight: 286.808
CAS RN: 10031-20-6
Properties: rose red, sl deliq cryst [MER06]
Solubility: s in 0.5 parts H_2O; s alcohol [MER06]
Melting Point, °C: 64, with some decomposition [MER06]

1997

Compound: Manganese(II) carbonate
Synonym: rhodochrosite
Formula: $MnCO_3$
Molecular Formula: $CMnO_3$
Molecular Weight: 114.947
CAS RN: 598-62-9
Properties: pink solid; trig; gradually turns light
brown in air; photoluminescent; hardness
3–4 Mohs; can be made by precipitation of
a water soluble Mn(II) salt with an alkali
carbonate [HAW93] [KIR81] [MER06]
Solubility: sl s H_2O; s dil acids [KIR81]
Density, g/cm³: 3.125 [KIR81]
Melting Point, °C: decomposes at >200 [KIR81]

1998

Compound: Manganese(II) chloride
Synonym: sacchite
Formula: $MnCl_2$
Molecular Formula: Cl_2Mn
Molecular Weight: 125.843
CAS RN: 7773-01-5
Properties: pink cryst or flakes; trig, hygr; enthalpy
of fusion 30.70 kJ/mol; obtained by reaction
of Mn, MnO, $Mn(OH)_2$, or $MnCO_3$ and HCl,
followed by crystallization and dehydration;
used as a chlorination catalyst for organic
materials, as a paint dryer, as a dietary
supplement [CRC10] [HAW93] [KIR81]
Solubility: s pyridine, ethanol; i ether [KIR78];
g/100 g soln, H_2O: 38.8 (0°C), 43.6 (25°C), 53.5
(100°C); equilibrium solid phases: $MnCl_2 \cdot 4H_2O$
(0°C, 25°C), $MnCl_2$ (100°C) [KRU93]
Density, g/cm³: 2.977 [KIR81]
Melting Point, °C: 652 [KIR81]
Boiling Point, °C: 1190 [KIR81]

1999

Compound: Manganese(II) chloride tetrahydrate
Formula: $MnCl_2 \cdot 4H_2O$
Molecular Formula: $Cl_2H_8MnO_4$
Molecular Weight: 197.905
CAS RN: 13446-34-9
Properties: reddish, sl deliq, monocl cryst; there is
a dihydrate, 20603-88-7 [KIR81] [MER06]
Solubility: s 0.7 parts H_2O; s alcohol; i ether [MER06]
Density, g/cm³: 1.913 [HAW93]
Melting Point, °C: 87.5 [HAW93]

2000

Compound: Manganese(II) citrate
Synonym: manganous citrate
Formula: $Mn_3(C_6H_5O_7)_2$
Molecular Formula: $C_{12}H_{10}Mn_3O_{14}$
Molecular Weight: 543.017
CAS RN: 71799-92-3
Properties: white powd; used as a food and feed
additive and as a dietary supplement [HAW93]
Solubility: s H_2O containing dissolved
sodium citrate [HAW93]

2001

Compound: Manganese(II) dihydrogen
phosphate dihydrate
Formula: $Mn(H_2PO_4)_2 \cdot 2H_2O$
Molecular Formula: $H_8MnO_{10}P_2$
Molecular Weight: 284.944
CAS RN: 18718-07-5
Properties: almost colorless cryst; deliq [KIR81]
Solubility: s H_2O; i ethanol [KIR81]
Reactions: minus H_2O at 100°C [KIR81]

2002

Compound: Manganese(II) dithionate
Formula: $Mn(SO_3)_2$
Molecular Formula: MnO_6S_2
Molecular Weight: 215.066
CAS RN: 13568-72-4
Properties: tricl cryst [CRC10]
Solubility: s H_2O [HAW93]
Density, g/cm³: 1.76 [HAW93]

2003

Compound: Manganese(II) fluoride
Synonym: manganous fluoride
Formula: MnF_2
Molecular Formula: F_2Mn
Molecular Weight: 92.935
CAS RN: 7782-64-1
Properties: hygr; −80 mesh with 99.5% purity;
reddish powd [HAW93] [CER91] [STR93]
Solubility: g/100 g soln, H_2O: 0.800 (0°C), 1.00
(23.5°C), 0.48 (100°C); equilibrium solid phase,
$MnF_2 \cdot 4H_2O$ (0°C), MnF_2 (25°C, 100°C) [KRU93];
s dil HF, conc HCl or HNO_3 [MER06]
Density, g/cm³: 3.98 [MER06]
Melting Point, °C: 856 [MER06]

2004
Compound: Manganese(II) hydrogen phosphate trihydrate
Formula: $MnHPO_4 \cdot 3H_2O$
Molecular Formula: H_7MnO_7P
Molecular Weight: 204.959
CAS RN: 10236-39-2
Properties: has two forms: gray, prepared by decomposition of $MnCO_3$, and pink, prepared by reaction between phosphoric acid and $Mn_3(PO_4)_2$ [MER06]
Solubility: pink form: v sl s H_2O; s dil acids; gray form: s only in hot conc HCl [MER06]
Reactions: minus H_2O >100°C [CRC10]

2005
Compound: Manganese(II) hydroxide
Synonym: pyrochroite
Formula: $Mn(OH)_2$
Molecular Formula: H_2MnO_2
Molecular Weight: 88.953
CAS RN: 18933-05-6
Properties: white to pink cryst; hardness is 2.5 Mohs [HAW93]
Solubility: mol/L soln, H_2O: 3.6×10^{-5} (25°C) [KRU93]
Density, g/cm³: 3.258 [HAW93]
Melting Point, °C: decomposes [HAW93]

2006
Compound: Manganese(II) hypophosphite monohydrate
Formula: $Mn(H_2PO_2)_2 \cdot H_2O$
Molecular Formula: $H_6MnO_5P_2$
Molecular Weight: 202.931
CAS RN: 10043-84-2
Properties: pink, odorless, almost tasteless cryst or powd; used as a food additive and dietary supplement [HAW93] [MER06]
Solubility: 1 g/6.5 mL H_2O, 1 g/6 mL boiling H_2O; i alcohol [MER06]
Reactions: evolves phosphine when heated [MER06]

2007
Compound: Manganese(II) iodide
Formula: MnI_2
Molecular Formula: I_2Mn
Molecular Weight: 308.747
CAS RN: 7790-33-2
Properties: red-brown powd; hygr [STR93]
Solubility: s H_2O, with gradual decomposition; s alcohol [HAW93]

Density, g/cm³: 5.01 [HAW93]
Melting Point, °C: 638 [HAW93]

2008
Compound: Manganese(II) iodide tetrahydrate
Formula: $MnI_2 \cdot 4H_2O$
Molecular Formula: $H_8I_2MnO_4$
Molecular Weight: 380.809
CAS RN: 7790-33-2
Properties: rose red cryst; rapidly turns brown when exposed to air and light due to liberation of iodine [MER06]
Solubility: v s H_2O, with gradual decomposition; s alcohol [MER06]
Melting Point, °C: decomposes [CRC10]

2009
Compound: Manganese(II) metasilicate
Formula: $MnSiO_3$
Molecular Formula: MnO_3Si
Molecular Weight: 131.022
CAS RN: 7759-00-4
Properties: red ortho cryst [CRC10]
Solubility: i H_2O [CRC10]
Density, g/cm³: 3.48 [CRC10]
Melting Point, °C: 1291

2010
Compound: Manganese(II) molybdate
Formula: $MnMoO_4$
Molecular Formula: $MnMoO_4$
Molecular Weight: 214.876
CAS RN: 14013-15-1
Properties: yellow or off-white powd; monocl [KIR81] [STR93]
Density, g/cm³: 4.05 [LID94]

2011
Compound: Manganese(II) nitrate
Formula: $Mn(NO_3)_2$
Molecular Formula: MnN_2O_6
Molecular Weight: 178.948
CAS RN: 10377-66-9
Properties: liq; available in dissolved form; forms pink solutions [ALF95] [STR93]
Solubility: g/100 g soln, H_2O: 50.49 (0°C), 61.74 (25°C); solid phase, $Mn(NO_3)_2 \cdot 6H_2O$ [KRU93]

2012
Compound: Manganese(II) nitrate hexahydrate
Formula: $Mn(NO_3)_2 \cdot 6H_2O$

Molecular Formula: $H_{12}MnN_2O_{12}$
Molecular Weight: 287.040
CAS RN: 17141-63-8
Properties: rose colored, deliq, monocl needles; used in ceramics, as a catalyst [HAW93] [MER06]
Solubility: v s H_2O, alcohol [MER06]
Density, g/cm³: 1.8 [MER06]

2013
Compound: Manganese(II) nitrate tetrahydrate
Formula: $Mn(NO_3)_2 \cdot 4H_2O$
Molecular Formula: $H_8MnN_2O_{10}$
Molecular Weight: 251.010
CAS RN: 20694-39-7
Properties: pink, deliq cryst masses below 20°C [MER06]
Solubility: v s H_2O, s alcohol [MER06]
Density, g/cm³: 2.129 [MER06]
Melting Point, °C: 37.1 [MER06]

2014
Compound: Manganese(II) oxalate dihydrate
Formula: $MnC_2O_4 \cdot 2H_2O$
Molecular Formula: $C_2H_4MnO_6$
Molecular Weight: 178.988
CAS RN: 6556-16-7
Properties: white, cryst powd; used as a paint and varnish drier [HAW93] [MER06]
Solubility: g/100 g soln in H_2O: 0.0198 (0°C), 0.0309 ± 0.0002 (25°C); solid phase $MnC_2O_4 \cdot 2H_2O$ [KRU93]
Density, g/cm³: 2.453 [HAW93]
Melting Point, °C: decomposes at 150 [MER06]
Reactions: minus $2H_2O$ at 100°C [HAW93]

2015
Compound: Manganese(II) oxide
Synonym: manganosite
Formula: MnO
Molecular Formula: MnO
Molecular Weight: 70.937
CAS RN: 1344-43-0
Properties: grass green powd; cub; enthalpy of fusion 54.40 kJ/mol; produced from MnO_2 ores by roasting in a reducing atm, e.g., methane; used in textile printing, in ceramics, in paints [HAW93] [CRC10]
Solubility: i H_2O; s acids [HAW93]
Density, g/cm³: 5.37 [KIR81]
Melting Point, °C: 1840 [LID94]

2016
Compound: Manganese(II) perchlorate hexahydrate
Formula: $Mn(ClO_4)_2 \cdot 6H_2O$
Molecular Formula: $Cl_2H_{12}MnO_{14}$
Molecular Weight: 361.930
CAS RN: 15364-94-0
Properties: pink cryst; hygr [STR93]
Density, g/cm³: 2.10 [LID94]
Melting Point, °C: decomposes at 165 [AES93]

2017
Compound: Manganese(II) phosphate heptahydrate
Formula: $Mn_3(PO_4)_2 \cdot 7H_2O$
Molecular Formula: $H_{14}Mn_3O_{15}P_2$
Molecular Weight: 480.864
CAS RN: 10236-39-2
Properties: reddish white powd; used in conversion coating of steel, aluminum, and other metals [HAW93]
Solubility: i H_2O; s mineral acids [HAW93]

2018
Compound: Manganese(II) pyrophosphate
Formula: $Mn_2P_2O_7$
Molecular Formula: $Mn_2O_7P_2$
Molecular Weight: 283.819
CAS RN: 53731-35-4
Properties: white powd [HAW93]
Solubility: i H_2O; s solutions of potassium or sodium pyrophosphate [HAW93]
Density, g/cm³: 3.71 [HAW93]
Melting Point, °C: 1196 [HAW93]

2019
Compound: Manganese(II) pyrophosphate trihydrate
Formula: $Mn_2P_2O_7 \cdot 3H_2O$
Molecular Formula: $H_6Mn_2O_{10}P_2$
Molecular Weight: 337.866
CAS RN: 53731-35-4
Properties: white or nearly white powd [MER06]
Solubility: i H_2O, s in excess of alkali pyrophosphate, acids [MER06]

2020
Compound: Manganese(II) selenide
Formula: MnSe
Molecular Formula: MnSe
Molecular Weight: 133.898
CAS RN: 1313-22-0
Properties: black cryst; −20 mesh with 99.9% purity [CER91] [STR93]

Density, g/cm³: 5.45 [LID94]
Melting Point, °C: 1460 [LID94]

2021
Compound: Manganese(II) sulfate
Formula: $MnSO_4$
Molecular Formula: MnO_4S
Molecular Weight: 151.002
CAS RN: 7785-87-7
Properties: almost white; ortho-rhomb [KIR81]
Solubility: g/100 g soln, H_2O: 34.6 (0°C), 39.2 (25°C), 26.1 (100.7°C); solid phase, $MnSO_4 \cdot 7H_2O$ (0°C), $MnSO_4 \cdot H_2O$ (25°C, 100.7°C) [KRU93]
Density, g/cm³: 3.25 [KIR81]
Melting Point, °C: 700 [HAW93]
Boiling Point, °C: decomposes at 850 [HAW93]

2022
Compound: Manganese(II) sulfate monohydrate
Formula: $MnSO_4 \cdot H_2O$
Molecular Formula: H_2MnO_5S
Molecular Weight: 169.017
CAS RN: 10034-96-5
Properties: pale red, sl efflorescent cryst; used in dyeing [MER06]
Solubility: s in about 1 part cold H_2O, 0.6 parts boiling; i alcohol [MER06]
Density, g/cm³: 2.95 [LID94]
Reactions: minus H_2O at 400°C–450°C [MER06]

2023
Compound: Manganese(II) sulfate tetrahydrate
Formula: $MnSO_4 \cdot 4H_2O$
Molecular Formula: H_8MnO_8S
Molecular Weight: 223.063
CAS RN: 10101-68-5
Properties: translucent, pale rose red; efflorescent prisms; used in fertilizers, as a feed additive, in ceramics [HAW93]
Solubility: s H_2O; i alcohol [HAW93]
Density, g/cm³: 2.107 [HAW93]
Melting Point, °C: 30 [HAW93]

2024
Compound: Manganese(II) sulfide
Synonyms: alabandite, manganblende
Formula: MnS
Molecular Formula: MnS
Molecular Weight: 87.004
CAS RN: 18820-29-6

Properties: pink, green, or brownish green powd; three cryst forms: α: green cub; β: red cub; γ: red hex; used as an additive in steel production [HAW93] [MER06]
Solubility: i H_2O; s dil acids [MER06]
Density, g/cm³: 3.99 [STR93]
Melting Point, °C: decomposes at 1610 [AES93]
Reactions: readily oxidized in moist air to sulfate [MER06]

2025
Compound: Manganese(II) telluride
Formula: MnTe
Molecular Formula: MnTe
Molecular Weight: 182.538
CAS RN: 12032-88-1
Properties: hex cryst; possibly contains some $MnTe_2$, 12032-89-2; 6 mm pieces and smaller with 99.9% purity [LID94] [CER91]
Density, g/cm³: 6.0 [LID94]

2026
Compound: Manganese(II) tetraborate octahydrate
Formula: $MnB_4O_7 \cdot 8H_2O$
Molecular Formula: $B_4H_{16}MnO_{15}$
Molecular Weight: 354.300
CAS RN: 12228-91-0
Properties: red solid [CRC10]
Solubility: i H_2O, EtOH; s dil acids [CRC10]

2027
Compound: Manganese(II) titanate
Synonym: pyrophanite
Formula: $MnTiO_3$
Molecular Formula: MnO_3Ti
Molecular Weight: 150.803
CAS RN: 12032-74-5
Properties: −100 mesh with 99.9% purity; red hex cryst [LID94] [CER91] [KIR83]
Density, g/cm³: 4.54 [KIR83]
Melting Point, °C: 1360 [KIR83]

2028
Compound: Manganese(II) tungstate
Formula: $MnWO_4$
Molecular Formula: MnO_4W
Molecular Weight: 302.776
CAS RN: 13918-22-4
Properties: −200 mesh with 99.9% purity; off-white powd [STR93] [CER91]
Density, g/cm³: 7.2 [LID94]

2029

Compound: Manganese(II) zirconate
Formula: $MnZrO_3$
Molecular Formula: MnO_3Zr
Molecular Weight: 194.160
CAS RN: 70692-94-3
Properties: reacted product; −200 mesh
  with 99.5% purity [CER91]

2030

Compound: Manganese(II,III) oxide
Synonym: hausmannite
Formula: Mn_3O_4
Molecular Formula: Mn_3O_4
Molecular Weight: 228.812
CAS RN: 1317-35-7
Properties: brownish powd; tetr, $a = 0.5762$ nm,
  $c = 0.9470$ nm [THA92] [HAW93]
Solubility: i H_2O [KIR81]; s HCl, releasing Cl_2 [MER06]
Density, g/cm³: 4.876 [HAW93]
Melting Point, °C: 1564 [HAW93]

2031

Compound: Manganese(III) acetate dihydrate
Formula: $Mn(CH_3COO)_3 \cdot 2H_2O$
Molecular Formula: $C_6H_{13}MnO_8$
Molecular Weight: 268.102
CAS RN: 19513-05-4
Properties: brown cryst; mild selective
  oxidizing agent [ALD94] [STR93]

2032

Compound: Manganese(III) acetylacetonate
Synonyms: 2,4-pentanedione,
  manganese(III) derivative
Formula: $Mn(CH_3COCH=C(O)CH_3)_3$
Molecular Formula: $C_{15}H_{21}MnO_6$
Molecular Weight: 352.266
CAS RN: 14284-89-0
Properties: black cryst; hygr [STR93]
Melting Point, °C: decomposes at 160 [ALD94]

2033

Compound: Manganese(III) fluoride
Formula: MnF_3
Molecular Formula: F_3Mn
Molecular Weight: 111.933
CAS RN: 7783-53-1
Properties: red cryst; monocl; hygr; used as a
  fluorinating agent [HAW93] [KIR81]
Solubility: decomposed by H_2O [KIR81]

Density, g/cm³: 3.54 [KIR81]
Melting Point, °C: decomposes at >600 [KIR81]

2034

Compound: Manganese(III) hydroxide
Synonym: manganite
Formula: γ-$MnO(OH)$
Molecular Formula: $HMnO_2$
Molecular Weight: 87.945
CAS RN: 1332-63-4
Properties: black solid; monocl [KIR81]
Solubility: i H_2O; disproportionates in dil acid [KIR81]
Density, g/cm³: 4.2–4.4 [KIR81]
Melting Point, °C: decomposes at 250 [KIR81]
Reactions: transformed to Mn_2O_3 at 250°C [KIR81]

2035

Compound: Manganese(III) oxide
Synonym: braunite
Formula: α-Mn_2O_3
Molecular Formula: Mn_2O_3
Molecular Weight: 157.874
CAS RN: 1317-34-6
Properties: black to brown solid; rhomb,
  cub; very hard [HAW93] [KIR81]
Solubility: i H_2O [KIR81]; s HCl, evolving Cl_2 [MER06]
Density, g/cm³: 4.50 [HAW93]
Melting Point, °C: decomposes at 1080 [HAW93]

2036

Compound: Manganese(IV) oxide
Synonyms: manganese dioxide, pyrolusite
Formula: MnO_2
Molecular Formula: MnO_2
Molecular Weight: 86.937
CAS RN: 1313-13-9
Properties: black cryst or powd; α-MnO_2:
  $a = 0.97876$ nm, $c = 0.28650$ nm; λ-MnO_2: cub,
  $a = 0.8029$ nm; used as an oxidizing agent, in dry
  cell batteries [HAW93] [THA92] [ROS92]
Solubility: i H_2O, HNO_3, cold H_2SO_4; slowly dissolves
  in HCl, evolving Cl_2; s dil H_2SO_4, dil HNO_3 in
  the presence of H_2O_2 or oxalic acid [MER06]
Density, g/cm³: 5.026 [STR93]; α: 4.21 [ROS92]
Melting Point, °C: decomposes at 535 [STR93]
Reactions: minus O at 535°C [CRC10]

2037

Compound: Manganese(IV) telluride
Synonym: manganese ditelluride
Formula: $MnTe_2$

Molecular Formula: MnTe$_2$
Molecular Weight: 310.14
CAS RN: 12032-89-2
Properties: 6 mm pieces and down [STR93]

2038
Compound: Manganese(VII) oxide
Synonym: manganese heptoxide
Formula: Mn$_2$O$_7$
Molecular Formula: Mn$_2$O$_7$
Molecular Weight: 221.872
CAS RN: 12057-92-0
Properties: dark red oil; hygr [KIR81]
Solubility: v s H$_2$O [KIR81]
Density, g/cm^3: 2.396 [KIR81]
Melting Point, °C: 5.9 [KIR81]
Boiling Point, °C: decomposes at 55 [KIR81]

2039
Compound: Mendelevium
Formula: Md
Molecular Formula: Md
Molecular Weight: 258
CAS RN: 7440-11-1
Properties: discovered in 1955 by Ghiorso and colleagues
at Lawrence Berkeley Laboratory; preparation:
bombardment of ^{253}Es with He ions [KIR78]
Melting Point, °C: 827 [LID94]

2040
Compound: Mercury
Synonym: quicksilver
Formula: Hg
Molecular Formula: Hg
Molecular Weight: 200.59
CAS RN: 7439-97-6
Properties: silvery white, heavy liq metal; surface
tension is 480 dynes/cm; enthalpy of vaporization
59.11 kJ/mol; enthalpy of fusion 2.29 kJ/mol;
resistivity (20°C) 95.8 μohm · cm; critical density
3.56 g/cm^3; critical pressure 74.2 MPa; critical
temp 1677°C; expansion coefficient of liq at 20°C
is 182×10^{-6}/°C; viscosity at 20°C is 1.55 mPa · s
[KIR81] [HAW93] [COT88] [CRC10]
Solubility: 20–30 μg/L in H$_2$O; s in boiling
H$_2$SO$_4$, HNO$_3$ [KIR81] [HAW93]
Density, g/cm^3: 13.534 [MER06]
Melting Point, °C: −38.87 [MER06]
Boiling Point, °C: 356.73 [CRC10]
Thermal Conductivity, W/(m · K): 9.2 (25°C) [KIR81]
Thermal Expansion Coefficient: liq: 100°C
(1.458), 200°C (3.307) [CLA66]

2041
Compound: Mercury(I) acetate
Synonym: mercurous acetate
Formula: Hg$_2$(CH$_3$COO)$_2$
Molecular Formula: C$_4$H$_6$Hg$_2$O$_4$
Molecular Weight: 519.370
CAS RN: 631-60-7
Properties: colorless scales; lustrous leaflets
or cryst powd; light sensitive, becomes
dark; aq solutions decomposed quickly
by light and heat; used in medicine as an
antibacterial agent [HAW93] [MER06]
Solubility: 0.75 g/100 mL H$_2$O (12°C) [CRC10]
Melting Point, °C: decomposes [CRC10]

2042
Compound: Mercury(I) bromate
Formula: Hg$_2$(BrO$_3$)$_2$
Molecular Formula: Br$_2$Hg$_2$O$_6$
Molecular Weight: 656.98
CAS RN: 13465-33-3
Properties: col cryst [CRC10]
Solubility: i H$_2$O; sl acid [CRC10]
Melting Point, °C: decomposes [CRC10]

2043
Compound: Mercury(I) bromide
Synonym: mercurous bromide
Formula: Hg$_2$Br$_2$
Molecular Formula: Br$_2$Hg$_2$
Molecular Weight: 560.988
CAS RN: 15385-58-7
Properties: white, odorless, tasteless powd; tetr; sensitive
to light (darkens); decomposed by hot HCl or alkali
bromides; becomes yellow when heated, returns to
white color when cooled; prepared by oxidation of
Hg with Br$_2$ or as a precipitate by addition of NaBr
to HgNO$_3$ solution [KIR81] [HAW93] [MER06]
Solubility: g/L soln, H$_2$O: 3.9×10^{-5} (25°C)
[KRU93]; i alcohol, ether [MER06]; s in
fuming HNO$_3$, hot conc H$_2$SO$_4$ [HAW93]
Density, g/cm^3: 7.307 [HAW93]
Melting Point, °C: 405 [HAW93]
Boiling Point, °C: sublimes at 340–350 [HAW93]

2044
Compound: Mercury(I) carbonate
Formula: Hg$_2$CO$_3$
Molecular Formula: CHg$_2$O$_3$
Molecular Weight: 461.189
CAS RN: 50968-00-8

Properties: yellow powd [LAN05]
Solubility: g/L soln, H_2O: 8.8×10^{-9} (25°C) [KRU93]
Melting Point, °C: decomposes at 130 [LAN05]

2045
Compound: Mercury(I) chlorate
Synonym: mercurous chlorate
Formula: $Hg_2(ClO_3)_2$
Molecular Formula: $Cl_2Hg_2O_6$
Molecular Weight: 568.081
CAS RN: 10294-44-7
Properties: white cryst; decomposes at
~250°C to O_2, HgO, $HgCl_2$ [MER06]
Solubility: sl s H_2O, hydrolyzed in hot
H_2O forming basic salt [MER06]; s
alcohol, acetic acid [HAW93]
Density, g/cm³: 6.409 [HAW93]
Melting Point, °C: decomposes at 250 [HAW93]

2046
Compound: Mercury(I) chloride
Synonyms: mercurous chloride, calomel
Formula: Hg_2Cl_2
Molecular Formula: Cl_2Hg_2
Molecular Weight: 472.085
CAS RN: 10112-91-1
Properties: white, odorless, tasteless powd; rhomb
cryst; slowly decomposed by sunlight to liq
Hg and $HgCl_2$; can be obtained by oxidation
of Hg with Cl_2; used as a fungicide, as a
reference electrode, in ceramic painting;
decomposed by alkalies [HAW93] [MER06]
Solubility: g/100 g soln, H_2O: 0.000140
(0.5°C); mol/L soln, H_2O: 7.5×10^{-6} (25°C)
[KRU93]; i alcohol, ether [MER06]
Density, g/cm³: 7.15 [MER06]
Melting Point, °C: 302 [HAW93]
Boiling Point, °C: 384 [HAW93]

2047
Compound: Mercury(I) chromate
Synonym: mercurous chromate
Formula: Hg_2CrO_4
Molecular Formula: $CrHg_2O_4$
Molecular Weight: 517.174
CAS RN: 13444-75-2
Properties: brick red powd; ortho-rhomb cryst;
can have a variable composition; used for
coloring ceramics green [HAW93] [KIR78]
Solubility: i H_2O, alcohol; s conc HNO_3 [HAW93]
Melting Point, °C: decomposes [CRC10]

2048
Compound: Mercury(I) fluoride
Synonym: mercurous fluoride
Formula: Hg_2F_2
Molecular Formula: F_2Hg_2
Molecular Weight: 439.177
CAS RN: 13967-25-4
Properties: small, yellow cub cryst; blackens
in light; preparation by reaction of
Hg_2CO_3 and HF [MER06]
Solubility: hydrolyzed in H_2O to
Hg(I), HgO, HF [MER06]
Density, g/cm³: 8.73 [MER06]
Melting Point, °C: sublimes at ~240 [MER06]
Boiling Point, °C: decomposes at 570 [MER06]

2049
Compound: Mercury(I) iodate
Formula: $Hg_2(IO_3)_2$
Molecular Formula: $Hg_2I_2O_6$
Molecular Weight: 750.985
CAS RN: 13465-35-9
Properties: yellowish [LAN05]
Solubility: g/L soln, H_2O: 6.0×10^{-7} (25°C) [KRU93]
Melting Point, °C: volatilizes at 250 [LAN05]

2050
Compound: Mercury(I) iodide
Synonym: mercurous iodide
Formula: Hg_2I_2
Molecular Formula: Hg_2I_2
Molecular Weight: 654.989
CAS RN: 15385-57-6
Properties: bright yellow, amorphous, odorless, tasteless
powd; darkens or becomes greenish in light forming
HgI and liq Hg; color changes to dark yellow,
orange, and orange-red when heated, reverse order
when cooled; enthalpy of fusion 27.00 kJ/mol;
can be made by precipitation in $HgNO_3$ solution
with KI; used as a topical antibacterial agent in
medicine [HAW93] [MER06] [KIR81] [CRC10]
Solubility: g/L soln, H_2O: 2.0×10^{-7} (25°C)
[KRU93]; i alcohol, ether [MER06]; s
castor oil, ammonia [HAW93]
Density, g/cm³: 7.65–7.75 [HAW93]
Melting Point, °C: 290, when rapidly heated,
decomposes to HgI, Hg(I) [MER06]
Boiling Point, °C: sublimes at 140 [HAW93]

2051
Compound: Mercury(I) nitrate dihydrate
Synonym: mercurous nitrate

Formula: $HgNO_3 \cdot 2H_2O$
Molecular Formula: H_4HgNO_5
Molecular Weight: 298.626
CAS RN: 10415-75-5
Properties: short prismatic cryst; effloresces, becoming anhydrous in dry air; light sensitive; used as an analytical reagent [HAW93]
Solubility: sensitive to H_2O: s in small quantities of warm H_2O, but hydrolyzes in larger amounts [HAW93]
Density, g/cm³: 4.785 (3.9°C) [HAW93]
Melting Point, °C: decomposes at 70 [HAW93]

2052
Compound: Mercury(I) nitrate monohydrate
Formula: $HgNO_3 \cdot H_2O$
Molecular Formula: H_2HgNO_4
Molecular Weight: 280.611
CAS RN: 7782-86-7
Properties: white cryst; anhydrous is white monocl, which can be prepared by dissolution of Hg in hot dil HNO_3 followed by crystallization, some $Hg(NO_3)_2$ is also formed [KIR81] [STR93]
Density, g/cm³: 4.79 [STR93]
Melting Point, °C: 70 [STR93]

2053
Compound: Mercury(I) nitrite
Formula: $Hg_2(NO_2)_2$
Molecular Formula: $Hg_2N_2O_4$
Molecular Weight: 493.19
CAS RN: 13492-25-6
Properties: yellow cryst [CRC10]
Solubility: reac H_2O [CRC10]
Density, g/cm³: 7.3 [CRC10]
Melting Point, °C: decomposes at 100 [CRC10]

2054
Compound: Mercury(I) oxalate
Formula: $Hg_2C_2O_4$
Molecular Formula: $C_2Hg_2O_4$
Molecular Weight: 489.20
CAS RN: 2949-11-3
Properties: cryst [CRC10]
Solubility: i H_2O; sl HNO_3 [CRC10]

2055
Compound: Mercury(I) oxide
Synonym: mercurous oxide
Formula: Hg_2O
Molecular Formula: Hg_2O
Molecular Weight: 417.179

CAS RN: 15829-53-5
Properties: black or brownish black powd; light sensitive; no evidence that Hg_2O has been isolated; on adding NaOH solution to $HgNO_3$, what x-ray data indicate could be intimate mixture of HgO and Hg [MER06]
Solubility: i H_2O; s HNO_3; reacts with HCl to form calomel, Hg_2Cl_2 [MER06]
Density, g/cm³: 9.8 [HAW93]
Melting Point, °C: decomposes at 100 [HAW93]

2056
Compound: Mercury(I) perchlorate tetrahydrate
Formula: $HgClO_4 \cdot 4H_2O$
Molecular Formula: ClH_8HgO_8
Molecular Weight: 372.102
CAS RN: 65202-12-2
Properties: white cryst [STR93]
Solubility: g/100 g soln in H_2O: 282 (0°C), 394 (25°C), 580 (99°C); solid phase: $Hg_2(ClO_4)_2 \cdot 4H_2O$ (0°C, 25°C), $Hg_2(ClO_4)_2 \cdot 2H_2O$ (99°C) [KRU93]

2057
Compound: Mercury(I) sulfate
Synonym: mercurous sulfate
Formula: Hg_2SO_4
Molecular Formula: Hg_2O_4S
Molecular Weight: 497.224
CAS RN: 7783-36-0
Properties: white to sl yellow cryst powd; becomes gray and forms Hg and $HgSO_4$ in light; decomposes on heating; forms as a precipitate by addition of dil H_2SO_4 to a solution of $HgNO_3$; used in Clark and Weston standard cells [KIR81] [HAW93] [MER06]
Solubility: 0.05 g/100 g H_2O [KIR81]; s dil HNO_3 [MER06]
Density, g/cm³: 7.56 [MER06]
Melting Point, °C: decomposes [AES93]

2058
Compound: Mercury(I) sulfide
Formula: Hg_2S
Molecular Formula: Hg_2S
Molecular Weight: 433.246
CAS RN: 51595-71-2
Properties: black [LAN05]
Solubility: g/L soln, H_2O: 2.8×10^{-23} (25°C) [KRU93]
Melting Point, °C: decomposes [LAN05]

2059
Compound: Mercury(I) thiocyanate
Formula: $Hg_2(SCN)_2$

Molecular Formula: $C_2Hg_2N_2S_2$
Molecular Weight: 517.34
CAS RN: 13465-37-7
Properties: col powd [CRC10]
Solubility: g/100 g H_2O: 0.03^{25} [CRC10]

2060
Compound: Mercury(I) tungstate
Formula: Hg_2WO_4
Molecular Formula: Hg_2O_4W
Molecular Weight: 649.02
CAS RN: 38705-19-0
Properties: yellow amorphous solid [CRC10]
Solubility: i H_2O, EtOH [CRC10]
Melting Point, °C: decomposes [CRC10]

2061
Compound: Mercury(II) acetate
Synonym: mercuric acetate
Formula: $Hg(CH_3COO)_2$
Molecular Formula: $C_4H_6HgO_4$
Molecular Weight: 318.680
CAS RN: 1600-27-7
Properties: white cryst or light yellow powd; −60 mesh with 99.9% purity; sensitive to light; prepared by dissolving HgO in warm 20% acetic acid; used in pharmaceuticals and as a catalyst in organic synthesis [HAW93] [STR93] [KIR81] [CER91]
Solubility: s H_2O, alcohol [HAW93]
Density, g/cm³: 3.25 [HAW93]
Melting Point, °C: 179–182 [ALD94]

2062
Compound: Mercury(II) amide chloride
Formula: $Hg(NH_2)Cl$
Molecular Formula: ClH_2HgN
Molecular Weight: 252.07
CAS RN: 10124-48-8
Properties: white solid [CRC10]
Solubility: i H_2O, EtOH; sol warm acid [CRC10]
Density, g/cm³: 5.38 [CRC10]
Boiling Point, °C: sublimes [CRC10]

2063
Compound: Mercury(II) arsenate
Synonym: mercuric arsenate
Formula: $HgHAsO_4$
Molecular Formula: $AsHHgO_4$
Molecular Weight: 340.518
CAS RN: 7784-37-4

Properties: yellow powd; used to waterproof paints and in antifouling paints [HAW93] [MER06]
Solubility: i H_2O; s HCl, HNO_3 [MER06]

2064
Compound: Mercury(II) basic carbonate
Formula: $HgCO_3 \cdot 3HgO$
Molecular Formula: CHg_4O_6
Molecular Weight: 910.367
CAS RN: 13004-83-6
Properties: brown precipitate formed by adding sodium carbonate solution to $HgCl_2$ solution, forming a slurry; refluxing the slurry decomposes carbonate to red HgO [KIR81]

2065
Compound: Mercury(II) benzoate monohydrate
Formula: $Hg(C_7H_5O_2)_2 \cdot H_2O$
Molecular Formula: $C_{14}H_{12}HgO_5$
Molecular Weight: 460.836
CAS RN: 583-15-3
Properties: white cryst; odorless cryst powd; sensitive to light [MER06] [HAW93]
Solubility: 1.2 g/100 mL H_2O (15°C), 2.5 g/100 mL H_2O (100°C) [CRC10]
Melting Point, °C: 165 [HAW93]
Reactions: hydrolyzed in boiling H_2O to a basic salt and benzoic acid [MER06]

2066
Compound: Mercury(II) bromate
Formula: $Hg(BrO_3)_2$
Molecular Formula: $Hg(BrO_3)_2$
Molecular Weight: 456.39
CAS RN: 26522-91-8
Properties: cryst [CRC10]
Solubility: g/100 g H_2O: 0.15; s acid [CRC10]
Melting Point, °C: decomposes at 130 [CRC10]

2067
Compound: Mercury(II) bromide
Formula: $HgBr_2$
Molecular Formula: Br_2Hg
Molecular Weight: 360.398
CAS RN: 7789-47-1
Properties: white rhomb cryst or powd; sensitive to light; enthalpy of vaporization 58.89 kJ/mol; enthalpy of fusion 17.90 kJ/mol; can be prepared by precipitation below 75°C from $Hg(NO_3)_2$ solution with NaBr, followed by drying; used in medicine [HAW93] [MER06] [KIR81] [CRC10]

Solubility: g/100 g soln, H_2O: 0.3 (0°C), 0.611 ± 0.002
  (25°C), 4.7 (100°C) [KRU93]; v s hot alcohol,
  methanol, HCl, HBr; sl s chloroform [MER06]
Density, g/cm³: 6.109 [STR93]
Melting Point, °C: 236 [CRC10]
Boiling Point, °C: 322 [CRC10]

2068
Compound: Mercury(II) chlorate
Formula: $Hg(ClO_3)_2$
Molecular Formula: Cl_2HgO_6
Molecular Weight: 367.49
CAS RN: 13465-30-0
Properties: white needles [CRC10]
Solubility: g/100 g H_2O: 25 [CRC10]
Density, g/cm³: 4.998 [CRC10]
Melting Point, °C: decomposes [CRC10]

2069
Compound: Mercury(II) chloride
Synonym: corrosive sublimate
Formula: $HgCl_2$
Molecular Formula: Cl_2Hg
Molecular Weight: 271.495
CAS RN: 7487-94-7
Properties: cryst or white granules or powd;
  odorless; volatilizes unchanged at ~300°C;
  vapor pressure 13 Pa (100°C), 400 Pa (150°C);
  enthalpy of vaporization 58.9 kJ/mol; enthalpy
  of fusion 19.41 kJ/mol; obtained by reaction
  of Hg with Cl_2; used to coagulate albumin, to
  manufacture mercury compounds, as a disinfectant
  [HAW93] [MER06] [KIR81] [JAN71]
Solubility: g/100 g soln, H_2O: 3.5 (0°C), 6.8 (25°C), 36.5
  (100°C) [KRU93]; 1 g dissolves in 3.8 mL alcohol,
  200 mL benzene, 22 mL ether, 12 mL glycerol,
  40 mL acetic acid; s other organics [MER06]
Density, g/cm³: 5.44 [HAW93]; vapor: 9.8 [KIR81]
Melting Point, °C: 276 [JAN71]
Boiling Point, °C: 304 [JAN71]

2070
Compound: Mercury(II) chloride ammoniated
Synonym: ammoniated mercuric chloride
Formula: $Hg(NH_2)Cl$
Molecular Formula: ClH_2HgN
Molecular Weight: 252.066
CAS RN: 10124-48-8
Properties: white lumps, odorless powd; stable in air,
  darkens when exposed to light; used in medicine
  as a topical anti-infective [HAW93] [MER06]

Solubility: i H_2O, alcohol; s warm HCl,
  HNO_3, acetic acid [MER06]
Density, g/cm³: 5.38 [MER06]

2071
Compound: Mercury(II) chromate
Synonym: mercuric chromate
Formula: $HgCrO_4$
Molecular Formula: $CrHgO_4$
Molecular Weight: 316.584
CAS RN: 13444-75-2
Properties: red ortho-rhomb; used in
  antifouling formulations [KIR78]
Solubility: sl s H_2O [KIR78]
Density, g/cm³: 6.06 [LID94]
Melting Point, °C: decomposes [CRC10]

2072
Compound: Mercury(II) cyanide
Synonym: mercuric cyanide
Formula: $Hg(CN)_2$
Molecular Formula: C_2HgN_2
Molecular Weight: 252.625
CAS RN: 592-04-1
Properties: colorless, odorless; tetr cryst; darkens
  if exposed to light; can be made by reaction
  of aq slurry of yellow HgO with excess HCN,
  followed by heating to 95°C and filtration; used
  as an antiseptic in medicine, in germicidal soaps,
  and in photography [HAW93] [MER06]
Solubility: g/100 g soln, H_2O: 6.31 (0°C),
  10.06 ± 0.06 (25°C), 35.05 (101.1°C) [KRU93];
  1 g dissolves in 13 mL alcohol, 4 mL methanol;
  sl s ether; slowly s glycerol [MER06]
Density, g/cm³: 3.996 [MER06]
Melting Point, °C: decomposes at 320 [MER06]

2073
Compound: Mercury(II) dichromate
Synonym: mercuric dichromate
Formula: $HgCr_2O_7$
Molecular Formula: Cr_2HgO_7
Molecular Weight: 416.578
CAS RN: 7789-10-8
Properties: red, heavy, cryst powd [MER06]
Solubility: i H_2O; s HCl, HNO_3 [MER06]

2074
Compound: Mercury(II) fluoride
Synonym: mercuric fluoride

Formula: HgF$_2$
Molecular Formula: F$_2$Hg
Molecular Weight: 238.587
CAS RN: 7783-39-3
Properties: transparent cryst; white powd or cub cryst; very sensitive to moisture; can be obtained from a reaction of F$_2$ with HgCl$_2$ or HgO; used in the synthesis of organic fluoride compounds [HAW93] [MER06] [KIR78]
Solubility: turns yellow and hydrolyzes in H$_2$O after prolonged time [MER06]; moderately s in alcohol [HAW93]
Density, g/cm³: 8.95 [MER06]
Melting Point, °C: decomposes at 645 [HAW93]

2075
Compound: Mercury(II) fulminate
Formula: Hg(CNO)$_2$
Molecular Formula: C$_2$HgN$_2$O$_2$
Molecular Weight: 284.624
CAS RN: 628-86-4
Properties: gray, cryst powd; explodes readily if dry; used to manufacture caps and detonators for explosives [HAW93]
Solubility: sl s cold H$_2$O, s hot H$_2$O; s alcohol, NH$_4$OH [HAW93]
Density, g/cm³: 4.42 [HAW93]
Melting Point, °C: explodes [HAW93]

2076
Compound: Mercury(II) hydrogen arsenate
Formula: HgHAsO$_4$
Molecular Formula: AsHHgO$_4$
Molecular Weight: 340.52
CAS RN: 7784-37-4
Properties: yellow powd [CRC10]
Solubility: i H$_2$O; s acid [CRC10]

2077
Compound: Mercury(II) iodate
Synonym: mercuric iodate
Formula: Hg(IO$_3$)$_2$
Molecular Formula: HgI$_2$O$_6$
Molecular Weight: 550.395
CAS RN: 7783-32-6
Properties: white powd [MER06]
Solubility: g/100 g H$_2$O: 0.002 (20°C) [MER06]
Melting Point, °C: decomposes at 175 [LID94]

2078
Compound: Mercury(II) iodide(α)
Synonym: mercuric iodide
Formula: α-HgI$_2$
Molecular Formula: HgI$_2$
Molecular Weight: 454.399
CAS RN: 7774-29-0
Properties: scarlet red, heavy odorless, almost tasteless powd; tetr; sensitive to light; enthalpy of vaporization 59.2 kJ/mol; enthalpy of fusion 18.90 kJ/mol; obtained as a precipitate from a solution of HgCl$_2$ and KI; used to treat skin diseases and as an analytical reagent [KIR81] [CRC10] [MER06]
Solubility: 0.006 g/100 g H$_2$O (25°C); Hg dissolves in: 115 mL alcohol, 20 mL boiling alcohol, about 120 mL ether, about 60 mL acetone, 910 mL chloroform, 75 mL ethyl acetate [MER06]
Density, g/cm³: 6.28 [MER06]
Melting Point, °C: 259 [MER06]
Boiling Point, °C: 354 [CRC10]
Reactions: transition red → yellow at 130°C [MER06]

2079
Compound: Mercury(II) iodide(β)
Synonym: coccinite
Formula: β-HgI$_2$
Molecular Formula: HgI$_2$
Molecular Weight: 454.399
CAS RN: 7774-29-0
Properties: rhomb; turns yellow from red α-form when α-form is heated to 130°C, then reverses color change when cooled; used as an antiseptic in medicine and as Nessler's reagent [HAW93] [STR93] [CRC10]
Solubility: g/L soln, H$_2$O: 0.050 ± 0.06 (25°C) [KRU93]; s boiling alcohol [HAW93]
Density, g/cm³: 6.094 [CRC10]
Melting Point, °C: 259 (under argon) [STR93]
Boiling Point, °C: 349 [HAW93]

2080
Compound: Mercury(II) nitrate dihydrate
Formula: Hg(NO$_3$)$_2$·2H$_2$O
Molecular Formula: H$_4$HgN$_2$O$_8$
Molecular Weight: 360.63
CAS RN: 22852-67-1
Properties: monocl cryst [CRC10]
Solubility: s H$_2$O [CRC10]
Density, g/cm³: 4.78 [CRC10]

2081
Compound: Mercury(II) nitrate
Formula: $Hg(NO_3)_2$
Molecular Formula: HgN_2O_6
Molecular Weight: 324.60
CAS RN: 10045-94-0
Properties: col hygr cryst [CRC10]
Solubility: s H_2O; i EtOH [CRC10]
Density, g/cm³: 4.3 [CRC10]
Melting Point, °C: 79 [CRC10]

2082
Compound: Mercury(II) nitrate hemihydrate
Synonym: mercuric nitrate
Formula: $Hg(NO_3)_2 \cdot 1/2H_2O$
Molecular Formula: $HHgN_2O_{6.5}$
Molecular Weight: 333.607
CAS RN: 10045-94-0
Properties: colorless cryst or white deliq powd; decomposes by heating; made by dissolution of Hg in hot conc HNO_3, then cooling to crystallize product; used in the nitration of aromatic organic compounds and in felt manufacture [HAW93]
Solubility: s H_2O, HNO_3; i alcohol [HAW93]
Density, g/cm³: 4.39 [CRC10]
Melting Point, °C: 79 [HAW93]
Boiling Point, °C: decomposes [CRC10]

2083
Compound: Mercury(II) nitrate monohydrate
Synonym: mercuric nitrate monohydrate
Formula: $Hg(NO_3)_2 \cdot H_2O$
Molecular Formula: $H_2HgN_2O_7$
Molecular Weight: 342.615
CAS RN: 7783-34-8
Properties: white or sl yellow, deliq cryst powd [MER06]
Solubility: s H_2O, decomposes; s dil acids [MER06]
Density, g/cm³: 4.3 [STR93]

2084
Compound: Mercury(II) oleate
Synonym: mercuric oleate
Formula: $Hg(C_{17}H_{33}COO)_2$
Molecular Formula: $C_{36}H_{66}HgO_4$
Molecular Weight: 763.508
CAS RN: 1191-80-6
Properties: yellowish brown; odor of oleic acid; sensitive to light; used as an antiseptic and in antifouling paints [HAW93] [MER06]
Solubility: i H_2O; sl s alcohol, ether; v s in fixed oils [MER06]

2085
Compound: Mercury(II) oxalate
Formula: HgC_2O_4
Molecular Formula: C_2HgO_4
Molecular Weight: 288.610
CAS RN: 3444-13-1
Properties: powd [LAN05]
Solubility: g/100g H_2O: 0.0107 (20°C) [KRU93]
Melting Point, °C: decomposes at 165 [LAN05]

2086
Compound: Mercury(II) oxide red
Synonym: red mercuric oxide
Formula: HgO
Molecular Formula: HgO
Molecular Weight: 216.589
CAS RN: 21908-53-2
Properties: bright red or orange red, odorless cryst; ortho-rhomb; sensitive to light, decomposes to Hg and O_2; obtained either by thermal decomposition of $Hg(NO_3)_2$ or by precipitation from a hot solution of HgCl by Na_2CO_3; used in paint pigments, perfumery and cosmetics, and as an antiseptic [HAW93] [MER06] [KIR81]
Solubility: i H_2O; s dil HCl, HNO_3; i alcohol [MER06]
Density, g/cm³: 11.14 [MER06]
Melting Point, °C: decomposes at 500 [STR93]
Reactions: decomposes to Hg(I) and O_2 at 500°C [MER06]

2087
Compound: Mercury(II) oxide yellow
Synonym: yellow mercuric oxide
Formula: HgO
Molecular Formula: HgO
Molecular Weight: 216.589
CAS RN: 21908-53-2
Properties: yellow or light orange yellow, odorless powd; ortho-rhomb; stable in air; darkens on exposure to light; can be prepared by precipitation from a water soluble mercuric salt by an alkali; used as an antiseptic [KIR81] [HAW93] [MER06]
Solubility: i H_2O; s dil HCl, HNO_3 [MER06]
Density, g/cm³: 11.14 [MER06]
Melting Point, °C: decomposes at 500 [STR93]

2088
Compound: Mercury(II) oxide sulfate
Formula: $(Hg_3O_2)SO_4$
Molecular Formula: Hg_3O_6S
Molecular Weight: 729.83

CAS RN: 1312-03-4
Properties: yellow powd [CRC10]
Solubility: i H_2O; s acid [CRC10]

2089
Compound: Mercury(II) oxycyanide
Synonym: mercuric oxycyanide
Formula: $Hg(CN)_2 \cdot HgO$
Molecular Formula: $C_2Hg_2N_2O$
Molecular Weight: 469.214
CAS RN: 1335-31-5
Properties: white, ortho-rhomb cryst or cryst powd; exploded by percussion or if in contact with flame; prepared from excess yellow HgO in a slurry with HCN [MER06] [KIR81]
Solubility: 1 g/80 mL cold H_2O, more soluble in hot H_2O [MER06]
Density, g/cm³: 4.44 [MER06]
Melting Point, °C: can explode [CRC10]

2090
Compound: Mercury(II) perchlorate trihydrate
Formula: $Hg(ClO_4)_2 \cdot 3H_2O$
Molecular Formula: $Cl_2H_6HgO_{11}$
Molecular Weight: 453.536
CAS RN: 73491-34-6
Properties: white cryst [STR93]
Density, g/cm³: ~4 [STR93]

2091
Compound: Mercury(II) phosphate
Formula: $Hg_3(PO_4)_2$
Molecular Formula: $Hg_3O_8P_2$
Molecular Weight: 791.713
CAS RN: 7782-66-3
Properties: heavy, white or yellowish powd [HAW93]
Solubility: i H_2O, alcohol; s in acids [HAW93]

2092
Compound: Mercury(II) selenide
Synonym: tiemannite
Formula: HgSe
Molecular Formula: HgSe
Molecular Weight: 279.550
CAS RN: 20601-83-6
Properties: gray powd; sublimes in vacuum [HAW93] [STR93]
Solubility: i H_2O [HAW93]
Density, g/cm³: 8.266 [HAW93]
Melting Point, °C: sublimes in vacuum [CRC10]

2093
Compound: Mercury(II) sulfate
Synonym: mercuric sulfate
Formula: $HgSO_4$
Molecular Formula: HgO_4S
Molecular Weight: 296.654
CAS RN: 7783-35-9
Properties: white, odorless granules or cryst powd; sensitive to light; decomposes at red heat; can be obtained by reacting yellow HgO with H_2SO_4 solution; used as a catalyst for converting acetylene to acetaldehyde and as a battery electrolyte [HAW93] [MER06] [KIR81]
Solubility: decomposed in H_2O to yellow insoluble basic sulfate and H_2SO_4; s HCl, hot dil H_2SO_4, conc NaCl solution [MER06]
Density, g/cm³: 6.47 [MER06]
Melting Point, °C: decomposes [AES93]

2094
Compound: Mercury(II) sulfide(α)
Synonym: cinnabar
Formula: α-HgS
Molecular Formula: HgS
Molecular Weight: 232.656
CAS RN: 1344-48-5
Properties: bright scarlet red powd, lumps, hex cryst; sensitive to light (blackens); not attacked by HNO_3 or cold HCl; decomposed by hot conc H_2SO_4; can be prepared by heating black HgS in a conc solution of alkali polysulfide [KIR81] [MER06]
Solubility: i H_2O; s aqua regia, warm HI [MER06]
Density, g/cm³: 8.10 [STR93]
Melting Point, °C: sublimes at 583 [STR93]

2095
Compound: Mercury(II) sulfide(β)
Synonym: metacinnabar
Formula: β-HgS
Molecular Formula: HgS
Molecular Weight: 232.656
CAS RN: 1344-48-5
Properties: black or grayish black, odorless, tasteless, cub; can exist indefinitely in metastable state at room temp; prepared by mixing soluble mercuric salts and sulfides [KIR81] [MER06]
Solubility: i H_2O, alcohol, dil mineral acids [MER06]
Density, g/cm³: 7.70 [LID94]
Melting Point, °C: sublimes at 583 [STR93]

2096
Compound: Mercury(II) telluride
Formula: HgTe
Molecular Formula: HgTe
Molecular Weight: 328.190
CAS RN: 12068-90-5
Properties: gray powd; used as a semiconductor in solar cells, in thin-film transistors and in infrared detectors [HAW93] [STR93]
Density, g/cm³: 8.17 [CRC10]
Melting Point, °C: 673 [CRC10]
Thermal Conductivity, W/(m·K): 2 (25°C) [CRC10]

2097
Compound: Mercury(II) tetrathiocyanatocobaltate(II)
Synonym: cobalt(II) tetrathiocyanatomercurate(II)
Formula: HgCo(SCN)$_4$
Molecular Formula: C$_4$CoHgN$_4$S$_4$
Molecular Weight: 491.845
CAS RN: 27685-51-4
Properties: blue cryst; high-purity magnetic susceptibility standard [ALD94] [ALF95]

2098
Compound: Mercury(II) thiocyanate
Synonym: mercuric thiocyante
Formula: Hg(SCN)$_2$
Molecular Formula: C$_2$HgN$_2$S$_2$
Molecular Weight: 316.757
CAS RN: 592-85-8
Properties: odorless powd; radially arranged cryst needles; if heated, swells to considerable volume; sensitive to light [MER06]
Solubility: 0.069 g/100 mL H$_2$O (25°C); more soluble in boiling water, but decomposes; s dil HCl [MER06]
Density, g/cm³: 3.71 [LID94]
Melting Point, °C: decomposes at 165 [ALD94]
Reactions: decomposes into Hg, N$_2$, other products at ~165°C [MER06]

2099
Compound: Mercury(II) tungstate
Formula: HgWO$_4$
Molecular Formula: HgO$_4$W
Molecular Weight: 448.43
CAS RN: 37913-38-5
Properties: yellow cryst [CRC10]
Solubility: i H$_2$O, EtOH [CRC10]
Melting Point, °C: decomposes [CRC10]

2100
Compound: Metaphosphoric acid
Synonyms: phosphoric acid, meta
Formula: (HPO$_3$)$_n$
Molecular Formula: HO$_3$P, n=1
Molecular Weight: 79.980, n=1
CAS RN: 10343-62-1
Properties: transparent, glass-like solid or soft silky masses; hygr; volatilizes at red heat (HPO$_3$)$_n$ [MER06]
Solubility: dissolves very slowly in cold H$_2$O to form H$_3$PO$_4$, formation of H$_3$PO$_4$ accelerated by boiling; s alcohol [MER06]

2101
Compound: Metavanadic acid
Formula: HVO$_3$
Molecular Formula: HO$_3$V
Molecular Weight: 99.948
CAS RN: 13470-24-1
Properties: yellow scales [KIR83]
Solubility: s acids and alkalies [KIR83]

2102
Compound: Methylcyclopentadienyl-manganese tricarbonyl
Formula: C$_5$H$_4$CH$_3$Mn(CO)$_3$
Molecular Formula: C$_9$H$_7$MnO$_3$
Molecular Weight: 218.091
CAS RN: 12108-13-3
Properties: yellow liq; metallocene derivative [STR93] [ALD94]
Density, g/cm³: 1.38 [ALD94]
Melting Point, °C: −1 [ALD94]
Boiling Point, °C: 232–233 [ALD94]

2103
Compound: Methylgermane
Formula: GeH$_3$CH$_3$
Molecular Formula: CH$_6$Ge
Molecular Weight: 90.70
CAS RN: 1449-65-6
Properties: col gas [CRC10]
Density, g/L: 3.706
Melting Point, °C: −158 [CRC10]
Boiling Point, °C: −23 [CRC10]

2104
Compound: Molybdenum
Formula: Mo
Molecular Formula: Mo
Molecular Weight: 95.94
CAS RN: 7439-98-7

Properties: dark gray or black powd, metallic luster or silver-white color; bcc; enthalpy of fusion 37.48 kJ/mol; enthalpy of vaporization 491 kJ/mol; electrical resistivity (0°C) 50 μohm·m, (1000°C) 320; optical reflectivity 46% at 500 nm, 93% at 10,000 nm; has excellent corrosion resistance; used in crucible form to melt reactive metals such as barium, strontium, and indium, and to evaporate fluorides, chlorides, and reactive sulfides [KIR81] [MER06] [CER91] [CRC10]

Solubility: i H_2O, dil acids, conc HCl, alkali hydroxides, fused alkalis; reacts with HNO_3, hot conc H_2SO_4 [MER06]

Density, g/cm³: 10.28 [MER06]

Melting Point, °C: 2622 [MER06]

Boiling Point, °C: 4639 [CRC10]

Reactions: oxidizes rapidly to MoO_3 at >600°C [KIR81]

Thermal Conductivity, W/(m·K): 138 (25°C), 122 (500°C), 101 (1000°C), 82 (1500°C) [ALD94] [KIR81]

Thermal Expansion Coefficient: 0°C–400°C: 0.23%; 0°C–800°C: 0.46%; 0°C–1200°C: 0.72% [KIR81]

2105
Compound: Molybdenum acetate dimer
Formula: $[Mo(CH_3COO)_2]_2$
Molecular Formula: $C_8H_{12}Mo_2O_8$
Molecular Weight: 428.058
CAS RN: 14221-06-8
Properties: yellow cryst; sensitive to air [STR93]

2106
Compound: Molybdenum aluminide
Formula: Mo_3Al
Molecular Formula: $AlMo_3$
Molecular Weight: 314.802
CAS RN: 12003-72-4
Properties: −150, +325, −325 mesh with 99.5% purity; compound is a cermet, which can be flame sprayed [CER91] [HAW93]

2107
Compound: Molybdenum boride
Formula: Mo_2B
Molecular Formula: BMo_2
Molecular Weight: 202.691
CAS RN: 12006-99-4
Properties: tetr; refractory material; one of several borides, other formulas: MoB, MoB_2, Mo_2B_2, Mo_2B_5; used in brazes to join molybdenum and tungsten and tantalum for electronic and corrosion protection applications, and as sputtering material with 99.5% purity to produce wear-resistant and semiconductive films [KIR78] [HAW93] [CER91]

Density, g/cm³: 9.26 [CRC10]
Melting Point, °C: 2280 [KIR78]

2108
Compound: Molybdenum carbide
Formula: MoC
Molecular Formula: CMo
Molecular Weight: 107.951
CAS RN: 12011-97-1
Properties: −325 mesh, 10 μm or less, 95% purity; fcc, a=0.42810 nm [CIC73] [CER91]
Density, g/cm³: 9.15 [KIR78]
Melting Point, °C: 2577 [CIC73]

2109
Compound: Molybdenum carbide
Synonym: β-Mo_2C
Formula: Mo_2C
Molecular Formula: CMo_2
Molecular Weight: 203.891
CAS RN: 12069-89-5
Properties: gray powd; ortho-rhomb, a=0.4733 nm, b=0.60344 nm, c=0.52056 nm; in 99.5% purity, used as a sputtering target to produce wear-resistant films and semiconducting films; there is also material with formula MoC, 12011-97-1 [CIC73] [STR93] [CER91]
Density, g/cm³: 9.18 [STR93]
Melting Point, °C: 2687 [STR93]
Thermal Expansion Coefficient: 7.8×10^{-6}/K [KIR78]

2110
Compound: Molybdenum carbonyl
Synonym: molybdenum hexacarbonyl
Formula: $Mo(CO)_6$
Molecular Formula: C_6MoO_6
Molecular Weight: 264.002
CAS RN: 13939-06-5
Properties: white, shiny cryst; ortho-rhomb, a=1.20 nm, b=0.64 nm, c=1.12 nm; vapor pressure ~0.1 mm Hg at 20°C, ~43 mm Hg at 101°C; can be prepared by reacting $MoCl_5$ with zinc dust and CO in ether at high pressure; used in depositing molybdenum, for example to form molybdenum mirror [HAW93] [KIR81]
Solubility: i H_2O; s ceresin, paraffin oil, benzene; sl s ether and other organic solvents [HAW93]
Density, g/cm³: 1.96 [HAW93]
Melting Point, °C: decomposes at 150–151 (sublimes) [HAW93]

2111
Compound: Molybdenum disulfide
Synonym: molybdenite
Formula: MoS_2
Molecular Formula: MoS_2
Molecular Weight: 160.072
CAS RN: 1317-33-5
Properties: lead gray; hex; can be prepared by direct
  reaction of the elements to form a black and
  lustrous powd; greasy to the touch with low
  coefficient of friction; hardness 1–1.5 Mohs;
  similar to graphite in appearance; principal ore
  for molybdenum; as a 99% pure material, used
  as a sputtering target to form lubricant film
  on bearings and other moving parts [HAW93]
  [KIR81] [MER06] [CER91] [JAN85]
Solubility: i H_2O, dil acids; s H_2SO_4, conc
  HNO_3 [HAW93] [MER06]
Density, g/cm³: 5.06 [MER06]; black
  powd: 4.80 [STR93]
Melting Point, °C: 1750 [JAN85]
Reactions: begins to sublime at 450°C [MER06]

2112
Compound: Molybdenum metaphosphate
Formula: $Mo(PO_3)_6$
Molecular Formula: $MoO_{18}P_6$
Molecular Weight: 569.772
CAS RN: 133863-98-6
Properties: yellow powd [HAW93]
Solubility: i H_2O, most acids; sl s hot
  aqua regia [HAW93]
Density, g/cm³: 3.28 (0°C) [HAW93]

2113
Compound: Molybdenum mononitride
Formula: MoN
Molecular Formula: MoN
Molecular Weight: 109.947
CAS RN: 12033-19-1
Properties: hex, a=0.5725 nm, c=0.5608 nm [CIC73]
Density, g/cm³: 9.20 [LID94]

2114
Compound: Molybdenum nitride
Formula: Mo_2N
Molecular Formula: Mo_2N
Molecular Weight: 205.887
CAS RN: 12033-31-7
Properties: gray; fcc, a=0.416 nm; microhardness
  1700; transition temp 5.0 K [KIR81]

Density, g/cm³: 9.46 [KIR81]
Melting Point, °C: decomposes at 790 [KIR81]
Thermal Expansion Coefficient: 6.7×10^{-6} [KIR81]

2115
Compound: Molybdenum oxytetrafluoride
Formula: $MoOF_4$
Molecular Formula: F_4MoO
Molecular Weight: 187.933
CAS RN: 14459-59-7
Properties: volatile at moderate temp; there
  is a tetrahydrate (77727-63-0) and the
  compound MoO_2F_2 (13824-57-2) [KIR81]
Density, g/cm³: 3.00 [CRC10]
Melting Point, °C: $MoOF_4$: 98; MoO_2F_2
  sublimes at 270 [KIR81]
Boiling Point, °C: 180 [CRC10]

2116
Compound: Molybdenum pentaboride
Formula: Mo_2B_5
Molecular Formula: B_5Mo_2
Molecular Weight: 245.935
CAS RN: 12007-97-5
Properties: refractory material; borides used as
  a braze to join molybdenum, tungsten, and
  tantalum, and niobium parts for electronic,
  corrosion, and abrasion protection, used
  as a sputtering target with 99.5% purity to
  produce films which may be wear-resistant and
  semiconducting [CER91] [HAW93] [KIR78]
Melting Point, °C: 1600 [HAW93]
Reactions: transformed to MoB_2 at 1600°C [HAW93]

2117
Compound: Molybdenum phosphide
Formula: MoP
Molecular Formula: MoP
Molecular Weight: 126.914
CAS RN: 12163-69-8
Properties: gray-green powd; −200 mesh with
  99.5% purity [CER91] [CRC10]
Density, g/cm³: 7.34 [LID94]

2118
Compound: Molybdenum silicide
Synonym: molybdenum disilicide
Formula: $MoSi_2$
Molecular Formula: $MoSi_2$
Molecular Weight: 152.111

CAS RN: 12136-78-6
Properties: a cermet; dark gray cryst powd; has high stress rupture strength; used in electrical resistors, as a high temp protective coating, and in crucible form for melting bismuth, gallium, lead, silicon, silver, tin, zinc and to contain mercury, also as sputtering targets of 99.5%–99.95% purity to fabricate electrodes for integrated circuits [HAW93] [CER91]
Solubility: s HF and HNO_3; i aqua regia, other acids [HAW93]
Density, g/cm³: 6.31 [HAW93]
Melting Point, °C: 1870–2030 [HAW93]
Thermal Expansion Coefficient: (volume) 100°C (0.166), 200°C (0.400), 400°C (0.899), 800°C (1.960), 1200°C (3.048) [CLA66]

2119
Compound: Molybdenum(II) bromide
Formula: $MoBr_2$
Molecular Formula: Br_2Mo
Molecular Weight: 255.75
CAS RN: 13446-56-5
Properties: yellow-red cryst [CRC10]
Solubility: i H_2O, EtOH [CRC10]
Density, g/cm³: 4.88 [CRC10]
Melting Point, °C: decomposes at 700 [CRC10]

2120
Compound: Molybdenum(II) chloride
Formula: $MoCl_2$
Molecular Formula: Cl_2Mo
Molecular Weight: 166.845
CAS RN: 13478-17-6
Properties: yellow amorphous solid; –100 mesh with 99.5% purity [CER91] [CRC10]
Density, g/cm³: 3.714 [CRC10]
Melting Point, °C: decomposes at 530 [LID94]

2121
Compound: Molybdenum(II) iodide
Formula: MoI_2
Molecular Formula: I_2Mo
Molecular Weight: 349.749
CAS RN: 14055-74-4
Properties: black cryst; sensitive to air and moisture; can be prepared by reduction of MoI_3 (14055-75-5) with Mo, H_2, or a hydrocarbon [KIR81] [STR93]
Density, g/cm³: 5.278 [STR93]

2122
Compound: Molybdenum(III) bromide
Formula: $MoBr_3$
Molecular Formula: Br_3Mo
Molecular Weight: 335.652
CAS RN: 13446-57-6
Properties: can be prepared from $MoBr_4$ by reduction with Mo, H_2, or a hydrocarbon; another bromide is $MoBr_2$, which can be prepared by reduction of $MoBr_4$ (13446-56-5); oxybromides are: $MoOBr_3$ (13596-04-8) and MoO_2Br_2 (13595-98-7) [KIR81]
Melting Point, °C: decomposes [KIR81]

2123
Compound: Molybdenum(III) chloride
Formula: $MoCl_3$
Molecular Formula: Cl_3Mo
Molecular Weight: 202.298
CAS RN: 13478-18-7
Properties: –100 mesh with 99.5% purity; purple cryst [STR93] [CER91]
Density, g/cm³: 3.59 [STR93]
Melting Point, °C: disproportionates at >410 [KIR81]

2124
Compound: Molybdenum(III) fluoride
Formula: MoF_3
Molecular Formula: F_3Mo
Molecular Weight: 152.94
CAS RN: 20193-58-2
Properties: yellow brown hex cryst [CRC10]
Solubility: i H_2O [CRC10]
Density, g/cm³: 4.64 [CRC10]
Melting Point, °C: >600 [CRC10]

2125
Compound: Molybdenum(III) iodide
Formula: MoI_3
Molecular Formula: I_3Mo
Molecular Weight: 476.65
CAS RN: 14055-75-5
Properties: black solid [CRC10]
Solubility: i H_2O [CRC10]
Melting Point, °C: 927 [CRC10]

2126
Compound: Molybdenum(III) oxide
Synonym: molybdenum sesquioxide
Formula: Mo_2O_3
Molecular Formula: Mo_2O_3

Molecular Weight: 239.878
CAS RN: 1313-29-7
Properties: known only in the hydrated form
    Mo(OH)$_3$, but generally given the formula
    Mo$_2$O$_3$; grayish-black powd; used as a catalyst
    for organic synthesis; feed additive [HAW93]
Solubility: i H$_2$O; in alkalies; sl s acids [HAW93]

2127
Compound: Molybdenum(III) sulfide
Formula: Mo$_2$S$_3$
Molecular Formula: Mo$_2$S$_3$
Molecular Weight: 288.078
CAS RN: 12033-33-9
Properties: enthalpy of fusion 129.7 kJ/mol; can be
    prepared by reacting Mo and S in a sealed tube,
    under vacuum at high temperatures; there is also
    MoS$_2$, 1317-33-5 [CER91] [KIR81] [JAN85]
Density, g/cm^3: 5.91 [CRC10]
Melting Point, °C: 1807 [JAN85]
Boiling Point, °C: decomposes at 1870 [JAN85]

2128
Compound: Molybdenum(IV) bromide
Formula: MoBr$_4$
Molecular Formula: Br$_4$Mo
Molecular Weight: 415.556
CAS RN: 13520-59-7
Properties: black needles; deliq; obtained by
    bromination of Mo [KIR81] [CRC10]
Solubility: reacts with H$_2$O [LID94]
Melting Point, °C: decomposes [KIR81]

2129
Compound: Molybdenum(IV) chloride
Synonym: molybdenum tetrachloride
Formula: MoCl$_4$
Molecular Formula: Cl$_4$Mo
Molecular Weight: 237.751
CAS RN: 13320-71-3
Properties: black cryst; sensitive to both
    air and moisture [STR93]
Solubility: reacts with H$_2$O [LID94]
Melting Point, °C: decomposes [STR93]

2130
Compound: Molybdenum(IV) fluoride
Formula: MoF$_4$
Molecular Formula: F$_4$Mo
Molecular Weight: 171.93

CAS RN: 23412-45-5
Properties: green cryst [CRC10]
Solubility: reac H$_2$O [CRC10]
Melting Point, °C: decomposes [CRC10]

2131
Compound: Molybdenum(IV) iodide
Formula: MoI$_4$
Molecular Formula: I$_4$Mo
Molecular Weight: 603.56
CAS RN: 14055-76-6
Properties: black cryst [CRC10]
Melting Point, °C: decomposes at 100 [CRC10]

2132
Compound: Molybdenum(IV) oxide
Synonym: molybdenum dioxide
Formula: MoO$_2$
Molecular Formula: MoO$_2$
Molecular Weight: 127.939
CAS RN: 18868-43-4
Properties: tetr; brownish violet or lead gray,
    nonvolatile powd; can be made by reduction of
    MoO$_3$ with H$_2$ at 300°C–400°C, at higher temp
    Mo obtained ~500°C [CER91] [HAW93] [KIR81]
Solubility: sl s H$_2$SO$_4$; i HCl, HF, and alkalies [HAW93]
Density, g/cm^3: 6.47 [STR93]
Melting Point, °C: decomposes [JAN85]

2133
Compound: Molybdenum(IV) selenide
Formula: MoSe$_2$
Molecular Formula: MoSe$_2$
Molecular Weight: 253.860
CAS RN: 12058-18-3
Properties: gray powd; used as a solid lubricant and
    lubricant film prepared by sputtering 99.9%
    pure material [HAW93] [STR93] [CER91]
Density, g/cm^3: 6.0 [STR93]
Melting Point, °C: >1200 [STR93]

2134
Compound: Molybdenum(IV) sulfide
Formula: MoS$_2$
Molecular Formula: MoS$_2$
Molecular Weight: 160.07
CAS RN: 1317-33-5
Properties: black powd or hex cryst [CRC10]
Density, g/cm^3: 5.06 [CRC10]
Solubility: i H$_2$O; s conc acid [CRC10]
Melting Point, °C: 1750 [CRC10]

2135
Compound: Molybdenum(IV) telluride
Formula: $MoTe_2$
Molecular Formula: $MoTe_2$
Molecular Weight: 351.140
CAS RN: 12058-20-7
Properties: gray hex and 40 μm powd; used as a solid
  lubricant and with 99.9% purity as a sputtering target
  to produce lubricant films [HAW93] [CER91] [LID94]
Density, g/cm³: 7.7 [LID94]

2136
Compound: Molybdenum(V) chloride
Synonym: molybdenum pentachloride
Formula: $MoCl_5$
Molecular Formula: Cl_5Mo
Molecular Weight: 273.204
CAS RN: 10241-05-1
Properties: greenish black solid, is dark red liq or
  vapor; hygr; reacts with atm oxygen; enthalpy
  of vaporization 62.8 kJ/mol; enthalpy of fusion
  19.00 kJ/mol; used as a chlorination catalyst for
  vapor deposition of molybdenum [HAW93] [CRC10]
Solubility: s in dry ether and dry alcohol, in
  other organic solvents [HAW93]
Density, g/cm³: 2.928 [ALD94]
Melting Point, °C: 194 [CRC10]
Boiling Point, °C: 268 [HAW93]

2137
Compound: Molybdenum(V) fluoride
Formula: MoF_5
Molecular Formula: F_5Mo
Molecular Weight: 190.93
CAS RN: 13819-84-6
Properties: yellow monocl cryst [CRC10]
Solubility: reac H_2O [CRC10]
Melting Point, °C: 67 [CRC10]
Density, g/cm³: 3.5 [CRC10]

2138
Compound: Molybdenum(V) oxytrichloride
Formula: $MoOCl_3$
Molecular Formula: Cl_3MoO
Molecular Weight: 218.297
CAS RN: 13814-74-9
Properties: green cryst; can be prepared by refluxing
  $MoOCl_4$ in benzene; other oxychlorides are:
  MoOCl (41004-72-2), MoO_2Cl (20770-33-6),
  $MoOCl_2$ (24989-40-0) sublimes, MoO_2Cl_2
  (13637-68-8) mp 184°C in a sealed tube,
  Mo_2OCl_8 (77727-64-1) [KIR81] [CRC10]

Solubility: reacts with H_2O [LID94]
Melting Point, °C: 302 [KIR81]
Boiling Point, °C: sublimes [LID94]

2139
Compound: Molybdenum(VI) acid monohydrate
Synonym: molybdic acid monohydrate
Formula: $H_2MoO_4 \cdot H_2O$
Molecular Formula: H_4MoO_5
Molecular Weight: 179.969
CAS RN: 7782-91-4
Properties: white powd [MER06]
Solubility: 0.133 g/100 mL H_2O (18°C),
  2.568 (70°C) [CRC10]
Density, g/cm³: 3.1 [MER06]
Melting Point, °C: decomposes [CRC10]
Reactions: minus H_2O at 70°C [CRC10]

2140
Compound: Molybdenum(VI) dioxydichloride
Synonym: molybdenum dichloride dioxide
Formula: MoO_2Cl_2
Molecular Formula: Cl_2MoO_2
Molecular Weight: 198.844
CAS RN: 13637-68-8
Properties: yellowish-orange solid [LID94]
Density, g/cm³: 3.31 [ALD94]
Melting Point, °C: 184 [KIR81]

2141
Compound: Molybdenum(VI) dioxydifluoride
Synonym: molybdenum dioxide difluoride
Formula: MoO_2F_2
Molecular Formula: F_2MoO_2
Molecular Weight: 165.936
CAS RN: 13824-57-2
Properties: white cryst; hygr [CRC10]
Density, g/cm³: 3.494 [CRC10]
Melting Point, °C: sublimes at 270 [KIR81]

2142
Compound: Molybdenum(VI) fluoride
Synonym: molybdenum hexafluoride
Formula: MoF_6
Molecular Formula: F_6Mo
Molecular Weight: 209.930
CAS RN: 7783-77-9

Properties: volatile, white, cub cryst; very hygr;
enthalpy of fusion 4.33 kJ/mol; enthalpy
of vaporization 27.7 kJ/mol; evolves blue-
white clouds in moist air; can be prepared
by reacting molybdenum metal with F_2 at
elevated temperatures; other fluorides are:
MoF_5, 13819-84-6, mp 64°C, MoF_4, 23412-
45-5, MoF_3, 20193-58-2, mp decomposes, MoF_2,
20205-60-1 [KIR81] [MER06] [CRC10]
Solubility: hydrolyzed in H_2O; s anhydrous
HF: 1.5 moles/1000 g HF [MER06]
Density, g/cm³: 2.30 [ALD94]
Melting Point, °C: 17.5 [MER06]
Boiling Point, °C: 35.0 [MER06]

2143
Compound: Molybdenum(VI) oxide
Synonyms: molybdenum trioxide, molybdic anhydride
Formula: MoO_3
Molecular Formula: MoO_3
Molecular Weight: 143.938
CAS RN: 1313-27-5
Properties: white or sl yellow to sl bluish powd or
granules; ortho-rhomb, a=0.39628 nm, b=1.3855 nm;
c=0.36964 nm; enthalpy of vaporization 138 kJ/
mol; enthalpy of fusion 48.00 kJ/mol; produced
by roasting MoS_2; used in 99.99% pure form
as a sputtering target for luminescent coatings
[KIR81] [MER06] [CER91] [CRC10]
Solubility: g/100 g H_2O: 0.134 (20°C), 0.285
(30°C), 1.74 (80°C) [LAN05]; s conc
mineral acids, alkali hydroxides; after
strongly ignited is v sl s acids [MER06]
Density, g/cm³: 4.696 [MER06]
Melting Point, °C: 801 [CRC10]
Boiling Point, °C: 1155 [MER06]
Reactions: can sublime at >795°C [MER06]

2144
Compound: Molybdenum(VI) oxytetrachloride
Formula: $MoOCl_4$
Molecular Formula: Cl_4MoO
Molecular Weight: 253.750
CAS RN: 13814-75-0
Properties: green powd; sensitive to moisture;
volatile at ordinary temp [KIR81] [STR93]
Melting Point, °C: 100–101 [STR93]

2145
Compound: Molybdenum(VI) oxytetrafluoride
Formula: $MoOF_4$

Molecular Formula: F_4MoO
Molecular Weight: 187.93
CAS RN: 14459-59-7
Properties: volatile solid [CRC10]
Density, g/cm³: 3.00
Melting Point, °C: 97.2 [CRC10]
Boiling Point, °C: 186.0 [CRC10]

2146
Compound: Molybdenum(VI) sulfide
Synonym: molybdenum trisulfide
Formula: MoS_3
Molecular Formula: MoS_3
Molecular Weight: 192.138
CAS RN: 12033-29-3
Properties: brownish black amorphous
powd; formed by acidifying a solution of
ammonium tetrathiomolybdate [KIR81]
Melting Point, °C: decomposes [CRC10]

2147
Compound: Molybdic silicic acid hydrate
Formula: $H_4SiMo_{12}O_{40} \cdot xH_2O$
Molecular Formula: $H_4Mo_{12}O_{40}Si$ (anhydrous)
Molecular Weight: 1823.373 (anhydrous)
CAS RN: 11089-20-6
Properties: x is commonly 6–8; yellow
cryst powd; thermally stable; used in
photography, as a catalyst [HAW93]
Solubility: s H_2O, alcohol, acetone; i benzene;
decomposes in strongly basic solutions [HAW93]
Density, g/cm³: 2.82 [HAW93]

2148
Compound: Molybdophosphoric acid
Formula: $H_3P(Mo_3O_{10})_4$
Molecular Formula: $H_3Mo_{12}O_{40}P$
Molecular Weight: 1825.25
CAS RN: 51429-74-4
Properties: bright yellow cryst [CRC10]

2149
Compound: Monofluorophosphoric acid
Formula: $(HO)_2P(O)F$
Molecular Formula: FH_2O_3P
Molecular Weight: 99.986
CAS RN: 13537-32-1

Properties: colorless or yellow viscous liq; used in metal cleaners, electrolytic or chemical polishing agents [HAW93] [STR93]
Solubility: miscible with H_2O [HAW93]
Density, g/cm³: 1.818 [HAW93]

2150
Compound: Monoiodosilane
Formula: SiH_3I
Molecular Formula: H_3ISi
Molecular Weight: 158.014
CAS RN: 13598-42-0
Properties: enthalpy of vaporization 28.9 kJ/mol; entropy of vaporization 90.4 kJ/(mol·K) [CIC73]
Melting Point, °C: −57 [CIC73]
Boiling Point, °C: 45.6 [CIC73]

2151
Compound: Neodymium
Formula: Nd
Molecular Formula: Nd
Molecular Weight: 144.24
CAS RN: 7440-00-8
Properties: soft, silver white metal; yellowish in air; hex, at room temp; bcc >868°C; enthalpy of fusion 7.14 kJ/mol; enthalpy of sublimation 327.6 kJ/mol; electrical resistivity (20°C) 64.0 µohm·cm; radius of atom is 0.1821 nm; radius of ion 0.0995 nm for Nd^{+++}, rose-colored solution [MER06] [KIR82] [CRC10] [ALD94]
Solubility: slowly reacts with cold H_2O; rapidly with hot H_2O [MER06]
Density, g/cm³: 7.003 [MER06]
Melting Point, °C: 1021 [ALD94]
Boiling Point, °C: 3074 [KIR82]
Thermal Conductivity, W/(m·K): 16.5 (25°C) [CRC10]
Thermal Expansion Coefficient: 9.6×10^{-6}/K [CRC10]

2152
Compound: Neodymium acetate
Formula: $Nd(C_2H_3O_2)_3$
Molecular Formula: $C_6H_9NdO_6$
Molecular Weight: 321.373
CAS RN: 6192-13-6
Properties: red-purple cryst [CRC10]
Solubility: s H_2O [CRC10]

2153
Compound: Neodymium acetate monohydrate
Synonyms: acetic acid, neodymium salt monohydrate

Formula: $Nd(CH_3COO)_3 \cdot H_2O$
Molecular Formula: $C_6H_{11}NdO_7$
Molecular Weight: 339.389
CAS RN: 6192-13-8
Properties: light purple cryst [STR93]
Solubility: 26.2 g/100 mL H_2O [CRC10]

2154
Compound: Neodymium boride
Formula: NdB_6
Molecular Formula: B_6Nd
Molecular Weight: 209.106
CAS RN: 12008-23-0
Properties: −325 mesh 10 µm or less with 99.9% purity; refractory material [KIR78] [CER91]
Density, g/cm³: 4.93 [LID94]
Melting Point, °C: 2540 [KIR78]

2155
Compound: Neodymium bromate nonahydrate
Formula: $Nd(BrO_3)_3 \cdot 9H_2O$
Molecular Formula: $Br_3H_{18}NdO_{18}$
Molecular Weight: 690.084
CAS RN: 15162-92-2
Properties: red; hex [CRC10]
Solubility: g/100 g H_2O: 43.9 (0°C), 75.6 (20°C), 116 (40°C) [LAN05]
Melting Point, °C: 66.7 [CRC10]
Reactions: minus $9H_2O$ at 150°C [CRC10]

2156
Compound: Neodymium bromide
Formula: $NdBr_3$
Molecular Formula: Br_3Nd
Molecular Weight: 383.952
CAS RN: 13536-80-6
Properties: −20 mesh with 99.9% purity; green powd; hygr [STR93] [CER91]
Density, g/cm³: 5.3 [LID94]
Melting Point, °C: 684 [STR93]
Boiling Point, °C: 1540 [STR93]

2157
Compound: Neodymium carbonate hydrate
Formula: $Nd_2(CO_3)_3 \cdot xH_2O$
Molecular Formula: $C_3Nd_2O_9$ (anhydrous)
Molecular Weight: 468.508 (anhydrous)

CAS RN: 38245-38-4
Properties: light purple powd [STR93]
Solubility: i H_2O; s acids [HAW93]

2158
Compound: Neodymium cerium copper oxide
Formula: $Nd_{1.85}Ce_{0.15}CuO_4$
Molecular Formula: $Ce_{0.15}CuNd_{1.85}O_4$
Molecular Weight: 415.405
CAS RN: 119800-94-1
Properties: superconductor; general formula is $Nd_{2-x}Ce_xCuO_4$; material with formula where x=0.15 has T_c 20 K; cuprate semiconductors are metastable at low temperatures; oriented thin films can be prepared from bulk cuprates or from metal oxides by sputtering or by laser ablation; potential uses of superconductors include frictionless bearings, microwave, and electronic devices [CEN92]

2159
Compound: Neodymium chloride
Formula: $NdCl_2$
Molecular Formula: Cl_2Nd
Molecular Weight: 215.148
CAS RN: 25469-93-6
Properties: green hygr solid [CRC10]
Melting Point, °C: 841 [CRC10]

2160
Compound: Neodymium chloride
Formula: $NdCl_3$
Molecular Formula: Cl_3Nd
Molecular Weight: 250.598
CAS RN: 10024-93-8
Properties: −20 mesh with 99.9% purity; violet powd [ALD94] [STR93] [CER91]
Solubility: g/100 g H_2O: 96.7 (10°C), 98.0 (20°C), 105 (60°C) [LAN05]; v s alcohol; i ether, chloroform [HAW93] [MER06]
Density, g/cm³: 4.134 [STR93]
Melting Point, °C: 784 [STR93]
Boiling Point, °C: 1600 [HAW93]

2161
Compound: Neodymium chloride hexahydrate
Formula: $NdCl_3 \cdot 6H_2O$
Molecular Formula: $Cl_3H_{12}NdO_6$
Molecular Weight: 358.689
CAS RN: 13477-89-9

Properties: −4 mesh with 99.9% purity; purple cryst [STR93] [CER91]
Solubility: 2.46 parts per 1 part H_2O [MER06]; s alcohol [HAW93]
Density, g/cm³: 2.282 [STR93]
Melting Point, °C: 124 [MER06]
Reactions: minus $6H_2O$ at 160°C [HAW93]

2162
Compound: Neodymium fluoride
Formula: NdF_3
Molecular Formula: F_3Nd
Molecular Weight: 201.235
CAS RN: 13709-42-7
Properties: purple powd or 99.9% pure sintered tablets; hygr; tablets used as evaporation and sputtering material for multilayers, used with ZnS [STR93] [CER91]
Solubility: i H_2O [HAW93]
Density, g/cm³: 6.506 [STR93]
Melting Point, °C: 1410 [HAW93]
Boiling Point, °C: 2300 [HAW93]

2163
Compound: Neodymium hexafluoroacetylacetonate dihydrate
Synonyms: 1,1,1,5,5,5-hexafluoro-2,4-pentanedione, Nd
Formula: $Nd(CF_3COCHCOCF_3)_3 \cdot 2H_2O$
Molecular Formula: $C_{15}H_7F_{18}NdO_8$
Molecular Weight: 801.427
CAS RN: 47814-18-6
Properties: cryst [STR93]

2164
Compound: Neodymium hydride
Formula: NdH_{2-3}
Molecular Formula: H_2Nd; H_3Nd
Molecular Weight: NdH_2: 146.256; NdH_3: 147.264
CAS RN: 13864-04-5
Properties: lumps under argon atm; −60 mesh with 99.9% purity [CER91] [ALF95]

2165
Compound: Neodymium hydroxide
Formula: $Nd(OH)_3$
Molecular Formula: H_3NdO_3
Molecular Weight: 195.262
CAS RN: 16469-17-3

Properties: bluish or pink precipitate; heating at 300°C–350°C converts compound to grayish-brown $2Nd_2O_3 \cdot H_2O$, further heating converts hydroxide to $Nd_2O_3 \cdot H_2O$ [MER06] [ALF95]

2166
Compound: Neodymium iodide
Formula: NdI_3
Molecular Formula: I_3Nd
Molecular Weight: 524.953
CAS RN: 13813-24-6
Properties: −20 mesh with 99.9% purity; green powd; hygr [STR93] [CER91]
Melting Point, °C: 775 [STR93]

2167
Compound: Neodymium nitrate
Formula: $Nd(NO_3)_3$
Molecular Formula: N_3NdO_9
Molecular Weight: 330.257
CAS RN: 10045-95-1
Properties: violet hygr cryst [CRC10]
Solubility: g/100 g H_2O: 152^{25}; s EtOH [CRC10]

2168
Compound: Neodymium nitrate hexahydrate
Formula: $Nd(NO_3)_3 \cdot 6H_2O$
Molecular Formula: $H_{12}N_3NdO_{15}$
Molecular Weight: 438.346
CAS RN: 14517-29-4
Properties: purple cryst; hygr [STR93]
Solubility: g/100 g H_2O: 127 (0°C), 142 (20°C), 211 (60°C) [LAN05]

2169
Compound: Neodymium nitride
Formula: NdN
Molecular Formula: NNd
Molecular Weight: 158.247
CAS RN: 25764-11-8
Properties: black powd; −60 mesh with 99.9% purity; NaCl cryst system, a = 0.514 nm [CIC73] [CER91] [CRC10]
Solubility: decomposed by H_2O [CRC10]
Density, g/cm³: 7.69 [LID94]

2170
Compound: Neodymium oxalate decahydrate
Formula: $Nd_2(C_2O_4)_3 \cdot 10H_2O$
Molecular Formula: $C_6H_{20}Nd_2O_{22}$

Molecular Weight: 732.692
CAS RN: 14551-74-7
Properties: rose cryst [STR93]
Solubility: 0.000074 g/100 mL H_2O (25°C) [CRC10]
Reactions: minus H_2O 40°C–50°C [AES93]

2171
Compound: Neodymium oxide
Synonym: neodymia
Formula: Nd_2O_3
Molecular Formula: Nd_2O_3
Molecular Weight: 336.478
CAS RN: 1313-97-9
Properties: pure material is a blue powd; technical material has a brown color; hygr; absorbs atm CO_2; hex; has sl red fluorescence; used in ceramic capacitors, in coloring glass, and in television tubes, and as an evaporated material of 99.9% purity, it is possibly reactive to radio frequencies [HAW93] [MER06] [CER91]
Solubility: 5.7×10^{-6} g mol/L H_2O (29°C); s dil acids [MER06]
Density, g/cm³: 7.24 [STR93]
Melting Point, °C: 2272 [STR93]

2172
Compound: Neodymium perchlorate hexahydrate
Formula: $Nd(ClO_4)_3 \cdot 6H_2O$
Molecular Formula: $Cl_3H_{12}NdO_{18}$
Molecular Weight: 550.683
CAS RN: 17522-69-9
Properties: cryst [ALF95]

2173
Compound: Neodymium phosphate hydrate
Formula: $NdPO_4 \cdot xH_2O$
Molecular Formula: NdO_4P (anhydrous)
Molecular Weight: 239.211 (anhydrous)
CAS RN: 14298-32-9
Properties: powd [ALF95]

2174
Compound: Neodymium silicide
Formula: $NdSi_2$
Molecular Formula: $NdSi_2$
Molecular Weight: 200.411
CAS RN: 12137-04-1
Properties: 10 mm and down lump [ALF93]

2175
Compound: Neodymium sulfate
Formula: $Nd_2(SO_4)_3$
Molecular Formula: $Nd_2O_{12}S_3$
Molecular Weight: 576.671
CAS RN: 101509-27-7
Properties: pinkish needles [MER06]
Solubility: g/100 g H_2O: 13.0 (0°C), 7.1 (20°C), 1.2 (90°C) [LAN05]
Melting Point, °C: decomposes at 700–800 [MER06]

2176
Compound: Neodymium sulfate octahydrate
Formula: $Nd_2(SO_4)_3 \cdot 8H_2O$
Molecular Formula: $H_{16}Nd_2O_{20}S_3$
Molecular Weight: 720.793
CAS RN: 13477-91-3
Properties: purple cryst [STR93]
Solubility: s cold H_2O, less soluble in hot H_2O [HAW93]
Density, g/cm³: 2.85 [HAW93]
Melting Point, °C: decomposes at 800 [HAW93]

2177
Compound: Neodymium sulfide
Formula: Nd_2S_3
Molecular Formula: Nd_2S_3
Molecular Weight: 384.678
CAS RN: 12035-32-4
Properties: −200 mesh with 99.9% purity [CER91]
Density, g/cm³: 5.46 [LID94]
Melting Point, °C: 2207 [LID94]

2178
Compound: Neodymium telluride
Formula: Nd_2Te_3
Molecular Formula: Nd_2Te_3
Molecular Weight: 671.280
CAS RN: 12035-35-7
Properties: −20 mesh with 99.9% purity gray powd [STR93] [CER91]
Density, g/cm³: 7.0 [LID94]
Melting Point, °C: 1377 [LID94]

2179
Compound: Neodymium trifluoroacetylacetonate
Synonyms: 1,1,1-trifluoro-2,4-pentanedione, Nd derivative
Formula: $Nd(CF_3COCH=C(O)CH_3)_3$

Molecular Formula: $C_{15}H_{12}F_9NdO_6$
Molecular Weight: 603.482
CAS RN: 37473-67-9
Properties: blue-pink cryst [STR93]
Melting Point, °C: 140–142 [STR93]

2180
Compound: Neodymium tris(cyclopentadienyl)
Synonym: tris(cyclopentadienyl)neodymium
Formula: $Nd(C_5H_5)_3$
Molecular Formula: $C_{15}H_{15}Nd$
Molecular Weight: 339.53
CAS RN: 1273-98-9
Properties: powd; sensitive to air and moisture [STR93]
Melting Point, °C: 417 [STR93]
Boiling Point, °C: sublimes at 220 (0.01 mm Hg) [STR93]

2181
Compound: Neon
Formula: Ne
Molecular Formula: Ne
Molecular Weight: 20.1797
CAS RN: 7440-01-9
Properties: inert, odorless gas; critical temp −228.7°C; critical pressure 2.65 MPa; enthalpy of vaporization 1.71 kJ/mol; enthalpy of fusion 0.34 kJ/mol; heat capacity (25°C) 20.79; sonic velocity at 0°C is 433 m/s; viscosity (25°C) 31.73 Pa·s [KIR78] [CRC10] [AIR87]
Solubility: 10.5 mL/100 g H_2O (20°C, 101.32 kPa) [KIR78]; Henry's law constants, $k \times 10^{-4}$: 13.023 (70.0°C), 12.022 (124.5°C), 9.805 (174.5°C), 7.166 (226.4°C), 4.160 (283.7°C) [POT78]
Density, g/cm³: gas (101.3 kPa, 0°C) 0.0009000 [KIR78]
Melting Point, °C: −248.59 [CRC10]
Boiling Point, °C: −246.08 [CRC10]
Thermal Conductivity, W/(m·K): gas: (101.32 kPa, 0°C) 0.04607 [KIR78]

2182
Compound: Neptunium
Formula: Np
Molecular Formula: Np
Molecular Weight: 237
CAS RN: 7439-99-8

Properties: silvery metal; α: ortho-rhomb, a=0.4721 nm, b=0.4888 nm, c=0.6661 nm, stable from room temp to 280°C; β: tetra, a=0.4895 nm, c=0.3386 nm, stable from 280–577°C; γ: a=0.3518 nm; bcc, stable from 577–637°C; enthalpy of vaporization 418 kJ/mol; enthalpy of fusion 3.20 kJ/mol; ^{237}Np, $t_{1/2}$=2.14×10^{+6} years, $t_{1/2}$ of ^{236}Np 1.29×10^{+6} years; discovered in 1940; produced in kg amounts as a by-product of plutonium production [MER06] [KIR78] [CRC10]
Solubility: s HCl [HAW93]
Density, g/cm³: 20.45 [KIR78]
Melting Point, °C: 637 [KIR91]
Boiling Point, °C: 3900 [KIR91]
Thermal Conductivity, W/(m·K): 6.3 [CRC10]

2183
Compound: Neptunium(IV) oxide
Synonym: neptunium dioxide
Formula: NpO_2
Molecular Formula: NpO_2
Molecular Weight: 269
CAS RN: 12035-79-9
Properties: cub; dark olive powd [HAW93] [CRC10]
Density, g/cm³: 11.11 [CRC10]
Melting Point, °C: 2547 [LID94]

2184
Compound: Nickel
Formula: Ni
Molecular Formula: Ni
Molecular Weight: 58.6934
CAS RN: 7440-02-0
Properties: white, ferromagnetic metal; fcc, a=0.35238 nm; decomposes steam at red heat; electrical resistivity (20°C) 6.844 μohm·cm; Curie temp 353°C; Poisson's ratio 0.30; saturation magnetization 0.617 T; residual magnetization 0.300 T; coercive force 239 A/m; initial permeability 0.251 mH/m, maximum permeability 2.51–3.77 mH/m; enthalpy of vaporization 377.5 kJ/mol; enthalpy of fusion 17.48 kJ/mol [KIR81] [MER06] [CRC10] [ALD94]
Solubility: i H_2O; slowly attacked by dil HCl, H_2SO_4, readily by HNO_3 [MER06]
Density, g/cm³: 8.908 [HAW93]
Melting Point, °C: 1453 [ALD94]
Boiling Point, °C: 2732 (extrapolated) [KIR81]
Thermal Conductivity, W/(m·K): 90.0 (25°C), 82.8 (100°C), 63.6 (300°C), 61.9 (500°C) [KIR81] [ALD94]

Thermal Expansion Coefficient: 500°C: 15.2×10^{-6}/°C [KIR81]

2185
Compound: Nickel acetate tetrahydrate
Synonyms: acetic acid, nickel(II) salt
Formula: $Ni(CH_3COO)_2 \cdot 4H_2O$
Molecular Formula: $C_4H_{14}NiO_8$
Molecular Weight: 248.843
CAS RN: 6018-89-9
Properties: green monocl cryst; hygr; used in textiles as a mordant [HAW93]
Solubility: s in 6 parts H_2O; s alcohol [MER06]
Density, g/cm³: 1.744 [MER06]
Melting Point, °C: decomposes at 250 [HAW93]

2186
Compound: Nickel acetylacetonate
Synonyms: 2,4-pentanedione, Ni(II) derivative
Formula: $Ni(CH_3COCH=C(O)CH_3)_2$
Molecular Formula: $C_{10}H_{14}NiO_4$
Molecular Weight: 256.912
CAS RN: 3264-82-2
Properties: light green powd; hygr; trimer; there is a hydrate, CAS RN 120156-44-7 [STR93] [ALD94] [COT88]
Melting Point, °C: decomposes at 238 [STR93]

2187
Compound: Nickel aluminide
Formula: $NiAl_3$
Molecular Formula: Al_3Ni
Molecular Weight: 139.638
CAS RN: 12004-71-6
Properties: −20 mesh with 99.9% purity; there is also NiAl, 12003-78-0; a cermet that can be flame sprayed [CER91] [HAW93]

2188
Compound: Nickel ammonium chloride hexahydrate
Formula: $NH_4NiCl_3 \cdot 6H_2O$
Molecular Formula: $Cl_3H_{16}NNiO_6$
Molecular Weight: 291.182
CAS RN: 16122-03-5 (anhydrous)
Properties: green hygr cryst [CRC10]
Solubility: s H_2O [CRC10]
Density, g/cm³: 1.65 [CRC10]

2189
Compound: Nickel ammonium sulfate
Formula: $Fe_2(SO_4)_3 \cdot 9H_2O$
Molecular Formula: $Fe_2H_{18}O_{21}S_3$
Molecular Weight: 562.015
CAS RN: 13520-56-4
Properties: yellow hex cryst [CRC10]
Solubility: g/100 g H_2O: 440^{20} [CRC10]
Density, g/cm³: 2.1 [CRC10]
Melting Point, °C: decomposes at 400 [CRC10]

2190
Compound: Nickel ammonium sulfate hexahydrate
Formula: $Ni(NH_4)_2(SO_4)_2 \cdot 6H_2O$
Molecular Formula: $N_4NiH_{20}O_{14}S_2$
Molecular Weight: 394.987
CAS RN: 7785-20-8
Properties: blue-green cryst [CRC10]
Solubility: g/100 g H_2O: 6.5^{20}; s H_2O; i EtOH [CRC10]
Density, g/cm³: 1.92 [CRC10]
Melting Point, °C: decomposes at 130 [CRC10]

2191
Compound: Nickel antimonide
Synonym: breithauptite
Formula: NiSb
Molecular Formula: NiSb
Molecular Weight: 180.453
CAS RN: 12035-52-8
Properties: reddish hex; 6 mm pieces and smaller with 99.5% purity; there is also Ni_3Sb, 12503-49-0 [CER91] [CRC10]
Density, g/cm³: 8.74 [LID94]
Melting Point, °C: 1158 [CRC10]
Boiling Point, °C: decomposes at 1400 [CRC10]

2192
Compound: Nickel arsenate octahydrate
Formula: $Ni_3(AsO_4)_2 \cdot 8H_2O$
Molecular Formula: $As_2H_{16}Ni_{13}O_{16}$
Molecular Weight: 598.040
CAS RN: 7784-48-7
Properties: yellowish green powd; formed when an aq solution of arsenic anhydride is reacted with nickel carbonate; used as a catalyst for hardening fats, which are used in soap [HAW93] [KIR81]
Solubility: i H_2O; s acids [HAW93]
Density, g/cm³: 4.98 [HAW93]
Melting Point, °C: decomposes to NiO and As_2O_5 [KIR81]

2193
Compound: Nickel arsenide
Synonym: niccolite
Formula: NiAs
Molecular Formula: AsNi
Molecular Weight: 133.615
CAS RN: 27016-75-7
Properties: hex; 6 mm pieces and smaller of 99.5% purity; there is also $NiAs_2$, 12068-61-0, of the same purity [CER91] [CRC10]
Density, g/cm³: 7.57 [CRC10]
Melting Point, °C: 968 [CRC10]

2194
Compound: Nickel basic carbonate tetrahydrate
Synonym: zaratite
Formula: $NiCO_3 \cdot 2Ni(OH)_2 \cdot 4H_2O$
Molecular Formula: $CH_{12}Ni_3O_{11}$
Molecular Weight: 376.179
CAS RN: 3333-67-3
Properties: light green rhomb cryst or brown powd; can be obtained by adding sodium carbonate to a nickel salt solution; used in the manufacture of catalyst, electroplating, ceramic colors, and glazes [HAW93] [KIR81]
Solubility: i H_2O; s in dil acids and in ammonia [HAW93]
Density, g/cm³: 2.6 [HAW93]

2195
Compound: Nickel bis(cyclopentadienyl)
Synonym: bis(cyclopentadienyl)nickel
Formula: $Ni(C_5H_5)_2$
Molecular Formula: $C_{10}H_{10}Ni$
Molecular Weight: 188.883
CAS RN: 1271-28-9
Properties: bright green cryst; fairly stable in air [COT88] [ALF95]
Melting Point, °C: 171–173 [ALF95]
Reactions: forms CpNiNO with NO [COT88]

2196
Compound: Nickel boride
Formula: Ni_2B
Molecular Formula: BNi_2
Molecular Weight: 128.198
CAS RN: 12007-01-1
Properties: −35 mesh with 99% purity; refractory material [CER91] [KIR78]
Density, g/cm³: 7.90 [LID94]
Melting Point, °C: 1230 [KIR78]

2197

Compound: Nickel boride
Formula: NiB
Molecular Formula: BNi
Molecular Weight: 69.504
CAS RN: 12007-00-0
Properties: −35 mesh with 99% purity; refractory material; silver green [CER91] [STR93] [KIR78]
Density, g/cm³: 7.39 [STR93]
Melting Point, °C: 1080 [KIR78]

2198

Compound: Nickel boride
Formula: Ni₃B
Molecular Formula: BNi₃
Molecular Weight: 186.891
CAS RN: 12007-02-2
Properties: −35 mesh with 99% purity; refractory material [KIR78]
Density, g/cm³: 8.17 [LID94]
Melting Point, °C: 1155 [KIR78]

2199

Compound: Nickel bromide
Formula: NiBr₂
Molecular Formula: Br₂Ni
Molecular Weight: 218.501
CAS RN: 13462-88-9
Properties: brownish yellow or yellow, lustrous scales [HAW93]
Solubility: g/100 g soln, H_2O: 53.0 (0°C), 57.3 (25°C), 60.8 (100°C); solid phase, NiBr₂·6H₂O [KRU93]
Density, g/cm³: 5.098 [STR93]
Melting Point, °C: 963 [STR93]

2200

Compound: Nickel bromide trihydrate
Formula: NiBr₂·3H₂O
Molecular Formula: Br₂H₆NiO₃
Molecular Weight: 272.547
CAS RN: 13462-88-9
Properties: yellowish green, very deliq cryst; hexahydrate obtained from a reaction of black NiO and HBr [KIR81] [MER06]
Solubility: 199 g/100 mL H_2O (0°C), 316 g/100 mL H_2O (100°C) [CRC10]
Melting Point, °C: 300, decomposes [STR93]
Reactions: minus 3H₂O ~200°C [MER06]

2201

Compound: Nickel carbonate
Formula: NiCO₃
Molecular Formula: CNiO₃
Molecular Weight: 118.702
CAS RN: 3333-67-3
Properties: light green rhomb; −325 mesh 10 μm and smaller; can be prepared by adding Na₂CO₃ solution to a nickel salt solution or by heating Ni powd in NH₃ and CO₂, followed by boiling of NH₃ to obtain purer material; used in manufacture of catalysts, colored glass, and certain nickel pigments [KIR81] [CER91]
Solubility: 0.0093 g/100 mL H_2O (25°C) [CRC10]
Density, g/cm³: 4.39 [LID94]
Melting Point, °C: decomposes [CRC10]

2202

Compound: Nickel carbonate hydroxide tetrahydrate
Formula: 2NiCO₃·3Ni(OH)₂·4H₂O
Molecular Formula: C₂H₁₄Ni₁₅O₁₆
Molecular Weight: 587.591
CAS RN: 12244-51-8
Properties: green powd [AES93] [STR93] [ALD94]
Melting Point, °C: decomposes [CRC10]

2203

Compound: Nickel carbonyl
Formula: Ni(CO)₄
Molecular Formula: C₄NiO₄
Molecular Weight: 170.735
CAS RN: 13463-39-3
Properties: colorless, volatile liq; oxidizes in air; explodes at ~60°C; vapor pressure, kPa: 19.2 (0°C), 28.7 (10°C), 44.0 (20°C); critical temp 200°C; vapor density is four times greater than air; manufactured by reacting CO with Ni; used to prepare highly pure nickel and in alkaline nickel batteries [KIR81] [HAW93] [MER06]
Solubility: s in ~5000 parts air free H_2O; s alcohol, benzene, chloroform, acetone, CCl₄ [MER06]
Density, g/cm³: 1.3185 [HAW93]
Melting Point, °C: −19.3 [MER06]
Boiling Point, °C: 42.6 [KIR81]
Reactions: decomposed by heat to Ni and 4CO [DOU83]

2204

Compound: Nickel chlorate hexahydrate
Formula: Ni(ClO₃)₂·6H₂O
Molecular Formula: Cl₂H₁₂NiO₁₂
Molecular Weight: 333.687
CAS RN: 67952-43-6

Properties: dark red [CRC10]
Solubility: mol/100 mol H_2O: 8.88 (0°C), 12.02 (25°C); solid phase, $Ni(ClO_3)_2 \cdot 6H_2O$ (0°C), $Ni(ClO_3)_2 \cdot 4H_2O$ (25°C) [KRU93]; g/100 g H_2O: 111 (0°C), 133 (20°C), 308 (80°C) [LAN05]
Density, g/cm³: 2.07 [CRC10]
Melting Point, °C: decomposes at 80 [CRC10]

2205

Compound: Nickel chloride
Formula: $NiCl_2$
Molecular Formula: Cl_2Ni
Molecular Weight: 129.598
CAS RN: 7718-54-9
Properties: golden yellow powd; hygr; readily absorbs NH_3; enthalpy of fusion 71.20 kJ/mol [CRC10] [STR93] [MER06]
Solubility: g/100 g soln, H_2O: 35.04 (0°C), 39.6 (25°C), 46.7 (100.2°C); solid phase, $NiCl_2 \cdot 6H_2O$ (0°C, 25°C), $NiCl_2 \cdot 2H_2O$ (100.2°C) [KRU93]
Density, g/cm³: 3.55 [HAW93]
Melting Point, °C: 1031 [CRC10]
Boiling Point, °C: sublimes [LID94]
Reactions: sublimable in absence of air [MER06]

2206

Compound: Nickel chloride hexahydrate
Formula: $NiCl_2 \cdot 6H_2O$
Molecular Formula: $Cl_2H_{12}NiO_6$
Molecular Weight: 237.689
CAS RN: 7791-20-0
Properties: green cryst; monocl; formed when Ni or NiO is reacted with hot aq HCl solution; used in nickel electroplating [KIR81] [MER06]
Solubility: 4.9208 ± 0.0028 mol/kg, and 4.9172 ± 0.0049 mol/kg, in H_2O (25°C) [RAR87b], [RAR92]; s alcohol [MER06]
Reactions: sublimes at 973°C [STR93]

2207

Compound: Nickel chromate
Formula: $NiCrO_4$
Molecular Formula: $CrNiO_4$
Molecular Weight: 174.687
CAS RN: 14721-18-7
Properties: maroon; used in catalysts; gives dark red aq solutions; there is also nickel chromium oxide, $NiCr_2O_4$, 12018-18-7, −100 mesh with 99% purity [KIR78] [CER91]
Solubility: v sl s H_2O [KIR78]

2208

Compound: Nickel cyanide tetrahydrate
Formula: $Ni(CN)_2 \cdot 4H_2O$
Molecular Formula: $C_2H_8N_2NiO_4$
Molecular Weight: 182.790
CAS RN: 13477-95-7
Properties: apple green plates or powd; prepared by reacting KCN and $NiSO_4$; used in metallurgy and in electroplating; anhydrous $Ni(CN)_2$, 557-19-7 [STR93] [MER06] [KIR81]
Solubility: i H_2O; sl s dil acids; v s alkali cyanides, ammonia, ammonium carbonate [MER06]
Melting Point, °C: decomposes above 200 [KIR81]
Reactions: minus $4H_2O$ at 200°C [HAW93]

2209

Compound: Nickel disilicide
Formula: $NiSi_2$
Molecular Formula: $NiSi_2$
Molecular Weight: 114.864
CAS RN: 12201-89-7
Properties: −80 mesh powd [ALF93]
Density, g/cm³: 4.83 [LID94]
Melting Point, °C: 1200 [ALF93]

2210

Compound: Nickel fluoride
Formula: NiF_2
Molecular Formula: F_2Ni
Molecular Weight: 96.690
CAS RN: 10028-18-9
Properties: yellowish to green tetr cryst; hygr [MER06] [STR93]
Solubility: g/100 g soln, H_2O: 2.50 (25°C), 2.52 (90°C); solid phase, $NiF_2 \cdot 4H_2O$ [KRU93]; i alcohol, ether [MER06]
Density, g/cm³: 4.72 [MER06]
Melting Point, °C: ~1100 [KIR78]
Reactions: sublimes under HF at >1000°C [MER06]

2211

Compound: Nickel fluoride tetrahydrate
Formula: $NiF_2 \cdot 4H_2O$
Molecular Formula: $F_2H_8NiO_4$
Molecular Weight: 168.752
CAS RN: 13940-83-5
Properties: green powd; can be prepared by reacting $NiCO_3$ with aq HF [KIR78] [STR93]

2212

Compound: Nickel hexafluoroacetylacetonate hydrate
Synonyms: 1,1,1,5,5,5-hexafluoro-2,4-
   pentanedione, Ni derivative
Formula: $Ni(CF_3COCH=C(O)CF_3)_2 \cdot xH_2O$
Molecular Formula: $C_{10}H_2F_{12}NiO_4$ (anhydrous)
Molecular Weight: 472.798
CAS RN: 14949-69-0
Properties: green cryst [STR94] [ALD94]
Melting Point, °C: 211–213 [STR93]

2213

Compound: Nickel hydroxide
Formula: $Ni(OH)_2$
Molecular Formula: H_2NiO_2
Molecular Weight: 92.708
CAS RN: 12054-48-7
Properties: fine green powd; formed when a nickel
   sulfate solution is treated with sodium hydroxide,
   then neutralized and filtered; used in the
   manufacture of Ni-Cd batteries [KIR81] [HAW93]
Solubility: mol/L soln, H_2O: 1×10^{-4} (20°C)
   [KRU93]; s in acids and in NH_4OH [HAW93]
Density, g/cm³: 4.15 [HAW93]
Melting Point, °C: decomposes at 230 [STR93]
Reactions: forms NiO and H_2O on
   decomposing [MER06]

2214

Compound: Nickel iodate
Formula: $Ni(IO_3)_2$
Molecular Formula: I_2NiO_6
Molecular Weight: 408.498
CAS RN: 13477-98-0
Properties: yellow needles [LAN05]
Solubility: mol/100 mol H_2O: 0.023 (0°C), 0.035
   (25°C), 0.044 (90°C); solid phase, $Ni(IO_3)_2 \cdot 2H_2O$
   (0°C, 25°C), $Ni(IO_3)_2$ (90°C) [KRU93]
Density, g/cm³: 5.07 [LAN05]

2215

Compound: Nickel iodate tetrahydrate
Formula: $Ni(IO_3)_2 \cdot 4H_2O$
Molecular Formula: $H_8I_2NiO_{10}$
Molecular Weight: 480.560
CAS RN: 13477-99-1
Properties: hex cryst; deliq [CRC10] [LAN05]
Solubility: g/100 g H_2O: 0.74 (0°C), 1.09
   (20°C), 1.43 (30°C) [LAN05]
Melting Point, °C: decomposes at 100 [LAN05]

2216

Compound: Nickel iodide
Formula: NiI_2
Molecular Formula: I_2Ni
Molecular Weight: 312.502
CAS RN: 13462-90-3
Properties: iron black color; hygr; sublimes
   in absence of air [MER06] [STR93]
Solubility: g/100 g soln, H_2O: 55.4 (0°C),
   60.7 (25°C), 65.3 (90°C) [KRU93]
Density, g/cm³: 5.83 [KIR81]
Melting Point, °C: 797 [KIR81]
Boiling Point, °C: sublimes [LID94]

2217

Compound: Nickel iodide hexahydrate
Formula: $NiI_2 \cdot 6H_2O$
Molecular Formula: $H_{12}I_2NiO_6$
Molecular Weight: 420.593
CAS RN: 7790-34-3
Properties: bluish green, very deliq cryst;
   obtained when nickel carbonate is reacted
   with hydriodic acid [KIR81]
Solubility: v s H_2O, alcohol [MER06]
Density, g/cm³: 5.83 [KIR81]
Melting Point, °C: 797 [KIR81]

2218

Compound: Nickel molybdate
Formula: $NiMoO_4$
Molecular Formula: $MoNiO_4$
Molecular Weight: 218.631
CAS RN: 14177-55-0
Properties: green; three forms; α monocl, β and γ;
   obtained by reacting the metal oxides [KIR81]
Density, g/cm³: α: 3.5; β: 4.9 [KIR81]
Melting Point, °C: 970 [KIR78]

2219

Compound: Nickel nitrate
Formula: $Ni(NO_3)_2$
Molecular Formula: N_2NiO_6
Molecular Weight: 182.702
CAS RN: 13138-45-9
Properties: prepared by reacting red fuming nitric
   acid with nickel nitrate hexahydrate [KIR81]
Solubility: g/100 g soln, H_2O: 44.2 (0°C), 50.0 (25°C),
   69.2 (99.5°C); solid phase, $Ni(NO_3)_2 \cdot 6H_2O$ (0°C,
   25°C), $Ni(NO_3)_2 \cdot 2H_2O$ (99.5°C) [KRU93]

2220
Compound: Nickel nitrate hexahydrate
Formula: Ni(NO$_3$)$_2\cdot$6H$_2$O
Molecular Formula: H$_{12}$N$_2$NiO$_{12}$
Molecular Weight: 290.794
CAS RN: 13478-00-7
Properties: green monocl deliq cryst; manufactured by several methods for example slowly adding nickel powd to a stirred mixture of nitric acid and water; used as an intermediate in the manufacture of nickel catalysts [KIR81] [MER06]
Solubility: s in 0.4 parts H$_2$O; s alcohol [MER06]
Density, g/cm³: 2.065 [HAW93]
Melting Point, °C: 56.7 [MER06]
Boiling Point, °C: 137 [MER06]
Reactions: minus H$_2$O on heating [KIR81]

2221
Compound: Nickel oxalate dihydrate
Formula: NiC$_2$O$_4\cdot$2H$_2$O
Molecular Formula: C$_2$H$_4$NiO$_6$
Molecular Weight: 182.743
CAS RN: 6018-94-6
Properties: light green powd [MER06]
Solubility: g/L soln, H$_2$O: 0.0118 (25°C); solid phase is NiC$_2$O$_4\cdot$2H$_2$O [KRU93]

2222
Compound: Nickel oxide
Synonym: bunsenite
Formula: NiO
Molecular Formula: NiO
Molecular Weight: 74.692
CAS RN: 1313-99-1
Properties: green powd, yellow when hot; cub; prepared by reaction of pure nickel powd and water in air at 1000°C; inert and refractory material; there is a black form with 76%–77% Ni, whereas green form has 78.5% Ni; black form is microcrystalline and prepared by calcination of the NiCO$_3$ or Ni(NO$_3$)$_2$ at 600°C, green form prepared by heating a mixture of H$_2$O and Ni powd in air at 1000°C; used in the manufacture of steels [KIR81] [MER06]
Solubility: in 0.00054 m alkaline phosphate solution: 0.0157×10^{-6} m (26.8°C), 0.0169×10^{-6} m (93.3°C), 0.0102×10^{-6} m (262.2°C) [ZIE89]; s acids [MER06]
Density, g/cm³: 7.45 [KIR81]
Melting Point, °C: 2090 [KIR81]
Reactions: β-Ni(OH)$_2$=NiO+H$_2$O at 195°C [ZIE89]

2223
Compound: Nickel oxide
Formula: Ni$_2$O$_3$
Molecular Formula: Ni$_2$O$_3$
Molecular Weight: 165.385
CAS RN: 1314-06-3
Properties: gray-black cub cryst [CRC10]
Solubility: i H$_2$O; s hot acid [CRC10]
Melting Point, °C: decomposes at ~600 [CRC10]

2224
Compound: Nickel perchlorate hexahydrate
Formula: Ni(ClO$_4$)$_2\cdot$6H$_2$O
Molecular Formula: Cl$_2$H$_{12}$NiO$_{14}$
Molecular Weight: 365.686
CAS RN: 13520-61-1
Properties: green cryst; hygr [STR93]
Solubility: g/100 g soln, H$_2$O: 104.6 (0°C), 112.2 (26°C); solid phase Ni(ClO$_4$)$_2\cdot$5H$_2$O (0°C) [KRU93]
Melting Point, °C: 140 [STR93]

2225
Compound: Nickel phosphate heptahydrate
Formula: Ni$_3$(PO$_4$)$_2\cdot$7H$_2$O
Molecular Formula: H$_{14}$Ni$_3$O$_{15}$P$_2$
Molecular Weight: 492.130
CAS RN: 10381-36-9
Properties: light green powd; can be prepared by reaction of nickel with hot dil phosphoric acid solution; used in electroplating [KIR81] [HAW93]
Solubility: i H$_2$O; s acids and NH$_4$OH [HAW93]
Melting Point, °C: decomposes on heating [KIR81]

2226
Compound: Nickel phosphate octahydrate
Formula: Ni$_3$(PO$_4$)$_2\cdot$8H$_2$O
Molecular Formula: H$_{16}$Ni$_3$O$_{16}$P$_2$
Molecular Weight: 510.145
CAS RN: 10381-36-9
Properties: light green powd [MER06]
Solubility: i H$_2$O; s acids, ammonia [MER06]
Melting Point, °C: decomposes [ALF95]

2227
Compound: Nickel phosphide
Formula: Ni$_2$P
Molecular Formula: Ni$_2$P
Molecular Weight: 148.361
CAS RN: 12035-64-2
Properties: gray cryst; −100 mesh with 99.5% purity [CER91] [CRC10]

Density, g/cm³: 7.33 [LID94]
Melting Point, °C: 1112 [AES93]

2228
Compound: Nickel selenate hexahydrate
Formula: NiSeO$_4$·6H$_2$O
Molecular Formula: H$_{12}$NiO$_{10}$Se
Molecular Weight: 309.743
CAS RN: 75060-62-5
Properties: green; tetr [CRC10]
Solubility: g/100 g H$_2$O: 27.36 (0°C), 36.20 (21.6°C), 83.99 (100°C); solid phase, NiSeO$_4$·6H$_2$O (0°C, 25°C), NiSeO$_4$·4H$_2$O (100°C) [KRU93]
Density, g/cm³: 2.314 [CRC10]

2229
Compound: Nickel selenide
Formula: NiSe
Molecular Formula: NiSe
Molecular Weight: 137.653
CAS RN: 1314-05-2
Properties: cub gray powd [CRC10]
Density, g/cm³: 7.2 [LID94]

2230
Compound: Nickel silicide
Formula: Ni$_2$Si
Molecular Formula: Ni$_2$Si
Molecular Weight: 145.473
CAS RN: 12059-14-2
Properties: −20 mesh powd [ALF93]
Density, g/cm³: 7.4 [LID94]
Melting Point, °C: 1255 [LID94]

2231
Compound: Nickel stannate dihydrate
Formula: NiSnO$_3$·2H$_2$O
Molecular Formula: H$_4$NiO$_5$Sn
Molecular Weight: 261.432
CAS RN: 12035-38-0
Properties: green powd; light colored cryst powd; used as an additive in ceramic capacitors [HAW93]
Reactions: minus 2H$_2$O at ~120°C [HAW93]

2232
Compound: Nickel stearate
Formula: Ni[CH$_3$(CH$_2$)$_{16}$COO]$_2$
Molecular Formula: C$_{36}$H$_{70}$NiO$_4$
Molecular Weight: 625.643

CAS RN: 2223-95-2
Properties: waxy green solid [STR93]
Density, g/cm³: 1.13 [STR93]
Melting Point, °C: 180 [KIR78]

2233
Compound: Nickel subsulfide
Synonym: heazlewoodite
Formula: Ni$_3$S$_2$
Molecular Formula: Ni$_3$S$_2$
Molecular Weight: 240.212
CAS RN: 12035-72-2
Properties: yellowish; naturally occurring nickel mineral [KIR81] [CRC10]
Density, g/cm³: 5.82 [CRC10]
Melting Point, °C: 790 [CRC10]

2234
Compound: Nickel sulfate
Formula: NiSO$_4$
Molecular Formula: NiO$_4$S
Molecular Weight: 154.757
CAS RN: 7786-81-4
Properties: greenish yellow cryst; used in manufacture of nickel catalysts; nickel plating [HAW93]
Solubility: g/100 g H$_2$O: 27.6 (0°C), 40.8 (25°C), 78.0 (100°C); solid phase, NiSO$_4$·7H$_2$O (0°C, 25°C), β-NiSO$_4$·6H$_2$O (100°C) [KRU93]; i alcohol and ether [HAW93]
Density, g/cm³: 3.68 [HAW93]
Melting Point, °C: 840, minus SO$_3$ [HAW93]

2235
Compound: Nickel sulfate heptahydrate
Formula: NiSO$_4$·7H$_2$O
Molecular Formula: H$_{14}$NiO$_{11}$S
Molecular Weight: 280.863
CAS RN: 10101-98-1
Properties: green cryst [HAW93]
Solubility: g/100 g H$_2$O: 26.2 (0°C), 37.7 (20°C), 50.4 (40°C) [LAN05]; s alcohol [HAW93]
Density, g/cm³: 1.98 [HAW93]
Reactions: minus H$_2$O at 99°C, minus 6H$_2$O at 103°C [CRC10] [HAW93]

2236
Compound: Nickel sulfate hexahydrate
Formula: NiSO$_4$·6H$_2$O
Molecular Formula: H$_{12}$NiO$_{10}$S
Molecular Weight: 262.848
CAS RN: 10101-97-0

Properties: two phases: α: blue to bluish green tetr
cryst; β: green transparent cryst, stable at 40°C;
sweet, astringent taste; somewhat efflorescent;
produced by adding Ni powd to hot dil H_2SO_4;
used as an electrolyte in metal finishing and
in electroless plating [KIR81] [MER06]

Solubility: g/100 g H_2O: (blue green) 40.1 (20°C),
43.6 (30°C), 47.6 (40°C); (green) 44.4 (20°C),
46.6 (30°C), 76.7 (100°C) [LAN05]; sl s
alcohol, more s methanol [MER06]

Density, g/cm³: 2.07 [STR93]

Melting Point, °C: decomposes at ~100 [LID94]

Reactions: transition α to β at 53.3°C; minus
$5H_2O$ at ~100°C; decomposes to NiO and
SO_3 at >800°C [KIR81] [MER06]

2237

Compound: Nickel sulfide

Synonym: millerite

Formula: NiS

Molecular Formula: NiS

Molecular Weight: 90.759

CAS RN: 16812-54-7

Properties: black powd; trig, yellow metallic
luster; enthalpy of fusion 30.10 kJ/mol; can
be formed by fusing Ni powd and molten
sulfur; other sulfides include Ni_2S, 12137-08-
5, Ni_3S_2, 12035-72-2 (heazlewoodite), NiS_2,
12035-51-7, and Ni_3S, 12137-12-1 (polydymite)
[KIR81] [STR93] [CRC10] [JAN85]

Solubility: 0.00036 g/100 mL H_2O (18°C) [CRC10]

Density, g/cm³: 5.3–5.65 [STR93]

Melting Point, °C: 976 [LID94]

2238

Compound: Nickel telluride

Formula: NiTe

Molecular Formula: NiTe

Molecular Weight: 186.293

CAS RN: 12142-88-0

Properties: 6 mm pieces and smaller
with 99.9% purity [CER91]

2239

Compound: Nickel tetrafluoroborate hexahydrate

Formula: $Ni(BF_4)_2 \cdot 6H_2O$

Molecular Formula: $B_2F_8H_{12}NiO_6$

Molecular Weight: 340.394

CAS RN: 15684-36-3

Properties: green cryst [STR93]

Density, g/cm³: 1.47 [STR93]

2240

Compound: Nickel thiocyanate

Formula: $Ni(SCN)_2$

Molecular Formula: $C_2N_2NiS_2$

Molecular Weight: 174.860

CAS RN: 13689-92-4

Properties: green powd [STR93]

Solubility: g/100 g soln, H_2O: 35.48 (25°C) [KRU93]

2241

Compound: Nickel titanate

Formula: $NiTiO_3$

Molecular Formula: NiO_3Ti

Molecular Weight: 154.558

CAS RN: 12035-39-1

Properties: brown powd or canary yellow with
rhomb structure; −325 mesh 10 μm or less with
99.9% purity [CER91] [KIR83] [STR93]

Density, g/cm³: 4.56 [STR93]

2242

Compound: Nickel trifluoroacetylacetonate dihydrate

Synonym: 1,1,1-trifluoro-2,4-
pentandione nickel derivative

Formula: $Ni(CF_3COCH=C(O)CH_3)_2 \cdot 2H_2O$

Molecular Formula: $C_{10}H_{12}F_6NiO_6$

Molecular Weight: 400.885

CAS RN: 14324-83-5

Properties: powd [ALF95]

2243

Compound: Nickel tungstate

Formula: $NiWO_4$

Molecular Formula: NiO_4W

Molecular Weight: 306.531

CAS RN: 14177-51-6

Properties: −325 mesh 10 μm or less
with 99.9% purity [CER91]

2244

Compound: Nickel vanadate

Formula: NiV_2O_6

Molecular Formula: NiO_6V_2

Molecular Weight: 256.572

CAS RN: 52502-12-2

Properties: −200 mesh with 99.9%
purity [CER91] [ALF95]

2245
Compound: Nickel(II,III) sulfide
Synonym: polydymite
Formula: Ni_3S_4
Molecular Formula: Ni_3S_4
Molecular Weight: 304.344
CAS RN: 12137-12-1
Properties: gray-black; naturally occurring nickel mineral [KIR81] [CRC10]
Density, g/cm³: 4.77 [LID94]

2246
Compound: Nickel(III) oxide
Synonym: nickel sesquioxide
Formula: Ni_2O_3
Molecular Formula: Ni_2O_3
Molecular Weight: 165.385
CAS RN: 1314-06-3
Properties: gray black powd [MER06]
Solubility: i H_2O; v sl s cold acid; dissolves in hot HCl releasing Cl_2; dissolves in hot H_2SO_4 or HNO_3 evolving O_2 [MER06]
Reactions: decomposes at ~600°C to give NiO and O_2 [MER06]

2247
Compound: Niobium
Formula: Nb
Molecular Formula: Nb
Molecular Weight: 92.90638
CAS RN: 7440-03-1
Properties: steel gray, lustrous metal; pure metal is ductile and malleable; bcc, lattice constant 0.33004 nm; enthalpy of fusion 30.0 kJ/mol; enthalpy of vaporization 697 kJ/mol; enthalpy of combustion 949 kJ/mol; vapor pressure (2300°C) 22 MPa; evaporation rate (2300°C) 1.9 μg/(cm² s); electrical resistivity 13–16 μohm·cm; produced from Nb_2O_5 and carbon at 1800°C–2000°C; used as an anodic film for rectification [KIR81] [MER06] [CER91] [CRC10]
Solubility: inert to HCl, HNO_3, aqua regia; attacked by fusion with alkali hydroxides [MER06]
Density, g/cm³: 8.57 [MER06]
Melting Point, °C: 2477 [CRC10]
Boiling Point, °C: 4944 [CRC10]
Thermal Conductivity, W/(m·K): 52.3 (25°C) [KIR81]
Thermal Expansion Coefficient: coefficient of linear expansion 7.1×10^{-6}/°C from 18°C–100°C [KIR81]

2248
Compound: Niobium boride
Formula: NbB
Molecular Formula: BNb
Molecular Weight: 103.717
CAS RN: 12045-19-1
Properties: gray ortho cryst [CRC10]
Density, g/cm³: 7.5 [CRC10]

2249
Compound: Niobium boride
Formula: NbB_2
Molecular Formula: B_2Nb
Molecular Weight: 114.528
CAS RN: 12007-29-3
Properties: gray hex cryst [CRC10]
Density, g/cm³: 6.97 [CRC10]
Melting Point, °C: 3050 [CRC10]

2250
Compound: Niobium carbide
Formula: Nb_2C
Molecular Formula: CNb_2
Molecular Weight: 197.824
CAS RN: 12011-99-3
Properties: −325 mesh, 10 μm or less, 99.5% purity; hex, a = 0.3127 nm, c = 0.4972 nm [CER91] [KIR81]
Density, g/cm³: 7.80 [KIR81]
Melting Point, °C: 3090 [KIR81]

2251
Compound: Niobium diboride
Formula: NbB_2
Molecular Formula: B_2Nb
Molecular Weight: 114.528
CAS RN: 12007-29-3
Properties: gray powd; hex, a = 0.3089 nm, c = 0.3303 nm; hardness 8+ Mohs; resistivity 65 μohm·cm at 25°C; refractory material; used as a sputtering target with 99.5% purity to produce thermionic conductor film [KIR81] [CER91] [ALF93]
Density, g/cm³: 6.97 [ALD94]
Melting Point, °C: 2900 [KIR78]
Thermal Conductivity, W/(m·K): 17 at 296 K [KIR81]

2252
Compound: Niobium disilicide
Formula: $NbSi_2$
Molecular Formula: $NbSi_2$
Molecular Weight: 149.077

CAS RN: 12034-80-9
Properties: −325 mesh powd; cryst solid; used as a refractory and as 99.5%–99.95% pure material as a sputtering target in the fabrication of integrated circuits [HAW93] [ALF93] [CER91]
Density, g/cm³: 5.7 [LID94]
Melting Point, °C: 1950 [HAW93]

2253
Compound: Niobium hydride
Formula: NbH
Molecular Formula: HNb
Molecular Weight: 93.914
CAS RN: 13981-86-7
Properties: gray powd; bcc; sensitive to moisture; reaction of H_2 with Nb at 300°C–1500°C can result in the formation of $NbH_{0.85}$ [STR93] [KIR81]
Density, g/cm³: 6.6 [STR93]

2254
Compound: Niobium monoboride
Formula: NbB
Molecular Formula: BNb
Molecular Weight: 103.717
CAS RN: 12045-19-1
Properties: gray powd; ortho-rhomb, a = 0.3298 nm, b = 0.872 nm, c = 0.3166 nm; resistivity 64.5 μohm·cm at 25°C; refractory material; commonly prepared by hot-pressing boron with niobium or niobium hydride; used as sputtering target with 99.5% purity to produce wear-resistant and semiconductor films, and can provide neutron absorbing layers on nuclear fuel pellets [KIR78] [KIR81] [CER91] [ALF93]
Density, g/cm³: 7.5 [KIR81]
Melting Point, °C: 2270 [KIR78]

2255
Compound: Niobium nitride
Formula: NbN
Molecular Formula: NNb
Molecular Weight: 106.913
CAS RN: 24621-21-4
Properties: dark gray; fcc, a = 0.4388 nm; hardness, 8+ Mohs; electrical resistivity 78 μohm·cm; transition temp 15.2 K; can be prepared by heating Nb metal in excess N_2 or NH_3 to 700°C–1100°C; used in the form of 99.5% pure sputtering target for increasing electrical stability of diodes, transistors, and integrated circuits [KIR81] [CIC73] [CER91]
Solubility: i HCl, HNO_3, H_2SO_4; attacked by hot caustic, lime or strong alkali, evolving NH_3 [KIR81]
Density, g/cm³: 8.47 [KIR81]

Melting Point, °C: 2575 [STR93]
Thermal Conductivity, W/(m·K): 3.8 [KIR81]
Thermal Expansion Coefficient: 10.1×10^{-6} [KIR81]

2256
Compound: Niobium phosphide
Formula: NbP
Molecular Formula: NbP
Molecular Weight: 123.880
CAS RN: 12034-66-1
Properties: tetr cryst; −200 mesh with 99.5% purity [LID94] [CER91]
Density, g/cm³: 6.5 [LID94]

2257
Compound: Niobium(II) oxide
Formula: NbO
Molecular Formula: NbO
Molecular Weight: 108.905
CAS RN: 12034-57-0
Properties: gray metallic appearance; can be obtained by reduction of Nb_2O_5 in H_2 at 1300°C–1700°C; −100 mesh with 99.9% purity; cub, a = 0.42108 nm; enthalpy of fusion 85.00 kJ/mol [CRC10] [CER91] [KIR81]
Density, g/cm³: 7.30 [KIR81]
Melting Point, °C: 1937 [CRC10]

2258
Compound: Niobium(IV) bromide
Formula: $NbBr_4$
Molecular Formula: Br_4Nb
Molecular Weight: 412.522
CAS RN: 13842-75-6
Properties: dark brown cryst [CRC10]
Density, g/cm³: 4.72 [CRC10]
Solubility: reac H_2O [CRC10]
Melting Point, °C: sublimes at 300 [CRC10]

2259
Compound: Niobium(IV) carbide
Formula: NbC
Molecular Formula: CNb
Molecular Weight: 104.917
CAS RN: 12069-94-2
Properties: lavender gray powd; −325 mesh, 10 μm or less, 99.9% purity; fcc, a = 0.4471 nm; hardness 9+ Mohs; resistivity 180 μohm·cm maximum; used in special steels, coating graphite for nuclear reactors; as a 99.5% pure material, used as a sputtering target to produce wear-resistant and semiconducting films [HAW93] [KIR81] [CER91]

Solubility: i H_2O, acids; s in a mixture of HNO_3 and HF [HAW93]
Density, g/cm³: 7.82 [STR93]
Melting Point, °C: 3500 [STR93]
Boiling Point, °C: 4300 [KIR81]
Thermal Conductivity, W/(m·K): 14 at 23°C [KIR81]
Thermal Expansion Coefficient: (volume) 100°C (0.141), 200°C (0.329), 400°C (0.740), 800°C (1.626), 1200°C (2.565) [CLA66]

2260
Compound: Niobium(IV) chloride
Formula: $NbCl_4$
Molecular Formula: Cl_4Nb
Molecular Weight: 234.718
CAS RN: 13569-70-5
Properties: violet-black monocl cryst [CRC10]
Solubility: reac H_2O [CRC10]
Density, g/cm³: 3.2 [CRC10]
Melting Point, °C: decomposes at 800 [CRC10]
Boiling Point, °C: sublimes at 275 [CRC10]

2261
Compound: Niobium(IV) oxide
Synonym: niobium dioxide
Formula: NbO_2
Molecular Formula: NbO_2
Molecular Weight: 124.905
CAS RN: 12034-59-2
Properties: white powd; −200 mesh with 99.9% purity; tetr, a=0.371 nm, c=0.5985 nm; enthalpy of fusion 92.00 kJ/mol; can be prepared by reduction of Nb_2O_5 with H_2 at 800°C–1300°C [KIR81] [STR93] [CER91] [CRC10]
Density, g/cm³: 5.9 [STR93]
Melting Point, °C: 1902 [CRC10]

2262
Compound: Niobium(IV) selenide
Formula: $NbSe_2$
Molecular Formula: $NbSe_2$
Molecular Weight: 250.826
CAS RN: 12034-77-4
Properties: gray black solid; has a higher electrical conductivity than graphite; used as a lubricant and conductor at high temperatures and in high vacuum, and as a 99.8% pure material, used as a sputtering target to produce electrically conductive lubricant film [HAW93] [CER91]
Density, g/cm³: 6.3 [LID94]
Melting Point, °C: >1316 [HAW93]

2263
Compound: Niobium(IV) sulfide
Formula: NbS_2
Molecular Formula: NbS_2
Molecular Weight: 157.038
CAS RN: 12136-97-9
Properties: black powd; formula also given as $NbS_{1.75}$; as a 99.8% pure material, used as sputtering target to produce lubricant film on bearings and other moving parts [STR93] [CER91]
Density, g/cm³: 4.4 [LID94]

2264
Compound: Niobium(IV) telluride
Formula: $NbTe_2$
Molecular Formula: $NbTe_2$
Molecular Weight: 348.106
CAS RN: 12034-83-2
Properties: hex cryst; −325 mesh with 10 μm average or less; 99.8% pure material used as a sputtering target for lubricant film [CER91] [LID94]
Density, g/cm³: 7.6 [LID94]

2265
Compound: Niobium(V) bromide
Synonym: niobium pentabromide
Formula: $NbBr_5$
Molecular Formula: Br_5Nb
Molecular Weight: 492.426
CAS RN: 13478-45-0
Properties: yellow powd; ortho-rhomb, a=0.6127 nm, b=1.2198 nm, c=1.855 nm; sensitive to moisture; can be formed by reacting Br_2 and niobium at ~500°C [STR93] [KIR81]
Solubility: s H_2O, alcohol, ethyl bromide [KIR81]
Density, g/cm³: 4.36 [KIR81]
Melting Point, °C: 150 [STR93]; 254 [KIR81]
Boiling Point, °C: 361.6 [STR93]

2266
Compound: Niobium(V) chloride
Synonym: niobium pentachloride
Formula: $NbCl_5$
Molecular Formula: Cl_5Nb
Molecular Weight: 270.170
CAS RN: 10026-12-7
Properties: yellow, very deliq; monocl, a=0.1830 nm, b=1.798 nm, c=0.5888 nm; melts to a reddish orange liq; decomposes in moist air evolving HCl; enthalpy of vaporization 52.7 kJ/mol; enthalpy of fusion 33.90 kJ/mol; can be obtained by chlorination of Nb metal at 300°C–350°C [MER06] [KIR81] [CRC10]

Solubility: hydrolyzes in H_2O; s HCl, CCl_4 [MER06] [KIR81]
Density, g/cm³: 2.75 [MER06]
Melting Point, °C: 204.7 [CRC10]
Boiling Point, °C: 254.05 [CRC10]
Reactions: starts to sublime at 125°C [MER06]

2267
Compound: Niobium(V) ethoxide
Formula: $Nb(OC_2H_5)_5$
Molecular Formula: $C_{10}H_{25}NbO_5$
Molecular Weight: 318.212
CAS RN: 3236-82-6
Properties: liq; flammable; moisture sensitive [ALD94]
Solubility: decomposed by H_2O [ALF95]
Density, g/cm³: 1.258 [ALD94]
Melting Point, °C: 6 [ALD94]
Boiling Point, °C: 140–142 (0.1 mm Hg) [ALD94]

2268
Compound: Niobium(V) fluoride
Synonym: niobium pentafluoride
Formula: NbF_5
Molecular Formula: F_5Nb
Molecular Weight: 187.898
CAS RN: 7783-68-8
Properties: strongly refractive, deliq, colorless monocl cryst; very hygr; lattice parameters a=0.963 nm, b=1.443 nm, c=0.512 nm; enthalpy of vaporization 52.3 kJ/mol; enthalpy of fusion 36.00 kJ/mol; obtained from Nb and F_2 or anhydrous HF at 250°C–300°C [KIR81] [MER06] [CRC10]
Solubility: hydrolyzes in H_2O; sl s CS_2, $CHCl_3$ [MER06]
Density, g/cm³: 2.6955 [MER06]; 3.92 [STR93]
Melting Point, °C: 80 [MER06]
Boiling Point, °C: 229 [CRC10]

2269
Compound: Niobium(V) fluorodioxide
Synonym: niobium dioxide fluoride
Formula: NbO_2F
Molecular Formula: $FNbO_2$
Molecular Weight: 143.903
CAS RN: 15195-33-2
Properties: cub, a=0.3902 nm; can be prepared by dissolution of Nb_2O_5 in 48% HF, followed by evaporation to dryness and heating to 250°C [KIR81]
Density, g/cm³: 4.0 [LID94]

2270
Compound: Niobium(V) iodide
Synonym: niobium pentiodide

Formula: NbI_5
Molecular Formula: I_5Nb
Molecular Weight: 727.428
CAS RN: 13779-92-5
Properties: black powd; monocl, a=1.058 nm, b=0.658 nm, c=1.388 nm; sensitive to moisture; obtained by reaction of excess I_2 with Nb metal in a sealed tube [STR93] [KIR81]
Density, g/cm³: 5.32 [LID94]
Melting Point, °C: decomposes at ~200 [KIR81]
Reactions: decomposes to NbI_4, 13870-21-8, at 206°C–270°C in vacuum [KIR81]

2271
Compound: Niobium(V) oxide
Synonym: niobium pentoxide
Formula: Nb_2O_5
Molecular Formula: Nb_2O_5
Molecular Weight: 265.810
CAS RN: 1313-96-8
Properties: white, ortho-rhomb cryst; becomes yellow when heated; α form is monocl, a=2.116 nm, b=0.3822 nm, c= 1.935 nm; enthalpy of fusion 104.3 kJ/mol; used as an evaporated material and sputtering target with 99.95% and 99.5% purity to prepare dielectric coatings and multilayers [KIR81] [MER06] [CER91] [CRC10]
Solubility: i H_2O; s HF, hot H_2SO_4 [MER06]
Density, g/cm³: 4.47 [ALD94]
Melting Point, °C: 1520 [MER06]

2272
Compound: Niobium(V) oxybromide
Formula: $NbOBr_3$
Molecular Formula: Br_3NbO
Molecular Weight: 348.617
CAS RN: 14459-75-7
Properties: yellowish brown solid; hydrolyzes readily in moist air, prepared by reacting Br_2 with a mixture of Nb_2O_5 and carbon at 550°C [KIR81]
Melting Point, °C: sublimes in vacuum at 180 [KIR81]
Boiling Point, °C: decomposes at ~320 [KIR81]
Reactions: decomposes in vacuum to $NbBr_5$ and Nb_2O_5 [KIR81]

2273
Compound: Niobium(V) oxychloride
Formula: $NbOCl_3$
Molecular Formula: Cl_3NbO
Molecular Weight: 215.263
CAS RN: 13597-20-1

Properties: white solid; tetr, a = 1.087 nm, c = 0.396 nm; can be prepared by air oxidation of $NbCl_5$ [KIR81]
Density, g/cm³: 3.72 [KIR81]
Melting Point, °C: sublimes in vacuum at ~200 [KIR81]

2274
Compound: Niobocene dichloride
Synonym: bis(cyclopentadienyl)niobium
Formula: $(C_5H_5)_2NbCl_2$
Molecular Formula: $C_{10}H_{10}Cl_2Nb$
Molecular Weight: 294.00
CAS RN: 12793-14-5
Properties: sensitive to moisture; uses: synthesis of transition metal complexes and organometallic compounds [ALD93]

2275
Compound: Nitric acid
Synonym: aqua fortis
Formula: HNO_3
Molecular Formula: HNO_3
Molecular Weight: 63.013
CAS RN: 7697-37-2
Properties: transparent, colorless or yellowish liq; fuming; hygr; corrosive; attacks almost all metals; yellowish color is due to formation of nitrogen dioxide when exposed to light; strong oxidizing agent; viscosity (25°C) is 0.761 cp; vapor pressure (25°C) is 62 mm Hg; specific conductance 3.77×10^{-2} (ohm·cm)$^{-1}$ at 25°C; enthalpy of vaporization 39.1 kJ/mol at 25°C; enthalpy of fusion 10.50 kJ/mol; [CRC10] [HAW93] [COT88]
Solubility: miscible with H_2O; decomposed by alcohol [HAW93]
Density, g/cm³: 1.504 [HAW93]
Melting Point, °C: −41.6 [CRC10]
Boiling Point, °C: decomposes at 78 [HAW93]

2276
Compound: Nitric oxide
Synonym: nitrogen monoxide
Formula: NO
Molecular Formula: NO
Molecular Weight: 30.006
CAS RN: 10102-43-9
Properties: colorless gas; reacts readily with oxygen at room temp to form NO_2; enthalpy of vaporization 13.83 kJ/mol; enthalpy of fusion 2.30 kJ/mol [CRC10] [HAW93]

Solubility: sl s H_2O [HAW93]
Density, g/cm³: 1.317 g/L [LID94]
Melting Point, °C: −163.6 [COT88]
Boiling Point, °C: −151.8 [COT88]

2277
Compound: Nitrogen
Formula: N_2
Molecular Formula: N_2
Molecular Weight: 28.013 (atomic weight 14.00674)
CAS RN: 7727-37-9
Properties: colorless, tasteless, odorless gas; chemically unreactive; specific volume (21.1°C, 1 atm) 0.86 m³/kg; critical temp −147.1°C; critical pressure 33.5 atm; critical density 0.311 g/cm³; enthalpy of vaporization 5.57 kJ/mol; enthalpy of fusion 0.71 kJ/mol [CRC10] [AIR87] [HAW93] [MER06]
Solubility: sl s H_2O; i alcohol [HAW93]
Density, g/cm³: gas: 1.25046 g/L [MER06]
Melting Point, °C: −210.01 [MER06]
Boiling Point, °C: −195.79 [MER06]
Thermal Conductivity, W/(m·K): 0.02583 (25°C) [ALD94]

2278
Compound: Nitrogen dioxide
Formula: NO_2
Molecular Formula: NO_2
Molecular Weight: 46.006
CAS RN: 10102-44-0
Properties: reddish brown gas; brown liq below 21.15°C; colorless solid at about −11°C; used in the production of nitric acid, as an oxidizing agent [HAW93] [MER06]
Solubility: decomposes in H_2O to HNO_3 and releases NO; s conc H_2SO_4, HNO_3 [MER06]
Density, g/cm³: gas: 2.0198 g/L [LID94]; liq: 1.448 [MER06]
Melting Point, °C: −9.3 [MER06]
Boiling Point, °C: 21.15 [MER06]

2279
Compound: Nitrogen pentoxide
Formula: N_2O_5
Molecular Formula: N_2O_5
Molecular Weight: 108.010
CAS RN: 10102-03-1
Properties: col hex cryst [CRC10]
Solubility: s chl
Density, g/cm³: 2.0 [CRC10]
Boiling Point, °C: 33 [CRC10]

2280
Compound: Nitrogen selenide
Synonym: selenium nitride
Formula: N_4Se_4
Molecular Formula: N_4Se_4
Molecular Weight: 371.867
CAS RN: 12033-88-4
Properties: orange red, amorphous powd or
    monocl cryst; preparation: by reaction of
    dry NH_3 and $SeCl_4$; explosive [MER06]
Solubility: i H_2O, ether, absolute alcohol; sl s
    CS_2, benzene, acetic acid [MER06]
Density, g/cm³: 4.2 [MER06]
Melting Point, °C: explodes [LID94]

2281
Compound: Nitrogen tetroxide
Formula: N_2O_4
Molecular Formula: N_2O_4
Molecular Weight: 92.011
CAS RN: 10544-72-6
Properties: col liq [CRC10]
Solubility: reac H_2O [CRC10]
Density, g/cm³: 1.45^{20} [CRC10]
Melting Point, °C: −9.3 [CRC10]
Boiling Point, °C: 21.15 [CRC10]

2282
Compound: Nitrogen trichloride
Formula: NCl_3
Molecular Formula: Cl_3N
Molecular Weight: 120.365
CAS RN: 10025-85-1
Properties: yellowish, thick, oily, liq; pungent odor;
    evaporates rapidly in air; very unstable; decomposes
    in light; explodes when heated to 93°C or when
    subjected to a flash of direct sunlight [MER06]
Solubility: i H_2O, decomposes in H_2O after
    24 h; s CS_2, phosphorus trichloride,
    benzene, CCl_4, $CHCl_3$ [MER06]
Density, g/cm³: 1.653 [MER06]
Melting Point, °C: <40 [HAW93]
Boiling Point, °C: <71 [HAW93]

2283
Compound: Nitrogen trifluoride
Formula: NF_3
Molecular Formula: F_3N
Molecular Weight: 71.002
CAS RN: 7783-54-2

Properties: colorless gas; oxidizing agent; moldy
    odor; critical temp −39.25°C; critical pressure
    4.53 MPa; critical volume 123.8 cm³/mol; enthalpy
    of vaporization 11.59 kJ/mol; enthalpy of fusion
    398 J/mol; chemically inert, does not attack glass,
    mercury; decomposed by electric sparks; used
    in electronics industry [MER06] [AIR87]
Solubility: 1.4×10^{-5} mol/mol H_2O at 25°C [KIR78]
Density, g/cm³: gas: 3.116 g/L [LID94];
    liq, at bp: 1.885 [MER06]
Melting Point, °C: −208.5 [MER06]
Boiling Point, °C: −129 [MER06]

2284
Compound: Nitrogen triiodide
Formula: NI_3
Molecular Formula: I_3N
Molecular Weight: 394.720
CAS RN: 13444-85-4
Properties: black cryst; unstable; can explode if
    touched; more stable if kept wet [HAW93]
Reactions: can explode [CRC10]

2285
Compound: Nitrogen trioxide
Synonym: dinitrogen trioxide
Formula: N_2O_3
Molecular Formula: N_2O_3
Molecular Weight: 76.011
CAS RN: 10544-73-7
Properties: blue liq; used as an oxidant in
    special fuel systems [HAW93]
Density, g/cm³: 1.447 (2°C) [HAW93]
Melting Point, °C: −102 [COT88]
Boiling Point, °C: decomposes at 3.5 [COT88]

2286
Compound: Nitrogen(V) oxide
Synonym: dinitrogen pentoxide
Formula: N_2O_5
Molecular Formula: N_2O_5
Molecular Weight: 108.010
CAS RN: 10102-03-1
Properties: colorless, hex cryst [MER06]
Solubility: v s chloroform without appreciable
    decomposition; less s CCl_4 [MER06]
Density, g/cm³: 2.0 [LID94]
Melting Point, °C: 30 [COT88]
Boiling Point, °C: decomposes at 47 [COT88]
Reactions: sublimes at 32.4°C [MER06]

2287
Compound: Nitronium hexafluoroantimonate
Formula: NO_2SbF_6
Molecular Formula: F_6NO_2Sb
Molecular Weight: 281.756
CAS RN: 17856-92-7
Properties: white cryst; sensitive to moisture [STR93]

2288
Compound: Nitronium hexafluorophosphate
Formula: NO_2PF_6
Molecular Formula: F_6NO_2P
Molecular Weight: 190.970
CAS RN: 19200-21-6
Properties: white cryst; sensitive to moisture [STR93]

2289
Compound: Nitronium tetrafluoroborate
Formula: NO_2BF_4
Molecular Formula: BF_4NO_2
Molecular Weight: 132.811
CAS RN: 13826-86-3
Properties: white cryst; sensitive to moisture [ALD94]

2290
Compound: Nitrosyl chloride
Formula: NOCl
Molecular Formula: ClNO
Molecular Weight: 65.459
CAS RN: 2696-92-6
Properties: yellowish red liq or yellow gas; forms nitric oxide and chlorine when heated; enthalpy of vaporization 25.78 kJ/mol; used as a catalyst [HAW93] [CRC10]
Solubility: decomposes in H_2O; s fuming H_2SO_4 [HAW93]
Density, g/cm³: gas: 2.872 g/L [LID94]; liq: 1.273 [HAW93]
Melting Point, °C: −61.5 [HAW93]
Boiling Point, °C: −5.55 [CRC10]

2291
Compound: Nitrosyl fluoride
Formula: FNO
Molecular Formula: FNO
Molecular Weight: 49.005
CAS RN: 7789-25-5
Properties: colorless gas, has bluish color if impure; reacts rapidly with glass; used as an oxidizer in rocket propellants and as a reagent for fluorination [MER06]

Solubility: reacts with H_2O to form NO, HNO_3, and HF [MER06]
Density, g/cm³: gas: 2.150 g/L [LID94]; liq: 1.326; solid: 1.719 [MER06]
Melting Point, °C: −132.5 [MER06]
Boiling Point, °C: −59.9 [MER06]

2292
Compound: Nitrosylsulfuric acid
Synonym: nitrosyl sulfate
Formula: $HOSO_2ONO$
Molecular Formula: HNO_5S
Molecular Weight: 127.078
CAS RN: 7782-78-7
Properties: prisms; oxidizing agent; forms an intermediate in the Chamber process for fuming sulfuric acid; obtained by reaction of SO_3, nitrogen oxides, and H_2O; uses: preparation of cryst diazonium sulfates, bleaching cereal milling products [MER06] [ALD94]
Solubility: decomposes in H_2O; s H_2SO_4 [MER06]
Melting Point, °C: decomposes at 73.5 [MER06]
Reactions: decomposes in moist air to form H_2SO_4 and HNO_3 [MER06]

2293
Compound: Nitrous acid
Formula: HNO_2
Molecular Formula: HNO_2
Molecular Weight: 47.014
CAS RN: 7782-77-6
Properties: weak acid; stable only in solution; light blue [HAW93]

2294
Compound: Nitrous oxide
Formula: N_2O
Molecular Formula: N_2O
Molecular Weight: 44.012
CAS RN: 10024-97-2
Properties: colorless gas; sweet taste; critical temp 36.5°C; critical pressure 7.26 MPa; enthalpy of vaporization 16.53 kJ/mol; enthalpy of fusion 6.54 kJ/mol; used in dentistry and medicine as an anesthetic and in electronics industry [CRC10] [AIR87] [HAW93]
Solubility: sl s H_2O; s alcohol, ether, conc H_2SO_4 [HAW93]
Density, g/cm³: gas: 1.931 g/L [LID94]; liq: 1.22 (−89°C) [HAW93]
Melting Point, °C: −90.8 [HAW93]
Boiling Point, °C: −88.5 [COT88]

2295
Compound: Nitryl chloride
Formula: NO_2Cl
Molecular Formula: $ClNO_2$
Molecular Weight: 81.459
CAS RN: 13444-90-1
Properties: colorless gas; odor of chlorine; yellow tinge to its solutions; enthalpy of vaporization 25.7 kJ/mol [CRC10] [HAW93]
Density, g/cm³: gas: 3.574 g/L [LID94]; liq: 1.33 [HAW93]
Melting Point, °C: −145 [HAW93]
Boiling Point, °C: −14.3 [MER06]

2296
Compound: Nitryl fluoride
Formula: NO_2F
Molecular Formula: FNO_2
Molecular Weight: 65.004
CAS RN: 10022-50-1
Properties: colorless gas; strong oxidizing agent; enthalpy of vaporization 18.05 kJ/mol; used in rocket propellants and as a fluorinating agent [HAW93] [CRC10]
Solubility: hydrolyzes with HNO_3 and HF as products [HAW93]
Density, g/cm³: gas: 2.852 g/L [LID94]; liq: 1.80 [HAW93]
Melting Point, °C: −165 [HAW93]
Boiling Point, °C: −72.4 [MER06]

2297
Compound: Nobelium
Formula: No
Molecular Formula: No
Molecular Weight: 259
CAS RN: 10028-14-5
Properties: synthetic radioactive element; one of the actinides; has nine very short-lived isotopes; discovered in 1958 by Ghiorso and colleagues [HAW93]
Melting Point, °C: 827 [LID94]

2298
Compound: Octadecaborane(22)
Formula: $B_{18}H_{22}$
Molecular Formula: $B_{18}H_{22}$
Molecular Weight: 216.733
CAS RN: 11071-61-7
Properties: yellow cryst [CRC10]
Solubility: org solvents [CRC10]
Melting Point, °C: 180 [CRC10]

2299
Compound: Nonaborane(15)
Formula: B_9H_{15}
Molecular Formula: B_9H_{15}
Molecular Weight: 112.418
CAS RN: 19465-30-6
Properties: col liq [CRC10]
Melting Point, °C: 2.7 [CRC10]

2300
Compound: Osmium
Formula: Os
Molecular Formula: Os
Molecular Weight: 190.230
CAS RN: 7440-04-2
Properties: bluish white, lustrous metal; has ten oxidation states, −2 to +8, with higher oxidation states being the most stable; closed-packed hex, a = 0.27341 nm; stable in cold air; hardness is 7.0 Mohs; electrical resistivity, μohm · cm: 8.12 (0°C), 9.66 (20°C); vapor pressure at mp 1.8 Pa; Young's modulus 558.6 GPa; enthalpy of fusion 57.85 kJ/mol; enthalpy of vaporization 738 kJ/mol [KIR82] [MER06] [ALD94]
Solubility: attacked by aqua regia; barely affected by HCl, H_2SO_4 [MER06]
Density, g/cm³: 22.61 [MER06]
Melting Point, °C: 3045 [ALD94]
Boiling Point, °C: 5027 [ALD94]
Thermal Conductivity, W/(m · K): 87.6 (25°C) [CRC10]
Thermal Expansion Coefficient: at 20°C is 6.1×10^{-6}/°C [KIR82]

2301
Compound: Osmium bis(cyclopentadienyl)
Synonym: bis(cyclopentadienyl)osmium
Formula: $(C_5H_5)_2Os$
Molecular Formula: $C_{10}H_{10}Os$
Molecular Weight: 320.419
CAS RN: 1273-81-0
Properties: cryst [ALF95]
Melting Point, °C: 226–228 [ALF95]

2302
Compound: Osmium carbonyl
Synonym: triosmium dodecacarbonyl
Formula: $Os_3(CO)_{12}$
Molecular Formula: $C_{12}O_{12}Os_3$
Molecular Weight: 906.815
CAS RN: 15696-40-9
Properties: yellow cryst; stable in air [DOU83] [STR93]

Density, g/cm³: 3.48 [STR93]
Melting Point, °C: 224 [ALD94]

2303

Compound: Osmium(II) chloride
Synonym: osmium dichloride
Formula: $OsCl_2$
Molecular Formula: Cl_2Os
Molecular Weight: 261.135
CAS RN: 13444-92-3
Properties: dark green needles; hygr; unstable with respect to atm oxygen [HAW93]
Solubility: i H_2O; s alcohol, ether [HAW93]

2304

Compound: Osmium(III) chloride
Synonym: osmium trichloride
Formula: $OsCl_3$
Molecular Formula: Cl_3Os
Molecular Weight: 296.588
CAS RN: 13444-93-4
Properties: dark gray cub solid [KIR82]
Solubility: i H_2O; s HNO_3 [KIR82]
Melting Point, °C: decomposes at >450 [KIR82]

2305

Compound: Osmium(III) chloride hydrate
Formula: $OsCl_3 \cdot xH_2O$
Molecular Formula: Cl_3Os (anhydrous)
Molecular Weight: 296.588 (anhydrous)
CAS RN: 14996-60-2
Properties: black cryst [STR93]
Melting Point, °C: decomposes at >500 [STR93]

2306

Compound: Osmium(IV) chloride
Synonym: osmium tetrachloride
Formula: $OsCl_4$
Molecular Formula: Cl_4Os
Molecular Weight: 332.041
CAS RN: 10026-01-4
Properties: red cryst [MER06]
Solubility: s H_2O resulting in a yellow solution, hydrolysis occurs after standing [MER06]
Density, g/cm³: 4.38 [MER06]
Reactions: sublimes at 450°C [MER06]

2307

Compound: Osmium(IV) oxide
Synonym: osmium dioxide

Formula: OsO_2
Molecular Formula: O_2Os
Molecular Weight: 222.229
CAS RN: 12036-02-1
Properties: dark bluish black powd, with rutile cryst form [KIR82] [ALD94]
Solubility: i H_2O, acids [KIR82]
Density, g/cm³: 11.4 [KIR82]

2308

Compound: Osmium(VI) fluoride
Synonym: osmium hexafluoride
Formula: OsF_6
Molecular Formula: F_6Os
Molecular Weight: 304.220
CAS RN: 13768-38-2
Properties: pale yellow, volatile solid; hydrolyzed when exposed to moisture [MER06]
Density, g/cm³: 4.1 [LID94]
Melting Point, °C: 32.1 [MER06]
Boiling Point, °C: 45.9 [MER06]

2309

Compound: Osmium(VIII) oxide
Synonym: osmium tetroxide
Formula: OsO_4
Molecular Formula: O_4Os
Molecular Weight: 254.228
CAS RN: 20816-12-0
Properties: pale yellow solid; monocl cryst; acrid, chlorine-like odor; volatile; vapor pressure 11 mm Hg (27°C); critical temp 405°C; enthalpy of fusion 9.80 kJ/mol; obtained by heating finely divided Os in air or O_2; used as an oxidation agent and catalyst [MER06] [CRC10]
Solubility: g/100 g H_2O: 5.26 (0°C), 5.75 (10°C), 6.43 (20°C) [LAN05]; 375 g/100 CCl_4; s benzene [MER06]
Density, g/cm³: 5.10 [MER06]
Melting Point, °C: 40.6 [MER06]
Boiling Point, °C: 130.0 [MER06]
Reactions: begins to sublime and distill below the boiling point [MER06]

2310

Compound: Oxalic acid
Synonym: ethanedioic acid
Formula: $(COOH)_2$
Molecular Formula: $C_2H_2O_4$
Molecular Weight: 90.035
CAS RN: 144-62-7

Properties: colorless, odorless, hygr solid; has two forms: rhomb α and monocl β; α prepared from sublimation of the dihydrate, β prepared by crystallization from acetic acid; rhomb is thermodynamically stable form at room temp; enthalpy of sublimation 90.58 kJ/mol; enthalpy of solution in water −9.58 kJ/mol; enthalpy of combustion −245.61 kJ/mol; enthalpy of decomposition 826.78 kJ/mol; vapor pressure (P), kPa, (57°C–107°C) log P = −(4726.95/T) + 11.3478 [KIR81]
Solubility: g/100 g H_2O: 3.54 (0°C), 9.52 (20°C), 120 (90°C) [LAN05]; g/100 g: 23.7 in ethanol (15.6°C), 1.5 in ethyl ether (25°C) [KIR81]
Density, g/cm³: α: 1.900; β: 1.896 [KIR81]
Melting Point, °C: α: 189.5; β: 182 [KIR81]
Reactions: sublimes starting at 100°C, rapidly by 125°C [KIR81]
Thermal Conductivity, W/(m·K): 0.9 at 0°C [KIR81]

2311
Compound: Oxalic acid dihydrate
Formula: HOOCCOOH·$2H_2O$
Molecular Formula: $C_2H_6O_6$
Molecular Weight: 126.066
CAS RN: 6153-56-6
Properties: transparent, colorless cryst; enthalpy of solution in water −35.5 kJ/mol; crystallizes from water; used to clean automobile radiators [HAW93] [KIR81]
Solubility: s H_2O, ether [KIR81]
Density, g/cm³: 1.653 [KIR81]
Melting Point, °C: 101.5 [HAW93]
Reactions: minus $2H_2O$ at 98°C–100°C [KIR81]

2312
Compound: Oxalyl chloride
Formula: ClCOCOCl
Molecular Formula: $C_2Cl_2O_2$
Molecular Weight: 126.926
CAS RN: 79-37-8
Properties: liq [ALF95]
Density, g/cm³: 1.455 [ALF95]
Boiling Point, °C: 63–64 [ALF95]

2313
Compound: Oxygen
Formula: O_2
Molecular Formula: O_2
Molecular Weight: 31.999 (atomic weight: 15.9994)
CAS RN: 7782-44-7

Properties: colorless, odorless, tasteless gas; diatomic; can be liquefied at −183°C to sl bluish liq; solidifies at −218°C; critical temp −118.95°C; critical pressure 50.14 atm; enthalpy of vaporization 6.820 kJ/mol; enthalpy of fusion 0.44 kJ/mol; specific volume (21.1°C, 101.3 kPa) 0.75 m³/kg [HAW93] [MER06] [AIR87] [CRC10] [ALD94]
Solubility: one vol gas dissolves in 32 volumes H_2O (20°C), in 7 volumes alcohol (20°C); s other organic liq, usually higher solubility than in H_2O [MER06]
Density, g/cm³: gas: 1.404 g/L [LID94]; liq: 1.14 g/mL [MER06]
Melting Point, °C: −218.79 [CRC10]
Boiling Point, °C: −182.96 [CRC10]
Thermal Conductivity, W/(m·K): 0.02658 (25°C) [ALD94]

2314
Compound: Ozone
Synonym: triatomic oxygen
Formula: O_3
Molecular Formula: O_3
Molecular Weight: 47.998
CAS RN: 10028-15-6
Properties: blue gas; unstable; pungent odor; oxidizing agent; can be liquefied at −12.1°C; solid is black violet; prepared by silent electric discharge in oxygen; used in drinking water purification, in industrial waste treatment [MER06] [COT88] [DOU83]
Density, g/cm³: gas: 2.144; liq: 1.614 [MER06]
Melting Point, °C: −193 [MER06]
Boiling Point, °C: −111.9 [MER06]
Reactions: reacts with most compounds at 25°C [COT88]

2315
Compound: Palladium
Formula: Pd
Molecular Formula: Pd
Molecular Weight: 106.42
CAS RN: 7440-05-3
Properties: silvery white metal; fcc, a = 0.389 nm; also occurs as black powd, spongy compressible mass; hardness, 4.8 Mohs; electrical resistivity 10.0 μohm·cm; Poisson's ratio 0.39; appreciably volatile at high temperatures; absorbs up to 800 times its own volume of $H_2(g)$; enthalpy of fusion 16.74 kJ/mol; enthalpy of vaporization 362 kJ/mol [HAW93] [MER06] [KIR82] [CRC10] [ALD94]
Solubility: s aqua regia, fused alkalies [HAW93]
Density, g/cm³: 12.02 [MER06]
Melting Point, °C: 1555 [MER06]
Boiling Point, °C: 3167 [MER06]

Reactions: forms dihalides at red heat with
    fluorine, chlorine [MER06]
Thermal Conductivity, W/(m·K): 75.3 (25°C) [KIR82]
Thermal Expansion Coefficient: 11.1×10^{-6}/°C [KIR82]

2316
Compound: Palladium(II) acetate
Synonym: palladous acetate
Formula: $Pd(CH_3COO)_2$
Molecular Formula: $C_4H_6O_4Pd$
Molecular Weight: 224.510
CAS RN: 3375-31-3
Properties: orange brown cryst; there is a trimer
    $[Pd(CH_3COO)_2]_3$ [MER06] [AES93]
Solubility: i H_2O; s with decomposition, HCl;
    s $CHCl_3$, methylene dichloride, acetone,
    acetonitrile, diethyl ether [MER06]
Melting Point, °C: decomposes at 205 [KIR82]

2317
Compound: Palladium(II) acetylacetonate
Synonyms: 2,4-pentanedione, palladium(II) derivative
Formula: $Pd(CH_3C(O)CH=COCH_3)_2$
Molecular Formula: $C_{10}H_{14}O_4Pd$
Molecular Weight: 304.639
CAS RN: 14024-61-4
Properties: yellow cryst [STR93]
Melting Point, °C: decomposes at 205 [STR93]

2318
Compound: Palladium(II) bromide
Formula: $PdBr_2$
Molecular Formula: Br_2Pd
Molecular Weight: 266.228
CAS RN: 13444-94-5
Properties: black cryst; hygr [STR93]
Density, g/cm³: 5.173 [STR93]
Melting Point, °C: decomposes [AES93]

2319
Compound: Palladium(II) chloride
Synonym: palladous chloride
Formula: $PdCl_2$
Molecular Formula: Cl_2Pd
Molecular Weight: 177.325
CAS RN: 7647-10-1
Properties: dark brown powd or cryst; red rhomb;
    hygr; used in the electroless deposition process
    [KIR82] [HAW93] [STR93] [MER06]
Solubility: s H_2O, HCl, alcohol, and acetone [HAW93]

Density, g/cm³: 4.0 [STR93]
Melting Point, °C: decomposes at 675 [HAW93]

2320
Compound: Palladium(II) chloride dihydrate
Formula: $PdCl_2 \cdot 2H_2O$
Molecular Formula: $Cl_2H_4O_2Pd$
Molecular Weight: 213.356
CAS RN: 7647-10-1
Properties: dark brown cryst; reduced by H_2 or
    CO in solution to the metal [MER06]
Solubility: s H_2O, alcohol, acetone [MER06]

2321
Compound: Palladium(II) cyanide
Formula: $Pd(CN)_2$
Molecular Formula: C_2N_2Pd
Molecular Weight: 158.455
CAS RN: 2035-66-7
Properties: yellow powd [STR93]
Melting Point, °C: decomposes [AES93]

2322
Compound: Palladium(II) fluoride
Formula: PdF_2
Molecular Formula: F_2Pd
Molecular Weight: 144.417
CAS RN: 13444-96-7
Properties: violet hygr cryst; paramagnetic
    [LID94] [KIR82]
Solubility: reacts with H_2O [LID94]
Density, g/cm³: 5.76 [LID94]
Melting Point, °C: 952 [LID94]

2323
Compound: Palladium(II) hexafluoroacetylacetonate
Synonyms: 1,1,1,5,5,5-hexafluoro-2,4-
    pentanedione Pd(II) derivative
Formula: $Pd(CF_3COCHCOCF_3)_2$
Molecular Formula: $C_{10}H_2F_{12}O_4Pd$
Molecular Weight: 520.524
CAS RN: 64916-48-9
Properties: cryst [ALF95] [ALD94]

2324
Compound: Palladium(II) iodide
Formula: PdI_2
Molecular Formula: I_2Pd

Molecular Weight: 360.229
CAS RN: 7790-38-7
Properties: black powd [HAW93]
Solubility: s KI soln; i H_2O, alcohol, ether [HAW93]
Density, g/cm³: 6.003 [HAW93]
Melting Point, °C: decomposes at 350 [HAW93]

2325
Compound: Palladium(II) nitrate
Synonym: palladous nitrate
Formula: $Pd(NO_3)_2$
Molecular Formula: N_2O_6Pd
Molecular Weight: 230.429
CAS RN: 10102-05-3
Properties: brown deliq cryst; heating
   causes decomposition; used as a
   catalyst [MER06] [HAW93]
Solubility: s H_2O giving a turbid solution, may
   form precipitate; s dil HNO_3 [MER06]
Density, g/cm³: 1.118 [ALD94]

2326
Compound: Palladium(II) oxalate
Formula: $Pd(C_2O_4)_2$
Molecular Formula: C_4O_8Pd
Molecular Weight: 194.42
CAS RN: 57592-57-1
Properties: powd [ALF95]

2327
Compound: Palladium(II) oxide
Synonym: palladium monoxide
Formula: PdO
Molecular Formula: OPd
Molecular Weight: 122.419
CAS RN: 1314-08-5
Properties: −20 mesh with 99.95% purity; black
   green or amber solid [HAW93] [CER91]
Solubility: i H_2O, acids; sl s aqua
   regia, 48% HBr [MER06]
Density, g/cm³: 8.70 [STR93]
Melting Point, °C: 870 [ALD94]

2328
Compound: Palladium(II) sulfate dihydrate
Formula: $PdSO_4 \cdot 2H_2O$
Molecular Formula: H_4O_6PdS
Molecular Weight: 238.514
CAS RN: 13566-03-5
Properties: brown cryst [STR93]

2329
Compound: Palladium(II) sulfide
Formula: PdS
Molecular Formula: PdS
Molecular Weight: 138.486
CAS RN: 12125-22-3
Properties: gray, tetr cryst [LID94]
Density, g/cm³: 6.60 [ALD94]
Melting Point, °C: 950 [LAN05]

2330
Compound: Palladium(II) tetraammine
   chloride monohydrate
Formula: $Pd(NH_3)_4Cl_2 \cdot H_2O$
Molecular Formula: $Cl_2H_{14}N_4OPd$
Molecular Weight: 263.463
CAS RN: 13933-31-8
Properties: hygr [ALD94]

2331
Compound: Palladium(III) fluoride
Formula: PdF_3
Molecular Formula: F_3Pd
Molecular Weight: 163.415
CAS RN: 12021-58-8
Properties: consists of Pd(II) and Pd(IV) with
   the formula $Pd(II)[PdIV)]F_6$ [KIR82]

2332
Compound: Pentaborane(11)
Synonym: dihydropentaborane(9)
Formula: B_5H_{11}
Molecular Formula: B_5H_{11}
Molecular Weight: 65.142
CAS RN: 18433-84-6
Properties: unstable liq; prepared from diborane;
   decomposes when heated or when allowed to
   stand for long periods of time, producing various
   products including diborane, tetraborane, hydrogen;
   spontaneously flammable in air; enthalpy of
   vaporization 31.8 kJ/mol [MER06] [CRC10]
Solubility: hydrolyzes in H_2O [MER06]
Melting Point, °C: −122 [COT88]
Boiling Point, °C: 63 [MER06]

2333
Compound: Pentaborane(9)
Synonym: pentaboron nonahydride
Formula: B_5H_9
Molecular Formula: B_5H_9

Molecular Weight: 63.126
CAS RN: 19624-22-7
Properties: liq; vapor pressure (0°C) 66 mm Hg; spontaneously flammable in air; can be prepared from diborane; forms diammine by reaction with NH_3 [MER06]
Solubility: hydrolyzed if heated [COT88]
Density, g/cm³: 0.61 [MER06]
Melting Point, °C: −46.6 [KIR78]
Boiling Point, °C: 60 [COT88]
Reactions: decomposes slowly at 150°C [MER06]

2334
Compound: Pentagermane
Formula: Ge_5H_{12}
Molecular Formula: Ge_5H_{12}
Molecular Weight: 375.30
CAS RN: 15587-39-0
Properties: col liq [CRC10]
Solubility: i H_2O [CRC10]
Boiling Point, °C: 234 [CRC10]

2335
Compound: Pentamethylcyclopentadienyltantalum tetrachloride
Formula: $[C_5(CH_3)_5]TaCl_4$
Molecular Formula: $C_{10}H_{15}Cl_4Ta$
Molecular Weight: 457.988
CAS RN: 71414-47-6
Properties: orange powd; sensitive to atm moisture and oxygen [STR93]
Melting Point, °C: 220 [STR93]

2336
Compound: Pentamminechlorocobalt(III) chloride
Formula: $[Co(NH_3)_5Cl]Cl_2$
Molecular Formula: $Cl_3CoH_{15}N_5$
Molecular Weight: 250.444
CAS RN: 13859-51-3
Properties: brick red cryst [KIR79] [ALD94]
Solubility: 24.87 g/100 mL cold H_2O; sl s HCl; i alcohol [KIR79]

2337
Compound: Perbromyl fluoride
Formula: BrO_3F
Molecular Formula: $BrFO_3$
Molecular Weight: 146.900
CAS RN: 37265-91-1

Properties: col gas [CRC10]
Solubility: reac H_2O [CRC10]
Melting Point, °C: −110 [CRC10]
Boiling Point, °C: decomposes at 20 [CRC10]

2338
Compound: Perchloric acid
Formula: $HClO_4$
Molecular Formula: $ClHO_4$
Molecular Weight: 100.459
CAS RN: 7601-90-3
Properties: colorless, fuming liq; hygr; conc acid is unstable, e.g., sensitive to shock; commercial aq acid contains 65%–70% $HClO_4$; some hydrates are: $HClO_4 \cdot H_2O$, 60477-26-1, a colorless oily liq, mp 50°C, bp decomposes; $HClO_4 \cdot 2H_2O$, 13445-00-6, colorless, mp −17.5°C, bp 203°C; $HClO_4 \cdot 3H_2O$, 35468-32-7, two forms: α, mp −37°C, β, mp −43.2°C [HAW93] [KIR79]
Solubility: s H_2O with evolution of heat [HAW93]
Density, g/cm³: 1.77 [KIR79]
Melting Point, °C: −112 [HAW93]
Boiling Point, °C: 110 (extrapolated) [KIR79]

2339
Compound: Perchloryl fluoride
Formula: ClO_3F
Molecular Formula: $ClFO_3$
Molecular Weight: 102.45
CAS RN: 7616-94-6
Properties: col gas [CRC10]
Density, g/L: 4.187 [CRC10]
Melting Point, °C: −147 [CRC10]
Boiling Point, °C: decomposes at −46.75 [CRC10]

2340
Compound: Performic acid
Formula: HCOOOH
Molecular Formula: CH_2O_3
Molecular Weight: 62.024
CAS RN: 107-32-4
Properties: colorless liq; solutions are unstable; used in epoxidation and hydroxylation reactions [HAW93]
Solubility: miscible with H_2O, alcohol, ether; s benzene, chloroform [HAW93]

2341
Compound: Periodic acid
Formula: $HIO_4 \cdot 2H_2O$

Molecular Formula: H_5IO_6
Molecular Weight: 227.940
CAS RN: 10450-60-9
Properties: white cryst; oxidizing agent; preparation: oxidation of an iodate by Cl_2 in basic solution; uses: quantitative measurement of α,β-dihydroxyorganic compounds, to increase the wet strength of paper [HAW93] [DOU83]
Solubility: s H_2O, alcohol; sl s ether [HAW93]
Melting Point, °C: 122 [HAW93]
Boiling Point, °C: decomposes at 130 [HAW93]
Reactions: minus $2H_2O$ at 100°C [HAW93]

2342
Compound: Periodyl fluoride
Formula: IO_3F
Molecular Formula: FIO_3
Molecular Weight: 193.900
CAS RN: 30708-86-2
Properties: col cryst [CRC10]
Boiling Point, °C: decomposes at >100 [CRC10]

2343
Compound: Periodic acid dihydrate
Formula: $HIO_4 \cdot 2H_2O$
Molecular Formula: H_5IO_6
Molecular Weight: 227.940
CAS RN: 10450-60-9
Properties: monocl hygr cryst [CRC10]
Solubility: s H_2O, EtOH; sl eth [CRC10]
Melting Point, °C: decomposes at 122 [CRC10]

2344
Compound: Peroxysulfuric acid
Synonym: Caro's acid
Formula: H_2SO_5
Molecular Formula: H_2O_5S
Molecular Weight: 114.079
CAS RN: 7722-86-3
Properties: white cryst; oxidant; used for testing aniline; in dye manufacturing [HAW93]
Melting Point, °C: 45, decomposing [HAW93]

2345
Compound: Perrhenic acid
Formula: $HReO_4$
Molecular Formula: HO_4Re
Molecular Weight: 251.213
CAS RN: 13768-11-1

Properties: colorless liq; only exists in solution; strong, very stable acid [HAW93] [STR93]
Solubility: v s H_2O and in organic solvents [HAW93]

2346
Compound: Phenylmercuric acetate
Synonym: PMA
Formula: $CH_3COOHgC_6H_5$
Molecular Formula: $C_8H_8HgO_2$
Molecular Weight: 336.740
CAS RN: 62-38-4
Properties: cryst prisms; preparation: heating mercuric acetate with benzene; uses: herbicide, fungicide [KIR81] [MER06]
Solubility: s ~600 parts H_2O, s CH_3COONH_4 aq solutions, s alcohol, benzene, acetone [MER06]
Melting Point, °C: 150–152 [ALD94]

2347
Compound: Phenylmercuric chloride
Synonym: chlorophenylmercury
Formula: C_6H_5HgCl
Molecular Formula: C_6H_5ClHg
Molecular Weight: 313.15
CAS RN: 100-56-1
Properties: white leaflets; uses: fungicide [MER06]
Solubility: ~20,000 parts H_2O; s benzene, ether [MER06]
Melting Point, °C: decomposes at 248–250 [ALD94]

2348
Compound: Phenylmercuric nitrate, basic
Formula: $C_6H_5HgNO_3 \cdot C_6H_5HgOH$
Molecular Formula: $C_{12}H_{11}Hg_2NO_4$
Molecular Weight: 634.404
CAS RN: 8003-05-2
Properties: pearly scales; preparation: boiling benzene and mercuric acetate, followed by treatment with an alkali nitrate; uses: antimicrobial, fungicide for tree treatment [MER06]
Solubility: s ~1250 parts H_2O; sl s alcohol [MER06]
Melting Point, °C: decomposes at 176–186 [ALD94]

2349
Compound: Phosphine
Formula: PH_3
Molecular Formula: H_3P
Molecular Weight: 33.998
CAS RN: 7803-51-2

Properties: colorless gas; spontaneously flammable in air; decaying fish odor; reacts violently with halogens, O_2; forms phosphonium salts with halogen acids; critical temp 51.6°C; critical pressure 6.53 MPa; enthalpy of vaporization 14.6 kJ/mol; formed from white phosphorus and an aq alkali hydroxide [AIR87] [MER06] [CRC10]
Solubility: 0.26 volumes in H_2O (20°C); i hot H_2O; sl s alcohol, ether, cuprous chloride solutions [HAW93] [MER06]
Density, g/cm³: 1.492 g/L [LID94]
Melting Point, °C: −133 [MER06]
Boiling Point, °C: −87.75 [CRC10]
Reactions: decomposed to H_2 and metal phosphide by hot metal [MER06]

2350
Compound: Phosphomolybdic acid hydrate
Formula: $H_3[P(Mo_3O_{10})_4] \cdot xH_2O$
Molecular Formula: $H_3Mo_{12}O_{40}P$ (anhydrous)
Molecular Weight: 1825.254 (anhydrous)
CAS RN: 11104-88-4
Properties: yellow cryst; oxidizing agent; used as a reagent for alkaloids, as a pigment; imparts water resistance to plastics [HAW93]
Solubility: s in less than 0.4 parts H_2O, alcohol, ether [MER06]
Density, g/cm³: 3.15 [HAW93]
Melting Point, °C: 78–90 [HAW93]

2351
Compound: Phosphonitrilic chloride trimer
Formula: $(PNCl_2)_3$
Molecular Formula: $Cl_6N_3P_3$
Molecular Weight: 347.657
CAS RN: 940-71-6
Properties: white cryst; sensitive to moisture [STR93]
Density, g/cm³: 1.98 [STR93]
Melting Point, °C: 128.8 [STR93]
Boiling Point, °C: 127 (12 mm Hg) [STR93]

2352
Compound: Phosphonium iodide
Formula: PH_4I
Molecular Formula: H_4IP
Molecular Weight: 161.910
CAS RN: 12125-09-6
Properties: large, transparent, colorless cryst; tetr; sublimes at room temp; decomposes to PH_3 and HI when heated or in presence of alcohol or water; rapid heating causes detonation [MER06]

Solubility: decomposed by H_2O, alcohol, evolving phosphine gas [HAW93]
Density, g/cm³: 2.86 [HAW93]
Melting Point, °C: 18.5 [HAW93]
Boiling Point, °C: 80 [HAW93]
Reactions: sublimes at 61.8°C [HAW93]

2353
Compound: Phosphoric acid
Synonym: orthophosphoric acid
Formula: H_3PO_4
Molecular Formula: H_3O_4P
Molecular Weight: 97.995
CAS RN: 7664-38-2
Properties: colorless, odorless, sparkling, syrupy liq or unstable ortho-rhomb cryst; acid dissociation constants: $K_1 = 7.107 \times 10^{-3}$, $K_2 = 7.99 \times 10^{-8}$, $K_3 = 4.8 \times 10^{-13}$; enthalpy of fusion 13.40 kJ/mol [MER06] [CRC10]
Solubility: s H_2O, alcohol [MER06]
Density, g/cm³: cryst: 1.834 [HAW93]
Melting Point, °C: cryst: 42.35 [MER06]
Boiling Point, °C: 407 [LID94]
Reactions: minus $1/2H_2O$ at 213°C, forming pyrophosphoric acid [HAW93]

2354
Compound: Phosphorous acid
Synonym: orthophosphorus acid
Formula: H_3PO_3
Molecular Formula: H_3O_3P
Molecular Weight: 81.996
CAS RN: 13598-36-2
Properties: white, very hygr and deliq, cryst mass; slowly oxidized in air to H_3PO_4; enthalpy of fusion 12.80 kJ/mol [MER06] [DOU83]
Solubility: v s H_2O, alcohol [MER06]
Density, g/cm³: 1.65; liq: 1.597 [MER06]
Melting Point, °C: 74.4 [CRC10]
Boiling Point, °C: decomposes to PH_3, H_3PO_4 above 180 [MER06]

2355
Compound: Phosphorus (black)
Synonym: black phosphorus
Formula: P
Molecular Formula: P
Molecular Weight: 30.973762
CAS RN: 7723-14-0

Properties: black solid, resembling graphite; obtained by heating white phosphorus under pressure; ortho-rhomb cryst; stable in air; not spontaneously flammable in air; under high pressure, transformed reversibly to a second rhomb cryst, density 3.56, and cub cryst, density 3.83; conducts electricity [HAW93] [MER06]
Solubility: i organic solvents [MER06]
Density, g/cm³: 2.691 [MER06]

2356
Compound: Phosphorus (red)
Synonym: red phosphorus
Formula: P
Molecular Formula: P
Molecular Weight: 30.973762
CAS RN: 7723-14-0
Properties: red to violet amorphous powd; 6 mm pieces and smaller with 99.995% purity; obtained from white phosphorus using catalysts by heating at 240°; less active than white phosphorus; has high electrical resistivity [HAW93] [MER06] [CER91]
Solubility: i organic solvents; s PBr_3 [MER06]
Density, g/cm³: 2.34 [MER06]
Melting Point, °C: sublimes at 416 [MER06]
Reactions: burns in air to form P_2O_5 at 260°C [MER06]

2357
Compound: Phosphorus (white)
Synonyms: white, yellow phosphorus
Formula: P_4
Molecular Formula: P_4
Molecular Weight: 123.895 (atomic weight: 30.974)
CAS RN: 12185-10-3
Properties: wax-like transparent cryst; darkens when exposed to light; impurities cause yellow color; α-form: exists at room temp; β-form: hex cryst; prepared from α-form at −79.6°C; volatile; sublimes in vacuum at ordinary temperatures; vapor density corresponds to formula P_4; hardness is 0.5 Mohs; electronegativity 2.06; enthalpy of vaporization 12.4 kJ/mol; enthalpy of fusion 0.66 kJ/mol; electrical resistivity (20°C) 10 μohm·cm [CRC10] [ALD94] [HAW93] [MER06] [COT88]
Solubility: 1 part/300,000 parts H_2O; 1 g/400 mL absolute alcohol; 1 g/200 mL $CHCl_2$; 1 g/40 mL benzene [MER06]
Density, g/cm³: α: 1.83; β. 1.88 [MER06]
Melting Point, °C: 44.1 (0.181 mm) [MER06]
Boiling Point, °C: 277 [CRC10]
Reactions: ignites in moist at ~30°C [MER06]
Thermal Conductivity, W/(m·K): 0.236 (25°C) [ALD94]

2358
Compound: Phosphorus heptasulfide
Formula: P_4S_7
Molecular Formula: P_4S_7
Molecular Weight: 348.357
CAS RN: 12037-82-0
Properties: light yellow cryst [HAW93]
Solubility: sl s carbon disulfide [HAW93]
Density, g/cm³: 2.19 [HAW93]
Melting Point, °C: 310 [HAW93]
Boiling Point, °C: 523 [HAW93]

2359
Compound: Phosphorus nitride
Formula: P_3N_5
Molecular Formula: N_5P_3
Molecular Weight: 162.955
CAS RN: 17739-47-8
Properties: white, amorphous solid; stable in air [HAW93]
Solubility: i cold H_2O, decomposed by hot H_2O (HAW93]; s in common organic solvents [HAW93]
Melting Point, °C: decomposes at 800 [HAW93]

2360
Compound: Phosphorus oxybromide
Synonym: phosphoryl bromide
Formula: $POBr_3$
Molecular Formula: Br_3OP
Molecular Weight: 286.685
CAS RN: 7789-59-5
Properties: colorless cryst; thin plates with faint orange tint; enthalpy of vaporization 38 kJ/mol; used as an intermediate in chemical processes [HAW93] [MER06] [CRC10]
Solubility: hydrolyzes slowly in H_2O to H_3PO_4 and HBr; s ether, benzene, chloroform, CS_2, conc H_2SO_4 [MER06]
Density, g/cm³: 2.822 [MER06]
Melting Point, °C: 56 [MER06]
Boiling Point, °C: 191.7 [CRC10]

2361
Compound: Phosphorus oxychloride
Synonym: phosphoryl chloride
Formula: $POCl_3$
Molecular Formula: Cl_3OP
Molecular Weight: 153.331
CAS RN: 10025-87-3

Properties: colorless, clear liq; 99.9% purity; fumes strongly; liberates heat by reaction with H_2O, alcohol; pungent odor; enthalpy of vaporization 34.35 at bp, 38.6 at 25°C; enthalpy of fusion 13.10 kJ/mol; used to manufacture esters for plasticizers and gasoline additives [HAW93] [MER06] [CER91] [CRC10]
Solubility: reacts with H_2O, alcohol [HAW93]
Density, g/cm³: 1.645 [MER06]
Melting Point, °C: 1.25 [ALD94]
Boiling Point, °C: 105.8 [MER06]

2362
Compound: Phosphorus oxyfluoride
Formula: POF_3
Molecular Formula: F_3OP
Molecular Weight: 103.968
CAS RN: 13478-20-1
Properties: colorless gas; critical temp 73.3°C; critical pressure 4.23 MPa; enthalpy of fusion 14.9 kJ/mol; enthalpy of vaporization 23.2 kJ/mol [KIR78]
Solubility: hydrolyzes [KIR78]
Density, g/cm³: 4.562 g/L [LID94]
Melting Point, °C: −39.1 [KIR78]
Boiling Point, °C: −39.7 [KIR78]

2363
Compound: Phosphorus triselenide
Formula: P_2Se_3
Molecular Formula: P_2Se_3
Molecular Weight: 298.828
CAS RN: 1314-86-9
Properties: dark red mass; heating causes decomposition; decomposed by moist air and in H_2O [MER06]
Solubility: s KOH; i CS_2 [MER06]
Density, g/cm³: 1.31 [LID94]
Melting Point, °C: 245 [LID94]
Boiling Point, °C: ~380 [LID94]

2364
Compound: Phosphorus(III) bromide
Synonym: phosphorus tribromide
Formula: PBr_3
Molecular Formula: Br_3P
Molecular Weight: 270.686
CAS RN: 7789-60-8
Properties: colorless, fuming liq; 99.9% purity; penetrating odor; enthalpy of vaporization 38.8 kJ/mol [CRC10] [CER91] [HAW93] [MER06]
Solubility: s with decomposition, H_2O and alcohol; s acetone, CS_2 [MER06]

Density, g/cm³: 2.852 (15°C) [STR93]
Melting Point, °C: −41.5 [MER06]
Boiling Point, °C: 172.95 [CRC10]

2365
Compound: Phosphorus(III) chloride
Synonym: phosphorus trichloride
Formula: PCl_3
Molecular Formula: Cl_3P
Molecular Weight: 137.332
CAS RN: 7719-12-2
Properties: 99.9% purity; colorless, clear, fuming liq; enthalpy of vaporization 30.5 kJ/mol at bp, 32.1 kJ/mol at 25°C; enthalpy of fusion 7.10 kJ/mol [MER06] [CRC10] [CER91]
Solubility: decomposed by H_2O, alcohol; s benzene, chloroform, ether, CS_2 [MER06]
Density, g/cm³: 1.574 [MER06]
Melting Point, °C: −112 [CRC10]
Boiling Point, °C: 75.95 [CRC10]

2366
Compound: Phosphorus(III) fluoride
Synonym: phosphorus trifluoride
Formula: PF_3
Molecular Formula: F_3P
Molecular Weight: 87.969
CAS RN: 7783-55-3
Properties: colorless gas; sensitive to air and moisture; critical temp −2.05°C; critical pressure 4.33 MPa; enthalpy of vaporization 16.5 kJ/mol [KIR78] [STR93]
Solubility: slowly hydrolyzed by H_2O [MER06]
Density, g/cm³: gas: 3.907 g/L [MER06]
Melting Point, °C: −151.30 [MER06]
Boiling Point, °C: −101.38 [MER06]

2367
Compound: Phosphorus(III) iodide
Synonym: phosphorus triiodide
Formula: PI_3
Molecular Formula: I_3P
Molecular Weight: 411.687
CAS RN: 13455-01-1
Properties: red or orange cryst; hygr; enthalpy of vaporization 43.9 kJ/mol [HAW93] [CRC10]
Solubility: decomposed by H_2O; s alcohol, CS_2 [HAW93]
Density, g/cm³: 4.18 [STR93]
Melting Point, °C: 61.5 [LID94]
Boiling Point, °C: decomposes at 227 [CRC10]

2368
Compound: Phosphorus(III) oxide
Synonym: phosphorus trioxide
Formula: P_2O_3
Molecular Formula: O_3P_2
Molecular Weight: 109.946
CAS RN: 1314-24-5
Properties: transparent cryst; monocl or colorless liq; disproportionates to red P and P_2O_4 if heated above 210°C [MER06]
Solubility: slowly forms H_3PO_3 in cold H_2O; reacts violently in hot H_2O forming red P, PH_3, and H_3PO_4 [MER06]
Density, g/cm³: 2.135 [MER06]
Melting Point, °C: 23.8 [MER06]
Boiling Point, °C: 173.1 (under N_2) [MER06]

2369
Compound: Phosphorus(III) sulfide
Synonym: phosphorus trisulfide
Formula: P_2S_3
Molecular Formula: P_2S_3
Molecular Weight: 158.146
CAS RN: 81129-00-2
Properties: grayish yellow mass; tasteless and odorless; decomposed by atm moisture [HAW93]
Solubility: s alcohol, CS_2, ether [HAW93]
Melting Point, °C: 290 [HAW93]
Boiling Point, °C: 490 [HAW93]

2370
Compound: Phosphorus(V) bromide
Synonym: phosphorus pentabromide
Formula: PBr_5
Molecular Formula: Br_5P
Molecular Weight: 430.494
CAS RN: 7789-69-7
Properties: yellow cryst; −60 mesh with 99.9% purity; decomposed by H_2O or alcohol; used as a brominating agent [HAW93] [MER06] [CER91]
Solubility: s CS_2, CCl_4 [MER06]
Melting Point, °C: decomposes at 106 [HAW93]

2371
Compound: Phosphorus(V) chloride
Synonym: phosphorus pentachloride
Formula: PCl_5
Molecular Formula: Cl_5P
Molecular Weight: 208.238
CAS RN: 10026-13-8

Properties: white to pale yellow; −60 mesh with 99.9% purity; deliq; fumes; sublimes without melting from 160°C–165°C; irritating odor; used as a chlorinating agent, catalyst, and dehydrating agent [HAW93] [MER06] [CER91]
Solubility: hydrolyzed in H_2O to H_3PO_4, HCl; s CS_2, CCl_4 [MER06]
Density, g/cm³: 3.60 [LID94]
Melting Point, °C: 148 (under pressure) [MER06]
Boiling Point, °C: 160 [MER06]

2372
Compound: Phosphorus(V) fluoride
Synonym: phosphorus pentafluoride
Formula: PF_5
Molecular Formula: F_5P
Molecular Weight: 125.966
CAS RN: 7647-19-0
Properties: colorless gas; nonflammable; fumes strongly in air; high thermal stability; critical temp 144.5°C; critical pressure 3.39 MPa; enthalpy of fusion 12.1 kJ/mol; enthalpy of vaporization 17.2 kJ/mol; can be prepared by reaction of PF_3 with F_2; used as a polymerization catalyst and in electronics industry [AIR87] [HAW93] [MER06] [CRC10] [KIR78]
Solubility: hydrolyzed in H_2O, eventually to H_3PO_4 [MER06]
Density, g/cm³: gas: 5.527 g/L [LID94]
Melting Point, °C: −93.8 [MER06]
Boiling Point, °C: −84.6 [MER06]

2373
Compound: Phosphorus(V) oxide
Synonym: phosphorus pentoxide
Formula: P_2O_5
Molecular Formula: O_5P_2
Molecular Weight: 141.945
CAS RN: 1314-56-3
Properties: soft, white powd; −100 mesh with 99.9% purity; several cryst and amorphous forms; very deliq; corrosive; not flammable; readily absorbs H_2O from air; forms H_3PO_4 in water, releasing heat; enthalpy of fusion 27.20 kJ/mol [CRC10] [CER91] [HAW93] [MER06]
Density, g/cm³: 2.39 [STR93]
Melting Point, °C: 420 [CRC10]
Boiling Point, °C: sublimes at 360 [MER06]

2374
Compound: Phosphorus(V) selenide
Synonym: phosphorus pentaselenide
Formula: P_2Se_5
Molecular Formula: P_2Se_5

Molecular Weight: 456.748
CAS RN: 1314-82-5
Properties: amorphous glass; blackish purple solid; decomposes in steam and boiling water [MER06]
Solubility: reacts with CCl_4; i CS_2 [MER06]

2375
Compound: Phosphorus(V) sulfide
Synonym: phosphorus pentasulfide
Formula: P_2S_5
Molecular Formula: P_2S_5
Molecular Weight: 222.278
CAS RN: 1314-80-3
Properties: light yellow or greenish yellow cryst; H_2S like odor; very hygr; forms P_2O_5 and SO_2 when ignited in air; decomposed by moist air; used in production of lubricating oils, in flotation agents, and safety matches [HAW93]
Solubility: s alkali hydroxides, CS_2 [HAW93]
Density, g/cm³: 2.03 [HAW93]
Melting Point, °C: 286–290 [HAW93]
Boiling Point, °C: 515 [HAW93]

2376
Compound: Phosphotungstic acid 24-hydrate
Synonym: tungstophosphoric acid
Formula: $H_3PW_{12}O_{40} \cdot 24H_2O$
Molecular Formula: $H_{51}O_{64}PW_{12}$
Molecular Weight: 3312.420
CAS RN: 12067-99-1
Properties: H_2O content can vary appreciably; white or sl yellowish green cryst or powd; efflorescent; used in analytical chemistry to detect many organic compounds such as phenols, alkaloids, and albumin [MER06]
Solubility: s ~0.5 parts H_2O; s alcohol, ether [MER06]
Melting Point, °C: 89 [HAW93]

2377
Compound: Platinum
Formula: Pt
Molecular Formula: Pt
Molecular Weight: 195.08
CAS RN: 7440-06-4
Properties: silvery gray, lustrous, ductile metal; fcc, a = 0.39231 nm; also has black powd and spongy mass forms; vapor pressure at mp 0.0187 Pa; electrical resistivity, μohm·cm: 10.6 (20°C), 9.85 (0°C); Brinell hardness 97; does not corrode or tarnish; attacked by Cl_2 at high temperatures; Poisson's ratio 0.39; enthalpy of fusion 22.17 kJ/mol; used as a catalyst for chemical, automotive, and petroleum industries [KIR82] [HAW93] [MER06] [CRC10]

Solubility: i H_2O, mineral acids; s aqua regia [HAW93] [MER06]
Density, g/cm³: 21.447 (calc) [MER06]
Melting Point, °C: 1768.4 [CRC10]
Boiling Point, °C: 3825 [CRC10]
Thermal Conductivity, W/(m·K): 71.1 (25°C) [KIR82]
Thermal Expansion Coefficient: (volume) 100°C (0.216), 200°C (0.494), 400°C (1.074), 800°C (2.339), 1200°C (3.750) [CLA66]

2378
Compound: Platinum acetylacetonate
Synonyms: 2,4-pentanedione, platinum(II) derivative
Formula: $Pt(CH_3C(O)CH=COCH_3)_2$
Molecular Formula: $C_{10}H_{14}O_4Pt$
Molecular Weight: 393.299
CAS RN: 15170-57-7
Properties: pale yellow cryst [STR93]
Melting Point, °C: 250–252 [ALD94]

2379
Compound: Platinum hexafluoride
Formula: PtF_6
Molecular Formula: F_6Pt
Molecular Weight: 309.070
CAS RN: 13693-05-5
Properties: dark brick red, rhomb solid; strong oxidizing agent; there are also PtF_4, 13455-15-7, and PtF_5, 13782-84-8 [KIR82]
Density, g/cm³: 3.83 [KIR82]
Melting Point, °C: 61.3 [KIR82]
Boiling Point, °C: 69.14 [KIR82]

2380
Compound: Platinum silicide
Formula: PtSi
Molecular Formula: PtSi
Molecular Weight: 223.166
CAS RN: 12137-83-6
Properties: ortho-rhomb cryst; −100 mesh powd [LID94] [ALF93]
Density, g/cm³: 12.4 [LID94]
Melting Point, °C: 1229 [ALF93]

2381
Compound: Platinum(II) bromide
Synonym: platinum dibromide
Formula: $PtBr_2$
Molecular Formula: Br_2Pt
Molecular Weight: 354.888

CAS RN: 13455-12-4
Properties: brown powd; PtBr₄, 13455-11-3,
    exists [KIR82] [STR93]
Density, g/cm³: 6.65 [STR93]
Melting Point, °C: decomposes at 250 [STR93]

2382
Compound: Platinum(II) chloride
Synonym: platinum dichloride
Formula: PtCl₂
Molecular Formula: Cl₂Pt
Molecular Weight: 265.985
CAS RN: 10025-65-7
Properties: grayish green to brown
    powd; hex [KIR82] [MER06]
Solubility: i H₂O, alcohol, ether; s HCl [MER06]
Density, g/cm³: 5.87 [MER06]; 6.05 [STR93]
Melting Point, °C: decomposes at 581 [STR93]
Reactions: decomposes at red heat
    yielding platinum [HAW93]

2383
Compound: Platinum(II) cyanide
Formula: Pt(CN)₂
Molecular Formula: C₂N₂Pt
Molecular Weight: 247.115
CAS RN: 592-06-3
Properties: yellowish green cryst [STR93]

2384
Compound: Platinum(II) hexafluoroacetylacetonate
Synonyms: 1,1,1,5,5,5-hexafluoro-2,4-
    pentanedione Pt(II) derivative
Formula: Pt(CF₃COCHCOCF₃)₂
Molecular Formula: C₁₀H₂F₁₂O₄Pt
Molecular Weight: 609.185
CAS RN: 65353-51-7
Properties: orange cryst [STR93]
Melting Point, °C: sublimes at 65 [STR93]

2385
Compound: Platinum(II) iodide
Synonym: platinum diiodide
Formula: PtI₂
Molecular Formula: I₂Pt
Molecular Weight: 448.889
CAS RN: 7790-39-8
Properties: heavy, black powd [MER06]
Solubility: i H₂O, alkali iodides [MER06]
Density, g/cm³: 6.40 [ALD94]
Melting Point, °C: decomposes at 325 [MER06]

2386
Compound: Platinum(II) oxide
Synonym: platinum monoxide
Formula: PtO
Molecular Formula: OPt
Molecular Weight: 211.079
CAS RN: 12035-82-4
Properties: tetr black cryst [LID94] [KIR82]
Solubility: i H₂O, alcohol; s aqua regia [KIR82]
Density, g/cm³: 14.9 [KIR82]
Melting Point, °C: decomposes at 500 [KIR82]

2387
Compound: Platinum(IV) chloride
Synonym: platinum tetrachloride
Formula: PtCl₄
Molecular Formula: Cl₄Pt
Molecular Weight: 336.891
CAS RN: 37773-49-2
Properties: reddish brown cryst; sensitive
    to moisture [STR93]
Density, g/cm³: 4.303 [STR93]
Melting Point, °C: 370, decomposes [STR93]

2388
Compound: Platinum(IV) chloride pentahydrate
Formula: PtCl₄·5H₂O
Molecular Formula: Cl₄H₁₀O₅Pt
Molecular Weight: 426.967
CAS RN: 13454-96-1
Properties: red cryst [HAW93]
Solubility: s H₂O and alcohol [HAW93]
Density, g/cm³: 2.43 [HAW93]
Reactions: minus 4H₂O at 100°C [HAW93]

2389
Compound: Platinum(IV) iodide
Synonym: platinic iodide
Formula: PtI₄
Molecular Formula: I₄Pt
Molecular Weight: 702.698
CAS RN: 7790-46-7
Properties: brownish black powd; PtI₃, 58782-
    50-6, exists [KIR82] [MER06]
Solubility: s H₂O [MER06]
Melting Point, °C: decomposes at 130 [LID94]

2390
Compound: Platinum(IV) oxide
Synonym: Adams' catalyst

Formula: PtO_2
Molecular Formula: O_2Pt
Molecular Weight: 227.079
CAS RN: 1314-15-4
Properties: −100 mesh with 99.9% purity; black powd [HAW93] [CER91]
Solubility: s conc acids, s dil KOH solutions [HAW93]
Density, g/cm³: 11.8 [LID94]
Melting Point, °C: 450 [AES93]

2391
Compound: Plutonium
Formula: α-Pu
Molecular Formula: Pu
Molecular Weight: 244
CAS RN: 7440-07-5
Properties: silvery white metal; highly reactive; α form: monocl, a=0.6183 nm, b=0.4822 nm, c=1.0963 nm; ionic radius of Pu^{++++} is 0.0887 nm; stable form from room temp to 115°C; enthalpy of vaporization 333.5 kJ/mol; enthalpy of fusion 2.82 kJ/mol; discovered in 1940–1941; prepared in ton quantities in nuclear reactors; ^{238}Pu produced in kg amounts from ^{237}Np; important fuel for producing power for terrestrial and extraterrestrial applications [MER06] [KIR78] [CRC10]
Density, g/cm³: 19.86 [KIR78]
Melting Point, °C: 646 [KIR91]
Boiling Point, °C: 3235 [KIR91]
Reactions: transitions: α→β, 115°C; β→γ, 185°C; γ→δ, 310°C; δ→δ', 452°C; δ'→ε, 480°C [KIR78]

2392
Compound: Plutonium nitride
Formula: PuN
Molecular Formula: NPu
Molecular Weight: 258
CAS RN: 12033-54-4
Properties: dark gray; fcc, a=0.4907 nm [KIR81]
Density, g/cm³: 14.4 [KIR81]
Melting Point, °C: 2550 [KIR81]

2393
Compound: Plutonium(III) chloride
Formula: $PuCl_3$
Molecular Formula: Cl_3Pu
Molecular Weight: 350
CAS RN: 13569-62-5
Properties: emerald green; hex, a=0.7394 nm, c=0.4243 nm [KIR78]
Density, g/cm³: 5.71 [KIR78]
Melting Point, °C: 760 [KIR91]

2394
Compound: Plutonium(III) fluoride
Formula: PuF_3
Molecular Formula: F_3Pu
Molecular Weight: 301
CAS RN: 13842-83-6
Properties: purple; hex, a=0.7092 nm, c=0.7254 nm [KIR78]
Density, g/cm³: 9.33 [KIR78]
Melting Point, °C: 1425 [KIR91]

2395
Compound: Plutonium(III) iodide
Formula: PuI_3
Molecular Formula: I_3Pu
Molecular Weight: 625
CAS RN: 13813-46-2
Properties: green; ortho-rhomb, a=0.4326 nm, b=1.3962 nm, c=0.9974 nm [KIR78]
Density, g/cm³: 6.92 [KIR78]

2396
Compound: Plutonium(IV) chloride
Formula: $PuCl_4$
Molecular Formula: Cl_4Pu
Molecular Weight: 386
CAS RN: 13536-92-0
Properties: greenish yellow; tetr, a=0.8377 nm, c=0.7481 nm [KIR78]
Density, g/cm³: 4.72 [KIR78]

2397
Compound: Plutonium(IV) fluoride
Formula: PuF_4
Molecular Formula: F_4Pu
Molecular Weight: 320
CAS RN: 13709-56-3
Density, g/cm³: 7.1 [CRC10]
Melting Point: 1037

2398
Compound: Plutonium(IV) oxide
Synonym: plutonium dioxide
Formula: PuO_2
Molecular Formula: O_2Pu
Molecular Weight: 276
CAS RN: 11116-03-3
Properties: yellowish green to brown; cub, a=0.53960 nm [KIR78]

Density, g/cm³: 11.46 [KIR78]
Melting Point, °C: 2400 [KIR91]

2399
Compound: Plutonium(VI) hexafluoride
Formula: PuF_6
Molecular Formula: F_6Pu
Molecular Weight: 358
CAS RN: 13693-06-6
Properties: reddish brown; ortho-rhomb, a=0.9912 nm, b=0.8942 nm, c=0.5206 nm [KIR78]
Density, g/cm³: 5.081 [KIR78]
Melting Point, °C: 52 [KIR91]

2400
Compound: Polonium
Formula: α-Po
Molecular Formula: Po
Molecular Weight: 209
CAS RN: 7440-08-6
Properties: radioactive solid, α-emitter; resistivity at 0°C is 42 μohm·cm; resembles Te and Bi in chemical properties; coexists with β-form over temp range 18 to 54°C [MER06]
Density, g/cm³: 9.196 [MER06]
Melting Point, °C: 254 [MER06]
Boiling Point, °C: 962 [MER06]
Thermal Conductivity, W/(m·K): 20 [CRC10]

2401
Compound: Polonium
Formula: β-Po
Molecular Formula: Po
Molecular Weight: 209
CAS RN: 7440-08-6
Properties: radioactive solid; α-emitter; resistivity (0°C) 44 μohm·cm; chemical behavior resembles Te, Bi; coexists with α-form over temp range 18°C–54°C; electronegativity 1.76 [MER06] [COT88]
Density, g/cm³: 9.398 [MER06]
Melting Point, °C: 254 [MER06]
Boiling Point, °C: 962 [MER06]

2402
Compound: Polonium(IV) chloride
Synonym: polonium tetrachloride
Formula: $PoCl_4$
Molecular Formula: Cl_4Po
Molecular Weight: 351
CAS RN: 10026-02-5

Properties: hygr; bright yellow cryst; monocl or tricl; hydrolyzed in moist air to form a white solid; vapors are purplish brown, becoming bluish green at >500°C [MER06]
Solubility: s H_2O, slowly hydrolyzing; v s HCl, thionyl chloride; s ethanol, acetone; decomposed by dil HNO_3 [MER06]
Melting Point, °C: ~300 (in chlorine) [MER06]
Boiling Point, °C: 390 [LID94]
Reactions: turns scarlet red at 350°C [MER06]

2403
Compound: Polonium(IV) oxide
Synonym: polonium dioxide
Formula: PoO_2
Molecular Formula: O_2Po
Molecular Weight: 241
CAS RN: 7446-06-2
Properties: two cryst forms: low temp, yellow fcc; high temp, red tetr [MER06]
Solubility: s phosphoric acid, ammonium carbonate solution [MER06]
Density, g/cm³: 8.9 [LID94]
Melting Point, °C: sublimes at 885, color darkens to chocolate brown [MER06]
Reactions: decomposes into elements at 500°C under vacuum [MER06]

2404
Compound: Potassium
Synonym: kalium
Formula: K
Molecular Formula: K
Molecular Weight: 39.0983
CAS RN: 7440-09-7
Properties: soft, silvery white metal; tarnishes in air; bcc; brittle at low temperatures; reacts vigorously with O_2, H_2O, acids, hydroxides, halogens; enthalpy of fusion 2.32 kJ/mol; enthalpy of vaporization 81.13 kJ/mol; electrical resistivity (20°C) 6.1 μohm·cm; surface tension (100°C) 86 mN/m; viscosity (25°C) 0.258 mPa·s; ionic radius 0.133 nm; Pauling electronegativity 0.8 [KIR82] [MER06] [ALD94]
Solubility: s liq ammonia, ethylenediamine, aniline; some metals [MER06]
Density, g/cm³: 0.856 [MER06]
Melting Point, °C: 63.7 [CAB85]
Boiling Point, °C: 760 [CAB85]
Thermal Conductivity, W/(m·K): 102.5 (25°C) [ALD94]

2405
Compound: Potassium acetate
Synonyms: acetic acid, potassium salt
Formula: CH_3COOK
Molecular Formula: $C_2H_3KO_2$
Molecular Weight: 98.143
CAS RN: 127-08-2
Properties: white, lustrous cryst; very deliq; saline taste; usually prepared by reacting potassium carbonate with acetic acid; used as a dehydrating agent and as a textile conditioner [HAW93] [MER06] [KIR82]
Solubility: g/100 g, H_2O: 216 (0°C), 256 (20°C), 398 (90°C); solid phase, CH_3COOK [LAN05]; s alcohol, i ether [HAW93]
Density, g/cm³: 1.57 [MER06]
Melting Point, °C: 292 [MER06]

2406
Compound: Potassium acetylacetonate hemihydrate
Synonyms: 2,4-pentanedione, potassium derivative hemihdyrate
Formula: $K(CH_3COCH=C(O)CH_3) \cdot 1/2H_2O$
Molecular Formula: $C_5H_8KO_{2.5}$
Molecular Weight: 147.215
CAS RN: 57402-46-7
Properties: off-white powd; hygr [ALD94] [STR93]
Melting Point, °C: decomposes at 215 [STR93]

2407
Compound: Potassium aluminate trihydrate
Formula: $K_2Al_2O_4 \cdot 3H_2O$
Molecular Formula: $Al_2H_6K_2O_7$
Molecular Weight: 250.202
CAS RN: 12003-63-1
Properties: hard, lustrous cryst; using in dyeing, printing, paper sizing [HAW93] [MER06]
Solubility: v s H_2O, hydrolyzes giving alkaline solution; i alcohol [HAW93] [MER06]

2408
Compound: Potassium aluminum sulfate
Synonym: burnt alum
Formula: $KAl(SO_4)_2$
Molecular Formula: $AlKO_8S_2$
Molecular Weight: 258.207
CAS RN: 10043-67-1
Properties: white powd; absorbs atm moisture [MER06]
Solubility: g/100 g H_2O: 3.00 (0°C), 5.90 (20°C), 109 (90°C) [LAN05]

2409
Compound: Potassium aluminum sulfate dodecahydrate
Synonym: kalinite
Formula: $KAl(SO_4)_2 \cdot 12H_2O$
Molecular Formula: $AlH_{24}KO_{20}S_2$
Molecular Weight: 474.391
CAS RN: 7784-24-9
Properties: colorless, trans cryst [MER06]
Solubility: 1 g/7.2 mL H_2O, 1 g/0.3 mL boiling H_2O; s glyercol [MER06]
Density, g/cm³: 1.725 [MER06]
Melting Point, °C: 92.5 [MER06]
Reactions: minus $12H_2O$ at ~200°C [MER06]

2410
Compound: Potassium antimony oxalate trihydrate
Synonym: antimony potassium oxalate
Formula: $K_3[Sb(OOCCOO)_3] \cdot 3H_2O$
Molecular Formula: $C_6H_6K_3O_{15}Sb$
Molecular Weight: 557.160
CAS RN: 5965-33-3
Properties: cryst powd [MER06]
Solubility: s H_2O [MER06]

2411
Compound: Potassium antimony tartrate hemihydrate
Synonym: tartar emetic
Formula: $K(SbO)C_4O_6 \cdot 1/2H_2O$
Molecular Formula: $C_4HKO_{7.5}Sb$
Molecular Weight: 329.903
CAS RN: 28300-74-5
Properties: transparent cryst; effloresces in air; sweetish metallic taste; used in medicine, textiles and, leather, as an insecticide [HAW93] [MER06]
Solubility: 1 g/12 mL H_2O, 1 g/3 mL boiling H_2O [MER06]
Density, g/cm³: 2.6 [HAW93]
Reactions: dehydrates at 100°C [HAW93]

2412
Compound: Potassium azide
Formula: KN_3
Molecular Formula: KN_3
Molecular Weight: 81.118
CAS RN: 20762-60-1
Properties: colorless; body-center; tetr, a = 0.6091 nm, c = 0.7056 nm [CIC73] [CRC10]
Solubility: g/100 g H_2O: 41.4 (0°C), 50.8 (20°C), 61.0 (40°C), 106 (100°C) [LAN05]
Density, g/cm³: 2.04 [CRC10]
Melting Point, °C: 350 (vacuum) [CRC10]

2413
Compound: Potassium bis(oxalato) platinate(II) dihydrate
Formula: $K_2Pt(C_2O_4)_2 \cdot 2H_2O$
Molecular Formula: $C_4H_4K_2O_{10}Pt$
Molecular Weight: 485.346
CAS RN: 14224-64-5
Properties: cryst [ALF95]

2414
Compound: Potassium borohydride
Synonym: potassium tetrahydroborate
Formula: KBH_4
Molecular Formula: BH_4K
Molecular Weight: 53.941
CAS RN: 13762-51-1
Properties: white, cryst powd; nonhygr cryst; supports combustion; decomposes without melting at ~500°C; thermally more stable and less reactive than sodium borohydride; used as a reducing agent for aldehydes, ketones, and acid chlorides [MER06]
Solubility: w/w H_2O, 19% (25°C); alkaline solutions stable [MER06]
Density, g/cm³: 1.11 [LID94]
Melting Point, °C: decomposes at 500 [KIR80]

2415
Compound: Potassium bromate
Formula: $KBrO_3$
Molecular Formula: $BrKO_3$
Molecular Weight: 167.000
CAS RN: 7758-01-2
Properties: white cryst or granules; oxidizing agent; used as a laboratory reagent, in permanent wave formulations, as a food additive [HAW93] [MER06]
Solubility: g/100 g soln, H_2O: 2.96 (0°C), 7.53 (25°C), 33.3 (100°C) [KRU93]; i alcohol [MER06]
Density, g/cm³: 3.27 [MER06]
Melting Point, °C: ~350 [MER06]
Reactions: ~370°C, decomposes with evolution of O_2 [MER06]

2416
Compound: Potassium bromide
Formula: KBr
Molecular Formula: BrK
Molecular Weight: 119.002
CAS RN: 7758-02-3
Properties: colorless cub cryst or white granules or powd; hygr; strong, bitter, saline taste; enthalpy of fusion 25.50 kJ/mol; can be prepared by reacting bromine with potassium carbonate; used in photography, engraving and lithography, spectroscopy [HAW93] [CRC10] [STR93] [MER06] [KIR82]
Solubility: g/100 g soln, H_2O: 40.61 (25°C); solid phase, KBr [KRU93]; 1 g/250 mL alcohol; 1 g/4.6 mL glycerol [MER06]
Density, g/cm³: 2.75 [MER06]
Melting Point, °C: 734 [CRC10]
Boiling Point, °C: 1435 [STR93]
Thermal Expansion Coefficient: (volume) 100°C (0.951), 200°C (2.265), 400°C (5.116) [CLA66]

2417
Compound: Potassium carbonate
Synonyms: salt of tartar, pearl ash
Formula: K_2CO_3
Molecular Formula: CK_2O_3
Molecular Weight: 138.206
CAS RN: 584-08-7
Properties: white monocl; hygr, odorless granules or translucent powd; enthalpy of fusion 27.60 kJ/mol; commonly produced by the carbonation of KOH; used in special glasses, e.g., optical and color TV tubes, pigments; general-purpose food additive [HAW93] [MER06] [KIR82] [CRC10]
Solubility: g/100 g soln, H_2O: 51.25 (0°C), 52.85 (25°C), 60.90 (100°C); solid phase, $K_2CO_3 \cdot 1\text{-}1/2H_2O$ [KRU93]
Density, g/cm³: 2.428 [HAW93]
Melting Point, °C: 891 [MER06]
Boiling Point, °C: decomposes [STR93]

2418
Compound: Potassium carbonate hemitrihydrate
Synonym: potassium carbonate sesquihydrate
Formula: $K_2CO_3 \cdot 1\text{-}1/2H_2O$
Molecular Formula: $CH_3K_2O_{4.5}$
Molecular Weight: 165.229
CAS RN: 6381-79-9
Properties: small, granular cryst; not hygr, if fully hydrated with 1.5 H_2O; formula also given as $2 K_2CO_3 \cdot 3H_2O$ [MER06]
Solubility: 129.4 g/100 mL H_2O (0°C), 268 g/100 mL H_2O (100°C) [LAN05]
Density, g/cm³: 2.043 [CRC10]
Melting Point, °C: 891 [ALD94]

2419

Compound: Potassium chlorate
Synonym: potcrate
Formula: $KClO_3$
Molecular Formula: $ClKO_3$
Molecular Weight: 122.549
CAS RN: 3811-04-9
Properties: monocl lustrous cryst or white granules or powd; cooling, saline taste; used as an oxidizing agent, in explosives, in matches, is a source of oxygen; can react violently with organic matter; explodes with H_2SO_4 [MER06] [HAW93]
Solubility: g/100 g soln, H_2O: 3.2 (0°C), 7.9 (25°C), 36.0 (100°C) [KRU93]; ~1 g/50 mL glycerol; i alcohol [MER06]
Density, g/cm³: 2.32 [MER06]
Melting Point, °C: 368 [MER06]
Boiling Point, °C: decomposes to perchlorate and O_2 at >368 [MER06]

2420

Compound: Potassium chloride
Synonym: sylvite
Formula: KCl
Molecular Formula: ClK
Molecular Weight: 74.551
CAS RN: 7447-40-7
Properties: white, cub cryst or powd; −40 mesh with 99.999% purity; strong saline taste; enthalpy of fusion 26.53 kJ/mol; used in fertilizers, pharmaceutical preparations [MER06] [HAW93] [CER91] [CRC10]
Solubility: g/100 g soln, H_2O: 21.92 (0°C), 26.4 (25°C), 36.0 (100°C); equilibrium solid phase KCl at 25°C [KRU93]; 1 g/14 mL glycerol, 1 g/~250 mL alcohol [MER06]; 4.8088 ± 0.0024 mol/(kg H_2O) at 25°C [RAR85b]
Density, g/cm³: 1.984 [STR93]
Melting Point, °C: 776 [DOU83]
Boiling Point, °C: sublimes at 1500 [STR93]
Thermal Expansion Coefficient: (volume) 100°C (0.887), 200°C (1.994), 400°C (4.850), 600°C (8.805) [CLA66]

2421

Compound: Potassium chlorochromate
Synonym: Peligot's salt
Formula: $KCrO_3Cl$
Molecular Formula: $ClCrKO_3$
Molecular Weight: 174.545
CAS RN: 16037-50-6
Properties: red or orange cryst; monocl; evolves chlorine when heated; used as an oxidizing agent [HAW93] [KIR78]
Solubility: s H_2O, hydrolyzes [KIR78]
Density, g/cm³: 2.497 [KIR78]
Melting Point, °C: decomposes [KIR78]

2422

Compound: Potassium chromate
Synonym: tarapacaite
Formula: K_2CrO_4
Molecular Formula: CrK_2O_4
Molecular Weight: 194.191
CAS RN: 7789-00-6
Properties: yellow cryst; ortho-rhomb; used in analytical chemistry as a reagent and in inks [HAW93] [KIR78]
Solubility: g/100 g H_2O: 58.8 (0°C), 65.1 (25°C), 80.1 (100°C) [KRU93]
Density, g/cm³: 2.732 [KIR78]
Melting Point, °C: 971 [KIR78]

2423

Compound: Potassium chromium(III) oxalate trihydrate
Synonym: potassium tris(oxalato) chromate
Formula: $K_3Cr(C_2O_4)_3 \cdot 3H_2O$
Molecular Formula: $C_6H_6CrK_3O_{15}$
Molecular Weight: 487.396
CAS RN: 15275-09-9
Properties: black-green, monocl; hygr; prepared from reaction of oxalic acid, potassium oxalate, and potassium dichromate; used in tanning and dyeing wool [MER06]
Solubility: s H_2O [MER06]

2424

Compound: Potassium chromium(III) sulfate dodecahydrate
Synonym: chrome alum
Formula: $CrK(SO_4)_2 \cdot 12H_2O$
Molecular Formula: $CrH_{24}KO_{20}S_2$
Molecular Weight: 499.405
CAS RN: 7789-99-0
Properties: large, reddish violet to black; efflorescent; octahedral cub cryst; ruby red under transmitted light; aq solution is violet when cold, green when hot, color returns to violet on cooling; used in tanning, textile dyeing, ceramics [HAW93] [MER06]
Solubility: 4 parts cold, 2 parts boiling H_2O [MER06]
Density, g/cm³: 1.813 [HAW93]

Melting Point, °C: 89 [HAW93]
Reactions: minus $10H_2O$ at 100°C [HAW93]

2425
Compound: Potassium citrate
Formula: $K_3C_6H_5O_7$
Molecular Formula: $C_6H_5K_3O_7$
Molecular Weight: 306.397
CAS RN: 866-84-2
Properties: used to modify the burning
    rate of papers [KIR78]
Solubility: 60.91 g/100 g saturated solution
    in water (25°C) [MER06]
Melting Point, °C: decomposes at 230 [KIR78]

2426
Compound: Potassium citrate monohydrate
Synonyms: citric acid, tripotassium salt monohydrate
Formula: $KOOCCH_2C(OH)(COOK)CH_2COOK \cdot H_2O$
Molecular Formula: $C_6H_7K_3O_8$
Molecular Weight: 324.412
CAS RN: 6100-05-6
Properties: colorless or white cryst or powd;
    deliq; cooling, saline taste, odorless; used
    as an antacid, as a sequestrant for metals
    and as a buffer in food [HAW93]
Solubility: g anhydrous/100 g H_2O: 153 (10°C), 172
    (20°C), 194 (30°C) [LAN05]; 1 g/2.5 mL glycerol
    (dissolves slowly); sl s alcohol [MER06] [HAW93]
Density, g/cm³: 1.98 [HAW93]
Melting Point, °C: decomposes at 230 [HAW93]
Reactions: minus H_2O at 180°C [HAW93]

2427
Compound: Potassium cobalt(II) selenate hexahydrate
Formula: $K_2Co(SeO_4)_2 \cdot 6H_2O$
Molecular Formula: $CoH_{12}K_2O_{14}Se_2$
Molecular Weight: 531.137
CAS RN: 28041-86-3
Properties: garnet red; monocl cryst;
    stable in air [MER06]
Density, g/cm³: 2.514 [MER06]

2428
Compound: Potassium copper(I) cyanide
Formula: $KCu(CN)_2$
Molecular Formula: C_2CuKN_2
Molecular Weight: 150.679
CAS RN: 13682-73-0
Properties: white, cryst salt; used in
    copper plating baths [HAW93]

2429
Compound: Potassium cyanate
Formula: KCNO
Molecular Formula: CKNO
Molecular Weight: 81.115
CAS RN: 590-28-3
Properties: white; cryst powd; used in herbicides,
    treatment of sickle cell anemia [MER06] [HAW93]
Solubility: s H_2O; v sl s alcohol [MER06]
Density, g/cm³: 2.05 [MER06]
Melting Point, °C: decomposes at 700–900 [HAW93]

2430
Compound: Potassium cyanide
Formula: KCN
Molecular Formula: CKN
Molecular Weight: 65.116
CAS RN: 151-50-8
Properties: white, deliq, granular powd or fused pieces;
    odor of hydrogen cyanide; gradually decomposed
    in moist air; enthalpy of fusion 1470 J/mol; enthalpy
    of solution 11700 J/mol [MER06] [KIR78]
Solubility: s in: 2 parts cold H_2O, 1 part
    boiling H_2O, 2 parts glycerol, 100 parts
    alcohol, 25 parts methanol [MER06]
Density, g/cm³: cub: 1.553 (20°C); ortho-
    rhomb 1.62 (−60°C) [KIR78]
Melting Point, °C: 634 [MER06]

2431
Compound: Potassium cyanoaurite
Synonym: potassium dicyanoaurate(I)
Formula: $KAu(CN)_2$
Molecular Formula: C_2AuKN_2
Molecular Weight: 288.104
CAS RN: 13967-50-5
Properties: white, cryst powd; used in
    electrogilding [HAW93]
Solubility: 1 g/7 mL H_2O, 1 g/0.5 mL boiling
    H_2O; sl s alcohol; i ether [MER06]
Density, g/cm³: 3.45 [ALD94]

2432
Compound: Potassium dichromate
Synonym: potassium bichromate
Formula: $K_2Cr_2O_7$
Molecular Formula: $Cr_2K_2O_7$
Molecular Weight: 294.185
CAS RN: 7778-50-9

Properties: bright reddish orange cryst; tricl; not hygr or deliq; decomposes at ~500°C; oxidizing agent [KIR78] [MER06]
Solubility: g/100 g H_2O: 4.6 (0°C), 15.0 (25°C), 97 (100°C); solid phase, $K_2Cr_2O_7$ [KRU93]
Density, g/cm³: 2.676 [KIR78]
Melting Point, °C: 398 [KIR78]
Boiling Point, °C: decomposes at −500 [LID94]

2433
Compound: Potassium dihydrogen arsenate
Synonym: Macquer's salt
Formula: KH_2AsO_4
Molecular Formula: AsH_2KO_4
Molecular Weight: 180.034
CAS RN: 7784-41-0
Properties: colorless cryst or white cryst mass or powd; used in preserving hides, to print textiles, in insecticides [HAW93] [MER06]
Solubility: g/100 g soln, H_2O: 15.7 (0°C), 23.6 (25°C), 48.0 (100°C); solid phase, $KH_2AsO_4 \cdot H_2O$ (0°C, 25°C), KH_2AsO_4 (100°C) [KRU93]; slowly s in 1.6 parts glycerol; i alcohol [MER06]
Density, g/cm³: 2.867 [HAW93]
Melting Point, °C: 288 [HAW93]

2434
Compound: Potassium dihydrogen hypophosphite
Formula: KH_2PO_2
Molecular Formula: H_2KO_2P
Molecular Weight: 104.087
CAS RN: 7782-87-8
Properties: white cryst or granules; deliq; decomposes when strongly heated in air, evolving PH_3; can explode when mixed with oxidizing agents, e.g., chlorates [MER06]
Solubility: 200 g/100 mL H_2O (25°C) [CRC10]
Melting Point, °C: decomposes [CRC10]

2435
Compound: Potassium dihydrogen phosphate
Synonym: potassium phosphate monobasic
Formula: KH_2PO_4
Molecular Formula: H_2KO_4P
Molecular Weight: 136.085
CAS RN: 7778-77-0
Properties: tetr cryst or white granular powd; used in baking powd, in yeast foods, as a buffer, and sequestrant for metals [HAW93] [MER06] [KIR82]
Solubility: g/100 g soln, H_2O: 12.4 (0°C), 20.0 (25°C),

45.5 (90°C) [KRU93]; i alcohol [MER06]
Density, g/cm³: 2.338 [HAW93]
Melting Point, °C: 253 [HAW93]
Reactions: minus H_2O forming metaphosphate at 400°C [MER06]

2436
Compound: Potassium dihydrogen phosphite
Synonym: potassium monobasic phosphite
Formula: KH_2PO_3
Molecular Formula: H_2KO_3P
Molecular Weight: 120.086
CAS RN: 13598-36-2
Properties: white powd; hygr; slowly oxidized in air to the phosphate [HAW93]
Solubility: 220 g/100 mL H_2O (20°C) [CRC10]
Melting Point, °C: decomposes [CRC10]

2437
Compound: Potassium dithionate
Synonym: potassium hyposulfate
Formula: $K_2(SO_3)_2$
Molecular Formula: $K_2O_6S_2$
Molecular Weight: 238.325
CAS RN: 13455-20-4
Properties: colorless cryst; used as an analytical reagent [HAW93]
Solubility: g/100 g H_2O: 2.6 (0°C), 6.6 (20°C), 9.3 (30°C) [LAN05]; i alcohol [HAW93]
Density, g/cm³: 2.27 [HAW93]
Melting Point, °C: decomposes [CRC10]

2438
Compound: Potassium ferricyanide
Synonym: potassium hexacyanoferrate(III)
Formula: $K_3Fe(CN)_6$
Molecular Formula: $C_6FeK_3N_6$
Molecular Weight: 329.248
CAS RN: 13746-66-2
Properties: ruby red cryst or powd; lustrous; used to temper steel, as an etching liq, in electroplating [HAW93] [MER06]
Solubility: g/100 g H_2O: 30.2 (0°C), 46 (20°C), 70 (60°C) [LAN05]; s alcohol; decomposed by acids [MER06]
Density, g/cm³: 1.89 [MER06]
Melting Point, °C: decomposes [STR93]

2439
Compound: Potassium ferrocyanide trihydrate
Synonym: potassium hexacyanoferrate(II) trihydrate
Formula: $K_4Fe(CN)_6 \cdot 3H_2O$

Molecular Formula: $C_6H_6FeK_4N_6O_3$
Molecular Weight: 422.390
CAS RN: 14459-95-1
Properties: lemon yellow; soft, sl efflorescent cryst; used in dyeing, in tempering steel [HAW93] [MER06]
Solubility: g anhydrous/100 g H_2O: 14.3 (0°C), 28.2 (20°C), 74.2 (100°C) [LAN05]; i alcohol [HAW93]
Density, g/cm³: 1.85 [MER06]
Melting Point, °C: decomposes at 70 [STR93]
Reactions: begins to give up H_2O at 60°C; anhydrous by 100°C [MER06]

2440
Compound: Potassium fluoride
Formula: KF
Molecular Formula: FK
Molecular Weight: 58.096
CAS RN: 7789-23-3
Properties: cub cryst; usually white, deliq powd; sharp saline taste; enthalpy of fusion 27.2 kJ/mol; enthalpy of vaporization 173 kJ/mol; commercial preparation is from KOH and HF, followed by drying; used to etch glass, as a preservative [HAW93] [MER06] [CRC10]
Solubility: g/100 g soln, H_2O: 30.90 (0°C), 50.41 (25°C), 60.01 (80°C); solid phase, $KF \cdot 4H_2O$ (0°C), $KF \cdot 2H_2O$ (25°C), KF (80°C) [KRU93]; s acids, HF, liq NH_3 [MER06]
Density, g/cm³: 2.481 [MER06]
Melting Point, °C: 858 [CRC10]
Boiling Point, °C: 1505 [MER06]

2441
Compound: Potassium fluoride dihydrate
Formula: $KF \cdot 2H_2O$
Molecular Formula: FH_4KO_2
Molecular Weight: 94.127
CAS RN: 13455-21-5
Properties: white monocl cryst; has been used as stationary phase in gas chromatography [ALD94] [MER06]
Solubility: 349.3 g/100 mL H_2O (18°C) [MER06]
Density, g/cm³: 2.454 [STR93]
Melting Point, °C: decomposes at 41 [LID94]

2442
Compound: Potassium fullerene
Formula: K_3C_{60}
Molecular Formula: $C_{60}K_3$
Molecular Weight: 837.955
CAS RN: 137232-17-8

Properties: fcc, lattice constant 1.4253 nm; superconductor, T_c 19.3 K; bulk modulus 28 GPa; cohesive energy 24.2 eV; enthalpy of formation 4.9 eV; density of states 25/(eV/C_{60}); electron effective mass 1.3; hole effective mass 1.5, 3.4; potential uses include optical limiter to protect materials from damage due to high light intensities and for the fabrication of industrial diamonds [DRE93]

2443
Compound: Potassium gold(III) oxide trihydrate
Synonym: potassium aurate
Formula: $KAuO_2 \cdot 3H_2O$
Molecular Formula: AuH_6KO_5
Molecular Weight: 322.110
CAS RN: 12446-76-3
Properties: yellow cryst; used to prepare other gold compounds [HAW93]
Solubility: s H_2O, alcohol [HAW93]
Melting Point, °C: decomposes [CRC10]

2444
Compound: Potassium heptafluoroniobate
Formula: K_2NbF_7
Molecular Formula: F_7K_2Nb
Molecular Weight: 304.092
CAS RN: 16924-03-1
Properties: lump [ALF95]

2445
Compound: Potassium heptafluorotantalate
Formula: K_2TaF_7
Molecular Formula: F_7K_2Ta
Molecular Weight: 392.134
CAS RN: 16924-00-8
Properties: colorless, rhomb needles when crystallized from solution containing HF, KF, and tantalum fluoride [KIR83]
Solubility: hydrolyzes in H_2O [KIR83]
Density, g/cm³: 5.24 [KIR83]
Melting Point, °C: 740 [KIR83]

2446
Compound: Potassium heptaiodobismuthate
Synonym: bismuth potassium iodide
Formula: K_4BiI_7
Molecular Formula: BiI_7K_4
Molecular Weight: 1253.704

CAS RN: 41944-01-8
Properties: red cryst; used to precipitate vitamins and antibiotics from aq solutions [HAW93] [MER06]
Solubility: decomposes in H_2O; s alkali iodide soln [MER06]

2447
Compound: Potassium hexabromoplatinate(IV)
Formula: K_2PtBr_6
Molecular Formula: Br_6K_2Pt
Molecular Weight: 752.701
CAS RN: 16920-93-7
Properties: reddish brown powd; hygr [STR93]
Density, g/cm³: 4.66 [STR93]
Melting Point, °C: decomposes at 400 [STR93]

2448
Compound: Potassium hexachloroiridate(IV)
Formula: K_2IrCl_6
Molecular Formula: Cl_6IrK_2
Molecular Weight: 483.130
CAS RN: 16920-56-2
Properties: black powd; hygr; there is also K_3IrCl_6, CAS RN 14024-41-0 [STR93] [ALD94]
Density, g/cm³: 3.546 [STR93]
Melting Point, °C: decomposes [STR93]

2449
Compound: Potassium hexachloroosmiate(IV)
Formula: K_2OsCl_6
Molecular Formula: Cl_6K_2Os
Molecular Weight: 481.143
CAS RN: 16871-60-6
Properties: dark red to almost black; cub cryst; hygr [STR93] [MER06]
Solubility: v s H_2O; sl s alcohol [MER06]
Density, g/cm³: 3.42 [ALD94]
Melting Point, °C: decomposes at 600 [ALD94]

2450
Compound: Potassium hexachloropalladate(IV)
Formula: K_2PdCl_6
Molecular Formula: Cl_6K_2Pd
Molecular Weight: 397.333
CAS RN: 16919-73-6
Properties: red powd; hygr [STR93]
Density, g/cm³: 2.738 [STR93]

2451
Compound: Potassium hexachloroplatinate(IV)
Formula: K_2PtCl_6
Molecular Formula: Cl_6K_2Pt
Molecular Weight: 485.993
CAS RN: 16921-30-5
Properties: small, yellowish orange cryst or powd; used in photography and as a reagent [HAW93]
Solubility: g/100 g H_2O: 0.48 (0°C), 0.78 (20°C), 5.03 (100°C) [LAN05]; i alcohol [HAW93]
Density, g/cm³: 3.50 [HAW93]
Melting Point, °C: decomposes at 250 [HAW93]

2452
Compound: Potassium hexachlororhenate(IV)
Formula: K_2ReCl_6
Molecular Formula: Cl_6K_2Re
Molecular Weight: 477.120
CAS RN: 16940-97-9
Properties: green powd; hygr [STR93]
Density, g/cm³: 3.34 [STR93]

2453
Compound: Potassium tetracyanocadmium
Synonym: cadmium potassium cyanide
Formula: $K_2Cd(CN)_4$
Molecular Formula: $C_4CdK_2N_4$
Molecular Weight: 294.679
CAS RN: 14402-75-6
Properties: highly refractive, cub cryst; when heated, melts to colorless liq, solidifying to a gray, cryst mass on cooling [MER06]
Solubility: s 3 parts cold H_2O, 1 part hot H_2O [MER06]
Density, g/cm³: 1.846 [MER06]
Melting Point, °C: ~450 [MER06]

2454
Compound: Potassium hexacyanocobalt(III)
Synonym: potassium cobalticyanide
Formula: $K_3Co(CN)_6$
Molecular Formula: $C_6CoK_3N_6$
Molecular Weight: 332.334
CAS RN: 13963-58-1
Properties: faintly yellow, monocl cryst when obtained from H_2O; sensitive to light; unstable if stored, generating HCN [MER06]
Solubility: v s H_2O, acetic acid solutions; i alcohol [MER06]
Density, g/cm³: 1.906 [MER06]
Melting Point, °C: decomposes, forming olive green mass [MER06]

2455
Compound: Potassium hexacyanoplatinate(IV)
Formula: $K_2Pt(CN)_6$
Molecular Formula: $C_6K_2N_6Pt$
Molecular Weight: 429.383
CAS RN: 16920-94-8
Properties: white powd [STR93]

2456
Compound: Potassium hexafluoroantimonate
Formula: $KSbF_6$
Molecular Formula: F_6KSb
Molecular Weight: 274.846
CAS RN: 16893-92-8
Properties: −6 mesh of 99.9% purity [CER91]

2457
Compound: Potassium hexafluoroarsenate(V)
Formula: $KAsF_6$
Molecular Formula: AsF_6K
Molecular Weight: 228.010
CAS RN: 17029-22-0
Properties: grayish white cryst [STR93]
Melting Point, °C: 400 [STR93]

2458
Compound: Potassium hexafluorogermanate
Formula: K_2GeF_6
Molecular Formula: F_6GeK_2
Molecular Weight: 264.797
CAS RN: 7783-73-5
Properties: white cryst; stable up to 500°C [HAW93]
Solubility: g/100 g H_2O: 0.25 (0°C), 0.50 (20°C), 0.96 (40°C) [LAN05]; s alcohol [HAW93]
Melting Point, °C: 730 [CRC10]
Boiling Point, °C: ~835 [CRC10]

2459
Compound: Potassium hexafluoromanganate(IV)
Formula: K_2MnF_6
Molecular Formula: F_6K_2Mn
Molecular Weight: 247.125
CAS RN: 16962-31-5
Properties: golden yellow; hex platelets; turns brown when heated, but returns to original color when cooled [MER06]
Solubility: hydrolyzed in H_2O, precipitates MnO_4 [MER06]
Melting Point, °C: decomposes [CRC10]

2460
Compound: Potassium hexafluoronickelate(IV)
Synonym: potassium nickel(IV) fluoride
Formula: K_2NiF_6
Molecular Formula: F_6K_2Ni
Molecular Weight: 250.880
CAS RN: 17218-47-2
Properties: powd; can be a source of F_2 because F_2 is evolved when the compound is heated [STR93]
Melting Point, °C: decomposes at 400 [STR93]

2461
Compound: Potassium hexafluorophosphate
Formula: KPF_6
Molecular Formula: F_6KP
Molecular Weight: 184.062
CAS RN: 17084-13-8
Properties: white cryst; hygr [ALD94] [STR93]
Solubility: 9.3 g/100 mL H_2O (25°C), 20.6 g/100 mL H_2O (50°C) [CRC10]
Density, g/cm³: 2.55 [STR93]
Melting Point, °C: ~575 [STR93]
Boiling Point, °C: decomposes [STR93]

2462
Compound: Potassium hexafluorosilicate
Synonym: hieratite
Formula: K_2SiF_6
Molecular Formula: F_6K_2Si
Molecular Weight: 220.273
CAS RN: 16871-90-2
Properties: white fine powd or cryst; used to manufacture opalescent glass, used in porcelain enamel and as an insecticide [MER06]
Solubility: g/100 g H_2O: 0.077 (0°C), 0.151 (20°C), 0.253 (40°C) [LAN05]; hydrolyzes in hot water to KF, HF, and silicic acid; i alcohol [MER06]
Density, g/cm³: 2.27 [MER06]
Melting Point, °C: decomposes [STR93]

2463
Compound: Potassium hexafluorotitanate monohydrate
Formula: $K_2TiF_6 \cdot H_2O$
Molecular Formula: $F_6H_2K_2OTi$
Molecular Weight: 258.082
CAS RN: 16919-27-0
Properties: colorless; monocl [CRC10]
Solubility: g/100 g H_2O: 0.55 (0°C), 0.91 (10°C), 1.28 (20°C) [LAN05]
Melting Point, °C: 780 [CRC10]
Reactions: minus H_2O at 32°C [CRC10]

2464

Compound: Potassium hexafluorozirconate
Formula: K_2ZrF_6
Molecular Formula: F_6K_2Zr
Molecular Weight: 283.411
CAS RN: 16923-95-8
Properties: colorless, monocl cryst [MER06] [CRC10]
Solubility: 0.781 g/100 mL H_2O (2°C),
25 g/100 mL H_2O (100°C) [CRC10]
Density, g/cm³: 3.48 [CRC10]

2465

Compound: Potassium hexametaphosphite
Formula: $(KPO_3)_6$
Molecular Formula: $K_6O_{18}P_6$
Molecular Weight: 708.420
CAS RN: 7790-53-6
Properties: white powd; hygr [CRC10] [STR93]
Solubility: g/100 mL soln, H_2O: 0.0041 (25°C) [KRU93]
Density, g/cm³: 1.207 [CRC10]
Melting Point, °C: 807 [STR93]
Boiling Point, °C: 1320 [CRC10]

2466

Compound: Potassium hexanitritocobalt(III)
Synonym: Fischer's salt
Formula: $K_3Co(NO_2)_6$
Molecular Formula: $CoK_3N_6O_{12}$
Molecular Weight: 452.261
CAS RN: 66942-97-0
Properties: yellow cryst powd [HAW93]
Solubility: 0.9 g/100 mL H_2O (17°C),
decomposed by hot H_2O [CRC10]
Melting Point, °C: decomposes at 200 [HAW93]

2467

Compound: Potassium hexanitritorhodate(III)
Formula: $K_3Rh(NO_2)_6$
Molecular Formula: $K_3N_6O_{12}Rh$
Molecular Weight: 496.234
CAS RN: 17712-66-2
Properties: white powd [STR93]

2468

Compound: Potassium hexathiocyanoplatinate(IV)
Formula: $K_2Pt(SCN)_6$
Molecular Formula: $C_6K_2N_6PtS_6$
Molecular Weight: 621.779
CAS RN: 17069-38-4
Properties: carmine red cryst [MER06]
Solubility: s H_2O [MER06]

2469

Compound: Potassium hydride
Formula: KH
Molecular Formula: HK
Molecular Weight: 40.106
CAS RN: 7693-26-7
Properties: white needles; slurry of gray
powd in oil; sensitive to atm oxygen
and moisture [STR93] [CRC10]
Solubility: decomposed by H_2O [CRC10]
Density, g/cm³: 1.47 [CRC10]
Melting Point, °C: decomposes [CRC10]

2470

Compound: Potassium hydrogen arsenite
Formula: $KAsO_2 \cdot HAsO_2$
Molecular Formula: As_2HKO_4
Molecular Weight: 253.947
CAS RN: 10124-50-2
Properties: white hygr powd; gradually decomposed by
atm CO_2; used in the manufacture of silver mirrors
to reduce silver salts to metallic silver [MER06]
Solubility: s H_2O [MER06]

2471

Compound: Potassium hydrogen carbonate
Synonym: potassium bicarbonate
Formula: $KHCO_3$
Molecular Formula: $CHKO_3$
Molecular Weight: 100.115
CAS RN: 298-14-6
Properties: monocl transparent cryst, white granules
or powd; sl alkaline or salty taste; obtained
by addition of CO_2 to a solution of potassium
carbonate; used in baking powd, soft drinks,
as an antacid [HAW93] [MER06] [KIR82]
Solubility: g/100 g soln, H_2O: 18.6 (0°C), 26.6 (25°C);
solid phase, $KHCO_3$ [KRU93]; i alcohol [MER06]
Density, g/cm³: 2.17 [HAW93]
Melting Point, °C: decomposes at 100–120 [HAW93]

2472

Compound: Potassium hydrogen fluoride
Formula: KHF_2
Molecular Formula: F_2HK
Molecular Weight: 78.103
CAS RN: 7789-29-9
Properties: colorless tetr cryst; decomposed by heat;
enthalpy of fusion 6.62 kJ/mol; produced by reaction
of KOH or K_2CO_3 with HF; used to etch glass, as a
flux for silver solders [HAW93] [MER06] [CRC10]

Solubility: g/100 mL H_2O: 30.1 (10°C), 39.2
   (20°C), 114.0 (80°C); s dil alcohol;
   i absolute alcohol [MER06]
Density, g/cm³: 2.37 [MER06]
Melting Point, °C: 238.7 [MER06]

2473
Compound: Potassium hydrogen iodate
Synonym: potassium acid iodate
Formula: $KH(IO_3)_2$
Molecular Formula: HI_2KO_6
Molecular Weight: 389.911
CAS RN: 13455-24-8
Properties: colorless monocl or rhomb cryst;
   used as an alkalimetric standard [KIR81]
Solubility: 9.15 parts/100 parts H_2O at 50°C [KIR81]

2474
Compound: Potassium hydrogen oxalate hemihydrate
Synonyms: potassium binoxalate, sorrel salt
Formula: $KHC_2O_4 \cdot 1/2H_2O$
Molecular Formula: $C_2H_2KO_{4.5}$
Molecular Weight: 137.133
CAS RN: 127-95-7
Properties: white, odorless cryst; bitter sharp
   taste; somewhat hygr; used to remove ink
   stains, in scouring metals, cleaning wood;
   decomposes when heated [HAW93] [MER06]
Solubility: s 40 parts cold H_2O, 6 parts
   boiling H_2O [MER06]
Density, g/cm³: 2.088 [HAW93]

2475
Compound: Potassium hydrogen phosphite
Formula: K_2HPO_3
Molecular Formula: HK_2O_3P
Molecular Weight: 158.177
CAS RN: 13492-26-7
Properties: white; deliq powd; slowly oxidized in air to
   the phosphate; decomposed by heating [MER06]
Solubility: v s H_2O; i alcohol [MER06]

2476
Compound: Potassium hydrogen selenite
Synonym: potassium biselenite
Formula: $KHSeO_3$
Molecular Formula: HKO_3Se
Molecular Weight: 167.064
CAS RN: 7782-70-9

Properties: ortho-rhomb prisms; very deliq; selenium
   oxide liberated by heating at >100°C [MER06]
Solubility: s H_2O; sl s alcohol [MER06]
Reactions: slowly loses H_2O at 100°C [MER06]

2477
Compound: Potassium hydrogen sulfate
Synonyms: mercallite, misenite
Formula: $KHSO_4$
Molecular Formula: HKO_4S
Molecular Weight: 136.170
CAS RN: 7646-93-7
Properties: white deliq monocl or rhomb cryst;
   used as a flux and in the manufacture of mixed
   fertilizers [HAW93] [MER06] [KIR82]
Solubility: g/100 g soln, H_2O: 26.6 (0°C),
   34.0 (25°C), 54.9 (100°C) [KRU93]
Density, g/cm³: 2.322 [STR93]
Melting Point, °C: 210 [KIR82]
Boiling Point, °C: decomposes [STR93]
Reactions: gives up H_2O at high temp to
   form pyrosulfate [MER06]

2478
Compound: Potassium hydrogen sulfide hemihydrate
Formula: $KHS \cdot 1/2H_2O$
Molecular Formula: $H_2KO_{0.5}S$
Molecular Weight: 81.180
CAS RN: 1310-61-8
Properties: tetr; turns yellow rapidly in air,
   forming polysulfides and H_2S; forms
   dark red liq when melted [MER06]
Solubility: v s H_2O, alcohol [MER06]
Density, g/cm³: 1.70 [MER06]
Melting Point, °C: 450–510 [MER06]
Reactions: loses H_2O at 175°C–200°C [MER06]

2479
Compound: Potassium hydrogen sulfite
Synonym: potassium bisulfite
Formula: $KHSO_3$
Molecular Formula: HKO_3S
Molecular Weight: 120.170
CAS RN: 7773-03-7
Properties: white, cryst powd; odor of sulfur dioxide;
   preparation: passing SO_2 through aq solution of
   K_2CO_3; uses: antiseptic, bleaching straw and textiles,
   reduce various organic compounds [HAW93]
Solubility: s H_2O; i alcohol [HAW93]
Melting Point, °C: decomposes at 190 [HAW93]

2480

Compound: Potassium hydrogen tartrate
Synonyms: tartaric acid, monopotassium salt
Formula: KOOCCH(OH)CH(OH)COOH
Molecular Formula: $C_4H_5KO_6$
Molecular Weight: 188.178
CAS RN: 868-14-4
Properties: colorless cryst or white, cryst powd;
pleasant acidic taste; used in baking powd,
medicine, as a food additive [HAW93] [MER06]
Solubility: g/100 g H_2O: 0.231 (0°C), 0.523 (20°C),
0.762 (30°C) [LAN05]; v s dil mineral
acids, solutions of alkalies [MER06]
Density, g/cm³: 1.984 [HAW93]

2481

Compound: Potassium hydroxide
Formula: KOH
Molecular Formula: HKO
Molecular Weight: 56.105
CAS RN: 1310-58-3
Properties: white or sl yellow lumps, rods,
pellets; rhomb; deliq; absorbs H_2O and
CO_2 from air; enthalpy of fusion 8.60 kJ/
mol; manufactured by electrolysis of KCl
solutions; used in soap manufacture, for
bleaching [HAW93] [MER06] [KIR82]
Solubility: g/100 g soln, H_2O: 49.00 (0°C), 54.27
(25°C), 64.59 (100°C); solid phase, KOH·$2H_2O$
(0°C, 25°C), KOH·H_2O (100°C) [KRU93]; s 3
parts alcohol, 2.5 parts glycerol [MER06]
Density, g/cm³: 2.044 [STR93]
Melting Point, °C: 406 [LID94]
Boiling Point, °C: 1324 [STR93]

2482

Compound: Potassium iodate
Formula: KIO_3
Molecular Formula: IKO_3
Molecular Weight: 214.001
CAS RN: 7758-05-6
Properties: white, odorless, monocl cryst or
powd; −80 mesh with 99.9% purity; can
be prepared by electrochemical oxidation
of KI; forms KIO_2F_2 when reacted with
HF [KIR81] [MER06] [CER91]
Solubility: g/100 g soln, H_2O: 4.4 (0°C), 8.4 (25°C),
24.4 (100°C) [KRU93]; i alcohol [MER06]
Density, g/cm³: 3.979 [KIR81]
Melting Point, °C: partially decomposes
at 560 [MER06]

2483

Compound: Potassium iodide
Formula: KI
Molecular Formula: IK
Molecular Weight: 166.003
CAS RN: 7681-11-0
Properties: colorless or white; cub cryst, white granules,
or powd; −20 mesh with 99.999% purity; sl deliq
in moist air; liberates iodine if exposed to air for
a lengthy time; sensitive to light; strong, bitter,
saline taste; enthalpy of fusion 24.00 kJ/mol;
produced by dissolution of I_2 in KOH solutions;
used in infrared transmission instrumentation,
as a dietary supplement [HAW93]
[MER06] [KIR82] [CER91] [CRC10]
Solubility: g/100 g H_2O: 127.5 (0°C), 148 (25°C),
207.2 ± 0.8 (100°C); solid phase, KI [KRU93];
1 g/22 mL alcohol, 1 g/8 mL boiling alcohol,
1 g/51 mL absolute alcohol [MER06]
Density, g/cm³: 3.123 [HAW93]
Melting Point, °C: 681 [CRC10]
Boiling Point, °C: 1330 [HAW93]
Thermal Expansion Coefficient: (volume) 100°C
(0.90), 200°C (2.04), 400°C (4.92) [CLA66]

2484

Compound: Potassium magnesium chloride sulfate
Synonym: kainite
Formula: $MgSO_4 \cdot KCl \cdot 3H_2O$
Molecular Formula: ClH_6KMgO_7S
Molecular Weight: 248.966
CAS RN: 1318-72-5
Properties: natural hydrated double salt; white, gray,
reddish, or colorless; vitreous luster; hardness
is 2.5–3; used as a fertilizer [HAW93]
Solubility: 79.56 g/100 mL H_2O (18°C) [CRC10]
Density, g/cm³: 2.131 [CRC10]

2485

Compound: Potassium magnesium sulfate
Synonym: langbeinite
Formula: $K_2SO_4 \cdot 2MgSO_4$
Molecular Formula: $K_2Mg_2O_{12}S_3$
Molecular Weight: 414.998
CAS RN: 13826-56-7
Properties: white, tetr cryst; used
in fertilizers [HAW93]
Solubility: g/100 g H_2O: 14.0 (0°C), 25.0
(20°C), 63.4 (80°C) [LAN05]
Density, g/cm³: 2.829 [HAW93]
Melting Point, °C: 927 [HAW93]

2486
Compound: Potassium manganate
Formula: K_2MnO_4
Molecular Formula: K_2MnO_4
Molecular Weight: 197.133
CAS RN: 10294-64-1
Properties: dark green cryst; oxidizing agent, liberates
Cl$_2$ from HCl; used in bleaching skins, fibers,
oils, as a disinfectant [HAW93] [MER06]
Solubility: s H_2O; s and stable in
KOH solutions [MER06]
Melting Point, °C: decomposes at 190 [MER06]

2487
Compound: Potassium metaarsenite monohydrate
Formula: $KH(AsO_2)_2 \cdot H_2O$
Molecular Formula: $As_2H_3KO_5$
Molecular Weight: 271.962
CAS RN: 10124-50-2
Properties: white powd; hygr; decomposes slowly in air;
used as a reducing agent in silvering mirrors [HAW93]
Solubility: s H_2O, sl s in alcohol [HAW93]

2488
Compound: Potassium molybdate
Formula: K_2MoO_4
Molecular Formula: K_2MoO_4
Molecular Weight: 238.135
CAS RN: 13446-49-6
Properties: white, deliq, microcryst powd; −200
mesh with 99.9% purity; isomorphous with
K_2SO_4, K_2CrO_4 [KIR81] [HAW93] [CER91]
Solubility: g/100 g soln, H_2O: 64.57 ± 0.1 (25°C),
66.54 (89.96°C); solid phase, K_2MoO_4 [KRU93]
Density, g/cm³: 2.342 [KIR81]
Melting Point, °C: 919 [HAW93]

2489
Compound: Potassium monohydrogen phosphate
Synonym: potassium dibasic phosphate
Formula: K_2HPO_4
Molecular Formula: HK_2O_4P
Molecular Weight: 174.176
CAS RN: 7758-11-4
Properties: white somewhat hygr powd;
ignited to pyrophosphate [MER06]
Solubility: g/100 g soln, H_2O: 45.8 ± 0.3 (0°C),
62.4 ± 0.4 (25°C), 73.8 (99.4°C); solid phase,
$K_2HPO_4 \cdot 6H_2O$ (0°C), $K_2HPO_4 \cdot 3H_2O$ (25°C),
K_2HPO_4 (99.4°C) [KRU93]; sl s alcohol [MER06]
Melting Point, °C: decomposes [CRC10]

2490
Compound: Potassium monoxide
Synonym: potassium oxide
Formula: K_2O
Molecular Formula: K_2O
Molecular Weight: 94.196
CAS RN: 12136-45-7
Properties: gray cryst mass; hygr [CRC10] [HAW93]
Solubility: s H_2O, forming KOH;
s alcohol and ether [HAW93]
Density, g/cm³: 2.32 (0°C) [HAW93]
Melting Point, °C: decomposes at 350 [HAW93]

2491
Compound: Potassium nickel sulfate hexahydrate
Synonym: nickel potassium sulfate
Formula: $K_2SO_4 \cdot NiSO_4 \cdot 6H_2O$
Molecular Formula: $H_{12}K_2NiO_{14}S_2$
Molecular Weight: 437.109
CAS RN: 10294-65-2
Properties: blueish green cryst; prepared
by crystallization from an aq solution;
has limited use as a dye mordant and in
metal finishing [KIR81] [HAW93]
Solubility: g/100 g H_2O: 3.37 (0°C), 5.94
(20°C), 33.4 (100°C) [LAN05]
Density, g/cm³: 2.124 [HAW93]
Melting Point, °C: decomposes at <100 [CRC10]

2492
Compound: Potassium niobate
Formula: $KNbO_3$
Molecular Formula: $KNbO_3$
Molecular Weight: 180.002
CAS RN: 12030-85-2
Properties: −100 mesh with 99.9% purity;
white powd [STR93] [CER91]

2493
Compound: Potassium niobate hexadecahydrate
Formula: $K_8Nb_6O_{19} \cdot 16H_2O$
Molecular Formula: $H_{32}K_8Nb_6O_{35}$
Molecular Weight: 1462.457
CAS RN: 12502-31-7
Properties: large monocl cryst; forms
supersaturated solutions in water [KIR81]
Solubility: saturated soln is 425 g/100 g H_2O at room
temp, more soluble in hot H_2O [KIR81]

2494

Compound: Potassium nitrate
Synonym: saltpeter
Formula: KNO$_3$
Molecular Formula: KNO$_3$
Molecular Weight: 101.103
CAS RN: 7757-79-1
Properties: colorless, transparent prisms, white granular, or rhomb and trig cryst powd; cools when dissolved in H$_2$O; enthalpy of fusion 10.10 kJ/mol; manufactured by reaction of KCl and HNO$_3$, with Cl$_2$ as a by-product [MER06] [KIR82] [CRC10]
Solubility: g/100 g soln, H$_2$O: 11.7 (0°C), 27.5 (25°C), 71.0 (100°C) [KRU93]; 1 g/620 mL alcohol; s glycerol; i absolute alcohol [MER06]
Density, g/cm^3: 2.109 [STR93]
Melting Point, °C: 333 [MER06]
Boiling Point, °C: decomposes at 400, evolving oxygen [MER06]

2495

Compound: Potassium nitrite
Formula: KNO$_2$
Molecular Formula: KNO$_2$
Molecular Weight: 85.104
CAS RN: 7758-09-0
Properties: white or sl yellow; deliq granules or rods; decomposed by acids, evolving brown fumes of nitrous anhydride [MER06]
Solubility: g/100 g soln, H$_2$O: 73.7 (0°C), 75.75 (25°C), 80.2 (100°C); solid phase, KNO$_2$ [KRU93]; sl s alcohol [MER06]
Density, g/cm^3: 1.915 [MER06]
Melting Point, °C: decomposes at 350 [ALD94]
Boiling Point, °C: explodes at 537 [HAW93]

2496

Compound: Potassium nitroprusside dihydrate
Formula: K$_2$[Fe(CN)$_5$(NO)]·2H$_2$O
Molecular Formula: C$_5$H$_4$FeK$_2$N$_6$O$_3$
Molecular Weight: 330.167
CAS RN: 14709-57-0
Properties: garnet red; hygr cryst [MER06]
Solubility: 100 g/100 mL H$_2$O (16°C) [CRC10]

2497

Compound: Potassium osmate dihydrate
Formula: K$_2$OsO$_4$·2H$_2$O
Molecular Formula: H$_4$K$_2$O$_6$Os
Molecular Weight: 368.455
CAS RN: 19718-36-6

Properties: purple powd; hygr rhomb cryst; slowly decomposes in aq solution, forming tetroxide [STR93] [MER06] [KIR82]
Solubility: s H$_2$O; i alcohol, ether [MER06]
Reactions: minus H$_2$O at 200°C [KIR82]

2498

Compound: Potassium oxalate monohydrate
Formula: K$_2$C$_2$O$_4$·H$_2$O
Molecular Formula: C$_2$H$_2$K$_2$O$_5$
Molecular Weight: 184.232
CAS RN: 6487-48-5
Properties: colorless, odorless cryst; efflorescent in warm, dry air; converted to carbonate by ignition [MER06]
Solubility: g/100 g soln, H$_2$O: 29.32 ± 0.04 (0°C), 27.4 (25°C), 44.5 (100°C); solid phase K$_2$C$_2$O$_4$·H$_2$O [KRU93]
Density, g/cm^3: 2.13 [MER06]
Melting Point, °C: decomposes when heated [HAW93]
Reactions: minus H$_2$O at ~160°C [MER06]

2499

Compound: Potassium pentaborate octahydrate
Formula: K$_2$B$_{10}$O$_{16}$·8H$_2$O
Molecular Formula: B$_{10}$H$_{16}$K$_2$O$_{24}$
Molecular Weight: 586.420
CAS RN: 12229-13-9
Properties: ortho-rhomb, white powd; enthalpy of dehydration 110.8 kJ/mol from 106.5°C–134°C [STR93] [KIR78]
Solubility: % anhydrous by weight, H$_2$O: 1.56 (0°C), 3.28 (25°C), 6.88 (50°C), 22.3 (100°C) [KIR78]
Density, g/cm^3: 1.74 [KIR78]
Melting Point, °C: 780 [STR93]

2500

Compound: Potassium pentachloronitrosyl iridium(III) hydrate
Formula: KIr(NO)Cl$_5$·xH$_2$O
Molecular Formula: Cl$_5$IrKNO (anhydrous)
Molecular Weight: 438.588 (anhydrous)
CAS RN: 22594-86-1
Properties: brown cryst [STR93]

2501

Compound: Potassium pentachlororuthenate(III) hydrate
Formula: K$_2$RuCl$_5$·xH$_2$O
Molecular Formula: Cl$_5$K$_2$Ru (anhydrous)
Molecular Weight: 356.530 (anhydrous)
CAS RN: 14404-33-2
Properties: brown powd [STR93]

2502
Compound: Potassium perborate monohydrate
Synonym: potassium peroxyborate
Formula: $2KBO_3 \cdot H_2O$
Molecular Formula: $B_2H_2K_2O_7$
Molecular Weight: 213.831
CAS RN: 28876-88-2
Properties: white cryst; formula also given
    as a hemihydrate [CRC10]
Solubility: g/100 g H_2O: 1.25 (0°C) [KRU93]
Melting Point, °C: decomposes at 150 [CRC10]
Reactions: minus O_2 at 100°C [CRC10]

2503
Compound: Potassium percarbonate monohydrate
Synonym: potassium peroxycarbonate
Formula: $K_2C_2O_6 \cdot H_2O$
Molecular Formula: $C_2H_2K_2O_7$
Molecular Weight: 216.231
CAS RN: 589-97-9
Properties: white, granular mass;
    sensitive to light [MER06]
Solubility: 1 part/15 parts cold H_2O;
    evolves O_2 in hot H_2O [MER06]
Melting Point, °C: 200–300 [HAW93]

2504
Compound: Potassium perchlorate
Formula: $KClO_4$
Molecular Formula: $ClKO_4$
Molecular Weight: 138.549
CAS RN: 7778-74-7
Properties: hygr; colorless cryst or white powd;
    oxidizing agent, reacts with organic matter,
    other oxidizable material; evolves O_2 if
    heated; can decompose by concussion; used
    in explosives, photography, pyrotechnics,
    and flares [HAW93] [MER06] [STR93]
Solubility: g/100 g soln, H_2O: 0.75 (0°C), 2.03 (25°C),
    18.2 (100°C); solid phase, $KClO_4$ [KRU93]
Density, g/cm³: 2.52 [MER06]
Melting Point, °C: decomposes at 400
    [MER06]; 610 [STR93]

2505
Compound: Potassium periodate
Formula: KIO_4
Molecular Formula: IKO_4
Molecular Weight: 230.001
CAS RN: 7790-21-8

Properties: white powd or colorless, transparent,
    tetr cryst; strong oxidizing agent in aq
    solutions, e.g., can oxidize manganese
    compounds to permanganate [MER06]
Solubility: g/100 g soln, H_2O: 0.169 (0.2°C),
    0.51 (25°C), 6.83 (97°C) [KRU93]
Density, g/cm³: 3.618 [MER06]
Melting Point, °C: 582 [MER06]
Boiling Point, °C: explodes [HAW93]

2506
Compound: Potassium permanganate
Formula: $KMnO_4$
Molecular Formula: $KMnO_4$
Molecular Weight: 158.034
CAS RN: 7722-64-7
Properties: dark purple cryst; stable in air; decomposes
    evolving O_2 at ~240°C; good oxidizing agent in aq
    solutions, e.g., oxidizes HCl to Cl_2; astringent taste;
    odorless; used as an oxidizer, as a disinfectant,
    bleach, in tanning [HAW93] [MER06]
Solubility: g/100 g soln, H_2O: 2.75 (0°C), 7.09
    (25°C); solid phase, $KMnO_4$ [KRU93]
Density, g/cm³: 2.70 [HAW93]
Melting Point, °C: decomposes [HAW93]

2507
Compound: Potassium peroxide
Formula: K_2O_2
Molecular Formula: K_2O_2
Molecular Weight: 110.196
CAS RN: 17014-71-0
Properties: yellow amorphous mass;
    decomposes in water, evolving oxygen;
    oxidizing agent; used for bleaching, in
    oxygen generating gas masks [HAW93]
Solubility: decomposed by H_2O [HAW93]
Melting Point, °C: 490 [HAW93]

2508
Compound: Potassium perrhenate
Formula: $KReO_4$
Molecular Formula: KO_4Re
Molecular Weight: 289.303
CAS RN: 10466-65-6
Properties: white, tetr cryst; −40 mesh with
    99.9% purity [STR93] [KIR82] [CER91]
Solubility: g/100 g H_2O: 0.34 (0°C), 0.99
    (20°C), 8.7 (80°C) [LAN05]
Density, g/cm³: 4.887 [STR93]
Melting Point, °C: 555 [STR93]
Boiling Point, °C: ~1365 [STR93]

2509

Compound: Potassium perruthenate
Formula: $KRuO_4$
Molecular Formula: KO_4Ru
Molecular Weight: 204.166
CAS RN: 10378-50-4
Properties: black tetr [KIR82]
Solubility: 1.1 g/100 mL H_2O (60°C) [CRC10]
Melting Point, °C: decomposes at 400 [KIR82]

2510

Compound: Potassium persulfate
Synonym: potassium peroxydisulfate
Formula: $K_2S_2O_8$
Molecular Formula: $K_2O_8S_2$
Molecular Weight: 270.324
CAS RN: 7727-21-1
Properties: colorless or white cryst; −100 mesh with 99.9% purity; unstable, gradually evolving O_2 with rate of evolution increasing with temp; strong oxidizing agent in aq solutions [MER06] [CER91]
Solubility: g/100 mL soln, H_2O: 1.620 (0°C), 5.840 (25°C) [KRU93]; i alcohol [MER06]
Density, g/cm³: 2.477 [HAW93]
Melting Point, °C: completely decomposed at ~100 [MER06]

2511

Compound: Potassium phosphate
Formula: K_3PO_4
Molecular Formula: K_3O_4P
Molecular Weight: 212.266
CAS RN: 7778-53-2
Properties: granular, white powd; deliq; orthorhomb cryst; used in gasoline purification, to soften water, in liq soaps, and as an emulsifier in foods [HAW93] [MER06]
Solubility: g/100 g soln, H_2O: 44.26 (0°C), 51.42 (25°C); solid phase, $K_3PO_4 \cdot 7H_2O$ [KRU93]; i alcohol [MER06]
Density, g/cm³: 2.564 [MER06]
Melting Point, °C: 1340 [MER06]

2512

Compound: Potassium pyrophosphate trihydrate
Formula: $K_4P_2O_7 \cdot 3H_2O$
Molecular Formula: $H_6K_4O_{10}P_2$
Molecular Weight: 384.374
CAS RN: 7320-34-5
Properties: colorless; deliq granules or cryst; somewhat hygr; used as a builder in soaps and detergents and as a sequestering agent; also anhydrous form [HAW93] [MER06] [STR93] [ALD94]

Solubility: v s H_2O; i alcohol [MER06]
Density, g/cm³: 2.33 [HAW93]
Melting Point, °C: 1090 [HAW93]
Reactions: minus $2H_2O$ at 180°C, minus $3H_2O$ at ~300°C [CRC10] [HAW93]

2513

Compound: Potassium pyrosulfate
Formula: $K_2S_2O_7$
Molecular Formula: $K_2O_7S_2$
Molecular Weight: 254.325
CAS RN: 7790-62-7
Properties: colorless needles; fused pieces or white cryst powd; used as a laboratory reagent [HAW93] [MER06]
Solubility: s H_2O [MER06]
Density, g/cm³: 2.28 [MER06]
Melting Point, °C: ~325 [MER06]

2514

Compound: Potassium pyrosulfite
Formula: $K_2S_2O_5$
Molecular Formula: $K_2O_5S_2$
Molecular Weight: 222.326
CAS RN: 16731-55-8
Properties: white cryst or cryst powd; odor of sulfur dioxide; SO_2 evolved from acidic solutions; oxidizes to sulfate in air [MER06]
Solubility: g/100 g soln, H_2O: 22.1 (0°C), 32.8 (25°C), 55.5 (94.0°C), solid phase $K_2S_2O_5$ [KRU93]
Density, g/cm³: 2.3 [HAW93]
Melting Point, °C: decomposes at 150–190 [HAW93]

2515

Compound: Potassium ruthenate(VI)
Formula: K_2RuO_4
Molecular Formula: K_2O_4Ru
Molecular Weight: 243.265
CAS RN: 31111-21-4
Properties: black with green luster; rhomb [KIR82]
Solubility: v s H_2O [KIR82]

2516

Compound: Potassium selenate
Formula: K_2SeO_4
Molecular Formula: K_2O_4Se
Molecular Weight: 221.155
CAS RN: 7790-59-2
Properties: −100 mesh with 99.5% purity; colorless cryst or white powd [MER06] [CER91]
Solubility: g/100 g soln, H_2O: 52.7 ± 0.9 (0°C), 53.3 ± 0.3 (25°C), 55.6 ± 0.6 (100°C) [KRU93]
Density, g/cm³: 3.07 [MER06]

2517
Compound: Potassium selenide
Formula: K_2Se
Molecular Formula: K_2Se
Molecular Weight: 157.157
CAS RN: 1312-74-9
Properties: cryst; reddens in air; color changes to brownish black when heated; deliq [MER06]
Solubility: s H_2O; i ammonia [MER06]
Density, g/cm³: 2.29 [LID94]
Melting Point, °C: 800 [LID94]

2518
Compound: Potassium selenite
Formula: K_2SeO_3
Molecular Formula: K_2O_3Se
Molecular Weight: 205.155
CAS RN: 10431-47-7
Properties: −100 mesh with 99.5% purity; white powd; hygr [STR93] [CER91]
Solubility: g/100 g soln, H_2O: 68.45 (0°C), 68.5 (24.3°C), 68.53 (100.6°C); solid phase, K_2SeO_3 (0°C, 100.6°C), $K_2SeO_3 \cdot 4H_2O + K_2SeO_3$ (24.3°C) [KRU93]
Melting Point, °C: 875 [STR93]
Boiling Point, °C: decomposes [STR93]

2519
Compound: Potassium silver cyanide
Synonym: silver potassium cyanide
Formula: $KAg(CN)_2$
Molecular Formula: C_2AgKN_2
Molecular Weight: 199.001
CAS RN: 506-61-6
Properties: white cryst; sensitive to light [MER06]
Solubility: s H_2O; silver cyanide precipitated from acid solutions [MER06]

2520
Compound: Potassium sodium carbonate hexahydrate
Synonym: sodium potassium carbonate hexahydrate
Formula: $KNaCO_3 \cdot 6H_2O$
Molecular Formula: $CH_{12}KNaO_9$
Molecular Weight: 230.189
CAS RN: 64399-16-2
Properties: colorless cryst; this double salt fuses more readily than the single salts; used as a flux in analysis [HAW93]
Solubility: 185.2 g/100 mL H_2O (15°C) [CRC10]
Density, g/cm³: 1.6344 [HAW93]
Melting Point, °C: decomposes at 135 [HAW93]
Reactions: minus $6H_2O$ at 100°C [CRC10]

2521
Compound: Potassium stannate trihydrate
Synonym: potassium hydroxystannate(IV)
Formula: $K_2SnO_3 \cdot 3H_2O$
Molecular Formula: $H_6K_2O_6Sn$
Molecular Weight: 298.951
CAS RN: 12142-33-5
Properties: white to light tan cryst; used to dye and print textiles, in alkaline tin plating baths; anhydrous available as −100 mesh with 99.9% purity [HAW93] [CER91]
Solubility: 110.5 g/100 mL H_2O at 15°C; i alcohol [MER06] [KIR83]
Density, g/cm³: 3.197 [MER06]
Melting Point, °C: decomposes at 140 [AES93]

2522
Compound: Potassium stannosulfate
Formula: $K_2Sn(SO_4)_2$
Molecular Formula: $K_2O_8S_2Sn$
Molecular Weight: 389.034
CAS RN: 27790-37-0
Properties: white cryst; partially decomposed by H_2O [MER06]
Solubility: s dil alkaline hydroxide solutions [MER06]

2523
Compound: Potassium stearate
Synonyms: stearic acid, potassium salt
Formula: $CH_3(CH_2)_{16}COOK$
Molecular Formula: $C_{18}H_{35}KO_2$
Molecular Weight: 322.573
CAS RN: 593-29-3
Properties: white, cryst powd; slight odor of fat; used in softening textiles [HAW93]
Solubility: slowly s cold H_2O, more readily s hot H_2O, alcohol [MER06]

2524
Compound: Potassium sulfate
Synonym: arcanite
Formula: K_2SO_4
Molecular Formula: K_2O_4S
Molecular Weight: 174.261
CAS RN: 7778-80-5
Properties: colorless or white, hard, rhomb or hex cryst, granules or powd; bitter saline taste; enthalpy of fusion 36.40 kJ/mol; also used in glass manufacture [MER06] [KIR82] [CRC10]

Solubility: g/100 g soln, H_2O: 6.9 (0°C), 10.75 (25°C), 19.4 (100°C); solid phase, K_2SO_4 [KRU93]; 1 g/75 mL glycerol; i alcohol [MER06]
Density, g/cm³: 2.66 [MER06]
Melting Point, °C: 1069 [CRC10]
Thermal Expansion Coefficient: (volume) 100°C (0.544), 200°C (2.118), 400°C (4.935) [CLA66]

2525
Compound: Potassium sulfide
Formula: K_2S
Molecular Formula: K_2S
Molecular Weight: 110.263
CAS RN: 1312-73-8
Properties: may contain polysulfides; red or yellowish red; cub cryst or fused plates; discolored in air; very hygr; unstable; may explode when struck or if rapidly heated; enthalpy of fusion 16.15 kJ/mol [STR93] [MER06] [HAW93] [CRC10]
Solubility: s H_2O, alcohol, glycerol; i ether [HAW93]
Density, g/cm³: 1.74 [MER06]
Melting Point, °C: 948 [CRC10]

2526
Compound: Potassium sulfide pentahydrate
Formula: $K_2S \cdot 5H_2O$
Molecular Formula: $H_{10}K_2O_5S$
Molecular Weight: 200.339
CAS RN: 1312-73-8
Properties: colorless; rhomb; odor of H_2S; turns yellow to yellowish red when exposed to air and light; aq solutions are alkaline and unstable [MER06]
Solubility: v s H_2O, alcohol, glycerol; i ether [MER06]
Melting Point, °C: 60 [MER06]
Reactions: minus $3H_2O$ at 150°C [CRC10]

2527
Compound: Potassium sulfite dihydrate
Formula: $K_2SO_3 \cdot 2H_2O$
Molecular Formula: $H_4K_2O_5S$
Molecular Weight: 194.292
CAS RN: 7790-56-9
Properties: white, monocl cryst or cryst powd; oxidizes gradually to sulfate in air; used as a photographic developer, as a food and wine preservative [HAW93] [MER06] [KIR82]
Solubility: g/100 g soln, H_2O: 47.52 (0°C), 49.01 (25°C), 55.53 (100°C); solid phase, K_2SO_3 [KRU93]
Melting Point, °C: decomposes [KIR82]

2528
Compound: Potassium superoxide
Synonym: potassium dioxide
Formula: KO_2
Molecular Formula: KO_2
Molecular Weight: 71.097
CAS RN: 12030-88-5
Properties: yellow powd; sensitive to moisture [STR93]
Solubility: decomposes in H_2O [CRC10]
Density, g/cm³: 2.14 [STR93]
Melting Point, °C: 380 [LID94]

2529
Compound: Potassium tantalate
Formula: $KTaO_3$
Molecular Formula: KO_3Ta
Molecular Weight: 268.044
CAS RN: 12030-91-0
Properties: −100 mesh with 99.9% purity [CER91]

2530
Compound: Potassium tellurate(VI) trihydrate
Formula: $K_2TeO_4 \cdot 3H_2O$
Molecular Formula: $H_6K_2O_7Te$
Molecular Weight: 323.841
CAS RN: 15571-91-2
Properties: white, cryst powd [MER06]
Solubility: s in 4 parts H_2O [MER06]

2531
Compound: Potassium tellurite
Formula: K_2TeO_3
Molecular Formula: K_2O_3Te
Molecular Weight: 253.795
CAS RN: 7790-58-1
Properties: white powd; granular; hygr; used in chemical analysis to test for bacteria [HAW93]
Solubility: g/100 g H_2O: 8.8 (0°C), 27.5 (20°C), 50.4 (30°C) [LAN05]
Melting Point, °C: decomposes at 460–470 [HAW93]

2532
Compound: Potassium tellurite(IV) hydrate
Formula: $K_2TeO_3 \cdot xH_2O$
Molecular Formula: K_2O_3Te (anhydrous)
Molecular Weight: 253.795 (anhydrous)
CAS RN: 123333-66-4
Properties: white, hygr powd [ALD94] [STR93]
Melting Point, °C: decomposes at 465 [STR93]

2533
Compound: Potassium tetraborate pentahydrate
Formula: $K_2B_4O_7 \cdot 5H_2O$
Molecular Formula: $B_4H_{10}K_2O_{12}$
Molecular Weight: 323.513
CAS RN: 1332-77-0
Properties: white, cryst powd [MER06]
Solubility: s 4 parts H_2O; sl s alcohol [MER06]

2534
Compound: Potassium tetraborate tetrahydrate
Formula: $K_2B_4O_7 \cdot 4H_2O$
Molecular Formula: $B_4H_8K_2O_{11}$
Molecular Weight: 305.498
CAS RN: 12045-78-2
Properties: −6 mesh with 99.9% purity; ortho-rhomb, white powd; dehydration temp depends on partial pressure of H_2O [CER91] [STR93] [KIR78]
Solubility: % anhydrous by weight, H_2O: 9.02 (10°C), 13.6 (25°C), 24.0 (50°C), 48.4 (100°C) [KIR78]
Density, g/cm³: 1.92 [KIR78]
Melting Point, °C: decomposes [STR93]
Reactions: reversible dehydration from 85°C–111°C [KIR78]

2535
Compound: Potassium tetrabromoaurate(III) dihydrate
Formula: $KAuBr_4 \cdot 2H_2O$
Molecular Formula: $AuBr_4H_4KO_2$
Molecular Weight: 591.712
CAS RN: 14323-32-1
Properties: violet cryst; sensitive to light [MER06]
Solubility: s H_2O, alcohol [MER06]
Melting Point, °C: decomposes at 120 [CRC10]

2536
Compound: Potassium tetrabromopalladate(II)
Formula: K_2PdBr_4
Molecular Formula: Br_4K_2Pd
Molecular Weight: 504.233
CAS RN: 13826-93-2
Properties: reddish brown powd; hygr [STR93]

2537
Compound: Potassium tetrabromoplatinate(II)
Formula: K_2PtBr_4
Molecular Formula: Br_4K_2Pt
Molecular Weight: 592.893

CAS RN: 13826-94-3
Properties: red powd [STR93]

2538
Compound: Potassium tetrachloroaurate(III)
Formula: $KAuCl_4$
Molecular Formula: $AuCl_4K$
Molecular Weight: 377.876
CAS RN: 13682-61-6
Properties: yellowish orange cryst [STR93]
Solubility: g/100 g H_2O: 38.3 (10°C), 61.8 (20°C), 405 (60°C) [LAN05]
Melting Point, °C: decomposes at 357 [AES93]

2539
Compound: Potassium tetrachloroaurate(III) dihydrate
Formula: $KAuCl_4 \cdot 2H_2O$
Molecular Formula: $AuCl_4H_4KO_2$
Molecular Weight: 413.907
CAS RN: 13005-39-5
Properties: yellow, monocl cryst; light sensitive [MER06] [HAW93]
Solubility: s H_2O, alcohol, ether [HAW93]

2540
Compound: Potassium tetrachloropalladate(II)
Formula: K_2PdCl_4
Molecular Formula: Cl_4K_2Pd
Molecular Weight: 326.428
CAS RN: 10025-98-6
Properties: brown powd; hygr [STR93]
Density, g/cm³: 2.67 [STR93]
Melting Point, °C: decomposes at 105 [ALD94]

2541
Compound: Potassium tetrachloroplatinate(II)
Synonym: potassium chloroplatinate
Formula: K_2PtCl_4
Molecular Formula: Cl_4K_2Pt
Molecular Weight: 415.088
CAS RN: 10025-99-7
Properties: pink or ruby red; tetr; hygr [MER06] [STR93] [KIR82]
Solubility: s H_2O; i alcohol [MER06] [KIR82]
Density, g/cm³: 3.38 [STR93]
Melting Point, °C: decomposes at >500 [KIR82]

2542
Compound: Potassium tetracyanomercurate(II)
Formula: $K_2Hg(CN)_4$

Molecular Formula: $C_4HgK_2N_4$
Molecular Weight: 382.858
CAS RN: 591-89-9
Properties: colorless or white cryst [MER06]
Solubility: s H_2O [MER06]

2543
Compound: Potassium tetracyanonickelate(II) monohydrate
Formula: $K_2Ni(CN)_4 \cdot H_2O$
Molecular Formula: $C_4H_2K_2N_4NiO$
Molecular Weight: 258.976
CAS RN: 14220-17-8
Properties: yellowish orange; cryst powd [MER06]
Solubility: s H_2O [MER06]
Density, g/cm³: 1.875 [CRC10]
Reactions: minus H_2O at ~100°C [MER06]

2544
Compound: Potassium tetracyanoplatinate(II)
Formula: $K_2Pt(CN)_4$
Molecular Formula: $C_4K_2N_4Pt$
Molecular Weight: 377.348
CAS RN: 562-76-5
Properties: yellow powd; hygr [STR93]
Solubility: g/100 g H_2O: 11.6 (0°C), 33.9 (20°C), 194 (90°C) [LAN05]

2545
Compound: Potassium tetracyanoplatinate(II) trihydrate
Formula: $K_2Pt(CN)_4 \cdot 3H_2O$
Molecular Formula: $C_4H_6K_2N_4O_3Pt$
Molecular Weight: 431.394
CAS RN: 14323-36-5
Properties: almost colorless; rhomb prisms; blue color in direction of principal axis [MER06]
Solubility: s H_2O [MER06]
Density, g/cm³: 2.455 [CRC10]
Melting Point, °C: decomposes at 400–600 [CRC10]
Reactions: minus $3H_2O$ at 100°C [CRC10]

2546
Compound: Potassium tetracyanozincate
Formula: $K_2Zn(CN)_4$
Molecular Formula: $C_4K_2N_4Zn$
Molecular Weight: 247.658
CAS RN: 14244-62-3
Properties: cryst powd [MER06]
Solubility: v s H_2O [MER06]

2547
Compound: Potassium tetrafluoroberyllate dihydrate
Synonym: beryllium potassium fluoride
Formula: $K_2BeF_4 \cdot 2H_2O$
Molecular Formula: $BeF_4H_4K_2O_2$
Molecular Weight: 199.234
CAS RN: 7787-50-0
Properties: brilliant cryst [MER06]
Solubility: s H_2O, conc K_2SO_4 [MER06]

2548
Compound: Potassium tetrafluoroborate
Synonym: avogadrite
Formula: KBF_4
Molecular Formula: BF_4K
Molecular Weight: 125.903
CAS RN: 14075-53-7
Properties: colorless; ortho-rhomb below 283°C, a=0.7032 nm, b=0.8674 nm, c=0.5496 nm; bipyrimidal or cub cryst; enthalpy of fusion 17.7 kJ/mol; can be prepared as a gelatinous precipitate from fluoroboric acid and KOH or K_2CO_3; used in plating, in flux for soldering [HAW93] [MER06] [KIR78] [JAN71]
Solubility: g/100 g H_2O: 0.3 (3°C), 0.448 (20°C), 0.55 (25°C), 1.4 (40°C), 6.27 (100°C); sl s boiling alcohol [MER06]
Density, g/cm³: 2.505 [MER06]
Melting Point, °C: 530 [MER06]

2549
Compound: Potassium tetraiodoaurate(III)
Synonym: gold potassium iodide
Formula: $KAuI_4$
Molecular Formula: AuI_4K
Molecular Weight: 743.683
CAS RN: 7791-29-9
Properties: black, lustrous cryst; sensitive to light [MER06]
Solubility: s H_2O, decomposes and liberates iodine [MER06]
Melting Point, °C: decomposes at 150 [CRC10]

2550
Compound: Potassium tetraiodocadmium dihydrate
Synonym: cadmium potassium iodide dihydrate
Formula: $K_2CdI_4 \cdot 2H_2O$
Molecular Formula: $CdH_4I_4K_2O_2$
Molecular Weight: 734.256
CAS RN: 584-10-1

Properties: large, water clear, somewhat
  distorted octahedra; deliq; turns yellow when
  aging due to release of I_2 [MER06]
Solubility: 1 part w/w dissolves at 15°C, in: 0.73
  parts H_2O, 1.4 parts alcohol; 24.5 parts
  ether; 4.5 parts of a 1:1 mixture of ether
  and alcohol; s ethyl acetate [MER06]
Density, g/cm³: 3.359 [MER06]

2551

Compound: Potassium tetraiodomercurate(II)
Formula: K_2HgI_4
Molecular Formula: HgI_4K_2
Molecular Weight: 786.404
CAS RN: 7783-33-7
Properties: sulfur yellow color; cryst; deliq; dihydrate
  known as Mayers reagent; dihydrate prepared by
  dissolving stoichiometric amounts of HgI_2 and KI
  in distilled water, which is used as an antiseptic and
  as a precipitant for alkaloids; in strongly alkaline
  solutions, known as Nessler's reagent, which is
  used for ammonia detection [KIR81] [MER06]
Solubility: v s H_2O; s alcohol, ether, acetone [MER06]
Density, g/cm³: 4.29 [LID94]
Melting Point, °C: 105 [CRC10]

2552

Compound: Potassium tetranitritoplatinate(II)
Formula: $K_2Pt(NO_2)_4$
Molecular Formula: $K_2N_4O_8Pt$
Molecular Weight: 457.299
CAS RN: 13815-39-9
Properties: white powd [STR93]

2553

Compound: Potassium tetraoxalate dihydrate
Formula: $KHC_2O_4 \cdot H_2C_2O_4 \cdot 2H_2O$
Molecular Formula: $C_4H_7KO_{10}$
Molecular Weight: 254.192
CAS RN: 127-96-8
Properties: colorless or white cryst [MER06]
Solubility: s 60 parts cold H_2O, 12 parts
  boiling H_2O; sl s alcohol [MER06]
Melting Point, °C: decomposes [CRC10]

2554

Compound: Potassium thioantimonate
  heminonahydrate
Formula: $K_3SbS_4 \cdot 4\text{-}1/2H_2O$
Molecular Formula: $H_9K_3O_{4.5}S_4Sb$

Molecular Weight: 448.385
CAS RN: 14693-02-8
Properties: colorless to yellowish cryst; formula also
  given as $2K_3SbS_4 \cdot 4\text{-}1/2H_2O$ [MER06] [CRC10]
Solubility: g anhydrous/100 g H_2O: 306 (0°C), 302
  (30°C), 381 (80°C) [LAN05]; i alcohol [MER06]

2555

Compound: Potassium thiocarbonate
Formula: K_2CS_3
Molecular Formula: CK_2S_3
Molecular Weight: 186.406
CAS RN: 26750-66-3
Properties: yellowish red; deliq granules or
  cryst; very hygr; used in medicine, as a
  soil fumigant [HAW93] [MER06]
Solubility: v s H_2O, giving strongly
  akaline solution [MER06]
Melting Point, °C: decomposes [CRC10]

2556

Compound: Potassium thiocyanate
Formula: KSCN
Molecular Formula: CKNS
Molecular Weight: 97.182
CAS RN: 333-20-0
Properties: colorless; deliq cryst; temp drops to ~30°C
  when dissolved in its own weight of H_2O [MER06]
Solubility: g/100 g H_2O: 177 (0°C), 239 (25°C),
  673.6 (99°C); solid phase, KSCN [KRU93];
  1 g dissolves in: 0.5 mL acetone, 12 mL
  alcohol, 8 mL boiling alcohol [MER06]
Density, g/cm³: 1.88 [HAW93]
Melting Point, °C: 173 [HAW93]
Boiling Point, °C: decomposes at 500 [HAW93]
Reactions: turns brown, green, blue when
  fused, white when cooled [HAW93]

2557

Compound: Potassium thiosulfate
Synonym: potassium hyposulfite
Formula: $K_2S_2O_3$
Molecular Formula: $K_2O_3S_2$
Molecular Weight: 190.327
CAS RN: 10294-66-3
Properties: colorless cryst; hygr [MER06]
Solubility: g/100 g H_2O: 96.1 (0°C), 165.0 (25°C),
  312.0 (90°C); solid phase, $K_2S_2O_3 \cdot 2H_2O$
  (0°C), $3K_2S_2O_3 \cdot 5H_2O$ (25°C), $K_2S_2O_3$
  (90°C) [KRU93]; i alcohol [MER06]
Density, g/cm³: 2.23 [CRC10]

2558
Compound: Potassium titanate
Formula: K_2TiO_3
Molecular Formula: K_2O_3Ti
Molecular Weight: 174.062
CAS RN: 12030-97-6
Properties: white; grayish brown powd; as a fiber, has a high index of refraction, can diffuse and reflect infrared radiation; used in fiber form in rockets, missiles, and in nuclear power as an insulator [HAW93] [STR93]
Solubility: hydrolyzes in H_2O to give a strongly alkaline solution [HAW93]
Density, g/cm³: 3.1 [LID94]
Melting Point, °C: 1515 [LID94]

2559
Compound: Potassium titanium oxalate dihydrate
Formula: $K_2TiO(C_2O_4)_2 \cdot 2H_2O$
Molecular Formula: $C_4H_4K_2O_{11}Ti$
Molecular Weight: 354.133
CAS RN: 14402-67-6
Properties: colorless, lustrous cryst; used as a mordant in cotton and leather dyeing [HAW93]
Solubility: v s H_2O [MER06]

2560
Compound: Potassium triiodide monohydrate
Formula: $KI_3 \cdot H_2O$
Molecular Formula: H_2I_3KO
Molecular Weight: 437.827
CAS RN: 7790-42-3
Properties: dark brown, hygr; monocl prisms; reasonably stable only in the monohydrate form; formula also given as hemihydrate [CRC10] [MER06]
Solubility: s H_2O; s with partial decomposition in alcohol and ether [MER06]
Density, g/cm³: 3.50 [MER06]
Melting Point, °C: 38 (closed tube) [MER06]
Reactions: decomposes, evolving I_2 with KI residue at 225°C [MER06]

2561
Compound: Potassium triiodozincate
Synonym: potassium zinc iodide
Formula: $KZnI_3$
Molecular Formula: I_3KZn
Molecular Weight: 485.202
CAS RN: 7790-43-4

Properties: cryst; very hygr [MER06]
Solubility: v s H_2O [MER06]

2562
Compound: Potassium triphosphate
Synonym: potassium tripolyphosphate
Formula: $K_5P_3O_{10}$
Molecular Formula: $K_5O_{10}P_3$
Molecular Weight: 448.407
CAS RN: 13845-36-8
Properties: white powd; prepared by dehydration of an equimolar mixture of mono- and dipotassium phosphates; used in detergents [KIR82] [STR93]
Density, g/cm³: 2.54 [STR93]
Melting Point, °C: 620 [STR93]

2563
Compound: Potassium tungstate
Formula: K_2WO_4
Molecular Formula: K_2O_4W
Molecular Weight: 326.035
CAS RN: 7790-60-5
Properties: −100 mesh with 99.5% purity; heavy; deliq; cryst powd [MER06] [CER91]
Solubility: s ~2 parts cold H_2O, ~0.7 parts boiling H_2O; i alcohol [MER06]
Density, g/cm³: 3.12 [MER06]
Melting Point, °C: 921 [MER06]

2564
Compound: Potassium tungstate dihydrate
Formula: $K_2WO_4 \cdot 2H_2O$
Molecular Formula: $H_4K_2O_6W$
Molecular Weight: 362.065
CAS RN: 7790-60-5
Properties: heavy, cryst powd; hygr [ALD94] [HAW93]
Solubility: 51.5 g/100 mL H_2O [CRC10]
Density, g/cm³: 3.1 [HAW93]
Melting Point, °C: 921 [HAW93]

2565
Compound: Potassium uranate
Formula: $K_2U_2O_7$
Molecular Formula: $K_2O_7U_2$
Molecular Weight: 666.251
CAS RN: 7790-63-8
Properties: cub orange powd [LID94] [MER06]
Solubility: i H_2O; s acids [MER06]
Density, g/cm³: 6.12 [LID94]

2566
Compound: Potassium uranyl nitrate
Formula: $K(UO_2)(NO_3)_3$
Molecular Formula: $KN_3O_{11}U$
Molecular Weight: 495.140
CAS RN: 18078-40-5
Properties: greenish yellow; cryst powd [MER06]
Solubility: s ~1 part H_2O [MER06]

2567
Compound: Potassium uranyl sulfate dihydrate
Formula: $K_2(UO_2)(SO_4)_2 \cdot 2H_2O$
Molecular Formula: $H_4K_2O_{12}S_2U$
Molecular Weight: 576.383
CAS RN: 27709-53-1
Properties: greenish yellow; cryst powd [MER06]
Solubility: v s H_2O [MER06]
Density, g/cm³: 3.363 [CRC10]
Reactions: minus $2H_2O$ at 120°C [CRC10]

2568
Compound: Potassium vanadate
Synonym: potassium metavanadate
Formula: KVO_3
Molecular Formula: KO_3V
Molecular Weight: 138.038
CAS RN: 13769-43-2
Properties: colorless cryst; −200 mesh with 99.9% purity; there are also K_3VO_4, 14293-78-8, and $K_4V_2O_7$, 14638-93-8 [CRC10] [CER91]

2569
Compound: Potassium zinc sulfate hexahydrate
Formula: $K_2Zn(SO_4)_2 \cdot 6H_2O$
Molecular Formula: $H_{12}K_2O_{14}S_2Zn$
Molecular Weight: 443.806
CAS RN: 13932-17-7
Properties: cryst [MER06]
Solubility: g/100 g H_2O: 13.0 (0°C), 25.9 (20°C), 72.1 (60°C) [LAN05]

2570
Compound: Potassium zirconate
Formula: K_2ZrO_3
Molecular Formula: K_2O_3Zr
Molecular Weight: 217.419
CAS RN: 12030-98-7
Properties: −200 mesh with 99.9% purity [CER91]

2571
Compound: Potassium zirconium sulfate trihydrate
Formula: $K_4Zr(SO_4)_4 \cdot 3H_2O$
Molecular Formula: $H_6K_4O_{19}S_4Zr$
Molecular Weight: 685.918
CAS RN: 53608-79-0
Properties: white, cryst powd [HAW93]
Solubility: sl s H_2O [MER06]

2572
Compound: Praseodymium acetate hydrate
Formula: $Pr(CH_3COO)_3 \cdot xH_2O$
Molecular Formula: $C_6H_9O_6Pr$ (anhydrous)
Molecular Weight: 318.041 (anhydrous)
CAS RN: 6192-12-7
Properties: green cryst; hygr [AES93] [STR93]

2573
Compound: Praseodymium acetylacetonate
Synonyms: 2,4-pentanedione, praseodymiun(III) derivative
Formula: $Pr(CH_3COCH=C(O)CH_3)_3$
Molecular Formula: $C_{15}H_{21}O_6Pr$
Molecular Weight: 438.236
CAS RN: 14553-09-4
Properties: powd [STR93]
Melting Point, °C: 146 [CRC10]

2574
Compound: Praseodymium barium copper oxide
Formula: $PrBa_2Cu_3O_7$
Molecular Formula: $Ba_2Cu_3O_7Pr$
Molecular Weight: 718.196
CAS RN: 126284-91-1
Properties: a=0.3878 nm, b=0.3940 nm, c=1.1761 nm; can be prepared by heating Pr_6O_{11}, which is free of both carbonate and hydroxide with stoichiometric amounts of $BaCO_3$ and CuO, then grinding the reactants and heating the resulting powd at 950°C for 12 h in air; CAS RN 126284-91-1 is for the compound with $PrBa_2CuO_{6.97}$; CAS RN 120309-22-8 refers to compound with $PrBa_2Cu_3O_{7.05}$ [CON87]

2575
Compound: Praseodymium boride
Formula: PrB_6
Molecular Formula: B_6Pr
Molecular Weight: 205.774
CAS RN: 12008-27-4

Properties: −325 mesh with 10 μm average or less [CER91]
Density, g/cm³: 4.84 [LID94]
Melting Point, °C: 2610 [LID94]

2576

Compound: Praseodymium bromate nonahydrate
Formula: $Pr(BrO_3)_3 \cdot 9H_2O$
Molecular Formula: $Br_3H_{18}O_{18}Pr$
Molecular Weight: 686.752
CAS RN: 15162-93-3
Properties: green; hex [CRC10]
Solubility: g/100 g H_2O: 55.9 (0°C), 91.8 (20°C), 144 (80°C) [LAN05]
Melting Point, °C: 56.5 [CRC10]
Reactions: minus $7H_2O$ at 170°C [CRC10]

2577

Compound: Praseodymium bromide
Formula: $PrBr_3$
Molecular Formula: Br_3Pr
Molecular Weight: 380.620
CAS RN: 13536-53-3
Properties: green cryst; −20 mesh with 99.9% purity [CER91] [CRC10]
Solubility: decomposed by H_2O [CRC10]
Density, g/cm³: 5.28 [LID94]
Melting Point, °C: 691 [AES93]
Boiling Point, °C: 1547 [CRC10]

2578

Compound: Praseodymium carbonate octahydrate
Formula: $Pr_2(CO_3)_3 \cdot 8H_2O$
Molecular Formula: $C_3H_{16}O_{17}Pr_2$
Molecular Weight: 605.965
CAS RN: 14948-62-0
Properties: light green powd [AES93] [STR93]
Reactions: minus $6H_2O$ at 100°C [CRC10]

2579

Compound: Praseodymium chloride
Formula: $PrCl_3$
Molecular Formula: Cl_3Pr
Molecular Weight: 247.266
CAS RN: 10361-79-2
Properties: −20 mesh of 99.9% purity; bluish green powd; hygr [STR93] [CER91]
Solubility: s H_2O, alcohol [MER06]
Density, g/cm³: 4.02 [STR93]
Melting Point, °C: 786 (under argon) [STR93]
Boiling Point, °C: 1700 [STR93]

2580

Compound: Praseodymium chloride heptahydrate
Formula: $PrCl_3 \cdot 7H_2O$
Molecular Formula: $Cl_3H_{14}O_7Pr$
Molecular Weight: 373.373
CAS RN: 10025-90-8
Properties: −4 mesh with 99.9% purity; hygr; green cryst [MER06] [STR93] [CER91]
Solubility: 334 g/100 mL H_2O (13°C) [CRC10]
Density, g/cm³: 2.250 [STR93]
Melting Point, °C: 115 [STR93]
Reactions: minus $7H_2O$ at 180°C–200°C if heated in HCl stream [MER06]

2581

Compound: Praseodymium fluoride
Formula: PrF_3
Molecular Formula: F_3Pr
Molecular Weight: 197.903
CAS RN: 13709-46-1
Properties: hygr, light blue powd and 99.9% pure melted pieces of 3–6 mm; hygr; melted pieces used as evaporation material for infrared multilayers, low and high ends only [STR93] [CER91]
Density, g/cm³: 6.3 [LID94]
Melting Point, °C: 1395 [LID94]

2582

Compound: Praseodymium hexafluoroacetylacetonate
Synonyms: 1,1,1,5,5,5-hexafluoro-2,4-pentanedione, Pr derivative
Formula: $Pr(CF_3COCHCOCF_3)_3$
Molecular Formula: $C_{15}H_3F_{18}O_6Pr$
Molecular Weight: 762.067
CAS RN: 47814-20-0
Properties: light green powd [STR93]

2583

Compound: Praseodymium hydride
Formula: PrH_3
Molecular Formula: H_3Pr
Molecular Weight: 143.931
CAS RN: 13864-03-4
Properties: lump under argon atm [ALF95]

2584

Compound: Praseodymium hydroxide
Formula: $Pr(OH)_3$
Molecular Formula: H_3O_3Pr

Molecular Weight: 191.929
CAS RN: 16469-16-2
Properties: gelatinous; pale green; obtained by reacting hydroxide with Pr salt; purple powd obtained by reacting Pr carbide with H_2O [MER06]

2585
Compound: Praseodymium iodide
Formula: PrI_3
Molecular Formula: I_3Pr
Molecular Weight: 521.621
CAS RN: 13813-23-5
Properties: green hygr cryst; −20 mesh with 99.9% purity [CER91] [CRC10]
Density, g/cm³: ~5.8 [LID94]
Melting Point, °C: 737 [CRC10]

2586
Compound: Praseodymium nitrate hexahydrate
Formula: $Pr(NO_3)_3 \cdot 6H_2O$
Molecular Formula: $H_{12}N_3O_{15}Pr$
Molecular Weight: 435.014
CAS RN: 15878-77-0
Properties: light green cryst; there is a pentahydrate, 14483-17-1 [AES93] [STR93]
Solubility: 5.0257 ± 0.0087 mol/kg (25°C), hexahydrate is the stable phase [RAR85b]; g anhydrous/100 g H_2O: 112 (20°C), 162 (30°C), 178 (40°C) [LAN05]

2587
Compound: Praseodymium nitride
Formula: PrN
Molecular Formula: NPr
Molecular Weight: 154.915
CAS RN: 25764-09-4
Properties: −60 mesh with 99.9% purity; NaCl crystal system, a = 0.515 nm [CER91] [CIC73]
Density, g/cm³: 7.46 [LID94]

2588
Compound: Praseodymium oxalate decahydrate
Formula: $Pr_2(C_2O_4)_3 \cdot 10H_2O$
Molecular Formula: $C_6H_{20}O_{22}Pr_2$
Molecular Weight: 726.027
CAS RN: 24992-60-7
Properties: light green powd [ALF95] [STR93]

2589
Compound: Praseodymium perchlorate hexahydrate
Formula: $Pr(ClO_4)_3 \cdot 6H_2O$

Molecular Formula: $Cl_3H_{12}O_{18}Pr$
Molecular Weight: 547.351
CAS RN: 13498-07-2
Properties: green cryst; hygr [ALF95] [STR93]

2590
Compound: Praseodymium phosphate
Formula: $PrPO_4$
Molecular Formula: O_4PPr
Molecular Weight: 235.879
CAS RN: 14298-31-8
Properties: green powd [STR93]

2591
Compound: Praseodymium silicide
Formula: $PrSi_2$
Molecular Formula: $PrSi_2$
Molecular Weight: 197.079
CAS RN: 12066-83-0
Properties: tetr cryst; 10 mm and down lump [LID94] [ALF93]
Density, g/cm³: 5.46 [LID94]
Melting Point, °C: 1712 [LID94]

2592
Compound: Praseodymium sulfate
Formula: $Pr_2(SO_4)_3$
Molecular Formula: $O_{12}Pr_2S_3$
Molecular Weight: 570.006
CAS RN: 10277-44-8
Properties: light green cryst [MER06]
Solubility: 0.1545 ± 0.0019 mol/kg (25°C), octahydrate is the equilibrium phase [RAR88]; g/100 g H_2O: 19.8 (0°C), 12.6 (20°C), 0.91 (100°C) [LAN05]
Density, g/cm³: 3.72 [CRC10]

2593
Compound: Praseodymium sulfate octahydrate
Formula: $Pr_2(SO_4)_3 \cdot 8H_2O$
Molecular Formula: $H_{16}O_{20}Pr_2S_3$
Molecular Weight: 714.128
CAS RN: 13510-41-3
Properties: light green cryst; monocl [CRC10] [AES93] [STR93]
Solubility: 17.4 g/100 mL H_2O (20°C) [CRC10]
Density, g/cm³: 2.827 [CRC10]

2594
Compound: Praseodymium sulfide
Formula: Pr_2S_3

Molecular Formula: Pr_2S_3
Molecular Weight: 378.013
CAS RN: 12038-13-0
Properties: cub cryst; −200 mesh with 99.9% purity [LID94] [CER91]
Density, g/cm³: 5.1 [LID94]
Melting Point, °C: 1765 [LID94]

2595

Compound: Praseodymium telluride
Formula: Pr_2Te_3
Molecular Formula: Pr_2Te_3
Molecular Weight: 664.615
CAS RN: 12038-12-9
Properties: cub cryst; −20 mesh with 99.9% purity; probably also contains $PrTe_3$, as well as other Pr-Te phases [LID94] [CER91]
Density, g/cm³: ~7.0 [LID94]
Melting Point, °C: 1500 [LID94]

2596

Compound: Praseodymium(α)
Formula: α-Pr
Molecular Formula: Pr
Molecular Weight: 140.90765
CAS RN: 7440-10-0
Properties: yellowish metal; hex; tarnishes in moist air, forming oxide film; enthalpy of fusion is 6.912 kJ/mol; enthalpy of sublimation is 355.6 kJ/mol; electrical resistivity 68 μohm·cm; radius of atom is 0.1828 nm; radius of ion is 0.1013 nm, Pr^{+++}; forms green solutions [MER06] [KIR82] [ALD94]
Density, g/cm³: 6.773 [KIR82]
Melting Point, °C: 935 [DOU83]
Reactions: α to β transition at 800°C [MER06]

2597

Compound: Praseodymium(β)
Formula: β-Pr
Molecular Formula: Pr
Molecular Weight: 140.90765
CAS RN: 7440-10-0
Properties: 12 mm pieces and smaller (under oil) of 99.9% purity; yellowish metal; forms oxide film in moist air; bcc; electrical resistivity (20°C) 68 μohm·cm; enthalpy of fusion 6.89 kJ/mol; enthalpy of vaporization 331 kJ/mol [MER06] [CER91] [CRC10] [ALD94]
Density, g/cm³: 6.64 [MER06]
Melting Point, °C: 935 [MER06]
Boiling Point, °C: 3510 [LID94]

Reactions: α to β transition at 800°C [MER06]
Thermal Conductivity, W/(m·K): 12.5 (25°C) [CRC10]
Thermal Expansion Coefficient: 6.7×10^{-6}/K [CRC10]

2598

Compound: Praseodymium(III) oxide
Formula: Pr_2O_3
Molecular Formula: O_3Pr_2
Molecular Weight: 329.813
CAS RN: 12036-32-7
Properties: yellow-green amorphous; 3–12 mm pieces with 10 μm average or less with 99.9% purity [CER91] [CRC10]
Solubility: 0.000020 g/100 mL H_2O (29°C) [CRC10]
Density, g/cm³: 7.07 [CRC10]
Melting Point, °C: 2300 [LID94]

2599

Compound: Praseodymium(III,IV) oxide
Formula: Pr_6O_{11}
Molecular Formula: $O_{11}Pr_6$
Molecular Weight: 1021.439
CAS RN: 12037-29-5
Properties: black powd or brown sintered pieces; used as an evaporation material, with interest in its reactivity to radio frequencies [STR93] [CER91]

2600

Compound: Promethium
Formula: Pm
Molecular Formula: Pm
Molecular Weight: 145
CAS RN: 7440-12-2
Properties: silvery metal; enthalpy of sublimation is 348 kJ/mol (estimated); radius of atom is 0.1811 nm; radius of ion is 0.0979 nm; Pr^{+++} has yellow-colored solutions [KIR82]
Density, g/cm³: 7.22 [MER06]
Melting Point, °C: 1042 [KIR82]
Boiling Point, °C: ~3000 [KIR82]
Thermal Conductivity, W/(m·K): 15 (25°C) [CRC10]

2601

Compound: 2,2-Bis(ethylferrocenyl)propane
Formula: $(C_2H_5C_5H_3FeC_5H_5C)_2C(CH_3)_2$
Molecular Formula: $C_{27}H_{32}Fe_2$
Molecular Weight: 468.245
CAS RN: 81579-74-0
Properties: red-brown viscous liq [STR93]
Density, g/cm³: 1.275 [STR93]

2602
Compound: Protactinium
Formula: Pa
Molecular Formula: Pa
Molecular Weight: 231.03588
CAS RN: 7440-13-3
Properties: shiny, malleable metal; readily tarnishes in air; α: tetr, a = 0.3929 nm, c = 0.3241 nm, stable from room temp up to 1170°C; β: bcc, a = 0.381 nm, stable from 1170°C–1575°C; first discovered in 1917; enthalpy of fusion 12.34 kJ/mol; can be produced by nuclear reaction $^{230}Th + n \rightarrow ^{231}Th + \gamma$ $^{231}Th \rightarrow ^{231}Pa$, however, Pa is also found in natural sources; $t_{1/2}$ ^{231}Pa is $3.25 \times 10^{+4}$ years [KIR78] [MER06] [CRC10]
Density, g/cm³: 15.37 [KIR78]
Melting Point, °C: 1575 [KIR91]
Reactions: reacts with H_2 at 250°C–300°C forming PaH_3 [MER06]

2603
Compound: Pyrophosphoric acid
Synonym: diphosphoric acid
Formula: $[P(O)(OH)_2]_2O$
Molecular Formula: $H_4O_7P_2$
Molecular Weight: 177.98
CAS RN: 2466-09-3
Properties: hygr; syrupy liq, glassy cryst; preparation: reaction of H_3PO_4 and $POCl_3$; uses: catalyst, metal treatment, stabilizer for organic peroxides [HAW93] [MER06]
Solubility: 709 g/100 mL H_2O (23°C); s alcohol, ether [MER06]
Melting Point, °C: 61–63 [ALD94]
Reactions: forms phosphoric acid on dissolution in hot H_2O [MER06]

2604
Compound: Radium
Formula: Ra
Molecular Formula: Ra
Molecular Weight: 226
CAS RN: 7440-14-4
Properties: brilliant white metal; bcc; blackens when exposed to air; spontaneously disintegrates, forming radon; compounds closely resemble those of barium; $t_{1/2} = 1600$ years; enthalpy of sublimation 130 kJ/mol [DOU83] [MER06] [GME77]
Solubility: evolves H_2 in H_2O [CRC10]
Density, g/cm³: 5.5 [MER06]
Melting Point, °C: 700 [MER06]
Boiling Point, °C: 1737 [MER06]

2605
Compound: Radium bromide
Formula: $RaBr_2$
Molecular Formula: Br_2Ra
Molecular Weight: 386
CAS RN: 10031-23-9
Properties: white or sl brownish cryst [MER06]
Solubility: s H_2O [MER06], s alcohol [HAW93]
Density, g/cm³: 5.79 [MER06]
Melting Point, °C: 728 [MER06]
Boiling Point, °C: sublimes at 900 [MER06]

2606
Compound: Radium carbonate
Formula: $RaCO_3$
Molecular Formula: CO_3Ra
Molecular Weight: 286
CAS RN: 7116-98-5
Properties: amorphous, radioactive powd; pure material is white, but impurities impart yellow, orange, or pink color [HAW93]
Solubility: i H_2O [HAW93]

2607
Compound: Radium chloride
Formula: $RaCl_2$
Molecular Formula: Cl_2Ra
Molecular Weight: 297
CAS RN: 10025-66-8
Properties: radioactive; white or sl brownish cryst [MER06]
Solubility: s H_2O [MER06]; s alcohol [HAW93]
Density, g/cm³: 4.91 [MER06]
Melting Point, °C: 1000 [MER06]

2608
Compound: Radium sulfate
Formula: $RaSO_4$
Molecular Formula: O_4RaS
Molecular Weight: 322
CAS RN: 7446-16-4
Properties: white, rhomb cryst when pure, else colored yellow, orange, pink; radioactive [HAW93] [CRC10]
Solubility: i H_2O, acids [HAW93]

2609
Compound: Radon
Formula: Rn
Molecular Formula: Rn
Molecular Weight: 222
CAS RN: 10043-92-2

Properties: colorless, inert gas; strongly radioactive; enthalpy of vaporization 18.10 kJ/mol; heat capacity (101.32 kPa, 25°C) ~21 J/(mol · K); viscosity (25°C, 101.32 kPa) 23.3 Pa · s; ^{222}Rn $t_{1/2} = 3.82$ days; can be condensed to a colorless, transparent liq [MER06] [HAW93] [KIR78]

Solubility: at 101.32 kPa, 230 mL/L H_2O (20°C) [KIR78]; s organic solvents [MER06]

Density, g/cm³: gas: (101.32 kPa, 0°C) 9.741 g/L [LID94]; liq: 4.4 [MER06]

Melting Point, °C: −71 [CRC10]

Boiling Point, °C: −61.7 [LID94]

2610

Compound: Rhenium

Formula: Re

Molecular Formula: Re

Molecular Weight: 186.207

CAS RN: 7440-15-5

Properties: 3–6 mm pieces of 99.99% purity; black to silvery gray metal; hex close-packed cryst, a = 0.2760 nm, c = 0.4458 nm; enthalpy of sublimation 791 kJ/mol; enthalpy of fusion 60.43 kJ/mol; electrical resistivity 19.3 μohm · cm at 20°C; tensile strength is 80,000 psi; modulus of elasticity is 0.46 Pa [KIR82] [HAW93] [MER06] [CER91] [CRC10]

Solubility: i HCl; attacked by HNO_3, H_2SO_4 [HAW93]

Density, g/cm³: 21.02 [MER06]

Melting Point, °C: 3180 [MER06]

Boiling Point, °C: 5596 [CRC10]

Thermal Conductivity, W/(m · K): 47.9 (25°C) [CRC10]

Thermal Expansion Coefficient: 6.2×10^{-6}/K [CRC10]

2611

Compound: Rhenium boride

Formula: Re_7B_3

Molecular Formula: B_3Re_7

Molecular Weight: 1335.882

CAS RN: 12355-99-6

Properties: formula is generally Re_7B_3 with possible traces of ReB_2; −100 mesh with 99.5% purity [CER91]

2612

Compound: Rhenium carbonyl

Synonym: dirhenium pentacarbonyl

Formula: $Re_2(CO)_{10}$

Molecular Formula: $C_{10}O_{10}Re_2$

Molecular Weight: 652.518

CAS RN: 14285-68-8

Properties: yellowish white, volatile cryst; can be prepared by reacting Re_2O_7, H_2, and CO under high pressure; used to produce highly pure Re metal [KIR82] [STR93]

Solubility: s in most organic solvents [KIR82]

Density, g/cm³: 2.87 [STR93]

Melting Point, °C: 170, decomposes [STR93]

2613

Compound: Rhenium pentacarbonyl bromide

Formula: $Re(CO)_5Br$

Molecular Formula: C_5BrO_5Re

Molecular Weight: 406.163

CAS RN: 14220-21-4

Properties: off-white cryst; used to study π-electron complexes [ALD94] [STR93]

Melting Point, °C: sublimes at 90 [STR93]

2614

Compound: Rhenium pentacarbonyl chloride

Formula: $Re(CO)_5Cl$

Molecular Formula: C_5ClO_5Re

Molecular Weight: 361.712

CAS RN: 14099-01-5

Properties: octahedral cryst; off-white; used to study π-electron-deficient compounds [STR93] [ALD94]

2615

Compound: Rhenium(III) bromide

Synonym: rhenium tribromide

Formula: $ReBr_3$

Molecular Formula: Br_3Re

Molecular Weight: 425.919

CAS RN: 13569-49-8

Properties: green-black cryst; −100 mesh with 99.9% purity structure consists of Fe_3Br_9 units, linked by bridging [CER91] [COT88] [CRC10]

Density, g/cm³: 6.10 [LID94]

Melting Point, °C: sublimes at 500 (vacuum) [CRC10]

2616

Compound: Rhenium(III) chloride

Formula: $ReCl_3$

Molecular Formula: Cl_3Re

Molecular Weight: 292.565

CAS RN: 13569-63-6

Properties: reddish black powd; hygr; sensitive to light; emits green vapor when heated from which the metal can be deposited [HAW93] [STR93]

Solubility: s H_2O and glacial acetic acid [HAW93]

Density, g/cm³: 4.81 [LID94]

Melting Point, °C: decomposes at 500 [LID94]

2617
Compound: Rhenium(III) iodide
Synonym: rhenium triiodide
Formula: ReI_3
Molecular Formula: I_3Re
Molecular Weight: 566.920
CAS RN: 15622-42-1
Properties: black; −80 mesh of 99.9% purity; structure consists of Re_3I_9 molecules linked by bridging; preparation: careful evaporation of $HReO_4$ solution containing excess HI [COT88] [CER91]
Melting Point, °C: decomposes [LID94]

2618
Compound: Rhenium(IV) chloride
Formula: $ReCl_4$
Molecular Formula: Cl_4Re
Molecular Weight: 328.018
CAS RN: 13569-71-6
Properties: greenish black cryst; structure consists of zigzag chains of Fe_2Co_9; −80 mesh; sensitive to moisture; forms when a solution of $HReO_4$ and HCl is carefully evaporated [KIR82] [STR93] [CER91] [COT88]
Density, g/cm³: 4.9 [STR93]
Melting Point, °C: decomposes at 300 [LID94]

2619
Compound: Rhenium(IV) fluoride
Formula: ReF_4
Molecular Formula: F_4Re
Molecular Weight: 262.201
CAS RN: 15192-42-4
Properties: dark green; blue [COT88] [CRC10]
Density, g/cm³: 7.49 [LID94]
Melting Point, °C: sublimes at >300 [COT88]

2620
Compound: Rhenium(IV) oxide
Formula: ReO_2
Molecular Formula: O_2Re
Molecular Weight: 218.206
CAS RN: 12036-09-8
Properties: black powd; distorted rutile structure; anhydrous available as −100 mesh with 99.95% purity [CER91] [STR93] [COT88]
Density, g/cm³: 11.4 [CRC10]
Melting Point, °C: decomposes at 1000 [STR93]

2621
Compound: Rhenium(IV) selenide
Formula: $ReSe_2$
Molecular Formula: $ReSe_2$
Molecular Weight: 344.127
CAS RN: 12038-64-1
Properties: −100 mesh with 99.9% purity; layer structures with considerable Re–Re bonding [COT88] [CER91]

2622
Compound: Rhenium(IV) silicide
Formula: $ReSi_2$
Molecular Formula: $ReSi_2$
Molecular Weight: 242.378
CAS RN: 12038-66-3
Properties: −80 mesh powd of 99.9% purity [CER91] [ALF93]

2623
Compound: Rhenium(IV) sulfide
Formula: ReS_2
Molecular Formula: ReS_2
Molecular Weight: 250.339
CAS RN: 12038-63-0
Properties: tricl cryst; commonly nonstoichiometric; −80 mesh with 99.9% purity; preparation: by heating Re_2S_7 with sulfur in vacuum; uses: effective catalyst for hydrogenation of organic compounds, not poisoned by sulfur compounds, catalyzes reduction of NO to N_2O at 100°C [CER91] [COT88] [LID94]
Density, g/cm³: 7.6 [LID94]

2624
Compound: Rhenium(IV) telluride
Formula: $ReTe_2$
Molecular Formula: $ReTe_2$
Molecular Weight: 441.407
CAS RN: 12067-00-4
Properties: well-characterized compound, which does not have a layer structure; −60 mesh with 99.9% purity [COT88] [CER91]

2625
Compound: Rhenium(V) bromide
Synonym: rhenium pentabromide
Formula: $ReBr_5$
Molecular Formula: Br_5Re
Molecular Weight: 585.727

CAS RN: 30937-53-2
Properties: dark brown; −100 mesh with
99.9% purity [COT88] [CER91]
Melting Point, °C: decomposes at 110 [LID94]

2626
Compound: Rhenium(V) chloride
Formula: ReCl$_5$
Molecular Formula: Cl$_5$Re
Molecular Weight: 363.471
CAS RN: 39368-69-9
Properties: sensitive to moisture; formula is also given
as the dimer Re$_2$Cl$_{10}$; dark green to black solid;
decomposed by heating; formed by reacting Re
with Cl$_2$ at ~600°C, yielding a dark red-brown
liq; starting material for rhenium porphyrin
complexes [ALD94] [HAW93] [KIR82] [COT88]
Solubility: decomposes in H$_2$O; s HCl
and alkalies [HAW93]
Density, g/cm^3: 4.9 [HAW93]
Melting Point, °C: decomposes [AES93]
Reactions: rapidly hydrolyzed to ReO$_4^-$
by aq basic solution [COT88]

2627
Compound: Rhenium(VI) fluoride
Formula: ReF$_6$
Molecular Formula: F$_6$Re
Molecular Weight: 300.197
CAS RN: 10049-17-9
Properties: yellow cub cryst; extremely hygr; bluish
vapors given off in air; transition point −1.9
or −3.45°C; enthalpy of sublimation 32.657 kJ/
mol (above transition), 41.407 kJ/mol (below
transition); enthalpy of fusion 4.635 kJ/mol;
enthalpy of vaporization 28.75 kJ/mol; entropy of
fusion 60.46 J/(mol · K); entropy of vaporization
93.4 J/(mol · K); can be produced by reacting Re
powd with F$_2$ at ~100°C; used in the chemical
vapor deposition of Re [KIR78] [MER06]
Solubility: s HNO$_3$; s 1.75 mol/1000 g
anhydrous HF [MER06]
Density, g/cm^3: 3.58 [KIR78]
Melting Point, °C: 18.5 [MER06]
Boiling Point, °C: 48 [CER91]

2628
Compound: Rhenium(VI) oxide
Synonym: rhenium trioxide
Formula: ReO$_3$
Molecular Formula: O$_3$Re

Molecular Weight: 234.205
CAS RN: 1314-28-9
Properties: −100 mesh with 99.95% purity;
red cub cryst; green luster imparted by
transmitted light; disproportionates in vacuum
to Re$_2$O$_7$ and ReO$_2$ [MER06] [CER91]
Solubility: i H$_2$O, alkalies, nonoxidizing acids [MER06]
Density, g/cm^3: 6.9–7.4 [MER06]
Melting Point, °C: decomposes at 400 [STR93]
Reactions: oxidized to HReO$_4$ by HNO$_3$ [MER06]

2629
Compound: Rhenium(VI) oxytetrachloride
Formula: ReOCl$_4$
Molecular Formula: Cl$_4$ORe
Molecular Weight: 344.017
CAS RN: 13814-76-1
Properties: orange square pyramid; preparation:
by reaction between Re and SO$_2$Cl$_2$ at
350°C [COT88] [CRC10] [KIR82]
Melting Point, °C: 29.3 [CRC10]
Boiling Point, °C: 223 [CRC10]

2630
Compound: Rhenium(VI) trioxychloride
Formula: ReO$_3$Cl
Molecular Formula: ClO$_3$Re
Molecular Weight: 269.658
CAS RN: 7791-09-5
Properties: clear, colorless liq; can be prepared
by chlorination of Re$_2$O$_7$ [MER06]
Solubility: hydrolyzed to HReO$_4$ in
H$_2$O; s CCl$_4$ [MER06]
Density, g/cm^3: 3.867 [HAW93]
Melting Point, °C: 4.5 [MER06]
Boiling Point, °C: 128 [MER06]

2631
Compound: Rhenium(VII) oxide
Synonym: rhenium heptaoxide
Formula: Re$_2$O$_7$
Molecular Formula: O$_7$Re$_2$
Molecular Weight: 484.410
CAS RN: 1314-68-7
Properties: −6 mesh with 99.99% purity; canary
yellow; very deliq cryst; absorbs water readily,
forming perrhenic acid, HReO$_4$; enthalpy of
fusion 64.20 kJ/mol; structure has infinite array of
alternating ReO$_4$ tetrahedra and ReO$_6$ octahedra;
preparation: reaction with O$_2$ and heated Re
metal [COT88] [CRC10] [MER06] [CER91]

Solubility: v s H_2O, alcohol, ether, ethyl
acetate, dioxane, pyridine [MER06]
Density, g/cm³: 6.103 [STR93]
Melting Point, °C: 220 [COT88]
Boiling Point, °C: 360 [ALD94]
Reactions: sublimes at 250°C [MER06]

2632
Compound: Rhenium(VII) sulfide
Formula: Re_2S_7
Molecular Formula: Re_2S_7
Molecular Weight: 596.876
CAS RN: 12038-67-4
Properties: brownish black solid; preparation:
precipitation from a saturated solution of
2–6 M HCl solution of ReO_4^- with H_2S;
used as a catalyst [HAW93] [COT88]
Solubility: i H_2O; s alkali sulfide solutions [HAW93]
Density, g/cm³: 4.87 [HAW93]
Melting Point, °C: decomposes to ReS at 600 [HAW93]

2633
Compound: Rhodium
Formula: Rh
Molecular Formula: Rh
Molecular Weight: 102.90550
CAS RN: 7440-16-6
Properties: silvery white, ductile metal; fcc,
a = 0.3803 nm; electrical resistivity at 0°C
4.51 μohm·cm; Vicker's hardness 120; Poisson's
ratio 0.26; resistant to acids; reacts with aqua
regia if finely divided; enthalpy of fusion
26.59 kJ/mol; enthalpy of vaporization 494 kJ/
mol [ALD94] [MER06] [KIR82] [CRC10]
Solubility: i acids, aqua regia; s fused $KHSO_4$ [HAW93]
Density, g/cm³: 12.41 [MER06]
Melting Point, °C: 1966 [MER06]
Boiling Point, °C: 3695 [LID94]
Thermal Conductivity, W/(m·K):
150.6 (25°C) [KIR82]
Thermal Expansion Coefficient: 8.3×10^{-6}/°C [KIR82]

2634
Compound: Rhodium carbonyl
Synonym: hexarhodium hexadecacarbonyl
Formula: $Rh_6(CO)_{16}$
Molecular Formula: $C_{16}O_{16}Rh_6$
Molecular Weight: 1065.61
CAS RN: 28407-51-4
Properties: black cryst; cluster structure; used as
a catalyst [COT88] [ALD94] [ALF95]

2635
Compound: Rhodium carbonyl chloride
Formula: $[Rh(CO)_2Cl]_2$
Molecular Formula: $C_4Cl_2O_4Rh_2$
Molecular Weight: 388.758
CAS RN: 14523-22-9
Properties: reddish orange cryst, planar structure;
stable in dry air; solutions in organic solvents
decompose in air; preparation: by passing ethanol
saturated CO over $RhCl_3 \cdot 3H_2O$ at ~100°C, followed
by sublimation of needles of the carbonyl; uses:
homogeneous catalyst [ALD94] [MER06] [COT88]
Solubility: s in most organic solvents;
i aliphatic hydrocarbons [MER06]
Melting Point, °C: 124–125 [MER06]

2636
Compound: Rhodium dodecacarbonyl
Formula: $Rh_4(CO)_{12}$
Molecular Formula: $C_{12}O_{12}Rh_4$
Molecular Weight: 747.747
CAS RN: 19584-30-6
Properties: dark red cryst; sensitive to light, atm
oxygen, and moisture [STR93] [ALD94]
Density, g/cm³: 2.52 [STR93]
Melting Point, °C: 150 [DOU83]

2637
Compound: Rhodium(II) acetate dimer
Formula: $Rh_2(CH_3COO)_4$
Molecular Formula: $C_8H_{12}O_8Rh_2$
Molecular Weight: 441.989
CAS RN: 15956-28-2
Properties: greenish black cryst; preparation: by heating
sodium acetate and $RhCl_3 \cdot 3H_2O$ in methanol; uses:
homogeneous catalyst [ALD94] [STR93] [COT88]

2638
Compound: Rhodium(III) acetylacetonate
Synonyms: 2,4-pentanedione, rhodium(III) derivative
Formula: $Rh(CH_3COCH=C(O)CH_3)_3$
Molecular Formula: $C_{15}H_{21}O_6Rh$
Molecular Weight: 400.234
CAS RN: 14284-92-5
Properties: yellow cryst [STR93]
Melting Point, °C: 263–264 [ALD94]
Boiling Point, °C: decomposes at 280 [STR93]

2639
Compound: Rhodium(III) bromide dihydrate
Synonym: rhodium tribromide dihydrate

Formula: $RhBr_3 \cdot 2H_2O$
Molecular Formula: $Br_3H_4O_2Rh$
Molecular Weight: 378.649
CAS RN: 15608-29-4
Properties: hygr, brownish black cryst [STR93] [ALD94]

2640
Compound: Rhodium(III) chloride
Synonym: rhodium trichloride
Formula: $RhCl_3$
Molecular Formula: Cl_3Rh
Molecular Weight: 209.264
CAS RN: 10049-07-7
Properties: reddish brown, monocl powd [KIR82] [HAW93]
Solubility: i H_2O; s alkali hydroxide or cyanide solutions [MER06]
Density, g/cm³: 5.38 [KIR82]
Melting Point, °C: decomposes at 450 [STR93]
Boiling Point, °C: sublimes at 800 [HAW93]

2641
Compound: Rhodium(III) chloride hydrate
Synonym: rhodium trichloride hydrate
Formula: $RhCl_3 \cdot xH_2O$
Molecular Formula: Cl_3Rh (anhydrous)
Molecular Weight: 209.264 (anhydrous)
CAS RN: 20765-98-4
Properties: dark red powd; hygr [STR93]
Melting Point, °C: decomposes at 100 [ALD94]

2642
Compound: Rhodium(III) iodide
Synonym: rhodium triiodide
Formula: RhI_3
Molecular Formula: I_3Rh
Molecular Weight: 483.619
CAS RN: 15492-38-3
Properties: monocl cryst; black powd; hygr [LID94] [STR93]
Density, g/cm³: 6.4 [LID94]

2643
Compound: Rhodium(III) nitrate
Formula: $Rh(NO_3)_3$
Molecular Formula: N_3O_9Rh
Molecular Weight: 288.921
CAS RN: 10139-58-9
Properties: available as amber-colored 10% soln [STR93]
Density, g/cm³: soln: 1.410 [ALD94]

2644
Compound: Rhodium(III) nitrate dihydrate
Formula: $Rh(NO_3)_3 \cdot 2H_2O$
Molecular Formula: $H_4N_3O_{11}Rh$
Molecular Weight: 324.951
CAS RN: 13465-43-5
Properties: deliq red cryst [AES93] [CRC10]

2645
Compound: Rhodium(III) oxide
Synonym: rhodium trioxide
Formula: Rh_2O_3
Molecular Formula: O_3Rh_2
Molecular Weight: 253.809
CAS RN: 12036-35-0
Properties: hex cryst; gray powd [LID94] [STR93]
Density, g/cm³: 8.2 [STR93]
Melting Point, °C: decomposes at 1100 [STR93]

2646
Compound: Rhodium(III) oxide pentahydrate
Formula: $Rh_2O_3 \cdot 5H_2O$
Molecular Formula: $H_{10}O_8Rh_2$
Molecular Weight: 343.885
CAS RN: 39373-27-8
Properties: yellow powd [STR93]
Melting Point, °C: decomposes [STR93]

2647
Compound: Rhodium(III) sulfate
Formula: $Rh_2(SO_4)_3$
Molecular Formula: $O_{12}Rh_2S_3$
Molecular Weight: 494.002
CAS RN: 10489-46-0
Properties: reddish yellow solid; ~8% aq soln [ALD94] [KIR82]
Density, g/cm³: soln: 1.217 [ALD94]
Melting Point, °C: decomposes at >500 [KIR82]

2648
Compound: Rhodium(IV) oxide dihydrate
Formula: $RhO_2 \cdot 2H_2O$
Molecular Formula: H_4O_4Rh
Molecular Weight: 170.935
CAS RN: 12137-27-8
Properties: olive-green powd [CRC10] [AES93]
Density, g/cm³: 8.20 [CRC10]
Melting Point, °C: decomposes at 1100–1150 [CRC10]

2649
Compound: Rubidium
Formula: Rb
Molecular Formula: Rb
Molecular Weight: 85.4678
CAS RN: 7440-17-7
Properties: lustrous, silvery white, soft metal; bcc; tarnishes readily in air; evolves H_2 in water; ignites spontaneously in oxygen; reacts with halogens, mercury; ionic radius 0.148 nm; viscosity (39°C) 0.6713 mPa·s; surface tension (39°C) 75 mN/m; electrical resistivity 11.0 μohm·cm; enthalpy of fusion 2.19 kJ/mol; enthalpy of vaporization 75.77 kJ/mol; used in photocells, as a catalyst or catalyst promoter [HAW93] [KIR82] [MER06] [CRC10] [ALD94]
Solubility: s in acids and alcohol [HAW93]
Density, g/cm³: solid: 1.532 [MER06]; liq: 1.472 at 39°C [KIR82]
Melting Point, °C: 38.5 [CAB85]
Boiling Point, °C: 688 [MER06]
Thermal Conductivity, W/(m·K): 58.2 (25°C) [CRC10]; 29.3 for liq [KIR82]

2650
Compound: Rubidium acetate
Synonyms: acetic acid, Rb salt
Formula: CH_3COORb
Molecular Formula: $C_2H_3O_2Rb$
Molecular Weight: 144.513
CAS RN: 563-67-7
Properties: −4 mesh with 99.9% purity; white, hygr cryst [CER91] [STR93]
Solubility: 86 g/100 mL H_2O (45°C), 89.3 g/100 mL (99.4°C) [CRC10]
Melting Point, °C: 246 [STR93]

2651
Compound: Rubidium acetylacetonate
Synonyms: 2,4-pentanedione, rubidium derivative
Formula: $[CH_3COCH=C(O)CH_3]Rb$
Molecular Formula: $C_5H_7O_2Rb$
Molecular Weight: 184.578
CAS RN: 66169-93-5
Properties: white cryst [CRC10]
Melting Point, °C: decomposes at 200 [ALD94]

2652
Compound: Rubidium aluminum sulfate dodecahydrate
Formula: $RbAl(SO_4)_2 \cdot 12H_2O$
Molecular Formula: $AlH_{24}O_{20}RbS_2$

Molecular Weight: 520.760
CAS RN: 7488-54-2
Properties: colorless cryst [MER06] [HAW93]
Solubility: g/100 g H_2O: 0.72 (0°C), 1.50 (20°C), 21.6 (80°C) [LAN05]; i alcohol [MER06]
Density, g/cm³: 1.867 [HAW93]
Melting Point, °C: dodecahydrate, 99–109 [MER06]

2653
Compound: Rubidium azide
Formula: RbN_3
Molecular Formula: N_3Rb
Molecular Weight: 127.488
CAS RN: 22756-36-1
Properties: colorless needles; tetr, a=0.636 nm, c=0.741 nm [CRC10] [CIC73]
Solubility: 107 g/100 mL H_2O (16°C) [CRC10]
Density, g/cm³: 2.79 [CRC10]
Melting Point, °C: decomposes at ~310 [CRC10]

2654
Compound: Rubidium bromate
Formula: $RbBrO_3$
Molecular Formula: BrO_3Rb
Molecular Weight: 213.370
CAS RN: 13446-70-3
Properties: cub [CRC10]
Solubility: g/100 g H_2O: 2.93 (25°C) [KRU93]
Density, g/cm³: 3.68 [LAN05]
Melting Point, °C: 430 [LAN05]

2655
Compound: Rubidium bromide
Formula: RbBr
Molecular Formula: BrRb
Molecular Weight: 165.372
CAS RN: 7789-39-1
Properties: −4 mesh with 99.9% purity; white, cryst powd; enthalpy of fusion 15.50 kJ/mol [MER06] [CRC10] [CER91]
Solubility: g/100 g soln, H_2O: 47.26 (0.5°C), 53.69 (25°C), 67.24 (113.5°C) [KRU93]
Density, g/cm³: 3.35 [MER06]
Melting Point, °C: 682 [CRC10]
Boiling Point, °C: 1340 [MER06]
Thermal Expansion Coefficient: (volume) 100°C (0.925), 200°C (2.171) [CLA66]

2656
Compound: Rubidium carbonate
Formula: Rb_2CO_3

Molecular Formula: CO_3Rb_2
Molecular Weight: 230.945
CAS RN: 584-09-8
Properties: −20 mesh with 99.9% purity; white, monocl cryst; extremely hygr; dissociates above 900°C; used in special glass formulations [LID94] [HAW93] [CER91]
Solubility: 450 g/100 mL H_2O at 20°C [KIR82]
Melting Point, °C: 837 [STR93]
Boiling Point, °C: decomposes at 740 [KIR82]

2657

Compound: Rubidium chlorate
Formula: $RbClO_3$
Molecular Formula: ClO_3Rb
Molecular Weight: 168.919
CAS RN: 13446-71-4
Properties: trimetric [CRC10]
Solubility: g/100 g H_2O: 2.138 (0°C), 5.36 (19.8°C), 62.80 (99°C) [KRU93]
Density, g/cm³: 3.19 [LAN05]

2658

Compound: Rubidium chloride
Formula: RbCl
Molecular Formula: ClRb
Molecular Weight: 120.921
CAS RN: 7791-11-9
Properties: −4 mesh with 99.9% purity; white, cryst powd; hygr; enthalpy of fusion 18.40 kJ/mol; used in testing for perchloric acid and as a source of Rb metal [HAW93] [CRC10] [CER91]
Solubility: g/100 g soln H_2O: 43.5 (0°), 48.4 ± 0.21 (25°), 58.9 (100°); equilibrium solid phase RbCl at 25° [KRU93]; 7.7832 ± 0.0083 mol/(kg·H_2O) at 25°C [RAR85b]
Density, g/cm³: 2.76 [MER06]
Melting Point, °C: 715 [CRC10]
Boiling Point, °C: 1390 [MER06]
Thermal Expansion Coefficient: (volume) 100°C (0.891), 200°C (2.079) [CLA66]

2659

Compound: Rubidium chromate
Formula: Rb_2CrO_4
Molecular Formula: CrO_4Rb_2
Molecular Weight: 286.930
CAS RN: 13446-72-5
Properties: −20 mesh with 99.9% purity; yellow cryst [STR93] [CER91]

Solubility: g/100 g soln, H_2O: 38.27 (0°), 43.265 (25°C) [KRU93]; g/100 g H_2O: 62.0 (0°C), 73.6 (20°C), 95.7 (60°C) [LAN05]
Density, g/cm³: 3.518 [STR93]

2660

Compound: Rubidium cobalt(II) sulfate hexahydrate
Formula: $Rb_2Co(SO_4)_2 \cdot 6H_2O$
Molecular Formula: $CoH_{12}O_{14}Rb_2S_2$
Molecular Weight: 530.088
CAS RN: 28038-39-3
Properties: ruby-red, monocl [CRC10]
Solubility: g/100 g H_2O: 5.10 (0°C), 10.8 (20°C), 70.1 (100°C) [LAN05]
Density, g/cm³: 2.56 [CRC10]

2661

Compound: Rubidium cyanide
Formula: RbCN
Molecular Formula: CNRb
Molecular Weight: 111.486
CAS RN: 19073-56-4
Properties: cub; white or colorless [LID94] [KIR78]
Density, g/cm³: 2.32 [CRC10]

2662

Compound: Rubidium dichromate
Formula: $Rb_2Cr_2O_7$
Molecular Formula: $Cr_2O_7Rb_2$
Molecular Weight: 386.924
CAS RN: 13446-73-6
Properties: red tricl or yellow monocl [LAN05]
Solubility: monocl: g/100 g H_2O: 5.9 (20°C), 10.0 (30°C), 15.2 (40°C), 32.3 (60°C); tricl: 5.8 (20°C), 9.5 (30°C), 14.8 (40°C), 32.4 (60°C) [LAN05]
Density, g/cm³: monocl: 3.021; tricl: 3.125 [CRC10] [LAN05]

2663

Compound: Rubidium fluoride
Formula: RbF
Molecular Formula: FRb
Molecular Weight: 104.466
CAS RN: 13446-74-7
Properties: −4 mesh with 99.9% purity; white cub; hygr; enthalpy of fusion 17.30 kJ/mol [STR93] [CRC10] [CER91] [LID94]
Solubility: g/100 g soln, H_2O: 75.06 (18°C); solid phase, RbF·1-1/2H_2O [KRU93]; i alcohol [HAW93]
Density, g/cm³: 3.557 [HAW93]
Melting Point, °C: 833 [CRC10]
Boiling Point, °C: 1410 [STR93]

2664
Compound: Rubidium fluoroborate
Synonym: rubidium borofluoride
Formula: RbBF$_4$
Molecular Formula: BF$_4$Rb
Molecular Weight: 172.273
CAS RN: 18909-68-7
Properties: ortho-rhomb below 245°C,
    a = 0.7296 nm, b = 0.9108 nm, c = 0.5636 nm;
    cub above 245°C [KIR78]
Solubility: 0.6 g/100 mL H$_2$O (17°C) [KIR78]
Density, g/cm^3: 2.820 [KIR78]
Melting Point, °C: decomposes at 612 [KIR78]

2665
Compound: Rubidium fullerene
Formula: Rb$_3$C$_{60}$
Molecular Formula: C$_{60}$Rb$_3$ ·
Molecular Weight: 977.063
CAS RN: 137926-73-9
Properties: fcc, lattice constant 1.4436 nm; bulk
    modulus 22 GPa; density of states 35 states/(eV/C$_{60}$);
    superconducting temp 29.4 K [DRE93]

2666
Compound: Rubidium hexafluorogermanate
Formula: Rb$_2$GeF$_6$
Molecular Formula: F$_6$GeRb$_2$
Molecular Weight: 357.536
CAS RN: 16962-48-4
Properties: white, cryst solid [HAW93]
Solubility: sl s cold H$_2$O, v s hot H$_2$O [HAW93]
Melting Point, °C: 696 [HAW93]

2667
Compound: Rubidium hydroxide
Formula: RbOH
Molecular Formula: HORb
Molecular Weight: 102.475
CAS RN: 1310-82-3
Properties: grayish white; deliq; stronger base than
    KOH; absorbs atm CO$_2$; possible use in low
    temp storage batteries; there is a hydrate form
    [HAW93] [MER06] [STR93] [CER91]
Solubility: g/100 g soln, H$_2$O: 63.39 (30°C)
    [KRU93]; s alcohol [MER06]
Density, g/cm^3: 3.203 [MER06]
Melting Point, °C: 300 [HAW93]

2668
Compound: Rubidium iodide
Formula: RbI
Molecular Formula: IRb
Molecular Weight: 212.372
CAS RN: 7790-29-6
Properties: −4 mesh with 99.9% purity; white
    cryst or cryst powd; discolors if exposed to
    light, air; enthalpy of fusion 12.50 kJ/mol
    [CRC10] [MER06] [CER91]
Solubility: g/100 g soln, H$_2$O: 55.50 (0°C),
    61.99 (25°C), 73.01 (93°C); solid phase,
    RbI [KRU93]; s alcohol [MER06]
Density, g/cm^3: 3.55 [MER06]
Melting Point, °C: 642 [MER06]
Boiling Point, °C: 1300 [MER06]

2669
Compound: Rubidium iron(III) sulfate dodecahydrate
Formula: RbFe(SO$_4$)$_2$ · 12H$_2$O
Molecular Formula: FeH$_{24}$O$_{20}$RbS$_2$
Molecular Weight: 549.624
CAS RN: 30622-97-0
Properties: cub [CRC10]
Solubility: g/100 g H$_2$O: 8.0 (10°C),
    20 (20°C), 52 (40°C) [LAN05]
Density, g/cm^3: 1.91–1.95 [CRC10]
Melting Point, °C: 45–53 [CRC10]

2670
Compound: Rubidium metavanadate
Formula: RbVO$_3$
Molecular Formula: O$_3$RbV
Molecular Weight: 184.408
CAS RN: 13597-45-0
Properties: −100 mesh with 99.9% purity [CER91]

2671
Compound: Rubidium molybdate
Formula: Rb$_2$MoO$_4$
Molecular Formula: MoO$_4$Rb$_2$
Molecular Weight: 330.874
CAS RN: 13718-22-4
Properties: −200 mesh with 99.9% purity;
    white [KIR81] [CER91]
Solubility: g/100 g soln, H$_2$O: 67.88 (18°C) [KRU93]
Melting Point, °C: 958, 919 [KIR81]

2672
Compound: Rubidium niobate
Formula: RbNbO₃
Molecular Formula: NbO₃Rb
Molecular Weight: 226.372
CAS RN: 12059-51-7
Properties: −200 mesh with 99.9% purity [CER91]

2673
Compound: Rubidium nitrate
Formula: RbNO₃
Molecular Formula: NO₃Rb
Molecular Weight: 147.473
CAS RN: 13126-12-0
Properties: −80 mesh with 99.9% purity; white cryst; hygr; enthalpy of fusion 5.60 kJ/mol [STR93] [CRC10] [CER91]
Solubility: g/100 g H₂O: 19.5 (0°C), 67.3 (25°C), 452 (100°C) [KRU93]
Density, g/cm³: 3.11 [STR93]
Melting Point, °C: 305 [CRC10]

2674
Compound: Rubidium orthovanadate
Formula: Rb₃VO₄
Molecular Formula: O₄Rb₃V
Molecular Weight: 371.343
CAS RN: 13566-05-7
Properties: −100 mesh with 99.9% purity [CER91]

2675
Compound: Rubidium oxide
Formula: Rb₂O
Molecular Formula: ORb₂
Molecular Weight: 186.935
CAS RN: 18088-11-4
Properties: cub; yellowish brown; sensitive to atm oxygen and moisture [STR93] [CRC10]
Density, g/cm³: 4.0 [LID94]
Melting Point, °C: decomposes at 400 [STR93]

2676
Compound: Rubidium perchlorate
Formula: RbClO₄
Molecular Formula: ClO₄Rb
Molecular Weight: 184.919
CAS RN: 13510-42-4
Properties: −4 mesh with 99.9% purity; white, rhomb cryst; hygr [STR93] [CRC10] [CER91]

Solubility: g/100 g H₂O: 1.1 (0°C), 1.8 (25°C), 22 (100°C) [KRU93]
Density, g/cm³: 2.80 [STR93]
Melting Point, °C: 281 [KIR79]
Boiling Point, °C: decomposes at 606 [KIR79]
Reactions: transition from ortho-rhomb to cub at 551–534 K [KIR79]

2677
Compound: Rubidium permanganate
Formula: RbMnO₄
Molecular Formula: MnO₄Rb
Molecular Weight: 204.404
CAS RN: 13465-49-1
Properties: cryst [LAN05]
Solubility: g/100 mL soln, H₂O: 1.06 (19°C) [KRU93]
Density, g/cm³: 3.325 [LAN05]
Melting Point, °C: decomposes at 295 [LAN05]

2678
Compound: Rubidium pyrovanadate
Formula: Rb₄V₂O₇
Molecular Formula: O₇Rb₄V₂
Molecular Weight: 555.750
CAS RN: 13597-61-0
Properties: −100 mesh with 99.9% purity [CER91]

2679
Compound: Rubidium selenide
Formula: Rb₂Se
Molecular Formula: Rb₂Se
Molecular Weight: 249.896
CAS RN: 31052-43-4
Properties: white cub cryst; −60 mesh with 99.5% purity [LID94] [CER91]
Density, g/cm³: 3.22 [LID94]
Melting Point, °C: 733 [LID94]

2680
Compound: Rubidium sulfate
Formula: Rb₂SO₄
Molecular Formula: O₄Rb₂S
Molecular Weight: 267.000
CAS RN: 7488-54-2
Properties: −20 mesh with 99.9% purity; white cryst [STR93] [CER91]
Solubility: g/100 g H₂O: 36.4 (0°C), 50.8 (25°C), 81.8 (100°C) [KRU93]
Density, g/cm³: 3.613 [STR93]
Melting Point, °C: 1050 [STR93]
Boiling Point, °C: ~1700 [STR93]

2793
Compound: Silver diethyldithiocarbamate
Synonyms: diethyldithiocarbamic acid, silver(I) salt
Formula: $(C_2H_5)_2NCS_2Ag$
Molecular Formula: $C_5H_{10}AgNS_2$
Molecular Weight: 256.141
CAS RN: 1470-61-7
Properties: light-sensitive; used as a reagent to detect arsenic [ALD94]
Melting Point, °C: 172–175 [ALD94]

2794
Compound: Silver difluoride
Synonym: Ag(II) fluoride
Formula: AgF_2
Molecular Formula: AgF_2
Molecular Weight: 145.865
CAS RN: 7783-95-1
Properties: white, when pure; usually grayish black or brownish solid; light sensitive; very hygr; reacts violently with H_2O; very strong oxidizing agent, e.g. can evolve ozone from dil acids and oxidizes iodides to iodine; obtained by reaction of F_2 on AgCl at 200°C; fluorinating agent [KIR83] [MER06]
Solubility: decomposed by H_2O [CRC10]
Density, g/cm³: 4.58 [STR93]
Melting Point, °C: 700, decomposes [STR93]

2795
Compound: Silver fluoride
Formula: AgF
Molecular Formula: AgF
Molecular Weight: 126.866
CAS RN: 7775-41-9
Properties: yellow or brownish cryst; cub; very hygr; darkens if exposed to light; forms basic fluoride in moist air; can form several hydrates; used as an antiseptic [HAW93] [MER06]
Solubility: mol/kg H_2O: 6.76 (0°C), 13.97 (25°C), 16.15 (108°C); solid phase, $AgF \cdot 4H_2O$ (0°C), $AgF \cdot 2H_2O$ (25°C), AgF (108°C) [KRU93]; s HF, NH_3, CH_3CN [MER06]
Density, g/cm³: 5.852 [MER06]
Melting Point, °C: 435 [MER06]
Boiling Point, °C: 1159 [HAW93]

2796
Compound: Silver hexafluoroantimonate(V)
Formula: $AgSbF_6$
Molecular Formula: AgF_6Sb
Molecular Weight: 343.616
CAS RN: 26042-64-8
Properties: −6 mesh with 99.9% purity; white powd; hygr; used as acidic catalyst in epoxide reactions [STR93] [ALD94] [CER91]

2797
Compound: Silver hexafluoroarsenate
Formula: $AgAsF_6$
Molecular Formula: $AgAsF_6$
Molecular Weight: 296.780
CAS RN: 12005-82-2
Properties: hygr powd [STR93]
Melting Point, °C: decomposes [STR93]

2798
Compound: Silver hexafluorophosphate
Formula: $AgPF_6$
Molecular Formula: AgF_6P
Molecular Weight: 252.832
CAS RN: 26042-63-7
Properties: −6 mesh with 99.9% purity; white cryst; hygr; sensitive to light; uses: acidic catalyst to synthesize some sulfides and vinyl fluorides [ALD94] [STR93] [CER91]

2799
Compound: Silver hydrogen fluoride
Formula: $AgHF_2$
Molecular Formula: AgF_2H
Molecular Weight: 146.873
CAS RN: 12249-52-4
Properties: light sensitive; hygr [STR93]
Melting Point, °C: decomposes [STR93]

2800
Compound: Silver iodate
Formula: $AgIO_3$
Molecular Formula: $AgIO_3$
Molecular Weight: 282.770
CAS RN: 7783-97-3
Properties: white, cryst powd; light sensitive [MER06]
Solubility: g/L soln, H_2O: 0.0505 (25°C) [KRU93]; s ~1000 parts 35% HNO_3 (25°C), 2.5 parts 10% ammonia [MER06]
Density, g/cm³: 5.53 [MER06]
Melting Point, °C: >200 [MER06]

2801
Compound: Silver iodide
Synonym: iodyrite
Formula: AgI
Molecular Formula: AgI
Molecular Weight: 234.772
CAS RN: 7783-96-2
Properties: −20 mesh with 99.999% purity;
    light yellow powd; darkens when exposed
    to light; hex or cub cryst; enthalpy of
    vaporization 143.9 kJ/mol; enthalpy of fusion
    9.41 kJ/mol [MER06] [CER91] [CRC10]
Solubility: 0.03 mg/L H_2O; i acids, except HI [MER06];
    s KI, KCN, NH_4OH, NaCl solutions [HAW93]
Density, g/cm³: 5.68 [LID94]
Melting Point, °C: 558 [CRC10]
Boiling Point, °C: 1506 [CRC10]
Thermal Expansion Coefficient: -2.5×10^{-6}/K [CRC10]

2802
Compound: Silver lactate monohydrate
Formula: $AgC_3H_5O_3 \cdot H_2O$
Molecular Formula: $C_3H_7AgO_4$
Molecular Weight: 214.955
CAS RN: 128-00-7
Properties: white or sl gray cryst powd;
    sensitive to light [MER06]
Solubility: s in ~15 parts H_2O; sl s alcohol [MER06]
Melting Point, °C: 212 [CRC10]

2803
Compound: Silver molybdate
Formula: Ag_2MoO_4
Molecular Formula: Ag_2MoO_4
Molecular Weight: 375.674
CAS RN: 13765-74-7
Properties: white, pale yellow (if fused) [KIR81]
Solubility: 3.86 mg/100 g soln in H_2O (25°C) [KRU93]
Density, g/cm³: 6.18 [LID94]
Melting Point, °C: 483 [KIR81]

2804
Compound: Silver nitrate
Formula: $AgNO_3$
Molecular Formula: $AgNO_3$
Molecular Weight: 169.873
CAS RN: 7761-88-8
Properties: −10 mesh with 99.999% purity;
    colorless, transparent rhomb cryst; pure
    material not light sensitive; enthalpy of fusion
    11.50 kJ/mol [CRC10] [MER06] [CER91]

Solubility: g/100 g soln, H_2O: 54.8 (0°C), 70.7 (25°C),
    88.0 (100°C) [KRU93]; 1 g dissolves in: 30 mL
    alcohol, 6.5 mL boiling alcohol; 253 mL acetone;
    v s ammonia solution; sl s ether [MER06]
Density, g/cm³: 4.352 [STR93]
Melting Point, °C: 212, forming yellowish liq [MER06]
Boiling Point, °C: decomposes, 440, into
    Ag, nitrogen, oxygen [MER06]

2805
Compound: Silver nitrite
Formula: $AgNO_2$
Molecular Formula: $AgNO_2$
Molecular Weight: 153.874
CAS RN: 7783-99-5
Properties: pale yellow needles; light sensitive, turning
    gray; used as a reagent for alcohols, as a standard
    solution for water analysis [HAW93] [MER06]
Solubility: g/100 g soln, H_2O: 0.155 (0°C),
    0.4135 (25°C) [KRU93]; i alcohol;
    decomposed by dil acids [MER06]
Density, g/cm³: 4.453 [STR93]
Melting Point, °C: decomposes, 140 [MER06]

2806
Compound: Silver oxalate
Formula: $Ag_2C_2O_4$
Molecular Formula: $C_2Ag_2O_4$
Molecular Weight: 303.756
CAS RN: 533-51-7
Properties: white, cryst powd [MER06]
Solubility: g/L soln, H_2O: 0.041 (25°C) [KRU93]; s
    moderately conc HNO_3, ammonia [MER06]
Density, g/cm³: 5.03 [MER06]
Melting Point, °C: decomposes [STR93]
Reactions: can explode [CRC10]

2807
Compound: Silver oxide
Formula: Ag_2O
Molecular Formula: Ag_2O
Molecular Weight: 231.735
CAS RN: 20667-12-3
Properties: brownish black powd; reduced by
    hydrogen, carbon monoxide, and most metals;
    light sensitive; used to polish glass, to purify
    drinking water, as a catalyst [HAW93] [MER06]
Solubility: 0.0013 g/100 mL H_2O (20°C),
    0.0053 g/100 mL H_2O (80°C) [CRC10]; v s
    dil HNO_3, ammonia; i alcohol [MER06]
Density, g/cm³: 7.2 [LID94]

Melting Point, °C: 300, decomposes [STR93]
Reactions: begins to decompose at ~200°C, rapidly at 250°C–300°C [MER06]

2808

Compound: Silver perchlorate
Formula: $AgClO_4$
Molecular Formula: $AgClO_4$
Molecular Weight: 207.319
CAS RN: 7783-93-9
Properties: colorless deliq cryst; used in the manufacture of explosives [MER06] [HAW93]
Solubility: g/100 g soln, H_2O: 81.7 ± 0.4 (0°C), 84.6 ± 0.1 (25°C), 88.8 (99°C); solid phase, $AgClO_4 \cdot H_2O$ (0°C, 25°C), $AgClO_4$ (99°C) [KRU93]; s in many organic solvents [MER06]
Density, g/cm³: 2.806 [MER06]
Melting Point, °C: 486, decomposes [MER06]

2809

Compound: Silver perchlorate monohydrate
Formula: $AgClO_4 \cdot H_2O$
Molecular Formula: $AgClH_2O_5$
Molecular Weight: 225.334
CAS RN: 14242-05-8
Properties: white cryst; hygr; stable up to 43°C [MER06] [STR93]
Solubility: 84.5 g/100 g saturated soln of H_2O (25°C) [MER06]

2810

Compound: Silver permanganate
Formula: $AgMnO_4$
Molecular Formula: $AgMnO_4$
Molecular Weight: 226.804
CAS RN: 7783-98-4
Properties: violet; light sensitive; cryst powd; used in gas masks and as an antiseptic [HAW93] [MER06]
Solubility: ~9 g/L H_2O, room temp, more s in hot H_2O [MER06]
Density, g/cm³: 4.49 [MER06]
Melting Point, °C: decomposes [HAW93]

2811

Compound: Silver peroxide
Synonym: silver(II) oxide
Formula: Ag_2O_2
Molecular Formula: Ag_2O_2
Molecular Weight: 247.735
CAS RN: 1301-96-8

Properties: charcoal gray powd; oxidant; malleable; ortho-rhomb or cub cryst; can be obtained by persulfate oxidation of Ag_2O in alkaline medium at 90°C; strongly oxidizing; used in the manufacture of silver-zinc batteries [HAW93] [MER06] [KIR83]
Solubility: 27 mg/L H_2O, decomposes (25°C); s alkalies, NH_4OH (evolving N_2); s dil acids, evolving O_2 [MER06]
Density, g/cm³: 7.483 [MER06]
Melting Point, °C: decomposes >100 to Ag and O_2 [MER06]

2812

Compound: Silver perrhenate
Formula: AgO_4Re
Molecular Formula: AgO_4Re
Molecular Weight: 358.07
CAS RN: 20654-56-2
Properties: white solid [STR08]
Density, g/cm³: 7.05 (25°C) [STR08]
Melting Point, °C: 430 [STR08]

2813

Compound: Silver phosphate
Synonyms: silver phosphate, tribasic
Formula: Ag_3PO_4
Molecular Formula: Ag_3O_4P
Molecular Weight: 418.574
CAS RN: 7784-09-0
Properties: yellow powd; darkened by light; reduced by hydrogen; used in photographic emulsions, in pharmaceuticals [HAW93] [MER06]
Solubility: s 15,550 parts H_2O; sl s dil acids; v s dil HNO_3, ammonia [MER06]
Density, g/cm³: 6.37 [MER06]
Melting Point, °C: 849 [MER06]

2814

Compound: Silver picrate monohydrate
Formula: $AgOC_6H_2(NO_2)_3 \cdot H_2O$
Molecular Formula: $C_6H_4AgN_3O_8$
Molecular Weight: 353.981
CAS RN: 146-84-9
Properties: yellow cryst; used as an antimicrobial agent [HAW93] [MER06]
Solubility: s ~50 parts H_2O; sl s alcohol; i $CHCl_3$, ether [MER06]
Reactions: can explode [HAW93]

2815

Compound: Silver selenate
Formula: Ag_2SeO_4

Molecular Formula: Ag_2O_4Se
Molecular Weight: 358.692
CAS RN: 7784-07-8
Properties: ortho-rhomb cryst; prepared from silver carbonate and sodium selenate [KIR83] [MER06]
Solubility: g/1000 g H_2O: 0.870 (25°C), 0.053 (100°C) [KRU93]
Density, g/cm³: 5.72 [MER06]

2816
Compound: Silver selenide
Formula: Ag_2Se
Molecular Formula: Ag_2Se
Molecular Weight: 294.696
CAS RN: 1302-09-6
Properties: gray hex microscopic needles; exists in two forms, transition temp 133°C; oxidized to Ag and selenium oxide when heated in O_2 [MER06]
Solubility: i H_2O [MER06]
Density, g/cm³: 8.216 [MER06]
Melting Point, °C: 880 [MER06]

2817
Compound: Silver selenite
Formula: Ag_2SeO_3
Molecular Formula: Ag_2O_3Se
Molecular Weight: 342.694
CAS RN: 7784-05-6
Properties: needles; decomposes >530°C to Ag, selenium oxide, oxygen [MER06]
Solubility: sl s cold H_2O, v s hot H_2O; s HNO_3 [MER06]
Density, g/cm³: 5.9297 [MER06]
Melting Point, °C: 530, decomposes [MER06]

2818
Compound: Silver subfluoride
Formula: Ag_2F
Molecular Formula: Ag_2F
Molecular Weight: 234.734
CAS RN: 1302-01-8
Properties: hex; bronze colored cryst with green luster; becomes grayish black as a result of prolonged exposure to air; good electrical conductor; quickly hydrolyzes in H_2O precipitating Ag powd; can be prepared by heating a conc solution of AgF with Ag powd [KIR78] [MER06]
Solubility: decomposes in H_2O [KIR78]
Density, g/cm³: 8.57 [MER06]
Melting Point, °C: decomposes to Ag, AgF 100–200 [MER06]

2819
Compound: Silver sulfate
Formula: Ag_2SO_4
Molecular Formula: Ag_2O_4S
Molecular Weight: 311.798
CAS RN: 10294-26-5
Properties: small colorless cryst or cryst powd; light sensitive, slowly darkening; can be made by reaction of metallic Ag and hot H_2SO_4 [KIR84] [MER06]
Solubility: g/100 g soln, H_2O: 0.56 ± 0.01 (0°C), 0.834 (25°C), 1.39 (100°C) [KRU93]; s HNO_3, ammonia, conc H_2SO_4 [MER06]
Density, g/cm³: 5.45 [MER06]
Melting Point, °C: 657 [MER06]
Boiling Point, °C: decomposes, 1085 [MER06]

2820
Compound: Silver sulfide
Synonyms: acanthite, argentite
Formula: Ag_2S
Molecular Formula: Ag_2S
Molecular Weight: 247.802
CAS RN: 21548-73-2
Properties: −100 mesh with 99.9% purity; grayish black powd; orthro-rhomb, changes to cub >179°C; enthalpy of fusion 14.10 kJ/mol; used in ceramics [HAW93] [MER06] [CER91]
Solubility: i H_2O; s conc H_2SO_4, HNO_3 [HAW93]
Density, g/cm³: 7.32 [HAW93]
Melting Point, °C: 825 [HAW93]
Boiling Point, °C: decomposes [HAW93]

2821
Compound: Silver telluride
Synonym: hessite
Formula: Ag_2Te
Molecular Formula: Ag_2Te
Molecular Weight: 343.336
CAS RN: 12002-99-2
Properties: black cryst [STR93]
Density, g/cm³: 8.5 [STR93]
Melting Point, °C: 955 [STR93]

2822
Compound: Silver tetraiodomercurate(II) (α-form)
Synonym: mercury(II) silver iodide
Formula: Ag_2HgI_4
Molecular Formula: Ag_2HgI_4
Molecular Weight: 923.942

CAS RN: 7784-03-4
Properties: deep yellow; thermochromic powd; becomes blood red at 40°C–50°C, β-form; can be prepared by precipitation from a solution of $AgNO_3$ and K_2HgI_4 [KIR81] [CRC10] [MER06]
Solubility: i H_2O, dil acids; s in solutions of alkali iodides or cyanides [MER06]
Density, g/cm³: 6.02 [CRC10]
Melting Point, °C: β: decomposes 158 [CRC10]
Reactions: α to β transition, 50.7°C [CRC10]

2823
Compound: Silver thiocyanate
Formula: AgSCN
Molecular Formula: CAgNS
Molecular Weight: 165.952
CAS RN: 1701-93-5
Properties: white powd [STR93]
Solubility: g/L soln, H_2O: 0.00018 ± 0.00002 (25°C), 0.0064 (100°C) [KRU93]
Melting Point, °C: decomposes [STR93]

2824
Compound: Silver trifluoroacetate
Formula: AgO_2CCF_3
Molecular Formula: $C_2AgF_3O_2$
Molecular Weight: 220.88
CAS RN: 2966-50-9
Properties: white to off-white solid [STR08]

2825
Compound: Silver tungstate
Formula: Ag_2WO_4
Molecular Formula: Ag_2O_4W
Molecular Weight: 463.574
CAS RN: 13465-93-5
Properties: white powd; formula also written as $Ag_8W_4O_{16}$ [ALD94] [STR93]
Solubility: g/L soln, H_2O: 0.235 (25°C) [KRU93]

2826
Compound: Sodium
Synonym: natrium
Formula: Na
Molecular Formula: Na
Molecular Weight: 22.989768
CAS RN: 7440-23-5

Properties: soft silvery white metal; bcc, a = 0.4282 nm; lustrous, but tarnishes in air; enthalpy of fusion 2.60 kJ/mol; enthalpy of vaporization 97.4 kJ/mol; electrical resistivity (20°C) 4.69 μohm·cm; viscosity at 100°C is 0.680 mPa·s; surface tension at 400°C 161 mN/m; decomposes alcohol; burns with a yellow flame; reduces most oxides to elements [MER06] [KIR82] [ALD94]
Solubility: reacts violently with H_2O evolving H_2 and forming NaOH soln; s mercury, liq NH_3 [KIR82] [MER06]
Density, g/cm³: 0.968 at 20°C [MER06]
Melting Point, °C: 97.82 [MER06]
Boiling Point, °C: 881 [MER06]
Thermal Conductivity, W/(m·K): 142 at 25°C [ALD94]
Thermal Expansion Coefficient: 71×10^{-6}/K [CRC10]

2827
Compound: Sodium β-aluminum oxide
Formula: $β-Na_2O \cdot 11Al_2O_3$
Molecular Formula: $Al_{22}Na_2O_{34}$
Molecular Weight: 1183.554
CAS RN: 11138-49-1
Properties: hex, a = 0.558 nm, c = 2.245 nm [KIR78]
Density, g/cm³: 3.24 [KIR78]

2828
Compound: Sodium acetate
Synonyms: acetic acid, sodium salt
Formula: CH_3COONa
Molecular Formula: $C_2H_3NaO_2$
Molecular Weight: 82.035
CAS RN: 127-09-3
Properties: white powd; odorless; efflorescent [HAW93]
Solubility: g/100 g H_2O: 36.3 (0°C), 50.5 (25°C), 170 (100°C); solid phase, $CH_3COONa \cdot 3H_2O$ (0°C, 25°C) [KRU93]; s alcohol [MER06]
Density, g/cm³: 1.528 [STR93]
Melting Point, °C: 324 [STR93]

2829
Compound: Sodium acetate trihydrate
Synonyms: acetic acid, sodium salt trihydrate
Formula: $NaCH_3COO \cdot 3H_2O$
Molecular Formula: $C_2H_9NaO_5$
Molecular Weight: 136.080
CAS RN: 6131-90-4
Properties: transparent cryst or granules; efflorescent in warm air [MER06]

Solubility: 1 g/0.8 mL H_2O, 1 g/0.6 mL boiling
  H_2O, 1 g/19 mL alcohol [MER06]
Density, g/cm³: 1.45 [MER06]
Melting Point, °C: 58 [MER06]
Reactions: minus $3H_2O$ 120°C; decomposes
  at higher temp [MER06]

2830

Compound: Sodium acetylacetonate
Synonyms: 2,4-pentanedione, sodium derivative
Formula: $NaCH_3COCH=C(O)CH_3$
Molecular Formula: $C_5H_7NaO_2$
Molecular Weight: 122.100
CAS RN: 15435-71-9
Properties: off-white powd [STR93]
Melting Point, °C: 210 decomposes [STR93]

2831

Compound: Sodium acetylide
Synonym: sodium carbide
Formula: $NaC{\equiv}CH$
Molecular Formula: C_2HNa
Molecular Weight: 48.020
CAS RN: 1066-26-8
Properties: 18% suspension in xylene with gray
  color; sensitive to atm oxygen and moisture;
  freezing point of suspension is 29°C [STR93]

2832

Compound: Sodium aluminate
Formula: $NaAlO_2$
Molecular Formula: $AlNaO_2$
Molecular Weight: 81.971
CAS RN: 1302-42-7
Properties: white powd; hygr; can be prepared by
  fusing sodium carbonate and aluminum acetate
  in stoichiometric amounts at 800°C; used as a
  mordant, in water purification, in sizing paper,
  in cleaning compounds [HAW93] [KIR78]
Solubility: v s H_2O; i alcohol [MER06]
Density, g/cm³: 4.63 [LID94]
Melting Point, °C: 1650 [MER06]

2833

Compound: Sodium aluminum sulfate dodecahydrate
Synonym: sodium alum
Formula: $AlNa(SO_4)_2 \cdot 12H_2O$
Molecular Formula: $AlH_{24}NaO_{20}S_2$
Molecular Weight: 458.282
CAS RN: 10102-71-3
Properties: colorless cryst [MER06]

Solubility: g anhydrous/100 g H_2O: 37.4 (0°C), 39.7
  (20°C), 43.8 (40°C) [LAN05]; i alcohol [MER06]
Density, g/cm³: 1.61 [MER06]
Melting Point, °C: ~60 [MER06]

2834

Compound: Sodium amide
Synonym: sodamide
Formula: $NaNH_2$
Molecular Formula: H_2NNa
Molecular Weight: 39.013
CAS RN: 7782-92-5
Properties: −40 mesh protected from atm under
  hexane, with 96% purity; commercial
  product white to olive green; ortho-
  rhomb cryst; enthalpy of formation is
  −118.8 kJ/mol [CIC73] [CER91] [MER06]
Solubility: reacts violently with H_2O, forming NaOH
  and NH_3 [MER06]; 0.17 g/100 g liq NH_3 [CIC73]
Density, g/cm³: 1.39 [CIC73]
Melting Point, °C: 208 [CIC73]
Boiling Point, °C: 400 [HAW93]

2835

Compound: Sodium ammonium hydrogen
  phosphate tetrahydrate
Synonym: microcosmic salt
Formula: $NaNH_4HPO_4 \cdot 4H_2O$
Molecular Formula: $H_{13}NNaO_8P$
Molecular Weight: 209.069
CAS RN: 51750-73-3
Properties: odorless; monocl; efflorescent
  in air, evolving NH_3 [MER06]
Solubility: s about 5 parts cold, 1 part boiling
  H_2O [MER06]; i alcohol [HAW93]
Density, g/cm³: 1.544 [MER06] [ALD94]
Melting Point, °C: ~80, when rapidly heated [MER06]
Reactions: prolonged heating produces
  $NaPO_3$ [MER06]

2836

Compound: Sodium antimonate monohydrate
Synonym: sodium pyroantimonate monohydrate
Formula: $Na_2O \cdot Sb_2O_5 \cdot H_2O$
Molecular Formula: $H_2Na_2O_7Sb_2$
Molecular Weight: 403.512
CAS RN: 33908-66-6
Properties: white, granular powd; formula
  also given as $NaSb(OH)_6$, −200 mesh with
  99.9% purity [MER06] [CER91]
Solubility: sl s H_2O [MER06]

2837

Compound: Sodium arsenate dodecahydrate
Formula: $Na_3AsO_4 \cdot 12H_2O$
Molecular Formula: $AsH_{24}Na_3O_{16}$
Molecular Weight: 424.072
CAS RN: 7778-43-0
Properties: colorless cryst [HAW93]
Solubility: 38.9 g/100 mL H_2O (15.5°C) [CRC10]; sl
s in alcohol and glycerol; i ether [HAW93]
Density, g/cm³: 1.7539 [HAW93]
Melting Point, °C: 86 [HAW93]

2838

Compound: Sodium arsenite
Synonym: sodium metaarsenite
Formula: $NaAsO_2$
Molecular Formula: AsO_2Na
Molecular Weight: 129.911
CAS RN: 7784-46-5
Properties: white or grayish white powd;
somewhat hygr; absorbs atm CO_2; used in
soaps for taxidermists, in insecticides, as a
hide preservative [HAW93] [MER06]
Solubility: v s H_2O; sl s alcohol [MER06]
Density, g/cm³: 1.87 [HAW93]

2839

Compound: Sodium azide
Synonym: smite
Formula: NaN_3
Molecular Formula: N_3Na
Molecular Weight: 65.010
CAS RN: 26628-22-8
Properties: white powd; hex cryst; β-NaN_3: body-center
rhomb, with a=0.5488 nm; decomposes when
heated into Na, N_2 [CIC73] [MER06] [STR93]
Solubility: g/100 g H_2O: 38.9 (0°C), 40.8
(20°C), 55.3 (100°C) [LAN05]
Density, g/cm³: 1.846 [MER06]
Melting Point, °C: 300, decomposes [STR93]

2840

Compound: Sodium borodeuteride
Formula: $NaBD_4$
Molecular Formula: BD_4Na
Molecular Weight: 41.861
CAS RN: 15681-89-7
Properties: white powd; sensitive to moisture [STR93]
Density, g/cm³: 1.074 [STR93]
Melting Point, °C: ~400 [STR93]

2841

Compound: Sodium borohydride
Formula: $NaBH_4$
Molecular Formula: BH_4Na
Molecular Weight: 37.833
CAS RN: 16940-66-2
Properties: white cub; hygr; stable in dry air
up to 300°C, decomposes 400°C–500°C;
strong reducing agent [MER06]
Solubility: w/w H_2O: 55% (25°C), 88.5%
(60°C); s liq ammonia, ethylenediamine,
other organic solvents [MER06]
Density, g/cm³: 1.074 [MER06]
Melting Point, °C: ~400, decomposes [MER06]

2842

Compound: Sodium bromate
Formula: $NaBrO_3$
Molecular Formula: $BrNaO_3$
Molecular Weight: 150.892
CAS RN: 7789-38-0
Properties: colorless; cub cryst, granules or powd;
enthalpy of fusion 28.11 kJ/mol [CRC10] [MER06]
Solubility: g/100 g soln, H_2O: 28.29 (25°C), 47.6
(100°C); solid phase, $NaBrO_3$ [KRU93]
Density, g/cm³: 3.34 [MER06]
Melting Point, °C: 381, decomposes and
evolves oxygen [MER06]

2843

Compound: Sodium bromide
Formula: $NaBr$
Molecular Formula: $BrNa$
Molecular Weight: 102.894
CAS RN: 7647-15-6
Properties: white cub cryst, a=0.5977 nm; granules,
powd; absorbs moisture from air; enthalpy of
fusion 26.11 kJ/mol; preparation: by the addition
of stoichometric amount of HBr to NaOH
or Na_2CO_3 [CRC10] [KIR82] [MER06]
Solubility: g/100 g soln, H_2O: 44.47 (0°C), 48.61 (25°C),
53.8–54.8 (100°C); solid phase, $NaBr$-$2H_2O$ (0°C,
25°C), $NaBr$ (100°C) [KRU93]; s alcohol [HAW93]
Density, g/cm³: 3.203 [STR93]
Melting Point, °C: 747 [CRC10]
Boiling Point, °C: 1390 [STR93]

2844

Compound: Sodium bromide dihydrate
Formula: $NaBr \cdot 2H_2O$
Molecular Formula: BrH_4NaO_2
Molecular Weight: 138.925

CAS RN: 13466-08-5
Properties: white cryst powd [HAW93]
Solubility: 79.5 g/100 mL H_2O (0°C), 118.6 g/100 mL H_2O (80.5°C) [CRC10]; moderately s in alcohol [HAW93]
Density, g/cm³: 2.176 [HAW93]
Reactions: minus $2H_2O$ at 51°C [CRC10]

2845
Compound: Sodium *t*-butoxide
Formula: C_4H_9NaO
Molecular Formula: C_4H_9NaO
Molecular Weight: 96.11
CAS RN: 865-48-5
Properties: white to off-white solid; moisture-sensitive [STR08]

2846
Compound: Sodium cacodylate hydrate
Formula: $(CH_3)_2As(O)ONa \cdot xH_2O$
Molecular Formula: $C_2H_8AsNaO_3$
Molecular Weight: 159.91
CAS RN: 124-65-2

2847
Compound: Sodium carbonate
Synonym: soda ash
Formula: Na_2CO_3
Molecular Formula: CNa_2O_3
Molecular Weight: 105.989
CAS RN: 497-19-8
Properties: −100 mesh with 99.999% purity; hygr powd; gradually absorbs one mole H_2O from air; enthalpy of fusion 29.70 kJ/mol [MER06] [CER91] [CRC10]
Solubility: mol/kg H_2O: 0.66 (0°C), 2.77 (25°C), 4.22 (100°C); solid phase, $Na_2CO_3 \cdot 10H_2O$ (0°C, 25°C), $Na_2CO_3 \cdot H_2O$ (100°C) [KRU93]; s glycerol; i alcohol [MER06]
Density, g/cm³: 2.54 [LID94]
Melting Point, °C: 858.1 [LID94]
Reactions: minus CO_2 starting at 400°C [MER06]

2848
Compound: Sodium carbonate bicarbonate dihydrate
Synonyms: trona, sodium sesquicarbonate
Formula: $Na_2CO_3 \cdot NaHCO_3 \cdot 2H_2O$
Molecular Formula: $C_2H_5Na_3O_8$
Molecular Weight: 226.026
CAS RN: 497-19-8
Properties: monocl needles; stable in air [MER06]
Solubility: g/100 mL H_2O: 13 (0°C), 42 (100°C) [MER06]

Density, g/cm³: 2.112 [MER06]
Melting Point, °C: decomposes [CRC10]

2849
Compound: Sodium carbonate decahydrate
Synonyms: soda, washing soda
Formula: $Na_2CO_3 \cdot 10H_2O$
Molecular Formula: $CH_{20}Na_2O_{13}$
Molecular Weight: 286.142
CAS RN: 6132-02-1
Properties: transparent cryst; effloresces in air [MER06]
Solubility: s 2 parts cold H_2O, 0.25 parts boiling H_2O; glycerol; i alcohol [MER06]
Density, g/cm³: 1.46 [MER06]
Melting Point, °C: 34 [MER06]
Reactions: minus H_2O at 33.5°C [CRC10]

2850
Compound: Sodium carbonate monohydrate
Synonym: thermonatrite
Formula: $Na_2CO_3 \cdot H_2O$
Molecular Formula: $CH_2Na_2O_4$
Molecular Weight: 124.005
CAS RN: 5968-11-6
Properties: colorless; small cryst or powd; stable under ordinary atm conditions of temp and moisture [MER06]
Solubility: s 3 parts H_2O, 1.8 parts boiling H_2O, 7 parts glycerol; i alcohol [MER06]
Density, g/cm³: 2.25 [MER06]
Melting Point, °C: 109 [HAW93]
Reactions: forms anhydride at 100°C [MER06]

2851
Compound: Sodium carbonate peroxohydrate
Synonym: sodium percarbonate
Formula: $Na_2CO_3 \cdot 1\text{-}1/2H_2O_2$
Molecular Formula: $CH_3Na_2O_6$
Molecular Weight: 157.011
CAS RN: 15630-39-4
Properties: stable, microcryst powd; forms Na_2O_2 and Na_2CO_3 in water; used as a bleaching agent, mild antiseptic and a cleaner for dentures [HAW93]
Solubility: 120 g/kg H_2O (20°C) [HAW93]

2852
Compound: Sodium chlorate
Formula: $NaClO_3$

Molecular Formula: $ClNaO_3$
Molecular Weight: 106.441
CAS RN: 7775-09-9
Properties: cub cryst or white granules; sl hygr; strong oxidizing agent; enthalpy of fusion 22.10 kJ/mol [CRC10] [MER06] [KIR78]
Solubility: g/100 g soln, H_2O: 44.3 (0°C), 50.0 (25°C), 66.8 (100°C); solid phase, $NaClO_3$ [KRU93]; s alcohol [MER06]
Density, g/cm³: 2.490 [HAW93]
Melting Point, °C: 248–261, decomposes [STR93]
Boiling Point, °C: decomposes 300 [MER06]
Reactions: decomposes at 300°C, with evolution of O_2 [MER06]

2853

Compound: Sodium chloride
Synonym: halite
Formula: NaCl
Molecular Formula: ClNa
Molecular Weight: 58.443
CAS RN: 7647-14-5
Properties: colorless cub white cryst, granules or powd; hygr; hardness 2.5; enthalpy of fusion 28.16 kJ/mol; enthalpy of solution 3.757 kJ/mol; saturated brine has vapor pressure, kPa: 1.76 (20°C), 2.39 (25°C), 9.26 (50°C), 23.27 (70°C), 52.20 (90°C); bp of saturated brine is 108.7°C; specific gravity of saturated brine is 1.1978 [KIR82] [MER06] [STR93] [CRC10]
Solubility: g/100 g soln, H_2O: 35.63 (0°C), 35.92 (25°C), 39.4 (100°C); solid phase, NaCl [KRU93]; 1 g/10 mL glycerol; v sl s alcohol [MER06]; 6.1581 ± 0.0058 mol/(kg·H_2O) at 25°C [RAR85b]
Density, g/cm³: 2.165 [STR93]
Melting Point, °C: 800.7 [LID94]
Boiling Point, °C: 1465 [LID94]
Thermal Conductivity, W/(m·K): data for aq solutions of NaCl from 20°C to 330°C are found in [OZB80]
Thermal Expansion Coefficient: (volume) 100°C (0.963), 200°C (2.288), 400°C (5.256), 600°C (8.932) [CLA66]

2854

Compound: Sodium chlorite
Formula: $NaClO_2$
Molecular Formula: $ClNaO_2$
Molecular Weight: 90.442
CAS RN: 7758-19-2
Properties: white cryst or powd; sl hygr; strong oxidizing agent; used to improve the taste and odor of potable water, and as a bleaching agent for textiles and wood pulp [HAW93] [MER06]

Solubility: mol/mol soln, H_2O: 6.51 (25°C), 4.1 (60°C); solid phase, $NaClO_2 \cdot 3H_2O$ (25°C), $NaClO_2$ (60°C) [KRU93]
Reactions: decomposes at 180°C–200°C [MER06]

2855

Compound: Sodium chromate
Formula: Na_2CrO_4
Molecular Formula: $CrNa_2O_4$
Molecular Weight: 161.974
CAS RN: 7775-11-3
Properties: yellow cryst; ortho-rhomb [KIR78]
Solubility: g/100 g soln, H_2O: 24.2 (0°C), 45.8 (25°C), 56.1 (100°C); solid phase, $Na_2CrO_4 \cdot 10H_2O$ (0°C), $Na_2CrO_4 \cdot 6H_2O$ (25°C), Na_2CrO_4 (100°C) [KRU93]
Density, g/cm³: 2.723 [KIR78]
Melting Point, °C: 792 [KIR78]

2856

Compound: Sodium chromate decahydrate
Formula: $Na_2CrO_4 \cdot 10H_2O$
Molecular Formula: $CrH_{20}Na_2O_{14}$
Molecular Weight: 342.127
CAS RN: 7775-11-3
Properties: yellow, translucent, efflorescent cryst, water content can vary; used in inks, for dyeing, as a paint pigment, wood preservative and corrosion protectant [HAW93] [MER06]
Solubility: s H_2O; sl s in alcohol [HAW93]
Density, g/cm³: 1.483 [HAW93]
Melting Point, °C: 19.9 [HAW93]

2857

Compound: Sodium chromate tetrahydrate
Formula: $Na_2CrO_4 \cdot 4H_2O$
Molecular Formula: $CrH_8Na_2O_8$
Molecular Weight: 234.035
CAS RN: 10034-82-9
Properties: yellow; somewhat deliq cryst; used in pigment manufacture, leather tanning and as a corrosion inhibitor [MER06] [HAW93]
Solubility: s ~1 part H_2O; sl s alcohol [MER06]

2858

Compound: Sodium citrate dihydrate
Synonyms: citric acid, trisodium salt dihydrate
Formula: $NaOOCCH_2C(OH)(COONa)CH_2COONa \cdot 2H_2O$
Molecular Formula: $C_6H_9Na_3O_9$
Molecular Weight: 294.101
CAS RN: 6132-04-3

Properties: white cryst, granules or powd; odorless with cool saline taste; used in photography, as a sequestering agent for metals, as an anticoagulant for blood samples, and for emulsifying, acidifying and sequestering food [MER06]
Solubility: g/100 mL H_2O: 71 (25°C), 167 (100°C); i alcohol, ether [KIR78]
Reactions: minus $2H_2O$ at 150°C [MER06]

2859
Compound: Sodium citrate pentahydrate
Formula: $Na_3C_6H_5O_7 \cdot 5H_2O$
Molecular Formula: $C_6H_{15}Na_3O_{12}$
Molecular Weight: 348.147
CAS RN: 6858-44-2
Properties: large, colorless cryst or white granules; effloresces in air [MER06] [KIR78]
Solubility: g/100 mL H_2O: 92.6 (25°C), 250 (100°C); sl s alcohol, i ether [KIR78]
Density, g/cm³: 1.857 [CRC10]
Reactions: minus $5H_2O$ at 150°C [CRC10]

2860
Compound: Sodium copper chromate trihydrate
Formula: $Na_2O \cdot 4CuO \cdot 4CrO_3 \cdot 3H_2O$
Molecular Formula: $Cr_4Cu_4H_6Na_2O_{20}$
Molecular Weight: 834.184
CAS RN: 68399-60-0
Properties: maroon triclinic; used as an antifouling pigment [KIR78]
Solubility: v sl s H_2O [KIR78]
Density, g/cm³: 3.57 [KIR78]

2861
Compound: Sodium cyanate
Formula: NaOCN
Molecular Formula: CNNaO
Molecular Weight: 65.007
CAS RN: 917-61-3
Properties: colorless needles when prepared from alcohol medium; decomposes in H_2O to urea and Na_2CO_3 [MER06]
Solubility: 0.22 g/100 g alcohol (0°C); i ether [MER06]
Density, g/cm³: 1.893 [MER06]
Melting Point, °C: 550 [MER06]

2862
Compound: Sodium cyanide
Synonym: cyanogran
Formula: NaCN
Molecular Formula: CNNa
Molecular Weight: 49.008

CAS RN: 143-33-9
Properties: white granules or fused pieces; somewhat deliq; enthalpy of fusion 15.4 kJ/mol; enthalpy of vaporization 156.3 kJ/mol; enthalpy of solution 1510 J/mol; vapor pressure, kPa: 0.1013 (800°C), 1.652 (1000°C), 11.9 (1200°C), 41.8 (1360°C); used in electroplating of zinc, copper, brass gold and other metals; forms a dihydrate [KIR78] [MER06]
Solubility: g/100 g H_2O: 40.8 (0°C), 58.7 (20°C), 71.2 (30°C) [LAN05]; sl s alcohol [MER06]
Density, g/cm³: cub: 1.60, ortho-rhomb: 1.62–1.624 [KIR78]
Melting Point, °C: 563.7 [STR93]
Boiling Point, °C: 1496 [STR93]

2863
Compound: Sodium cyanoborohydride
Synonym: sodium cyanotrihydridoborate
Formula: $NaBH_3(CN)$
Molecular Formula: CH_3BNNa
Molecular Weight: 62.843
CAS RN: 25895-60-7
Properties: white powd; hygr; mild reducing agent, e.g. reduces aldehydes, ketones, oximes [MER06] [ALD94]
Solubility: 212 g/100 g H_2O (29°C); v s methanol; sl s ethanol [MER06]
Density, g/cm³: 1.199 [MER06]
Melting Point, °C: 240–242 [MER06]

2864
Compound: Sodium deuteride
Formula: NaD
Molecular Formula: DNa
Molecular Weight: 25.005
CAS RN: 15780-28-6
Properties: slurry of gray powd; sensitive to moisture; 20% in oil, 98% isotopic purity [STR93]

2865
Compound: Sodium diacetate
Synonym: sodium acid acetate
Formula: $CH_3COOH \cdot CH_3COONa$
Molecular Formula: $C_4H_7NaO_4$
Molecular Weight: 142.089
CAS RN: 126-96-5
Properties: white powd; can be used as a source of acetic acid; releases 42.25% of available CH_3COOH; used as a buffer, as a mold inhibitor, food preservative [HAW93] [MER06]
Solubility: s H_2O [MER06]; sl s alcohol; i ether [HAW93]
Melting Point, °C: decomposes >150 [MER06]

2866
Compound: Sodium dichromate dihydrate
Formula: $Na_2Cr_2O_7 \cdot 2H_2O$
Molecular Formula: $Cr_2H_4Na_2O_9$
Molecular Weight: 297.999
CAS RN: 7782-12-0
Properties: reddish orange cryst; monocl; somewhat deliq [KIR78] [MER06] [STR93]
Solubility: g/100 g soln in H_2O: 62.17 (0°C), 65.01 (25°C), 80.6 (100°C); solid phase, $Na_2Cr_2O_7 \cdot 2H_2O$ (0°C, 25°C), $Na_2Cr_2O_7 \cdot 6H_2O$ (100°C) [KRU93]
Density, g/cm³: 2.348 [KIR78]
Melting Point, °C: 84.6 (incongruent) [KIR78]
Reactions: minus $2H_2O$ ~100°C [MER06]

2867
Compound: Sodium dihydrogen phosphate dihydrate
Formula: $NaH_2PO_4 \cdot 2H_2O$
Molecular Formula: H_6NaO_6P
Molecular Weight: 156.008
CAS RN: 7558-80-7
Properties: ortho-rhomb, colorless cryst [MER06]
Solubility: g/100 g soln in H_2O: 36.5 (0°C), 48.5 (25°C), 71.0 (100°C); solid phase, $NaH_2PO_4 \cdot 2H_2O$ (0°C, 25°C), NaH_2PO_4 (100°C) [KRU93]
Density, g/cm³: 1.915 [MER06]
Melting Point, °C: 60 [MER06]

2868
Compound: Sodium dihydrogen phosphate monohydrate
Synonym: sodium monobasic phosphate
Formula: $NaH_2PO_4 \cdot H_2O$
Molecular Formula: H_4NaO_5P
Molecular Weight: 137.993
CAS RN: 10049-21-5
Properties: white; sl deliq cryst or granules [MER06] [STR93]
Solubility: 59.9 g/100 mL H_2O (0°C), 427 g/100 mL H_2O (100°C) [CRC10]; i alcohol [MER06]
Density, g/cm³: 2.04 [CRC10]
Melting Point, °C: decomposes, 204 [CRC10]
Reactions: minus H_2O at 100°C; forms metaphosphate when ignited [MER06]

2869
Compound: Sodium dihydrogen pyrophosphate
Synonym: sodium acid pyrophosphate
Formula: $Na_2H_2P_2O_7$
Molecular Formula: $H_2Na_2O_7P_2$
Molecular Weight: 221.939
CAS RN: 7758-16-9

Properties: white; fused masses or powd; used in baking powd [MER06]
Solubility: g/100 g H_2O: 4.47 (0°C), 12.0 (20°C), 18.4 (40°C) [LAN05]
Density, g/cm³: ~1.9 [LID94]
Melting Point, °C: decomposes 220 [MER06]

2870
Compound: Sodium dithionate
Synonym: sodium hyposulfate
Formula: $Na_2(SO_3)_2$
Molecular Formula: $Na_2O_6S_2$
Molecular Weight: 206.108
CAS RN: 7631-94-9
Properties: prepared by reacting MnO_2 with SO_2 to form the product MnS_2O_6, then reacting this product with Na_2CO_3 [MER06]
Solubility: g/100 g H_2O: 7.83 (0°C), 17.38 (25°C), 64.74 (100°C); solid phase, $Na_2S_2O_6 \cdot 2H_2O$ [KRU93]

2871
Compound: Sodium dithionate dihydrate
Formula: $Na_2S_2O_6 \cdot 2H_2O$
Molecular Formula: $H_4Na_2O_8S_2$
Molecular Weight: 242.139
CAS RN: 7631-94-9
Properties: colorless; ortho-rhomb cryst; stable in air [MER06]
Solubility: H_2O, w/w: 6.05% (0°C), 13.39% (20°C), 17.32% (30°C); i alcohol [MER06]
Density, g/cm³: 2.189 [MER06]
Reactions: minus $2H_2O$ at 110°C; decomposes to $Na_2SO_4 + SO_2$ at 267°C [MER06]

2872
Compound: Sodium ethoxide
Formula: $NaOC_2H_5$
Molecular Formula: C_2H_5NaO
Molecular Weight: 68.05
CAS RN: 141-52-6
Properties: white-to-yellow [STR08]

2873
Compound: Sodium ferricyanide monohydrate
Formula: $Na_3Fe(CN)_6 \cdot H_2O$
Molecular Formula: $C_6H_2FeN_6Na_3O$
Molecular Weight: 298.935
CAS RN: 14217-21-1
Properties: ruby red deliq cryst; used in the production of pigments, in dyeing and printing [HAW93] [MER06]
Solubility: s 5.5 parts H_2O, 1.5 parts boiling H_2O [MER06]; i alcohol [HAW93]

2874
Compound: Sodium ferrocyanide decahydrate
Synonym: yellow prussiate of soda
Formula: $Na_4Fe(CN)_6 \cdot 10H_2O$
Molecular Formula: $C_6H_{20}FeN_6Na_4O_{10}$
Molecular Weight: 484.063
CAS RN: 13601-19-9
Properties: pale yellow monocl cryst; somewhat
    efflorescent; steadily dehydrated >50°C [MER06]
Solubility: H_2O: 10.2% (1°C), 14.7% (17°C), 17.6%
    (25°C), 28.1% (53°C), 39% (85°C), 39.7%
    (96.6°C); i most organic solvents [MER06]
Density, g/cm³: 1.458 [HAW93]
Melting Point, °C: decomposes to NaCN,
    Fe, C, N_2 at 435 [MER06]
Reactions: minus $10H_2O$ at 81.5°C [MER06]

2875
Compound: Sodium fluoride
Synonym: villaumite
Formula: NaF
Molecular Formula: FNa
Molecular Weight: 41.988
CAS RN: 7681-49-4
Properties: clear lustrous white powd, or 99.9% pure
    melted pieces of 3–6 mm; cub or tetr cryst; enthalpy
    of fusion 33.35 kJ/mol; can be prepared by reacting
    hydrofluoric acid with soda ash or NaOH; used to
    fluoridate municipal drinking water, in toothpastes,
    for cryolite manufacture, as windows in ultraviolet and
    infrared detectors, in the form of melted pieces used as
    evaporation material for low index films, and reflection
    diminishing coatings [MER06] [CER91] [CRC10]
Solubility: g/100 g soln, H_2O: 3.53 (0°C), 3.98
    (25°C), 4.83 (100°C); solid phase, NaF
    (0°C, 25°C) [KRU93]; i alcohol [MER06]
Density, g/cm³: 2.78 [LID94]
Melting Point, °C: 996 [CRC10]
Boiling Point, °C: 1704 [MER06]

2876
Compound: Sodium fluoroborate
Synonym: sodium tetrafluoroborate
Formula: $NaBF_4$
Molecular Formula: BF_4Na
Molecular Weight: 109.795
CAS RN: 13755-29-8
Properties: white powd; ortho-rhomb below 240°C,
    a = 0.68358 nm, b = 0.62619 nm, c = 0.67916 nm;
    slowly decomposed by heat; can be prepared
    by reacting NaOH or Na_2CO_3 with fluoroboric
    acid; used in sand casting of aluminum and
    magnesium [HAW93] [MER06] [KIR78]

Solubility: g/100 mL H_2O: 108 (26°C), 210
    (100°C) [MER06]; s alcohol [KIR78]
Density, g/cm³: 2.47 [MER06]
Melting Point, °C: 384 [MER06]

2877
Compound: Sodium fluorophosphate
Formula: Na_2PO_3F
Molecular Formula: FNa_2O_3P
Molecular Weight: 143.939
CAS RN: 10163-15-2
Properties: white powd; used in the preparation of
    bactericides and fungicides [HAW93] [STR93]
Solubility: s H_2O [HAW93]
Melting Point, °C: ~625 [STR93]

2878
Compound: Sodium fluorosulfonate
Formula: $NaSO_3F$
Molecular Formula: $FNaO_3S$
Molecular Weight: 122.052
CAS RN: 14483-63-7
Properties: shiny leaflets; hygr [KIR78]
Solubility: s H_2O, alcohol, acetone; i ether [KIR78]
Melting Point, °C: decomposes [CRC10]

2879
Compound: Sodium formaldehyde sulfoxylate
Formula: $NaHSO_2 \cdot CH_2O \cdot 2H_2O$
Molecular Formula: CH_7NaO_5S
Molecular Weight: 154.120
CAS RN: 149-44-0
Properties: white solid; used as stripping and
    discharge agent for textiles [HAW93]
Solubility: s H_2O, alcohol [HAW93]
Melting Point, °C: 64 [HAW93]

2880
Compound: Sodium gold cyanide
Synonym: gold sodium cyanide
Formula: $NaAu(CN)_2$
Molecular Formula: C_2AuN_2Na
Molecular Weight: 271.992
CAS RN: 15280-09-8
Properties: white cryst yellow powd; used for gold
    plating electronic components [HAW93] [MER06]
Solubility: s H_2O [MER06]

2881
Compound: Sodium gold thiosulfate dihydrate
Synonym: gold sodium thiosulfate

Formula: $Na_3Au(S_2O_3)_2 \cdot 2H_2O$
Molecular Formula: $AuH_4Na_3O_8S_4$
Molecular Weight: 526.227
CAS RN: 10233-88-2
Properties: white; glistening cryst; needle like or prismatic; darkens slowly when exposed to light; aq solution decomposes on standing, and turns yellow [MER06]
Solubility: 1 gm/2 mL H_2O; i alcohol [MER06]
Density, g/cm³: 3.09 [MER06]
Reactions: minus $2H_2O$ at 150°C–160°C [MER06]

2882

Compound: Sodium hexachloroiridate(III) hydrate
Formula: $Na_3IrCl_6 \cdot xH_2O$
Molecular Formula: Cl_6IrNa_3 (anhydrous)
Molecular Weight: 473.905 (anhydrous)
CAS RN: 123334-23-6
Properties: greenish brown cryst; hygr [STR93] [ALD94]

2883

Compound: Sodium hexachloroiridate(IV) hexahydrate
Formula: $Na_2IrCl_6 \cdot 6H_2O$
Molecular Formula: $Cl_6H_{12}IrNa_2O_6$
Molecular Weight: 559.004
CAS RN: 19567-78-3
Properties: reddish black powd; hygr [STR93]
Solubility: g/100 g H_2O: 34.46 (15°C), 56.17 (30°C), 279.3 (80°C) [LAN05]
Melting Point, °C: 600, decomposes [STR93]

2884

Compound: Sodium hexachloroosmiate(IV) hydrate
Formula: $Na_2OsCl_6 \cdot xH_2O$
Molecular Formula: Cl_6Na_2Os (anhydrous)
Molecular Weight: 448.926 (anhydrous)
CAS RN: 1307-81-9
Properties: reddish orange powd; hygr; x<2 [STR93] [ALD94]

2885

Compound: Sodium hexachloropalladate(IV)
Formula: Na_2PdCl_6
Molecular Formula: Cl_6Na_2Pd
Molecular Weight: 365.116
CAS RN: 53823-60-2
Properties: reddish orange cryst; hygr [STR93]

2886

Compound: Sodium hexachloroplatinate(IV)
Formula: Na_2PtCl_6
Molecular Formula: Cl_6Na_2Pt
Molecular Weight: 453.776
CAS RN: 1307-82-0
Properties: yellow cryst; hygr; readily forms hexahydrate in moist air at 25°C when relative humidity >50% [MER06]
Solubility: s H_2O, alcohol [MER06]

2887

Compound: Sodium hexachloroplatinate(IV) hexahydrate
Formula: $Na_2PtCl_6 \cdot 6H_2O$
Molecular Formula: $Cl_6H_{12}Na_2O_6Pt$
Molecular Weight: 561.867
CAS RN: 19583-77-8
Properties: orange powd; hygr [STR93]

2888

Compound: Sodium hexachlororhodate(III) hydrate
Formula: $Na_3RhCl_6 \cdot xH_2O$
Molecular Formula: Cl_6Na_3Rh (anhydrous)
Molecular Weight: 384.591 (anhydrous)
CAS RN: 14972-70-4
Properties: red cryst; hygr [AES93] [STR93]
Melting Point, °C: 900 decomposes [AES93]

2889

Compound: Sodium hexafluoroaluminate
Synonym: cryolite
Formula: Na_3AlF_6
Molecular Formula: AlF_6Na_3
Molecular Weight: 209.941
CAS RN: 13775-53-6
Properties: white powd or melted pieces; liq vapor pressure 253 Pa (1012°C); monocl, a = 0.546 nm, b = 0.561 nm, c = 0.780 nm; enthalpy of vaporization 225 kJ/mol; hardness 2.5 Mohs; electrical conductivity of liq (1012°C) 2.82 (ohm·cm)$^{-1}$; viscosity of liq 6.7 mPa·s (1012°C); enthalpy of transition 8.21 kJ/mol (565°C); used in the molten form to dissolve Al_2O_3 in the manufacture of aluminum [KIR78] [CIC73] [STR93] [CER91]
Solubility: g/100 g H_2O: 0.042 (25°C), 0.135 (100°C) [CIC73]
Density, g/cm³: monocl: 2.97; cub (560°C): 2.77; liq: 2.087 [KIR78]
Melting Point, °C: 1012 [KIR78]
Reactions: transition from monocl to cub at 565°C [KIR78]

2890
Compound: Sodium hexafluoroantimonate(V)
Formula: $NaSbF_6$
Molecular Formula: F_6NaSb
Molecular Weight: 258.740
CAS RN: 16925-25-0
Properties: white powd [STR93]
Density, g/cm³: 3.375 [STR93]

2891
Compound: Sodium hexafluoroarsenate
Formula: $NaAsF_6$
Molecular Formula: AsF_6Na
Molecular Weight: 211.902
CAS RN: 12005-86-6
Properties: white powd [STR93]

2892
Compound: Sodium hexafluoroferrate(III)
Formula: Na_3FeF_6
Molecular Formula: F_6FeNa_3
Molecular Weight: 238.806
CAS RN: 20955-11-7
Properties: off-white to green [STR93]

2893
Compound: Sodium hexafluorogermanate
Formula: Na_2GeF_6
Molecular Formula: F_6GeNa_2
Molecular Weight: 232.580
CAS RN: 36470-39-0
Properties: white powd [STR93]
Solubility: g/100 g H_2O: 1.52 (0°C), 2.25
    (30°C), 3.36 (80°C) [LAN05]

2894
Compound: Sodium hexafluorophosphate
Formula: $NaPF_6$
Molecular Formula: F_6NaP
Molecular Weight: 167.954
CAS RN: 21324-39-0
Properties: white cryst; hygr [STR93]
Density, g/cm³: 2.369 [ALD94]

2895
Compound: Sodium hexafluorosilicate
Formula: Na_2SiF_6
Molecular Formula: F_6Na_2Si
Molecular Weight: 188.056

CAS RN: 16893-85-9
Properties: white granular powd; odorless, tasteless,
    free flowing; prepared from H_2SiF_6 and Na_2CO_3
    or NaCl; used in fluoridation, in laundry soaps and
    for mothproofing woolens [HAW93] [MER06]
Solubility: g/100 g H_2O: 4.35 (0°C), 7.2 (20°C),
    24.5 (100°C) [LAN05]; i alcohol [MER06]
Density, g/cm³: 2.679 [STR93]
Melting Point, °C: melts at red heat,
    decomposing [MER06]

2896
Compound: Sodium hexafluorostannate(IV)
Formula: Na_2SnF_6
Molecular Formula: F_6Na_2Sn
Molecular Weight: 278.680
CAS RN: 16924-51-9
Properties: white powd [STR93]

2897
Compound: Sodium hexafluorotitanate
Formula: Na_2TiF_6
Molecular Formula: F_6Na_2Ti
Molecular Weight: 207.837
CAS RN: 17116-13-1
Properties: white powd [STR93]

2898
Compound: Sodium hexafluorozirconate
Formula: Na_2ZrF_6
Molecular Formula: F_6Na_2Zr
Molecular Weight: 251.194
CAS RN: 16925-26-1
Properties: white powd [STR93]

2899
Compound: Sodium hexametaphosphate
Synonym: Graham's salt
Formula: $(NaPO_3)_6$
Molecular Formula: $Na_6O_{18}P_6$
Molecular Weight: 611.771
CAS RN: 10124-56-8
Properties: white powd [STR93]

2900
Compound: Sodium hexanitritocobalt(III)
Synonym: sodium cobaltinitrite
Formula: $Na_3Co(NO_2)_6$
Molecular Formula: $CoN_6Na_3O_{12}$
Molecular Weight: 403.935

CAS RN: 14649-73-1
Properties: yellow to brownish yellow; cryst powd; decomposed by mineral acids [MER06]
Solubility: v s H_2O; sl s alcohol [MER06]

2901
Compound: Sodium hydride
Formula: NaH
Molecular Formula: HNa
Molecular Weight: 23.998
CAS RN: 7646-69-7
Properties: silvery needles; commercial material is grayish white powd; reacts explosively with H_2O, vigorously with lower alcohols; ignites when standing in moist air; manufactured by reaction of H_2 with molten sodium metal which has been dispersed in mineral oil [KIR80] [MER06]
Solubility: s molten NaOH; i liq NH_3 [MER06]
Density, g/cm³: 1.396 [MER06]
Melting Point, °C: 425 with decomposition [MER06]

2902
Compound: Sodium hydrogen arsenate
Formula: Na_2HAsO_4
Molecular Formula: $AsHNa_2O_4$
Molecular Weight: 185.908
CAS RN: 7778-43-0
Properties: powd [MER06]
Solubility: g/100 g soln, H_2O: 5.59 (0.1°C), 29.33 (25°C), 66.5 (98.5°C); solid phase, $Na_2HAsO_4 \cdot 12H_2O$ (0.1°C), $Na_2HAsO_4 \cdot 7H_2O$ (25°C), Na_2HAsO_4 (98.5°C) [KRU93]; sl s alcohol [MER06]

2903
Compound: Sodium hydrogen arsenate heptahydrate
Formula: $Na_2HAsO_4 \cdot 7H_2O$
Molecular Formula: $AsH_{15}Na_2O_{11}$
Molecular Weight: 312.014
CAS RN: 10048-95-0
Properties: odorless cryst; effloresces in warm air; has been used as antimalarial dermatologic [MER06]
Solubility: s in 1.3 parts H_2O; s glycerol; sl s alcohol [MER06]
Density, g/cm³: 1.87 [MER06]
Melting Point, °C: 57 [MER06]
Reactions: minus $5H_2O$ ~50°C, becomes anhydrous at 100°C [MER06]

2904
Compound: Sodium hydrogen carbonate
Synonym: sodium bicarbonate

Formula: $NaHCO_3$
Molecular Formula: $CHNaO_3$
Molecular Weight: 84.007
CAS RN: 144-55-8
Properties: white cryst powd or granules; readily decomposed by weak acids [MER06]
Solubility: g/100 g soln, H_2O: 6.48 (0°C), 9.32 (25°C), 19.1 (100°C); solid phase, $NaHCO_3$ [KRU93]; i alcohol [MER06]
Density, g/cm³: 2.159 [ALD94]
Reactions: minus CO_2 ~50°C, forms Na_2CO_3 at 100°C [MER06]

2905
Compound: Sodium hydrogen fluoride
Synonym: sodium bifluoride
Formula: $NaHF_2$
Molecular Formula: F_2HNa
Molecular Weight: 61.995
CAS RN: 1333-83-1
Properties: white cryst powd [MER06]
Solubility: g/100 g soln in H_2O: 3.7 (20°C), 16.4 (80°C) [KIR78]
Density, g/cm³: 2.08 [LID94]
Melting Point, °C: decomposes above 160 [KIR78]

2906
Compound: Sodium hydrogen oxalate monohydrate
Formula: $NaHC_2O_4 \cdot H_2O$
Molecular Formula: $C_2H_3NaO_5$
Molecular Weight: 130.033
CAS RN: 1186-49-8
Properties: white powd [STR93]

2907
Compound: Sodium hydrogen phosphate
Formula: Na_2HPO_4
Molecular Formula: HNa_2O_4P
Molecular Weight: 141.959
CAS RN: 7558-79-4
Properties: hygr powd; absorbs from 2 to 7 moles H_2O from atm, depending on humidity and temp [MER06]
Solubility: g/100 g H_2O: 1.6 (0°C), 11.8 (25°C), 103.3 (100°C); solid phase, α-$Na_2HPO_4 \cdot 12H_2O$ (0°C, 25°C), Na_2HPO_4 (100°) [KRU93]
Density, g/cm³: 1.679 [STR93]

2908
Compound: Sodium hydrogen phosphate dodecahydrate
Formula: $Na_2HPO_4 \cdot 12H_2O$

Molecular Formula: $H_{25}Na_2O_{16}P$
Molecular Weight: 358.143
CAS RN: 10039-32-4
Properties: translucent cryst or granules; minus $5H_2O$ at ambient atm conditions [MER06]
Solubility: s 3 parts H_2O; i alcohol [MER06]
Density, g/cm³: ~1.5 [MER06]
Melting Point, °C: 34–35 [MER06]

2909
Compound: Sodium hydrogen phosphate heptahydrate
Formula: $Na_2HPO_4 \cdot 7H_2O$
Molecular Formula: $H_{15}Na_2O_{11}P$
Molecular Weight: 268.066
CAS RN: 7782-85-6
Properties: cryst or granular powd; stable in air [MER06]
Solubility: s 4 parts H_2O, more s in boiling H_2O; i alcohol [MER06]
Density, g/cm³: ~1.7 [MER06]

2910
Compound: Sodium hydrogen phosphite pentahydrate
Formula: $Na_2HPO_3 \cdot 5H_2O$
Molecular Formula: $H_{11}Na_2O_8P$
Molecular Weight: 216.036
CAS RN: 13708-85-5
Properties: white; hygr; cryst powd; used as an antidote to mercuric chloride poisoning [HAW93] [MER06]
Solubility: g anhydrous/100 g H_2O: 418 (0°C), 429 (20°C), 566 (30°C) [LAN05]; i alcohol [HAW93]
Melting Point, °C: 53 [HAW93]
Boiling Point, °C: 200–250, decomposes [HAW93]

2911
Compound: Sodium hydrogen sulfate
Synonym: niter cake
Formula: $NaHSO_4$
Molecular Formula: $HNaO_4S$
Molecular Weight: 120.062
CAS RN: 7681-38-1
Properties: fused hygr; tricl [KIR82] [MER06]
Solubility: s in 2 parts H_2O, 1 part boiling H_2O; decomposes in alcohol to sodium sulfate and H_2SO_4 [MER06]
Density, g/cm³: 2.435 [MER06]
Melting Point, °C: ~315 [MER06]

2912
Compound: Sodium hydrogen sulfate monohydrate
Synonym: sodium bisulfate monohydrate
Formula: $NaHSO_4 \cdot H_2O$

Molecular Formula: H_3NaO_5S
Molecular Weight: 138.077
CAS RN: 10034-88-5
Properties: cryst; forms pyrosulfate if heated strongly [MER06]
Solubility: s ~0.8 parts H_2O; decomposed by alcohol [MER06]
Density, g/cm³: 2.10 [LID94]

2913
Compound: Sodium hydrogen sulfide
Synonym: sodium hydrosulfide
Formula: NaHS
Molecular Formula: HNaS
Molecular Weight: 56.064
CAS RN: 16721-80-5
Properties: white to colorless; hydrogen sulfide odor; enthalpy of solution is 15.9 kJ/mol; cub cryst; very hygr; hydrolyzed in moist air to Na_2S and NaOH; color changed by heating in dry air: yellow, then orange as the temp increases [MER06] [KIR83]
Solubility: s H_2O, alcohol, ether [MER06]
Density, g/cm³: 1.79 [MER06]
Melting Point, °C: 350, becomes black liq [MER06]

2914
Compound: Sodium hydrogen sulfide dihydrate
Synonym: sodium bisulfide
Formula: $NaHS \cdot 2H_2O$
Molecular Formula: H_5NaO_2S
Molecular Weight: 92.095
CAS RN: 16721-80-5
Properties: lemon colored needle or flake forms; used in paper pulping and dyestuff processing [MER06] [HAW93]
Solubility: v s H_2O, alcohol, ether [MER06]
Melting Point, °C: 55 [MER06]

2915
Compound: Sodium hydrogen sulfide trihydrate
Formula: $NaHS \cdot 3H_2O$
Molecular Formula: H_7NaO_3S
Molecular Weight: 110.110
CAS RN: 16721-80-5
Properties: shiny; rhomb [MER06]
Melting Point, °C: 22 [MER06]

2916
Compound: Sodium hydrogen sulfite
Synonym: sodium bisulfite

Formula: NaHSO$_3$
Molecular Formula: HNaO$_3$S
Molecular Weight: 104.062
CAS RN: 7631-90-5
Properties: white; cryst powd; gradually oxidized to sulfate in air; has SO$_2$ odor [MER06]
Solubility: s 3.5 parts cold H$_2$O, 2 parts boiling water, in ~70 parts alcohol [MER06]
Density, g/cm^3: 1.48 [MER06]

2917
Compound: Sodium hydrogen tartrate monohydrate
Formula: NaHC$_4$H$_4$O$_6$·H$_2$O
Molecular Formula: C$_4$H$_7$NaO$_7$
Molecular Weight: 190.086
CAS RN: 868-18-8
Properties: white cryst [MER06]
Solubility: s in ~9 parts H$_2$O, 2 parts boiling H$_2$O; i alcohol [MER06]

2918
Compound: Sodium hydrosulfite
Synonym: sodium dithionite
Formula: Na$_2$S$_2$O$_4$
Molecular Formula: Na$_2$O$_4$S$_2$
Molecular Weight: 174.110
CAS RN: 7775-14-6
Properties: white or grayish white; cryst powd; oxidizes in air to bisulfite and bisulfate, more quickly oxidized in moist atm [MER06]
Solubility: v s H$_2$O; sl s alcohol [MER06]
Melting Point, °C: 55, decomposes [HAW93]

2919
Compound: Sodium hydroxide
Synonyms: caustic soda, soda lye
Formula: NaOH
Molecular Formula: HNaO
Molecular Weight: 39.997
CAS RN: 1310-73-2
Properties: white, deliq solid; fused solid; absorbs both CO$_2$ and H$_2$O from the atm; enthalpy of vaporization 175 kJ/mol; enthalpy of fusion 6.60 kJ/mol [CRC10] [HAW93] [MER06]
Solubility: g/100 g soln, H$_2$O: 29.6 (0°C), 53.3 (25°C), 77.6 (100°C) [KRU93]; 1 g dissolves in: 7.2 mL absolute alcohol, 4.2 mL methanol; s glycerol [MER06]
Density, g/cm^3: 2.13 [MER06]
Melting Point, °C: 323 [CRC10]
Boiling Point, °C: 1390 [STR93]

2920
Compound: Sodium hydroxide monohydrate
Formula: NaOH·H$_2$O
Molecular Formula: H$_3$NaO$_2$
Molecular Weight: 58.013
CAS RN: 12179-02-1
Properties: white powd; hygr [STR93]

2921
Compound: Sodium hypochlorite
Formula: NaClO
Molecular Formula: ClNaO
Molecular Weight: 74.442
CAS RN: 7681-52-9
Properties: very explosive; used to bleach wood pulp and textiles, is a disinfectant for municipal water and sewage, prevents the formation of fungi in oil production [KIR78] [MER06]
Solubility: g/100 g soln, H$_2$O: 22.7 (0°C), 44.0 (24.5°C); solid phase, NaClO·5H$_2$O [KRU93]
Density, g/cm^3: 1.097 [ALD94]

2922
Compound: Sodium hypochlorite pentahydrate
Formula: NaOCl·5H$_2$O
Molecular Formula: ClH$_{10}$NaO$_6$
Molecular Weight: 164.518
CAS RN: 7681-52-9
Properties: pale greenish color; cryst; very unstable; decomposed by atm CO$_2$; used to bleach paper pulp, textiles [HAW93] [MER06]
Solubility: 29.3 g/100 mL H$_2$O (0°C) [MER06], 94.2 g/100 mL H$_2$O (23°C) [CRC10]
Density, g/cm^3: 1.6 [LID94]
Melting Point, °C: 18 [MER06]

2923
Compound: Sodium hypophosphate decahydrate
Formula: Na$_4$P$_2$O$_6$·10H$_2$O
Molecular Formula: H$_{20}$Na$_4$O$_{16}$P$_2$
Molecular Weight: 430.056
CAS RN: 13721-43-2
Properties: colorless or white cryst [MER06]
Solubility: 1.49 g/100 mL H$_2$O (25°C), 5.46 g/100 mL H$_2$O (60°C) [CRC10]
Density, g/cm^3: 1.823 [CRC10]
Melting Point, °C: decomposes [CRC10]

2924
Compound: Sodium hypophosphite monohydrate
Formula: NaH$_2$PO$_2$·H$_2$O

Molecular Formula: H_4NaO_3P
Molecular Weight: 105.994
CAS RN: 123333-67-5
Properties: white, odorless; deliq; evolves
flammable phosphine when heated strongly;
strong reducing agent, e.g., explodes when
triturated with chlorates or other oxidizing
agents; used as reducing agent in the electroless
plating of nickel onto plastics [MER06]
Solubility: 100 g/100 g H_2O [KRU93];
s 1 part H_2O, 0.15 parts boiling H_2O;
s cold alcohol; i ether [MER06]
Melting Point, °C: decomposes [STR93]

2925
Compound: Sodium iodate
Formula: $NaIO_3$
Molecular Formula: $INaO_3$
Molecular Weight: 197.892
CAS RN: 7681-55-2
Properties: white, cryst powd; oxidizing agent;
can be prepared by electrochemical oxidation
of NaI; used as an antiseptic, disinfectant,
feed additive [HAW93] [MER06] [KIR81]
Solubility: g/100 g soln, H_2O: 2.42 (0°C),
8.62 (25°C), 24.8 (100°C); solid phase,
$NaIO_3 \cdot 5H_2O$ (0°C), $NaIO_3 \cdot H_2O$ (25°C), $NaIO_3$
(100°C) [KRU93]; i alcohol [MER06]
Density, g/cm³: 4.277 [STR93]
Melting Point, °C: decomposes [STR93]

2926
Compound: Sodium iodide
Formula: NaI
Molecular Formula: INa
Molecular Weight: 149.894
CAS RN: 7681-82-5
Properties: white deliq cub cryst or granules; gradually
absorbs up to ~5% H_2O from moist atm; iodine
slowly evolved, imparting brown coloration; enthalpy
of fusion 23.60 kJ/mol; preparation: from a reaction
between acidic iodide solution and NaOH or
Na_2CO_3; uses: in photography, as a solvent for iodine,
in cloud seeding [HAW93] [MER06] [CRC10]
Solubility: g/100 g soln, H_2O: 61.54 (0°C),
64.76 (25°C), 75.14 (100°C); solid phase,
$NaI \cdot 2H_2O$ (0°C, 25°C), NaI (100°C)
[KRU93]; 1 g dissolves in ~2 mL alcohol,
1 mL glycerol; s acetone [MER06]
Density, g/cm³: 3.667 [KIR82]
Melting Point, °C: 660 [CRC10]
Boiling Point, °C: 1304 [KIR82]

2927
Compound: Sodium iodide dihydrate
Formula: $NaI \cdot 2H_2O$
Molecular Formula: H_4INaO_2
Molecular Weight: 185.925
CAS RN: 13517-06-1
Properties: white, cub cryst or powd [HAW93]
Solubility: 318 g/100 mL H_2O (0°C),
1550 g/100 mL H_2O (100°C) [CRC10]
Density, g/cm³: 2.448 [HAW93]
Melting Point, °C: 752 [CRC10]

2928
Compound: Sodium metabismuthate
Synonym: sodium bismuthate
Formula: $NaBiO_3$
Molecular Formula: $BiNaO_3$
Molecular Weight: 279.968
CAS RN: 12232-99-4
Properties: yellow to yellowish brown powd;
somewhat hygr; slowly decomposes when
stored, decomposition accelerated by
moisture and high temp [MER06]
Solubility: i cold H_2O, decomposed in hot H_2O evolving
oxygen; evolves chlorine with HCl [MER06]

2929
Compound: Sodium metabisulfite
Synonym: sodium pyrosulfite
Formula: $Na_2S_2O_5$
Molecular Formula: $Na_2O_5S_2$
Molecular Weight: 190.109
CAS RN: 7681-57-4
Properties: white; cryst or powd; has odor of SO_2;
used as a food preservative [HAW93] [MER06]
Solubility: 54 g/100 mL H_2O (20°C),
82 g/100 mL H_2O (100°C) [CRC10]
Density, g/cm³: 1.48 [ALD94]
Melting Point, °C: decomposes >150 [CRC10]

2930
Compound: Sodium metaborate
Formula: $NaBO_2$
Molecular Formula: $BNaO_2$
Molecular Weight: 65.800
CAS RN: 7775-19-1
Properties: white pieces or powd; enthalpy of
fusion 36.20 kJ/mol; used as an herbicide
[HAW93] [CRC10] [MER06]
Solubility: g/100 g soln, H_2O: 14.10 (0°C), 22.00
(25°C), 55.60 (100°C); solid phase, $NaBO_2 \cdot 4H_2O$
(0°C, 25°C), $NaBO_2 \cdot 2H_2O$ (100°C) [KRU93]

Density, g/cm³: 2.464 [HAW93]
Melting Point, °C: 966 [CRC10]
Boiling Point, °C: 1434 [HAW93]

2931

Compound: Sodium metaborate dihydrate
Formula: $NaBO_2 \cdot 2H_2O$
Molecular Formula: BH_4NaO_4
Molecular Weight: 101.831
CAS RN: 35585-58-1
Properties: tricl; can be prepared by heating slurry of tetrahydrate above 54°C; stable phase for saturated solutions from 54 to 105°C; a hemihydrate forms at higher temperatures; absorbs atm CO_2 forming borax and sodium carbonate [KIR78]
Solubility: % anhydrous in H_2O: 14.82 (60°C), 19.88 (80°C), 28.22 (100°C); 0.3% in boiling alcohol [KIR78]
Density, g/cm³: 1.91 [KIR78]
Melting Point, °C: 90–95 [KIR78]

2932

Compound: Sodium metaborate tetrahydrate
Formula: $NaBO_2 \cdot 4H_2O$
Molecular Formula: BH_8NaO_6
Molecular Weight: 137.861
CAS RN: 10555-76-7
Properties: tricl; readily obtained by cooling solution of borax and a calculated amount of NaOH; absorbs atm CO_2, forming borax and sodium carbonate; stable phase in saturated solution from 11.5°C to 53.6°C, stable phase above 53.6°C is the dihydrate [KIR78]
Solubility: % anhydrous in H_2O soln: 14.5 (0°C), 21.7 (25°C), 34.1 (50°C); solubility in methanol is 26.4% (40°C) [KIR78]
Density, g/cm³: 1.74 [KIR78]
Melting Point, °C: ~54 melts in waters of hydration [KIR78]
Reactions: minus H_2O 120°C [CRC10]

2933

Compound: Sodium metagermanate
Formula: Na_2GeO_3
Molecular Formula: $GeNa_2O_3$
Molecular Weight: 166.588
CAS RN: 12025-19-3
Properties: white, monocl, deliq [CRC10]
Solubility: g/100 g H_2O: 14.4 (0°C), 23.8 (20°C), 116 (80°C) [LAN05]
Density, g/cm³: 3.31 [CRC10]
Melting Point, °C: 1083 [CRC10]

2934

Compound: Sodium metaniobate heptahydrate
Formula: $Na_2Nb_2O_6 \cdot 7H_2O$
Molecular Formula: $H_{14}Na_2Nb_2O_{13}$
Molecular Weight: 453.895
CAS RN: 67211-31-8
Properties: colorless tricl; obtained from niobium pentoxide and sodium hydroxide or sodium carbonate [KIR81]
Solubility: sl s H_2O [HAW93]
Density, g/cm³: 4.512–4.559 [CRC10]
Reactions: minus H_2O 100°C [CRC10]

2935

Compound: Sodium metasilicate
Formula: Na_2SiO_3
Molecular Formula: Na_2O_3Si
Molecular Weight: 122.064
CAS RN: 6834-92-0
Properties: white powd; bead shaped with uniform particle size; hygr; normally in the form of a glass or ortho-rhomb cryst; enthalpy of fusion 52.2 kJ/mol; the nonahydrate $Na_2SiO_3 \cdot 9H_2O$ is ortho-rhomb, efflorescent, melts at 48°C in its waters of cryst, has enthalpy of hydration of −101.04 kJ/mol; normally prepared by fusion of sand (SiO_2) and soda ash (Na_2CO_3); used for cleaning laundry, dairy and metals, for floor cleaning [HAW93] [MER06] [STR93] [OXY93]
Solubility: parts/100 parts soln, H_2O: 15.8 (20°C), 15.6 (35°C), 36.1 (45°C), 46.5 (55°C), 48.3 (60°C), 52.4 (70°C); i alcohol, acids, salt solutions [OXY93] [MER06]
Density, g/cm³: 2.614 [MER06]
Melting Point, °C: 1089 [MER06]

2936

Compound: Sodium metasilicate pentahydrate
Formula: $Na_2SiO_3 \cdot 5H_2O$
Molecular Formula: $H_{10}Na_2O_8Si$
Molecular Weight: 212.140
CAS RN: 13517-24-3
Properties: white powd; uniform particle size; enthalpy of fusion 30.5 kJ/mol; manufactured in brick furnace at 1300°C by fusing sand and soda ash, followed by removal of glass; used in the form of liq solutions for numerous applications such as in coatings, adhesives and cements, gels and catalysts [STR93] [OXY93]
Solubility: parts/100 parts soln in H_2O: 27.5 (20°C), 44.5 (35°C), 62.8 (45°C), 80.9 (55°C), 84.0 (60°C), 91.2 (70°C) [OXY93]
Melting Point, °C: 72.2 [OXY93]

2937
Compound: Sodium metatantalate
Formula: $NaTaO_3$
Molecular Formula: NaO_3Ta
Molecular Weight: 251.936
CAS RN: 12034-15-0
Properties: white powd, −100 mesh of
99.9% purity [CRC10] [STR93]

2938
Compound: Sodium metavanadate
Formula: $NaVO_3$
Molecular Formula: NaO_3V
Molecular Weight: 121.930
CAS RN: 13718-26-8
Properties: −200 mesh with 99.9% purity; colorless,
monocl, or pale green cryst powd; used in
inks, and in fur dyeing [CER91] [HAW93]
Solubility: 21.1 g/100 mL H_2O (25°C),
38.8 g/100 mL H_2O (75°C) [CRC10]
Melting Point, °C: 630 [STR93]

2939
Compound: Sodium molybdate
Formula: Na_2MoO_4
Molecular Formula: $MoNa_2O_4$
Molecular Weight: 205.918
CAS RN: 7631-95-0
Properties: white to off-white powd; small, lustrous,
cryst plates; prepared by evaporation of aq solution
of molybdic oxide and sodium hydroxide to form
the dihydrate, followed by heating at 100°C; used
in pigments and metal finishing, and to inhibit
corrosion in aq media [KIR81] [HAW93] [STR93]
Solubility: g/100 g soln, H_2O: 30.62 (0°C), 39.40 (25°C),
45.52 (100°C); solid phase, $Na_2MoO_4 \cdot 10H_2O$
(0°C), $Na_2MoO_4 \cdot 2H_2O$ (25°C, 100°C) [KRU93]
Density, g/cm³: 3.28 [STR93]
Melting Point, °C: 687 [STR93]

2940
Compound: Sodium molybdate dihydrate
Formula: $Na_2MoO_4 \cdot 2H_2O$
Molecular Formula: $H_4MoNa_2O_6$
Molecular Weight: 241.948
CAS RN: 10102-40-6
Properties: cryst powd [MER06]
Solubility: s 1.7 parts cold H_2O, ~0.7
parts boiling H_2O [MER06]
Density, g/cm³: 3.28 [STR93]
Reactions: minus $2H_2O$ 100°C [CRC10]

2941
Compound: Sodium molybdosilicate hydrate
Formula: $Na_4SiMo_{12}O_{40} \cdot xH_2O$
Molecular Formula: $Mo_{12}Na_4O_{40}Si$ (anhydrous)
Molecular Weight: 1911.301 (anhydrous)
CAS RN: 103443-51-2
Properties: yellow cryst; used as a catalyst [HAW93]
Solubility: s H_2O, acetone, alcohol, ethyl acetate;
i ether, benzene, cyclohexane [HAW93]
Density, g/cm³: 3.44 [HAW93]

2942
Compound: Sodium niobate
Formula: $NaNbO_3$
Molecular Formula: $NaNbO_3$
Molecular Weight: 163.894
CAS RN: 12034-09-2
Properties: −100 mesh with 99.9% purity;
white powd [CER91] [STR93]
Density, g/cm³: 4.55 [LID94]
Melting Point, °C: 1422 [AES93]

2943
Compound: Sodium nitrate
Synonym: niter
Formula: $NaNO_3$
Molecular Formula: $NNaO_3$
Molecular Weight: 84.995
CAS RN: 7631-99-4
Properties: colorless transparent trigonal cryst,
white granules or powd; deliq; enthalpy of
fusion 15.00 kJ/mol; preparation: extraction
from ore by brine at 70°C, followed by
crystallization [KIR82] [MER06] [CRC10]
Solubility: g/100 g H_2O: 73.0 (0°C), 87.6 (20°C),
180 (100°C) [LAN05]; 1 g dissolves in: 125 mL
alcohol, 52 mL boiling alcohol; 3470 mL absolute
alcohol; 300 mL absolute methanol [MER06]
Density, g/cm³: 2.261 [STR93]
Melting Point, °C: 307 [CRC10]
Boiling Point, °C: explodes at 537 [HAW93]
Thermal Expansion Coefficient: (volume)
100°C (1.076), 200°C (2.74) [CLA66]

2944
Compound: Sodium nitrite
Formula: $NaNO_2$
Molecular Formula: $NNaO_2$
Molecular Weight: 68.996
CAS RN: 7632-00-0

Properties: white or sl yellow hygr granules, rods or powd; body-centered ortho-rhomb, a=0.355 nm, b=0.556 nm, c=0.557 nm; enthalpy of transition at 158°C–165°C is 1192 kJ/mol; oxidizing agent; slowly oxidized to the nitrate by atm O_2; preparation: by the dissolution of nitrogen oxides in aq alkaline solutions [KIR82] [MER06]

Solubility: g/100 g soln, H_2O: 41.65 (0°C), 45.92 (25°C), 61.50 (99.9°C); solid phase, $NaNO_2$ [KRU93]; sl s alcohol; decomposed by weak acids, evolving brown fumes of N_2O_3 [MER06]

Density, g/cm³: 2.168 [STR93]

Melting Point, °C: 271 [MER06]

Boiling Point, °C: 320, decomposes [KIR82]

Reactions: forms Na_2O and N_2 or nitrogen oxide at 320°C [KIR82]

2945

Compound: Sodium nitroferricyanide(III) dihydrate

Synonym: sodium nitroprusside dihydrate

Formula: $Na_2[Fe(CN)_5NO] \cdot 2H_2O$

Molecular Formula: $C_5H_4FeN_6Na_2O_3$

Molecular Weight: 297.950

CAS RN: 13755-38-9

Properties: ruby red; transparent cryst; aq solutions decompose [MER06]

Solubility: s in ~2.3 parts H_2O; sl s alcohol [MER06]

Density, g/cm³: 1.72 [STR93]

2946

Compound: Sodium oleate

Formula: $C_{17}H_{33}COONa$

Molecular Formula: $C_{18}H_{33}NaO_2$

Molecular Weight: 304.449

CAS RN: 143-19-1

Properties: white powd; used in ore flotation, to waterproof textiles [HAW93]

Solubility: s H_2O, partially decomposes; s alcohol [HAW93]

Melting Point, °C: 232–235 [CRC10]

2947

Compound: Sodium orthosilicate

Formula: Na_4SiO_4

Molecular Formula: Na_4O_4Si

Molecular Weight: 184.043

CAS RN: 13472-30-5

Properties: white powd; used in laundries, for metal cleaning, in heavy duty cleaning [HAW93] [STR93]

Solubility: s H_2O [HAW93]

Melting Point, °C: 1018 [STR93]

2948

Compound: Sodium orthovanadate

Formula: Na_3VO_4

Molecular Formula: Na_3O_4V

Molecular Weight: 183.909

CAS RN: 13721-39-6

Properties: −200 mesh with 99.9% purity; colorless hex prisms [CER91] [KIR83]

Solubility: s H_2O [KIR83]

Melting Point, °C: 850–856 [KIR83]

2949

Compound: Sodium orthovanadate decahydrate

Formula: $Na_3VO_4 \cdot 10H_2O$

Molecular Formula: $H_{20}Na_3O_{14}V$

Molecular Weight: 364.062

CAS RN: 16519-60-1

Properties: white cryst [STR93]

Melting Point, °C: 850–866 [STR93]

2950

Compound: Sodium oxalate

Formula: $Na_2C_2O_4$

Molecular Formula: $C_2Na_2O_4$

Molecular Weight: 134.000

CAS RN: 62-76-0

Properties: white, odorless powd; used in textile and leather finishing, blueprinting [MER06] [HAW93]

Solubility: g/100 g soln, H_2O: 2.62 (0°C), 3.48 (25°C), 6.10 (100°C) [KRU93]; i alcohol [MER06]

Density, g/cm³: 2.34 [HAW93]

Melting Point, °C: 250–270, decomposes [HAW93]

2951

Compound: Sodium oxide

Synonym: sodium monoxide

Formula: Na_2O

Molecular Formula: Na_2O

Molecular Weight: 61.979

CAS RN: 1313-59-3

Properties: white; amorphous pieces or powd; reacts violently with H_2O, forming NaOH; enthalpy of fusion 48.00 kJ/mol [MER06] [CRC10]

Solubility: reacts with H_2O, forming NaOH [HAW93]

Density, g/cm³: 2.27 [MER06]

Melting Point, °C: 1132 [CRC10]

Boiling Point, °C: sublimes at 1274 [HAW93]

Reactions: decomposition begins >400°C to form Na_2O_2 and Na [MER06]

2952
Compound: Sodium paraperiodate
Formula: $Na_3H_2IO_6$
Molecular Formula: $H_2INa_3O_6$
Molecular Weight: 293.885
CAS RN: 13940-38-0
Properties: white cryst solid; used as a selective
    oxidizing agent for specific carbohydrates
    and amino acids [HAW93]
Solubility: v sl s H_2O; s in conc NaOH
    solutions [HAW93]

2953
Compound: Sodium pentaiodobismuthate tetrahydrate
Synonym: bismuth sodium iodide
Formula: $Na_2BiI_5 \cdot 4H_2O$
Molecular Formula: $BiH_8I_5Na_2O_4$
Molecular Weight: 961.544
CAS RN: 53778-50-0
Properties: odorless, red cryst, astringent taste [MER06]
Solubility: s H_2O, hydrolyzes [MER06]
Melting Point, °C: decomposes 93 [MER06]

2954
Compound: Sodium perborate monohydrate
Formula: $NaBO_3 \cdot H_2O$
Molecular Formula: BH_2NaO_4
Molecular Weight: 99.815
CAS RN: 10332-33-9
Properties: white, amorphous powd; used as
    a denture cleaner, as a bleaching agent
    in special detergents [HAW93]
Solubility: v s H_2O, reacting to give H_2O_2
    and sodium borate [HAW93]

2955
Compound: Sodium perborate tetrahydrate
Formula: $NaBO_3 \cdot 4H_2O$
Molecular Formula: BH_8NaO_7
Molecular Weight: 153.861
CAS RN: 10486-00-7
Properties: white, odorless, cryst powd; stable when cool
    and dry, else decomposes evolving O_2 [MER06]
Solubility: s ~40 parts H_2O, soln eventually decomposes
    in the sequence: $\rightarrow H_2O_2 \rightarrow O_2$ [MER06]
Melting Point, °C: 60, decomposes [ALD94]

2956
Compound: Sodium perchlorate
Formula: $NaClO_4$

Molecular Formula: $ClNaO_4$
Molecular Weight: 122.441
CAS RN: 7601-89-0
Properties: white powd; hygr; used in
    explosives, jet fuel [HAW93] [STR93]
Solubility: g/100 g soln, H_2O: 62.8 ± 0.1 (0°C), 67.7 ± 0.1
    (25°C), 76.75 (100°C); solid phase, $NaClO_4 \cdot H_2O$
    (0°C, 25°C), $NaClO_4$ (100°C) [KRU93]
Density, g/cm³: 2.499 [KIR79]
Melting Point, °C: 482 [STR93]
Boiling Point, °C: decomposes [HAW93]
Reactions: transition from ortho-rhomb
    to cub at 577–586 K [KIR79]

2957
Compound: Sodium perchlorate monohydrate
Formula: $NaClO_4 \cdot H_2O$
Molecular Formula: ClH_2NaO_5
Molecular Weight: 140.456
CAS RN: 7791-07-3
Properties: white cryst; deliq [MER06]
Solubility: 66 parts in 100 parts H_2O (0°C) [KIR79]
Density, g/cm³: 2.02 [MER06]
Melting Point, °C: decomposes ~130 [MER06]

2958
Compound: Sodium periodate
Synonym: sodium metaperiodate
Formula: $NaIO_4$
Molecular Formula: $INaO_4$
Molecular Weight: 213.892
CAS RN: 7790-28-5
Properties: white, tetr cryst; oxidant [ALD94] [MER06]
Solubility: g/100 g soln, H_2O: 12.62 (25°C);
    solid phase, $NaIO_4 \cdot 3H_2O$ [KRU93];
    s H_2SO_4, HNO_3, acetic acids [MER06]
Density, g/cm³: 3.865 [MER06]
Melting Point, °C: decomposes ~300 [MER06]

2959
Compound: Sodium periodate trihydrate
Synonym: sodium metaperiodate
Formula: $NaIO_4 \cdot 3H_2O$
Molecular Formula: H_6INaO_7
Molecular Weight: 267.938
CAS RN: 13472-31-6
Properties: stable phase below 34.5°C; white,
    efflorescent; trig cryst [MER06] [KIR81]
Solubility: 1 g/8 mL H_2O (20°C) [MER06]
Density, g/cm³: 3.219 (18°C) [HAW93]
Melting Point, °C: decomposes 175 [MER06]

2960
Compound: Sodium permanganate trihydrate
Formula: $NaMnO_4 \cdot 3H_2O$
Molecular Formula: H_6MnNaO_7
Molecular Weight: 195.972
CAS RN: 10101-50-5
Properties: purple to reddish black; very hygr; granules; used as an oxidizing agent, disinfectant; there is a monohydrate, CAS RN 79048-36-5 [MER06] [ALD94]
Solubility: v s H_2O; decomposed by alcohol [MER06]
Density, g/cm³: 2.47 [HAW93]
Melting Point, °C: 170, decomposes [HAW93]

2961
Compound: Sodium peroxide
Synonym: sodium dioxide
Formula: Na_2O_2
Molecular Formula: Na_2O_2
Molecular Weight: 77.979
CAS RN: 1313-60-6
Properties: yellowish white; granular powd; absorbs atm water and CO_2; reacts with dil acids to produce H_2O_2; strong oxidant, e.g. readily reacts with organic matter or other oxidizable materials [MER06]
Solubility: v s H_2O, producing $NaOH$, H_2O_2; H_2O_2 quickly decomposes evolving O_2 [MER06]
Density, g/cm³: 2.805 [STR93]
Melting Point, °C: 460, decomposes [STR93]

2962
Compound: Sodium perrhenate
Formula: $NaReO_4$
Molecular Formula: NaO_4Re
Molecular Weight: 273.195
CAS RN: 13472-33-8
Properties: white powd; hygr [STR93]
Solubility: 100 g/100 mL H_2O (20°C) [CRC10]
Density, g/cm³: 5.39 [STR93]
Melting Point, °C: 300 (in oxygen) [STR93]

2963
Compound: Sodium persulfate
Synonym: sodium peroxydisulfate
Formula: $Na_2S_2O_8$
Molecular Formula: $Na_2O_8S_2$
Molecular Weight: 238.107
CAS RN: 7775-27-1
Properties: white; cryst powd; gradually decomposes if standing, with rate of decomposition accelerated by moisture and high temp; strong oxidizing agent; used as bleaching agent for fats, oils, fabrics [MER06]

Solubility: 549 g/L H_2O (20°C); decomposed by alcohol [MER06]
Density, g/cm³: 2.400 [ALD94]

2964
Compound: Sodium phosphate
Synonyms: trisodium phosphate, sodium orthophosphate
Formula: Na_3PO_4
Molecular Formula: Na_3O_4P
Molecular Weight: 163.940
CAS RN: 7601-54-9
Properties: hygr; −100 mesh powd; uses: photographic developers, clarify sugar, clean boiler scale, soften water, paper manufacturing, laundering, detergents [ALD94] [MER06] [AES93]
Solubility: g/100 g H_2O: 5.38 (0°C), 14.53 (25°C), 94.6 (100°C); solid phase, $Na_3PO_4 \cdot 1/4NaOH \cdot 12H_2O$ (0°C, 25°C), $Na_3PO_4 \cdot 6H_2O$ (100°C) [KRU93]

2965
Compound: Sodium phosphate dodecahydrate
Formula: $Na_3PO_4 \cdot 12H_2O$
Molecular Formula: $H_{24}Na_3O_{16}P$
Molecular Weight: 380.124
CAS RN: 10101-89-0
Properties: colorless or white cryst; used to soften water, as a detergent and metal cleaner [MER06] [HAW93] [STR93]
Solubility: s 3.5 parts H_2O, 1 part boiling H_2O; i alcohol [MER06]
Density, g/cm³: 1.62 [HAW93]
Melting Point, °C: ~75, when heated rapidly [MER06]
Reactions: minus $12H_2O$ at 100°C [HAW93]

2966
Compound: Sodium phosphide
Formula: Na_3P
Molecular Formula: Na_3P
Molecular Weight: 99.943
CAS RN: 12058-85-4
Properties: red solid; decomposes when heated or if immersed in water, evolving phosphine; thermally stable up to 650°C; there are also Na_2P, 12439-14-4, and Na_3P_{11}, 39343-85-6; produced by reacting Na and P, then stored under oil [KIR82] [HAW93]
Solubility: decomposes in H_2O, evolving PH_3 [CRC10] [HAW93]
Melting Point, °C: stable up to 650 [KIR82]

2967
Compound: Sodium phosphomolybdate
Synonym: sodium 12-molybdophosphate
Formula: $Na_3PO_4 \cdot 12MoO_3$
Molecular Formula: $MO_{12}Na_3O_{40}P$
Molecular Weight: 1891.199
CAS RN: 1313-30-0
Properties: yellow cryst; used in chemical analysis, neuromicroscopy, imparts water resistance to plastics, adhesives, cements [HAW93] [MER06]
Solubility: v s H_2O [MER06]
Density, g/cm³: 2.83 [HAW93]

2968
Compound: Sodium phosphotungstate
Synonym: sodium 12-tungstophosphate
Formula: $2Na_2O \cdot P_2O_5 \cdot 12WO_3 \cdot 18H_2O$
Molecular Formula: $H_{36}Na_4O_{61}P_2W_{12}$
Molecular Weight: 3372.236
CAS RN: 51312-42-6
Properties: yellowish white; granular powd; used as a reagent, and in the manufacture of pigments [HAW93] [MER06]
Solubility: v s H_2O, alcohol [HAW93]

2969
Compound: Sodium polyphosphate
Synonym: sodium polymetaphosphate
Formula: $Na_{(n+2)}P_nO_{(3n+1)}$
Molecular Formula: $Na_5P_3O_{10}$
CAS RN: 50813-16-6
Properties: clear hygr glass; two most important of the polyphosphates are with n=2 and with n=3; preparation: rapid chilling of molten sodium metaphosphate; used in water treatment for sequestering metals [HAW93] [MER06] [ALD94]
Solubility: s H_2O [MER06]
Melting Point, °C: 628 [MER06]

2970
Compound: Sodium potassium tartrate tetrahydrate
Synonyms: Rochelle salt, potassium sodium tartrate
Formula: $NaKC_4H_4O_6 \cdot 4H_2O$
Molecular Formula: $C_4H_{12}KNaO_{10}$
Molecular Weight: 282.221
CAS RN: 304-59-6
Properties: translucent cryst or white cryst powd; sl efflorescent in warm air; has cool, saline taste; used in baking powd, as a cathartic in medicine, for silvering mirrors [HAW93] [MER06]
Solubility: g anhydrous/100 g H_2O: 31.9 (0°C), 67.8 (20°C), 102 (30°C) [LAN05]; sl s alcohol [MER06]

Density, g/cm³: 1.79 [MER06]
Melting Point, °C: 70–80 [MER06]
Boiling Point, °C: decomposition begins at 220 [MER06]
Reactions: minus $3H_2O$ at 100°C; anhydrous 130°C–140°C [MER06]

2971
Compound: Sodium pyrophosphate
Formula: $Na_4P_2O_7$
Molecular Formula: $Na_4O_7P_2$
Molecular Weight: 265.902
CAS RN: 7722-88-5
Properties: colorless, transparent cryst or white powd; used in water softening, as a metal cleaner [HAW93]
Solubility: g/100 g soln, H_2O: 2.236 (0°C), 6.618 (25°C), 31.15 (96°C); solid phase, $Na_4P_2O_7 \cdot 10H_2O$ (0°C, 25°C), $Na_4P_2O_7$ (100°C) [KRU93]
Density, g/cm³: 2.45 [HAW93]
Melting Point, °C: 880 [HAW93]

2972
Compound: Sodium pyrophosphate decahydrate
Formula: $Na_4P_2O_7 \cdot 10H_2O$
Molecular Formula: $H_{20}Na_4O_{17}P_2$
Molecular Weight: 446.055
CAS RN: 13472-36-1
Properties: white powd [STR93]
Solubility: s H_2O; decomposed by alcohol [HAW93]
Density, g/cm³: 1.815–1.836 [STR93]
Reactions: minus H_2O 94°C [HAW93]

2973
Compound: Sodium pyrovanadate
Formula: $Na_4V_2O_7$
Molecular Formula: $Na_4O_7V_2$
Molecular Weight: 305.838
CAS RN: 13517-26-5
Properties: −200 mesh with 99.9% purity; colorless hex prisms [KIR83] [CER91]
Solubility: s H_2O [KIR83]
Melting Point, °C: 632–654 [KIR83]

2974
Compound: Sodium selenate
Formula: Na_2SeO_4
Molecular Formula: Na_2O_4Se
Molecular Weight: 188.938
CAS RN: 13410-01-0

Properties: −100 mesh with 99.5% purity; white powd;
uses: insecticide [MER06] [STR93] [CER91]
Solubility: g/100 g soln, H_2O: 11.74 (0°C), 36.91 (25.2°C),
42.14 (100°C); solid phase, $Na_2SeO_4 \cdot 10H_2O$
(0°C, 25°C), Na_2SeO_4 (100°C) [KRU93]
Density, g/cm³: 3.213 [STR93]

2975
Compound: Sodium selenate decahydrate
Formula: $Na_2SeO_4 \cdot 10H_2O$
Molecular Formula: $H_{20}Na_2O_{14}Se$
Molecular Weight: 369.091
CAS RN: 10102-23-5
Properties: white cryst [MER06]
Solubility: v s H_2O [MER06]
Density, g/cm³: 1.603–1.620 [HAW93]

2976
Compound: Sodium selenide
Formula: Na_2Se
Molecular Formula: Na_2Se
Molecular Weight: 124.940
CAS RN: 1313-85-5
Properties: −60 mesh, dry under argon with 99.9%
purity; amorphous cryst: deliq; becomes red
if exposed to the atm; decahydrate: needles,
becomes red, then brown in air; hexadecahydrate:
prisms; decomposes in air to Na_2CO_3, Se
and some Na_2Se [MER06] [CER91]
Solubility: decomposes in H_2O [MER06]
Density, g/cm³: 2.625 [MER06]
Melting Point, °C: Na_2Se: >875;
hexadecahydrate: 40 [MER06]

2977
Compound: Sodium selenite
Formula: Na_2SeO_3
Molecular Formula: Na_2O_3Se
Molecular Weight: 172.938
CAS RN: 10102-18-8
Properties: −100 mesh with 99.5% purity; white powd;
tetr prisms; stable in air [MER06] [STR93] [CER91]
Solubility: g/100 g soln, H_2O: 47.28 (24.4°C), 45.3
(103.3°C); solid phase, $Na_2SeO_3 \cdot 5H_2O$ (24.4°C),
Na_2SeO_3 (103.3°C) [KRU93]; i alcohol [MER06]
Melting Point, °C: decomposes [AES93]

2978
Compound: Sodium selenite pentahydrate
Formula: $Na_2SeO_3 \cdot 5H_2O$

Molecular Formula: $H_{10}Na_2O_8$
Molecular Weight: 263.014
CAS RN: 26970-82-1
Properties: white acicular cryst; evolves $5H_2O$ in
dry air; used in glass manufacturing to control
color, in decorating porcelain, and for testing
seed germination [HAW93] [MER06] [ALD94]
Solubility: s H_2O; i alcohol [HAW93]

2979
Compound: Sodium silicate
Synonym: waterglass
Formula: $Na_2O \cdot xSiO_2$
Molecular Formula: Na_2SiO_3
CAS RN: 1344-09-8
Properties: colorless to white; usual compositions:
Na_2SiO_3, $Na_6Si_2O_7$, $Na_2Si_3O_7$; contains variable
amounts of H_2O, e.g. $Na_2SiO_3 \cdot 5H_2O$; produced
by fusion of sand and soda ash; used as a catalyst
and in silica gels, soaps [HAW93] [MER06]
Solubility: v sl s cold H_2O [MER06]

2980
Compound: Sodium stannate trihydrate
Formula: $Na_2SnO_3 \cdot 3H_2O$
Molecular Formula: $H_6Na_2O_6Sn$
Molecular Weight: 266.734
CAS RN: 12209-98-2
Properties: −100 mesh with 99.9% purity; white
or colorless; cryst; decomposed in air and by
weak acids; used as a mordant in dyeing, and
in ceramics [HAW93] [MER06] [CER91]
Solubility: g/100 g H_2O: 46.0 (0°C), 43.7 (20°C),
38.9 (40°C) [LAN05]; i alcohol [MER06]
Reactions: minus $3H_2O$ at 140°C [HAW93]

2981
Compound: Sodium stearate
Synonyms: stearic acid, sodium salt
Formula: $CH_3(CH_2)_{16}COONa$
Molecular Formula: $C_{18}H_{35}NaO_2$
Molecular Weight: 306.465
CAS RN: 822-16-2
Properties: usually contains sodium palmitate;
white powd; soapy feel; hydrolyzes in water
to given an alkaline solution; used as a
waterproofing and gelling agent; in toothpaste
and cosmetics [HAW93] [MER06]
Solubility: slowly s cold H_2O, alcohol;
v s hot H_2O, alcohol [MER06]

2982
Compound: Sodium sulfate
Synonyms: mirabilite, thenardite
Formula: Na$_2$SO$_4$
Molecular Formula: Na$_2$O$_4$S
Molecular Weight: 142.044
CAS RN: 7757-82-6
Properties: white odorless powd or ortho-rhomb cryst; enthalpy of fusion is 23.60 kJ/mol; used in the manufacture of kraft paper, paperboard and glass; also used as a filler in synthetic detergents [HAW93] [MER06] [KIR83] [CRC10]
Solubility: g/100 g H$_2$O: 4.5 (0°C), 28.0 (25°C), 42.2 (100°C); solid phase, Na$_2$SO$_4$·10H$_2$O (0°C, 25°C), Na$_2$SO$_4$ (100°C) [KRU93]; i alcohol [MER06]
Density, g/cm^3: 2.68 [STR93]
Melting Point, °C: 884 [CRC10]

2983
Compound: Sodium sulfate decahydrate
Synonyms: Glauber's salt, mirabilite
Formula: Na$_2$SO$_4$·10H$_2$O
Molecular Formula: H$_{20}$Na$_2$O$_{14}$S
Molecular Weight: 322.197
CAS RN: 7727-73-3
Properties: large, transparent monocl cryst or granules; effloresces; enthalpy of crystallization 74.98 J/mol at 25°C; energy storage capacity is more than seven times greater than that of water; used in solar energy to store heat, and in air conditioning [MER06] [HAW93]
Solubility: s in 3.3 parts H$_2$O (15°C), 1.5 parts (25°C); s glycerol; i alcohol [MER06]
Density, g/cm^3: 1.464 (cryst) [HAW93]
Melting Point, °C: 32.4 [MER06]
Reactions: gives up waters of hydration at 100°C [HAW93]

2984
Compound: Sodium sulfate heptahydrate
Formula: Na$_2$SO$_4$·7H$_2$O
Molecular Formula: H$_{14}$Na$_2$O$_{11}$S
Molecular Weight: 268.150
CAS RN: 7727-73-3
Properties: white rhomb; tetr [CRC10] [LAN05]
Solubility: g/100 g H$_2$O: 19.5 (°C), 30.0 (10°C), 44.1 (20°C) [LAN05]
Reactions: minus 7H$_2$O 24.4°C [CRC10]

2985
Compound: Sodium sulfide
Formula: Na$_2$S

Molecular Formula: Na$_2$S
Molecular Weight: 78.046
CAS RN: 1313-82-2
Properties: white cub cryst or granules; very hygr; discolors when exposed to the atm, slowly forming sodium carbonate and sodium thiosulfate; enthalpy of solution is −63.5 kJ/mol; enthalpy of fusion 19.00 kJ/mol; crystallzes from aq solution as the nonahydrate, 1313-84-4 [KIR82] [MER06] [CRC10]
Solubility: g/100 g soln, H$_2$O: 8.8 (0°C), 15.3 (25°C), 60.1 (95°C); solid phase, Na$_2$S·9H$_2$O (0°C, 25°C), Na$_2$S·H$_2$O (95°C) [KRU93]; sl s alcohol; i ether [MER06]
Density, g/cm^3: 1.856 [MER06]
Melting Point, °C: 1180 (vacuum) [MER06]

2986
Compound: Sodium sulfide nonahydrate
Formula: Na$_2$S·9H$_2$O
Molecular Formula: H$_{18}$Na$_2$O$_9$S
Molecular Weight: 240.184
CAS RN: 1313-84-4
Properties: tetr; deliq; cryst; H$_2$S odor; becomes yellow, then brownish black if subjected to atm exposure; decomposed by acids [MER06]
Solubility: 1 g/0.5 mL H$_2$O (25°C); sl s alcohol; i ether [MER06]
Density, g/cm^3: 1.427 [MER06]
Melting Point, °C: 920, decomposes [HAW93]

2987
Compound: Sodium sulfide pentahydrate
Formula: Na$_2$S·5H$_2$O
Molecular Formula: H$_{10}$Na$_2$O$_5$S
Molecular Weight: 168.122
CAS RN: 1313-83-3
Properties: flat, shiny cryst; flammable; evolves H$_2$S in acid solutions [MER06] [ALD94]
Solubility: v s H$_2$O, alcohol; i ether [MER06]
Density, g/cm^3: 1.58 [LID94]
Melting Point, °C: 120 [MER06]
Reactions: minus 3H$_2$O 100°C [CRC10], dehydrates at 120°C [MER06]

2988
Compound: Sodium sulfite
Formula: Na$_2$SO$_3$
Molecular Formula: Na$_2$O$_3$S
Molecular Weight: 126.044
CAS RN: 7757-83-7

Properties: white, small cryst or powd; fairly stable to oxidation; used in paper and dyes industries [HAW93] [MER06]
Solubility: g/100 g H_2O: 13.85 (0°C), 30.5 ± 0.4 (25°C), 26.3 (100°C); solid phase, $Na_2SO_3 \cdot 7H_2O$ (0°C, 25°C), Na_2SO_3 (100°C) [KRU93]
Density, g/cm³: 2.633 [STR93]
Melting Point, °C: decomposes [STR93]

2989
Compound: Sodium sulfite heptahydrate
Formula: $Na_2SO_3 \cdot 7H_2O$
Molecular Formula: $H_{14}Na_2O_{10}S$
Molecular Weight: 252.151
CAS RN: 10102-15-5
Properties: efflorescent cryst; oxidizing in air to sulfate [MER06]
Solubility: s 1.6 parts H_2O, ~30 parts glycerol; sl s alcohol [MER06]
Density, g/cm³: 1.539 [CRC10]
Reactions: minus $7H_2O$ 150°C [CRC10]

2990
Compound: Sodium tartrate dihydrate
Formula: $Na_2C_4H_4O_6 \cdot 2H_2O$
Molecular Formula: $C_4H_8Na_2O_8$
Molecular Weight: 230.083
CAS RN: 868-18-8
Properties: white cryst or granules; used as food additive, as a sequestrant and stabilizer [HAW93] [MER06]
Solubility: s in ~3 parts H_2O, 1.5 parts boiling H_2O; i alcohol [MER06]
Density, g/cm³: 1.794 [HAW93]
Reactions: minus $2H_2O$ at 150°C [HAW93]

2991
Compound: Sodium tellurate(VI)
Formula: Na_2TeO_4
Molecular Formula: Na_2O_4Te
Molecular Weight: 237.578
CAS RN: 10102-83-4
Properties: white powd [MER06]
Solubility: s in 130 parts cold H_2O, 50 parts boiling H_2O [MER06]

2992
Compound: Sodium tellurate(VI) dihydrate
Formula: $Na_2TeO_4 \cdot 2H_2O$
Molecular Formula: $H_4Na_2O_6Te$
Molecular Weight: 273.608
CAS RN: 26006-71-3

Properties: −100 mesh with 99.5% purity; white powd [STR93] [CER91]
Melting Point, °C: decomposes [STR93]

2993
Compound: Sodium tellurite(IV)
Formula: Na_2TeO_3
Molecular Formula: Na_2O_3Te
Molecular Weight: 221.578
CAS RN: 10102-20-2
Properties: −100 mesh with 99.5% purity; white powd; used in bacteriology, and in medicine [HAW93] [CER91]
Solubility: s H_2O [MER06]

2994
Compound: Sodium tetraborate
Synonym: sodium borate
Formula: $Na_2B_4O_7$
Molecular Formula: $B_4Na_2O_7$
Molecular Weight: 201.220
CAS RN: 1330-43-4
Properties: fused sodium borate; white powd or glassy plates; hygr; becomes opaque in air; partially hydrates in damp air; enthalpy of formation of glass form −3256.6 kJ/mol; has several cryst forms; enthalpy of fusion of cryst form 81.2 kJ/mol; used in the manufacture of glass, enamels and other ceramics [KIR78] [HAW93] [MER06]
Solubility: g/100 g soln, H_2O: 1.18 (0°C), 3.13 (25°C), 28.22 (100°C); solid phase, $Na_2B_4O_7 \cdot 10H_2O$ (0°C, 25°C), $Na_2B_4O_7 \cdot 4H_2O$ (100°C) [KRU93]
Density, g/cm³: 2.367 [STR93]
Melting Point, °C: 741 [STR93]
Boiling Point, °C: 1575 [STR93]

2995
Compound: Sodium tetraborate decahydrate
Synonym: borax
Formula: $Na_2B_4O_7 \cdot 10H_2O$
Molecular Formula: $B_4H_{20}Na_2O_{17}$
Molecular Weight: 381.373
CAS RN: 1303-96-4
Properties: hard, odorless cryst, granules or powd; effloresces in dry air; monocl; specific heat, 1.611 kJ/(kg · K); cryst habit may be changed by adding various substances, and by altering conditions [MER06] [KIR78]
Solubility: %w anhydrous salt: 1.18 (0°C); 3.13 (25°C); 15.90 (60°C) [KIR78]; 1 g/1 mL glycerol; i alcohol [MER06]

Density, g/cm³: 1.73 [MER06]
Melting Point, °C: 75, when heated rapidly [MER06]
Reactions: loses $5H_2O$ at 100°C; loses $9H_2O$ at 150°C, dehydrates at 320°C [MER06]

2996
Compound: Sodium tetraborate pentahydrate
Synonym: tincalconite
Formula: $Na_2B_4O_7 \cdot 5H_2O$
Molecular Formula: $B_4H_{10}Na_2O_{12}$
Molecular Weight: 291.296
CAS RN: 12045-88-4
Properties: free flowing powd; trig; specific heat 1.32 kJ/(kg·K); enthalpy of formation −4784.4 MJ/ mol; used as a weed killer, and to control fungus growth on citrus fruit [KIR78]
Solubility: %w of anhydrous in H_2O: 16.40 (60°C), 23.38 (80°C), 34.63 (100°C); %w of pentahydrate at 25°C: 16.9% in methanol, 31.1% in ethylene glycol, 10.0% in diethylene glycol [KIR78]
Density, g/cm³: 1.815 [HAW93]
Reactions: minus H_2O at 122°C [HAW93]

2997
Compound: Sodium tetraborate tetrahydrate
Synonym: ernite
Formula: $Na_2B_4O_7 \cdot 4H_2O$
Molecular Formula: $B_4H_8Na_2O_{11}$
Molecular Weight: 273.281
CAS RN: 12045-87-3
Properties: monocl; specific heat ~1.2 kJ/(kg·K); enthalpy of formation −4489.0 kJ/mol; absorbs water to form borax at relative humidities above 70% [KIR78]
Solubility: % anhydrous in H_2O: 14.82 (60°C), 17.12 (70°C), 19.88 (80°C), 23.31 (90°C), 28.22 (100°C) [KIR78]
Density, g/cm³: 1.95 [LID94]

2998
Compound: Sodium tetrabromoaurate(III)
Formula: $NaAuBr_4$
Molecular Formula: $AuBr_4Na$
Molecular Weight: 539.573
CAS RN: 52495-41-7
Properties: reddish black cryst [STR93]

2999
Compound: Sodium tetrachloroaluminate
Formula: $NaAlCl_4$
Molecular Formula: $AlCl_4Na$
Molecular Weight: 191.783

CAS RN: 7784-16-9
Properties: yellow hygr powd; used as a catalyst for organic reactions [CRC10] [HAW93]
Solubility: s H_2O [MER06]
Density, g/cm³: 2.01 [LID94]
Melting Point, °C: 185 [CRC10]

3000
Compound: Sodium tetrachloroaurate(III) dihydrate
Formula: $NaAuCl_4 \cdot 2H_2O$
Molecular Formula: $AuCl_4H_4NaO_2$
Molecular Weight: 397.799
CAS RN: 13874-02-7
Properties: yellowish orange cryst; rhomb; stable up to 100°C; used in photography, in staining fine glass, for decorating porcelain and in medicine [HAW93] [MER06]
Solubility: g anhydrous/100 g H_2O: 139 (10°C), 151 (20°C), 900 (60°C) [LAN05]; s alcohol, ether [MER06]
Melting Point, °C: 100 decomposes [AES93]

3001
Compound: Sodium tetrachloropalladate(II) trihydrate
Formula: $Na_2PdCl_4 \cdot 3H_2O$
Molecular Formula: $Cl_4H_6Na_2O_3Pd$
Molecular Weight: 384.256
CAS RN: 13820-53-6
Properties: reddish brown powd [STR93]

3002
Compound: Sodium tetrafluoroberyllate
Synonym: beryllium sodium fluoride
Formula: Na_2BeF_4
Molecular Formula: BeF_4Na_2
Molecular Weight: 130.986
CAS RN: 13871-27-7
Properties: ortho-rhomb or monocl cryst [MER06]
Solubility: g/100 g H_2O: 1.33 (0°C), 1.44 (20°C), 2.73 (90°C) [LAN05]
Density, g/cm³: 2.47 [LID94]
Melting Point, °C: 575 [LID94]

3003
Compound: Sodium tetrasulfide
Formula: Na_2S_4
Molecular Formula: Na_2S_4
Molecular Weight: 174.244
CAS RN: 12034-39-8

Properties: yellow, hygr cryst; can be clear, dark red liq; prepared by reacting Na_2S with S; used to reduce organic nitro compounds, for the manufacture of sulfur dyes, and in the preparation of metal sulfide finishes [HAW93]
Density, g/cm³: 1.335 at 15.5°C [KIR83]
Melting Point, °C: cryst: 275 [HAW93]; solidifies at −33 to 10 [KIR83]
Boiling Point, °C: 115 [KIR83]

3004
Compound: Sodium thioantimonate nonahydrate
Synonym: Schlippe's salt
Formula: $Na_3SbS_4 \cdot 9H_2O$
Molecular Formula: $H_{18}Na_3O_9S_4Sb$
Molecular Weight: 481.131
CAS RN: 13776-84-6
Properties: colorless or light yellow large cryst; becomes covered with reddish brown coating of antimony sulfide if exposed to air [MER06]
Solubility: g anhydrous/100 g H_2O: 13.4 (0°C), 27.9 (20°C), 88.3 (80°C) [LAN05]; i alcohol; decomposed by weak acids [MER06]
Density, g/cm³: 1.806 [CRC10]
Melting Point, °C: 87 [CRC10]
Boiling Point, °C: decomposes 234 [CRC10]

3005
Compound: Sodium thiocyanate
Formula: NaSCN
Molecular Formula: CNNaS
Molecular Weight: 81.074
CAS RN: 540-72-7
Properties: colorless cryst or white powd; hygr; sensitive to light; used as an analytical reagent, in dyeing and printing textiles [HAW93]
Solubility: g/100 g H_2O: 142.6 (25°C), 225.6 (101.4°C); solid phase, $NaSCN \cdot H_2O$ (25°C), NaSCN (100°C) [KRU93]; s alcohol [HAW93]
Melting Point, °C: 287 [HAW93]

3006
Compound: Sodium thiophosphate dodecahydrate
Formula: $Na_3PO_3S \cdot 12H_2O$
Molecular Formula: $H_{24}Na_3O_{15}PS$
Molecular Weight: 396.191
CAS RN: 51674-17-0
Properties: thin, six sided leaflets when prepared from water solvent; effloresces in dry air [MER06] [ALD94]
Solubility: v s warm H_2O [MER06]
Melting Point, °C: 60 [MER06]

3007
Compound: Sodium thiosulfate
Formula: $Na_2S_2O_3$
Molecular Formula: $Na_2O_3S_2$
Molecular Weight: 158.110
CAS RN: 7772-98-7
Properties: colorless monocl powd; hygr [CRC10] [ALD94] [MER06]
Solubility: s H_2O; i alcohol [MER06]
Density, g/cm³: 1.667 [ALD94]

3008
Compound: Sodium thiosulfate pentahydrate
Synonym: hypo
Formula: $Na_2S_2O_3 \cdot 5H_2O$
Molecular Formula: $H_{10}Na_2O_8S_2$
Molecular Weight: 248.186
CAS RN: 10102-17-7
Properties: colorless cryst or granules; effloresces in warm dry air; somewhat deliq in moist air [MER06]
Solubility: g/100 g H_2O: 50.2 (0°C), 70.1 (20°C), 104 (60°C) [LAN05]; i alcohol [MER06]
Density, g/cm³: 1.729 [STR93]
Melting Point, °C: 48, decomposes [STR93]
Reactions: minus $5H_2O$ at 100°C; decomposes at higher temperatures [MER06]

3009
Compound: Sodium titanate
Formula: $Na_2Ti_3O_7$
Molecular Formula: $Na_2O_7Ti_3$
Molecular Weight: 301.577
CAS RN: 12034-36-5
Properties: −200 mesh with 99.9% purity; white cryst; used in welding [HAW93] [CER91]
Solubility: i H_2O [HAW93]
Density, g/cm³: 3.35–3.50 [STR93]
Melting Point, °C: 1128 [STR93]

3010
Compound: Sodium trimetaphosphate hexahydrate
Synonym: Knorre's salt
Formula: $(NaPO_3)_3 \cdot 6H_2O$
Molecular Formula: $H_{12}Na_3O_{15}P_3$
Molecular Weight: 413.976
CAS RN: 7785-84-4
Properties: efflorescent; tricl-rhomb prisms [MER06]
Solubility: 1 g/4.5 mL H_2O; i alcohol [MER06]
Density, g/cm³: 1.786; anhydrous: 2.49 [MER06]
Melting Point, °C: 53 [MER06]
Reactions: minus H_2O when stored; anhydrous at 100°C [MER06]

3011
Compound: Sodium triphosphate
Synonym: sodium tripolyphosphate
Formula: $Na_5P_3O_{10}$
Molecular Formula: $Na_5O_{10}P_3$
Molecular Weight: 367.864
CAS RN: 7758-29-4
Properties: white cryst powd; has two cryst forms; sl hygr; granules; used to soften water, as a food additive, and as a sequestering agent [HAW93] [MER06]
Solubility: g/100 g soln, H_2O: 13.98 (0°C), 12.96 (25°C), 16.50 (70°C); solid phase, $Na_5P_3O_{10} \cdot 6H_2O$ [KRU93]
Melting Point, °C: 622 [HAW93]
Reactions: cryst form transition at 417°C [HAW93]

3012
Compound: Sodium tungstate
Formula: Na_2WO_4
Molecular Formula: Na_2O_4W
Molecular Weight: 293.818
CAS RN: 13472-45-2
Properties: −200 mesh with 99.9% purity; white rhomb powd [STR93] [KIR83] [CER91]
Solubility: g/100 g H_2O: 71.5 (0°C), 73.0 (20°C), 97.2 (100°C) [LAN05]
Density, g/cm³: 4.179 [KIR83]
Melting Point, °C: 698 [KIR83]

3013
Compound: Sodium tungstate dihydrate
Formula: $Na_2WO_4 \cdot 2H_2O$
Molecular Formula: $H_4Na_2O_6W$
Molecular Weight: 329.848
CAS RN: 10213-10-2
Properties: colorless cryst or white rhomb cryst powd; effloresces in dry air [KIR83] [MER06]
Solubility: s in ~1.1 parts H_2O; i alcohol [MER06]
Density, g/cm³: 3.245 [HAW93]
Melting Point, °C: 692 [STR93]
Reactions: minus $2H_2O$ at 100°C [MER06]

3014
Compound: Sodium uranate monohydrate
Synonym: sodium metauranate
Formula: $Na_2U_2O_7 \cdot H_2O$
Molecular Formula: $H_2Na_2O_8U_2$
Molecular Weight: 652.049
CAS RN: 13721-34-1
Properties: yellow powd [MER06]
Solubility: i H_2O; s acids [MER06]

3015
Compound: Sodium uranyl carbonate
Formula: $2Na_2CO_3 \cdot UO_2CO_3$
Molecular Formula: $C_3Na_4O_{11}U$
Molecular Weight: 542.014
CAS RN: 60897-40-7
Properties: yellow; obtained when uranium ores are leached with soda ash at high temperatures and high pressures [KIR83] [CRC10]
Melting Point, °C: decomposes at 400 [KIR83]

3016
Compound: Sodium zirconate
Formula: Na_2ZrO_3
Molecular Formula: Na_2O_3Zr
Molecular Weight: 185.202
CAS RN: 12201-48-8
Properties: −200 mesh with 99.5% purity [CER91]

3017
Compound: Stannic bromide
Synonym: tin(IV) bromide
Formula: $SnBr_4$
Molecular Formula: Br_4Sn
Molecular Weight: 438.326
CAS RN: 7789-67-5
Properties: white cryst mass; fumes strongly in air; enthalpy of vaporization 43.5 kJ/mol; enthalpy of fusion 12.00 kJ/mol; used in mineral separations [HAW93] [MER06] [CRC10]
Solubility: v s water, evolving heat; s alcohol [MER06]
Density, g/cm³: 3.34 [MER06]
Melting Point, °C: 31 [MER06]
Boiling Point, °C: 202 [MER06]

3018
Compound: Stannic chloride
Synonym: tin(IV) chloride
Formula: $SnCl_4$
Molecular Formula: Cl_4Sn
Molecular Weight: 260.521
CAS RN: 7646-78-8
Properties: colorless liq; fumes in air; enthalpy of vaporization 34.9 kJ/mol; enthalpy of fusion 9.20 kJ/mol [MER06] [STR93] [CRC10]
Solubility: s H_2O, evolving heat; s alcohol, CCl_4, benzene, toluene, acetone, kerosene, gasoline [MER06]
Density, g/cm³: 2.2788 [HAW93]
Melting Point, °C: −33 [MER06]
Boiling Point, °C: 114 [DOU83]

3019

Compound: Stannic chloride pentahydrate
Synonym: tin(IV) chloride pentahydrate
Formula: $SnCl_4 \cdot 5H_2O$
Molecular Formula: $Cl_4H_{10}O_5Sn$
Molecular Weight: 350.697
CAS RN: 10026-06-9
Properties: white or slighlty yellow cryst or fused small lumps; slight odor of HCl [MER06]
Solubility: v s H_2O, alcohol [MER06]
Density, g/cm³: 2.04 [KIR83]
Melting Point, °C: ~56 decomposes [KIR83]

3020

Compound: Stannic chromate
Formula: $Sn(CrO_4)_2$
Molecular Formula: Cr_2O_8Sn
Molecular Weight: 350.697
CAS RN: 38455-77-5
Properties: brownish yellow, cryst powd; used in coloring porcelain [HAW93] [MER06]
Solubility: s H_2O [MER06]
Melting Point, °C: decomposed by heating [MER06]

3021

Compound: Stannic fluoride
Synonym: tin(IV) fluoride
Formula: SnF_4
Molecular Formula: F_4Sn
Molecular Weight: 194.704
CAS RN: 7783-62-2
Properties: snow white, tetr cryst; very hygr; can be prepared by reacting F_2 with many stannous or stannic compounds [KIR78] [MER06]
Solubility: hydrolyzes in H_2O [MER06]
Density, g/cm³: 4.78 [MER06]
Melting Point, °C: sublimes 705 [STR93]

3022

Compound: Stannic iodide
Synonym: tin(IV) iodide
Formula: SnI_4
Molecular Formula: I_4Sn
Molecular Weight: 626.328
CAS RN: 7790-47-8
Properties: −6 mesh with 99.999% purity; yellow to reddish cryst; hydrolyzes in H_2O; enthalpy of vaporization 56.9 kJ/mol [CRC10] [MER06] [CER91]
Solubility: s alcohol, benzene, chloroform, ether, carbon disulfide [MER06]
Density, g/cm³: 4.473 [STR93]

Melting Point, °C: 144.5 [STR93]
Boiling Point, °C: 364.5 [CRC10]

3023

Compound: Stannic oxide
Synonym: cassiterite
Formula: SnO_2
Molecular Formula: O_2Sn
Molecular Weight: 150.709
CAS RN: 18282-10-5
Properties: white or sl gray powd or 3–12 mm sintered pieces of 99.9% purity; manufactured by blowing hot air over molten tin or by calcining the hydrated oxide; sintered pieces used as an evaporation material for anti static film and transparent heating elements, 99.9% pure material used as a sputtering target for transparent conductive films and in varistors [KIR83] [MER06] [CER91]
Solubility: i H_2O, alcohol, cold acids; slowly dissolves in hot, conc KOH or NaOH solutions [MER06]
Density, g/cm³: 6.95 [MER06]
Melting Point, °C: 1630 [STR93]
Reactions: sublimes at 1800°C–1900°C [HAW93]

3024

Compound: Stannic selenide
Synonym: tin diselenide
Formula: $SnSe_2$
Molecular Formula: Se_2Sn
Molecular Weight: 276.630
CAS RN: 20770-09-6
Properties: reddish brown cryst [MER06]
Solubility: s in alkali, conc acids; decomposed by HNO_3 [MER06]
Density, g/cm³: 4.85 [MER06]
Melting Point, °C: 650 [MER06]

3025

Compound: Stannic selenite
Synonym: tin selenite
Formula: $Sn(SeO_3)_2$
Molecular Formula: O_6Se_2Sn
Molecular Weight: 372.626
CAS RN: 7446-25-5
Properties: cryst powd [MER06]
Solubility: i H_2O; s in excess warm HCl [MER06]

3026

Compound: Stannic sulfide
Synonyms: mosaic gold, tin disulfide
Formula: SnS_2

Molecular Formula: S_2Sn
Molecular Weight: 182.842
CAS RN: 1315-01-1
Properties: yellow to brown powd; golden leaflets, metallic luster, has been used as a pigment [MER06] [HAW93]
Solubility: i H_2O, dil mineral acids; s aqua regia, alkali hydroxide solutions [MER06]
Density, g/cm³: 4.3 [MER06]
Melting Point, °C: decomposes at 600 [HAW93]

3027
Compound: Stannous acetate
Synonym: tin(II) acetate
Formula: $Sn(CH_3COO)_2$
Molecular Formula: $C_4H_6O_4Sn$
Molecular Weight: 236.800
CAS RN: 638-39-1
Properties: white, ortho-rhomb cryst; decomposed by water; used as a reducing agent [HAW93] [MER06]
Solubility: s dil HCl [MER06]
Density, g/cm³: 2.31 [MER06]
Melting Point, °C: 182.5–183 [MER06]
Boiling Point, °C: sublimes, 155 (0.1 mm Hg) [STR93]

3028
Compound: Stannous bromide
Synonym: tin(II) bromide
Formula: $SnBr_2$
Molecular Formula: Br_2Sn
Molecular Weight: 278.518
CAS RN: 10031-24-0
Properties: yellowish powd; oxidizes in air, turning brown; sensitive to moisture; enthalpy of vaporization 102 kJ/mol [CRC10] [MER06] [STR93] [HAW93]
Solubility: s in a small amount of H_2O, decomposed by a larger volume; s alcohol, ether, acetone [MER06]
Density, g/cm³: 5.12 [MER06]
Melting Point, °C: 215 [MER06]
Boiling Point, °C: 639 [CRC10]

3029
Compound: Stannous chloride
Synonym: tin(II) chloride
Formula: $SnCl_2$
Molecular Formula: Cl_2Sn
Molecular Weight: 189.615
CAS RN: 7772-99-8
Properties: white; ortho-rhomb cryst; mass of flakes; can absorb atm oxygen; enthalpy of vaporization 86.8 kJ/mol; enthalpy of fusion 12.80 kJ/mol; used as a reducing agent [HAW93] [MER06] [CRC10]

Solubility: s H_2O, ethanol, acetone, ether, methyl acetate [MER06]
Density, g/cm³: 3.95 [MER06]
Melting Point, °C: 246 [DOU83]
Boiling Point, °C: 606 [DOU83]

3030
Compound: Stannous chloride dihydrate
Synonym: tin(II) chloride dihydrate
Formula: $SnCl_2 \cdot 2H_2O$
Molecular Formula: $Cl_2H_4O_2Sn$
Molecular Weight: 225.646
CAS RN: 10025-69-1
Properties: white cryst; absorbs atm O_2, forming insoluble oxychloride; reducing agent [MER06] [STR93]
Solubility: s in less than its own weight of H_2O; forms insoluble material with more water; v s dil, conc HCl; s alcohol, NaOH soln, glacial acetic acid [MER06]
Density, g/cm³: 2.71 [MER06]
Melting Point, °C: 37–38 [MER06]
Boiling Point, °C: 652 [ALD94]
Reactions: decomposes when strongly heated [MER06]

3031
Compound: Stannous fluoride
Synonym: tin(II) fluoride
Formula: SnF_2
Molecular Formula: F_2Sn
Molecular Weight: 156.707
CAS RN: 7783-47-3
Properties: white powd; hygr; monocl; forms an oxyfluoride in air; can be prepared by reacting SnO with aq HF; used in dental preparations [KIR78] [MER06] [STR93]
Solubility: g/100 g in H_2O: 31 (0°), 78.5 (106°C); i alcohol, ether, chloroform [HAW93] [KIR83]
Density, g/cm³: 4.57 [MER06]
Melting Point, °C: 219 [STR93]
Boiling Point, °C: 850 [STR93]

3032
Compound: Stannous fluoroborate
Formula: $Sn(BF)_2$
Molecular Formula: B_2F_2Sn
Molecular Weight: 178.329
CAS RN: 13814-97-6
Properties: only available as a solution, e.g. 47% w; prepared by dissolution of SnO in fluoroboric acid; used in tin and tin-lead plating baths [KIR83]

3033

Compound: Stannous fluorophosphate
Synonym: tin(II) fluorophosphate
Formula: $SnPO_3F$
Molecular Formula: FO_3PSn
Molecular Weight: 216.680
CAS RN: 52262-58-5
Properties: white powd [STR93]

3034

Compound: Stannous iodide
Synonym: tin(II) iodide
Formula: SnI_2
Molecular Formula: I_2Sn
Molecular Weight: 372.519
CAS RN: 10294-70-9
Properties: reddish orange powd; moisture
   sensitive; enthalpy of vaporization
   105 kJ/mol [CRC10] [STR93]
Solubility: g/100 g H_2O: 0.99 (20°C), 1.42
   (40°C), 4.20 (100°C) [LAN05]; decomposes
   in H_2O; s alkali chloride or iodide
   solutions, benzene, $CHCl_3$ [MER06]
Density, g/cm³: 5.28 [MER06]
Melting Point, °C: 320 [MER06]
Boiling Point, °C: 714 [CRC10]

3035

Compound: Stannous oxalate
Synonym: tin(II) oxalate
Formula: SnC_2O_4
Molecular Formula: C_2O_4Sn
Molecular Weight: 206.730
CAS RN: 814-94-8
Properties: heavy, white, cryst powd; used in dyeing
   and printing textiles [HAW93] [MER06]
Solubility: i H_2O; s dil HCl [MER06]
Density, g/cm³: 3.56 [MER06]
Melting Point, °C: 280, decomposes [STR93]

3036

Compound: Stannous oxide
Synonym: tin(II) oxide
Formula: SnO
Molecular Formula: OSn
Molecular Weight: 134.709
CAS RN: 21651-19-4
Properties: brownish black powd;
   unstable in air [HAW93]
Solubility: i H_2O, alcohol; s acids, conc NaOH,
   KOH solutions [MER06]

Density, g/cm³: 6.45 [MER06]
Melting Point, °C: 1080 (600 mm Hg),
   decomposes [HAW93]

3037

Compound: Stannous pyrophosphate
Synonym: tin(II) pyrophosphate
Formula: $Sn_2P_2O_7$
Molecular Formula: $O_7P_2Sn_2$
Molecular Weight: 411.363
CAS RN: 15578-26-4
Properties: white free flowing cryst; prepared
   from stannous chloride and sodium
   pyrophosphate; used in toothpastes [HAW93]
Solubility: i H_2O; s conc acid [MER06]
Density, g/cm³: 4.009 [MER06]
Melting Point, °C: decomposes above 400 [KIR83]

3038

Compound: Stannous selenide
Synonym: tin(II) selenide
Formula: SnSe
Molecular Formula: SeSn
Molecular Weight: 197.670
CAS RN: 1315-06-6
Properties: 3–12 mm pieces with 99.999% purity;
   steel gray prisms [MER06] [CER91]
Solubility: i H_2O; s aqua regia, alkali sulfide
   and selenide solutions [MER06]
Density, g/cm³: 6.18 [MER06]
Melting Point, °C: 861 [MER06]

3039

Compound: Stannous stearate
Synonym: tin(II) stearate
Formula: $Sn[CH_3(CH_2)_{16}COO]_2$
Molecular Formula: $C_{36}H_{70}O_4Sn$
Molecular Weight: 685.660
CAS RN: 7637-13-0
Properties: off-white powd [STR93]

3040

Compound: Stannous sulfate
Synonym: tin(II) sulfate
Formula: $SnSO_4$
Molecular Formula: O_4SSn
Molecular Weight: 214.774
CAS RN: 7488-55-3

Properties: snow white ortho-rhomb cryst; can be prepared by reacting tin with excess sulfuric acid at 100°C for several days; principal use is in tin plating baths [KIR83] [MER06]

Solubility: 330 g/L H_2O at H_2O, hydrolyzes with precipitation of basic salt; s dil H_2SO_4 [MER06] [KIR83]

Melting Point, °C: decomposes at 378 to SnO_2 and SO_2 [MER06]

3041

Compound: Stannous sulfide

Synonym: tin(II) sulfide

Formula: SnS

Molecular Formula: SSn

Molecular Weight: 150.776

CAS RN: 1314-95-0

Properties: dark gray cryst, or black amorphous powd; has been used as a pigment [KIR83] [MER06]

Solubility: i H_2O, alkali hydroxides; s conc HCl, hot conc H_2SO_4 [MER06]

Density, g/cm³: 5.08 [MER06]

Melting Point, °C: 880 [HAW93]

Boiling Point, °C: 1230 [HAW93]

3042

Compound: Stannous tartrate

Synonym: tin(II) tartrate

Formula: $SnC_4H_4O_6$

Molecular Formula: $C_4H_4O_6Sn$

Molecular Weight: 266.782

CAS RN: 815-85-0

Properties: heavy, white, cryst powd; used in dyeing and printing fabrics [HAW93]

Solubility: s H_2O, dil HCl [MER06]

3043

Compound: Stannous telluride

Synonym: tin(II) telluride

Formula: SnTe

Molecular Formula: SnTe

Molecular Weight: 246.310

CAS RN: 12040-02-7

Properties: gray cryst; 3–12 mm pieces with 99.999% and 99.8% purity [CER91]

Density, g/cm³: 6.45 [CRC10]

Melting Point, °C: 807 (max) [CRC10]

Thermal Conductivity, W/(m·K): 9.1 [CRC10]

3044

Compound: Strontium

Formula: Sr

Molecular Formula: Sr

Molecular Weight: 87.62

CAS RN: 7440-24-6

Properties: silvery white metal; fcc; active metal, e.g. forms oxide film in air, electrical resistivity (20°C) 23 μohm·cm; enthalpy of fusion 7.43 kJ/mol; enthalpy of vaporization 136.9 kJ/mol; electronegativity 1.10 [CIC73] [MER06] [CRC10] [ALD94]

Solubility: reacts quickly with H_2O; s alcohol [HAW93]

Density, g/cm³: 2.63 [CIC73]

Melting Point, °C: 769 [CRC10]

Boiling Point, °C: 1384 [CRC10]

Thermal Conductivity, W/(m·K): 35.3 (25°C) [ALD94]

Thermal Expansion Coefficient: 22.5×10^{-6}/K [CRC10]

3045

Compound: Strontium acetate

Formula: $Sr(CH_3COO)_2$

Molecular Formula: $C_4H_6O_4Sr$

Molecular Weight: 205.710

CAS RN: 543-94-2

Properties: white powd [STR93]

Solubility: g/100 g H_2O: 36.93 (0.05°C), 40.19 (25°C), 36.36 (97°C); solid phase, $Sr(CH_3COO)_2 \cdot 4H_2O$ (0°C), $Sr(CH_3COO)_2 \cdot 1/2H_2O$ (25°C, 97°C) [KRU93]

Density, g/cm³: 2.099 [STR93]

Melting Point, °C: decomposes [CRC10]

3046

Compound: Strontium acetate hemihydrate

Formula: $Sr(CH_3COO)_2 \cdot 1/2H_2O$

Molecular Formula: $C_4H_7O_{4.5}Sr$

Molecular Weight: 214.717

CAS RN: 543-94-2

Properties: white, cryst powd; ignites to $SrCO_3$ [ALF95] [MER06]

Solubility: s in 2.5 parts H_2O; sl s alcohol [MER06]

Reactions: minus $1/2H_2O$ at 150°C [MER06]

3047

Compound: Strontium acetylacetonate

Synonyms: 2,4-pentanedione, strontium derivative

Formula: $Sr(CH_3COCH=C(O)CH_3)_2$

Molecular Formula: $C_{10}H_{14}O_4Sr$

Molecular Weight: 285.839

CAS RN: 12193-47-4

Properties: white powd [STR93]
Melting Point, °C: 220 decomposes [STR93]

3048

Compound: Strontium aluminate
Formula: $SrAl_2O_4$
Molecular Formula: Al_2O_4Sr
Molecular Weight: 205.581
CAS RN: 12004-37-4
Properties: −100 mesh with 99.5% purity [CER91]

3049

Compound: Strontium bromate monohydrate
Formula: $Sr(BrO_3)_2 \cdot H_2O$
Molecular Formula: $Br_2H_2O_7Sr$
Molecular Weight: 361.440
CAS RN: 14519-18-7
Properties: lustrous powd; white to sl
  yellow cryst; hygr [HAW93]
Solubility: g/100 g soln H_2O: 18.32 (0°C), 27.25 (25°C),
  41.00 (104°C); solid phases: $Sr(BrO_3)_2 \cdot H_2O$
  (0°C, 25°C), $Sr(BrO_3)_2$ (104°C) [KRU93]
Density, g/cm³: 3.773 [HAW93]
Melting Point, °C: decomposes at 240 [HAW93]
Reactions: minus H_2O at 120°C [HAW93]

3050

Compound: Strontium bromide
Formula: $SrBr_2$
Molecular Formula: Br_2Sr
Molecular Weight: 247.428
CAS RN: 10476-81-0
Properties: white powd; −20 mesh with 99.5% purity;
  enthalpy of fusion 10.12 kJ/mol [CRC10] [CER91]
Solubility: g/100 g H_2O: 85.2 (0°C), 107.0
  (25°C), 222.5 (100°C) [KRU93]
Density, g/cm³: 4.216 [STR93]
Melting Point, °C: 657 [CRC10]

3051

Compound: Strontium bromide hexahydrate
Formula: $SrBr_2 \cdot 6H_2O$
Molecular Formula: $Br_2H_{12}O_6Sr$
Molecular Weight: 355.519
CAS RN: 10476-81-0
Properties: colorless; deliq; cryst or white granules;
  used as a sedative in medicine [HAW93] [MER06]
Solubility: s in 0.35 parts H_2O; s alcohol; i ether [MER06]
Density, g/cm³: 2.386 [CRC10]
Reactions: minus $4H_2O$ at 89°C, minus
  $6H_2O$ by 180°C [HAW93]

3052

Compound: Strontium carbide
Formula: SrC_2
Molecular Formula: C_2Sr
Molecular Weight: 111.642
CAS RN: 12071-29-3
Properties: black tetr; −8 mesh with
  99% purity [CER91] [CRC10]
Solubility: decomposed by H_2O [CRC10]
Density, g/cm³: 3.2 [CRC10]
Melting Point, °C: >1700 [CRC10]

3053

Compound: Strontium carbonate
Synonym: strontianite
Formula: $SrCO_3$
Molecular Formula: CO_3Sr
Molecular Weight: 147.629
CAS RN: 1633-05-2
Properties: −20 mesh with 99.999% purity, <30 ppm
  Ba; white powd; hygr [STR93] [CER91]
Solubility: g/L H_2O: 0.00082 (8.8°C), 0.0109
  (24°C) [KRU93]; s dil acids [MER06]
Density, g/cm³: 3.70 [STR93]
Melting Point, °C: 1497 [STR93]
Reactions: minus CO_2 at 1340 [HAW93]
Thermal Expansion Coefficient: (volume)
  100°C (0.541), 200°C (1.168), 400°C
  (2.473), 800°C (5.726) [CLA66]

3054

Compound: Strontium chlorate
Formula: $Sr(ClO_3)_2$
Molecular Formula: Cl_2O_6Sr
Molecular Weight: 254.521
CAS RN: 7791-10-8
Properties: colorless or white cryst
  powd [HAW93] [MER06]
Solubility: g/100 g soln, H_2O: 61.40 (0°C),
  63.78 (25°C), 67.08 (95°C); solid phase,
  $Sr(ClO_3)_2 \cdot 3H_2O$ (0°C), $Sr(ClO_3)_2$ (25°C,
  95°C) [KRU93]; sl s alcohol [MER06]
Density, g/cm³: 3.15 [MER06]
Melting Point, °C: 120, decomposes [HAW93]

3055

Compound: Strontium chloride
Formula: $SrCl_2$
Molecular Formula: Cl_2Sr
Molecular Weight: 158.525
CAS RN: 10476-85-4

Properties: −40 mesh with 99.5% purity;
white powd; hygr; enthalpy of fusion
16.20 kJ/mol [STR93] [CRC10] [CER91]
Solubility: g/100 g H_2O: 43.5 (0°C), 55.8 (25°C),
100.8 (100°C); solid phase, $SrCl_2 \cdot 6H_2O$ (0°C,
25°C), $SrCl_2 \cdot 2H_2O$ (100°C) [KRU93]
Density, g/cm³: 3.052 [STR93]
Melting Point, °C: 874 [CRC10]
Boiling Point, °C: 1250 [STR93]

3056
Compound: Strontium chloride hexahydrate
Formula: $SrCl_2 \cdot 6H_2O$
Molecular Formula: $Cl_2H_{12}O_6Sr$
Molecular Weight: 266.617
CAS RN: 10025-70-4
Properties: colorless cryst; white granules; effloresces
in air; deliq in presence of moisture [MER06]
Solubility: 3.5195 ± 0.0026 mol/(kg · H_2O) at
25°C [RAR85b]; s alcohol [MER06]
Density, g/cm³: 1.96 [MER06]
Melting Point, °C: 115 [STR93]
Reactions: minus $5H_2O$ at 100°C; minus
$6H_2O$ at 150°C [MER06]

3057
Compound: Strontium chromate
Formula: $SrCrO_4$
Molecular Formula: CrO_4Sr
Molecular Weight: 203.614
CAS RN: 7789-06-2
Properties: −80 mesh with 99% purity;
light yellow; monocl; has been used in
metal coatings to protect the metal from
corrosion [HAW93] [KIR78] [CER91]
Solubility: s dil acids [KIR78]; g/L soln, H_2O:
0.91 (25°C), 0.43 (100°C) [KRU93]
Density, g/cm³: 3.895 [KIR78]
Melting Point, °C: decomposes [KIR78]

3058
Compound: Strontium dithionate tetrahydrate
Formula: $Sr(SO_3)_2 \cdot 4H_2O$
Molecular Formula: $H_8O_{10}S_2Sr$
Molecular Weight: 319.809
CAS RN: 13845-16-4
Properties: trig [CRC10]
Solubility: g/100 g soln, H_2O: 4.51
(0°C), 10.8 (20°C) [KRU93]
Density, g/cm³: 2.373 [CRC10]
Reactions: minus $4H_2O$, 78°C [CRC10]

3059
Compound: Strontium ferrite
Synonym: strontium dodecairon nonadecaoxide
Formula: $SrFe_{12}O_{19}$
Molecular Formula: $Fe_{12}O_{19}Sr$
Molecular Weight: 1061.773
CAS RN: 12023-91-5
Properties: powd; −325 mesh with 99.5%
purity [CER91] [ALF93]

3060
Compound: Strontium fluoride
Formula: SrF_2
Molecular Formula: F_2Sr
Molecular Weight: 125.617
CAS RN: 7783-48-4
Properties: white powd, and 99.9% pure melted
pieces of 3–6 mm; hygr; enthalpy of fusion
29.70 kJ/mol; pieces used as evaporation and
sputtering material for infrared transparent
films [STR93] [CER91] [CRC10]
Solubility: g/L soln, H_2O: 0.1135 (0.26°C),
0.21 ± 0.13 (25°C) [KRU93]; s dil acids;
decomposed by strong acids [MER06]
Density, g/cm³: 4.24 [MER06]
Melting Point, °C: 1477 [CRC10]
Boiling Point, °C: 2489 [STR93]
Reactions: oxidized to SrO >1000°C [MER06]

3061
Compound: Strontium hexaboride
Synonym: strontium boride
Formula: SrB_6
Molecular Formula: B_6Sr
Molecular Weight: 152.486
CAS RN: 12046-54-7
Properties: −325 mesh 10 μm or less with 99.5%
purity; refractory material [KIR78] [CER91]
Density, g/cm³: 3.39 [ALD94]
Melting Point, °C: 2235 [KIR78]

3062
Compound: Strontium hydride
Formula: SrH_2
Molecular Formula: H_2Sr
Molecular Weight: 89.636
CAS RN: 13598-33-9
Properties: −60 mesh with 99.5% purity;
resembles CaH_2 in both properties
and reactivity [KIR80] [CER91]
Density, g/cm³: 3.72 [KIR80]

3063
Compound: Strontium hydroxide
Formula: $Sr(OH)_2$
Molecular Formula: H_2O_2Sr
Molecular Weight: 121.635
CAS RN: 18480-07-4
Properties: colorless deliq cryst; absorbs H_2O from air; enthalpy of fusion 21.00 kJ/mol [HAW93] [CRC10]
Solubility: g/100 g soln in H_2O: 0.90 (0°C), 2.16 ± 0.41 (25°C), 47.71 (100°C); solid phase, $Sr(OH)_2 \cdot 8H_2O$ [KRU93]
Density, g/cm³: 3.625 [HAW93]
Melting Point, °C: 512 [JAN71]
Reactions: minus H_2O at 710°C [CRC10]

3064
Compound: Strontium hydroxide octahydrate
Formula: $Sr(OH)_2 \cdot 8H_2O$
Molecular Formula: $H_{18}O_{10}Sr$
Molecular Weight: 265.757
CAS RN: 1311-10-0
Properties: colorless; deliq cryst or white powd; forms $SrCO_3$ by reaction with atm CO_2 [MER06]
Solubility: s in 50 parts H_2O; 2.1 parts boiling H_2O [MER06]
Density, g/cm³: 1.90 [STR93]
Reactions: minus some H_2O ~100°C [MER06]

3065
Compound: Strontium iodate
Formula: $Sr(IO_3)_2$
Molecular Formula: I_2O_6Sr
Molecular Weight: 437.425
CAS RN: 13470-01-4
Properties: tricl; −80 mesh with 99.5% purity [CER91] [CRC10]
Solubility: g/100 g soln, H_2O: 0.098 (0°C), 0.165 (25°C), 0.350 (100°C); solid phase, $Sr(IO_3)_2$ [KRU93]
Density, g/cm³: 5.045 [CRC10]

3066
Compound: Strontium iodide
Formula: SrI_2
Molecular Formula: I_2Sr
Molecular Weight: 341.429
CAS RN: 10476-86-5
Properties: −80 mesh with 99.5% purity; white cryst; enthalpy of fusion 19.70 kJ/mol [CER91] [HAW93] [CRC10]

Solubility: g/100 g H_2O: 165.3 (0°C), 181.2 (25°C), 383.1 (100°C); solid phase, $SrI_2 \cdot 6H_2O$ (0°C, 25°C), $SrI_2 \cdot 2H_2O$ (100°C) [KRU93]
Density, g/cm³: 4.549 [HAW93]
Melting Point, °C: 515 [CRC10]
Boiling Point, °C: decomposes [HAW93]

3067
Compound: Strontium iodide hexahydrate
Formula: $SrI_2 \cdot 6H_2O$
Molecular Formula: $H_{12}I_2O_6Sr$
Molecular Weight: 449.520
CAS RN: 10476-86-5
Properties: colorless to yellowish; deliq; sensitive to light and atm O_2 with partial oxidation freeing I_2 [MER06]
Solubility: s in 0.2 parts H_2O; s alcohol [MER06]
Density, g/cm³: 2.67 [HAW93]
Melting Point, °C: ~120, when rapidly heated [MER06]

3068
Compound: Strontium lactate trihydrate
Formula: $Sr(C_3H_5O_3)_2 \cdot 3H_2O$
Molecular Formula: $C_6H_{16}O_9Sr$
Molecular Weight: 319.808
CAS RN: 29870-99-3
Properties: white, odorless, granular powd [MER06]
Solubility: s in 3 parts H_2O, 0.5 parts boiling H_2O; sl s alcohol [MER06]
Reactions: minus $3H_2O$ at 120°C [MER06]

3069
Compound: Strontium molybdate(VI)
Formula: $SrMoO_4$
Molecular Formula: MoO_4Sr
Molecular Weight: 247.558
CAS RN: 13470-04-7
Properties: −200 mesh with 99.9% purity; white cryst powd; scheelite structure, c/a = 2.23; used as an anticorrosion pigment, used in electronic and optical applications, in solid state lasers [HAW93] [KIR81] [CER91]
Solubility: ~0.003 g/100 g H_2O [KIR81]
Density, g/cm³: 4.662 [KIR81]
Melting Point, °C: ~1040 [KIR81]

3070
Compound: Strontium niobate
Formula: $SrNb_2O_6$

Molecular Formula: Nb_2O_6Sr
Molecular Weight: 369.429
CAS RN: 12034-89-8
Properties: monocl cryst; −200 mesh with
99.9% purity [LID94] [CER91]
Density, g/cm³: 5.11 [LID94]
Melting Point, °C: 1225 [LID94]

3071

Compound: Strontium nitrate
Formula: $Sr(NO_3)_2$
Molecular Formula: N_2O_6Sr
Molecular Weight: 211.629
CAS RN: 10042-76-9
Properties: −8 mesh with 99.995% purity; white cryst;
used in pyrotechnics, in marine signals, matches,
and railroad flares [HAW93] [STR93] [CER91]
Solubility: g/100 g soln, H_2O: 28.2 (0.1°C),
40.7 (20°C), 51.2 (105°C); solid phase,
$Sr(NO_3)_2 \cdot 4H_2O$ (0.1°C, 20°C), $Sr(NO_3)_2$ (105°C)
[KRU93]; sl s alcohol, acetone [MER06]
Density, g/cm³: 2.99 [MER06]
Melting Point, °C: 570 [MER06]

3072

Compound: Strontium nitride
Formula: Sr_3N_2
Molecular Formula: N_2Sr_3
Molecular Weight: 290.873
CAS RN: 12033-82-8
Properties: −60 mesh with 99.5% purity [CER91]
Melting Point, °C: 1030 [CIC73]

3073

Compound: Strontium nitrite
Formula: $Sr(NO_2)_2$
Molecular Formula: N_2O_4Sr
Molecular Weight: 179.631
CAS RN: 13470-06-9
Properties: white or yellowish powd;
hygr needles [HAW93]
Solubility: g/100 g soln, H_2O: 43.1 (35°C),
58.1 (98°C); solid phase, $Sr(NO_2)_2 \cdot H_2O$
[KRU93]; i alcohol [HAW93]
Density, g/cm³: 2.8 [HAW93]
Melting Point, °C: decomposes at 240 [HAW93]

3074

Compound: Strontium oxalate
Formula: SrC_2O_4
Molecular Formula: C_2O_4Sr
Molecular Weight: 175.640

CAS RN: 814-95-9
Properties: white powd [STR93]
Solubility: g/L soln, H_2O: 0.057 (0°C),
0.077 (20°C) [KRU93]

3075

Compound: Strontium oxalate monohydrate
Formula: $SrC_2O_4 \cdot H_2O$
Molecular Formula: $C_2H_2O_5Sr$
Molecular Weight: 193.655
CAS RN: 814-95-9
Properties: white, odorless; cryst powd;
used in pyrotechnics, tanning, catalyst
manufacturing [HAW93] [MER06]
Solubility: s in 20,000 parts H_2O; 1900 parts 3.5%
acetic acid; s dil HCl, HNO_3 [MER06]
Reactions: minus H_2O at 150°C [HAW93]

3076

Compound: Strontium oxide
Synonym: strontia
Formula: SrO
Molecular Formula: OSr
Molecular Weight: 103.619
CAS RN: 1314-11-0
Properties: white to grayish white; reacts with water,
forming $Sr(OH)_2$, with evolution of heat; enthalpy
of fusion 75.00 kJ/mol [MER06] [CRC10]
Solubility: g/100 g H_2O: 1.03 (30°C), 1.05
(40°C), 12.15 (100°C) [LAN05]; reacts with
H_2O to form the hydroxide [HAW93]
Density, g/cm³: 4.7 [MER06]
Melting Point, °C: 2430 [AES93]
Boiling Point, °C: ~3000 [HAW93]

3077

Compound: Strontium perchlorate
Formula: $Sr(ClO_4)_2$
Molecular Formula: Cl_2O_8Sr
Molecular Weight: 286.520
CAS RN: 13450-97-0
Properties: colorless cryst [HAW93]
Solubility: mol/kg H_2O: 8.16 (0°C), 10.67 (25°C),
12.70 (40°C); solid phase, $Sr(ClO_4)_2 \cdot 4H_2O$
(0°C), $Sr(ClO_4)_2 \cdot 2H_2O$ (25°C), $3Sr(ClO_4)_2 \cdot 2H_2O$
(above 40°C) [KRU93]; s alcohol [HAW93]

3078

Compound: Strontium perchlorate hexahydrate
Formula: $Sr(ClO_4)_2 \cdot 6H_2O$
Molecular Formula: $Cl_2H_{12}O_{14}Sr$

Molecular Weight: 394.612
CAS RN: 13450-97-0
Properties: white cryst; hygr [STR93]
Melting Point, °C: <100 [STR93]

3079
Compound: Strontium permanganate trihydrate
Formula: $Sr(MnO_4)_2 \cdot 3H_2O$
Molecular Formula: $H_6Mn_2O_{11}Sr$
Molecular Weight: 379.537
CAS RN: 14446-13-0
Properties: purple cub [CRC10]
Solubility: g/100 g soln, H_2O: 2.5 (0°C) [KRU93]
Density, g/cm³: 2.75 [CRC10]
Melting Point, °C: decomposes 175 [CRC10]

3080
Compound: Strontium peroxide
Formula: SrO_2
Molecular Formula: O_2Sr
Molecular Weight: 119.619
CAS RN: 1314-18-7
Properties: white powd; unstable if standing, decomposes in air under ambient conditions; decomposed in water, evolving O_2 [MER06]
Solubility: decomposed to H_2O_2 by dil acids [MER06]
Density, g/cm³: 4.56 [HAW93]
Melting Point, °C: 215, decomposes [HAW93]

3081
Compound: Strontium peroxide octahydrate
Formula: $SrO_2 \cdot 8H_2O$
Molecular Formula: $H_{16}O_{10}Sr$
Molecular Weight: 263.741
CAS RN: 1314-18-7
Properties: white powd [HAW93]
Solubility: sl s cold H_2O, decomposed by hot H_2O; s NH_4Cl solutions and alcohol [HAW93]
Density, g/cm³: 1.951 [HAW93]
Melting Point, °C: decomposes [CRC10]
Reactions: minus $8H_2O$ at 100°C [HAW93]

3082
Compound: Strontium salicylate dihydrate
Formula: $Sr(C_7H_5O_3)_2 \cdot 2H_2O$
Molecular Formula: $C_{14}H_{14}O_8Sr$
Molecular Weight: 397.880
CAS RN: 6160-38-9
Properties: white cryst or powd; odorless; light sensitive; heat causes decomposition; used in manufacturing of pharmaceuticals and fine chemicals [HAW93]

Solubility: s H_2O and alcohol [HAW93]
Melting Point, °C: decomposes [HAW93]

3083
Compound: Strontium selenate
Formula: $SrSeO_4$
Molecular Formula: O_4SeSr
Molecular Weight: 230.578
CAS RN: 7446-21-1
Properties: ortho-rhomb cryst; prepared by heating strontium carbonate with selenium or selenium oxide [MER06]
Solubility: i H_2O; s HCl [MER06]
Density, g/cm³: 4.25 [MER06]

3084
Compound: Strontium selenide
Formula: SrSe
Molecular Formula: SeSr
Molecular Weight: 166.580
CAS RN: 1315-07-7
Properties: white cub; −20 mesh with 99.5% purity [CER91] [CRC10]
Density, g/cm³: 4.38 [CRC10]
Melting Point, °C: 1600 [AES93]

3085
Compound: Strontium silicide
Formula: $SrSi_2$
Molecular Formula: Si_2Sr
Molecular Weight: 143.791
CAS RN: 12138-28-2
Properties: silver-gray cub cryst; 10 mm & down lump [LID94] [ALF93]
Density, g/cm³: 3.35 [LID94]
Melting Point, °C: 1100 [LID94]

3086
Compound: Strontium stannate
Formula: $SrSnO_3$
Molecular Formula: O_3SnSr
Molecular Weight: 254.328
CAS RN: 12143-34-9
Properties: −200 mesh with 99.5% purity [CER91]

3087
Compound: Strontium sulfate
Synonym: celestite
Formula: $SrSO_4$
Molecular Formula: O_4SSr

Molecular Weight: 183.684
CAS RN: 7759-02-6
Properties: hygr white cryst or precipitate; used in pyrotechnics, in ceramics and glass [HAW93] [STR93]
Solubility: g/100 mL soln, H_2O: 0.0121 (5°C), 0.0135 (25°C), 0.0113 (95°C); solid phase, $SrSO_4$ [KRU93]; s HCl, HNO_3 [MER06]
Density, g/cm³: 3.96 [MER06]
Melting Point, °C: 1600 [STR93]

3088
Compound: Strontium sulfide
Formula: SrS
Molecular Formula: SSr
Molecular Weight: 119.686
CAS RN: 1314-96-1
Properties: −200 mesh with 99.9% purity; gray powd; has odor of H_2S in moist atm [MER06] [CER91]
Solubility: sl s H_2O; s acids, decomposes [MER06]
Density, g/cm³: 3.70 [MER06]
Melting Point, °C: >2000 [STR93]

3089
Compound: Strontium tantalate
Formula: $SrTa_2O_6$
Molecular Formula: O_6SrTa_2
Molecular Weight: 545.512
CAS RN: 12065-74-6
Properties: reacted product; −200 mesh with 99.9% purity [CER91]

3090
Compound: Strontium tartrate tetrahydrate
Formula: $SrC_4H_4O_6 \cdot 4H_2O$
Molecular Formula: $C_4H_{12}O_{10}Sr$
Molecular Weight: 307.753
CAS RN: 6100-96-5
Properties: monocl white cryst; used in pyrotechnics [HAW93] [CRC10]
Solubility: sl s H_2O [HAW93]
Density, g/cm³: 1.966 [HAW93]

3091
Compound: Strontium thiosulfate pentahydrate
Synonym: strontium hyposulfite
Formula: $SrS_2O_3 \cdot 5H_2O$
Molecular Formula: $H_{10}O_8S_2Sr$
Molecular Weight: 289.826
CAS RN: 15123-90-7
Properties: monocl fine needles [CRC10] [HAW93]
Solubility: g/100 g solution H_2O: 8.78 (0°C), 21.10 (27.5°C), 26.80 (40°C); solid phase: $SrS_2O_3 \cdot 3H_2O$ [KRU93]; i alcohol [HAW93]

Density, g/cm³: 2.17 [HAW93]
Reactions: minus $4H_2O$ at 100°C [HAW93]

3092
Compound: Strontium titanate
Formula: $SrTiO_3$
Molecular Formula: O_3SrTi
Molecular Weight: 183.485
CAS RN: 12060-59-2
Properties: −200 mesh of 99.9% purity; white powd; cub; hardness 6–6.5 Mohs; has properties of refractive index, dispersion and optical transmission which are comparable to diamond; highly pure material can be obtained by calcining the double strontium titanate; used in electronics and in electrical insulation; as 99.9% pure material, used as sputtering target for thin film capacitors [HAW93] [KIR83] [CER91]
Solubility: i H_2O and in most solvents [HAW93]
Density, g/cm³: 4.81 (HAW93)
Melting Point, °C: 2060 [HAW93]

3093
Compound: Strontium tungstate
Formula: $SrWO_4$
Molecular Formula: O_4SrW
Molecular Weight: 335.458
CAS RN: 13451-05-3
Properties: −200 mesh with 99.9% purity; white tetr, a = 0.540 nm, c = 1.109 nm [KIR83] [CER91]
Solubility: 0.14 g/100 mL H_2O (15°C) [CRC10]
Density, g/cm³: 6.187 [KIR83]
Melting Point, °C: decomposes [CRC10]

3094
Compound: Strontium vanadate
Formula: SrV_2O_6
Molecular Formula: O_6SrV_2
Molecular Weight: 285.499
CAS RN: 12435-86-8
Properties: −200 mesh with 99.9% purity [CER91]

3095
Compound: Strontium zirconate
Formula: $SrZrO_3$
Molecular Formula: O_3SrZr
Molecular Weight: 226.842
CAS RN: 12036-39-4
Properties: white powd; used in electronics, and as the 99% pure material, is used as a sputtering target for thin film capacitors [HAW93] [STR93] [CER91]
Melting Point, °C: 2600 [HAW93]

3096

Compound: Sulfur chloride
Synonym: sulfur monochloride
Formula: S_2Cl_2
Molecular Formula: Cl_2S_2
Molecular Weight: 135.037
CAS RN: 10025-67-9
Properties: light amber to yellowish red; fuming, oily liq; penetrating odor; dielectric constant, 4.9 (22°C); dipole moment, 1.60 [MER06]
Solubility: s alcohol, benzene, ether, CS_2, CCl_4, oils; decomposed in water, forming sulfur, HCl, SO_2, H_2S, sulfite, thiosulfate [MER06]
Density, g/cm³: 1.6885 [MER06]
Melting Point, °C: −77 [MER06]
Boiling Point, °C: 138 [MER06]

3097

Compound: Sulfur dioxide
Formula: SO_2
Molecular Formula: O_2S
Molecular Weight: 64.065
CAS RN: 7446-09-5
Properties: colorless gas; not flammable; mild reducing agent, e.g. bleaches vegetable colors; vapor pressure is 3.2 atms at 20°C; critical temp 157.5°C; critical pressure 7.87 MPa; enthalpy of vaporization 24.94 kJ/mol [CRC10] [AIR87] [HAW93] [MER06]
Solubility: % H_2O: 17.7 (0°C), 11.9 (15°C), 8.5 (25°C), 6.4 (35°C); % other solvents: 25, alcohol; 32, methanol [MER06]
Density, g/cm³: liq: 1.5 [MER06]
Melting Point, °C: −72 [MER06]
Boiling Point, °C: −10.0 [CRC10]

3098

Compound: Sulfur hexafluoride
Formula: SF_6
Molecular Formula: F_6S
Molecular Weight: 146.056
CAS RN: 2551-62-4
Properties: colorless, odorless gas; very stable, e.g. to electrical discharge (in transformer oil); does not attack glass; enthalpy of sublimation 23.59 kJ/mol; triple point −50.52°C; enthalpy of vaporization 9.642 kJ/mol; enthalpy of fusion 5.02 kJ/mol; dielectric constant of gas 1.00204; viscosity of gas 0.01576 mPa·s; critical temp 45.55°C; critical pressure 3.759 MPa; critical density 0.737 g/cm³ [KIR78] [MER06] [CRC10]
Solubility: sl s H_2O; s alcohol [MER06]
Density, g/cm³: gas: 6.5 g/L; liq: 1.67 [HAW93]
Melting Point, °C: −50.8 [MER06]

Boiling Point, °C: sublimes −63.8 [MER06]
Thermal Conductivity, W/(m·K): liq is 0.0583; gas is 0.01415 [KIR78]

3099

Compound: Sulfur tetrafluoride
Formula: SF_4
Molecular Formula: F_4S
Molecular Weight: 108.060
CAS RN: 7783-60-0
Properties: colorless gas; stable up to 600°C; reacts violently with H_2O; decomposed by conc H_2SO_4; critical temp 90.9°C; enthalpy of vaporization 26.44 kJ/mol; used in electronics industry [AIR87] [MER06] [CRC10]
Solubility: decomposes in H_2O [HAW93]
Density, g/cm³: liq: 1.95 (−78°C); solid: 2.349 (−18.3°C) [MER06]
Melting Point, °C: −121.0 [MER06]
Boiling Point, °C: −40.5 [CRC10]
Reactions: attacks glass, but not quartz [MER06]

3100

Compound: Sulfur trioxide N,N-dimethylformamide complex
Formula: $HCON(CH_3)_2 \cdot SO_3$
Molecular Formula: $C_3H_7NO_4S$
Molecular Weight: 153.159
CAS RN: 29584-42-7
Properties: corrosive; uses: mild sulfating agent [ALD94]
Melting Point, °C: 115–158 [ALD94]

3101

Compound: Sulfur trioxide(α)
Formula: α-SO_3
Molecular Formula: O_3S
Molecular Weight: 80.064
CAS RN: 7446-11-9
Properties: solid; needles; vapor pressure (25°C) 73 mm; stable form is α; enthalpy of vaporization 43.14 kJ/mol at 25°C; enthalpy of fusion 8.60 kJ/mol [CRC10] [MER06] [HAW93]
Solubility: reacts vigorously with H_2O to form H_2SO_4 [MER06]
Density, g/cm³: 1.97 [CRC10]
Melting Point, °C: 16.8 [CRC10]
Boiling Point, °C: 45 [CRC10]

3102

Compound: Sulfur trioxide(β)
Formula: β-SO_3

Molecular Formula: O_3S
Molecular Weight: 80.064
CAS RN: 7446-11-9
Properties: dimer; needles; metastable; vapor pressure, 25°C, 344 mm [MER06] [CRC10]
Solubility: decomposed by H_2O [CRC10]
Melting Point, °C: 32.5 [MER06]

3103
Compound: Sulfur trioxide(γ)
Formula: γ-SO_3
Molecular Formula: O_3S
Molecular Weight: 80.064
CAS RN: 7446-11-9
Properties: form can be icy mass or liq; metastable [MER06] [DOU83]
Solubility: reacts violently with H_2O, forming H_2SO_4 [MER06]
Density, g/cm³: liq: 1.9224 [MER06]
Melting Point, °C: 16.8 [MER06]
Boiling Point, °C: 44.8 [MER06]

3104
Compound: Sulfur(α)
Synonym: brimstone
Formula: α-S
Molecular Formula: S
Molecular Weight: 32.066
CAS RN: 7704-34-9
Properties: yellow cryst; ortho-rhomb; stable form at usual temperatures; electrical resistivity (20°C) $2 \times 10^{+23}$ μohm·cm; electronegativity 2.44; enthalpy of vaporization 45 kJ/mol; enthalpy of fusion 1.72 kJ/mol [CRC10] [MER06] [COT88]
Solubility: i H_2O; sl s alcohol and ether; s CS_2, CCl_4, benzene [HAW93]
Density, g/cm³: 2.06 [MER06]
Melting Point, °C: 112.8 [ALD94]
Boiling Point, °C: 444.674 [ALD94]
Reactions: forms monocl S at 94.5°C [MER06]
Thermal Conductivity, W/(m·K): 0.205 (25°C) [ALD94]

3105
Compound: Sulfur(β)
Formula: β-S
Molecular Formula: S
Molecular Weight: 32.066
CAS RN: 7704-34-9
Properties: monocl; light yellow; opaque; brittle; stable form from 94.5°C to 120°C; transforms if left standing to ortho-rhomb at a slow rate [MER06]
Solubility: i H_2O; sl s alcohol, ether; s 1 g/2 mL CS_2; s benzene, toluene, acetone [MER06]

Density, g/cm³: 1.957 [ALD94]
Melting Point, °C: 119.0 [ALD94]
Boiling Point, °C: 444.674 [ALD94]

3106
Compound: Sulfur(γ)
Synonym: mother-of-pearl sulfur
Formula: γ-S
Molecular Formula: S
Molecular Weight: 32.066
CAS RN: 7704-34-9
Properties: yellow, amorphous [CRC10]
Solubility: i H_2O [CRC10]
Density, g/cm³: 1.92 [CRC10]
Melting Point, °C: 106.8 [MER06]
Boiling Point, °C: 444.6 [CRC10]

3107
Compound: Sulfuric acid
Synonym: oil of vitriol
Formula: H_2SO_4
Molecular Formula: H_2O_4S
Molecular Weight: 98.080
CAS RN: 7664-93-9
Properties: clear, colorless, oily liq; absorbs moisture from atm; can char organic materials, e.g. sugar; miscible with water, evolving heat; enthalpy of fusion 10.71 kJ/mol; specific conductance 1.044×10^{-2} at 25°C; dielectric constant 110 at 20°C [MER06] [COT88] [CRC10]
Solubility: miscible with H_2O [HAW93]
Density, g/cm³: ~1.84 [MER06]
Melting Point, °C: 10.31 [CRC10]
Boiling Point, °C: ~290 [MER06]
Reactions: decomposes to SO_3 and H_2O at 340°C [MER06]

3108
Compound: Sulfuric acid fuming
Synonym: oleum
Formula: $H_2SO_4 + SO_3$
Molecular Formula: $H_2S_2O_7$
Molecular Weight: 178.144
CAS RN: 8014-95-7
Properties: commercial acid contains up to 30% SO_3; colorless, or sl colored, viscous liq; choking fumes of SO_3 [MER06]

3109
Compound: Sulfurous acid
Formula: H_2SO_3

Molecular Formula: H_2O_3S
Molecular Weight: 82.080
CAS RN: 7782-99-2
Properties: solution of sulfur dioxide in water; colorless; clear liq; odor of SO_2; gradually oxidized to sulfate by atm O_2; mild reducing agent; e.g. dental bleach [MER06]
Solubility: s H_2O [HAW93]
Density, g/cm³: ~1.03 [MER06]

3110

Compound: Sulfuryl chloride
Formula: SO_2Cl_2
Molecular Formula: Cl_2O_2S
Molecular Weight: 134.970
CAS RN: 7791-25-5
Properties: colorless liq; pungent odor; turns yellow slowly when standing due to slight dissociation into Cl_2 and SO_2; violent reaction with alkalies; enthalpy of vaporization 31.4 kJ/mol [CRC10] [MER06]
Solubility: slowly decomposed in H_2O, forming H_2SO_4 and HCl [MER06]; miscible with benzene, toluene, ether, acetic acid, other organic materials [MER06]
Density, g/cm³: 1.664 [MER06]
Melting Point, °C: −54.1, −46 [MER06]
Boiling Point, °C: 69.3 [MER06]
Reactions: forms $SO_2Cl_2 \cdot 15H_2O$ with icy cold H_2O [MER06]

3111

Compound: Sulfuryl fluoride
Formula: SO_2F_2
Molecular Formula: F_2O_2S
Molecular Weight: 102.062
CAS RN: 2699-79-8
Properties: colorless odorless gas; not very reactive; stable up to 400°C; not hydrolyzed in H_2O, but hydrolyzes in NaOH solutions; used in insecticides, fumigants [MER06] [HAW93]
Solubility: mL SO_2F_2/100 mL solvent: 4–5, H_2O; 24–27, alcohol; 210–220, toluene; 136–138, CCl_4, [MER06]
Density, g/cm³: gas: 4.55 g/L; liq; 1.7 [CRC10]
Melting Point, °C: −135.8 [MER06]
Boiling Point, °C: −55.4 [MER06]

3112

Compound: Tantalum
Formula: Ta
Molecular Formula: Ta
Molecular Weight: 180.9479
CAS RN: 7440-25-7

Properties: gray, very hard, malleable metal; bcc, a = 0.33026 nm; electrical resistivity (18°C) 12.4 μohm · cm; Poisson's ratio 0.35; Young's modulus at room temp 186; slowly reacts with fused alkalies; reacts with F_2, Cl_2, O_2 when heated; absorbs H_2 at high temp; enthalpy of vaporization 732.8 kJ/mol; enthalpy of fusion 36.57 kJ/mol; used to contain molten metals such as sodium [KIR83] [ALD94] [MER06] [CER91] [CRC] [CAB93]
Solubility: very resistant to attack by acids except HF, resistant to alkali solutions [KIR83]
Density, g/cm³: 16.69 [MER06]
Melting Point, °C: 2996 [ALD94]
Boiling Point, °C: 5429 [MER06]
Thermal Conductivity, W/(m · K): 54.4 at 20°C [KIR83]
Thermal Expansion Coefficient: 8×10^{-6}/°C over the temp range 20°C–1500°C [HAW93]

3113

Compound: Tantalum aluminide
Formula: $TaAl_3$
Molecular Formula: Al_3Ta
Molecular Weight: 261.859
CAS RN: 12004-76-1
Properties: −80 mesh with 99.5% purity; gray refractory powd; oxidizes slowly in air above 500°C; formed by adding Al metal to potassium fluorotantalate at ~1000°C [KIR83] [CER91]
Solubility: i acids and alkalies [KIR83]
Density, g/cm³: 7.02 [KIR83]
Melting Point, °C: ~1400 [KIR83]

3114

Compound: Tantalum boride
Formula: TaB
Molecular Formula: BTa
Molecular Weight: 191.759
CAS RN: 12007-07-7
Properties: −325 mesh with 99.5% purity; refractory material; used as a sputtering target of 99.5% purity to produce wear-resistant and semiconductive films, and other uses [KIR78] [CER91]
Density, g/cm³: 14.2 [LID94]
Melting Point, °C: 2040 [KIR78]

3115

Compound: Tantalum carbide
Formula: Ta_2C
Molecular Formula: CTa_2
Molecular Weight: 373.907
CAS RN: 12070-07-4

Properties: hex, refractory; −325 mesh, 10 μm or less, 99.5% purity; hex, a=0.3106 nm [CER91] [CIC73]
Density, g/cm³: 15.1 [LID94]
Melting Point, °C: 3327 [LID94]

3116
Compound: Tantalum diboride
Synonym: tantalum boride
Formula: TaB₂
Molecular Formula: B₂Ta
Molecular Weight: 202.570
CAS RN: 12077-35-1
Properties: gray metallic powd; hardness >8 mohs; can be formed by heating tantalum and boron in vacuum at ~1800°C; 99.5% pure material used as a sputtering target to produce wear-resistant films and semiconductor films, and other applications [KIR83] [CER91]
Solubility: i acids and alkalies [KIR83]
Density, g/cm³: 11.15 [KIR83]
Melting Point, °C: ~3000 [KIR83]

3117
Compound: Tantalum disulfide
Formula: TaS₂
Molecular Formula: S₂Ta
Molecular Weight: 245.080
CAS RN: 12143-72-5
Properties: black powd or cryst; used as a solid lubricant, also as a 99% pure material used as a sputtering target to form lubricant film on bearings and other moving parts; there is TaS compound [HAW93] [CER91] [STR93]
Solubility: i H₂O [HAW93]
Density, g/cm³: 6.86 [LID94]
Melting Point, °C: for TaS: >1300 [STR93]

3118
Compound: Tantalum ethoxide
Formula: Ta(OC₂H₅)₅
Molecular Formula: C₁₀H₂₅O₅Ta
Molecular Weight: 406.254
CAS RN: 6074-84-6
Properties: yellow liq; moisture sensitive; 99.999% purity, <100 ppm Nb [CER91] [STR93]
Density, g/cm³: 1.566 [ALD94]
Melting Point, °C: 21 [CER91]
Boiling Point, °C: 145 at 0.1 mm Hg [CER91]

3119
Compound: Tantalum hydride
Formula: TaH

Molecular Formula: HTa
Molecular Weight: 181.956
CAS RN: 13981-95-8
Properties: gray, brittle with metallic luster; can form when H₂ is absorbed by Ta at 450°C; hydrogen is released when TaH is heated above 800°C; material is a superconductor [KIR83]
Solubility: resistant to attack by acids [KIR83]
Density, g/cm³: 15.1 [KIR83]

3120
Compound: Tantalum monocarbide
Formula: TaC
Molecular Formula: CTa
Molecular Weight: 192.959
CAS RN: 12070-06-3
Properties: cryst, very fine golden brown powd; fcc, a=0.44555 nm; refractory material; resistivity is 30 μohm·cm at room temp; hardness 9–10 Mohs prepared by reaction of tantalum powd and carbon black at ~1900°C in an inert atm; used in crucible form for melting zirconium oxide and similar oxides with high melting points, and as a sputtering target [CER91] [KIR83] [CIC73]
Solubility: i acids except mixture of HF and HNO₃ [KIR83]
Density, g/cm³: 13.9 [KIR83]
Melting Point, °C: 3880 [KIR83]
Boiling Point, °C: 5500 [HAW93]
Thermal Conductivity, W/(m·K): 22 [KIR78]
Thermal Expansion Coefficient: 6.29×10⁻⁶/K [KIR78]

3121
Compound: Tantalum nitride(δ)
Formula: δ-TaN
Molecular Formula: NTa
Molecular Weight: 194.955
CAS RN: 12033-62-4
Properties: yellowish gray; fcc, a=0.4336 nm; microhardness 3200; transition temp 17.8 K; used as a 99.5% pure sputtering target to increase electrical stability of diodes, transistors and integrated circuits [KIR81] [CER91]
Solubility: i acids; decomposed by KOH with evolution of NH₃ [KIR83]
Density, g/cm³: 15.6 [KIR81]
Melting Point, °C: 2950 [KIR81]

3122
Compound: Tantalum nitride(ε)
Formula: ε-TaN
Molecular Formula: NTa

Molecular Weight: 194.955
CAS RN: 12033-62-4
Properties: brown, bronze or black cryst; hex, a=0.5191 nm, c=0.2906 nm; electrical resistivity 128 μohm · m; hardness 1100 microhardness; transition temp 1.8 K; forms when tantalum is heated in pure nitrogen ~1100°C [KIR81] [KIR83] [HAW93] [CIC73]
Solubility: i H_2O; sl s in aqua regia, HNO_3 and HF [HAW93]
Density, g/cm³: 13.8 [KIR83]
Melting Point, °C: 2800 [KIR83]
Reactions: evolves N_2 if heated to 2000°C [KIR83]
Thermal Conductivity, W/(m · K): 9.54 [KIR81]

3123
Compound: Tantalum pentabromide
Synonym: tantalum(V) bromide
Formula: $TaBr_5$
Molecular Formula: Br_5Ta
Molecular Weight: 580.468
CAS RN: 13451-11-1
Properties: yellow cryst powd; sensitive to moisture; enthalpy of vaporization 62.3 kJ/mol; enthalpy of fusion 45.60 kJ/mol; can be prepared by heating tantalum metal in pure bromine gas above 300°C [STR93] [CRC10]
Density, g/cm³: 4.67 [STR93]
Melting Point, °C: 240 [ALD94]
Boiling Point, °C: 349 [CRC10]

3124
Compound: Tantalum pentachloride
Synonym: tantalum(V) chloride
Formula: $TaCl_5$
Molecular Formula: Cl_5Ta
Molecular Weight: 358.212
CAS RN: 7721-01-9
Properties: resublimed yellow cryst powd; monocl; decomposed in moist atm; enthalpy of fusion 35.10 kJ/mol; enthalpy of vaporization 54.8 kJ/mol; formed by reaction of Cl_2 with tantalum at 200°C; used in the chlorination of organic materials [HAW93] [MER06] [CRC10]
Solubility: decomposed in H_2O; s absolute alcohol [MER06]
Density, g/cm³: 3.68 [MER06]
Melting Point, °C: 216 [CRC10]
Boiling Point, °C: 242 [CRC10]

3125
Compound: Tantalum pentafluoride
Synonym: tantalum(V) fluoride

Formula: TaF_5
Molecular Formula: F_5Ta
Molecular Weight: 275.940
CAS RN: 7783-71-3
Properties: off-white deliq powd; enthalpy of vaporization is 56.9 kJ/mol; slowly etches glass; can be produced by fluorination of Ta metal; used as a catalyst in organic reactions [MER06] [STR93] [HAW93] [CRC10] [KIR78]
Solubility: s H_2O, ether, conc HNO_3; sl s hot CS_2, CCl_4 [MER06]
Density, g/cm³: 4.74 [MER06]
Melting Point, °C: 96.8 [KIR83]
Boiling Point, °C: 229.5 [MER06]

3126
Compound: Tantalum pentaiodide
Synonym: tantalum(V) iodide
Formula: TaI_5
Molecular Formula: I_5Ta
Molecular Weight: 815.470
CAS RN: 14693-81-3
Properties: hex black powd; sensitive to moisture [LID94] [STR93]
Density, g/cm³: 5.80 [LID94]
Melting Point, °C: 496 [KIR83]
Boiling Point, °C: 543 [KIR83]
Reactions: minus iodine above 1000°C [KIR83]

3127
Compound: Tantalum pentoxide
Synonym: tantalum(V) oxide
Formula: Ta_2O_5
Molecular Formula: O_5Ta_2
Molecular Weight: 441.893
CAS RN: 1314-61-0
Properties: white microcryst rhomb powd; decomposed by fusing with $KHSO_4$ or KOH; forms potassium tantalate when fused with KOH; enthalpy of fusion 120.00 kJ/mol; can be prepared by igniting Ta in air; used in optical glass, lasers and as a dielectric material, also as an evaporation material and sputtering target of 99.95% purity for dielectric films and multilayers [HAW93] [MER06] [KIR83] [CER91] [CRC10]
Solubility: i H_2O, alcohol, mineral acids; s HF [MER06]
Density, g/cm³: 8.2 [KIR83]
Melting Point, °C: 1800 [KIR83]
Reactions: reacts with carbon at 1900°C to form the carbide [KIR83]

3128
Compound: Tantalum pentoxide hydrate
Synonym: tantalic acid
Formula: $Ta_2O_5 \cdot xH_2O$
Molecular Formula: O_5Ta_2 (anhydrous)
Molecular Weight: 441.893 (anhydrous)
CAS RN: 75397-94-3
Properties: white insoluble precipitate formed by leaching a potassium pyrosulfate fusion of tantalum in H_2O; tantalic acid forms organic complexes with tannic, oxalic, tartaric, citric and pyrogallic acids; used in analytical chemistry [KIR83]

3129
Compound: Tantalum phosphide
Formula: TaP
Molecular Formula: PTa
Molecular Weight: 211.922
CAS RN: 12037-63-7
Properties: −100 mesh with 99.5% purity [CER91]

3130
Compound: Tantalum selenide
Formula: $TaSe_2$
Molecular Formula: Se_2Ta
Molecular Weight: 338.868
CAS RN: 12039-55-3
Properties: −325 mesh, $10\,\mu m$ or less, 99.8% pure material used as a sputtering target to produce lubricant films [CER91]

3131
Compound: Tantalum silicide
Formula: $TaSi_2$
Molecular Formula: Si_2Ta
Molecular Weight: 237.119
CAS RN: 12039-79-1
Properties: gray powd; used in the form of a 99.5%–99.95% pure material as a sputtering target in the fabrication of integrated circuits; there is also the compound Ta_5Si_3, 12067-56-0 [STR93] [CER91]
Density, g/cm^3: 9.14 [STR93]
Melting Point, °C: 2200 [STR93]

3132
Compound: Tantalum telluride
Formula: $TaTe_2$
Molecular Formula: $TaTe_2$
Molecular Weight: 436.148

CAS RN: 12067-66-2
Properties: −325 mesh $10\,\mu m$ or less with 99.8% purity [CER91]

3133
Compound: Tantalum tetroxide
Formula: Ta_2O_4
Molecular Formula: O_4Ta_2
Molecular Weight: 425.894
CAS RN: 12035-90-4
Properties: dark gray powd; probably forms when the pentoxide is partially reduced by carbon at 1900°C [KIR83] [CRC10]
Reactions: oxidizes on heating [CRC10]

3134
Compound: Tantalum trisilicide
Formula: Ta_5Si_3
Molecular Formula: Si_3Ta_5
Molecular Weight: 988.996
CAS RN: 12067-56-0
Properties: −325 mesh powd; used as a 99.5 or 99.95% material as a sputtering target in the fabrication of integrated circuits [ALF93] [CER91]

3135
Compound: Technetium
Formula: Tc
Molecular Formula: Tc
Molecular Weight: 98
CAS RN: 7440-26-8
Properties: closed-packed hex, a=0.2741 nm, c=0.4399 nm; enthalpy of sublimation 650 kJ/mol; enthalpy of vaporization ~577 kJ/mol; enthalpy of fusion 33.29 kJ/mol; slowly tarnishes in moist air; when obtained from H_2 reduction of ammonium pertechnate, has silvery gray color, and a spongy mass; resembles rhenium in chemical behavior; Debye constant 455 K; used as a metallurgical tracer, in nuclear medicine, and to protect against corrosion [HAW93] [MER06] [RAR83] [CRC10]
Density, g/cm^3: 11.5 [HAW93]
Melting Point, °C: 2167 [RAR83]
Boiling Point, °C: ~4600 [RAR83]
Thermal Conductivity, $W/(m \cdot K)$: 50.6 [CRC10]
Thermal Expansion Coefficient: a-axis: $7.04 \times 10^{-6}/K$; c-axis: $7.06 \times 10^{-6}/K$ [RAR83]

3136
Compound: Technetium dioxide
Formula: TcO_2

Molecular Formula: O_2Tc
Molecular Weight: 130
CAS RN: 12036-16-7
Properties: there is a monohydrate, CAS RN 42861-23-4, and a dihydrate, CAS RN 60003-95-4 [ERI92]
Solubility: H_2O: $TcO_2 \cdot nH_2O = TcO(OH)_2(aq) + (n-1)H_2O$, log $K = -8.16$; $TcO_2 \cdot n2H_2O + H_2O = Tc(OH)_3^- + (n-1)H_2O + H^+$, log $K = -19.20$; reference also contains predominance diagram [ERI92]

3137
Compound: Telluric acid
Synonym: orthotelluric acid
Formula: H_6TeO_6
Molecular Formula: H_6O_6Te
Molecular Weight: 229.644
CAS RN: 7803-68-1
Properties: −40 mesh with 99.5% purity; white solid; monocl, cub; very weak acid, $K_1 = 2 \times 10^{-8}$, $K_2 = 1 \times 10^{-11}$ [MER06] [CER91]
Solubility: g H_2TeO_6/100 g H_2O: 16.2 (0°C), 41.6 (20°C), 155 (100°C) [LAN05]; tends to polymerize; s dil HNO_3 [MER06]
Density, g/cm³: monocl: 3.068; cub: 3.163 [MER06]
Melting Point, °C: 136 [HAW93]
Boiling Point, °C: 160, decomposes [STR93]

3138
Compound: Tellurium
Formula: Te
Molecular Formula: Te
Molecular Weight: 127.60
CAS RN: 13494-80-9
Properties: grayish white, lustrous, brittle; rhomb cryst; hardness, 2.3 Mohs; Poisson's ratio 0.33 at 30°C; enthalpy of fusion 17.87 kJ/mol; enthalpy of vaporization 114.1 kJ/mol; electrical resistivity (20°C) $(5.8-33) \times 10^{+3}$ μohm·cm; modulus of elasticity 4140 MPa; viscosity at mp 1.8–1.95 mPa·s; electronegativity 2.01; burns with greenish blue flame; p-type semiconductor; used in thin film devices as blocking contact [HAW93] [MER06] [CRC10] [KIR83] [COT88] [CER91] [ALD94]
Solubility: i H_2O, benzene, CS_2 [MER06]
Density, g/cm³: 6.24 (cryst) [KIR83]
Melting Point, °C: 449.8 [MER06]
Boiling Point, °C: 989.8 [ALD94]
Reactions: reacts with HNO_3, conc H_2SO_4, KOH, forming red solution [MER06]
Thermal Conductivity, W/(m·K): 1.97 to 3.38 (25°C) [ALD94]

3139
Compound: Tellurium decafluoride
Formula: Te_2F_{10}
Molecular Formula: $F_{10}Te_2$
Molecular Weight: 445.184
CAS RN: 53214-07-6
Properties: volatile, colorless liq; stable [KIR83]
Melting Point, °C: −33.7 [KIR83]
Boiling Point, °C: 59 [KIR83]

3140
Compound: Tellurium dibromide
Synonym: tellurous bromide
Formula: $TeBr_2$
Molecular Formula: Br_2Te_2
Molecular Weight: 287.408
CAS RN: 7789-54-0
Properties: greenish black cryst mass, or gray to black needles; very hygr; has a violet vapor [HAW93]
Solubility: decomposed by H_2O; s ether; sl s chloroform [MER06]
Melting Point, °C: 210 [HAW93]
Boiling Point, °C: 339 [HAW93]

3141
Compound: Tellurium dichloride
Synonym: tellurous chloride
Formula: $TeCl_2$
Molecular Formula: Cl_2Te
Molecular Weight: 198.505
CAS RN: 10025-71-5
Properties: black amorphous solid; hygr; melts to a black liq; purple vapor, disproportionates in ether, dioxane [MER06] [KIR83]
Solubility: decomposed by H_2O; i CCl_4 [MER06] [HAW93]
Density, g/cm³: 6.9 [HAW93]
Melting Point, °C: 208 [MER06]
Boiling Point, °C: 328 [MER06]

3142
Compound: Tellurium dioxide
Synonym: tellurite
Formula: TeO_2
Molecular Formula: O_2Te
Molecular Weight: 159.599
CAS RN: 7446-07-3
Properties: white cryst; tetr and ortho-rhomb; yellow when heated; made by dissolving Te in strong HNO_3 to form $2TeO_2 \cdot HNO_3$ at 400°C–430°C, then decomposing this product [KIR83] [MER06]

Solubility: s H_2O ~1:150,000; s NaOH,
  HCl solutions [MER06]
Density, g/cm³: tetr: 5.75; ortho-rhomb: 6.04 [MER06]
Melting Point, °C: 733, forming yellow liq [MER06]
Boiling Point, °C: 1245 [HAW93]

3143
Compound: Tellurium disulfide
Formula: TeS_2
Molecular Formula: S_2Te
Molecular Weight: 191.732
CAS RN: 7446-35-7
Properties: red powd; eventually turns to
  brown amorphous powd; fuses to a
  gray, lustrous mass [HAW93]
Solubility: i H_2O, acids; s in alkali sulfides [HAW93]

3144
Compound: Tellurium hexafluoride
Formula: TeF_6
Molecular Formula: F_6Te
Molecular Weight: 241.590
CAS RN: 7783-80-4
Properties: colorless gas; does not attack
  glass when pure [MER06]
Solubility: slowly mixes with H_2O, hydrolyzing
  to telluric acid [MER06]
Density, g/cm³: solid (−191°C): 4.006;
  liq (−10°C): 2.499 [MER06]
Melting Point, °C: −37.6 [MER06]
Boiling Point, °C: 35.5 [CRC10]
Reactions: reduced by Te to TeF_4 [KIR83]

3145
Compound: Tellurium nitrate
Synonym: basic tellurium nitrate
Formula: $TeO_2 \cdot NO_3$
Molecular Formula: NO_5Te
Molecular Weight: 221.604
CAS RN: 64535-94-0
Properties: prepared by dissolution
  of Te in HNO_3 [KIR83]
Melting Point, °C: decomposes from
  190 to 300 [KIR83]

3146
Compound: Tellurium nitride
Formula: Te_3N_4
Molecular Formula: N_4Te_3
Molecular Weight: 438.827
CAS RN: 12164-01-1

Properties: citron-yellow colored solid; unstable, and
  can detonate readily if struck or heated [KIR83]
Solubility: could explode on contact with H_2O [KIR83]

3147
Compound: Tellurium sulfate
Synonym: basic tellurium sulfate
Formula: $2TeO_2 \cdot SO_3$
Molecular Formula: O_7STe_2
Molecular Weight: 399.262
CAS RN: 12068-84-8
Properties: prepared by dissolution of TeO_2 solution in
  H_2SO_4, followed by slow evaporation to form the
  compound; stable up to 440°C–500°C [KIR83]
Solubility: slowly hydrolyzed by cold H_2O,
  rapidly by hot H_2O [KIR83]

3148
Compound: Tellurium tetrabromide
Formula: $TeBr_4$
Molecular Formula: Br_4Te
Molecular Weight: 447.216
CAS RN: 10031-27-3
Properties: −4 mesh with 99.9% purity; orange cryst,
  turning red when hot [MER06] [CER91]
Solubility: s in small volume of H_2O,
  hydrolyzes in larger volume [MER06]
Density, g/cm³: 4.3 [MER06]
Melting Point, °C: ~380 [MER06]
Boiling Point, °C: 420, decomposes to
  $TeBr_2$ and Br_2 [HAW93]

3149
Compound: Tellurium tetrachloride
Formula: $TeCl_4$
Molecular Formula: Cl_4Te
Molecular Weight: 269.411
CAS RN: 10026-07-0
Properties: white, cryst sold; very hygr; decomposed
  by water to TeO_2 and HCl; melts to a yellow
  liq, dark red at higher temperatures; enthalpy
  of vaporization 77 kJ/mol [MER06] [CRC10]
Solubility: s in absolute alcohol, toluene [MER06]
Density, g/cm³: 3.01 [MER06]
Melting Point, °C: 225 [MER06]
Boiling Point, °C: ~390 [KIR83]

3150
Compound: Tellurium tetrafluoride
Formula: TeF_4
Molecular Formula: F_4Te

Molecular Weight: 203.594
CAS RN: 15192-26-4
Properties: white hygr needles; attacks glass, silica, and copper at 200°C, does not attack platinum below 300°C [KIR83]
Solubility: readily hydrolyzed [KIR83]
Melting Point, °C: decomposes to TaF_6 at 194 [KIR83]

3151
Compound: Tellurium tetraiodide
Formula: TeI_4
Molecular Formula: I_4Te
Molecular Weight: 635.218
CAS RN: 7790-48-9
Properties: −4 mesh with 99.9% purity; grayish black volatile cryst; stable in moist air [MER06] [KIR83] [CER91]
Solubility: hydrolyzed by H_2O forming TeO_2, HI; s HI; sl s acetone [MER06]
Density, g/cm³: 5.05 [MER06]
Melting Point, °C: 280 [MER06]
Reactions: I_2 evolved by heating [MER06]

3152
Compound: Tellurium trioxide
Formula: TeO_3
Molecular Formula: O_3Te
Molecular Weight: 175.598
CAS RN: 13451-18-8
Properties: −60 mesh with 99.9% purity; two forms: yellowish orange α, and grayish β; α is a strong oxidant, e.g. reacting vigorously with Al, Sn, C, P and S [KIR83] [CER91]
Density, g/cm³: α: 5.07; β: 6.21 [KIR83]
Melting Point, °C: decomposes [CRC10]

3153
Compound: Tellurous acid
Formula: H_2TeO_3
Molecular Formula: H_2O_3Te
Molecular Weight: 177.614
CAS RN: 10049-23-7
Properties: unstable white cryst, or cryst powd; dehydrates readily to TeO_2 [KIR83] [MER06]
Solubility: sl s H_2O; s in dil acids, alkalies [MER06]
Density, g/cm³: 3.05 [HAW93]
Melting Point, °C: 40, decomposes [HAW93]

3154
Compound: Terbium
Formula: Tb

Molecular Formula: Tb
Molecular Weight: 158.92534
CAS RN: 7440-27-9
Properties: silvery gray metal; easily oxidized by atm O_2; hex close-packed; electrical resistivity (20°C) 116 µohm·cm; enthalpy of fusion is 10.80 kJ/mol; enthalpy of sublimation 288.7 kJ/mol; radius of atom is 0.17833 nm; radius of Tb^{+++} ion is 0.0923 nm; forms colorless solutions; used as a phosphor activator [HAW93] [MER06] [KIR82] [ALD94]
Solubility: slowly reacts with H_2O; s dil acids [HAW93]
Density, g/cm³: 8.27 [MER06]
Melting Point, °C: 1356 [MER06]
Boiling Point, °C: 3230 [ALD94]
Thermal Conductivity, W/(m·K): 11.1 (25°C) [CRC10]
Thermal Expansion Coefficient: 10.3×10^{-6}/K [CRC10]

3155
Compound: Terbium acetate hydrate
Formula: $Tb(CH_3COO)_3 \cdot xH_2O$
Molecular Formula: $C_6H_9O_6Tb$ (anhydrous)
Molecular Weight: 336.059 (anhydrous)
CAS RN: 100587-92-6
Properties: hygr [ALD94]

3156
Compound: Terbium acetylacetonate trihydrate
Synonyms: 2,4-pentanedione, terbium(III) derivative
Formula: $Tb(CH_3COCH=C(O)(CH_3)_3 \cdot 3H_2O$
Molecular Formula: $C_{15}H_{27}O_9Tb$
Molecular Weight: 510.299
CAS RN: 14284-95-8
Properties: white powd; hygr [STR93]
Melting Point, °C: 168–170 [STR93]

3157
Compound: Terbium bromide
Formula: $TbBr_3$
Molecular Formula: Br_3Tb
Molecular Weight: 398.637
CAS RN: 14456-47-4
Properties: hex, silvery gray; −20 mesh with 99.9% purity [CER91] [CRC10]
Melting Point, °C: 827 [AES93]
Boiling Point, °C: 1490 [CRC10]

3158
Compound: Terbium carbonate hydrate
Formula: $Tb_2(CO_3)_3 \cdot xH_2O$
Molecular Formula: $C_3O_9Tb_2$ (anhydrous)
Molecular Weight: 497.878 (anhydrous)

CAS RN: 100587-96-0
Properties: white powd [STR93]

3159
Compound: Terbium chloride
Formula: TbCl$_3$
Molecular Formula: Cl$_3$Tb
Molecular Weight: 265.283
CAS RN: 10042-88-3
Properties: −20 mesh with 99.9% purity;
off-white powd; hygr [STR93] [CER91]
Solubility: s H$_2$O [MER06]
Density, g/cm³: 4.35 [MER06]
Melting Point, °C: 588 [MER06]

3160
Compound: Terbium chloride hexahydrate
Formula: TbCl$_3$·6H$_2$O
Molecular Formula: Cl$_3$H$_{12}$O$_6$Tb
Molecular Weight: 373.374
CAS RN: 13798-24-8
Properties: −4 mesh with 99.9% purity;
transparent, colorless prismatic cryst;
very hygr [HAW93] [CER91]
Solubility: v s H$_2$O; forms supersaturated
solutions [MER06]
Density, g/cm³: 4.35 [HAW93]
Melting Point, °C: 588 (anhydrous) [HAW93]
Reactions: minus 6H$_2$O 180°C–200°C
(in HCl gas stream) [MER06]

3161
Compound: Terbium fluoride
Formula: TbF$_3$
Molecular Formula: F$_3$Tb
Molecular Weight: 215.920
CAS RN: 13708-63-9
Properties: white powd, and 99.9% pure melted
pieces of 3–12 mm; hygr; pieces used as
evaporation material for possible application
to multilayers [STR93] [CER91]
Melting Point, °C: 1172 [STR93]

3162
Compound: Terbium hydride
Formula: TbH$_{2-3}$
Molecular Formula: H$_2$Tb; H$_3$Tb
Molecular Weight: TbH$_2$: 160.941; TbH$_3$: 161.949
CAS RN: 13598-54-4
Properties: −60 mesh with 99.9% purity [CER91]

3163
Compound: Terbium iodide
Formula: TbI$_3$
Molecular Formula: I$_3$Tb
Molecular Weight: 539.638
CAS RN: 13813-40-6
Properties: −20 mesh with 99.9% purity [CER91]
Density, g/cm³: ~5.2 [LID94]
Melting Point, °C: 946 [AES93]
Boiling Point, °C: >1300 [CRC10]

3164
Compound: Terbium nitrate hexahydrate
Formula: Tb(NO$_3$)$_3$·6H$_2$O
Molecular Formula: H$_{12}$N$_3$O$_{15}$Tb
Molecular Weight: 453.031
CAS RN: 13451-19-9
Properties: colorless; monocl cryst or
white powd [HAW93] [MER06]
Solubility: s H$_2$O [HAW93]
Melting Point, °C: 893 [AES93]

3165
Compound: Terbium nitride
Formula: TbN
Molecular Formula: NTb
Molecular Weight: 172.932
CAS RN: 12033-64-6
Properties: cub cryst; −40 mesh with
99.9% purity [LID94] [CER91]
Density, g/cm³: 9.55 [LID94]

3166
Compound: Terbium oxalate hydrate
Formula: Tb$_2$(C$_2$O$_4$)$_3$·xH$_2$O
Molecular Formula: C$_6$O$_{12}$Tb$_2$ (anhydrous)
Molecular Weight: 581.909 (anhydrous)
CAS RN: 24670-06-2
Properties: white powd, x = 10 [CRC10] [STR93]
Density, g/cm³: x = 10: 2.60 [STR93]
Reactions: minus H$_2$O at 40°C [CRC10]

3167
Compound: Terbium perchlorate hexahydrate
Formula: Tb(ClO$_4$)$_3$·6H$_2$O
Molecular Formula: Cl$_3$H$_{12}$O$_{18}$Tb
Molecular Weight: 565.367
CAS RN: 14014-09-6
Properties: white cryst; hygr [STR93]

3168

Compound: Terbium silicide
Formula: $TbSi_2$
Molecular Formula: Si_2Tb
Molecular Weight: 215.096
CAS RN: 12039-80-4
Properties: ortho-rhomb cryst; 10 mm & down lump [LID94] [ALF93]
Density, g/cm³: 6.66 [LID94]

3169

Compound: Terbium sulfate octahydrate
Formula: $Tb_2(SO_4)_3 \cdot 8H_2O$
Molecular Formula: $H_{16}O_{20}S_3Tb_2$
Molecular Weight: 750.164
CAS RN: 13842-67-6
Properties: white cryst [STR93]
Solubility: s H_2O [HAW93]
Reactions: minus $8H_2O$ at 360°C [HAW93]

3170

Compound: Terbium sulfide
Formula: Tb_2S_3
Molecular Formula: S_3Tb_2
Molecular Weight: 414.049
CAS RN: 12138-11-3
Properties: cub cryst; −200 mesh with 99.9% purity [LID94] [CER91]
Density, g/cm³: 6.35 [LID94]

3171

Compound: Terbium(III,IV) oxide
Formula: Tb_4O_7
Molecular Formula: O_7Tb_4
Molecular Weight: 747.697
CAS RN: 12037-01-3
Properties: dark brown powd, or sintered pieces 3–12 mm; sl hygr; absorbs atm CO_2; used as an evaporation material of 99.9% purity; possibly reactive to radio frequencies [HAW93] [CER91]
Solubility: s dil acids [HAW93]
Reactions: minus O_2 on heating [CRC10]

3172

Compound: Tetraborane(10)
Formula: B_4H_{10}
Molecular Formula: B_4H_{10}
Molecular Weight: 53.323
CAS RN: 18283-93-7

Properties: gas with disagreeable odor; vapor pressure, mm Hg: (0°C) 388, (6°C) 580; enthalpy of vaporization 27.1 kJ/mol; decomposes in a few hours at room temp, more rapidly at 100°C; spontaneously flammable in air, unless pure [MER06] [COT88] [CRC10]
Solubility: hydrolyzes in H_2O to boric acid with evolution of hydrogen [MER06]
Density, g/cm³: 2.34 g/L [LID94]
Melting Point, °C: −120 [KIR78]
Boiling Point, °C: 18 [KIR78]

3173

Compound: Tetrabromodiborane
Formula: B_2Br_4
Molecular Formula: B_2Br_4
Molecular Weight: 341.238
CAS RN: 14355-29-4
Properties: col liq [CRC10]
Melting Point, °C: ~1 [CRC10]
Boiling Point, °C: decomposes at 20 [CRC10]

3174

Compound: Tetrachlorodiborane
Formula: B_2Cl_4
Molecular Formula: B_2Cl_4
Molecular Weight: 163.434
CAS RN: 13701-67-2
Properties: colorless liq; flam [CRC10]
Melting Point, °C: −92.6 [CRC10]
Boiling Point, °C: 66.5 [CRC10]

3175

Compound: Tetradecaborane(18)
Formula: $B_{14}H_{18}$
Molecular Formula: $B_{14}H_{18}$
Molecular Weight: 169.497
CAS RN: 55606-55-8
Properties: visc yellow oil [CRC10]
Solubility: s cychex, CS_2 [CRC10]
Boiling Point, °C: decomposes at 100 [CRC10]

3176

Compound: Tetraethyl lead
Synonym: TEL
Formula: $(C_2H_5)_4Pb$
Molecular Formula: $C_8H_{20}Pb$
Molecular Weight: 323.447
CAS RN: 78-00-2

Properties: colorless liq; burns with orange-colored flame; obtained by reacting $PbCl_2$ and zinc-ethyl; uses: formerly used as gasoline additive [ALD94] [MER06]
Solubility: i H_2O; s benzene, petroleum ether, gasoline [MER06]
Density, g/cm³: 1.653 [ALD94]
Melting Point, °C: −136 [ALD94]
Boiling Point, °C: 84–85 at 15 mm Hg [ALD94]

3177
Compound: Tetraethyl silane
Formula: $Si(C_2H_5)_4$
Molecular Formula: $C_8H_{20}Si$
Molecular Weight: 144.332
CAS RN: 631-36-7
Properties: hygr [ALD94]
Density, g/cm³: 0.766 [ALD94]
Melting Point, °C: −82.5 [ALD94]
Boiling Point, °C: 153–154 [ALD94]

3178
Compound: Tetraethylammonium bromide
Synonym: TEAB
Formula: $(C_2H_5)_4NBr$
Molecular Formula: $C_8H_{20}BrN$
Molecular Weight: 210.158
CAS RN: 71-91-0
Properties: hygr cryst; prepared from triethylamine and ethyl bromide; used as ganglion blocking agent [MER06] [ALD94]
Solubility: v s H_2O, alcohol, chloroform [MER06]
Melting Point, °C: 285, decomposes [ALD94]
Thermal Expansion Coefficient: from 25°C to 100°C (0.18), 200°C (0.42), 400°C (0.90), 600°C (1.38), 800°C (1.86), 1000°C (2.34), 1200°C (2.82) [TAY91b]

3179
Compound: Tetraethylammonium chloride
Synonym: T.E.A. chloride
Formula: $(C_2H_5)_4NCl$
Molecular Formula: $C_8H_{20}ClN$
Molecular Weight: 165.706
CAS RN: 56-34-8
Properties: deliq cryst, tetrahydrate is monocl; ganglion blocking agent [MER06]
Solubility: s H_2O [MER06]
Density, g/cm³: 1.0801 [MER06]
Melting Point, °C: 37.5 (tetrahydrate) [MER06]

3180
Compound: Tetraethylorthosilicate
Synonym: ethyl silicate
Formula: $Si(OC_2H_5)_4$
Molecular Formula: $C_8H_{20}O_4Si$
Molecular Weight: 208.329
CAS RN: 78-10-4
Properties: colorless, flammable liq; flash point 52°C; prepared by reaction of absolute ethanol with $SiCl_4$; sensitive to moisture; used in the sol-gel preparation of zircon, in weatherproofing and hardening stone [MER06] [ALD94]
Solubility: miscible with ethanol; reacts with H_2O [MER06]
Density, g/cm³: 0.934 [ALD94]
Melting Point, °C: −77 [MER06]
Boiling Point, °C: 165–166 [MER06]

3181
Compound: Tetragermane
Formula: Ge_4H_{10}
Molecular Formula: Ge_4H_{10}
Molecular Weight: 300.64
CAS RN: 14691-47-5
Properties: col liq [CRC10]
Solubility: i H_2O [CRC10]
Boiling Point, °C: 176.9 [CRC10]

3182
Compound: Tetrafluoroboric acid
Formula: HBF_4
Molecular Formula: HBF_4
Molecular Weight: 87.813
CAS RN: 16872-11-0
Properties: col liq [CRC10]
Solubility: v s H_2O, EtOH [CRC10]
Density, g/cm³: ~1.8 [CRC10]
Boiling Point: decomposes at 130 [CRC10]

3183
Compound: Tetrafluorodiborane
Formula: B_2F_4
Molecular Formula: B_2F_4
Molecular Weight: 97.616
CAS RN: 13965-73-6
Properties: col gas; flam [CRC10]
Solubility: reac H_2O [CRC10]
Density, g/L: 3.990 [CRC10]
Melting Point, °C: −56 [CRC10]
Boiling Point, °C: −34 [CRC10]

3184

Compound: Tetracarbonyldihydroiron
Formula: $Fe(CO)_4H_2$
Molecular Formula: $C_4H_2FeO_4$
Molecular Weight: 169.902
CAS RN: 12002-28-7
Properties: col liq; stab low temp [CRC10]
Solubility: s alk [CRC10]
Melting Point, °C: −70 [CRC10]
Boiling Point, °C: decomposes at −20 [CRC10]

3185

Compound: Tetramethylgermane
Formula: $(CH_3)_4Ge$
Molecular Formula: $C_4H_{12}Ge$
Molecular Weight: 132.749
CAS RN: 865-52-1
Properties: liq; flammable [ALD94]
Density, g/cm³: 0.978 [ALD94]
Melting Point, °C: −88 [ALD94]
Boiling Point, °C: 43–44 [ALD94]

3186

Compound: Tetramethyltin
Formula: $(CH_3)_4Sn$
Molecular Formula: $C_4H_{12}Sn$
Molecular Weight: 178.849
CAS RN: 594-27-4
Properties: liq; flammable [ALD94]
Density, g/cm³: 1.291 [ALD94]
Melting Point, °C: −54 [ALD94]
Boiling Point, °C: 74–75 [ALD94]

3187

Compound: Tetrapropylammonium perruthenate(VII)
Synonym: TPAP
Formula: $(CH_3CH_2CH_2)_4NRuO_4$
Molecular Formula: $C_{12}H_{28}NO_4Ru$
Molecular Weight: 351.428
CAS RN: 114615-82-6
Properties: sold; mild catalytic oxidant [ALD94]
Melting Point, °C: 165, decomposes [ALD94]
Reactions: explodes when heated [ALD94]

3188

Compound: Thallium
Formula: Tl
Molecular Formula: Tl
Molecular Weight: 204.3833
CAS RN: 7440-28-0

Properties: α-Tl: hex; β-Tl: cub; bluish white, very
soft; enthalpy of fusion 4.14 kJ/mol; vaporization
enthalpy 165 kJ/mol; Brinell hardness is 2;
resistivity 16.6 μohm · cm; oxidizes in air, forming
oxide film on surface; used in low melting
alloys as electrode to measure the dissolved
oxygen content of waters; photoelectric [CIC73]
[HAW93] [ALD94] [MER06] [KIR83] [CRC10]
Solubility: i H_2O; reacts with HNO_3, H_2SO_4 [MER06]
Density, g/cm³: α: 11.85; β: 11.86–11.87 [CIC73]
Melting Point, °C: 303.5 [MER06]
Boiling Point, °C: 1553 [COT88]
Reactions: volatilization begins at 174°C [MER06]
Thermal Conductivity, W/(m · K): 46.1 (25°C) [CRC10]
Thermal Expansion Coefficient: (volume) × 10^{-6}/°C:
90 (20°C) α, 124 β, 140 liq [CIC73]

3189

Compound: Thallium barium calcium copper oxide
Formula: $Tl_4Ba_3Ca_3Cu_4O_{13}$
Molecular Formula: $Ba_3Ca_3Cu_4O_{13}Tl_4$
Molecular Weight: 1808.924
CAS RN: 119000-19-0
Properties: superconductor; 99.99% and
99.9% purity 20 μm powd [ALF93]

3190

Compound: Thallium barium calcium copper oxide
Formula: $Tl_2Ba_2Ca_2Cu_3O_{10}$
Molecular Formula: $Ba_2Ca_2Cu_3O_{10}Tl_2$
Molecular Weight: 1114.209
CAS RN: 127241-75-2
Properties: superconductor (2223 phase); 99.999%
and 99.9% purity, 20 μm powd; T_c 115–127 K;
for material with formula $Tl_2Ba_2Ca_3Cu_4O_{12}$, T_c
is 113–119 K [CEN92] [ALF93] [ASM93]

3191

Compound: Thallium barium calcium copper oxide
Formula: $Tl_2Ba_2CaCu_2O_8$
Molecular Formula: $Ba_2CaCu_2O_8Tl_2$
Molecular Weight: 1018.664
CAS RN: 125720-69-1
Properties: superconductor; 20 μm powd, 99.999%
purity and 99.9%; T_c 95–110 K [ALF93] [ASM93]

3192

Compound: Thallium(I) acetate
Synonym: thallous acetate
Formula: CH_3COOTl
Molecular Formula: $C_2H_3O_2Tl$

Molecular Weight: 263.428
CAS RN: 563-68-8
Properties: −4 mesh with 99.9% purity; white cryst; deliq; used in ore flotation separation; there is a hemitrihydrate, 2570-63-0 [HAW93] [MER06] [CER91]
Solubility: s H_2O, alcohol [MER06]
Density, g/cm³: 3.68 [HAW93]
Melting Point, °C: 131 [STR93]

3193

Compound: Thallium(I) acetylacetonate
Synonyms: 2,4-pentanedione, thallium(I) derivative
Formula: $TlCH_3COCH=C(O)CH_3$
Molecular Formula: $C_5H_7O_2Tl$
Molecular Weight: 303.493
CAS RN: 25955-51-5
Properties: white powd [STR93]

3194

Compound: Thallium(I) azide
Formula: TlN_3
Molecular Formula: N_3Tl
Molecular Weight: 246.403
CAS RN: 13847-66-0
Properties: yellow; body-center tetr, a = 0.623 nm, c = 0.675 nm [CIC73] [CRC10]
Solubility: g/100 g H_2O: 0.171 (0°C), 0.236 (10°C), 0.364 (20°°C) [LAN05]
Melting Point, °C: 330 (vacuum) [CRC10]

3195

Compound: Thallium(I) bromide
Synonym: thallous bromide
Formula: TlBr
Molecular Formula: BrTl
Molecular Weight: 284.287
CAS RN: 7789-40-4
Properties: −20 mesh with 99.999% purity; pale yellow; cryst powd; enthalpy of vaporization 99.56 kJ/mol; enthalpy of fusion 25.10 kJ/mol; used in a mixture with TlI in infrared transmitters [HAW93] [MER06] [CER91] [CRC10]
Solubility: g/100 g H_2O: 0.022 (0°C), 0.048 (20°C), 0.177 (60°C) [LAN05]
Density, g/cm³: 7.557 [HAW93]
Melting Point, °C: 460 [CRC10]
Boiling Point, °C: 819 [CRC10]

3196

Compound: Thallium(I) carbonate
Synonym: thallous carbonate

Formula: Tl_2CO_3
Molecular Formula: CO_3Tl_2
Molecular Weight: 468.776
CAS RN: 6533-73-9
Properties: shiny white monocl cryst; highly refractive; melts to a dark gray mass; enthalpy of fusion 18.40 kJ/mol; used in testing for carbon disulfide and in artificial diamonds [HAW93] [KIR83] [CRC10]
Solubility: g/100 g H_2O: 5.3 (20°C), 12.2 (60°C), 27.2 (100°C) [LAN05]; i alcohol [MER06]
Density, g/cm³: 7.11 [STR93]
Melting Point, °C: 272 [MER06]

3197

Compound: Thallium(I) chlorate
Formula: $TlClO_3$
Molecular Formula: ClO_3Tl
Molecular Weight: 287.834
CAS RN: 13453-30-0
Properties: needles [LAN05]
Solubility: g/100 g H_2O: 2.00 (0°C), 3.92 (20°C), 57.3 (100°C) [LAN05]
Density, g/cm³: 5.047 [LAN05]

3198

Compound: Thallium(I) chloride
Synonym: thallous chloride
Formula: TlCl
Molecular Formula: ClTl
Molecular Weight: 239.836
CAS RN: 7791-12-0
Properties: white cryst powd, and 99.9% pure melted pieces of 3–12 mm; turns violet when exposed to light; enthalpy of vaporization 102.2 kJ/mol; enthalpy of fusion 17.80 kJ/mol; used as a chlorination catalyst and in suntan lamp monitors, and as an evaporation material to deposit long wavelength coatings up to >50 μm, has index of ~1.90 at 20 μm [HAW93] [CER91] [CRC10]
Solubility: g/100 g H_2O: 0.21 (0°C), 0.33 (20°C), 1.80 (100°C) [LAN05]; i alcohol [MER06]
Density, g/cm³: 7.004 [HAW93]
Melting Point, °C: 430 [MER06]
Boiling Point, °C: 807 [CRC10]

3199

Compound: Thallium(I) cyanide
Synonym: thallous cyanide
Formula: TlCN
Molecular Formula: CNTl
Molecular Weight: 230.401
CAS RN: 13453-34-4

Properties: white; hex platelets [MER06]
Solubility: 16.8 g/100 mL H_2O, s a, alcohol [MER06]
Density, g/cm³: 6.523 [MER06]
Melting Point, °C: decomposes [CRC10]

3200
Compound: Thallium(I) ethoxide
Formula: $TlOC_2H_5$
Molecular Formula: C_2H_5OTl
Molecular Weight: 249.444
CAS RN: 20398-06-5
Properties: cloudy, dense liq; sensitive
  to moisture [STR93]
Density, g/cm³: 3.493 (20°C) [STR93]
Melting Point, °C: −3 [STR93]
Boiling Point, °C: decomposes at 130 [STR93]

3201
Compound: Thallium(I) fluoride
Synonym: thallous fluoride
Formula: TlF
Molecular Formula: FTl
Molecular Weight: 223.381
CAS RN: 7789-27-7
Properties: white powd; ortho-rhomb; hard,
  shiny cryst; can deliq, e.g., if breathed upon,
  however reverts to anhydrous form in dry
  air; not typically hygr; enthalpy of fusion
  14.00 kJ/mol [MER06] [STR93] [CRC10]
Solubility: 78.6 g/100 g H_2O at 15°C;
  s alcohols, HF [KIR83]
Density, g/cm³: 8.36 [MER06]
Melting Point, °C: 322 [MER06]
Boiling Point, °C: 655 [STR93]

3202
Compound: Thallium(I) formate
Formula: $HCOOTl$
Molecular Formula: CHO_2Tl
Molecular Weight: 249.401
CAS RN: 992-98-3
Properties: colorless needles or white
  powd; hygr [STR93][KIR83]
Solubility: 500 g/100 g H_2O at 10°C;
  s methanol [KIR83]
Density, g/cm³: 4.967 [STR93]
Melting Point, °C: 101 [STR93]

3203
Compound: Thallium(I) hexafluoroacetylacetonate
Synonyms: 1,1,1,5,5,5-hexafluoro-2,4-
  pentanedione, Tl(I) derivative

Formula: $TlCF_3COCHCOCF_3$
Molecular Formula: $C_5HF_6O_2Tl$
Molecular Weight: 411.435
CAS RN: 15444-43-6
Properties: yellow cryst [STR93]
Melting Point, °C: 126–128 [STR93]
Boiling Point, °C: decomposes [STR93]
Reactions: sublimes at 140°C (0.1 mm Hg) [STR94]

3204
Compound: Thallium(I) hexafluorophosphate
Formula: $TlPF_6$
Molecular Formula: FePTl
Molecular Weight: 349.347
CAS RN: 60969-19-9
Properties: white cryst [STR93]

3205
Compound: Thallium(I) hydroxide
Synonym: thallous hydroxide
Formula: TlOH
Molecular Formula: HOTl
Molecular Weight: 221.390
CAS RN: 1310-83-4
Properties: yellow needles; gives strongly alkaline
  solution when dissolved in H_2O, which turns
  tumeric paper brown color [MER06]
Solubility: g/100 g H_2O: 25.4 (0°C), 35.0
  (20°C), 150 (100°C) [LAN05]
Density, g/cm³: 7.44 [KIR83]
Melting Point, °C: decomposes at 139 [KIR83]

3206
Compound: Thallium(I) iodide
Synonym: thallous iodide
Formula: α-TlI
Molecular Formula: ITl
Molecular Weight: 331.287
CAS RN: 7790-30-9
Properties: yellow; rhomb cryst; turns red at
  170°C; enthalpy of vaporization 104.7 kJ/mol;
  enthalpy of fusion 13.10 kJ/mol; used mixed
  with TlBr in infrared transmitters [HAW93]
  [MER06] [CRC10] [KIR83]
Solubility: g/100 g H_2O: 0.002 (0°C), 0.006
  (20°C), 0.120 (100°C) [LAN05]
Density, g/cm³: 7.29 [KIR83]
Melting Point, °C: 440 [MER06]
Boiling Point, °C: 824 [MER06]

3207
Compound: Thallium(I) molybdate
Formula: Tl_2MoO_4
Molecular Formula: MoO_4Tl_2
Molecular Weight: 568.705
CAS RN: 34128-09-1
Properties: white when precipitated, yellow
  during fusion; cub, a=0.926 nm [KIR81]
Solubility: i H_2O [KIR81]
Melting Point, °C: red heat [KIR81]

3208
Compound: Thallium(I) nitrate
Synonym: thallous nitrate
Formula: $TlNO_3$
Molecular Formula: NO_3Tl
Molecular Weight: 266.388
CAS RN: 10102-45-1
Properties: white cryst; enthalpy of fusion
  9.60 kJ/mol; used in analysis and as a pyrotechnic
  (green fire) [HAW93] [MER06] [CRC10]
Solubility: g/100 g H_2O: 3.90 (0°C), 9.55 (20°C),
  414 (100°C) [LAN05]; i alcohol [MER06]
Density, g/cm³: 5.55 [MER06]
Melting Point, °C: 206 [MER06]
Boiling Point, °C: decomposes at 450 [MER06]

3209
Compound: Thallium(I) nitrite
Formula: $TlNO_2$
Molecular Formula: NO_2Tl
Molecular Weight: 250.389
CAS RN: 13824-63-6
Properties: yellow cryst [CRC10]
Solubility: g/100 g H_2O: 17.9 (0°C), 40.3
  (20°C), 750 (90°C) [LAN05]
Melting Point, °C: 182 [CRC10]

3210
Compound: Thallium(I) oxalate
Formula: $Tl_2C_2O_4$
Molecular Formula: $C_2O_4Tl_2$
Molecular Weight: 496.786
CAS RN: 30737-24-7
Properties: white powd [STR93]
Solubility: 1.48 g/100 mL H_2O (15°C),
  9.02 g/100 mL H_2O (100°C) [CRC10]
Density, g/cm³: 6.31 [STR93]

3211
Compound: Thallium(I) oxide
Synonym: thallous oxide
Formula: Tl_2O
Molecular Formula: OTl_2
Molecular Weight: 424.766
CAS RN: 1314-12-1
Properties: black powd; gradually oxides to
  Tl_2O_3 in air; used in artificial gems, to test
  for ozone, in optical glass [MER06]
Solubility: s H_2O, hydrolyzes to
  hydroxide; s alcohol [MER06]
Density, g/cm³: 9.52 (16°C) [HAW93]
Melting Point, °C: 300 [HAW93]
Boiling Point, °C: 1080 [HAW93]

3212
Compound: Thallium(I) perchlorate
Formula: $TlClO_4$
Molecular Formula: ClO_4Tl
Molecular Weight: 303.834
CAS RN: 13453-40-2
Properties: colorless rhomb [CRC10] [LAN05]
Solubility: g/100 g H_2O: 6.00 (0°C), 13.1
  (20°C), 81.5 (80°C) [LAN05]
Density, g/cm³: 4.89 [LAN05]
Melting Point, °C: 501 [LAN05]
Boiling Point, °C: decomposes [LAN05]

3213
Compound: Thallium(I) picrate
Formula: $TlOC_6H_2(NO_2)_3$
Molecular Formula: $C_6H_2N_3O_7Tl$
Molecular Weight: 432.481
CAS RN: 23293-27-8
Properties: red monocl or yellow tricl [CRC10]
Solubility: g/100 g H_2O: 0.135 (0°C), 0.40
  (20°C), 1.73 (60°C) [LAN05]
Density, g/cm³: red: 3.164; yellow: 2.993 [CRC10]
Reactions: explodes, 723°C–725°C [CRC10]

3214
Compound: Thallium(I) selenate
Synonym: thallous selenate
Formula: Tl_2SeO_4
Molecular Formula: O_4SeTl_2
Molecular Weight: 551.725
CAS RN: 7446-22-2
Properties: ortho-rhomb cryst [MER06]

Solubility: g/100 g H_2O: 2.13 (9.3°C), 2.4 (12°C), 10.86 (100°C); i alcohol, ether [MER06]
Density, g/cm³: 6.875 [MER06]
Melting Point, °C: >400 [MER06]

3215
Compound: Thallium(I) selenide
Synonym: thallous selenide
Formula: Tl_2Se
Molecular Formula: $SeTl_2$
Molecular Weight: 487.727
CAS RN: 15572-25-5
Properties: dark gray plates with metallic luster [MER06]
Solubility: i H_2O, acids [MER06]
Density, g/cm³: 9.05 [CRC10]
Melting Point, °C: 340 [MER06]

3216
Compound: Thallium(I) sulfate
Synonym: thallous sulfate
Formula: Tl_2SO_4
Molecular Formula: O_4STl_2
Molecular Weight: 504.831
CAS RN: 7446-18-6
Properties: white, rhomb prisms; enthalpy of fusion 23.00 kJ/mol; used to analyze for iodine in the presence of chlorine, in ozonometry and as a pesticide [HAW93] [MER06] [CRC10]
Solubility: g/100 mL H_2O: 2.70 (0°C), 4.87 (20°C), 18.45 (100°C) [MER06]
Density, g/cm³: 6.77 [MER06]
Melting Point, °C: 632 [MER06]
Boiling Point, °C: decomposes [STR93]

3217
Compound: Thallium(I) sulfide
Synonym: thallous sulfide
Formula: Tl_2S
Molecular Formula: STl_2
Molecular Weight: 440.833
CAS RN: 1314-97-2
Properties: −20 mesh with 99.9% purity; bluish black cryst powd; enthalpy of vaporization 154 kJ/mol; enthalpy of fusion 12.00 kJ/mol; used in infrared photocells [HAW93] [MER06] [CER91] [CRC10]
Solubility: sl s H_2O, alkali hydroxides, sulfides, cyanides; s mineral acids [MER06]
Density, g/cm³: 8.39 [MER06]
Melting Point, °C: 448.5 [MER06]
Boiling Point, °C: 1367 [CRC10]

3218
Compound: Thallium(I) trifluoroacetylacetonate
Synonyms: 1,1,1-trifluoro-2,4-pentanedione, thallium(I) derivative
Formula: $TlCF_3COCH=C(O)CH_3$
Molecular Formula: $C_5H_4F_3O_2Tl$
Molecular Weight: 357.464
CAS RN: 54412-40-7
Properties: white powd [ALD94]
Melting Point, °C: 110 decomposes [ALD94]

3219
Compound: Thallium(III) acetate
Synonyms: acetic acid, thallium(III) salt
Formula: $Tl(CH_3COO)_3$
Molecular Formula: $C_6H_9O_6Tl$
Molecular Weight: 381.517
CAS RN: 2570-63-0
Properties: light sensitive; uses: together with bromine, selective electrophilic aromatic brominator [ALD94]
Melting Point, °C: decomposes at 182 [ALD94]

3220
Compound: Thallium(III) bromide
Synonym: thallic bromide
Formula: $TlBr_3$
Molecular Formula: Br_3Tl
Molecular Weight: 444.095
CAS RN: 13701-90-1
Properties: yellow [KIR83]
Solubility: s H_2O, alcohols [KIR83]

3221
Compound: Thallium(III) chloride
Synonym: thallic chloride
Formula: $TlCl_3$
Molecular Formula: Cl_3Tl
Molecular Weight: 310.741
CAS RN: 13453-32-2
Properties: hexagonal plate [KIR83]
Solubility: v s H_2O, alcohols, ether [KIR83]
Density, g/cm³: 4.7 [LID94]
Melting Point, °C: 155 [KIR83]

3222
Compound: Thallium(III) chloride hydrate
Synonym: thallic chloride hydrate
Formula: $TlCl_3 \cdot xH_2O$

Molecular Formula: Cl_3Tl (anhydrous)
Molecular Weight: 310.741 (anhydrous)
CAS RN: 13453-33-3
Properties: white cryst [STR93]
Melting Point, °C: 37 [STR93]

3223
Compound: Thallium(III) fluoride
Synonym: thallic fluoride
Formula: TlF_3
Molecular Formula: F_3Tl
Molecular Weight: 261.378
CAS RN: 7783-57-5
Properties: olive green ortho-rhomb cryst; very
   sensitive to moisture; quickly decomposed by water;
   decomposed if heated in air [MER06] [KIR83]
Solubility: i H_2O [KIR83]
Density, g/cm³: 8.65 [MER06]
Melting Point, °C: decomposes at 550 in air [MER06]

3224
Compound: Thallium(III) nitrate
Synonym: thallic nitrate
Formula: $Tl(NO_3)_3$
Molecular Formula: N_3O_9Tl
Molecular Weight: 390.398
CAS RN: 13746-98-0
Properties: colorless; there is a trihydrate, CAS
   RN 13453-38-8 [KIR83] [ALD94]
Solubility: decomposes in H_2O [KIR83]
Melting Point, °C: 102–105 (trihydrate) [ALD94]

3225
Compound: Thallium(III) oxide
Synonym: thallic oxide
Formula: Tl_2O_3
Molecular Formula: O_3Tl_2
Molecular Weight: 456.765
CAS RN: 1314-32-5
Properties: brown powd; oxidant, e.g., reacts with HCl
   evolving Cl_2, and with H_2SO_4 evolving O_2 [MER06]
Solubility: i H_2O [MER06]
Density, g/cm³: 10.11 [KIR83]
Melting Point, °C: ~717 [STR93]

3226
Compound: Thallium(III) perchlorate hexahydrate
Formula: $Tl(ClO_4)_3 \cdot 6H_2O$
Molecular Formula: $Cl_3H_{12}O_{18}Tl$
Molecular Weight: 610.825

CAS RN: 15596-83-5
Properties: white cryst; hygr [STR93]

3227
Compound: Thallium(III) trifluoroacetate
Synonyms: trifluoroacetic acid, thallium(III) salt
Formula: $(CF_3COO)_3Tl$
Molecular Weight: 543.42
Molecular Formula: $C_6F_9O_6Tl$
CAS RN: 23586-53-0
Properties: hygr; oxidizing agent; uses: organic
   sulfide bond formation [ALD94]
Melting Point, °C: 213, decomposes [ALD94]

3228
Compound: Thionyl bromide
Formula: $SOBr_2$
Molecular Formula: Br_2OS
Molecular Weight: 207.873
CAS RN: 507-16-4
Properties: orange-yellow liq; slowly
   decomposes on standing; prepared by
   reaction of $SOCl_2$ and HBr [MER06]
Solubility: hydrolyzed by H_2O; miscible with
   benzene, chloroform, CCl_4 [MER06]
Density, g/cm³: 2.688 [MER06]
Melting Point, °C: −52 [MER06]
Boiling Point, °C: 138 [MER06]

3229
Compound: Thionyl chloride
Synonym: sulfurous oxychloride
Formula: $SOCl_2$
Molecular Formula: Cl_2OS
Molecular Weight: 118.970
CAS RN: 7719-09-7
Properties: pale yellow to red liq; suffocating odor;
   sensitive to moisture; enthalpy of vaporization
   31.7 kJ/mol at bp, 31 kJ/mol at 25°C; produced
   by oxidation of SCl_2 by SO_3; used in pesticides,
   engineering plastics [HAW93] [STR93] [MER06]
Solubility: decomposes in H_2O;
   s benzene, CCl_4 [HAW93]
Density, g/cm³: 1.638 [HAW93]
Melting Point, °C: −105 [HAW93]
Boiling Point, °C: 79 [ALD94]
Reactions: decomposes at 140°C [HAW93]

3230
Compound: Thionyl fluoride
Formula: SOF_2

Molecular Formula: F_2OS
Molecular Weight: 86.062
CAS RN: 7783-42-8
Properties: colorless gas with suffocating odor; does not attack glass; obtained from reaction between SbF_3 and $SOCl_2$ in presence of SbF_5 [MER06]
Solubility: hydrolyzed by H_2O; s ether, benzene [MER06]
Density, g/cm³: gas: 3.84 g/L [CRC10]; liq: 1.780 (−100°C) [MER06]
Melting Point, °C: −129.5 [MER06]
Boiling Point, °C: −43.8 [MER06]

3231

Compound: Thiophosphoryl chloride
Synonym: phosphorus sulfochloride
Formula: $PSCl_3$
Molecular Formula: Cl_3PS
Molecular Weight: 169.398
CAS RN: 3982-91-0
Properties: fuming liq; crystallizes to α-form at −40.8°C, to β-form at −36.2°C [MER06] [ALD94]
Solubility: hydrolyzes in H_2O, forming H_3PO_4, HCl, H_2S; hydrolyzes rapidly in alkaline solutions; s benzene, CCl_4, CS_2, chloroform [MER06]
Density, g/cm³: 1.635 [STR93]
Melting Point, °C: −35 [STR93]
Boiling Point, °C: 125 [STR93]

3232

Compound: Thorium
Formula: Th
Molecular Formula: Th
Molecular Weight: 232.0381
CAS RN: 7440-29-1
Properties: soft; grayish white, lustrous metal; somewhat ductile; α: fcc up to 1400°C, a=0.5086 nm; β: bcc, a=0.411 nm, stable 1400°C–1750°C; enthalpy of vaporization 564 kJ/mol; enthalpy of fusion 13.81 kJ/mol; electrical resistivity 14 μohm·cm; Poisson's ratio 0.27; thermal diffusivity 0.28 cm²/s at 200°C; $t_{1/2}$ ^{232}Th is $1.41 \times 10^{+10}$ years; ionic radius of Th^{++++} is 0.0972 nm [KIR78] [KIR83] [MER06] [CRC10]
Solubility: s acids; i H_2O, alkalies [HAW93]
Density, g/cm³: α: 11.724 [KIR91]
Melting Point, °C: 1750 [KIR91]
Boiling Point, °C: 3800 [ALD94]
Reactions: phase change from fcc to bcc ~1345°C [KIR83]
Thermal Conductivity, W/(m·K): 54.0 (25°C) [CRC10]
Thermal Expansion Coefficient: 11.0×10^{-6}/K [CRC10]

3233

Compound: Thorium acetylacetonate
Synonyms: 2,4-pentanedione, thorium(IV) derivative
Formula: $Th(CH_3COCH=C(O)CH_3)_4$
Molecular Formula: $C_{20}H_{28}O_8Th$
Molecular Weight: 628.475
CAS RN: 102192-40-5
Properties: cryst powd [HAW93]
Solubility: sl s H_2O, not readily hydrolyzed [HAW93]

3234

Compound: Thorium bromide
Formula: $ThBr_4$
Molecular Formula: Br_4Th
Molecular Weight: 551.654
CAS RN: 13453-49-1
Properties: colorless hygr; −8 mesh with 99.5% purity; can be prepared by reacting Th with Br_2; light sensitive and easily hydrolyzed [KIR83] [CER91] [CRC10]
Density, g/cm³: 5.67 [CRC10]
Melting Point, °C: sublimes at 610 [CRC10]

3235

Compound: Thorium carbide
Formula: ThC
Molecular Formula: CTh
Molecular Weight: 244.049
CAS RN: 12012-16-7
Properties: −40 mesh with 99.5% purity; fcc, a=0.5346 nm; prepared by reacting stoichiometric amounts of Th and C; reactive, e.g., burns in air to form ThO_2; there is a ThC_2, 12071-31-7 [KIR83] [CIC73] [CER91]
Solubility: readily hydrolyzes in H_2O evolving methane [KIR83]
Melting Point, °C: 2655 [KIR83]

3236

Compound: Thorium chloride
Formula: $ThCl_4$
Molecular Formula: Cl_4Th
Molecular Weight: 373.849
CAS RN: 10026-08-1
Properties: tetr; grayish white powd; hygr lustrous needles; enthalpy of vaporization 146.4 kJ/mol; enthalpy of fusion 40.20 kJ/mol; can be prepared by reacting Th with Cl_2; used in incandescent lamps [HAW93] [CRC10] [STR93] [KIR83] [COT88]
Solubility: s H_2O, alcohol [MER06]

Density, g/cm³: 4.59 [MER06]
Melting Point, °C: 770 [MER06]
Boiling Point, °C: 921 [MER06]

3237
Compound: Thorium dicarbide
Formula: ThC_2
Molecular Formula: C_2Th
Molecular Weight: 256.060
CAS RN: 12071-31-7
Properties: yellow solid; α form: monocl, a = 1.0555 nm, b = 0.8233 nm, c = 0.4201 nm; β form: tetr; γ form: cub, a = 0.5808 nm; decomposed by water; formed by heating ThO_2 and excess carbon; used as a nuclear fuel [HAW93] [CIC73] [KIR83]
Solubility: decomposed in H_2O, with evolution of ethane [KIR83]
Density, g/cm³: 8.96 (18°C) [HAW93]
Melting Point, °C: 2630–2680 [HAW93]
Boiling Point, °C: –5000 [HAW93]
Reactions: α to β transition at 1427°C, β to γ at 1497°C [CIC73]

3238
Compound: Thorium fluoride
Formula: ThF_4
Molecular Formula: F_4Th
Molecular Weight: 308.032
CAS RN: 13709-59-6
Properties: white powd; hygr; reacts with atm moisture to form $ThOF_2$ at temperatures >500°C; enthalpy of vaporization 258 kJ/mol; can be prepared by reaction of Th and F_2; used to produce thorium metal and in high temp ceramics; as 99.99% pure material, used as a sputtering target to produce low-index film with no absorption in visible and ultraviolet [HAW93] [STR93] [KIR83] [CER91] [CRC10]
Density, g/cm³: 6.32 [STR93]
Melting Point, °C: 1068 [KIR91]

3239
Compound: Thorium hexaboride
Formula: ThB_6
Molecular Formula: B_6Th
Molecular Weight: 296.904
CAS RN: 12229-63-9
Properties: –100 mesh with 99.8% purity, there is a ThB_4, 12007-83-9; refractory material [KIR78]
Density, g/cm³: 6.4 [CRC10]
Melting Point, °C: 2195; tetraboride 2500 [KIR78]

3240
Compound: Thorium hexafluoroacetylacetonate
Synonyms: 1,1,1,5,5,5-hexafluoro-2,4-pentanedione, thorium derivative
Formula: $Th(CF_3COCHCOCF_3)_4$
Molecular Formula: $C_{20}H_4F_{24}O_8Th$
Molecular Weight: 1060.247
CAS RN: 18865-75-3
Properties: white powd [STR93]
Melting Point, °C: 100–101 [STR93]

3241
Compound: Thorium hydride
Formula: ThH_2
Molecular Formula: H_2Th
Molecular Weight: 234.054
CAS RN: 16689-88-6
Properties: tetr cryst; here are also ThH_3, 40004-84-0, –60 mesh with 99.5% purity, and $ThH_{3.75}$ (Th_4H_{15}), 12055-07-1; the third hydride exhibits superconductivity [LID94] [KIR83] [CER91]
Density, g/cm³: 9.5 [LID94]

3242
Compound: Thorium hydroxide
Formula: $Th(OH)_4$
Molecular Formula: H_4O_4Th
Molecular Weight: 300.068
CAS RN: 13825-36-0
Properties: prepared by addition of alkali to a solution of Th^{++++} salt, yielding a gelatinous precipitate which is subsequently dehydrated; absorbs CO_2 to form $ThOCO_3$ [KIR83]
Melting Point, °C: decomposes [CRC10]
Reactions: minus water >470°C to form ThO_2 [KIR83]

3243
Compound: Thorium iodide
Formula: ThI_4
Molecular Formula: I_4Th
Molecular Weight: 739.656
CAS RN: 7790-49-0
Properties: pale yellow cryst; obtained from a reaction between Th and I_2; decomposed by light or heat [MER06] [KIR83]
Melting Point, °C: 556 [KIR91]
Boiling Point, °C: 837 [MER06]

3244
Compound: Thorium nitrate
Formula: $Th(NO_3)_4$

Molecular Formula: $N_4O_{12}Th$
Molecular Weight: 480.058
CAS RN: 13823-29-5
Properties: plates, deliq; obtained when thorium hydroxide is dissolved in a nitric acid solution [KIR83] [CRC10]
Solubility: g/100 g H_2O: 186 (0°C), 187 (10°C), 191 (20°C) [LAN05]; additional solubility data are in [SIE94]
Melting Point, °C: decomposes at 500 [CRC10]

3245
Compound: Thorium nitrate tetrahydrate
Formula: $Th(NO_3)_4 \cdot 4H_2O$
Molecular Formula: $H_8N_4O_{16}Th$
Molecular Weight: 552.119
CAS RN: 33088-16-3
Properties: white cryst; sl deliq; used in thoriated tungsten filaments and as a reagent for fluoride determination [HAW93] [MER06]
Solubility: v s H_2O, alcohol [MER06]
Melting Point, °C: decomposes at 500 [HAW93]

3246
Compound: Thorium nitride
Formula: ThN
Molecular Formula: NTh
Molecular Weight: 246.045
CAS RN: 12033-65-7
Properties: gray refractory solid; fcc, a=0.5159 nm; electrical resistivity 20 μohm·cm; microhardness 600; there is also the compound Th_3N_4, CAS RN 12033-90-8, −100 mesh with 99.5% purity [CER91] [KIR81]
Solubility: slowly hydrolyzed by H_2O [COT88]
Density, g/cm³: 11.9 [KIR81]
Melting Point, °C: 2820 [KIR81]

3247
Compound: Thorium orthosilicate
Synonym: thorite
Formula: $ThSiO_4$
Molecular Formula: O_4SiTh
Molecular Weight: 324.122
CAS RN: 51184-23-7
Properties: black to orange; zircon structure, a=0.71328 nm, c=0.63188 nm; hardness is 4.5–5 [HAW93] [SUB90]
Density, g/cm³: 4.4–5.2 [HAW93]
Thermal Expansion Coefficient: (25°C–500°C) 2.5×10^{-6}/°C [SUB90]

3248
Compound: Thorium oxalate dihydrate
Formula: $Th(C_2O_4)_2 \cdot 2H_2O$
Molecular Formula: $C_4H_4O_{10}Th$
Molecular Weight: 444.108
CAS RN: 24012-17-7
Properties: white powd; there is a hexahydrate, white cryst, which is precipitated from up to 2 M HNO_3; used in ceramics [HAW93] [COT88]
Solubility: i H_2O and most acids [HAW93]
Density, g/cm³: anhydrous: 4.637 (16°C) [HAW93]
Melting Point, °C: decomposes at >300–400 to ThO_2 [HAW93]

3249
Compound: Thorium oxide
Synonyms: thoria, thorianite
Formula: ThO_2
Molecular Formula: O_2Th
Molecular Weight: 264.037
CAS RN: 1314-20-1
Properties: white, heavy, cryst, cub powd, or 3–12 mm sintered pieces; hardness 6.5 Mohs; used in ceramics, gas mantles, crucibles, thoriated tungsten filaments, and in crucible form for melting hafnium, iridium, iron, manganese, silicon, thorium, titanium, uranium, and zirconium; used as an evaporation material and sputtering target of 99.99% and 99.9% purity for highly durable beam splitter [HAW93] [MER06] [KIR83] [CER91]
Solubility: i H_2O, alkalies [MER06]
Density, g/cm³: 9.86 [STR93]; 10.01 [KIR80]
Melting Point, °C: ~3050 [KIR91]
Boiling Point, °C: 4400 [HAW93]
Thermal Conductivity, W/(m·K): 5.1 (500°C), 3.0 (1000°C) [KIR80]
Thermal Expansion Coefficient: (volume) 100°C (0.234), 200°C (0.517), 400°C (1.100), 800°C (2.249), 1000°C (2.833) [CLA66]

3250
Compound: Thorium oxyfluoride
Formula: $ThOF_2$
Molecular Formula: F_2OTh
Molecular Weight: 286.034
CAS RN: 13597-30-3
Properties: 3–6 mm pieces (sintered) with 99.9% purity [CER91]

3251
Compound: Thorium perchlorate
Formula: $Th(ClO_4)_4$

Molecular Formula: $Cl_4O_{16}Th$
Molecular Weight: 629.839
CAS RN: 16045-17-3
Properties: obtained by dissolution of thorium hydroxide in perchloric acid solution; the tetrahydrate can be crystallized from an acidic solution which is then dehydrated to $Th(ClO_4)_4$ at ~280°C [KIR83]
Solubility: v s H_2O [KIR83]
Melting Point, °C: ~355 decomposes to ThO_2 [KIR83]

3252
Compound: Thorium selenide
Formula: $ThSe_2$
Molecular Formula: Se_2Th
Molecular Weight: 389.958
CAS RN: 60763-24-8
Properties: ortho-rhomb cryst; −80 mesh with 99.5 purity [LID94] [CER91]
Density, g/cm³: 805 [LID94]

3253
Compound: Thorium silicide
Formula: $ThSi_2$
Molecular Formula: Si_2Th
Molecular Weight: 288.209
CAS RN: 12067-54-8
Properties: black tetr; −80 mesh with 99.5% purity [CER91] [CRC10]
Density, g/cm³: 7.96 [CRC10]

3254
Compound: Thorium sulfate nonahydrate
Formula: $Th(SO_4)_2 \cdot 9H_2O$
Molecular Formula: $H_{18}O_{17}S_2Th$
Molecular Weight: 586.303
CAS RN: 10381-37-0
Properties: colorless or white; monocl cryst; decomposes when heated strongly [MER06]
Solubility: g/100 g H_2O: 0.74 (0°C), 1.38 (20°C), 3.00 (40°C) [LAN05]
Density, g/cm³: 2.8 [MER06]
Reactions: minus $9H_2O$ at 400°C [CRC10]

3255
Compound: Thorium sulfate octahydrate
Formula: $Th(SO_4)_2 \cdot 8H_2O$
Molecular Formula: $H_{16}O_{16}S_2Th$
Molecular Weight: 568.287
CAS RN: 10381-37-0
Properties: monocl white cryst powd [HAW93] [CRC10]

Solubility: sl s water; s ice water [HAW93]
Density, g/cm³: 2.8 [HAW93]
Reactions: minus $4H_2O$ at 42°C, minus $8H_2O$ at 400°C [HAW93]

3256
Compound: Thorium sulfate tetrahydrate
Formula: $Th(SO_4)_2 \cdot 4H_2O$
Molecular Formula: $H_8O_{12}S_2Th$
Molecular Weight: 496.227
CAS RN: 10381-37-0
Properties: white needles [CRC10]
Solubility: g/100 g H_2O: 4.04 (40°C), 1.63 (60°C) [LAN05]
Reactions: minus $4H_2O$, 400°C [CRC10]

3257
Compound: Thorium sulfide
Formula: ThS_2
Molecular Formula: S_2Th
Molecular Weight: 296.170
CAS RN: 12138-07-7
Properties: dark brown cryst; begins to decompose above 1500°C; used as a solid lubricant [HAW93] [KIR83]
Solubility: i H_2O; s acids [HAW93] [COT88]
Density, g/cm³: 7.30 [HAW93]
Melting Point, °C: 1875–1975 (in vacuum) [HAW93]

3258
Compound: Thorium tetracyanoplatinate(II) hexadecahydrate
Formula: $Th[Pt(CN)_4]_2 \cdot 16H_2O$
Molecular Formula: $C_8H_{32}N_8O_{16}Pt_2Th$
Molecular Weight: 1118.594
CAS RN: 14481-33-5
Properties: yellow cryst [MER06]
Solubility: sl s H_2O [MER06]

3259
Compound: Thulium
Formula: Tm
Molecular Formula: Tm
Molecular Weight: 168.9342
CAS RN: 7440-30-4
Properties: silvery white metal; easily worked; hex close-packed; electrical resistivity (20°C) 90 µohm · cm; enthalpy of fusion 16.84 kJ/mol; enthalpy of sublimation 232.2 kJ/mol; radius of atom 0.17462 nm; radius of Tm^{+++} ion 0.0870 nm; forms light green colored solutions [MER06] [KIR82] [ALD94]

Solubility: slowly reacts with H_2O;
s in dil acids [HAW93]
Density, g/cm³: 9.32 [KIR82]
Melting Point, °C: 1545 [KIR82]
Boiling Point, °C: 1950 [KIR82]
Thermal Conductivity, W/(m·K): 16.9 (25°C) [CRC10]
Thermal Expansion Coefficient: 13.3×10^{-6}/K [CRC10]

3260
Compound: Thulium acetate monohydrate
Formula: $Tm(CH_3COO)_3 \cdot H_2O$
Molecular Formula: $C_6H_{11}O_7Tm$
Molecular Weight: 364.083
CAS RN: 39156-80-4
Properties: white powd [STR93]

3261
Compound: Thulium acetylacetonate trihydrate
Synonyms: 2,4-pentanedione, thulium(III) derivative
Formula: $Tm(CH_3COCH=C(O)CH_3)_3 \cdot 3H_2O$
Molecular Formula: $C_{15}H_{27}O_9Tm$
Molecular Weight: 520.308
CAS RN: 14589-44-7
Properties: white powd [STR93]

3262
Compound: Thulium bromide
Formula: $TmBr_3$
Molecular Formula: Br_3Tm
Molecular Weight: 408.646
CAS RN: 14456-51-0
Properties: −20 mesh of 99.9% purity [CER91]
Melting Point, °C: 952 [AES93]
Boiling Point, °C: 1440 [CRC10]

3263
Compound: Thulium chloride
Formula: $TmCl_3$
Molecular Formula: Cl_3Tm
Molecular Weight: 275.292
CAS RN: 13537-18-3
Properties: −20 mesh with 99.9% purity;
off-white powd; hygr [STR93] [CER91]
Melting Point, °C: 821 [STR93]

3264
Compound: Thulium chloride heptahydrate
Formula: $TmCl_3 \cdot 7H_2O$
Molecular Formula: $Cl_3H_{14}O_7Tm$
Molecular Weight: 401.399

CAS RN: 13778-39-7
Properties: −4 mesh with 99.9% purity; light green
cryst; deliq; there is a hexahydrate, CAS RN
1331-74-4 [HAW93] [STR93] [CER91] [ALD94]
Solubility: s H_2O, alcohol [MER06]
Melting Point, °C: 824 [HAW93]
Boiling Point, °C: 1440 [HAW93]

3265
Compound: Thulium fluoride
Formula: TmF_3
Molecular Formula: F_3Tm
Molecular Weight: 225.929
CAS RN: 13760-79-7
Properties: off-white powd; hygr [STR93]
Melting Point, °C: 1158 [STR93]
Boiling Point, °C: >2200 [CRC10]

3266
Compound: Thulium hydroxide
Formula: $Tm(OH)_3$
Molecular Formula: H_3O_3Tm
Molecular Weight: 219.956
CAS RN: 1311-33-7
Properties: white precipitate [MER06]

3267
Compound: Thulium iodide
Formula: TmI_3
Molecular Formula: I_3Tm
Molecular Weight: 549.553
CAS RN: 13813-43-9
Properties: yellow cryst; −20 mesh with
99.9% purity [CRC10] [CER91]
Melting Point, °C: 1015 [CRC10]
Boiling Point, °C: 1260 [CRC10]

3268
Compound: Thulium nitrate hexahydrate
Formula: $Tm(NO_3)_3 \cdot 6H_2O$
Molecular Formula: $H_{12}N_3O_{15}Tm$
Molecular Weight: 463.040
CAS RN: 36548-87-5
Properties: off-white cryst [STR93]

3269
Compound: Thulium oxalate hexahydrate
Formula: $Tm_2(C_2O_4)_3 \cdot 6H_2O$

Molecular Formula: $C_6H_{12}O_{18}Tm_2$
Molecular Weight: 710.018
CAS RN: 26677-68-9
Properties: cryst; greenish white precipitate [STR93] [MER06]
Solubility: s aq solutions of alkali oxalates forming double oxalates [MER06]
Reactions: minus H_2O at 50°C [HAW93]

3270
Compound: Thulium oxide
Synonym: thulia
Formula: Tm_2O_3
Molecular Formula: O_3Tm_2
Molecular Weight: 385.866
CAS RN: 12036-44-1
Properties: dense white powd, with greenish tinge, or 3–12 mm sintered pieces; sl hygr; absorbs atm H_2O and CO_2; has reddish incandescence when heated, which changes to yellow and then white if heating is prolonged; used as an evaporation material of 99.9% purity to deposit films possibly reactive to radio frequencies [HAW93] [CER91]
Solubility: slowly dissolves in strong acids [HAW93]
Density, g/cm³: 8.6 [HAW93]

3271
Compound: Thulium silicide
Formula: $TmSi_2$
Molecular Formula: Si_2Tm
Molecular Weight: 225.105
CAS RN: 12039-84-8
Properties: 10 mm & down lump [ALF93]

3272
Compound: Thulium sulfate octahydrate
Formula: $Tm_2(SO_4)_3 \cdot 8H_2O$
Molecular Formula: $H_{16}O_{20}S_3Tm_2$
Molecular Weight: 770.181
CAS RN: 13778-40-0
Properties: white cryst; obtained from an aq solution of $TlCl_3$ and H_2SO_4 by precipitating with alcohol [MER06] [STR93]

3273
Compound: Thulium sulfide
Formula: Tm_2S_3
Molecular Formula: S_3Tm_2
Molecular Weight: 434.001

CAS RN: 12166-30-2
Properties: −200 mesh with 99.9% purity [CER91]

3274
Compound: Tin (gray)
Synonym: gray tin
Formula: Sn
Molecular Formula: Sn
Molecular Weight: 118.710
CAS RN: 7440-31-5
Properties: amorphous; unstable, brittle; formed by white tin at −40°C, but slowly reverts back to white form >20°C [MER06]
Density, g/cm³: 5.77 [KIR83]
Reactions: gray to white transformation at 13.2°C [KIR83]

3275
Compound: Tin (white)
Synonym: white tin
Formula: Sn
Molecular Formula: Sn
Molecular Weight: 118.710
CAS RN: 7440-31-5
Properties: almost silvery-white, lustrous, very malleable; easily powdered; brittle at 200°C; has thin oxide film; enthalpy of fusion 7.03 kJ/mol; enthalpy of vaporization 296.4 kJ/mol; electrical resistivity (0°C) 11.0 µohm·cm, (100°C) 15.5 µohm·cm; Brinell hardness at 20°C is 3.9; tensile strength at 15°C is 14.5 MPa; electronegativity 1.8–1.9; uses include cryogenic switching devices; band gap 0.082 eV (0 K) [MER06] [KIR83] [KIR82] [COT88] [CER91] [CRC10]
Solubility: reacts slowly with cold dil HCl, dil HNO_3, hot dil H_2SO_4; readily with conc HCl, aqua regia [MER06]
Density, g/cm³: 7.31 [MER06]
Melting Point, °C: 231.9 [KIR83]
Boiling Point, °C: 2270 [ALD94]
Reactions: crumbles to gray amorphous powd at −40°C ("gray tin") [MER06]
Thermal Conductivity, W/(m·K): 66.6 (25°C) [CRC10]
Thermal Expansion Coefficient: (volume) 100°C (0.54), 200°C (1.27) [CLA66]

3276
Compound: Tin hydride
Synonym: stannane

Formula: SnH$_4$
Molecular Formula: H$_4$Sn
Molecular Weight: 122.742
CAS RN: 2406-52-2
Properties: colorless poisonous gas; decomposes rapidly at room temp; enthalpy of vaporization 19.05 kJ/mol [KIR80] [CRC10]
Melting Point, °C: −150 [KIR80]
Boiling Point, °C: −51.8 [CRC10]

3277
Compound: Tin monophosphide
Formula: SnP
Molecular Formula: PSn
Molecular Weight: 149.684
CAS RN: 25324-56-5
Properties: white powd; −100 mesh of 99.5% purity [CER91] [AES93]
Density, g/cm^3: 6.56 [CRC10]
Melting Point, °C: decomposes [AES93]

3278
Compound: Tin triphosphide
Formula: Sn$_4$P$_3$
Molecular Formula: P$_3$Sn$_4$
Molecular Weight: 567.761
CAS RN: 12286-33-8
Properties: white cryst; other phosphides are Sn$_2$P$_3$ [53095-87-7] and SnP$_3$ [37367-13-8] [KIR82] [CRC10]
Density, g/cm^3: 5.181 [CRC10]
Melting Point, °C: decomposes at <480 [CRC10]

3279
Compound: Titanic acid
Synonym: orthotitanic acid
Formula: Ti(OH)$_4$
Molecular Formula: H$_4$O$_4$Ti
Molecular Weight: 115.897
CAS RN: 20338-08-3
Properties: white powd; variable water content; can be obtained as a precipitate by adding NaOH solution to a solution of a Ti(IV) salt; used as a mordant [HAW93] [KIR83]
Solubility: i H$_2$O; s dil HCl [KIR83]

3280
Compound: Titanium
Formula: Ti

Molecular Formula: Ti
Molecular Weight: 47.867
CAS RN: 7440-32-6
Properties: dark gray lustrous metal; two phases: α-form, hex, stable below 882.5°C; β-form, bcc, stable above 882.5°C; brittle when cold, else ductile; Vickers hardness is 80–100; Poisson's ratio ~0.41; enthalpy of fusion 14.15 kJ/mol; enthalpy of vaporization 425 kJ/mol; electrical resistivity 42.0 µohm·cm at 20°C [HAW93] [MER06] [KIR83] [CRC10] [ALD94]
Solubility: i H$_2$O [HAW93]
Density, g/cm^3: α: 4.506; β: 4.400 (885°C) [KIR83] [MER06]
Melting Point, °C: 1660 [ALD94]
Boiling Point, °C: 3277 [MER06]
Reactions: reacts with: F$_2$ (150°C); Cl$_2$ (300°C); Br$_2$ (360°C) [MER06]
Thermal Conductivity, W/(m·K): 21.9 at 25°C [KIR83]
Thermal Expansion Coefficient: (volume) 100°C (0.240), 200°C (0.567), 400°C (1.316), 600°C (2.095) [CLA66]

3281
Compound: Titanium boride
Formula: TiB$_2$
Molecular Formula: B$_2$Ti
Molecular Weight: 69.489
CAS RN: 12045-63-5
Properties: gray cryst refractory material; hex, a=0.3028 nm, c=0.3228 nm; hardness 9+ Mohs; electrical resistivity 28.4 µohm·cm (20°C); superconducting at 1.26 K; can be prepared by direct reaction of Ti with β at 2000°C; used as a high temp electrical conductor, as a cermet component, in crucibles to melt metals such as aluminum and tin, and as a 99.5% pure sputtering target to form films which increase cutting tool life [HAW93] [KIR78] [KIR83] [CER91]
Density, g/cm^3: 4.50 [STR93]
Melting Point, °C: 2900 [STR93]

3282
Compound: Titanium carbide
Formula: TiC
Molecular Formula: CTi
Molecular Weight: 59.878
CAS RN: 12070-08-5

Properties: gray cub, a=0.4328 nm; extremely hard; resistivity at room temp 60 μohm·cm; enthalpy of fusion 71 kJ/mol; hardness 9–10 Mohs; superconducting at 1.1 K; made by reaction of titanium with carbon at high temperatures; used as an additive in cutting tools, in crucible form for melting metals such as bismuth, zinc, and cadmium, and as 99.5% pure sputtering target to prepare wear-resistant semiconducting films [HAW93] [STR93] [CER91] [JAN71]
Solubility: i H_2O; s nitric acid, aqua regia [HAW93]
Density, g/cm³: 4.93 [STR93]
Melting Point, °C: ~3140 [STR93]
Boiling Point, °C: 4820 [STR93]
Thermal Conductivity, W/(m·K): 21 [KIR78]
Thermal Expansion Coefficient: (volume) 100°C (0.125), 200°C (0.326), 400°C (0.771), 800°C (1.736), 1200°C (2.869) [CLA66]

3283
Compound: Titanium dibromide
Formula: $TiBr_2$
Molecular Formula: Br_2Ti
Molecular Weight: 207.675
CAS RN: 13783-04-5
Properties: black powd; strong reducing agent; ignites spontaneously in air; can be made by disporportionation of $TiBr_4$ at 400°C [KIR83]
Solubility: reacts with H_2O evolving hydrogen and forming Ti^{+++} [KIR83]
Density, g/cm³: 4.31 [KIR83]
Melting Point, °C: decomposes at >500 [CRC10]

3284
Compound: Titanium dichloride
Formula: $TiCl_2$
Molecular Formula: Cl_2Ti
Molecular Weight: 118.772
CAS RN: 10049-06-6
Properties: black hex cryst; burns in air if heated; enthalpy of vaporization 232 kJ/mol; can be made by heating $TiCl_3$ in vacuum at 475°C [MER06] [KIR83] [CRC10]
Solubility: decomposed by H_2O evolving H_2 and forming $TiCl_3$ [KIR83]; s alcohol; i chloroform, ether, CS_2 [MER06]
Density, g/cm³: 3.13 [MER06]
Melting Point, °C: 1035 [MER06]
Boiling Point, °C: 1500 [KIR83]

3285
Compound: Titanium diiodide
Synonym: titanium(II) iodide

Formula: TiI_2
Molecular Formula: I_2Ti
Molecular Weight: 301.676
CAS RN: 13783-07-8
Properties: hygr black powd, hex; can be prepared by reacting Ti and I_2 at 440°C [KIR82] [CRC10]
Solubility: reacts quickly with H_2O, evolving hydrogen and forming TiI_3 solution [KIR83]
Density, g/cm³: 4.99 [CRC10]
Melting Point, °C: 600 [CRC10]
Boiling Point, °C: 1000 [CRC10]

3286
Compound: Titanium dioxide
Synonym: anatase
Formula: TiO_2
Molecular Formula: O_2Ti
Molecular Weight: 79.866
CAS RN: 1317-70-0
Properties: white; tetr, a=0.3758 nm, c=0.9514 nm; hardness 5.5–6 Mohs; enthalpy of transition to rutile 12.6 kJ/mol; can be prepared by hydrolysis of titanium tetraethoxide solutions in ethanol solution to yield amorphous hydrous powd, followed by hydrothermal treatment at 200°C–282°C for 5 h and refluxing [KIR83] [OGU88]
Density, g/cm³: 3.9 [KIR83]
Reactions: transforms to rutile ~700°C [KIR83]

3287
Compound: Titanium dioxide
Synonym: brookite
Formula: TiO_2
Molecular Formula: O_2Ti
Molecular Weight: 79.866
CAS RN: 13463-67-7
Properties: ortho-rhomb, a=0.9166 nm, b=0.5436 nm, c=0.5135 nm; hardness 5.5–6 Mohs; produced by heating amorphous TiO_2 with NaOH or KOH for several days at 200°C–600°C [KIR83]
Density, g/cm³: 4.0 [KIR83]

3288
Compound: Titanium dioxide
Synonym: rutile
Formula: TiO_2
Molecular Formula: O_2Ti
Molecular Weight: 79.866
CAS RN: 1317-80-2

Properties: white powd or 99.9% pure gold sintered tablets; tetr, a=0.533 nm, c=0.6645 nm; thermally stable form of TiO$_2$; hardness 7–7.5 Mohs; dielectric constant 114; used as a white pigment in paints, paper, rubber, etc., as an opacifying agent, and with 99.99% and 99.9% purity as a sputtering target to prepare high index films, and multilayer interference filters [HAW93] [MER06] [CER91] [KIR83]

Solubility: in 0.00058 mol/kg alkaline phosphate solutions: in units of 10^{-6} mol/kg·H$_2$O: 0.00125 (20°C), 0.00109 (120.5°C), 0.00244 (218.3°C), 0.0232 (287.2°C) [ZIE93]; s hot conc H$_2$SO$_4$, HF [MER06]

Density, g/cm³: 4.23 [MER06]

Melting Point, °C: 1855 [MER06]

Boiling Point, °C: 2500–3000 [STR93]

Thermal Conductivity, W/(m·K): 3.8 (500°C), 3.3 (1000°C) [KIR80]

Thermal Expansion Coefficient: (volume) 100°C (0.182), 200°C (0.434), 400°C (0.968), 800°C (2.063), 1000°C (2.861) [CLA66]

3289

Compound: Titanium diselenide

Synonym: Ti(IV) selenide

Formula: TiSe$_2$

Molecular Formula: Se$_2$Ti

Molecular Weight: 205.800

CAS RN: 12067-45-7

Properties: −325 mesh white powd [AES93]

3290

Compound: Titanium disulfide

Synonym: titanium(IV) sulfide

Formula: TiS$_2$

Molecular Formula: S$_2$Ti

Molecular Weight: 111.999

CAS RN: 12039-13-3

Properties: yellowish brown powd; hex; sensitive to moisture, forming H$_2$S and TiO$_2$; decomposed by steam; is obtained as a product of the reaction of H$_2$S and TiCl$_4$ at 600°C; used as a solid lubricant [HAW93] [STR93] [KIR83]

Solubility: stable to HCl, s cold or hot H$_2$SO$_4$; decomposed by hot NaOH [KIR83]

Density, g/cm³: 3.22 [STR93]

3291

Compound: Titanium ditelluride

Synonym: Ti(IV) telluride

Formula: TiTe$_2$

Molecular Formula: Te$_2$Ti

Molecular Weight: 303.080

CAS RN: 12067-15-3

Properties: −325 mesh black powd [AES93]

3292

Compound: Titanium hydride

Formula: TiH$_2$

Molecular Formula: H$_2$Ti

Molecular Weight: 49.883

CAS RN: 7704-98-5

Properties: grayish black metallic powd; stable in air; dissociates at 450°C; can produce 448 mL H$_2$/g; stable at room temp; burns quietly when ignited but violent reaction if oxidizing agents are present; industrial preparation by reaction of titanium sponge with H$_2$ at 200°C–600°C, then cooling in H$_2$; used as a source for Ti powd, as a getter in electronic tubes, and to seal metals [KIR80] [MER06]

Solubility: i H$_2$O [KIR80]

Density, g/cm³: 3.9 [STR93]

Melting Point, °C: decomposes at 400 [STR93]

Reactions: dissociates from 300°C to 600°C [KIR80]

3293

Compound: Titanium isopropoxide

Synonym: titanium isopropylate

Formula: Ti[OCH(CH$_3$)$_2$]$_4$

Molecular Formula: C$_{12}$H$_{28}$O$_4$Ti

Molecular Weight: 284.232

CAS RN: 546-68-9

Properties: colorless liq; fumes in air; used as a polymerization catalyst; used to prepare barium titanate, aluminum titanate, and TiO$_2$–CeO$_2$ coatings by the sol–gel process [MER06] [MAK90] [RIT86] [YAM89] [STR93]

Solubility: decomposes rapidly in H$_2$O; s absolute ethanol, ether, benzene, chloroform [MER06]

Density, g/cm³: 0.955 [STR93]

Melting Point, °C: ~20 [MER06]

Boiling Point, °C: 220 [MER06]

3294

Compound: Titanium monosulfide

Formula: TiS

Molecular Formula: STi

Molecular Weight: 79.933

CAS RN: 12039-07-5

Properties: hex, dark brown solid; can be obtained by direct reaction of Ti and S [KIR83]

Solubility: attacked by conc HCl and HNO_3
  but not by alkalies [KIR83]
Density, g/cm³: 4.05 [KIR83]

3295
Compound: Titanium monoxide
Synonym: titanium(II) oxide
Formula: TiO
Molecular Formula: OTi
Molecular Weight: 63.866
CAS RN: 12137-20-1
Properties: bronze pellets; fcc; weakly basic
  oxide with no important industrial
  uses [HAW93] [STR93] [KIR83]
Solubility: s hot 40% HF and 30% H_2O_2 [KIR83]
Density, g/cm³: 4.95 [STR93]
Melting Point, °C: 1700 [STR93]
Boiling Point, °C: >3000 [STR93]

3296
Compound: Titanium nitride
Formula: TiN
Molecular Formula: NTi
Molecular Weight: 61.874
CAS RN: 25583-20-4
Properties: bronze powd; fcc, a=0.4246 nm; hardness,
  8–9 Mohs; electrical resistivity 21.7 μohm·cm
  (20°C); transition temp 4.8 K; can be prepared by
  reaction between finely divided Ti and nitrogen at
  1000°C–1400°C; used in cermets, rectifiers and in
  crucible form for melting metals such as aluminum,
  bismuth, cadmium, lead, steel, tin; used as 99.5%
  pure sputtering target to increase life of cutting
  tools [HAW93] [CIC73] [KIR81] [KIR83] [CER91]
Solubility: attacked by boiling aqua regia,
  decomposed by boiling alkalies evolving
  ammonia; otherwise highly stable [KIR83]
Density, g/cm³: 5.22 [STR93]
Melting Point, °C: 2930 [STR93]
Thermal Conductivity, W/(m·K): 29.1 [KIR81]
Thermal Expansion Coefficient:
  9.35×10^{-6}/°C [KIR81]

3297
Compound: Titanium oxalate decahydrate
Formula: $Ti_2(C_2O_4)_3 \cdot 10H_2O$
Molecular Formula: $C_6H_{20}O_{22}Ti_2$
Molecular Weight: 539.946
CAS RN: 28212-09-1
Properties: yellow prisms; prepared from
  oxalic acid and $TiCl_3$ [HAW93]
Solubility: s H_2O; i alcohol, ether [HAW93]

3298
Compound: Titanium oxysulfate
Formula: $TiOSO_4$
Molecular Formula: O_5STi
Molecular Weight: 159.930
CAS RN: 13825-75-6
Properties: white or sl yellow powd;
  decomposed by water [MER06]

3299
Compound: Titanium phosphide
Formula: TiP
Molecular Formula: PTi
Molecular Weight: 78.841
CAS RN: 12037-65-9
Properties: hard, gray metallic powd with hex
  structure; prepared by heating phosphine
  with $TiCl_4$; stable up to 1400°C; finds use as
  a catalyst in organic reactions [KIR83]
Solubility: not attacked by common acids [KIR83]
Density, g/cm³: 4.08 [KIR83]

3300
Compound: Titanium silicide
Synonym: titanium disilicide
Formula: $TiSi_2$
Molecular Formula: Si_2Ti
Molecular Weight: 104.051
CAS RN: 12039-83-7
Properties: ortho-rhomb black powd, a=0.8236 nm,
  b=0.4773 nm, c=0.8523 nm; hardness 4.5 Mohs;
  resistivity 123 μohm·cm can be prepared
  by reaction of the elements; used in special
  alloy applications, as a flame-resistant coating
  material, also as 99.5 or 99.9% pure material,
  as a sputtering target in the fabrication of
  integrated circuits [HAW93] [STR93] [CER91]
Solubility: s HF, resistant to mineral acids
  and alkali solutions [KIR83]
Density, g/cm³: 4.39 [STR93]
Melting Point, °C: 1540 [STR93]

3301
Compound: Titanium sulfate
Synonym: titanous sulfate
Formula: $Ti_2(SO_4)_3$
Molecular Formula: $O_{12}S_3Ti_2$
Molecular Weight: 383.925
CAS RN: 10343-61-0

Properties: green, cryst powd; produced by reduction of Ti(IV) in sulfuric acid solution; used as a reducing agent in the textile industry [HAW93] [MER06] [KIR83]

Solubility: i H_2O, alcohol, conc H_2SO_4; s dil HCl, dil H_2SO_4 both giving violet solutions [MER06]

3302

Compound: Titanium tetrabromide

Synonym: titanium(IV) bromide

Formula: $TiBr_4$

Molecular Formula: Br_4Ti

Molecular Weight: 367.483

CAS RN: 7789-68-6

Properties: amber yellow or orange cub; very hygr; enthalpy of vaporization 44.37 kJ/mol; enthalpy of fusion 12.90 kJ/mol; can be made by reaction between $TiCl_4$ and HBr [MER06] [KIR83] [CRC10]

Solubility: dissolves with hydrolysis in H_2O [KIR83]

Density, g/cm³: 3.25 [MER06]

Melting Point, °C: 39 [CRC10]

Boiling Point, °C: 230 [CRC10]

3303

Compound: Titanium tetrachloride

Formula: $TiCl_4$

Molecular Formula: Cl_4Ti

Molecular Weight: 189.678

CAS RN: 7550-45-0

Properties: pale yellow or colorless liq; absorbs atm moisture and emits dense white cloud; vapor pressure (20°C) 1.33 kPa; enthalpy of vaporization 36.2 kJ/mol; enthalpy of fusion 9.97 kJ/mol; critical temp 358°C; viscosity 0.079 mPa s; dielectric constant (20°C) 2.79; manufactured by the chlorination of titanium compounds, for example rutile [MER06] [STR93] [KIR83] [CRC10]

Solubility: s cold H_2O, alcohol; decomposed in hot H_2O [MER06]; s dil HCl [HAW93]

Density, g/cm³: 1.726 [MER06]

Melting Point, °C: −45 [ALD94]

Boiling Point, °C: 136.4 [ALD94]

3304

Compound: Titanium tetrafluoride

Formula: TiF_4

Molecular Formula: F_4Ti

Molecular Weight: 123.861

CAS RN: 7783-63-3

Properties: powd or white mass; very hygr; can be produced by the reaction between F_2 and $TiCl_4$ at 250°C [MER06]

Solubility: hydrolyzes in H_2O; s alcohol, pyridine [MER06]

Density, g/cm³: 2.798 [MER06]

Boiling Point, °C: sublimes at 284 [MER06]

3305

Compound: Titanium tetraiodide

Formula: TiI_4

Molecular Formula: I_4Ti

Molecular Weight: 555.485

CAS RN: 7720-83-4

Properties: red powd; sensitive to moisture; enthalpy of vaporization 58.4 kJ/mol; enthalpy of fusion 19.80 kJ/mol; prepared by reacting Ti and I_2 under controlled conditions; used extensively as a catalyst in organic reactions [KIR83] [STR93] [CRC10]

Solubility: dissolves and hydrolyzes in H_2O [KIR82]

Density, g/cm³: 4.3 [STR93]

Melting Point, °C: 150 [CRC10]

Boiling Point, °C: 377 [CRC10]

3306

Compound: Titanium tribromide

Formula: $TiBr_3$

Molecular Formula: Br_3Ti

Molecular Weight: 287.579

CAS RN: 13135-31-4

Properties: bluish black cryst powd; hex plates or needles; disproportinates at 400°C to di- and tetrabromides [KIR83]

Solubility: s H_2O resulting in dark violet solution [KIR83]

Melting Point, °C: 115, hexahydrate [CRC10]

3307

Compound: Titanium trichloride

Formula: $TiCl_3$

Molecular Formula: Cl_3Ti

Molecular Weight: 154.225

CAS RN: 7705-07-9

Properties: dark reddish violet cryst; deliq; unstable, decomposes above 500°C; strong reducing agent; reaction between hydrogen and $TiCl_4$ produces α-$TiCl_3$ (violet); β (brown), and γ (violet) forms are produced by reacting $TiCl_4$ with aluminum alkyls; a τ (violet) form is made by grinding the α or γ forms; enthalpy of vaporization 124 kJ/mol; used extensively as a catalyst for the polymerization of hydrocarbons [KIR83] [MER06] [CRC10]

Solubility: s H_2O (exothermic), alcohol [MER06]
Density, g/cm³: 2.640 [STR93]
Melting Point, °C: decomposes at 440 [STR93]
Boiling Point, °C: 960 [CRC10]

3308

Compound: Titanium trifluoride
Formula: TiF_3
Molecular Formula: F_3Ti
Molecular Weight: 104.862
CAS RN: 13470-08-1
Properties: violet powd; sensitive to moisture; can be obtained by dissolving Ti metal in aq HF [STR93]
Solubility: i H_2O, dil acids and alkalies [KIR82]
Density, g/cm³: 3.40 [STR93]
Reactions: disproportionates >950°C [CER91]

3309

Compound: Titanium trioxide
Formula: Ti_2O_3
Molecular Formula: O_3Ti_2
Molecular Weight: 143.732
CAS RN: 1344-54-3
Properties: violet sintered tablets; hexagonal; oxidizes to TiO_2; sintered material of 99.9% purity used as an evaporation material for interference films, thin film resistors and capacitors [KIR83] [CER91]
Solubility: 40% hot HF [KIR83]
Density, g/cm³: 4.486 [KIR83]
Melting Point, °C: 1900 [KIR83]

3310

Compound: Titanium trisilicide
Formula: Ti_5Si_3
Molecular Formula: Si_3Ti_5
Molecular Weight: 323.657
CAS RN: 12067-57-1
Properties: −325 mesh gray powd [AES93]
Melting Point, °C: 2130 [AES93]

3311

Compound: Titanium trisulfide
Synonym: titanium sesquisulfide
Formula: Ti_2S_3
Molecular Formula: S_3Ti_2
Molecular Weight: 191.932
CAS RN: 12039-16-6
Properties: black cryst hex solid; can be prepared by direct reaction of Ti and S at 800°C [KIR83]
Density, g/cm³: 3.52 [KIR83]

3312

Compound: Titanium(IV) oxide acetylacetonate
Synonyms: 2,4-pentanedione, Ti(IV) derivative
Formula: $[CH_3COCH=C(O)CH_3]_2TiO$
Molecular Formula: $C_{10}H_{14}O_5Ti$
Molecular Weight: 262.098
CAS RN: 14024-64-7
Properties: cryst powd; hydrolysis resistant; preparation: reaction of titanium oxychloride with acetylacetone and sodium carbonate; uses: cross-linking agent for cellulosic fibers [HAW93]
Solubility: sl s H_2O [HAW93]
Melting Point, °C: 200, decomposes [ALD94]

3313

Compound: Titanocene dichloride
Synonym: bis(cyclopentadienyl)titanium dichloride
Formula: $(C_5H_5)_2TiCl_2$
Molecular Formula: $C_{10}H_{10}Cl_2Ti$
Molecular Weight: 248.975
CAS RN: 1271-19-8
Properties: bright red acidular cryst from toluene; sensitive to moisture; uses: synthesis of many transition metal complexes and organometallic compounds, catalyst [MER06] [ALF95]
Solubility: sl s H_2O [MER06]
Density, g/cm³: 1.6 [ALD94]
Melting Point, °C: 289–290 [ALF95]
Reactions: with H_2O forms Cp_2TiOH+ [COT88]

3314

Compound: Tribromogermane
Formula: $GeHBr_3$
Molecular Formula: Br_3GeH
Molecular Weight: 313.36
CAS RN: 14779-70-5
Properties: col liq [CRC10]
Solubility: reac H_2O [CRC10]
Melting Point, °C: −25 [CRC10]
Boiling Point, °C: decomposes [CRC10]

3315

Compound: Trichlorogermane
Formula: $GeHCl_3$
Molecular Formula: Cl_3GeH
Molecular Weight: 180.01
CAS RN: 1184-65-2
Properties: liq [CRC10]
Solubility: reac H_2O [CRC10]
Density, g/cm³: 1.93 [CRC10]

Melting Point, °C: −71 [CRC10]
Boiling Point, °C: 75.3 [CRC10]

3316

Compound: Tribromosilane
Formula: SiHBr$_3$
Molecular Formula: Br$_3$HSi
Molecular Weight: 268.806
CAS RN: 7789-57-3
Properties: enthalpy of vaporization
34.8 kJ/mol; entropy of vaporization
87.9 kJ/(mol · K) [CRC10] [CIC73]
Melting Point, °C: −73 [CIC73]
Boiling Point, °C: 109 [CRC10]

3317

Compound: Trichlorofluorogermane
Formula: GeCl$_3$F
Molecular Formula: Cl$_3$FGe
Molecular Weight: 198.00
CAS RN: 24422-20-6
Properties: liq [CRC10]
Melting Point, °C: −49.8 [CRC10]
Boiling Point, °C: 37.5 [CRC10]

3318

Compound: Trichlorosilane
Synonym: silicochloroform
Formula: SiHCl$_3$
Molecular Formula: Cl$_3$HSi
Molecular Weight: 135.452
CAS RN: 10025-78-2
Properties: colorless volatile mobile liq; supports
combustion; enthalpy of vaporization
25.7 kJ/mol at 25°C; entropy of vaporization
82.8 kJ/(mol · K); used in organic synthesis
[CIC73] [CRC10] [MER06] [STR93]
Solubility: decomposed by H$_2$O [MER06]
Density, g/cm^3: (25°C) 1.3313 [MER06]
Melting Point, °C: −118 [CIC73]
Boiling Point, °C: 33 [CRC10]

3319

Compound: Tridecaborane(19)
Formula: B$_{13}$H$_{19}$
Molecular Formula: B$_{13}$H$_{19}$
Molecular Weight: 159.694
CAS RN: 43093-20-5
Properties: yellow cryst [CRC10]
Melting Point, °C: 44 [CRC10]

3320

Compound: Triethylphosphine
Formula: (C$_2$H$_5$)$_3$P
Molecular Formula: C$_6$H$_{15}$P
Molecular Weight: 118.16
CAS RN: 554-70-1
Properties: colorless liq; preparation from white
phosphorus, ethylene, and hydrogen under pressure;
used in organic synthesis [MER06] [ALD94]
Solubility: i H$_2$O; miscible with alcohol, ether [MER06]
Density, g/cm^3: 0.80 [MER06]
Melting Point, °C: −17 [ALD94]
Boiling Point, °C: 127–128 [MER06]

3321

Compound: Trifluoromethane
Synonym: halocarbon-23
Formula: CHF$_3$
Molecular Formula: CHF$_3$
Molecular Weight: 70.013
CAS RN: 75-46-7
Properties: colorless gas with ethereal odor; critical
temp 25.9°C; critical pressure 4.84 MPa;
enthalpy of vaporization 170.5 kJ/mol;
used in electronics industry [AIR87]
Melting Point, °C: −155.2 [AIR87]
Boiling Point, °C: −82.2 [AIR87]

3322

Compound: Trifluorosilane
Synonym: silicofluoroform
Formula: SiHF$_3$
Molecular Formula: F$_3$HSi
Molecular Weight: 86.089
CAS RN: 13465-71-9
Properties: colorless gas; enthalpy of vaporization
16.1 kJ/mol; entropy of vaporization
87.9 kJ/(mol · K) [CIC73] [CRC10]
Density, g/cm^3: 1.86 (0°C) [CRC10]
Melting Point, °C: −131 [CIC73]
Boiling Point, °C: −94.4 [CIC73]

3323

Compound: Trigermane
Formula: Ge$_3$H$_8$
Molecular Formula: Ge$_3$H$_8$
Molecular Weight: 225.98
CAS RN: 14691-44-2
Properties: col liq [CRC10]
Solubility: i H$_2$O
Density, g/cm^3: 2.20^{-105} [CRC10]

Melting Point, °C: −105.6 [CRC10]
Boiling Point, °C: 110.5 [CRC10]

3324
Compound: Triiodosilane
Synonym: silicoiodoform
Formula: $SiHI_3$
Molecular Formula: HI_3Si
Molecular Weight: 409.807
CAS RN: 13465-72-0
Properties: colorless liq; enthalpy of vaporization 62.8 kJ/mol; entropy of vaporization 159 kJ/(mol · K) [CIC73] [CRC10]
Density, g/cm³: 3.314 (20°C) [CRC10]
Melting Point, °C: 8 [CIC73]
Boiling Point, °C: decomposes at 220 [CIC73]

3325
Compound: Tris(ethylenediammine) cobalt(III) chloride trihydrate
Formula: $Co[H_2NCH_2CH_2NH_2]_3Cl_3 \cdot 3H_2O$
Molecular Formula: $C_6H_{28}Cl_3CoN_6O_2$
Molecular Weight: 399.629
CAS RN: 14883-80-8
Properties: brown prisms [KIR79]
Solubility: v s H_2O [KIR79]
Density, g/cm³: 1.542 [KIR79]
Melting Point, °C: decomposes at 275 [ALD94]
Reactions: minus $3H_2O$ at 100°C [KIR79]

3326
Compound: Tris(triphenylphosphine)rhodium(I) chloride
Synonym: Wilkinson's catalyst
Formula: $Rh[P(C_6H_5)_3]_3$
Molecular Formula: $C_{54}H_{45}ClP_3Rh$
Molecular Weight: 925.231
CAS RN: 14694-95-2
Properties: burgundy-red cryst when prepared in ethanol by reaction between excess triphenylphosphine and $RhCl_3 \cdot 3H_2O$; used as a homogeneous catalyst [MER06] [ALD94]
Solubility: ~20 g/L (25°C) chloroform [MER06]
Melting Point, °C: 157–158 [MER06]

3327
Compound: Tritium
Formula: T_2
Molecular Formula: T_2
Molecular Weight: 6.032

CAS RN: 10028-17-8
Properties: gas with $t_{1/2}$ of 12.26 years; low β-emitter; critical temp −232.56°C; critical pressure 18.317 atm; enthalpy of sublimation 1640 J/mol; enthalpy of vaporization 1390 J/mol; used in hydrogen bomb and in radioactive tracers [MER06] [KIR78]
Density, g/cm³: liq: 45.35 mol/L [KIR78]
Melting Point, °C: −254.54 [MER06]
Boiling Point, °C: −248.12 [MER06]

3328
Compound: Tritium dioxide
Formula: T_2O
Molecular Formula: OT_2
Molecular Weight: 22.032
CAS RN: 14940-65-9
Properties: liq; triple point 4.49°C; liq vapor pressure at 25°C 2.64 kPa; enthalpy of vaporization ~45.81 kJ/mol; ionization constant ~6×10^{16}; temp of maximum density 13.4°C [KIR78]
Density, g/cm³: 1.2138 [KIR78]
Boiling Point, °C: 101.51 [KIR78]

3329
Compound: Tungsten
Synonym: wolfram
Formula: W
Molecular Formula: W
Molecular Weight: 183.84
CAS RN: 7440-33-7
Properties: steel gray to tin white metal; bcc, a = 0.316524 nm; enthalpy of fusion 52.31 kJ/mol; enthalpy of sublimation (25°C) 859.8 kJ/mol; electrical resistivity (20°C) 5.5 μohm · cm; stable in dry air, unless heated to red heat, then forms WO_3; used in crucible form for growing single cryst from reactive melts, and to evaporate metals and compounds for thin films [MER06] [KIR83] [CER91] [CRC10]
Solubility: s slowly in fused KOH or Na_2CO_3 in the presence of air [MER06]
Density, g/cm³: 19.254 [KIR83]
Melting Point, °C: 3410 [ALD94]
Boiling Point, °C: 5660 [ALD94]
Thermal Conductivity, W/(m · K): 146 (227°C), 118 (727°C), 100 (1727°C), 90 (3127°C) [KIR83]; 173 (25°C) [ALD94]
Thermal Expansion Coefficient: (volume) 100°C (0.095), 200°C (0.228), 400°C (0.513), 800°C (1.104), 1200 (1.774) [CLA66]

3330

Compound: Tungsten boride
Formula: WB
Molecular Formula: BW
Molecular Weight: 194.651
CAS RN: 12007-09-9
Properties: refractory material; black powd; in −325 mesh 99.5% pure form, used as a sputtering target for producing wear-resistant and semiconducting films, and other applications [KIR78] [STR93] [CER91]
Density, g/cm³: 15.2 [LID94]
Melting Point, °C: 2660 [KIR78]

3331

Compound: Tungsten boride
Synonym: ditungsten boride
Formula: W_2B
Molecular Formula: BW_2
Molecular Weight: 378.491
CAS RN: 12007-10-2
Properties: refractory material; black powd; forms when tungsten and boron are hot pressed; extremely hard and has almost metallic electrical conductivity; −325 mesh in 99.5% pure form used as a sputtering target for fabricating wear-resistant and semiconductor films [KIR83] [STR93] [CER91]
Density, g/cm³: 16.0 [LID94]
Melting Point, °C: 2670 [KIR78]

3332

Compound: Tungsten carbide
Formula: W_2C
Molecular Formula: CW_2
Molecular Weight: 379.691
CAS RN: 12070-13-2
Properties: black hex, a = 0.29982 nm, c = 0.4722 nm; can be prepared by heating tungsten and carbon in the presence of hydrogen at high temperatures; hardness approaches that of diamond; brittle; used in hard metals, and as a 99.5% pure sputtering target to produce wear-resistant films and semiconductor films [KIR83] [CIC73] [CER91] [CRC10]
Density, g/cm³: 17.15 [KIR83]
Melting Point, °C: ~2800 [KIR83]

3333

Compound: Tungsten carbide
Formula: WC
Molecular Formula: CW
Molecular Weight: 195.851
CAS RN: 12070-12-1
Properties: gray powd; hex, a = 0.29063 nm, c = 0.28386 nm; hardness 9+ Mohs; can be prepared by heating tungsten and carbon at high temperatures, sometimes in the presence of hydrogen; used in dies and cutting tools, wear-resistant parts, electrical resistors, as an abrasive, in crucible form used to melt copper, tin, bismuth and cobalt, and as a 99.5% pure sputtering target to prepare wear-resistant semiconductor films [HAW93] [CIC73] [STR93] [KIR83] [CER91] [GEI92]
Solubility: i H_2O; attacked by a mixture of HNO_3, HF acids [HAW93]
Density, g/cm³: 15.63 [STR93]
Melting Point, °C: ~2870 [STR93]
Boiling Point, °C: 6000 [STR93]
Thermal Conductivity, W/(m·K): 121 [KIR78]
Thermal Expansion Coefficient: (volume) 100°C (0.104), 200°C (0.236), 400°C (0.507), 800°C (1.103), 1200°C (1.768) [CLA66]

3334

Compound: Tungsten carbonyl
Synonym: tungsten hexacarbonyl
Formula: $W(CO)_6$
Molecular Formula: C_6O_6W
Molecular Weight: 351.902
CAS RN: 14040-11-0
Properties: white, volatile, highly refractive cryst; very stable; vapor pressure is 13.3 Pa (20°C) and 160 Pa (67°C); may be prepared by reducing WCl_6 with Al in anhydrous ether under 10 MPa of CO at 70°C [KIR83] [HAW93]
Solubility: i H_2O; s in organic solvents [HAW93]
Density, g/cm³: 2.65 [HAW93]
Melting Point, °C: decomposes at 169–170 [STR93]

3335

Compound: Tungsten dibromide
Formula: WBr_2
Molecular Formula: Br_2W
Molecular Weight: 343.648
CAS RN: 13470-10-5
Properties: black powd; prepared by the reduction of WBr_5 with H_2 [KIR83]
Melting Point, °C: decomposes at 400 [KIR83]

3336

Compound: Tungsten dichloride
Formula: WCl_2
Molecular Formula: Cl_2W
Molecular Weight: 254.745

CAS RN: 13470-12-7
Properties: gray amorphous powd; prepared by reduction of WCl_6 by Al in molten $NaAlCl_6$ [KIR83] [CRC10]
Density, g/cm³: 5.436 [CRC10]

3337
Compound: Tungsten diiodide
Formula: WI_2
Molecular Formula: I_2W
Molecular Weight: 437.649
CAS RN: 13470-17-2
Properties: brown powd [KIR83]
Density, g/cm³: 6.79 [KIR83]
Melting Point, °C: decomposes [CRC10]

3338
Compound: Tungsten dinitride
Formula: WN_2
Molecular Formula: N_2W
Molecular Weight: 211.853
CAS RN: 60922-26-1
Properties: brown; hex, a = 0.2893 nm, c = 0.2826 nm [CIC73] [CRC10]
Density, g/cm³: 7.7 [LID94]
Melting Point, °C: decomposes 600 [CIC73]

3339
Compound: Tungsten dioxide
Synonym: tungsten(IV) oxide
Formula: WO_2
Molecular Formula: O_2W
Molecular Weight: 215.839
CAS RN: 12036-22-5
Properties: brown powd, may become purple on standing; formed when WO_3 is reduced by H_2 at 575°C–600°C [KIR83] [STR93]
Density, g/cm³: 12.11 [STR93]
Melting Point, °C: 1500–1600 [STR93]

3340
Compound: Tungsten dioxydibromide
Formula: WO_2Br_2
Molecular Formula: Br_2O_2W
Molecular Weight: 375.647
CAS RN: 13520-75-7
Properties: light red cryst that forms when a mixture of bromine and oxygen is passed over tungsten at 300°C [KIR83]
Melting Point, °C: decomposes [CRC10]

3341
Compound: Tungsten diselenide
Formula: WSe_2
Molecular Formula: Se_2W
Molecular Weight: 341.760
CAS RN: 12067-46-8
Properties: −325 mesh black powd; dry, solid lubricant with exceptional stability at high temperatures, and in high vacuum; used in the form of 99.8% pure material as a sputtering target to produce lubricant films [HAW93] [CER91] [AES93]

3342
Compound: Tungsten disilicide
Formula: WSi_2
Molecular Formula: Si_2W
Molecular Weight: 240.011
CAS RN: 12039-88-2
Properties: bluish gray tetr powd, a = 0.3212 nm, c = 0.7880 nm; attacked by fluorine, chlorine, and fused alkalies; can be used in high temp thermocouples in combination with $MoSi_2$, in an oxidizing atm; as a 99.95% and 99.5% pure material, used as a sputtering target in the fabrication of integrated circuits [KIR83] [STR93] [CER91]
Solubility: i H_2O [KIR83]
Density, g/cm³: 9.4 [STR93]
Melting Point, °C: 2165 [STR93]
Thermal Expansion Coefficient: (volume) 100°C (0.164), 200°C (0.364), 400°C (0.868), 800°C (1.886), 1200°C (3.047) [CLA66]

3343
Compound: Tungsten disulfide
Synonym: tungstenite
Formula: WS_2
Molecular Formula: S_2W
Molecular Weight: 247.972
CAS RN: 12138-09-9
Properties: soft, grayish black powd which is relatively inert and unreactive; exists in mineral form, or can be prepared by heating tungsten powd with sulfur at 900°C; used as a solid lubricant, and as a 99.8% pure material as a sputtering target for lubricant films on bearings and other moving parts [HAW93] [STR93] [KIR83] [CER91]
Solubility: i H_2O, HCl, alkali [KIR83]
Density, g/cm³: 7.5 [STR93]
Melting Point, °C: decomposes at 1250 [STR93]

3344

Compound: Tungsten hexabromide
Formula: WBr_6
Molecular Formula: Br_6W
Molecular Weight: 663.264
CAS RN: 13701-86-5
Properties: bluish black cryst; made by reaction of BBr_3 with WCl_6 [KIR83]
Density, g/cm³: 6.9 [CRC10]
Melting Point, °C: 232 [KIR83]

3345

Compound: Tungsten hexachloride
Synonym: tungsten(VI) chloride
Formula: WCl_6
Molecular Formula: Cl_6W
Molecular Weight: 396.556
CAS RN: 13283-01-7
Properties: purple hex cryst; sensitive to moisture; vapor pressure is 43 mm Hg (215°C); enthalpy of vaporization 52.7 kJ/mol; enthalpy of fusion 6.60 kJ/mol; decomposed by moist air, and water; can be obtained by direct reaction of Cl_2 and W at 600°C [HAW93] [STR93] [KIR83] [CRC10]
Solubility: s organic solvents such as ethanol [HAW93]
Density, g/cm³: 3.52 [HAW93]
Melting Point, °C: 275 [CRC10]
Boiling Point, °C: 346.75 [CRC10]

3346

Compound: Tungsten hexafluoride
Synonym: tungsten(VI) fluoride
Formula: WF_6
Molecular Formula: F_6W
Molecular Weight: 297.830
CAS RN: 7783-82-6
Properties: colorless gas or pale yellow liq when condensed; critical temp 169.8°C; critical pressure 4.27 MPa; triple point 2.0°C at 55.1 kPa; transition point −8.4°C; enthalpy of vaporization 27 kJ/mol; enthalpy of fusion 4.10 kJ/mol; vapor pressure (17°C) 100.49 kPa; produced by reacting tungsten powd with gaseous fluorine above 350°C; used in electronics industry [KIR78] [MER06] [AIR87] [CRC10]
Solubility: decomposes in H_2O; s anhydrous HF: 3.14 moles/100 g [HAW93] [MER06]
Density, g/cm³: gas: 12.9 g/L [STR93]; liq: 3.441 (15°C) [KIR78]
Melting Point, °C: 2.3 [MER06]
Boiling Point, °C: 17.5 [ALD94]

3347

Compound: Tungsten nitride
Formula: W_2N
Molecular Formula: NW_2
Molecular Weight: 381.687
CAS RN: 12033-72-6
Properties: gray; fcc, a=0.412 nm; can be prepared by heating tungsten in ammonia; there is also a WN phase, 12058-38-7 [KIR83] [KIR81]
Density, g/cm³: 17.7 [KIR81]
Melting Point, °C: decomposes [KIR81]

3348

Compound: Tungsten oxychloride
Synonym: tungsten(VI) tetrachloride monoxide
Formula: $WOCl_4$
Molecular Formula: Cl_4OW
Molecular Weight: 341.650
CAS RN: 13520-78-0
Properties: dark red, acicular cryst; enthalpy of vaporization 67.8 kJ/mol; enthalpy of fusion 45.00 kJ/mol; obtained by refluxing $SOCl_2$ with tungsten trioxide; used in incandescent lamps [HAW93] [CRC10]
Solubility: decomposed by H_2O; s CS_2, benzene [KIR83] [HAW93]
Density, g/cm³: 11.92 [HAW93]
Melting Point, °C: 211 [KIR83]
Boiling Point, °C: 227.5 [CRC10]

3349

Compound: Tungsten oxydichloride
Synonym: tungsten(VI) dichloride dioxide
Formula: WO_2Cl_2
Molecular Formula: Cl_2O_2W
Molecular Weight: 286.744
CAS RN: 13520-76-8
Properties: pale yellow cryst solid; obtained by reacting CCl_4 and WO_2 at 250°C [KIR83]
Solubility: i cold H_2O, partially decomposed by hot H_2O; i alkaline solutions [KIR83]
Melting Point, °C: 266 [KIR83]

3350

Compound: Tungsten oxydiiodide
Formula: WOI_2
Molecular Formula: I_2OW
Molecular Weight: 453.648
CAS RN: 14447-89-3
Properties: obtained by heating tungsten, tungsten trioxide, and iodine in a 500°C–700°C temp gradient for 36 h [KIR83]

3351
Compound: Tungsten oxytetrabromide
Formula: $WOBr_4$
Molecular Formula: Br_4OW
Molecular Weight: 519.455
CAS RN: 13520-77-9
Properties: black deliq needles; made by reacting CBr_4 and WO_2 at 250°C [KIR83]
Melting Point, °C: 277 [KIR83]
Boiling Point, °C: 327 [KIR83]

3352
Compound: Tungsten oxytetrafluoride
Formula: WOF_4
Molecular Formula: F_4OW
Molecular Weight: 275.833
CAS RN: 13520-79-1
Properties: colorless plates; can be prepared by reacting W metal with an O_2–F_2 mixture at high temperatures; hygr [KIR83]
Solubility: decomposes to tungstic acid in H_2O [KIR83]
Melting Point, °C: 110 [CRC10]
Boiling Point, °C: 187 [CRC10]

3353
Compound: Tungsten oxytrichloride
Formula: $WOCl_3$
Molecular Formula: Cl_3OW
Molecular Weight: 306.197
CAS RN: 14249-98-0
Properties: green solid; can be obtained by reduction of $WOCl_4$ with Al in a sealed system at 100°C–140°C [KIR83]

3354
Compound: Tungsten pentaboride
Formula: W_2B_5
Molecular Formula: B_5W_2
Molecular Weight: 421.735
CAS RN: 12007-98-6
Properties: black powd, –325 mesh; refractory material [KIR78] [STR93]
Melting Point, °C: 2365 [KIR78]

3355
Compound: Tungsten pentabromide
Synonym: tungsten(V) bromide
Formula: WBr_5
Molecular Formula: Br_5W

Molecular Weight: 583.360
CAS RN: 13470-11-6
Properties: brownish violet cryst; high moisture sensitivity; prepared by reaction of bromine vapor on metallic tungsten at 450°C–500°C [KIR83]
Melting Point, °C: 276 [KIR83]
Boiling Point, °C: 333 [KIR83]

3356
Compound: Tungsten pentachloride
Formula: WCl_5
Molecular Formula: Cl_5W
Molecular Weight: 361.104
CAS RN: 13470-14-9
Properties: black cryst deliq solid; can be prepared by reducing WCl_6 with red phosphorus [KIR83]
Solubility: decomposes in H_2O to a blue oxide; v sl s CS_2 [KIR83]
Density, g/cm³: 3.875 [CRC10]
Melting Point, °C: 243 [KIR83]
Boiling Point, °C: 275.6 [KIR83]

3357
Compound: Tungsten telluride
Synonym: tungsten(IV) telluride
Formula: WTe_2
Molecular Formula: Te_2W
Molecular Weight: 439.040
CAS RN: 12067-26-4
Properties: gray powd; used in the form of a –325 mesh, 99.8% pure material as a sputtering target to form lubricant film [STR93] [CER91]

3358
Compound: Tungsten tetrabromide
Synonym: tungsten(IV) bromide
Formula: WBr_4
Molecular Formula: Br_4W
Molecular Weight: 503.456
CAS RN: 12045-94-2
Properties: black ortho-rhomb cryst; prepared by reduction of WBr_5 with Al [KIR83]

3359
Compound: Tungsten tetrachloride
Synonym: tungsten(IV) chloride
Formula: WCl_4

Molecular Formula: Cl_4W
Molecular Weight: 325.651
CAS RN: 13470-13-8
Properties: gray powd; sensitive to moisture; decomposes if heated; diamagnetic; obtained by reduction of WCl_6 with Al [KIR83] [STR93]
Density, g/cm³: 4.624 [STR93]
Melting Point, °C: decomposes [CRC10]

3360
Compound: Tungsten tetraiodide
Formula: WI_4
Molecular Formula: I_4W
Molecular Weight: 691.458
CAS RN: 14055-84-6
Properties: black powd; decomposed by air; prepared by reacting conc hydriodic acid and WCl_6 at 100°C [KIR83]
Density, g/cm³: 5.2 [CRC10]
Melting Point, °C: decomposes [CRC10]

3361
Compound: Tungsten tribromide
Formula: WBr_3
Molecular Formula: Br_3W
Molecular Weight: 423.552
CAS RN: 15163-24-3
Properties: black powd; thermally unstable; prepared by reaction of bromine and WBr_2 at 50°C in a sealed system [KIR83]
Solubility: i H_2O [KIR83]

3362
Compound: Tungsten triiodide
Formula: WI_3
Molecular Formula: I_3W
Molecular Weight: 564.553
CAS RN: 15513-69-6
Properties: can be prepared by reacting iodine and $W(CO)_6$ in a sealed system at 120°C [KIR83]

3363
Compound: Tungsten trioxide
Synonym: tungsten(VI) oxide
Formula: WO_3
Molecular Formula: O_3W
Molecular Weight: 231.838
CAS RN: 1314-35-8

Properties: canary yellow, heavy powd which becomes dark orange when heated and reverts to original color when cooled, also greenish yellow 3–12 mm sintered pieces; enthalpy of fusion 73.00 kJ/mol; can be prepared from tungstic acid; used as starting material to produce tungsten powd, sintered pieces used as evaporation material and sputtering target for shadow casting in electron microscopy [KIR83] [MER06] [CER91] [CRC10]
Solubility: i H_2O; s caustic alkalies; v sl s in acids [MER06]
Density, g/cm³: 7.16 [STR93]
Melting Point, °C: 1472 [CRC10]
Reactions: phase change from pseudorhomb to tetr above 700°C [KIR83]

3364
Compound: Tungsten trisilicide
Formula: W_5Si_3
Molecular Formula: Si_3W_5
Molecular Weight: 1003.457
CAS RN: 12039-95-1
Properties: bluish gray; very hard solid; attacked by fused alkalies and mixtures of nitric and hydrofluoric acids [HAW93]
Solubility: i H_2O [HAW93]
Density, g/cm³: 9.4 [HAW93]
Melting Point, °C: >900 [HAW93]

3365
Compound: Tungsten trisulfide
Formula: WS_3
Molecular Formula: S_3W
Molecular Weight: 280.038
CAS RN: 12125-19-8
Properties: chocolate brown powd; can be obtained from an alkali metal thiotungstate by treating with HCl [KIR83]
Solubility: sl s cold H_2O, forms colloid in hot H_2O; s alkali carbonate and hydroxide solutions [KIR83]

3366
Compound: Tungstic acid
Formula: H_2WO_4
Molecular Formula: H_2O_4W
Molecular Weight: 249.854
CAS RN: 7783-03-1
Properties: amorphous yellow powd; prepared by precipitation from hot tungstate solutions with strong acids, followed by boiling in an acidic medium; used in textile and plastics industries [KIR83] [STR93]

Solubility: i H_2O, acids; s alkalies [KIR83]
Density, g/cm³: 5.5 [STR93]
Reactions: minus H_2O at 100°C [CRC10]

3367

Compound: Tungstophosphoric acid hydrate
Synonym: 12-tungstophosphate
Formula: $H_3PO_4 \cdot 12WO_3 \cdot xH_2O$
Molecular Formula: $H_3O_{40}PW_{12}$ (anhydrous)
Molecular Weight: 2880.053 (anhydrous)
CAS RN: 12501-23-4
Properties: yellowish white solid;
hygr [HAW93] [STR93]
Solubility: s H_2O, acetone, ether [HAW93]
Melting Point, °C: 89, for x = 24 [HAW93]

3368

Compound: Uranium
Formula: U
Molecular Formula: U
Molecular Weight: 238.0289
CAS RN: 7440-61-1
Properties: silvery white, lustrous; black powd, when
obtained by reduction of UF_4; three forms, α-form:
ortho-rhomb, a = 0.2854 nm, b = 0.5869 nm,
c = 0.4956 nm; β: tetr, a = 1.0763 nm, c = 0.5652 nm; γ:
bcc, a = 0.3524 nm; resistivity 29 μohm·cm; enthalpy
of fusion 9.14 kJ/mol; enthalpy of sublimation
1062.73 kJ/mol; flammable in air forming U_3O_8;
$t_{1/2}$ ^{238}U is 4.47 × 10^{+9} years; ionic radius of U^{++++}
0.0918 nm [MER06] [KIR78] [KIR83] [CRC10]
Density, g/cm³: α: 19.07; β: 18.11; γ: 18.06 [KIR83]
Melting Point, °C: 1132 [KIR91]
Boiling Point, °C: 3818 [ALD94]
Reactions: transition α to β at 667.7°C,
β to γ at 774.8°C [MER06]
Thermal Conductivity, W/(m·K): 25.1
(36°C), 26.3 (100°C), 29.7 (200°C), 31.4
(300°C), 32.6 (400°C) [KIR83]
Thermal Expansion Coefficient: 13.9 × 10^{-6}/K [CRC10]

3369

Compound: Uranium diboride
Formula: UB_2
Molecular Formula: B_2U
Molecular Weight: 259.651
CAS RN: 12007-36-2
Properties: hex; −8 mesh with 99.5% purity;
refractory material [KIR78] [CER91] [CRC10]
Density, g/cm³: 12.7 [CRC10]
Melting Point, °C: 2385 [KIR78]

3370

Compound: Uranium dicarbide
Formula: UC_2
Molecular Formula: C_2U
Molecular Weight: 262.051
CAS RN: 12071-33-9
Properties: gray tetr cryst, a = 0.35241 nm,
c = 0.59962 nm; used in the form of
pellets or microspheres to fuel nuclear
reactors [HAW93] [CIC73]
Solubility: decomposes in H_2O; sl s alcohol [HAW93]
Density, g/cm³: 11.28 [HAW93]
Melting Point, °C: 2350 [HAW93]
Boiling Point, °C: 4370 [HAW93]
Reactions: transition tetr to cub at 1765°C [CIC73]

3371

Compound: Uranium dioxide
Synonym: uraninite
Formula: UO_2
Molecular Formula: O_2U
Molecular Weight: 270.028
CAS RN: 1344-57-6
Properties: −100 mesh; brown to black powd; cub
cryst; widely used to manufacture fuel pellets
for power reactors [KIR83] [MER06] [STR94]
Solubility: i H_2O, dil acids; s conc acids [MER06]
Density, g/cm³: 10.97 [MER06]
Melting Point, °C: 2865 [MER06]
Thermal Conductivity, W/(m·K): 5.1
(500°C), 3.4 (1000°C) [KIR80]
Thermal Expansion Coefficient: (volume)
100°C (0.199), 200°C (0.468), 400°C (1.045),
800°C (2.638), 1000°C (3.115) [CLA66]

3372

Compound: Uranium hexafluoride
Formula: UF_6
Molecular Formula: F_6U
Molecular Weight: 352.019
CAS RN: 7783-81-5
Properties: white; volatile; monocl solid; reacts
vigorously with H_2O, forming mainly UO_2F_2
and HF; enthalpy of vaporization at 64.01°C
28.899 kJ/mol; enthalpy of fusion 19.19 kJ/
mol; enthalpy of sublimation 48.095 kJ/
mol; triple point 64.052°C at 151 kPa; can be
prepared by direct reaction of uranium metal
and fluorine [KIR83] [MER06] [CRC10]
Solubility: s liq Cl_2, Br_2; gives dark red fuming solution
with nitrobenzene; s CCl_4, CH_3Cl [MER06]

Density, g/cm³: solid: 5.09; liq: 3.595 [MER06]
Melting Point, °C: 64 [KIR91]

3373
Compound: Uranium monocarbide
Formula: UC
Molecular Formula: CU
Molecular Weight: 250.040
CAS RN: 12070-09-6
Properties: gray with metallic appearance; fcc, a=0.49605 nm; can be prepared by arc melting stoichiometric amounts of the elements in an inert atm; reacts with oxygen [CIC73]
Density, g/cm³: 13.63 [KIR78]
Melting Point, °C: 2790 [CIC73]
Thermal Conductivity, W/(m·K): 25 [KIR78]
Thermal Expansion Coefficient: 9.1×10^{-6}/K [KIR78]

3374
Compound: Uranium mononitride
Formula: UN
Molecular Formula: NU
Molecular Weight: 252.036
CAS RN: 25658-43-9
Properties: dark gray; fcc, a=0.4890 nm; electrical resistivity 176 μohm·cm; Knoop hardness 580; only stable uranium nitride above 1300°C; formed by reacting uranium and nitrogen; there are two other nitrides, $UN_{1.5}$, 12033-85-1, and $UN_{1.75}$, 12266-20-5 [KIR83] [KIR81]
Density, g/cm³: 14.4 [KIR81]
Melting Point, °C: 2800 [CIC73]
Thermal Conductivity, W/(m·K): 15.5 [KIR81]
Thermal Expansion Coefficient: 8.0×10^{-6} [KIR81]

3375
Compound: Uranium pentabromide
Formula: UBr_5
Molecular Formula: Br_5U
Molecular Weight: 637.549
CAS RN: 13775-16-1
Properties: dark brown hygr unstable solid; formed when UBr_4 is extracted with liq Br_2 at 55°C, followed by recrystallization from liq Br_2 [KIR83]

3376
Compound: Uranium pentachloride
Formula: UCl_5
Molecular Formula: Cl_5U
Molecular Weight: 415.293
CAS RN: 13470-21-8

Properties: reddish brown cryst, with metallic luster; can be formed by reaction of UO_3 with CCl_4 or Cl_2 [KIR83]
Solubility: s liq Cl_2 [KIR83]
Density, g/cm³: 3.81 [CRC10]
Melting Point, °C: decomposes 300 [CRC10]

3377
Compound: Uranium pentafluoride
Formula: UF_5
Molecular Formula: F_5U
Molecular Weight: 333.021
CAS RN: 13775-07-0
Properties: two forms: α-UF_5 is grayish white and can be obtained by reduction of UF_6 with HBr, β-UF_5 is yellowish white and is obtained by reacting UF_6 with UF_4 at 150°C–200°C; both forms are hygr; have blue solutions in anhydrous HF [KIR83]
Density, g/cm³: 5.81 [LID94]
Melting Point, °C: 348 [LID94]

3378
Compound: Uranium tetraboride
Formula: UB_4
Molecular Formula: B_4U
Molecular Weight: 281.273
CAS RN: 12007-84-0
Properties: brown; −8 mesh with 99.9% purity and −60 mesh with 99.7% purity; refractory material [KIR78] [CRC10] [CER91]
Density, g/cm³: 5.35 [CRC10]
Melting Point, °C: 2495 [KIR78]

3379
Compound: Uranium tetrabromide
Formula: UBr_4
Molecular Formula: Br_4U
Molecular Weight: 557.645
CAS RN: 13470-20-7
Properties: dark brown, very hygr cryst; prepared by heating uranium turnings in a nitrogen gas stream saturated with bromine vapor; can be purified by vacuum distillation in a similar nitrogen stream [KIR83]
Density, g/cm³: 5.35 [CRC10]
Melting Point, °C: 519 [KIR91]
Boiling Point, °C: 765 [KIR83]

3380
Compound: Uranium tetrachloride
Formula: UCl_4

Molecular Formula: Cl_4U
Molecular Weight: 379.840
CAS RN: 10026-10-5
Properties: dark green octahedral cryst; oxidizes in air; decomposes in water; enthalpy of fusion 45.00 kJ/mol; prepared by the reaction $UO_3 + 3CCl_3CCl=CCl_2 \rightarrow UCl_4 + Cl_2 + 3CCl_2=CClOCl$ [KIR83] [CRC10] [MER06]
Solubility: v s H_2O, with decomposition [MER06]
Density, g/cm³: 4.725 [MER06]
Melting Point, °C: 590 [KIR91]
Boiling Point, °C: 791 [MER06]

3381
Compound: Uranium tetrafluoride
Formula: UF_4
Molecular Formula: F_4U
Molecular Weight: 314.023
CAS RN: 10049-14-6
Properties: monocl green cryst; reacts when heated with atm O_2, forming U_3O_8; prepared by reacting UO_2 with excess gaseous HF at ~550°C; used to produce both uranium metal and UF_6 [MER06] [KIR83]
Solubility: i H_2O; s conc acids, alkalies, but decomposes [MER06]
Density, g/cm³: 6.70 [HAW93]
Melting Point, °C: 960 [KIR91]

3382
Compound: Uranium tetraiodide
Formula: UI_4
Molecular Formula: I_4U
Molecular Weight: 745.647
CAS RN: 13470-22-9
Properties: black lustrous cryst; can be made by reaction of iodine and uranium metal; similar properties for UI_3 and for preparation of UI_3, 13775-18-8 [KIR83]
Density, g/cm³: 5.6 [CRC10]
Melting Point, °C: 506 [CRC10]
Boiling Point, °C: 729 [CRC10]

3383
Compound: Uranium tribromide
Formula: UBr_3
Molecular Formula: Br_3U
Molecular Weight: 477.741
CAS RN: 13470-19-4
Properties: reddish brown cryst; prepared by reacting UH_3 and HBr, or directly from the two elements; black cryst obtained when purified by gas phase transport [KIR83]

Density, g/cm³: 6.53 [CRC10]
Melting Point, °C: 730 [KIR91]

3384
Compound: Uranium tricarbide
Formula: U_2C_3
Molecular Formula: C_3U_2
Molecular Weight: 512.091
CAS RN: 12612-73-6
Properties: −60 mesh; gray with metallic appearance; bcc, a=0.80889 nm; can be prepared by arc welding stoichiometric amounts of the two elements [CIC73] [CER91]
Density, g/cm³: 12.7 [LID94]
Melting Point, °C: decomposes at 1727 [CIC73]

3385
Compound: Uranium trichloride
Formula: UCl_3
Molecular Formula: Cl_3U
Molecular Weight: 344.387
CAS RN: 10025-93-1
Properties: dark purple cryst; somewhat hygr; obtained by reacting UH_3 with HCl at 250°C–300°C; used in molten salt electrolytes to refine uranium metal [KIR83] [MER06]
Solubility: v s H_2O, evolves H_2, solution changes color from purple to green due to oxidation of U^{+++} [MER06]
Density, g/cm³: 5.51 [MER06]
Melting Point, °C: 835 [KIR91]

3386
Compound: Uranium trifluoride
Formula: UF_3
Molecular Formula: F_3U
Molecular Weight: 295.024
CAS RN: 13775-06-9
Properties: black mass containing small deep purple cryst; has been used as a component in molten salt systems [KIR83]
Solubility: i H_2O; s HNO_3, hot H_2SO_4, hot $HClO_4$ [KIR83]
Density, g/cm³: 8.9 [LID94]
Melting Point, °C: decomposes at >1140 [KIR91]

3387
Compound: Uranium trihydride
Formula: UH_3
Molecular Formula: H_3U
Molecular Weight: 241.053

CAS RN: 13598-56-6
Properties: −100 mesh; brownish gray to black powd; conducts electricity; prepared by heating uranium metal in a hydrogen atm at 150°C–200°C; used to prepare finely divided uranium by decomposition reaction [HAW93] [CER91]
Density, g/cm³: 10.92 [HAW93]

3388
Compound: Uranium trinitride
Formula: U_2N_3
Molecular Formula: N_3U_2
Molecular Weight: 518.078
CAS RN: 12033-83-9
Properties: bcc, a = 1.0678 nm [CIC73]
Density, g/cm³: 11.24 [CIC73]
Melting Point, °C: decomposes [CIC73]

3389
Compound: Uranium trioxide
Synonym: uranium(VI) oxide
Formula: UO_3
Molecular Formula: O_3U
Molecular Weight: 286.027
CAS RN: 1344-58-7
Properties: −100 mesh; has six forms: α is hex brown, β is orange monocl, γ is bright yellow rhomb, δ is red cub, ε is brick red tricl, η is rhomb; UO_3 can be obtained by thermal decomposition of uranyl compounds, e.g, carbonates, oxalates nitrates [KIR83] [CER91]
Solubility: i H_2O; s acids [MER06]
Density, g/cm³: 7.29 [MER06]
Melting Point, °C: decomposes at 650 [KIR91]

3390
Compound: Uranium(IV) sulfate octahydrate
Formula: $U(SO_4)_2 \cdot 8H_2O$
Molecular Formula: $H_{16}O_{16}S_2U$
Molecular Weight: 574.278
CAS RN: 19086-22-7
Solubility: g/100 g H_2O: 11.9 (20°C), 17.9 (30°C), 29.2 (40°C), 55.8 (60°C) [LAN05]
Melting Point, °C: decomposes at 90 [CRC10]

3391
Compound: Uranium(IV) sulfate tetrahydrate
Formula: $U(SO_4)_2 \cdot 4H_2O$
Molecular Formula: $H_8O_{12}S_2U$
Molecular Weight: 502.218
CAS RN: 13470-23-0

Properties: green, rhomb [LAN05]
Solubility: g/100 g H_2O: 10.1 (30°C), 9.0 (40°C), 7.7 (60°C) [LAN05]
Melting Point, °C: decomposes at 90 [CRC10]
Reactions: minus $4H_2O$ at 300°C [LAN05]

3392
Compound: Uranium(V,VI) oxide
Formula: U_3O_8
Molecular Formula: O_8U_3
Molecular Weight: 842.082
CAS RN: 1344-59-8
Properties: greenish black powd [STR93]
Density, g/cm³: 8.30 [STR93]
Melting Point, °C: 1150, decomposes to UO_2 [CRC10] [KIR91]

3393
Compound: Uranyl acetate dihydrate
Formula: $UO_2(CH_3COO)_2 \cdot 2H_2O$
Molecular Formula: $C_4H_{10}O_8U$
Molecular Weight: 424.147
CAS RN: 6159-44-0
Properties: yellow; cryst powd; slight odor of acetic acid; used as bacterial oxidation activator and in copying inks [MER06] [HAW93]
Solubility: s in 10 parts H_2O; sl s alcohol [MER06]
Density, g/cm³: 2.89 [MER06]
Reactions: minus $2H_2O$ at 110°C [CRC10]

3394
Compound: Uranyl acetylacetonate
Synonyms: 2,4-pentanedione, uranyl derivative
Formula: $UO_2(CH_3COCH=C(O)CH_3)_2$
Molecular Formula: $C_{10}H_{14}O_6U$
Molecular Weight: 468.248
CAS RN: 18039-69-5
Properties: cryst [AES93]

3395
Compound: Uranyl carbonate
Synonym: rutherfordine
Formula: UO_2CO_3
Molecular Formula: CO_5U
Molecular Weight: 330.037
CAS RN: 12202-79-8
Properties: naturally occurring mineral [KIR83]

3396
Compound: Uranyl chloride
Formula: UO_2Cl_2

Molecular Formula: Cl_2O_2U
Molecular Weight: 340.933
CAS RN: 7791-26-6
Properties: bright yellow cryst; ortho-rhomb; very hygr; very volatile >775°C [MER06]
Solubility: v s H_2O; s acetone, alcohol; i benzene [MER06]
Melting Point, °C: 578 [CRC10]
Boiling Point, °C: decomposes [CRC10]

3397

Compound: Uranyl chloride trihydrate
Formula: $UO_2Cl_2 \cdot 3H_2O$
Molecular Formula: Cl_2HeO_5U
Molecular Weight: 394.979
CAS RN: 13867-67-9
Properties: yellow powd; hygr [STR93]

3398

Compound: Uranyl fluoride
Formula: F_2O_2U
Molecular Formula: UO_2F_2
Molecular Weight: 308.025
CAS RN: 13536-84-0
Properties: yellow hygr solid [CRC10]
Solubility, g/100 g H_2O: 64.4^{20}; i bz [CRC10]

3399

Compound: Uranyl hydrogen phosphate tetrahydrate
Formula: $UO_2HPO_4 \cdot 4H_2O$
Molecular Formula: $H_9O_{10}PU$
Molecular Weight: 438.068
CAS RN: 18433-48-2
Properties: yellow; microcryst powd [MER06]
Solubility: i H_2O; s acids [MER06]

3400

Compound: Uranyl nitrate
Formula: $UO_2(NO_3)_2$
Molecular Formula: N_2O_8U
Molecular Weight: 283.698
CAS RN: 10102-06-4
Properties: yellow cryst [CRC10]
Solubility, g/100 g H_2O: 127^{25} [CRC10]

3401

Compound: Uranyl nitrate hexahydrate
Formula: $UO_2(NO_3)_2 \cdot 6H_2O$
Molecular Formula: $H_{12}N_2O_{14}U$
Molecular Weight: 502.129

CAS RN: 13520-83-7
Properties: yellow ortho-rhomb cryst; hygr; greenish luster by reflected light; can be prepared by heating a solution of uranyl nitrate to 188°C, then cooling to room temp [MER06] [STR93] [KIR83]
Solubility: g/100 g H_2O: 98 (0°C), 122 (20°C), 474 (100°C) [LAN05]; v s alcohol, ether [MER06], data are in [SIE94]
Density, g/cm³: 2.807 [MER06]
Melting Point, °C: 60.2 [STR93]
Boiling Point, °C: 118 [HAW93]

3402

Compound: Uranyl oxalate trihydrate
Formula: $UO_2C_2O_4 \cdot 3H_2O$
Molecular Formula: $C_2H_6O_9U$
Molecular Weight: 412.094
CAS RN: 22429-50-1
Properties: yellow cryst [CRC10]
Solubility: g anhydrous/100 g H_2O: 0.45 (10°C), 0.50 (20°C), 3.16 (100°C) [LAN05]
Reactions: minus H_2O, 110°C [CRC10]

3403

Compound: Uranyl sulfate
Formula: UO_2SO_4
Molecular Formula: O_6SU
Molecular Weight: 366.090
CAS RN: 1314-64-3
Properties: yellow cryst [CRC10]

3404

Compound: Uranyl sulfate monohydrate
Formula: $UO_2SO_4 \cdot H_2O$
Molecular Formula: H_2O_7SU
Molecular Weight: 384.107
CAS RN: 19415-82-8
Properties: stable up to 600°C; used to leach uranyl sulfate at this temp from sulfates of iron and aluminum [KIR83]

3405

Compound: Uranyl sulfate trihydrate
Formula: $UO_2SO_4 \cdot 3H_2O$
Molecular Formula: H_6O_9SU
Molecular Weight: 420.138
CAS RN: 12384-63-3
Properties: lemon yellow; cryst mass; property of stability at 600°C can be used in order to leach uranium from iron and aluminum at 600°C [KIR83] [MER06]

Solubility: s in ~5 parts H_2O, 25 parts alcohol [MER06]
Density, g/cm³: 3.28 [MER06]
Melting Point, °C: decomposes at 100 [CRC10]

3406
Compound: Vanadium
Formula: V
Molecular Formula: V
Molecular Weight: 50.9415
CAS RN: 7440-62-2
Properties: light gray or white, lustrous powd, or fused hard lumps; bcc, a = 0.3026 nm; enthalpy of fusion 21.50 kJ/mol; enthalpy of evaporation 458.6 kJ/mol; stable to moist air under typical conditions; electrical resistivity 24.2 μohm·cm at 20°C; Poisson's ratio 0.36; superconductivity transition 5.13 K; inert towards hot or cold HCl, cold H_2SO_4; uses include film resistors [MER06] [KIR83] [CER91] [CRC10]
Solubility: i H_2O; reacts with hot H_2SO_4, HF, HNO_3, aqua regia [MER06]
Density, g/cm³: 6.11 [MER06]
Melting Point, °C: 1890 [ALD94]
Boiling Point, °C: 3380 [ALD94]
Thermal Conductivity, W/(m·K): 30.7 at 25°C [ALD94]; 31 at 100°C [KIR83]
Thermal Expansion Coefficient: 8.3×10^{-6}/°C (23°C–100°C) [KIR83]

3407
Compound: Vanadium bis(cyclopentadienyl) dichloride
Synonym: bis(cyclopentadienyl)vanadium dichloride
Formula: $V(C_5H_5)_2Cl_2$
Molecular Formula: $C_{10}H_{10}Cl_2V$
Molecular Weight: 252.036
CAS RN: 12086-48-6
Properties: cryst [ALF95]
Density, g/cm³: 1.6 [ALF95]
Melting Point, °C: decomposes at 250 [ALF95]

3408
Compound: Vanadium carbide
Formula: VC
Molecular Formula: CV
Molecular Weight: 62.953
CAS RN: 12070-10-9
Properties: black cub cryst, a = 0.41355 nm; hardness 2800 kg/mm²; resistivity at room temp is 150 μohm·cm; used in alloys for cutting tools, and in 99.5% pure form as a sputtering target for producing wear-resistant films and semiconductor films [HAW93] [KIR83] [CIC73] [CER91]
Solubility: i H_2O; s HNO_3 with decomposition [KIR83]

Density, g/cm³: 5.77 [KIR83]
Melting Point, °C: 2810 [KIR83]
Boiling Point, °C: 3900 [KIR83]
Thermal Expansion Coefficient: 7.2×10^{-6}/K [KIR78]

3409
Compound: Vanadium carbonyl
Synonym: vanadium hexacarbonyl
Formula: $V(CO)_6$
Molecular Formula: C_6O_6V
Molecular Weight: 219.004
CAS RN: 20644-87-5
Properties: bluish green cryst; sensitive to atm O_2, pyromorphic; paramagnetic [MER06] [HAW93]
Melting Point, °C: 60–70, decomposes without melting [HAW93]
Boiling Point, °C: sublimes at 50 (15 mm Hg) [HAW93]

3410
Compound: Vanadium diboride
Formula: VB_2
Molecular Formula: B_2V
Molecular Weight: 72.564
CAS RN: 12007-37-3
Properties: refractory material; used as a 99.5% pure sputtering target to produce wear-resistant and semiconductive films [KIR78] [CER91]
Density, g/cm³: 5.100 [ALD94]
Melting Point, °C: 2450 [KIR78]

3411
Compound: Vanadium dibromide
Formula: VBr_2
Molecular Formula: Br_2V
Molecular Weight: 210.750
CAS RN: 14890-41-6
Properties: hex brownish orange cryst [LID94] [KIR83]
Density, g/cm³: 4.58 [LID94]

3412
Compound: Vanadium dichloride
Formula: VCl_2
Molecular Formula: Cl_2V
Molecular Weight: 121.847
CAS RN: 10580-52-6
Properties: apple green; hex plates; strong reducing agent; preparation: heating VCl_3 in N_2 atm, followed by sublimation in N_2 atm; used to purify HCl by removing arsenic [HAW93]

Solubility: decomposed in hot H_2O;
    s alcohol, ether [HAW93]
Density, g/cm³: 3.23 (18°C) [HAW93]

3413

Compound: Vanadium diiodide
Synonym: vanadium(II) iodide
Formula: VI_2
Molecular Formula: I_2V
Molecular Weight: 304.751
CAS RN: 15513-84-5
Properties: red hex [CRC10] [KIR83]
Density, g/cm³: 5.44 [CRC10]
Melting Point, °C: sublimes at 750–800 [CRC10]

3414

Compound: Vanadium dioxide
Synonym: vanadium(IV) oxide
Formula: V_2O_4
Molecular Formula: O_4V_2
Molecular Weight: 165.881
CAS RN: 12036-21-4
Properties: formula also VO_2; bluish black powd;
    slowly oxidizes in air; can be prepared
    by reacting V_2O_5 at its melting point with
    reductants such as sugar or oxalic acid; used
    as a high temp catalyst [HAW93] [KIR83]
Solubility: i H_2O; s acids, alkalies [HAW93]
Density, g/cm³: 4.339 [STR93]
Melting Point, °C: 1967 [STR93]

3415

Compound: Vanadium disilicide
Formula: VSi_2
Molecular Formula: Si_2V
Molecular Weight: 336.817
CAS RN: 12039-87-1
Properties: metallic prisms; –325 mesh; as a 99.5% pure
    material, used as a sputtering target in the fabrication
    of integrated circuits and as an electrochemical
    cathode [ALD94] [KIR83] [ALF93] [CER91]
Solubility: s HF [KIR83]
Density, g/cm³: 4.42 [KIR83]

3416

Compound: Vanadium gallide
Formula: V_3Ga
Molecular Formula: GaV_3
Molecular Weight: 222.548
CAS RN: 12024-15-6
Properties: –100 mesh of 99.5% purity;
    superconducting material [KIR83] [CER91]

3417

Compound: Vanadium monoboride
Formula: VB
Molecular Formula: BV
Molecular Weight: 61.753
CAS RN: 12045-27-1
Properties: refractory material; in the form of
    a 99.5% pure material used as a sputtering
    target to produce semiconductor and wear-
    resistant films [KIR78] [CER91]
Melting Point, °C: 2250 [KIR78]

3418

Compound: Vanadium monocarbide
Synonym: divanadium carbide
Formula: V_2C
Molecular Formula: CV_2
Molecular Weight: 113.894
CAS RN: 12012-17-8
Properties: hex, a=0.41655 nm, b=0.29020 nm,
    c=0.4577 nm [CIC73]
Melting Point, °C: 2167 [CIC73]

3419

Compound: Vanadium monosilicide
Formula: V_3Si
Molecular Formula: SiV_3
Molecular Weight: 180.911
CAS RN: 12039-76-8
Properties: cub cryst; superconducting; –100
    mesh; as a 99.5% pure sputtering target, used
    to produce resistant and semiconducting
    films in the fabrication of integrated
    circuits [LID94] [ALF93] [CER91]
Density, g/cm³: 5.70 [LID94]
Melting Point, °C: 1935 [ALF93]

3420

Compound: Vanadium monoxide
Synonym: vanadium(II) oxide
Formula: VO
Molecular Formula: OV
Molecular Weight: 66.941
CAS RN: 12035-98-2
Properties: –80 mesh powd; light green
    cryst; enthalpy of fusion 63.00 kJ/
    mol [CRC10] [KIR83] [STR93]
Solubility: s acids [KIR83]
Density, g/cm³: 5.758 [KIR83]
Melting Point, °C: 1790 [CRC10]

3421
Compound: Vanadium nitride
Formula: VN
Molecular Formula: NV
Molecular Weight: 64.949
CAS RN: 24646-85-3
Properties: black powd; fcc, a=0.4140 nm; hardness 9–10 Mohs; electrical resistivity 85 μohm·cm; transition temp 7.5 K; used as a 99.5% pure sputtering target to produce films [KIR81] [CIC73] [STR93] [CER91]
Solubility: i H_2O; s aqua regia [HAW93]
Density, g/cm³: 6.13 [STR93]
Melting Point, °C: 2320 [STR93]
Thermal Conductivity, W/(m·K): 11.3 [KIR81]
Thermal Expansion Coefficient: 8.1×10^{-6} [KIR81]

3422
Compound: Vanadium oxytrichloride
Synonym: vanadium(V) oxytrichloride
Formula: $VOCl_3$
Molecular Formula: Cl_3OV
Molecular Weight: 173.299
CAS RN: 7727-18-6
Properties: yellow liq; evolves red fumes in moist atm; can be used as a nonionizing solvent dissolving most nonmetals; hydrolyzes in moisture; enthalpy of vaporization 36.78 kJ/mol [CRC10] [HAW93] [MER06]
Solubility: decomposes in H_2O to vanadic acid and HCl; s methanol, ether, acetone, acids [KIR83] [MER06]
Density, g/cm³: 1.829 [STR93]
Melting Point, °C: –77 [MER06]
Boiling Point, °C: 126–127 [ALD94]

3423
Compound: Vanadium oxytrifluoride
Synonym: vanadium(V) oxytrifluoride
Formula: VOF_3
Molecular Formula: F_3OV
Molecular Weight: 123.936
CAS RN: 13709-31-4
Properties: yellowish orange powd; sensitive to moisture [STR93]
Density, g/cm³: 2.459 [STR93]
Melting Point, °C: 300 [STR93]
Boiling Point, °C: 480 [STR93]

3424
Compound: Vanadium pentafluoride
Synonym: vanadium(V) fluoride

Formula: VF_5
Molecular Formula: F_5V
Molecular Weight: 145.934
CAS RN: 7783-72-4
Properties: liq; etches glass slowly at room temp; appreciable vapor pressure at room temp; enthalpy of vaporization 44.52 kJ/mol; enthalpy of fusion 49.96 kJ/mol [CRC10] [MER06]
Solubility: hydrolyzed in H_2O, dil alkali; s anhydrous HF [MER06]
Density, g/cm³: 2.502 [MER06]
Melting Point, °C: 19.5 [MER06]

3425
Compound: Vanadium pentasulfide
Synonym: vanadium(V) sulfide
Formula: V_2S_5
Molecular Formula: S_5V_2
Molecular Weight: 262.213
CAS RN: 12138-17-9
Properties: greenish black powd; decomposes when heated [HAW93]
Density, g/cm³: 3.0 [CRC10]
Melting Point, °C: decomposes [CRC10]

3426
Compound: Vanadium pentoxide
Synonym: vanadium(V) oxide
Formula: V_2O_5
Molecular Formula: O_5V_2
Molecular Weight: 181.880
CAS RN: 1314-62-1
Properties: yellow to rust brown ortho-rhomb cryst; reversibly evolves O_2, 700°C–1125°C; enthalpy of fusion 64.50 kJ/mol [CRC10] [MER06]
Solubility: ~1 g/125 mL H_2O; s conc acids, forming red to yellow solutions [MER06]
Density, g/cm³: 3.357 [STR93]
Melting Point, °C: 690 [ALD94]
Boiling Point, °C: 1750, decomposes [HAW93]

3427
Compound: Vanadium sulfide
Formula: V_2S_2
Molecular Formula: S_2V_2
Molecular Weight: 166.015
CAS RN: 12138-08-8
Properties: black; formula also VS; used as a solid lubricant and as an electrode in lithium based batteries [HAW93] [CRC10]
Density, g/cm³: 4.2 [CRC10]
Melting Point, °C: decomposes [CRC10]

3428
Compound: Vanadium tetrachloride
Synonym: vanadium(IV) chloride
Formula: VCl_4
Molecular Formula: Cl_4V
Molecular Weight: 192.753
CAS RN: 7632-51-1
Properties: red liq; decomposes slowly to VCl_3
    and Cl_2 below 63°C; enthalpy of vaporization
    41.4 kJ/mol (bp), 42.5 kJ/mol (25°C); enthalpy
    of fusion 2.30 kJ/mol [CRC10] [HAW93]
Solubility: s alcohol, ether [HAW93]
Density, g/cm³: 1.816 [HAW93]
Melting Point, °C: −28 [ALD94]
Boiling Point, °C: 154 [ALD94]

3429
Compound: Vanadium tetrafluoride
Synonym: vanadium(IV) fluoride
Formula: VF_4
Molecular Formula: F_4V
Molecular Weight: 126.936
CAS RN: 10049-16-8
Properties: bright lime green powd; very
    hygr; disproportionates in vacuum to VF_3
    and VF_5 at 100°C–120°C [MER06]
Solubility: v s H_2O imparting blue color [MER06]
Density, g/cm³: 3.15 [MER06]
Melting Point, °C: 325, decomposes [STR93]

3430
Compound: Vanadium tribromide
Synonym: vanadium(III) bromide
Formula: VBr_3
Molecular Formula: Br_3V
Molecular Weight: 290.654
CAS RN: 13470-26-3
Properties: −20 mesh; black powd;
    sensitive to moisture [STR93]
Density, g/cm³: 4.0 [STR93]

3431
Compound: Vanadium trichloride
Synonym: vanadium(III) chloride
Formula: VCl_3
Molecular Formula: Cl_3V
Molecular Weight: 157.300
CAS RN: 7718-98-1
Properties: purple powd; sensitive to moisture;
    decomposes if heated; used to prepare
    organovanadium compounds [HAW93] [STR93]

Solubility: decomposes in H_2O;
    s alcohol, ether [HAW93]
Density, g/cm³: 3.00 [STR93]
Melting Point, °C: decomposes [KIR83]

3432
Compound: Vanadium trifluoride
Synonym: vanadium(III) fluoride
Formula: VF_3
Molecular Formula: F_3V
Molecular Weight: 107.937
CAS RN: 10049-12-4
Properties: greenish yellow powd; sublimes
    at bright red heat [MER06]
Solubility: i H_2O, alcohol [MER06]
Density, g/cm³: 3.363 [MER06]
Melting Point, °C: decomposes at 1406 [STR93]
Boiling Point, °C: sublimes at 800 [STR93]

3433
Compound: Vanadium trifluoride trihydrate
Formula: $VF_3 \cdot 3H_2O$
Molecular Formula: $F_3H_6O_3V$
Molecular Weight: 161.983
CAS RN: 10049-12-4
Properties: dark green; rhomb cryst [MER06]
Solubility: sl s H_2O [MER06]
Reactions: minus H_2O at 100°C [MER06]

3434
Compound: Vanadium trioxide
Synonym: vanadium(III) oxide
Formula: V_2O_3
Molecular Formula: O_3V_2
Molecular Weight: 149.881
CAS RN: 1314-34-7
Properties: black powd; gradually forms indigo blue
    cryst, V_2O_4, in air; used as a catalyst to convert
    ethylene to ethanol [HAW93] [MER06]
Solubility: i H_2O; s with difficulty in acids [MER06]
Density, g/cm³: 4.87 [MER06]
Melting Point, °C: 1940 [MER06]

3435
Compound: Vanadium trisulfate
Synonym: vanadium(III) sulfate
Formula: $V_2(SO_4)_3$
Molecular Formula: $O_{12}S_3V_2$
Molecular Weight: 390.074
CAS RN: 13701-70-7

Properties: lemon yellow powd; decomposes to VOSO$_4$ and SO$_2$, when heated ~410°C in vacuum; stable in dry air; strong reducing agent [MER06]
Solubility: very slowly dissolves in H$_2$O at room temp; s HNO$_3$ [MER06]
Melting Point, °C: decomposes ~400 [MER06]

3436
Compound: Vanadium trisulfide
Synonym: vanadium(III) sulfide
Formula: V$_2$S$_3$
Molecular Formula: S$_3$V$_2$
Molecular Weight: 198.081
CAS RN: 1315-03-3
Properties: −325 mesh; greenish black powd; heating causes decomposition [MER06] [STR93]
Solubility: i H$_2$O, cold HCl, dil H$_2$SO$_4$; s hot HCl, hot dil H$_2$SO$_4$, HNO$_3$ [MER06]
Density, g/cm^3: 4.7 [MER06]
Melting Point, °C: decomposes at >600 [CRC10]

3437
Compound: Vanadium(II) sulfate heptahydrate
Formula: VSO$_4 \cdot$ 7H$_2$O
Molecular Formula: H$_{14}$O$_{11}$SV
Molecular Weight: 273.112
CAS RN: 36907-42-3
Properties: violet monocl [CRC10] [KIR83]
Reactions: decomposes on heating in air [CRC10]

3438
Compound: Vanadium(III) acetylacetonate
Synonyms: 2,4-pentanedione, vanadium(III) derivative
Formula: V(CH$_3$COCH=C(O)CH$_3$)$_3$
Molecular Formula: C$_{15}$H$_{21}$O$_6$V
Molecular Weight: 348.270
CAS RN: 13476-99-8
Properties: brown cryst; sensitive to air [KIR83] [STR93]
Solubility: s methanol, acetone, benzene, chloroform [KIR83]
Density, g/cm^3: 0.9–1.2 [KIR83]
Melting Point, °C: 178–190 [KIR83]
Boiling Point, °C: sublimes at 170 (0.05 mm Hg) [STR93]

3439
Compound: Vanadocene
Synonym: bis(cyclopentadienyl)vanadium
Formula: V(C$_5$H$_5$)$_2$
Molecular Formula: C$_{10}$H$_{10}$V
Molecular Weight: 181.131
CAS RN: 1277-47-0

Properties: purple cryst; air and moisture sensitive [STR93]
Reactions: sublimes at 200°C (0.1 mm Hg) [STR93]

3440
Compound: Vanadocene dichloride
Formula: V(C$_5$H$_5$)$_2$Cl$_2$
Molecular Formula: C$_{10}$H$_{10}$Cl$_2$V
Molecular Weight: 252.034
CAS RN: 12083-48-6
Properties: dark green cryst [CRC10]
Solubility: s H$_2$O, chl, EtOH [CRC10]
Melting Point, °C: decomposes at 205 [CRC10]

3441
Compound: Vanadyl bromide
Formula: VOBr
Molecular Formula: BrOV
Molecular Weight: 146.845
CAS RN: 13520-88-2
Properties: violet [KIR83]
Density, g/cm^3: 4.0 [CRC10]
Melting Point, °C: decomposes at 482 [CRC10]

3442
Compound: Vanadyl chloride
Formula: VOCl
Molecular Formula: ClOV
Molecular Weight: 102.394
CAS RN: 13520-87-1
Properties: yellow-brown powd [CRC10] [KIR83]
Density, g/cm^3: 2.824 [CRC10]

3443
Compound: Vanadyl dibromide
Formula: VOBr$_2$
Molecular Formula: Br$_2$OV
Molecular Weight: 226.749
CAS RN: 13520-89-3
Properties: yellowish brown powd; deliq [CRC10] [KIR83]
Melting Point, °C: decomposes at 180 [CRC10]

3444
Compound: Vanadyl dichloride
Formula: VOCl$_2$
Molecular Formula: Cl$_2$OV
Molecular Weight: 137.846
CAS RN: 10213-09-9

Properties: green cryst; very deliq; disproportionates at 384°C to VOCl and VOCl₃ [MER06]
Solubility: slowly decomposed in water; s absolute alcohol, glacial acetic acid [MER06]
Density, g/cm³: 2.88 [MER06]

3445
Compound: Vanadyl difluoride
Formula: VOF₂
Molecular Formula: F₂OV
Molecular Weight: 104.938
CAS RN: 13814-83-0
Properties: yellow [KIR83]
Density, g/cm³: 3.396 [CRC10]
Melting Point, °C: decomposes [CRC10]

3446
Compound: Vanadyl selenite monohydrate
Formula: VOSeO₃·H₂O
Molecular Formula: H₂O₅SeV
Molecular Weight: 211.915
CAS RN: 133578-89-9
Properties: green plates; tric, a = 0.5969 nm, b = 0.6155 nm, c = 0.6349 nm; has magnetic properties; can be prepared by heating H₂SeO₃ and V₂O₅ in an autoclave at 200°C for ~48 h; material selectively intercalates alcohols [HUA91]
Density, g/cm³: 3.506 [HUA91]
Reactions: minus H₂O at 240°C–280°C [HUA91]

3447
Compound: Vanadyl sulfate dihydrate
Formula: VOSO₄·2H₂O
Molecular Formula: H₄O₇SV
Molecular Weight: 199.036
CAS RN: 27774-13-6
Properties: blue cryst powd; used as a mordant, catalyst reducing agent, colorant in glasses and ceramics; there is a trihydrate, CAS RN 12210-47-8 [HAW93] [MER06] [ALD94]
Solubility: s H₂O [MER06]

3448
Compound: Vanadyl tribromide
Formula: VOBr₃
Molecular Formula: Br₃OV
Molecular Weight: 306.653
CAS RN: 13520-90-6
Properties: deep red liq [KIR83]
Density, g/cm³: 2.933 [CRC10]
Boiling Point, °C: 130 [CRC10]
Reactions: decomposes at 180°C [CRC10]

3449
Compound: Vitreous silica
Formula: SiO₂
Molecular Formula: O₂Si
Molecular Weight: 60.085
CAS RN: 60676-86-0
Properties: hardness 5.5 Mohs; velocity of sound 5730 m/s [CIC73]
Density, g/cm³: 2.1957 at 0°C [CIC73]
Melting Point, °C: ~1500 [CIC73]
Boiling Point, °C: 2950 [CIC73]
Thermal Conductivity, W/(m·K): 1.423 (100°C) [CIC73]

3450
Compound: Water
Synonym: hydrogen oxide
Formula: H₂O
Molecular Formula: H₂O
Molecular Weight: 18.015
CAS RN: 7732-18-5
Properties: colorless, odorless, tasteless liq; dielectric constant 78.54; viscosity 1.005 cp (20°C); vapor pressure 760 mm Hg (100°C); triple point 273.16 K at 4.6 mm Hg; surface tension 73 dyne/cm (20°C); enthalpy of fusion (ice) 6.008 kJ/mol; enthalpy of vaporization 40.65 kJ/mol; critical temp 374.2°C [DOU83] [HAW93] [MER06]
Density, g/cm³: 0.9970 [LID94]
Melting Point, °C: 0 [MER06]
Boiling Point, °C: 100 [MER06]
Thermal Conductivity, W/(m·K): values from 20°C to 330°C are found in [OZB80]

3451
Compound: Xenon
Formula: Xe
Molecular Formula: Xe
Molecular Weight: 131.29
CAS RN: 7440-63-3
Properties: colorless, odorless gas; enthalpy of vaporization 12.62 kJ/mol; enthalpy of fusion 1.81 kJ/mol; critical temp 16.6°C; critical pressure 5.84 MPa; sonic velocity (101.32 kPa, 0°C) 168 m/s; viscosity (101.32 kPa, 25°C) 23.1 Pa·s; dielectric constant 1.0012 at 25°C and 1 atm; used in flash lamps, lasers, anesthesia [HAW93] [KIR78] [AIR87] [CRC10]
Solubility: 101.32 kPa: 108.1 mL/1000 g H₂O (20°C) [KIR78]; Henry's law constants, k × 10⁻⁴: 2.558 (70.3°C), 2.586 (125.5°C), 2.485 (175.7), 2.048 (225.1°C), 1.308 (284.2°C) [POT78]
Density, g/cm³: gas: 101.3 kPa, 0°C, 0.0058971 [KIR78]

Melting Point, °C: −111.75 [CRC10]
Boiling Point, °C: −108.05 [CRC10]
Thermal Conductivity, W/(m·K): gas
(101.32 kPa, 0°C) 0.00565 [ALD94]

3452
Compound: Xenon difluoride
Synonym: xenon fluoride
Formula: XeF$_2$
Molecular Formula: F$_2$Xe
Molecular Weight: 169.287
CAS RN: 13709-36-9
Properties: white stable cryst; powerful oxidizing
agent; enthalpy of sublimation 55.73 kJ/mol; tetr,
a = 0.4315 nm, c = 0.6990 nm; obtained from F$_2$
and Xe under high pressure [DOU83] [KIR78]
Solubility: 25 g/L H$_2$O (0°C) [MER06]; hydrolyzes
to Xe + O$_2$; v s liq HF [COT88]
Density, g/cm³: 4.32 [KIR78]
Melting Point, °C: 129.03 [KIR78]
Boiling Point, °C: sublimes without
decomposition [MER06]

3453
Compound: Xenon dioxydifluoride
Formula: XeO$_2$F$_2$
Molecular Formula: F$_2$O$_2$Xe
Molecular Weight: 201.286
CAS RN: 13875-06-4
Properties: colorless; ortho-rhomb, a = 0.6443 nm,
b = 0.6288 nm, c = 0.8312 nm [KIR78]
Density, g/cm³: 4.10 [KIR78]
Melting Point, °C: 30.8 explodes [KIR78]

3454
Compound: Xenon fluoride hexafluoroantimonate
Formula: XeF$_3$SbF$_6$
Molecular Formula: F$_9$SbXe
Molecular Weight: 424.036
CAS RN: 39797-63-2
Properties: yellowish green; monocl, a = 0.5394 nm,
b = 1.5559 nm, c = 0.5394 nm [KIR78]
Density, g/cm³: 3.92 [KIR78]
Melting Point, °C: 109–113 [KIR78]

3455
Compound: Xenon fluoride hexafluoroarsenate
Formula: Xe$_2$F$_3$AsF$_6$
Molecular Formula: AsF$_9$Xe$_2$
Molecular Weight: 377.198
CAS RN: 50432-32-1

Properties: yellowish green; monocl, a = 1.5443 nm,
b = 0.8678 nm, c = 2.0888 nm [KIR78]
Density, g/cm³: 3.62 [KIR78]
Melting Point, °C: 99 [KIR78]

3456
Compound: Xenon fluoride hexafluororuthenate
Formula: XeFRuF$_6$
Molecular Formula: F$_7$RuXe
Molecular Weight: 365.349
CAS RN: 22527-13-5
Properties: yellowish green; monocl, a = 0.7991 nm,
b = 1.1086 nm, c = 0.7250 nm [KIR78]
Density, g/cm³: 3.78 [KIR78]
Melting Point, °C: 110–111 [KIR78]

3457
Compound: Xenon fluoride monodecafluoroantimonate
Formula: XeFSb$_2$F$_{11}$
Molecular Formula: F$_{12}$Sb$_2$Xe
Molecular Weight: 602.785
CAS RN: 15364-10-0
Properties: yellow; monocl, a = 0.807 nm,
b = 0.955 nm, c = 0.733 nm [KIR78]
Density, g/cm³: 3.69 [KIR78]
Melting Point, °C: 63 [KIR78]

3458
Compound: Xenon hexafluoride
Formula: XeF$_6$
Molecular Formula: F$_6$Xe
Molecular Weight: 245.280
CAS RN: 13693-09-9
Properties: colorless solid; has greenish yellow
vapor; monocl, a = 0.933 nm, b = 1.096 nm,
c = 0.895 nm; enthalpy of sublimation
59.12 kJ/mol; very strong oxidizing agent;
preparation: reaction between Xe and excess
F$_2$ at 250°C [DOU83] [MER06] [KIR78]
Solubility: hydrolyzed in H$_2$O, forming
XeOF$_4$ and XeO$_3$ [MER06]
Density, g/cm³: 3.56 [KIR78]
Melting Point, °C: 49.48 [KIR78]
Boiling Point, °C: 75.57 [MER06]

3459
Compound: Xenon oxydifluoride
Formula: XeOF$_2$
Molecular Formula: F$_2$OXe
Molecular Weight: 185.286

CAS RN: 13780-64-8
Properties: yellow; formed as an unstable product of the partial hydrolysis of XeF_4 [KIR78]
Melting Point, °C: explodes ~0 [KIR78]

3460
Compound: Xenon oxytetrafluoride
Formula: $XeOF_4$
Molecular Formula: F_4OXe
Molecular Weight: 223.283
CAS RN: 13774-85-1
Properties: colorless volatile stable liq at 25°C [KIR78]
Melting Point, °C: −46.2 [KIR78]

3461
Compound: Xenon pentafluoride hexafluoroarsenate
Formula: XeF_5AsF_6
Molecular Formula: $AsF_{11}Xe$
Molecular Weight: 415.194
CAS RN: 20328-94-3
Properties: white; monocl, a = 0.5886 nm, b = 1.6564 nm, c = 0.8051 nm [KIR78]
Density, g/cm³: 3.51 [KIR78]
Melting Point, °C: 130.5 [KIR78]

3462
Compound: Xenon pentafluoride hexafluororuthenate
Formula: XeF_5RuF_6
Molecular Formula: $F_{11}RuXe$
Molecular Weight: 441.342
CAS RN: 39796-98-0
Properties: green; ortho-rhomb, a = 1.6771 nm, b = 0.8206 nm, c = 0.5617 nm [KIR78]
Density, g/cm³: 3.79 [KIR78]
Melting Point, °C: 152 [KIR78]

3463
Compound: Xenon tetrafluoride
Formula: XeF_4
Molecular Formula: F_4Xe
Molecular Weight: 207.284
CAS RN: 13709-61-0
Properties: colorless cryst; readily prepared by mixing fluorine and xenon; enthalpy of sublimation 60.92 kJ/mol; monocl, a = 0.5050 nm, b = 0.5922 nm, c = 0.5771 nm [KIR78]
Solubility: reacts violently with H_2O, forming Xe, O_2, HF, and XeO_3 [DOU83]
Density, g/cm³: 4.04 [KIR78]
Melting Point, °C: 117.10 [KIR78]

3464
Compound: Xenon tetroxide
Formula: XeO_4
Molecular Formula: O_4Xe
Molecular Weight: 195.288
CAS RN: 12340-14-6
Properties: yellow solid; unstable, can explode [KIR78]
Melting Point, °C: decomposes at <0 [KIR78]

3465
Compound: Xenon trifluoride monodecafluoroantimonate
Formula: $XeF_3Sb_2F_{11}$
Molecular Formula: $F_{14}Sb_2Xe$
Molecular Weight: 640.776
CAS RN: 35718-37-7
Properties: yellowish green; tric, a = 0.8237 nm, b = 0.9984 nm, c = 0.8004 nm [KIR78]
Density, g/cm³: 3.98 [KIR78]
Melting Point, °C: 81–83 [KIR78]

3466
Compound: Xenon trioxide
Formula: XeO_3
Molecular Formula: O_3Xe
Molecular Weight: 179.288
CAS RN: 13776-58-4
Properties: colorless solid; hygr; strongly explosive; ortho-rhomb, a = 0.6163 nm, b = 0.8115 nm, c = 0.5234 nm [MER06] [KIR78]
Solubility: s H_2O [KIR78]
Density, g/cm³: 4.55 [KIR78]
Melting Point, °C: explodes ~25 [KIR78]
Reactions: forms $HXeO_4$ in basic solutions [DOU83]

3467
Compound: Ytterbium
Formula: Yb
Molecular Formula: Yb
Molecular Weight: 173.04
CAS RN: 7440-64-4
Properties: silvery, ductile metal; fcc, room temp; bcc >798°C; enthalpy of fusion 7.657 kJ/ mol; enthalpy of sublimation 152.1 kJ/mol; atom radius 0.19392 nm; ion radius 0.0858 nm, Yb^{+++}, colorless; electrical resistivity (20°C) 28 μohm·cm [MER06] [KIR82] [ALD94]
Solubility: slowly reacts with H_2O; s dil acids, ammonia [HAW93]
Density, g/cm³: fcc: 6.9654 [KIR82]; bcc: 6.54 [MER06]
Melting Point, °C: 819 [KIR82]
Boiling Point, °C: 1196 [KIR82]

Thermal Conductivity, W/(m·K): 34.9 (25°C) [CRC10]
Thermal Expansion Coefficient: 26.3×10^{-6}/K [CRC10]

3468
Compound: Ytterbium acetate tetrahydrate
Formula: $Yb(CH_3COO)_3 \cdot 4H_2O$
Molecular Formula: $C_6H_{17}O_{10}Yb$
Molecular Weight: 422.235
CAS RN: 15280-58-7
Properties: white hygr powd; cryst aggregates [AES93] [STR93] [ALD94]
Density, g/cm³: 2.09 [STR93]
Reactions: minus $4H_2O$ at 100°C [CRC10]

3469
Compound: Ytterbium acetylacetonate
Synonyms: 2,4-pentanedione, ytterbium(III) derivative
Formula: $Yb(CH_3COCH=C(O)CH_3)_3$
Molecular Formula: $C_{15}H_{21}O_6Yb$
Molecular Weight: 470.368
CAS RN: 14284-98-1
Properties: powd [STR93]

3470
Compound: Ytterbium bromide hydrate
Formula: $YbBr_3 \cdot xH_2O$
Molecular Formula: Br_3Yb (anhydrous)
Molecular Weight: 412.752 (anhydrous)
CAS RN: 15163-03-8
Properties: white cryst; anhydrous YBr_3, 13759-89-2, −20 mesh with 99.9% purity [STR93] [CER91]
Melting Point, °C: 956 [AES93] (anhydrous)

3471
Compound: Ytterbium carbonate hydrate
Formula: $Yb_2(CO_3)_3 \cdot xH_2O$
Molecular Formula: $C_3O_9Yb_2$ (anhydrous)
Molecular Weight: 526.112 (anhydrous)
CAS RN: 64360-98-1
Properties: cryst [AES93]
Melting Point, °C: decomposes [AES93]

3472
Compound: Ytterbium chloride
Formula: $YbCl_3$
Molecular Formula: Cl_3Yb
Molecular Weight: 279.398
CAS RN: 10361-91-8
Properties: −20 mesh with 99.9% purity; white powd; hygr [STR93] [CER91]
Melting Point, °C: 875 [LID94]

3473
Compound: Ytterbium chloride hexahydrate
Formula: $YbCl_3 \cdot 6H_2O$
Molecular Formula: $Cl_3H_{12}O_6Yb$
Molecular Weight: 387.489
CAS RN: 10035-01-5
Properties: −4 mesh with 99.9% purity; green cryst; hygr [HAW93] [CER91]
Solubility: v s H_2O [HAW93]
Density, g/cm³: 2.575 [MER06]
Melting Point, °C: 865 [HAW93]
Reactions: minus $6H_2O$ at 180°C [HAW93]

3474
Compound: Ytterbium fluoride
Formula: YbF_3
Molecular Formula: F_3Yb
Molecular Weight: 230.035
CAS RN: 13760-80-0
Properties: 3–12 mm fused pieces with 99.9% purity; white powd; hygr [STR93] [CER91]
Solubility: i H_2O [HAW93]
Density, g/cm³: 8.168 [STR93]
Melting Point, °C: 1157 [STR93]
Boiling Point, °C: 2200 [STR93]

3475
Compound: Ytterbium hydride
Formula: YbH_3
Molecular Formula: H_3Yb
Molecular Weight: 176.064
CAS RN: 32997-62-9
Properties: in the form of lumps, in ampoule under Ar [AES93]

3476
Compound: Ytterbium nitrate pentahydrate
Formula: $Yb(NO_3)_3 \cdot 5H_2O$
Molecular Formula: $H_{10}N_3O_{14}Yb$
Molecular Weight: 449.131
CAS RN: 35725-34-9
Properties: white cryst [STR93] [ALD94]

3477
Compound: Ytterbium oxalate decahydrate
Formula: $Yb_2(C_2O_4)_3 \cdot 10H_2O$
Molecular Formula: $C_6H_{20}O_{22}Yb_2$
Molecular Weight: 790.292
CAS RN: 51373-68-3
Properties: white cryst [STR93]

Solubility: 0.0001 g/100 mL H_2O [CRC10]
Density, g/cm³: 2.644 [CRC10]

3478

Compound: Ytterbium oxide
Synonym: ytterbia
Formula: Yb_2O_3
Molecular Formula: O_3Yb_2
Molecular Weight: 394.078
CAS RN: 1314-37-0
Properties: if pure, colorless mass; brownish or yellowish
    if thulia is present, also 3–12 mm sintered pieces
    of 99.9% purity; sl hygr; absorbs atm H_2O and
    NH_3; used in special alloys, dielectric ceramics,
    and special glasses, also sintered pieces used as
    evaporation material to form film with reactivity to
    radio frequencies [HAW93] [MER06] [CER91]
Solubility: s dil acids [MER06]
Density, g/cm³: 9.2 [HAW93]
Melting Point, °C: 2346 [HAW93]

3479

Compound: Ytterbium perchlorate
Formula: $Yb(ClO_4)_3$
Molecular Formula: $Cl_3O_{12}Yb$
Molecular Weight: 471.390
CAS RN: 13498-08-3
Properties: off-white cryst [STR93]

3480

Compound: Ytterbium silicide
Formula: $YbSi_2$
Molecular Formula: Si_2Yb
Molecular Weight: 229.211
CAS RN: 12039-89-3
Properties: hex cryst; 10 mm & down
    lump [LID94] [ALF93]
Density, g/cm³: 7.54 [LID94]

3481

Compound: Ytterbium sulfate
Formula: $Yb_2(SO_4)_3$
Molecular Formula: $O_{12}S_3Yb_2$
Molecular Weight: 634.271
CAS RN: 10034-98-7
Properties: colorless cryst [CRC10]
Solubility: g/100 g H_2O: 44.2 (0°C), 22.2
    (30°C), 4.7 (100°C) [LAN05]
Density, g/cm³: 3.793 [LAN05]
Melting Point, °C: decomposes at 900 [LAN05]

3482

Compound: Ytterbium sulfate octahydrate
Formula: $Yb_2(SO_4)_3 \cdot 8H_2O$
Molecular Formula: $H_{16}O_{20}S_3Yb_2$
Molecular Weight: 778.393
CAS RN: 10034-98-7
Properties: lustrous, colorless cryst [MER06]
Solubility: s H_2O, solubility decreases
    with temp increase [MER06]
Density, g/cm³: 3.286 [STR93]

3483

Compound: Yttrium
Formula: Y
Molecular Formula: Y
Molecular Weight: 88.90585
CAS RN: 7440-65-5
Properties: white; silvery metal; hex close-packed
    cryst; enthalpy of fusion 11.43 kJ/mol; enthalpy
    of sublimation 424.7 kJ/mol; radius of atom
    0.1801 nm; ion radius of Y^{+++} 0.0893 nm aq
    solutions are colorless; electrical resistivity
    (20°C) 57 μohm · cm [KIR82] [ALD94]
Solubility: decomposes in cold H_2O, more
    rapidly in hot H_2O [MER06]; s dil
    acids, KOH solutions [HAW93]
Density, g/cm³: 4.4689 [KIR82]
Melting Point, °C: 1522 [KIR82]
Boiling Point, °C: 3338 [KIR82]
Thermal Conductivity, W/(m · K): 17.2 (25°C) [CRC10]
Thermal Expansion Coefficient: 10.6×10^{-6}/K [CRC10]

3484

Compound: Yttrium acetate hydrate
Formula: $Y(CH_3COO)_3 \cdot xH_2O$
Molecular Formula: $C_6H_9O_6Y$ (anhydrous)
Molecular Weight: 266.039 (anhydrous)
CAS RN: 23363-14-6
Properties: white cryst; x = 4, Molecular
    Weight 338.10 [AES93] [STR93]
Melting Point, °C: decomposes [AES93]

3485

Compound: Yttrium acetylacetonate trihydrate
Synonyms: 2,4-pentanedione, yttrium(III) derivative
Formula: $Y(CH_3COCH=C(O)CH_3)_3 \cdot 3H_2O$
Molecular Formula: $C_{15}H_{27}O_9Y$
Molecular Weight: 440.280
CAS RN: 15554-47-9
Properties: yellowish white cryst [STR93]
Melting Point, °C: 138–140 [STR93]

3486
Compound: Yttrium aluminum oxide
Synonym: YAG
Formula: $Y_3Al_5O_{12}$
Molecular Formula: $Al_5O_{12}Y_3$
Molecular Weight: 593.619
CAS RN: 12005-21-9
Properties: green cub cryst; 3–12 mm fused pieces; used in the form of a 99.99% pure material as a sputtering target to prepare bubble memory devices [CER91] [LID94]
Density, g/cm³: ~4.5 [LID94]

3487
Compound: Yttrium antimonide
Formula: YSb
Molecular Formula: SbY
Molecular Weight: 210.666
CAS RN: 12186-97-9
Properties: cub cryst; high purity semiconductor [HAW93] [LID94]
Density, g/cm³: 5.97 [LID94]
Melting Point, °C: 2310 [LID94]

3488
Compound: Yttrium arsenide
Formula: YAs
Molecular Formula: AsY
Molecular Weight: 163.828
CAS RN: 12255-48-0
Properties: cub cryst; high purity semiconductor [LID94] [HAW93]
Density, g/cm³: 5.59 [LID94]

3489
Compound: Yttrium barium copper oxide
Synonyms: supercon N-124, O-124
Formula: $YBa_2Cu_4O_8$
Molecular Formula: $Ba_2Cu_4O_8Y$
Molecular Weight: 745.739
CAS RN: 107539-20-8
Properties: N-124: 0.2 μm powd; wet processed from ACS grade nitrates; O-124: 20 μm powd; dry processed from ACS grade oxides; T_c is 81 K [STR93] [ASM93]

3490
Compound: Yttrium barium copper oxide
Synonym: supercon N-123
Formula: $YBa_2Cu_3O_7$

Molecular Formula: $Ba_2Cu_3O_7Y$
Molecular Weight: 666.194
CAS RN: 107539-20-8
Properties: ortho-rhomb; superconductor; 0.2 μm powd; wet processed from 99.999% nitrates; a hard grayish black sintered material prepared by reacting carbonate and hydroxide free Y_2O_3 with stoichiometric amounts of $BaCO_3$ and CuO in air at 950°C for 12 h; sensitive to atm CO_2 and H_2O; a = 0.3835 nm, b = 0.3884 nm, c = 1.1681 nm; transition temp, T_c, is 92 K; for $YBa_2Cu_{3.5}O_{7.5}$, T_c is 94 K [STR93] [CON87] [CEN92] [ASM93] [ALF93]

3491
Compound: Yttrium barium copper oxide
Synonym: supercon O-123
Formula: $YBa_2Cu_3O_7$
Molecular Formula: $Ba_2Cu_3O_7Y$
Molecular Weight: 666.194
CAS RN: 107539-20-8
Properties: 20 μm powd; dry processed from ACS grade oxides; T_c is 93 K for $YBa_2Cu_3O_7$; interplanar spacing 0.58 nm; effective mass to free electron mass ratio 2–2.5; coherence length parallel and perpendicular to conduction plane 1.5 nm and 0.15–3 nm; mean free path of charge carrier above T_c is 10.0 nm; penetration depth 145.0 nm; carrier density $3.1 \times 10^{+21}$/cm³; for x = 7 – δ, T_c is 92 K [CEN92] [STR93] [ASM93]
Reactions: metastable at low temperatures [CEN92]

3492
Compound: Yttrium boride
Synonym: yttrium hexaboride
Formula: YB_6
Molecular Formula: B_6Y
Molecular Weight: 153.772
CAS RN: 12008-32-1
Properties: −325 mesh 10 μm or less with 99.9% purity; refractory material [CER91] [KIR78]
Density, g/cm³: 3.2 [LID94]
Melting Point, °C: decomposes at 2600 [KIR78]

3493
Compound: Yttrium bromide
Formula: YBr_3
Molecular Formula: Br_3Y
Molecular Weight: 328.618
CAS RN: 13469-98-2
Properties: deliq; −20 mesh with 99.9% purity [CRC10] [CER91]

Solubility: g/100 g H_2O: 63.9 (0°C), 75.1
    (20°C), 123 (90°C) [LAN05]
Melting Point, °C: 904 [HAW93]

3494
Compound: Yttrium bromide nonahydrate
Formula: $YBr_3 \cdot 9H_2O$
Molecular Formula: $Br_3H_{18}O_9Y$
Molecular Weight: 490.756
CAS RN: 13469-98-2
Properties: colorless cryst [HAW93]
Solubility: s H_2O, alcohol; i ether [HAW93]

3495
Compound: Yttrium carbide
Formula: YC_2
Molecular Formula: C_2Y
Molecular Weight: 112.928
CAS RN: 12071-35-1
Properties: yellow; 12 mm pieces and smaller
    of 99.5% purity [CER91] [CRC10]
Density, g/cm³: 4.13 [CRC10]
Melting Point, °C: ~2400 [LID94]

3496
Compound: Yttrium carbonate trihydrate
Formula: $Y_2(CO_3)_3 \cdot 3H_2O$
Molecular Formula: $C_3H_6O_{12}Y_2$
Molecular Weight: 411.886
CAS RN: 5970-44-5
Properties: white to reddish white
    powd [MER06] [STR93]
Solubility: i H_2O; s dil mineral acids [MER06]

3497
Compound: Yttrium chloride
Formula: YCl_3
Molecular Formula: Cl_3Y
Molecular Weight: 195.264
CAS RN: 10361-92-9
Properties: −20 mesh with 99.9% purity;
    white powd; hygr [STR93] [CER91]
Solubility: g/100 g H_2O: 77.3 (0°C), 78.8
    (20°C), 80.8 (40°C) [LAN05]
Density, g/cm³: 2.67 [STR93]
Melting Point, °C: 721 [STR93]
Boiling Point, °C: 1507 [CRC10]

3498
Compound: Yttrium chloride hexahydrate
Formula: $YCl_3 \cdot 6H_2O$

Molecular Formula: $Cl_3H_{12}O_6Y$
Molecular Weight: 303.355
CAS RN: 10025-94-2
Properties: −4 mesh with 99.9% purity; colorless
    cryst; deliq [MER06] [CER91]
Solubility: 217 g/100 mL H_2O (20°C),
    235 g/100 mL H_2O (50°C) [CRC10]
Density, g/cm³: 2.18 [STR93]
Reactions: minus $6H_2O$ by heating
    in HCl stream [MER06]

3499
Compound: Yttrium fluoride
Formula: YF_3
Molecular Formula: F_3Y
Molecular Weight: 145.901
CAS RN: 13709-49-4
Properties: white powd; hygr; in the form of a
    99.9% pure material, is used as a sputtering
    target for multilayers [STR93] [CER91]
Density, g/cm³: 4.01 [STR93]
Melting Point, °C: 1152 [STR93]

3500
Compound: Yttrium hexafluoroacetylacetonate
Synonyms: 1,1,1,5,5,5-hexafluoro-2,4-
    pentanedione, yttrium derivative
Formula: $Y(CF_3COCHCOCF_3)_3$
Molecular Formula: $C_{15}H_3F_{18}O_6Y$
Molecular Weight: 710.062
CAS RN: 18911-76-7
Properties: white cryst [STR93]
Melting Point, °C: 166–170 [STR93]
Boiling Point, °C: decomposes at 240 [STR93]
Reactions: sublimes at 100°C (0.2 mm Hg) [STR93]

3501
Compound: Yttrium hydride
Formula: YH_3
Molecular Formula: H_3Y
Molecular Weight: 91.929
CAS RN: 13598-57-7
Properties: −60 mesh of 99.9% purity
    and lumps [CER91] [AES93]

3502
Compound: Yttrium hydroxide
Formula: $Y(OH)_3$
Molecular Formula: H_3O_3Y
Molecular Weight: 139.928
CAS RN: 16469-22-0

Properties: white; gelatinous precipitate; dries to a white powd, which absorbs atm CO_2 [MER06]
Reactions: decomposes on heating [CRC10]

3503
Compound: Yttrium iodide
Formula: YI_3
Molecular Formula: I_3Y
Molecular Weight: 469.619
CAS RN: 13470-38-7
Properties: white deliq flakes [CRC10] [AES93]
Melting Point, °C: 1004 [AES93]

3504
Compound: Yttrium iron oxide
Synonym: yttrium garnet
Formula: $Y_3Fe_5O_{12}$
Molecular Formula: $Fe_5O_{12}Y_3$
Molecular Weight: 737.936
CAS RN: 12063-56-8
Properties: used as 99.99% and 99.9% pure sputtering target to prepare ferromagnetic films and in bubble memory devices [CER91]

3505
Compound: Yttrium nitrate hexahydrate
Formula: $Y(NO_3)_3 \cdot 6H_2O$
Molecular Formula: $H_{12}N_3O_{15}Y$
Molecular Weight: 383.012
CAS RN: 13494-98-9
Properties: white deliq cryst [MER06] [STR93]
Solubility: g anhydrous/100 g H_2O: 93.1 (0°C), 123 (20°C), 200 (60°C) [LAN05]; partially decomposed in water to basic nitrates [MER06]; 5.2759 ± 0.0009 mol/(kg $\cdot$ H_2O) at 25°C [RAR85b]
Density, g/cm³: 2.68 [CRC10]
Reactions: minus $3H_2O$ at 100°C [CRC10]

3506
Compound: Yttrium oxalate nonahydrate
Formula: $Y_2(C_2O_4)_3 \cdot 9H_2O$
Molecular Formula: $C_6H_{18}O_{21}Y_2$
Molecular Weight: 604.008
CAS RN: 13266-82-5
Properties: white cryst [STR93]
Solubility: 0.0001 g/100 mL H_2O [CRC10]
Melting Point, °C: decomposes [AES93]

3507
Compound: Yttrium oxide
Synonym: yttria

Formula: Y_2O_3
Molecular Formula: O_3Y_2
Molecular Weight: 225.810
CAS RN: 1314-36-9
Properties: white powd or sintered tablets and pieces of 99.9% purity; bcc; readily absorbs atm CO_2; enthalpy of fusion 105.00 kJ/mol; used in crucible form for experimental, proprietary melting, also sintered pieces used as evaporation material for hard film dielectric coating and thin film capacitors, and as 99.999%, 99.99%, 99.9% pure sputtering target for preparing hard films, dielectric coatings, and thin film capacitor [CER91] [MER06] [CRC10]
Solubility: s dil acids [MER06]
Density, g/cm³: 5.03 [MER06]
Melting Point, °C: 2439 [CRC10]

3508
Compound: Yttrium perchlorate hexahydrate
Formula: $Y(ClO_4)_3 \cdot 6H_2O$
Molecular Formula: $Cl_3H_{12}O_{18}Y$
Molecular Weight: 495.348
CAS RN: 14017-56-2
Properties: white cryst [STR93]

3509
Compound: Yttrium phosphide
Formula: YP
Molecular Formula: PY
Molecular Weight: 119.880
CAS RN: 12294-01-8
Properties: cub cryst; high purity semiconductor [LID94] [HAW93]
Density, g/cm³: ~4.4 [LID94]

3510
Compound: Yttrium sulfate octahydrate
Formula: $Y_2(SO_4)_3 \cdot 8H_2O$
Molecular Formula: $H_{16}O_{20}S_3Y_2$
Molecular Weight: 610.125
CAS RN: 7446-33-5
Properties: small, reddish white; monocl cryst [MER06] [HAW93]
Solubility: g/100 g H_2O: 8.05 (0°C), 7.30 (20°C), 2.2 (90°C) [LAN05]; s conc H_2SO_4 reacting to produce $Y(HSO_4)_3$ [MER06]
Density, g/cm³: 2.558 [STR93]
Reactions: minus $8H_2O$ at 120°C [HAW93]

3511
Compound: Yttrium sulfide
Formula: Y_2S_3

Molecular Formula: S_3Y_2
Molecular Weight: 274.010
CAS RN: 12039-19-9
Properties: −200 mesh with 99.9% purity;
    yellow powd [STR93] [CER91]
Density, g/cm³: 3.87 [LID94]
Melting Point, °C: 1925 [LID94]

3512
Compound: Yttrium vanadate
Formula: YVO_3
Molecular Formula: O_3VY
Molecular Weight: 187.845
CAS RN: 12143-39-4
Properties: white cryst; 99.9% pure, doped with
    Eu_2O_3; used in phosphorescent coating on
    special currency papers, and as red phosphor
    in television tubes [HAW93] [CER91]

3513
Compound: Zinc
Formula: Zn
Molecular Formula: Zn
Molecular Weight: 65.39
CAS RN: 7440-66-6
Properties: lustrous bluish white; hex closed-packed
    metal; reacts with atm moisture producing surface of
    basic zinc carbonate; hardness 2.5 Mohs; electrical
    resistivity, 20°C, 5.8 μohm·cm; malleable when
    heated to 100°C–150°C; brittle at 210°C; enthalpy
    of fusion 7.387 kJ/mol; enthalpy of vaporization
    114.8 kJ/mol; uses include evaporating metal
    for metallized paper and capacitor dielectric
    films [KIR84] [MER06] [CER91] [ALD94]
Solubility: i H_2O; s HCl, HNO_3, alkaline
    hydroxides [MER06]
Density, g/cm³: 7.14 [MER06]
Melting Point, °C: 419.5 [MER06]
Boiling Point, °C: 908 [MER06]
Thermal Conductivity, W/(m·K): 113.0
    (18°C), 96.0 (419.5°C); liq: 60.7
    (419.5°C), 56.5 (750°C) [KIR84]
Thermal Expansion Coefficient: (volume) 100°C
    (0.717), 200°C (1.656), 400°C (3.699) [CLA66]

3514
Compound: Zinc acetate
Formula: $Zn(CH_3COO)_2$
Molecular Formula: $C_4H_6O_4Zn$
Molecular Weight: 183.479
CAS RN: 557-34-6

Properties: prepared from zinc nitrate and acetic
    anhydride; used to preserve wood, as a mordant
    in dyeing, and a blood test reagent [MER06]
Density, g/cm³: 1.84 [ALD94]

3515
Compound: Zinc acetate dihydrate
Formula: $Zn(CH_3COO)_2 \cdot 2H_2O$
Molecular Formula: $C_4H_{10}O_6Zn$
Molecular Weight: 219.509
CAS RN: 5970-45-6
Properties: white powd; pearly luster; somewhat
    efflorescent; monocl cryst; used as wood
    preservative, mordant, antiseptic, and
    a catalyst; anhydrous form: 99.99%
    pure powd; CAS RN 557-34-6 [AES93]
    [KIR84] [MER06] [STR93] [HAW93]
Solubility: g/100 g H_2O: 40 (25°C), 67 (100°C);
    3 g/100 g alcohol at 25°C [KIR84]
Density, g/cm³: 1.735 [MER06]
Melting Point, °C: 237 [MER06]
Reactions: minus $2H_2O$ at 100°C [HAW93]

3516
Compound: Zinc acetylacetonate hydrate
Synonyms: 2,4-pentanedione, zinc(II) derivative
Formula: $Zn(CH_3COCH=C(O)CH_3)_2 \cdot xH_2O$
Molecular Formula: $C_{10}H_{14}O_4Zn$ (anhydrous)
Molecular Weight: 263.609 (anhydrous)
CAS RN: 108503-47-5
Properties: cryst solid; trimer; hydrate is a white
    powd; used as a catalyst in the synthesis of long
    chain alcohols and aldehydes, and as a textile
    weighting agent [HAW93] [STR93] [COT88]
Solubility: decomposed by H_2O;
    s benzene, acetone [HAW93]
Melting Point, °C: 138 [HAW93]
Boiling Point, °C: sublimes [HAW93]

3517
Compound: Zinc ammonium chloride
Formula: $ZnCl_2 \cdot 2NH_4Cl$
Molecular Formula: $Cl_4H_8N_2Zn$
Molecular Weight: 243.278
CAS RN: 52628-25-8
Properties: white powd or cryst; used in
    galvanizing, as a flux for solder, and
    in adhesives [KIR84] [HAW93]
Solubility: g/100 g H_2O: 66 (0°C), 69 (30°C) [KIR84]
Density, g/cm³: 1.88 [KIR84]
Melting Point, °C: 150 (decomposes) [KIR84]

3518
Compound: Zinc antimonide
Formula: ZnSb
Molecular Formula: SbZn
Molecular Weight: 187.150
CAS RN: 12039-35-9
Properties: silvery white cryst, 99.5% pure melted pieces of 6 mm and smaller; used in thermoelectric devices and thermionic studies [HAW93] [CER91]
Solubility: decomposed by H_2O [HAW93]
Density, g/cm³: 6.33 [HAW93]
Melting Point, °C: 570 [HAW93]

3519
Compound: Zinc arsenate octahydrate
Synonym: koettigite
Formula: $Zn_3(AsO_4)_2 \cdot 8H_2O$
Molecular Formula: $As_2H_{16}O_{16}Zn_3$
Molecular Weight: 618.130
CAS RN: 13464-44-4
Properties: white powd; used as an insecticide and wood preservative [HAW93]
Solubility: i H_2O; s in acids, alkalies [HAW93]
Density, g/cm³: 3.31 (15°C) [HAW93]
Reactions: minus H_2O at 100°C [HAW93]

3520
Compound: Zinc arsenide
Formula: Zn_3As_2
Molecular Formula: As_2Zn_3
Molecular Weight: 346.013
CAS RN: 12006-40-5
Properties: gray tetr; −20 mesh with 99% purity, for arsine generation; 99.9999% purity electronic doping grade [CER91] [CRC10]
Density, g/cm³: 5.528 [ALD94]
Melting Point, °C: 1015 [AES93]

3521
Compound: Zinc arsenite
Synonym: zinc metaarsenite
Formula: $Zn(AsO_2)_2$
Molecular Formula: As_2O_4Zn
Molecular Weight: 279.231
CAS RN: 10326-24-6
Properties: colorless powd; used as insecticide and wood preservative [HAW93]
Solubility: i H_2O; s acids [HAW93]

3522
Compound: Zinc borate
Synonym: zinc diborate
Formula: $3ZnO \cdot 2B_2O_3$
Molecular Formula: $B_4O_9Zn_3$
Molecular Weight: 383.409
CAS RN: 27043-84-1
Properties: white, amorphous powd; used in medicine, for fireproofing textiles, as an inhibitor of fungus and mildew, and as a ceramic flux; most common borate is $2ZnO \cdot 3B_2O_3 \cdot 7H_2O$, x-ray structure is given as $Zn[B_2O_3(OH)_5] \cdot H_2O$; this hydrate can be prepared by adding borax to a solution of a soluble zinc salt; there are other hydrated borates, for example $Zn_3B_4O_9 \cdot 5H_2O$, CAS RN 12536-65-1, −325 mesh white powd [AES93] [HAW93] [KIR78]
Solubility: g/100 g H_2O: 0.007 (25°C); s dil acids [KIR84]
Density, g/cm³: 3.64 [HAW93]
Melting Point, °C: 980 [HAW93]

3523
Compound: Zinc borate hemiheptahydrate
Formula: $2ZnO \cdot 3B_2O_3 \cdot 3\text{-}1/2H_2O$
Molecular Formula: $B_6H_7O_{14.5}Zn_2$
Molecular Weight: 434.690
CAS RN: 12513-27-8
Properties: white tricl or powd; has about 20% hydrate water; used as a fire retardant material [KIR84] [HAW93] [CRC10]
Solubility: i H_2O [KIR84]
Density, g/cm³: 4.22 [KIR84]
Melting Point, °C: 980 [KIR84]

3524
Compound: Zinc borate pentahydrate
Formula: $3ZnO \cdot 2B_2O_3 \cdot 5H_2O$
Molecular Formula: $B_4H_{10}O_{14}Zn_3$
Molecular Weight: 473.487
CAS RN: 12536-65-1
Properties: −325 mesh white powd; formula also given as a dihydrate; used for fireproofing, in ceramics and fungicides [KIR84] [AES93]
Solubility: 0.007 g/100 g H_2O at 25°C; sl s HCl [KIR84]
Density, g/cm³: 3.64 [KIR84]

3525
Compound: Zinc bromate hexahydrate
Formula: $Zn(BrO_3)_2 \cdot 6H_2O$
Molecular Formula: $Br_2H_{12}O_{12}Zn$
Molecular Weight: 429.286
CAS RN: 13517-27-6

Properties: white solid; deliq; oxidizing agent [HAW93]
Solubility: v s H_2O [HAW93]
Density, g/cm³: 2.566 [HAW93]
Melting Point, °C: 100 [HAW93]
Reactions: minus $6H_2O$ at 200°C [CRC10]

3526

Compound: Zinc bromide
Formula: $ZnBr_2$
Molecular Formula: Br_2Zn
Molecular Weight: 225.198
CAS RN: 7699-45-8
Properties: white granular powd; very hygr; enthalpy
  of vaporization 118 kJ/mol; enthalpy of fusion
  16.70 kJ/mol; used in photographic emulsions, to
  manufacture rayon [MER06] [HAW93] [CRC10]
Solubility: mol/100 mol soln, H_2O: 31.1 (0°C), 37.6
  (25°C), 53.8 (100°C); solid phase, $ZnBr_2 \cdot 2H_2O$ (0°C,
  25°C), $ZnBr_2$ (100°C) [KRU93]; 1 g/0.5 mL 90%
  alcohol; s ether, alkali hydroxide solutions [MER06]
Density, g/cm³: 4.5 [LID94]
Melting Point, °C: 394 [CRC10]
Boiling Point, °C: 650 [CRC10]

3527

Compound: Zinc caprylate
Formula: $Zn(C_8H_{15}O_2)_2$
Molecular Formula: $C_{16}H_{30}O_4Zn$
Molecular Weight: 351.802
CAS RN: 557-09-5
Properties: lustrous scales; decomposes in moist atm
  forming caprylic acid; used as a fungicide [HAW93]
Solubility: sl s boiling H_2O; s boiling alcohol [HAW93]
Melting Point, °C: 136 [HAW93]

3528

Compound: Zinc carbonate
Synonym: smithsonite
Formula: $ZnCO_3$
Molecular Formula: CO_3Zn
Molecular Weight: 125.399
CAS RN: 3486-35-9
Properties: white, cryst powd; rhombohedr cryst;
  used in ceramics, as fireproofing filler, in
  cosmetics and lotions [HAW93] [MER06]
Solubility: mol/L soln, H_2O: 1.64×10^{-4} (solubility
  is given at $Pco_2 = 0.00032$ bar) [KRU93]; s dil
  acids, alkalies, NH_3 solutions [MER06]
Density, g/cm³: 4.42–4.45 [HAW93]
Melting Point, °C: decomposes [KIR84]
Reactions: evolves CO_2 at 300°C [HAW93]

3529

Compound: Zinc carbonate hydroxide
Formula: $3Zn(OH)_2 \cdot 2ZnCO_3$
Molecular Formula: $C_2H_6O_{12}Zn_5$
Molecular Weight: 549.013
CAS RN: 3486-35-9
Properties: white powd [STR93]
Density, g/cm³: 4.398 [STR93]

3530

Compound: Zinc chlorate
Formula: $Zn(ClO_3)_2$
Molecular Formula: Cl_2O_6Zn
Molecular Weight: 232.291
CAS RN: 10361-95-2
Properties: colorless to yellowish cryst;
  deliq; oxidizing agent [HAW93]
Solubility: mol/100 mol H_2O: 59.19 (0°C), 66.52
  (18°C), 75.44 (55°C); solid phase, $Zn(ClO_3)_2 \cdot 6H_2O$
  (0°C), $Zn(ClO_3)_2 \cdot 4H_2O$ (18°C, 55°C) [KRU93];
  s alcohol, glycerol, ether [HAW93]
Density, g/cm³: 2.15 [HAW93]
Melting Point, °C: decomposes at 60 [HAW93]

3531

Compound: Zinc chloride
Formula: $ZnCl_2$
Molecular Formula: Cl_2Zn
Molecular Weight: 136.295
CAS RN: 7646-85-7
Properties: white granules; very deliq; enthalpy
  of vaporization 126 kJ/mol; used as a catalyst,
  dehydrating agent in organic synthesis,
  as an antiseptic, and for fireproofing
  [HAW93] [MER06] [CRC10]
Solubility: g/100 g H_2O: 342 (0°C), 432 (25°), 615
  (100°C); solid phase, $ZnCl_2 \cdot H_2O$ (0°C), $ZnCl_2$ (25°C,
  100°C) [KRU93]; 1 g soluble in: 0.25 mL 2% HCl,
  1.3 mL alcohol, 2 mL glycerol; v s acetone [MER06]
Density, g/cm³: 2.91 [STR93]
Melting Point, °C: 283 [STR93]
Boiling Point, °C: 732 [STR93]

3532

Compound: Zinc chromate heptahydrate
Formula: $ZnCrO_4 \cdot 7H_2O$
Molecular Formula: $CrH_{14}O_{11}Zn$
Molecular Weight: 307.490
CAS RN: 13530-65-9
Properties: yellow solid; used as a pigment [HAW93]
Density, g/cm³: 3.4 [CRC10] (anhydrous)

3533
Compound: Zinc chromite
Formula: $ZnCr_2O_4$
Molecular Formula: Cr_2O_4Zn
Molecular Weight: 233.380
CAS RN: 12018-19-8
Properties: green cub spinel; pellets; used in catalysts [KIR78] [STR93]
Density, g/cm³: 5.3 [CRC10]

3534
Compound: Zinc citrate dihydrate
Synonyms: citric acid, zinc salt dihydrate
Formula: $Zn_3(C_6H_5O_7)_2 \cdot 2H_2O$
Molecular Formula: $C_{12}H_{14}O_{16}Zn_3$
Molecular Weight: 610.403
CAS RN: 546-46-3
Properties: colorless powd; can be made from zinc carbonate and citric acid; used in toothpaste and as a mouthwash [MER06]
Solubility: sl s H_2O; s dil mineral acids and alkali hydroxides [MER06]

3535
Compound: Zinc cyanide
Formula: $Zn(CN)_2$
Molecular Formula: C_2N_2Zn
Molecular Weight: 117.425
CAS RN: 557-21-1
Properties: white powd; readily decomposed by dil mineral acids, evolving HCN; has been used in plating, and as an insecticide [MER06]
Solubility: mol/L soln, H_2O: 4.2×10^{-5} (room temp) [KRU93]; s dil mineral acids, evolving HCN; i alcohol [HAW93]
Density, g/cm³: 1.852 [HAW93]
Melting Point, °C: 800, decomposes [HAW93]

3536
Compound: Zinc dichromate trihydrate
Formula: $ZnCr_2O_7 \cdot 3H_2O$
Molecular Formula: $Cr_2H_6O_{10}Zn$
Molecular Weight: 335.424
CAS RN: 7789-12-0
Properties: yellowish orange powd; preparation: reaction between chromic acid and $Zn(OH)_2$; used as a pigment [HAW93]
Solubility: s hot H_2O, acids; i alcohol and ether [HAW93]

3537
Compound: Zinc dimethyldithiocarbamate
Synonyms: dimethyldithiocarbamic acid, zinc salt
Formula: $Zn[(CH_3)_2NCS_2]_2$
Molecular Formula: $C_6H_{12}N_2S_4Zn$
Molecular Weight: 305.828
CAS RN: 137-30-4
Properties: cryst when obtained by reaction from hot chloroform + alcohol; used as accelerator for rubber vulcanization [MER06]
Solubility: i H_2O [MER06]
Density, g/cm³: 1.66
Melting Point, °C: 250–252 [ALD94]

3538
Compound: Zinc fluoride
Formula: ZnF_2
Molecular Formula: F_2Zn
Molecular Weight: 103.387
CAS RN: 7783-49-5
Properties: hygr; tetr needles or white cryst; enthalpy of vaporization 190.1 kJ/mol; finds use as ceramic glaze and as a wood preservative [HAW93] [MER06] [CRC10] [STR93]
Solubility: g/100mL soln, H_2O: 1.516 (25°C); solid phase, $ZnF_2 \cdot 4H_2O$ [KRU93]; sl s HF solutions; s HCl, HNO_3, NH_4OH [MER06]; i alcohol [HAW93]
Density, g/cm³: 4.95 [STR93]
Melting Point, °C: 872 [MER06]
Boiling Point, °C: 1500 [MER06]

3539
Compound: Zinc fluoride tetrahydrate
Formula: $ZnF_2 \cdot 4H_2O$
Molecular Formula: $F_2H_8O_4Zn$
Molecular Weight: 175.449
CAS RN: 13986-18-0
Properties: white cryst; rhombohedr [MER06] [STR93]
Solubility: mol/L soln, H_2O: 0.151 ± 0.004 (25°C) [KRU93]
Density, g/cm³: 2.255 [STR93]
Reactions: releases $4H_2O$ at 100°C [MER06]

3540
Compound: Zinc fluoroborate hexahydrate
Formula: $Zn(BF_4)_2 \cdot 6H_2O$
Molecular Formula: $B_2F_8H_{12}O_6Zn$
Molecular Weight: 347.090
CAS RN: 27860-83-9
Properties: hex cryst; used in zinc plating, for bonding and in the textile industry as a resin cure [LID94] [KIR84]

Solubility: >100 g/100 g H_2O at 25°C; s alcohol [KIR84]
Density, g/cm³: 2.12 [LID94]
Reactions: minus H_2O at 60°C [KIR84]

3541
Compound: Zinc formaldehyde sulfoxylate
Formula: $Zn(HSO_2 \cdot CH_2O)_2$
Molecular Formula: $C_2H_6O_6S_2Zn$
Molecular Weight: 255.588
CAS RN: 24887-06-7
Properties: rhomb prisms; used as a stripping and discharging agent in the textile industry [HAW93]
Solubility: v s H_2O; i alcohol; decomposed by acids [HAW93]
Melting Point, °C: 90 decomposes [KIR84]

3542
Compound: Zinc formate
Synonyms: formic acid, zinc salt
Formula: $Zn(CHOO)_2$
Molecular Formula: $C_2H_2O_4Zn$
Molecular Weight: 155.426
CAS RN: 557-41-5
Properties: colorless; readily forms dihydrate; can be obtained from reaction between zinc carbonate and formic acid [MER06] [CRC10]
Solubility: g/100 g H_2O: 3.70 (0°C), 5.20 (20°C), 38.0 (100°C) [LAN05]
Density, g/cm³: 2.368 [CRC10]
Reactions: decomposes on heating [CRC10]

3543
Compound: Zinc formate dihydrate
Formula: $Zn(CHO_2)_2 \cdot 2H_2O$
Molecular Formula: $C_2H_6O_6Zn$
Molecular Weight: 191.456
CAS RN: 5970-62-7
Properties: white cryst; has been used as a catalyst for the production of methanol, and as a waterproofing agent in the textile industry [HAW93]
Solubility: 5.2 g anhydrous/100 g H_2O (20°C); i alcohol [MER06]
Density, g/cm³: 2.207 [MER06]
Reactions: minus $2H_2O$ at 140°C [HAW93]

3544
Compound: Zinc hexafluoroacetylacetonate dihydrate
Synonyms: 1,1,1,5,5,5-hexafluoro-2,4-pentanedione, zinc derivative
Formula: $Zn(CF_3COCHCOCF_3)_2 \cdot 2H_2O$
Molecular Formula: $C_{10}H_6F_{12}O_6Zn$

Molecular Weight: 515.525
CAS RN: 16743-33-2
Properties: white powd [STR93]
Melting Point, °C: 157–158 [STR93]

3545
Compound: Zinc hexafluorosilicate hexahydrate
Formula: $ZnSiF_6 \cdot 6H_2O$
Molecular Formula: $F_6H_{12}O_6SiZn$
Molecular Weight: 315.557
CAS RN: 16871-71-9
Properties: white cryst; uses: harden concrete, preservative [HAW93] [MER06]
Solubility: s H_2O [MER06]
Density, g/cm³: 2.104 [CRC10]
Reactions: decomposes at 100°C [CRC10]

3546
Compound: Zinc hydroxide
Formula: $Zn(OH)_2$
Molecular Formula: H_2O_2Zn
Molecular Weight: 99.405
CAS RN: 20427-58-1
Properties: colorless cryst; used as an absorbent in surgical dressings and in rubber compounding [HAW93]
Solubility: sl s H_2O [HAW93]
Density, g/cm³: 3.053 [HAW93]
Melting Point, °C: 125, decomposes [HAW93]

3547
Compound: Zinc hypophosphite monohydrate
Formula: $Zn(H_2PO_2)_2 \cdot H_2O$
Molecular Formula: $H_6O_5P_2Zn$
Molecular Weight: 213.383
CAS RN: 7783-14-4
Properties: white cryst; hygr [HAW93]
Solubility: s H_2O and alkalies [HAW93]

3548
Compound: Zinc iodate
Formula: $Zn(IO_3)_2$
Molecular Formula: I_2O_6Zn
Molecular Weight: 415.195
CAS RN: 7790-37-6
Properties: white, cryst powd [MER06]
Solubility: g/100 g soln, H_2O: 0.622 (25°C) [KRU93]; s in 77 parts hot H_2O [MER06]
Density, g/cm³: 5.063 [CRC10]
Reactions: decomposes on heating [CRC10]

3549

Compound: Zinc iodide
Formula: ZnI_2
Molecular Formula: I_2Zn
Molecular Weight: 319.199
CAS RN: 10139-47-6
Properties: white or almost white; hygr cryst, granular powd; sensitive to air and light, turning brown due to iodine; used as topical antiseptic [HAW93] [MER06]
Solubility: g/100 g soln, H_2O: 81.11 (0°C), 81.20 (18°C), 83.62 (100°C); solid phase, ZnI_2 [KRU93]; 1 g/2 mL glycerol; v s alcohol, ether [MER06]
Density, g/cm³: 4.74 [MER06]
Melting Point, °C: ~446 [MER06]
Boiling Point, °C: ~625, decomposing [MER06]

3550

Compound: Zinc laurate
Synonyms: lauric acid, zinc salt
Formula: $Zn(C_{12}H_{23}O_2)_2$
Molecular Formula: $C_{24}H_{46}O_4Zn$
Molecular Weight: 464.017
CAS RN: 2452-01-9
Properties: white powd; used in paints and varnishes, in rubber compounding [HAW93]
Solubility: 0.01 g/100 mL H_2O (15°C), 0.019 g/100 mL H_2O (100°C) [CRC10]
Melting Point, °C: 128 [HAW93]

3551

Compound: Zinc molybdate
Formula: $ZnMoO_4$
Molecular Formula: MoO_4Zn
Molecular Weight: 225.328
CAS RN: 13767-32-3
Properties: white tetr; can be prepared from the two metal oxides [KIR81]
Solubility: 0.5 g/100 g H_2O [KIR81]
Density, g/cm³: 4.3 [LID94]
Melting Point, °C: >700 [KIR81]

3552

Compound: Zinc nitrate hexahydrate
Formula: $Zn(NO_3)_2 \cdot 6H_2O$
Molecular Formula: $H_{12}N_2O_{12}Zn$
Molecular Weight: 297.491
CAS RN: 10196-18-6
Properties: colorless cryst or lumps; used as catalyst and in latex coagulation [HAW93]
Solubility: g/100 g soln in H_2O: 48.3 (0.4°C), 56.1 (25.1°C), 90.0 (70.7°C); solid phase: $Zn(NO_3)_2 \cdot 6H_2O$ (0°C, 25.1°C), $Zn(NO_3)_2 \cdot H_2O$ (70.7°C) [KRU93]

Density, g/cm³: 2.065 [MER06]
Melting Point, °C: 36.4 [STR93]
Reactions: minus $6H_2O$ between 105°C and 131°C [HAW93]

3553

Compound: Zinc nitride
Formula: Zn_3N_2
Molecular Formula: N_2Zn_3
Molecular Weight: 224.183
CAS RN: 1313-49-1
Properties: −200 mesh with 99.9% purity; bluish gray, cryst material; a=0.972 nm [CER91] [MER06] [CIC73]
Density, g/cm³: 6.22 [CRC10]

3554

Compound: Zinc nitrite
Formula: $Zn(NO_2)_2$
Molecular Formula: N_2O_4Zn
Molecular Weight: 157.401
CAS RN: 10102-02-0
Properties: rapidly hydrolyzes in water; prepared by reaction of sodium nitrite and zinc sulfate in alcohol [MER06]

3555

Compound: Zinc oleate
Formula: $Zn(C_{18}H_{33}O_2)_2$
Molecular Formula: $C_{36}H_{66}O_4Zn$
Molecular Weight: 628.308
CAS RN: 557-07-3
Properties: white to tan, greasy powd; used in paints, resins, and varnishes [HAW93]
Solubility: i H_2O; s alcohol, ether, CS_2, benzene, petroleum ether [MER06]
Melting Point, °C: 70 [HAW93]

3556

Compound: Zinc oxalate dihydrate
Formula: $ZnC_2O_4 \cdot 2H_2O$
Molecular Formula: $C_2H_4O_6Zn$
Molecular Weight: 189.440
CAS RN: 547-68-2
Properties: white powd; used in organic synthesis [HAW93]
Solubility: g/L soln, H_2O: 0.018 (0°C), 0.0256 (25°C) [KRU93]; s dil mineral acids, ammonia solutions [MER06]
Density, g/cm³: 2.562 [HAW93]
Melting Point, °C: decomposes at 100 [HAW93]

71414-47-6　　2335: Pentamethyl-
　　　　　　　　cyclopentadienyltantalum
　　　　　　　　tetrachloride
71595-75-0　　1152: Decaborane(16)
71595-75-0　　1188: Dodecaborane(16)
71626-98-7　　665: Calcium iodide hexahydrate
71799-92-3　　2000: Manganese(II) citrate
71963-57-0　　1025: Cobalt(III) sepulchrate
　　　　　　　　trichloride
71965-17-8　　3615: Zirconyl basic nitrate
72520-94-6　　774: Cerous carbonate
　　　　　　　　pentahydrate
73157-11-6　　1382: Gallium azide
73491-34-6　　2090: Mercury(II) perchlorate
　　　　　　　　trihydrate
73560-00-6　　1590: Iodine nonoxide
74507-64-5　　1880: Magnesium bis(pentameth
　　　　　　　　ylcyclopentadienyl)
75060-62-5　　2228: Nickel selenate
　　　　　　　　hexahydrate
75397-94-3　　3128: Tantalum pentoxide
　　　　　　　　hydrate
75426-28-2　　2750: Selenium sulfide
75535-11-4　　1916: Magnesium iodide
　　　　　　　　hexahydrate
75926-22-6　　2744: Selenium hexasulfide
79490-00-9　　586: Cadmium perchlorate
80529-93-7　　1377: Gadopentetic acid
81029-06-3　　73: Aluminum perchlorate
　　　　　　　　nonahydrate
81129-00-2　　2369: Phosphorus(III) sulfide
81579-74-0　　2601: 2,2-Bis(ethylferrocenyl)
　　　　　　　　propane
82045-86-1　　430: Barium zirconium
　　　　　　　　phosphate
82642-06-6　　346: Barium copper yttrium
　　　　　　　　oxide-I
84359-31-9　　906: Chromium(III) phosphate
　　　　　　　　hexahydrate
86546-99-8　　1252: Europium(III) carbonate
　　　　　　　　hydrate
92141-86-1　　820: Cesium metaborate

97126-35-7　　1010: Cobalt(II) thiocyanate
　　　　　　　　trihydrate
99685-96-8　　718: Carbon
99685-96-8　　1350: Fullerene
100587-90-4　　1649: Lanthanum acetate
　　　　　　　　hydrate
100587-92-6　　3155: Terbium acetate hydrate
100587-96-0　　3158: Terbium carbonate
　　　　　　　　hydrate
101509-27-7　　2175: Neodymium sulfate
102192-40-5　　3233: Thorium acetylacetonate
103443-51-2　　2941: Sodium molybdosilicate
　　　　　　　　hydrate
107539-20-8　　3489: Yttrium barium copper
　　　　　　　　oxide
107539-20-8　　3490: Yttrium barium copper
　　　　　　　　oxide
107539-20-8　　3491: Yttrium barium copper
　　　　　　　　oxide
107539-20-8　　348: Barium copper yttrium
　　　　　　　　oxide-III
107782-11-6　　1342: Fluorine tetroxide
108249-27-0　　1508: Hydrazine monooxalate
108503-47-5　　3516: Zinc acetylacetonate
　　　　　　　　hydrate
109064-29-1　　347: Barium copper yttrium
　　　　　　　　oxide-II
109457-23-0　　1214: Erbium barium copper
　　　　　　　　oxide
110802-84-1　　1527: Hydrogen
　　　　　　　　hexachloroiridate(IV) hydrate
111419-39-7　　1676: Lanthanum strontium
　　　　　　　　copper oxide
114615-82-6　　3187: Tetrapropylammonium
　　　　　　　　perruthenate(VII)
114901-61-0　　504: Bismuth strontium calcium
　　　　　　　　copper oxide (1112)
114901-61-0　　505: Bismuth strontium calcium
　　　　　　　　copper oxide (2212)
114901-61-0　　506: Bismuth strontium calcium
　　　　　　　　copper oxide (2223)
115383-22-7　　729: Carbon fullerenes

118448-18-3　　619: Calcium bis(2,2,6,6-
　　　　　　　　tetramethyl-3,5-
　　　　　　　　heptanedionate)
119000-19-0　　3189: Thallium barium calcium
　　　　　　　　copper oxide
119800-94-1　　2158: Neodymium cerium
　　　　　　　　copper oxide
123333-66-4　　2532: Potassium tellurite(IV)
　　　　　　　　hydrate
123333-67-5　　2924: Sodium hypophosphite
　　　　　　　　monohydrate
123333-85-7　　1840: Lithium thiocyanate
　　　　　　　　hydrate
123333-98-2　　896: Chromium(III) fluoride
　　　　　　　　tetrahydrate
123334-23-6　　2882: Sodium
　　　　　　　　hexachloroiridate(III)
　　　　　　　　hydrate
124365-83-9　　349: Barium copper yttrium
　　　　　　　　oxide-IV
125720-69-1　　3191: Thallium barium calcium
　　　　　　　　copper oxide
126284-91-1　　2574: Praseodymium barium
　　　　　　　　copper oxide
127241-75-2　　3190: Thallium barium calcium
　　　　　　　　copper oxide
127386-54-3　　283: Antimony(V) fluoride
131159-39-2　　1351: Fullerenes
133578-89-9　　3446: Vanadyl selenite
　　　　　　　　monohydrate
133863-98-6　　2112: Molybdenum
　　　　　　　　metaphosphate
134929-59-2　　1352: Fullerene fluoride
137232-17-8　　2442: Potassium fullerene
137926-73-9　　2665: Rubidium fullerene
141326-12-7　　831: Cesium rubidium fullerene
141572-90-9　　652: Calcium
　　　　　　　　hexafluoroacetylacetonate
　　　　　　　　dihydrate
142617-56-9　　481: Bismuth
　　　　　　　　hexafluoroacetylacetonate

$Al_2O_7Si_2$ 79: Aluminum silicate
$Al_2O_9Te_3$ 91: Aluminum tellurite
$Al_2O_9Zr_3$ 96: Aluminum zirconate
$Al_2O_{12}S_3$ 86: Aluminum sulfate
$Al_2O_{12}W_3$ 95: Aluminum tungstate
Al_2S_3 88: Aluminum sulfide
Al_2Se_3 78: Aluminum selenide
Al_2Te_3 90: Aluminum telluride
Al_2Zr 97: Aluminum zirconium
Al_3Mn 1971: Manganese aluminide
Al_3Ni 2187: Nickel aluminide
Al_3Ta 3113: Tantalum aluminide
Al_3Zr 3586: Zirconium aluminide
$Al_4B_2O_9$ 23: Aluminum borate
Al_4Ba 325: Barium aluminide
$Al_4Mg_2O_{18}Si_5$ 1872: Magnesium aluminum silicate
$Al_5O_{12}Y_3$ 3486: Yttrium aluminum oxide
$Al_6O_{13}Si_2$ 82: Aluminum silicate
$Al_{22}Na_2O_{34}$ 2827: Sodium β-aluminum oxide
Am   101: Americium
$AmBr_3$ 102: Americium bromide
AmClO   111: Americium oxychloride
$AmCl_3$ 104: Americium chloride
AmF_3 105: Americium fluoride
AmF_4 114: Americium(IV) fluoride
AmH_3 106: Americium hydride
AmH_3O_3 107: Americium hydroxide
AmI_3 108: Americium iodide
AmO_2 115: Americium(IV) oxide
AmO_4P 112: Americium phosphate
Am_2O_3 109: Americium oxide(α)
                   110: Americium oxide(β)
Am_2S_3 113: Americium sulfide
Ar   290: Argon
ArF   291: Argon fluoride
As   292: Arsenic(α)
                   293: Arsenic(β)
$AsBiO_4$ 472: Bismuth arsenate
$AsBr_3$ 301: Arsenic(III) bromide
$AsCaHO_3$ 618: Calcium arsenite
$AsCl_3$ 302: Arsenic(III) chloride
AsCo   930: Cobalt arsenide
AsCoS   929: Cobalt arsenic sulfide
$AsCr_2$ 863: Chromium arsenide
$AsCuHO_3$ 1062: Copper(II) arsenite
$AsCu_3$ 1031: Copper arsenide
AsF_3 303: Arsenic(III) fluoride
AsF_5 311: Arsenic(V) fluoride
AsF_6K 2457: Potassium hexafluoroarsenate(V)
AsF_6Li 1794: Lithium hexafluoroarsenate
AsF_6Na 2891: Sodium hexafluoroarsenate
AsF_9Xe_2 3455: Xenon fluoride hexafluoroarsenate
$AsF_{11}Xe$ 3461: Xenon pentafluoride hexafluoroarsenate
AsFe   1617: Iron arsenide
$AsFeH_4O_6$ 1265: Ferric arsenate dihydrate
AsGa   1381: Gallium arsenide
$AsHHgO_4$ 2063: Mercury(II) arsenate

$AsHNa_2O_4$ 2902: Sodium hydrogen arsenate
$AsHO_4Pb$ 1752: Lead(II) hydrogen arsenate
AsH_2KO_4 2433: Potassium dihydrogen arsenate
AsH_3 316: Arsine
AsH_3O_3 315: Arsenious acid
AsH_3O_4 294: Arsenic acid
$AsH_4O_{4.5}$ 295: Arsenic acid hemihydrate
                   310: Arsenic(V) acid hemihydrate
AsH_6NO_4 143: Ammonium dihydrogen arsenate
$AsH_9N_2O_4$ 177: Ammonium hydrogen arsenate
$AsH_{12}N_3O_4$ 121: Ammonium arsenate (anhydrous)
$AsH_{15}Na_2O_{11}$ 2903: Sodium hydrogen arsenate heptahydrate
$AsH_{24}Na_3O_{16}$ 2837: Sodium arsenate dodecahydrate
AsI_2 299: Arsenic(II) iodide
AsI_3 304: Arsenic(III) iodide
AsIn   1555: Indium arsenide
$AsLi_3O_4$ 1768: Lithium arsenate
AsNi   2193: Nickel arsenide
AsO_2Na 2838: Sodium arsenite
$AsSb_3$ 258: Antimony arsenide
AsY   3488: Yttrium arsenide
As_2Ba_3 327: Barium arsenide
$As_2Ca_3O_8$ 617: Calcium arsenate
As_2Cd_3 561: Cadmium arsenide
As_2Co 932: Cobalt arsenide
$As_2Co_3H_{16}O_{16}$ 954: Cobalt(II) arsenate octahydrate
$As_2Cu_3O_8$ 1061: Copper(II) arsenate
$As_2Fe_3H_{12}O_{14}$ 1306: Ferrous arsenate hexahydrate
$As_2Fe_3O_8$ 1635: Iron(II) arsenate
$As_2Fe_4H_{10}O_{14}$ 1267: Ferric basic arsenite
As_2HKO_4 2470: Potassium hydrogen arsenite
$As_2H_3KO_5$ 2487: Potassium metaarsenite monohydrate
$As_2H_{16}Ni_{13}O_{16}$ 2192: Nickel arsenate octahydrate
$As_2H_{16}O_{16}Zn_3$ 3519: Zinc arsenate octahydrate
As_2Mg_3 1878: Magnesium arsenide
$As_2Mg_3O_8$ 1877: Magnesium arsenate hydrate (anhydrous)
As_2O_3 305: Arsenic(III) oxide
                   306: Arsenic(III) oxide
As_2O_4Pb 1691: Lead arsenite
As_2O_4Zn 3521: Zinc arsenite
As_2O_5 312: Arsenic(V) oxide
$As_2O_8Pb_3$ 1690: Lead arsenate
As_2S_2 300: Arsenic(II) sulfide
As_2S_3 308: Arsenic(III) sulfide
As_2S_5 314: Arsenic(V) sulfide
As_2Se 298: Arsenic hemiselenide
As_2Se_3 307: Arsenic(III) selenide

As_2Se_5 313: Arsenic(V) selenide
As_2Te_3 309: Arsenic(III) telluride
As_2Zn_3 3520: Zinc arsenide
As_3Co 931: Cobalt arsenide
As_4S_4 296: Arsenic disulfide
At   317: Astatine
Au   1428: Gold
AuBr   1429: Gold(I) bromide
$AuBr_3$ 1435: Gold(III) bromide
$AuBr_4H_4KO_2$ 2535: Potassium tetrabromoaurate(III) dihydrate
$AuBr_4H_{11}O_5$ 1538: Hydrogen tetrabromoaurate(III) pentahydrate
$AuBr_4Na$ 2998: Sodium tetrabromoaurate(III)
AuCl   1431: Gold(I) chloride
$AuCl_3$ 1436: Gold(III) chloride
$AuCl_4H$ 1540: Hydrogen tetrachloroaurate(III) (anhydrous) hydrate
$AuCl_4H_4KO_2$ 2539: Potassium tetrachloroaurate(III) dihydrate
$AuCl_4H_4N$ 237: Ammonium tetrachloroaurate(III) (anhydrous) hydrate
$AuCl_4H_4NaO_2$ 3000: Sodium tetrachloroaurate(III) dihydrate
$AuCl_4H_9O_4$ 550: Chloroauric(III) acid tetrahydrate
                   1541: Hydrogen tetrachloroaurate(III) tetrahydrate
$AuCl_4K$ 2538: Potassium tetrachloroaurate(III)
AuF_3 1438: Gold(III) fluoride
AuH_3O_3 1439: Gold(III) hydroxide
$AuH_4Na_3O_8S_4$ 2881: Sodium gold thiosulfate dihydrate
AuH_6KO_5 2443: Potassium gold(III) oxide trihydrate
AuI   1433: Gold(I) iodide
AuI_3 1440: Gold(III) iodide
AuI_4K 2549: Potassium tetraiodoaurate(III)
Au_2O_3 1441: Gold(III) oxide
$Au_2O_{12}Se_3$ 1442: Gold(III) selenate
Au_2S 1434: Gold(I) sulfide
Au_2S_3 1444: Gold(III) sulfide
Au_2Se_3 1443: Gold(III) selenide
B   524: Boron
BAs   525: Boron arsenide
BBe_2 442: Beryllium boride-II
BBe_4 441: Beryllium boride-I
BBr_3 533: Boron tribromide
$BCd_5H_{36}O_{58}W_{12}$ 563: Cadmium borotungstate octadecahydrate
BCl_3 534: Boron trichloride
BCo   933: Cobalt boride
BCo_2 934: Cobalt boride
BCr   870: Chromium monoboride
BCr_2 864: Chromium boride

Br₃Sb	262: Antimony(III) bromide	CCaO₃	628: Calcium carbonate
Br₃Sc	2724: Scandium bromide		629: Calcium carbonate
Br₃Tb	3157: Terbium bromide		630: Calcium carbonate
Br₃Ti	3306: Titanium tribromide	CCdO₃	566: Cadmium carbonate
Br₃Tl	3220: Thallium(III) bromide	CClN	1143: Cyanogen chloride
Br₃Tm	3262: Thulium bromide	CCl₂F₂	1172:
Br₃U	3383: Uranium tribromide		Dichlorodifluoromethane
Br₃V	3430: Vanadium tribromide	CCl₂O	743: Carbonyl chloride
Br₃W	3361: Tungsten tribromide	CCl₄	739: Carbon tetrachloride
Br₃Y	3493: Yttrium bromide	CCoO₃	959: Cobalt(II) carbonate
Br₃Yb	3470: Ytterbium bromide	CCsN	807: Cesium cyanide
(anhydrous)	hydrate	CCs₂O₃	803: Cesium carbonate
Br₄Ge	1418: Germanium(IV)	CCuN	1043: Copper(I) cyanide
	bromide	CCuNS	1054: Copper(I) thiocyanate
Br₄Hf	1449: Hafnium bromide	CCuO₃	1072: Copper(II) carbonate
Br₄K₂Pd	2536: Potassium	CF₂O	744: Carbonyl fluoride
	tetrabromopalladate(II)	CF₄	740: Carbon tetrafluoride
Br₄K₂Pt	2537: Potassium	CFeO₃	1310: Ferrous carbonate
	tetrabromoplatinate(II)	CFe₃	1620: Iron carbide
Br₄Mo	2128: Molybdenum(IV)	CFN	1144: Cyanogen fluoride
	bromide	CH₂BrCl	551: Bromochloromethane
Br₄Nb	2258: Niobium(IV) bromide	CH₂Cu₂O₅	1073: Copper(II) carbonate
Br₄OW	3351: Tungsten		hydroxide
	oxytetrabromide	CH₂Na₂O₄	2850: Sodium carbonate
Br₄Pb	1756: Lead(IV) bromide		monohydrate
Br₄Se	2751: Selenium tetrabromide	CH₂O₃	2340: Performic acid
Br₄Si	2772: Silicon tetrabromide	CH₃AlCl₂	100: Dichloromethyl-
Br₄Sn	3017: Stannic bromide		aluminum
Br₄Te	3148: Tellurium tetrabromide	CH₃BNNa	2863: Sodium
Br₄Th	3234: Thorium bromide		cyanoborohydride
Br₄Ti	3302: Titanium tetrabromide	CH₃K₂O₄.₅	2418: Potassium carbonate
Br₄U	3379: Uranium tetrabromide		hemitrihydrate
Br₄W	3358: Tungsten tetrabromide	CH₃Na₂O₆	2851: Sodium carbonate
Br₄Zr	3588: Zirconium bromide		peroxohydrate
Br₅Nb	2265: Niobium(V) bromide	CH₄MgO₅	1888: Magnesium carbonate
Br₅P	2370: Phosphorus(V)		dihydrate
	bromide	CH₄N₂	141: Ammonium cyanide
Br₅Re	2625: Rhenium(V) bromide	CH₄N₂S	248: Ammonium
Br₅Ta	3123: Tantalum pentabromide		thiocyanate
Br₅U	3375: Uranium pentabromide	CH₅NO₂	156: Ammonium formate
Br₅W	3355: Tungsten pentabromide	CH₅NO₃	179: Ammonium hydrogen
Br₆H₈N₂Os	159: Ammonium		carbonate
	hexabromoosmiate(IV)	CH₆Ge	2103: Methylgermane
Br₆H₈N₂Pt	160: Ammonium	CH₆MgO₆	1892: Magnesium carbonate
	hexabromoplatinate(IV)		trihydrate
Br₆K₂Pt	2447: Potassium	CH₆N₂O₂	127: Ammonium carbamate
	hexabromoplatinate(IV)	CH₆N₂S₂	146: Ammonium
Br₆W	3344: Tungsten hexabromide		dithiocarbamate
C	721: Carbon	CH₇NaO₅S	2879: Sodium formaldehyde
	723: Carbon		sulfoxylate
	724: Carbon (amorphous)	CH₈BeO₇	448: Beryllium carbonate
	733: Carbon soot		tetrahydrate
CAgN	2791: Silver cyanide	CH₈Mg₂O₈	1890: Magnesium carbonate
CAgNS	2823: Silver thiocyanate		hydroxide trihydrate
CAg₂O₃	2785: Silver carbonate	CH₈N₂O₃	128: Ammonium carbonate
CAuClO	1430: Gold(I) carbonyl	CH₁₀MgO₈	1891: Magnesium carbonate
	chloride		pentahydrate
CAuN	1432: Gold(I) cyanide	CH₁₂KNaO₉	2520: Potassium sodium
CB₄	526: Boron carbide		carbonate hexahydrate
CBaO₃	337: Barium carbonate	CH₁₂Ni₃O₁₁	2194: Nickel basic carbonate
CBe₂	447: Beryllium carbide		tetrahydrate
CBH₃O	518: Borane carbonyl	CH₂₀Na₂O₁₃	2849: Sodium carbonate
CBrN	1142: Cyanogen bromide		decahydrate
CBr₂O	742: Carbonyl bromide	CHBi₂O₅.₅	473: Bismuth basic carbonate
CBr₄	738: Carbon tetrabromide		hemihydrate
CCaN₂	641: Calcium cyanamide	CHCsO₂	810: Cesium formate

CHCsO₃	813: Cesium hydrogen	CLi₂O₃	1776: Lithium carbonate
	carbonate	CMgO₃	1887: Magnesium carbonate
CHf	1450: Hafnium carbide	CMnO₃	1997: Manganese(II)
CHF₃	3321: Trifluoromethane		carbonate
CHg₂O₃	2044: Mercury(I) carbonate	CMn₃	1978: Manganese carbide
CHg₄O₆	2064: Mercury(II) basic	CMo	2108: Molybdenum carbide
	carbonate	CMo₂	2109: Molybdenum carbide
CHKO₃	2471: Potassium hydrogen	CN₄	1141: Cyanogen azide
	carbonate	CNa₂O₃	2847: Sodium carbonate
CHLiO₃	1792: Lithium formate	CNb	2259: Niobium(IV) carbide
	monohydrate	CNb₂	2250: Niobium carbide
	1799: Lithium hydrogen	CNiO₃	2201: Nickel carbonate
	carbonate	CNNa	2862: Sodium cyanide
CHN	1524: Hydrogen cyanide	CNNaO	2861: Sodium cyanate
CHNaO₃	2904: Sodium hydrogen	CNNaS	3005: Sodium thiocyanate
	carbonate	CNRb	2661: Rubidium cyanide
CHO₂T₁	3202: Thallium(I) formate	CNTl	3199: Thallium(I) cyanide
CI₃H₁₂LuO₁₈	1860: Lutetium perchlorate	CO	730: Carbon monoxide
	hexahydrate	CO₂	725: Carbon dioxide
CI₄	741: Carbon tetraiodide	CO₃Pb	1698: Lead carbonate
CIN	1145: Cyanogen iodide	CO₃Ra	2606: Radium carbonate
	1583: Iodine cyanide	CO₃Rb₂	2656: Rubidium carbonate
CK₂O₃	2417: Potassium carbonate	CO₃Sr	3053: Strontium carbonate
CK₂S₃	2555: Potassium	CO₃Tl₂	3196: Thallium(I) carbonate
	thiocarbonate	CO₃Zn	3528: Zinc carbonate
CKN	2430: Potassium cyanide	CO₅U	3395: Uranyl carbonate
CKNO	2429: Potassium cyanate	CO₈Zr₃	3590: Zirconium carbonate
CKNS	2556: Potassium thiocyanate	(anhydrous)	basic hydrate
CLiN	1784: Lithium cyanide	COS	732: Carbon oxysulfide
CLiNS	1836: Lithium thiocyanate	COSe	731: Carbon oxyselenide
CLiNS	1840: Lithium	CS₂	727: Carbon disulfide
(anhydrous)	thiocyanate hydrate	CSe₂	726: Carbon diselenide
		CSSe	736: Carbon sulfide selenide
		CSTe	737: Carbon sulfide telluride

C_4O_8Pd	2326: Palladium(II) oxalate	$C_6H_4AgN_3O_8$	2814: Silver picrate monohydrate
C_5BrMnO_5	1983: Manganese pentacarbonyl bromide	$C_6H_5Ag_3O_7$	2790: Silver citrate
C_5BrO_5Re	2613: Rhenium pentacarbonyl bromide	$C_6H_5AlO_7$	34: Aluminum citrate
		$C_6H_5BiO_7$	478: Bismuth citrate
C_5ClO_5Re	2614: Rhenium pentacarbonyl chloride	C_6H_5ClHg	2347: Phenylmercuric chloride
C_5FeO_5	1626: iron pentacarbonyl	$C_6H_5K_3O_7$	2425: Potassium citrate
$C_5HF_6O_2Tl$	3203: Thallium(I) hexafluoro-acetylacetonate	$C_6H_{5+y}Fe_xN_yO_7$	148: Ammonium ferric citrate
		$C_6H_6CrK_3O_{15}$	907: Chromium(III) potassium oxalate trihydrate
$C_5H_4F_3O_2Tl$	3218: Thallium(I) trifluoroacetylacetonate		2423: Potassium chromium(III) oxalate trihydrate
$C_5H_4FeK_2N_6O_3$	2496: Potassium nitroprusside dihydrate		
$C_5H_4FeN_6Na_2O_3$	2945: Sodium nitroferricyanide(III) dihydrate	$C_6H_6FeK_4N_6O_3$	2439: Potassium ferrocyanide trihydrate
		$C_6H_6K_3O_{15}Sb$	275: Antimony(III) potassium oxalate trihydrate
$C_5H_5Cl_3Zr$	3592: Zirconium cyclopentadienyl trichloride		2410: Potassium antimony oxalate trihydrate
$C_5H_5Cl_4Nb$	1150: Cyclopenta-dienylniobium tetrachloride	$C_6H_6N_4O_7$	221: Ammonium picrate
		$C_6H_7K_3O_8$	2426: Potassium citrate monohydrate
C_5H_5In	1148: Cyclopenta-dienylindium(I)	$C_6H_8FeO_8$	1315: Ferrous citrate monohydrate
C_5H_5Li	1785: Lithium cyclopentadienide	$C_6H_8GeN_2O_{12}$ (anhydrous)	157: Ammonium germanium oxalate hydrate
$C_5H_7AgO_2$	2779: Silver acetylacetonate	$C_6H_9AlO_6$	17: Aluminum acetate
$C_5H_7CsO_2$	795: Cesium acetylacetonate	$C_6H_9BiO_6$	470: Bismuth acetate
$C_5H_7LiO_2$	1763: Lithium acetylacetonate	$C_6H_9CoO_6$	1014: Cobalt(III) acetate
		$C_6H_9CrO_6$	860: Chromium(III) acetate
$C_5H_7NaO_2$	2830: Sodium acetylacetonate	$C_6H_9Cu_2O_{9.5}$	1032: Copper citrate hemipentahydrate
$C_5H_7O_2Rb$	2651: Rubidium acetylacetonate	$C_6H_9InO_6$	1552: Indium acetate
$C_5H_7O_2Tl$	3193: Thallium(I) acetylacetonate	$C_6H_9LaO_6$ (anhydrous)	1649: Lanthanum acetate hydrate
$C_5H_8FeN_8O$	204: Ammonium nitroferricyanide	$C_6H_9LuO_6$ (anhydrous)	1846: Lutetium acetate hydrate
$C_5H_8KO_{2.5}$	2406: Potassium acetylacetonate hemihydrate	$C_6H_9Na_3O_9$	2858: Sodium citrate dihydrate
		$C_6H_9NdO_6$	2152: Neodymium acetate
$C_5H_{10}AgNS_2$	2793: Silver diethyldithiocarbamate	$C_6H_9O_6Pr$ (anhydrous)	2572: Praseodymium acetate hydrate
$C_5H_{13}NO_2$	255: Ammonium valerate	$C_6H_9O_6Sb$	261: Antimony(III) acetate
$C_5H_{15}NSn$	1183: Dimethylaminotri-methyltin	$C_6H_9O_6Sc$ (anhydrous)	2722: Scandium acetate hydrate
$C_6Bi_2O_{12}$	491: Bismuth oxalate	$C_6H_9O_6Tb$ (anhydrous)	3155: Terbium acetate hydrate
$C_6CoK_3N_6$	2454: Potassium hexacyanocobalt(III)	$C_6H_9O_6Tl$	3219: Thallium(III) acetate
$C_6Co_2FeN_6$ (anhydrous)	972: Cobalt(II) ferrocyanide hydrate	$C_6H_9O_6Y$ (anhydrous)	3484: Yttrium acetate hydrate
C_6CrO_6	867: Chromium carbonyl	$C_6H_9O_{9.5}$	1079: Copper(II) citrate hemipentahydrate
$C_6Cu_2FeN_6$	1087: Copper(II) ferrocyanide	$C_6H_{10}CaO_4$	692: Calcium propionate
$C_6Eu_2O_{12}$	1258: Europium(III) oxalate	$C_6H_{10}O_4Zn$	3564: Zinc propionate
$C_6F_9O_6Tl$	3227: Thallium(III) trifluoroacetate	$C_6H_{10}O_6Pb$	1753: Lead(II) lactate
		$C_6H_{10}O_{17}Sc_2$	2730: Scandium oxalate pentahydrate
$C_6FeK_3N_6$	2438: Potassium ferricyanide	$C_6H_{11}CrO_7$	887: Chromium(III) acetate monohydrate
$C_6Fe_2O_{12}$	1284: Ferric oxalate		
$C_6H_2Al_2O_{13}$	63: Aluminum oxalate monohydrate	$C_6H_{11}HoO_7$	1479: Holmium acetate monohydrate
$C_6H_2FeN_6Na_3O$	2873: Sodium ferricyanide monohydrate	$C_6H_{11}NdO_7$	2153: Neodymium acetate monohydrate
$C_6H_2N_3O_7Tl$	3213: Thallium(I) picrate		
$C_6H_{11}O_7Tm$	3260: Thulium acetate monohydrate		
$C_6H_{12}Ba_2FeN_6O_6$	356: Barium ferrocyanide hexahydrate		
$C_6H_{12}CeO_{7.5}$	767: Cerous acetate hemitrihydrate		
$C_6H_{12}Lu_2O_{18}$	1858: Lutetium oxalate hexahydrate		
$C_6H_{12}N_2S_4Zn$	3537: Zinc dimethyldithiocarbamate		
$C_6H_{12}O_{18}Tm_2$	3269: Thulium oxalate hexahydrate		
$C_6H_{13}Li_3O_{11}$	1782: Lithium citrate tetrahydrate		
$C_6H_{13}MnO_8$	2031: Manganese(III) acetate dihydrate		
$C_6H_{14}CuO_8$	1102: Copper(II) lactate dihydrate		
$C_6H_{14}LiN$	1789: Lithium diisopropylamide		
$C_6H_{14}N_2O_7$	180: Ammonium hydrogen citrate		
$C_6H_{15}AlO_3$	39: Aluminum ethoxide		
$C_6H_{15}AsO_6$	297: Arsenic(III) ethoxide		
$C_6H_{15}FeO_{12}$	1272: Ferric citrate pentahydrate		
$C_6H_{15}Na_3O_{12}$	2859: Sodium citrate pentahydrate		
$C_6H_{15}O_9Sm$	2699: Samarium acetate trihydrate		
$C_6H_{15}P$	3320: Triethylphosphine		
$C_6H_{16}MgO_{12}$	1899: Magnesium citrate pentahydrate		
$C_6H_{16}O_9Sr$	3068: Strontium lactate trihydrate		
$C_6H_{17}DyO_{10}$	1190: Dysprosium acetate tetrahydrate		
$C_6H_{17}ErO_{10}$	1212: Erbium acetate tetrahydrate		
$C_6H_{17}GdO_{10}$	1354: Gadolinium acetate tetrahydrate		
$C_6H_{17}N_3O_7$	137: Ammonium citrate tribasic		
$C_6H_{17}O_{10}Yb$	3468: Ytterbium acetate tetrahydrate		
$C_6H_{18}AlCl$	99: Chlorodiisobutyl-aluminum		
$C_6H_{18}Ce_2O_{21}$	783: Cerous oxalate nonahydrate		
$C_6H_{18}FeN_3O_{15}$	149: Ammonium ferric oxalate trihydrate		
$C_6H_{18}FeN_9O_3$	151: Ammonium ferricyanide trihydrate		
$C_6H_{18}FeN_{10}O$	169: Ammonium hexacyanoferrate(II) monohydrate		
$C_6H_{18}O_{21}Y_2$	3506: Yttrium oxalate nonahydrate		
$C_6H_{20}Ac_2O_{22}$	9: Actinium oxalate decahydrate		
$C_6H_{20}Dy_2O_{22}$	1203: Dysprosium oxalate decahydrate		
$C_6H_{20}Er_2O_{22}$	1228: Erbium oxalate decahydrate		
$C_6H_{20}FeN_6Na_4O_{10}$	2874: Sodium ferrocyanide decahydrate		

$Cl_2H_{16}O_{17}Zr$ 3620: Zirconyl perchlorate octahydrate

Cl_2HeO_5U 3397: Uranyl chloride trihydrate

Cl_2Hf 1451: Hafnium(II) chloride

Cl_2Hg 2069: Mercury(II) chloride

Cl_2HgO_6 2068: Mercury(II) chlorate

Cl_2Hg_2 2046: Mercury(I) chloride

$Cl_2Hg_2O_6$ 2045: Mercury(I) chlorate

$Cl_2HI_2NiO_{14}$ 2224: Nickel perchlorate hexahydrate

Cl_2In 1562: Indium(II) chloride

Cl_2Mg 1894: Magnesium chloride

Cl_2MgO_8 1934: Magnesium perchlorate

Cl_2Mn 1998: Manganese(II) chloride

Cl_2Mo 2120: Molybdenum(II) chloride

Cl_2MoO_2 2140: Molybdenum(VI) dioxydichloride

Cl_2Nd 2159: Neodymium chloride

Cl_2Ni 2205: Nickel chloride

Cl_2O 847: Chlorine monoxide

Cl_2OS 3229: Thionyl chloride

Cl_2OSe 2747: Selenium oxychloride

Cl_2OV 3444: Vanadyl dichloride

Cl_2OZr (anhydrous) 3616: Zirconyl chloride hydrate

Cl_2O_2S 3110: Sulfuryl chloride

Cl_2O_2U 3396: Uranyl chloride

Cl_2O_2W 3349: Tungsten oxydichloride

Cl_2O_3 1166: Dichlorine trioxide

Cl_2O_4 849: Chlorine perchlorate

Cl_2O_4Pb 1701: Lead chlorite

Cl_2O_6 1165: Dichlorine hexoxide

Cl_2O_6Pb 1699: Lead chlorate

Cl_2O_6Sr 3054: Strontium chlorate

Cl_2O_6Zn 3530: Zinc chlorate

Cl_2O_7 845: Chlorine heptoxide; 1164: Dichlorine heptoxide

Cl_2O_8Pb 1722: Lead(II) perchlorate

Cl_2O_8Sr 3077: Strontium perchlorate

Cl_2Os 2303: Osmium(II) chloride

Cl_2Pb 1700: Lead chloride

Cl_2Pd 2319: Palladium(II) chloride

Cl_2Pt 2382: Platinum(II) chloride

Cl_2Ra 2607: Radium chloride

Cl_2S_2 3096: Sulfur chloride

Cl_2Se_2 2740: Selenium chloride

Cl_2Sm 2720: Samarium(II) chloride

Cl_2Sn 3029: Stannous chloride

Cl_2Sr 3055: Strontium chloride

Cl_2Te 3141: Tellurium dichloride

Cl_2Ti 3284: Titanium dichloride

Cl_2V 3412: Vanadium dichloride

Cl_2W 3336: Tungsten dichloride

Cl_2Zn 3531: Zinc chloride

$Cl_3CoH_{15}N_5$ 2336: Pentamminechlorocobalt(III) chloride

$Cl_3CoH_{18}N_6$ 1471: Hexaamminecobalt(III) chloride

Cl_3Cr 893: Chromium(III) chloride

$Cl_3CrH_{12}O_6$ 894: Chromium(III) chloride hexahydrate

Cl_3CrO_{12} 903: Chromium(III) perchlorate

Cl_3Dy 1195: Dysprosium chloride

$Cl_3DyH_{12}O_6$ 1196: Dysprosium chloride hexahydrate

Cl_3DyO_{12} (anhydrous) 1205: Dysprosium perchlorate hydrate

Cl_3Er 1220: Erbium chloride

$Cl_3ErH_{12}O_6$ 1221: Erbium chloride hexahydrate

Cl_3ErO_{12} (anhydrous) 1230: Erbium perchlorate hydrate

Cl_3Eu 1253: Europium(III) chloride

$Cl_3EuH_{12}O_6$ 1254: Europium(III) chloride hexahydrate

$Cl_3EuH_{12}O_{18}$ 1260: Europium(III) perchlorate hexahydrate

Cl_3FGe 3317: Trichlorofluorogermane

Cl_3Fe 1269: Ferric chloride

$Cl_3FeH_{12}O_6$ 1270: Ferric chloride hexahydrate

$Cl_3FeH_{12}O_{18}$ 1288: Ferric perchlorate hexahydrate

Cl_3FeO_{12} (anhydrous) 1289: Ferric perchlorate hydrate

Cl_3Ga 1391: Gallium(III) chloride

$Cl_3GaH_{12}O_{18}$ 1401: Gallium(III) perchlorate hexahydrate

Cl_3Gd 1358: Gadolinium chloride

$Cl_3GdH_{12}O_6$ 1359: Gadolinium chloride hexahydrate

Cl_3GdO_{12} (anhydrous) 1369: Gadolinium perchlorate hydrate

Cl_3GeH 3315: Trichlorogermane

Cl_3HSi 3318: Trichlorosilane

$Cl_3H_2NO_2Ru$ 2690: Ruthenium nitrosyl chloride monohydrate

$Cl_3H_6O_{15}Sb$ 273: Antimony(III) perchlorate trihydrate

$Cl_3H_8InO_4$ 1566: Indium(III) chloride tetrahydrate

$Cl_3H_{12}HoO_6$ 1483: Holmium chloride hexahydrate

$Cl_3H_{12}HoO_{18}$ 1491: Holmium perchlorate hexahydrate

$Cl_3H_{12}LaO_6$ 1660: Lanthanum chloride hexahydrate

$Cl_3H_{12}LaO_{18}$ 1673: Lanthanum perchlorate hexahydrate

$Cl_3H_{12}LuO_6$ 1850: Lutetium chloride hexahydrate

$Cl_3H_{12}NdO_6$ 2161: Neodymium chloride hexahydrate

$Cl_3H_{12}NdO_{18}$ 2172: Neodymium perchlorate hexahydrate

$Cl_3H_{12}O_6Sc$ 2727: Scandium chloride hexahydrate

$Cl_3H_{12}O_6Sm$ 2706: Samarium chloride hexahydrate

$Cl_3H_{12}O_6Tb$ 3160: Terbium chloride hexahydrate

$Cl_3H_{12}O_6Y$ 3498: Yttrium chloride hexahydrate

$Cl_3H_{12}O_6Yb$ 3473: Ytterbium chloride hexahydrate

$Cl_3H_{12}O_{18}Pr$ 2589: Praseodymium perchlorate hexahydrate

$Cl_3H_{12}O_{18}Tb$ 3167: Terbium perchlorate hexahydrate

$Cl_3H_{12}O_{18}Tl$ 3226: Thallium(III) perchlorate hexahydrate

$Cl_3H_{12}O_{18}Y$ 3508: Yttrium perchlorate hexahydrate

$Cl_3H_{14}LaO_7$ 1659: Lanthanum chloride heptahydrate

$Cl_3H_{14}O_7Pr$ 2580: Praseodymium chloride heptahydrate

$Cl_3H_{14}O_7Tm$ 3264: Thulium chloride heptahydrate

$Cl_3H_{16}InO_{20}$ 1573: Indium(III) perchlorate octahydrate

$Cl_3H_{16}MgNO_6$ 195: Ammonium magnesium chloride hexahydrate

$Cl_3H_{16}NNiO_6$ 200: Ammonium nickel chloride hexahydrate; 2188: Nickel ammonium chloride hexahydrate

$Cl_3H_{18}N_6Ru$ 1472: Hexaammineruthenium(III) chloride

Cl_3Ho 1482: Holmium chloride

Cl_3I 1594: Iodine trichloride

Cl_3In 1565: Indium(III) chloride

Cl_3Ir 1608: Iridium(III) chloride

Cl_3Ir (anhydrous) 1609: Iridium(III) chloride hydrate

Cl_3La 1658: Lanthanum chloride

Cl_3Lu 1849: Lutetium chloride

Cl_3Mo 2123: Molybdenum(III) chloride

Cl_3MoO 2138: Molybdenum(V) oxytrichloride

Cl_3N 2282: Nitrogen trichloride

Cl_3NbO 2273: Niobium(V) oxychloride

Cl_3Nd 2160: Neodymium chloride

Cl_3OP 2361: Phosphorus oxychloride

Cl_3OV 3422: Vanadium oxytrichloride

Cl_3OW 3353: Tungsten oxytrichloride

$Cl_3O_{12}Sm$ (anhydrous) 2714: Samarium perchlorate hydrate

$Cl_3O_{12}Yb$ 3479: Ytterbium perchlorate

Cl_3Os 2304: Osmium(III) chloride

Cl_3Os (anhydrous) 2305: Osmium(III) chloride hydrate

Cl_3P 2365: Phosphorus(III) chloride

Cl_3PS 3231: Thiophosphoryl chloride

Cl_3Pr 2579: Praseodymium chloride

Cl_3Pu 2393: Plutonium(III) chloride

Cl_3Rc 2616: Rhenium(III) chloride

Cl_3Rh 2640: Rhodium(III) chloride

Cl_3Rh (anhydrous) 2641: Rhodium(III) chloride hydrate

Cl_3Ru 2693: Ruthenium(III) chloride

Cl_3Ru (anhydrous) 2694: Ruthenium(III) chloride hydrate

Cl_3Sb 263: Antimony(III) chloride